HANDBOOK OF

Intelligent Scaffolds for Tissue Engineering and Regenerative Medicine

HANDBOOK OF

Intelligent Scaffolds for Tissue Engineering and Regenerative Medicine

edited by Gilson Khang

PAN STANFORD PUBLISHING

Published by

Pan Stanford Publishing Pte. Ltd.
Penthouse Level, Suntec Tower 3
8 Temasek Boulevard
Singapore 038988

Email: editorial@panstanford.com
Web: www.panstanford.com

British Library Cataloguing-in-Publication Data
A catalogue record for this book is available from the British Library.

Handbook of Intelligent Scaffolds for Tissue Engineering and Regenerative Medicine

ISBN 978-981-4267-85-4 (Hardcover)
ISBN 978-981-4267-86-1 (eBook)

Printed in the USA

I would like to dedicate this handbook to my wife, Isabella Seong Hee Koh; my children, Jerome Taeuk and Daniel Taehoon; and my mother.

—G.K.

Contents

Preface

It has been recognized that regenerative medicine and tissue engineering offer an alternative technique to whole-organ and tissue transplantation for diseased, failed, or malfunctioning organs. Millions of patients suffer from end-stage organ failure or tissue loss annually. The only way to solve this problem might be organ transplantation and biomaterials transplantation. However, in order to avoid the shortage of donor organs and other problems caused by poor biocompatibility of biomaterials, a new hybridized method combined with cells and biomaterials had been introduced as regenerative medicine and tissue engineering around 20 years ago.

The specialty of regenerative medicine and tissue engineering continues to grow and change rapidly. This area saw major advances in the past few years. This field for academic research and commercialization is needed in multidisciplinary areas such as adult, embryoinic, and induced pluripotent cells, genetic programming, nuclear transfer, cloning, genomics, proteomics, nanotechnology, biomaterials, etc. Thanks to the latest 20 years' endeavor, several tissue-engineered products (TEMPS) and regenerative medicinal products (RMP) are on the boundary of the translation of benchside discoveries to clinical therapies. For the reconstruction of a neo-tissue by regenerative medicine and tissue engineering, triad components such as (i) *cells* that are harvested and dissociated from the donor tissue, including nerve, liver, pancreas, cartilage, and bone, as well as embryonic stem cells, adult stem cells, induced pluripotent cells (iPS), or precursor cells; (ii) biomaterials as *scaffold substrates* whose cells are attached and cultured, resulting in the implantation at the desired site of the functioning tissue; and (iii) *growth factors* that are promoting and/or preventing cell adhesion, proliferation, migration, and differentiation by up-regulating or down-regulating the synthesis of protein; growth factors; and receptors must be needed.

This handbook has concentrated on all the things for scaffolds among triad components, especially "intelligent" scaffolds from basic science, industries, to clinical applications. This textbook is organized into seven major areas. Part I, Introduction, reveals some of fundamentals of the biomaterials, scaffolds, and manufacturing methods. Part II covers ceramic and metal scaffolds. Part III, Intelligent Hydrogel, deals with various types of hydrogels for tissue regenerations. In Part IV, topics of scaffolds from electrospinning nanofibers have been covered. In Part V, novel biomaterials for scaffolds have been introduced, especially to mimic Mother Nature. The sixth part covers the recent novel fabrication methods for smart scaffolds. The last part, Part VII, of this handbook deals with the recent clinical trial of specific target organs using intelligent scaffolds. The authors have tried to dedicate the 45 chapters to the whole area of the recent topic of smart scaffolds for regenerative medicine and tissue engineering. I am indebted to the authors for their willing acceptance, devotion, and contribution to each recent topic.

I express my thanks to my students Mrs. Yong Ki Kim, Jung Bo Shim, and Young Un Kim for editing all manuscripts. Finally, I really appreciate our publisher, Mr. Stanford Chong. Without his trust and guidance, this huge work could not have been accomplished. Also, I would like to give special appreciation to Mr. Sarabjeet Garcha and Ms. Archana Ziradkar for the hard work.

Gilson Khang, PhD

Part I

INTRODUCTION

Chapter 1

BIOMATERIALS AND MANUFACTURING METHODS FOR SCAFFOLD IN REGENERATIVE MEDICINE

Gilson Khang

Department of BIN Fusion Technology & Department of Polymer-Nano Science and Technology, Chunbuk National University, 664-14, Dukjin, Jeonju 561-756, Korea

gskhang@jbnu.ac.kr

It has been recognized that regenerative medicine and tissue engineering offer an alternative technique to whole-organ and tissue transplantation for diseased, failed, or malfunctioning organs. In order to reconstitute a new tissue by regenerative medicine and tissue engineering techniques, three factors—(1) cells that are dissociated and harvested from the donor tissue, (2) biomaterials as scaffold substrates in which cells are attached and cultured, resulting in the implantation at the desired site of the functioning tissue, and (3) growth factors that are promoting and/or preventing cell adhesion, proliferation, migration, and differentiation for stem cells—are required. Among these three key components, scaffolds might play a very critical role in regenerative medicine and tissue engineering. The role of scaffolds is to induce and stimulate the growth of cells seeded within the porous structure of the scaffolds or of cells migrating from surrounding tissue, eventually

Handbook of Intelligent Scaffolds for Tissue Engineering and Regenerative Medicine
Edited by Gilson Khang

www.panstanford.com

mimicking a "Mother Nature" extracellular matrix (ECM). This handbook introduces the recent trends in the development of biomaterials and the fabrication methods of regenerative medicinal and tissue-engineered "intelligent" scaffolds.

In summary, in order to approach a more natural, three-dimensional (3D) environment, researches combined biology, biochemistry, material engineering, clinical benchside work, etc., with multidisciplinary networks and have attempted to redesign scaffolds that will support biological signals for tissue growth and reorganization.

1.1 Introduction

It has been recognized that regenerative medicine and tissue engineering offer an alternative technique to whole-organ and tissue transplantation for diseased, failed, or malfunctioning organs. Millions of patients suffer from end-stage organ failure or tissue loss annually. In the United States alone, at least eight million surgical operations have been carried out each year, requiring a total national health care cost exceeding $400 billion annually.[1–3] In case of cartilage disease, each year in the United States, surgery is performed on more than three million knees, hips, and other joints. Around 69% of American adults suffer from arthritis, which is frequently caused by cartilage damage because of sports injury, other trauma, and/or simply overuse. These degenerations of cartilage combined with its inability to self-repair leads to further degradation of the joints. In the head and the neck, similarly with cartilage disease, cartilage replacement or repair is needed for degenerative disease, traumatic injury, or agenesis. Over 500,000 patients need a surgical procedure as cartilage replacement. Similarly, septal reconstruction, auricular reconstruction, and laryngotracheal reconstruction require cartilage parts in patients.

Current clinical procedures for cartilage repair are not enough at restoring form and function. Cartilage autografts suffer from many problems such as limited donor tissue availability, donor site injury, scarring, and pain. Allogenic and alloplastic implants have a high risk of infection, graft resorption, and structural failure. Metallic, glass, and polymeric biomaterials alone do not easily integrate into the host tissue and have a limited lifetime. Furthermore, implantation generally requires invasive surgery.[4] In the past decade, much effort has been made in engineering ideal scaffolds for bone tissue regeneration (high porosity, proper pore size, biocompatibility, biodegradability, osteoinductivity, etc.), but none of the current materials fulfills all demands. Thus, a broad range of solutions have been developed for each particular function, for instance, devices with high mechanical stability for large bone defects in load-bearing long bones and moldable or injectable materials for craniofacial surgery.[1–4]

In order to avoid the shortage of donor organ and other problems caused from poor biocompatibility of only biomaterials, a new hybridized method combined with cells and biomaterials has been introduced as regenerative medicine and tissue engineering.[2] To reconstruct a new tissue by regenerative medicine and tissue engineering, triad components such as (i) cells that are harvested and dissociated from the donor tissue, including nerve, liver, pancreas, cartilage, and bone, as well as embryonic stem cells, adult stem cells, induced pluripotent cells (iPS), or precursor cells; (ii) biomaterials as scaffold substrates whose cells are attached and cultured, resulting in the implantation at the desired site of the functioning tissue; and (iii) growth factors that are promoting and/or preventing cell adhesion, proliferation, migration, and differentiation by up-regulating or down-regulating the synthesis of protein, growth factors, and receptors are needed, as shown in Fig. 1.1.[1,5] In a typical application for cartilage regeneration, donor cartilage or bone marrow–derived stem cells are harvested from the patient and dissociated into individual chondrocyte cells using enzymes such as collagenase and then mass-cultured *in vitro*. The chondrocyte cells or chondrogenesis stem cells using differentiation-induced molecules are then seeded onto a porous and synthetic biodegradable scaffold. This cell/polymer structure also is massively cultured in a bioreactor. The malfunctioned tissue is removed, and the cell/polymer structure is then implanted in the patient. Finally, the synthetic biodegradable scaffold bioresorbs into the body, and the chondrocyte cells produce collagen and glycosaminoglycan as their own natural ECM, resulting in regenerated cartilage. This approach

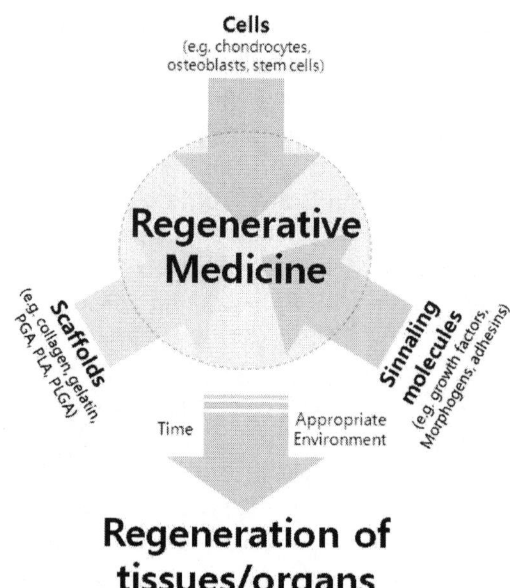

Figure 1.1. Tissue engineering triad. Combining three key elements such as cells, biomaterials, and signaling molecules, tissue-engineered neo-organs would be regenerated.

can be applied to manufacture theoretically almost all organs and tissues except several kinds of organs, like the brain.

1.2 Biomaterials for Regenerative Medicine and Tissue Engineering

1.2.1 *Importance of Scaffold Matrices in Regenerative Medicine and Tissue Engineering*

Scaffolds might play a very critical role in regenerative medicine and tissue engineering. The function of scaffolds is to direct the growth of cells seeded within the porous structure of the scaffolds or of cells migrating from surrounding tissue. The majority of mammalian cell types are anchorage dependent, resulting in dying if an adhesion substrate is not provided. Scaffold matrices can be used to achieve cell delivery with high loading and efficiency to specific sites. Therefore, the scaffold must provide a suitable substrate for cell attachment, cell proliferation, differentiated function, and cell migration. The prerequisite physicochemical properties of scaffolds are (i) to support and deliver for cells, (ii) to induce and differentiate and to be a conduit for tissue growth, (iii) to target the cell adhesion substrate, and (iv) to stimulate a cellular response. Other properties include (i) a wound-healing barrier, (ii) biocompatibility and biodegradability, (iii) relative ease of processability and malleability into desired shapes, (iv) high porosity with a large surface/volume, (v) mechanical strength and dimensional stability, (vi) sterilizability,

$$-[CH_2-CH_2]_n- \quad (1)$$

$$-[CH_2-CF_2]_n- \quad (2)$$

$$-[CF_2-CF_2]_n- \quad (3)$$

$$-[CH_2-CH_2-O]_n- \quad (4)$$

$$-[CH(OH)-CH_2]_n- \quad (5)$$

$$-[O-CH_2-CH_2-O-C(=O)-C_6H_4-C(=O)]_n- \quad (6)$$

$$-[O-CH_2-CH_2-CH_2-CH_2-O-C(=O)-C_6H_4-C(=O)]_n- \quad (7)$$

$$-[CH_2-C(CH_3)(C(=O)-O-CH_3)]_n- \quad (8)$$

$$-[CH_2-C(CH_3)(C(=O)-O-CH_2-CH_2-OH)]_n- \quad (9)$$

$$-[CH_2-CH(C(=O)-NH-CH(CH_3)_2)]_n- \quad (10)$$

11 (polypyrrole repeat unit, N in ring)

$$-[Si(CH_3)_2-O]_n- \quad (12)$$

13 (pyromellitimide repeat unit, $-[N(imide)-R]_n-$)

(A)

(A) Synthetic nondegradable polymers: 1. polyethylene, 2. poly(vinylidene fluoride), 3. polytetrafluoroethylene, 4. poly(ethylene oxide), 5. poly(vinyl alcohol), 6. poly(ethyleneterephthalate), 7. poly(butyleneterethphalate), 8. poly(methylmethacrylate), 9. poly(hydroxymethylmetacrylate), 10. poly(N-isopropylacrylamide), 11. polypyrrole, 12. poly(dimethyl siloxane), and 13. polyimides.

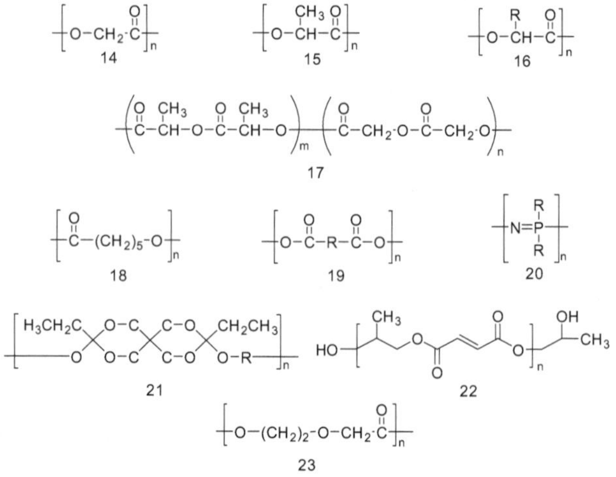

(B)

(B) Synthetic biodegradable: 14. poly(glycolicacid), 15. poly(lactic acid), 16. poly(hydroxyalkanoate), 17. poly(lactide-*co*-glycolide), 18. poly (ε-caprolactone), 19. polyanhydride, 20. polyphsphazene, 21. poly(ortho-ester), 22. poly(propylene fumarate), and 23. poly(dioxanone).

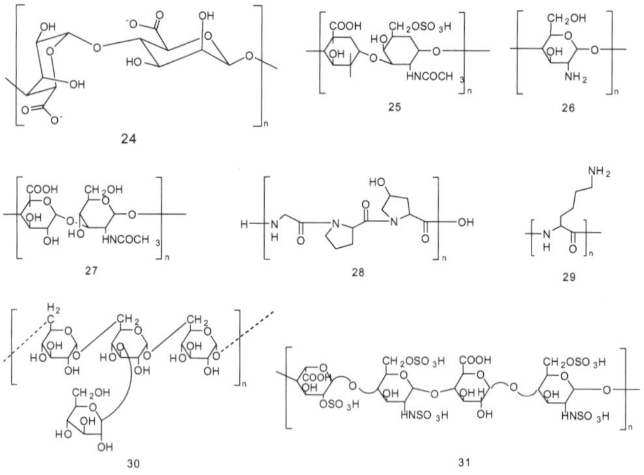

(C)

(C) Natural polymers: 24. alginate, 25. chindroitin-6-sulfate, 26. chitosan, 27. hyarunonan, 28. collagen, 29. polylysine, 30. dextran and 31. heparin.

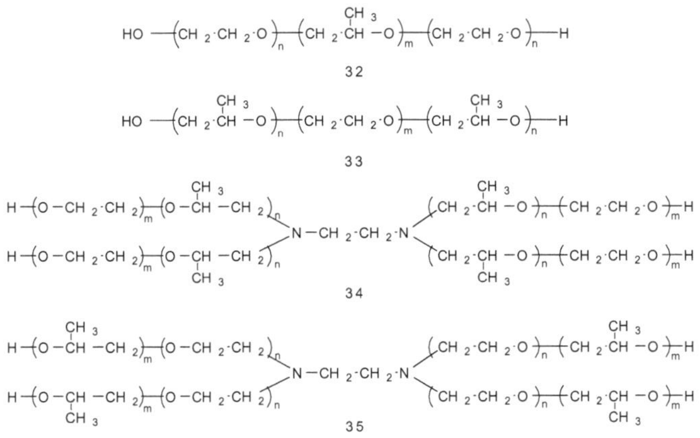

$$HO-(CH_2\cdot CH_2\cdot O)_n-(CH_2\cdot CH(CH_3)-O)_m-(CH_2\cdot CH_2\cdot O)_n-H$$

32

$$HO-(CH_2\cdot CH(CH_3)-O)_n-(CH_2\cdot CH_2\cdot O)_m-(CH_2\cdot CH(CH_3)-O)_n-H$$

33

$$[H-(O-CH_2\cdot CH_2)_m-(O-CH(CH_3)-CH_2)_n]_2N-CH_2\cdot CH_2\cdot N[(CH_2\cdot CH(CH_3)-O)_n-(CH_2\cdot CH_2\cdot O)_m-H]_2$$

34

$$[H-(O-CH(CH_3)-CH_2)_m-(O-CH_2\cdot CH_2)_n]_2N-CH_2\cdot CH_2\cdot N[(CH_2\cdot CH_2\cdot O)_n-(CH_2\cdot CH(CH_3)-O)_m-H]_2$$

35

(D)

(D) PEO-based hydrogel: 32. Pluronic, 33. Pluronic R, 34. Tetronic, and 35. Tetronic R.

Figure 1.2. Chemical structures of some commonly used biodegradable and nondegradable polymers in tissue engineering.[1]

etc.[1,5,7] Generally, 3D porous scaffolds can be fabricated from natural and synthetic polymers (Fig. 1.2 shows these chemical structures), ceramics, metal in very few cases, composite biomaterials, and cytokine release materials. Very recently, "intelligent" scaffolds are being extensively tested to mimic the human body's environment as the ECM to Mother Nature.

1.2.2 *Bioceramic Scaffolds*

"Bioceramic" is a term introduced for biomaterials that are produced by sintering or melting inorganic raw materials to create an amorphous or a crystalline solid body that can be used as an implant. Porous final products have been mainly used as scaffolds. The components of ceramics are calcium, silica, phosphorous, magnesium, potassium, and sodium. Bioceramic used in the fabrication for tissue engineering might be classified as nonresorbable (relatively inert), bioactive, or surface active (semi-inert) and biodegradable or resorbable (noninert). Alumina, zirconia, silicone nitride, and carbon are inert bioceramics. Certain glass ceramics are dense hydroxyapatites ($9CaO{\cdot}Ca(OH)_2{\cdot}3P_2O_5$) semi-inert (bioactive), and calcium phosphates, aluminum-calcium phosphates, coralline, tri-calcium phosphates ($3CaO{\cdot}P_2O_5$), zinc-calcium-phosphorous oxides, zinc-sulfate-calcium phosphates, ferric-calcium-phosphorous oxides, and calcium aluminates are resorbable ceramics.[8] Among these bioceramics, synthetic apatite and calcium phosphate minerals, coral-derived apatite, bioactive glass, and demineralized bone particle (DBP) will be introduced in this section since they are widely used in the hard-tissue engineering area.

1.2.2.1 Calcium phosphate

Synthetic, crystalline calcium phosphate can be crystallized into salts such as hydroxyapatite and β-whitlockite, depending on the Ca:P ratio, which are very tissue compatible and are used as bone substitutes in a granular, sponge form or a solid block. The apatite formed with calcium phosphate is considered closely related to the mineral phase of bone and teeth. Chemical composition of crystalline calcium phosphate is a mixture of $3CaO{\cdot}P_2O_5$, $9CaO{\cdot}Ca(OH)_2{\cdot}3P_2O_5$ and calcium pyrophosphate ($4CaO{\cdot}P_2O_5$). The active exchange of ions that occurs on the surface leads to the exchanging composition of mineral.[9] Also, the delivery of some elements to the new bone will form at the interface between the materials and the osteogenic cells when the porous ceramic scaffolds are implanted in the body with or without cells for the tissue-engineered bone.

1.2.2.2 Tricalcium phosphate

Tricalcium phosphate is the rapidly resorbable calcium phosphate ceramic resulting in 10 to 20 times resorption, faster than hydroxyapatite.[10] Porous tricalcium phosphate may stimulate local osteoblasts for new bone formation. Injectable calcium phosphate cement containing β-tricalcium phosphate, dibasic dicalcium phosphate, and tricalcium phosphate monoxide was investigated for the treatment of distal radius fractures. Calcium sulfate hemihydrate (plaster of Paris) as a synthetic graft material was also tested for the tissue-engineered bone.

1.2.2.3 Hydroxyapatite

Coral-derived apatite is a natural substance made by marine vertebrates.[11] The porous structure of coral is its unique physico-chemical property for the scaffold matrix as bone substitutes due to the structural similarity to bone. The main component of natural coral is calcium carbonate or aragonite, the metastable form of calcium carbonate, and it also can be converted to hydroxyapatite by a hydrothermal exchange process, resulting in the mixture of hydroxyapatite, $9CaO{\cdot}Ca(OH)_2{\cdot}3P_2O_5$, and fluoroapatite, $Ca_5(PO_4)_3F$. For the tissue-engineered bone, the hybrid structure of porous coral-derived scaffold and mesenchymal stem cell were demonstrated *in vitro*. This result showed the differentiation of bone marrow-derived stem cell to osteoblast, and successive mineralization was successfully accomplished.

1.2.2.4 Bioglass

Glass ceramics are polycrystalline ceramics manufactured by controlled crystallization of glasses using nucleating agents such as small amounts of metallic agents Pt groups, TiO_2, ZrO_2, and P_2O_5, resulting in a fine-grained ceramic that possesses excellent mechanical and thermal properties.[9] Typical bioglass ceramics developed for implantations are SiO_2-CaO-Na_2O-P_2O_5 and Li_2O-ZnO-SiO_2 systems. These bioglass scaffolds are suitable for the regeneration of inducing direct bonding with bone. The bonding to bone is related to the composition of each component.

1.2.2.5 Demineralized bone particle

One of the significant natural bioactive materials is DBP, which is a powerful inducer of new bone growth.[12] It has been recognized that DBP contains many kinds of osteogenic and chondrogenic cytokines as bone morphogenetic protein and widely uses a filling agent for bony defects in the clinic due to improved availability through the growing tissue bank industry. For the optimization of the application of DBP to tissue engineering, research studies such as nanohybrdization with synthetic (PLGA/DBP hybrid scaffolds) and natural organic compounds (collagen/DBP hybrid scaffolds) have been done.

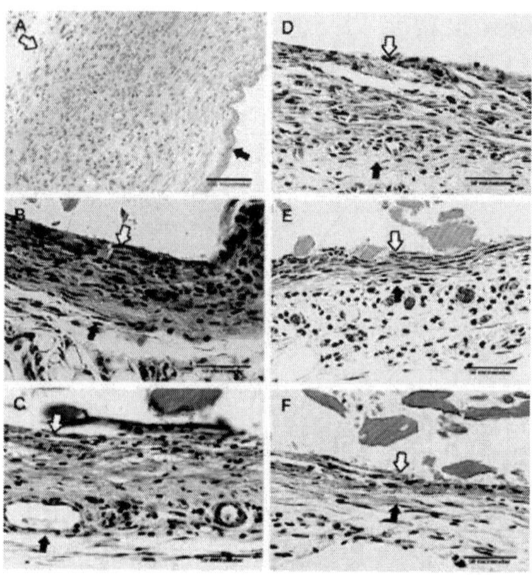

Figure 1.3. Photomicrographs of H&E stain sections of the directly bordering tissue after PLGA, hybrid PLGA/DBP films, and DBP. (A) The tissue implanted with PLGA, bar length = 100 μm (x100), (B–F) The tissue implanted with 10% DBP/PLGA, 20% DBP/PLGA, 40% DBP/PLGA, 80% DBP/PLGA, and DBP, respectively, bar length = 50 μm (x400). Note that the number of inflammatory cell and fibrous band thickness in vicinity to tissue-implanted samples was decreased as DBP content in PLGA film was increased. Polymer-tissue interface surfaces are indicated by a white arrow. The fibrous wall thickness was represented by black and white arrows. *Abbreviations*: H&E, hematoxylin and eosin.

Figure 1.3 shows the effect of DBP on the reduction of inflammatory reaction from the host tissue for a DBP/poly(lactide-*co*-glycolide) (PLGA) hybrid scaffold. After the hybridization of DBP into PLGA, the number of mononuclear phagocyte cells and the thickness of fibrotic wall thickness were dramatically decreased. It offers good information to design a natural scaffold (explained in chapter 16).[13]

The porosity, such as the size of the mean diameter and the surface area, is a critical factor for the growth and migration of a tissue into the bioceramic scaffolds.[9] Several methods were introduced to optimize the fabrication porous ceramics, such as dip casting, starch consolidation, polymeric sponge method, foaming method, organic additives, gel casting, slip casting, direct coagulation consolidation, hydrolysis-assisted solidification, and freezing methods (see sections 1.4 and 1.6). Therefore, it is very important to choose the appropriate preparation methods for the physical properties of desired organs.

1.2.3 *Synthetic Polymers*

One of the most significant shortages of natural polymers, as will be discussed in section 1.2.4, is typically expensive, suffering from batch-to-batch variation, and the possibility of cross-contamination from unknown viruses or unwanted diseases due to the isolation from plant, animal, and human tissue. On the contrary, synthetic polymeric biomaterials might have easily controlled physicochemical properties and quality and no immunogenecity. Also, they can be processed with various techniques and supplied consistently in large quantities.

In order to adjust the physical and mechanical properties of tissue-engineered scaffolds at the desired place in the human body, the molecular structure, molecular weight, etc., are easily adjusted during the synthetic process. There are largely divided two categories: (i) biodegradable and (ii) nonbiodegradable. Some nondegradable polymers are polyvinylalcohol (PVA), poly(hydroxylethylmethacryalte), and poly(N-isopropylacryamide). Some synthetic degradable polymers are the family of poly(α-hydroxy ester)s such as polyglycolide (PGA), polylactide (PLA) and its copolymer poly(lactide-*co*-glycolide) (PLGA), polyphosphazene, polyanhydride, poly(propylene fumarate), polycyanoacrylate, polycaprolactone, polydioxanone, biodegradable polyurethanes, etc.[1,5,7] (Chemical structures are shown in Fig. 1.2A,B,D.)

Among these two polymers, the synthetic biodegradable polymers were preferred for the application of regenerative medicine and tissue-engineered scaffolds to minimize the chronic foreign-body reaction and lead to the formation of the completely natural tissue. That is to say, they can form a temporary scaffold for mechanical and biochemical support. This section mainly focuses on the biodegradable polymers and then more detailed fabrication methods for polymers will be discussed in sections 1.4 and 1.6.

1.2.3.1 Poly(α-hydroxy ester)s

The family of poly(α-hydroxy acid)s such as PGA, PLA, and its copolymer PLGA that are among the few synthetic polymers approved for human clinical use by the Food and Drug Administration (FDA) are extensively used or tested for the scaffold materials as bioerodible materials due to good biocompatibility, controllable biodegradability, and relatively good processability.[14] It has been used for three decades as sutures of PGA, bone plate, screw, and reinforced materials for PLA and drug delivery devices of PLGA in surgical operation and whose safety has been proved in many medical applications.[15]

These polymers degrade by nonspecific hydrolytic scission of their ester bonds. PGA biodegrades by a combination of hydrolytic scission and enzymatic (esterase) action, producing glycolic acid, which can either enter the tricarboxylic acid (TCA) cycle or be excreted in urine and be eliminated as carbon dioxide and water. The hydrolysis of PLA yields lactic acid that is a normal by-product of anaerobic metabolism in the human body and is incorporated in the TCA cycle to be finally excreted by the body as carbon dioxide and water. With an additional methyl group to glycolide, PLA is much more hydrophobic than the highly crystalline PGA. As a result, PLA has a much slower rate degradation rate for over one year. The degradation time of PLGA as copolymers of these two polymers can be controlled from weeks to over a year by varying the ratio of monomers, its molecular weight, and the processing conditions. The synthetic methods and physicochemical properties such as melting temperature, glass transition temperature, tensile strength, Young's modulus, and elongation were reviewed elsewhere.[16]

The mechanism of biodegradation of poly(α-hydroxy acid)s is bulk degradation, which is characterized by a loss in the polymer molecular weight, while mass is maintained. Mass maintenance is useful for tissue engineering applications of those specific shapes. However, loss in molecular weight causes a significant decrease in mechanical properties. Degradation is depending on chemical history, porosity, crystallinity, steric hindrance, molecular weight, water uptake, and pH. Degradable products such as lactic acid and

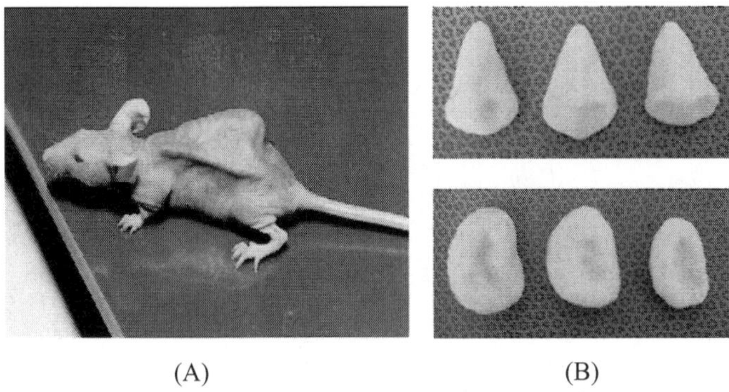

Figure 1.4. PGA nonwoven fiber scaffold with chondrocyte (A) and PGA scaffold for ear and nose form (B).

glycolic acid decrease the pH in the surrounding tissue, resulting in inflammation and potentially poor tissue development. PGA, PLA, and PLGA scaffolds were applied for regeneration of all tissue such as skin, cartilage, blood vessel, nerve, liver, dura mater, bone, and other tissue.[1,5,7] For the application of these polymers for the scaffolds, the development of fabrication methods for a porous structure is also important. Figure 1.4 reveals the PGA/chondrocyte hybrid constructs for the nose and PLGA/DBP hybrid constructs for the disc and spinal cord regenerations using tissue engineering made from our laboratory.

But, the significant drawback of the family of poly(α-hydroxy acid)s might be the induction of inflammatory cells and fibrotic wall thickness creating a hurdle for the clinical applications to patients and for launching in the market, as shown in Fig. 1.3. This was caused by the host response and acidic by-products of degradation, such as glycolic acid and lactic acid. It could be prolonged to three to four months as the end of biodegradation, as shown in Fig. 1.5. In order to solve this serious problem, we tried to hybridize PLGA and a natural polymer such as DBP and SIS. After the hybridization, the fibrotic wall thickness and macrophage reaction were dramatically decreased, as shown in Fig. 1.4, and will be discussed more in detail in chapter 16.

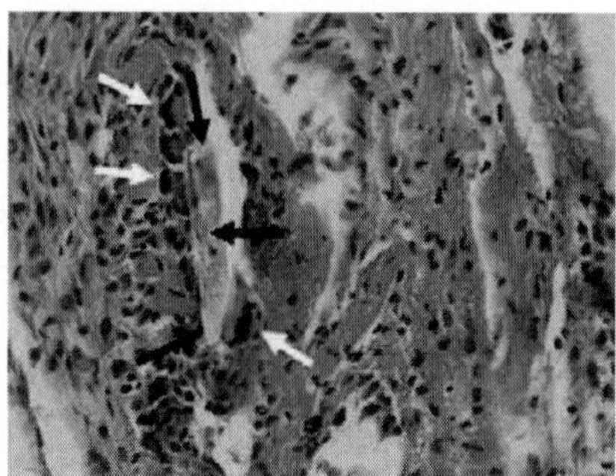

Figure 1.5. Foreign-body granuloma. Small PLGA debris (black arrows) broken off from the PLGA film and surrounded by macrophages and multinucleated giant cells (white arrows). These induced macrophages, and multinucleated giant cells were remaining over 2 months.

1.2.3.2 Polyanhydride

Polyanhydride is synthesized by the reaction of diacids with anhydride to form acetyl anhydride prepolymers. High-molecular-weight anhydrides are synthesized from the anhydride prepolymer in melt condensation. Polyanhydrides were modified with the reaction of imides to increase the physical properties. A typical example is copolymerization with aromatic imide monomers to increase the mechanical properties resulting in the polyanhydride-*co*-imide for the application of hard-tissue engineering. To control degradability and enhance mechanical properties, photocrosslinkable functional groups were introduced by the substituted methacrylate groups on polyanhydrides for the application of orthopedic tissue engineering.[16] The degradation mechanism of polyanhydrides appears as surface erosion with a highly predictable and controlled manner, whereas that of poly(α-hydroxy ester) is bulk erosion. To optimize the degradation behavior of anhydride-based copolymers, the controlling of polymer backbone chemistry, the ratio of monomer, and molecular weight can be performed.

1.2.3.3 Poly(propylene fumarate)

Poly(propylene fumarate) and its copolymer, biodegradable and unsaturated linear polyester, were synthesized as potential scaffold biomaterials. The degradation mechanism is hydrolytic chain scission, similar to poly(α-hydroxy ester). The mechanical strength and degradable behaviors were controlled by the reaction of cross-linking with vinyl monomer to unsaturated double bonds. Also, the physical property will be enhanced by the composite with degradable bioceramic β-tricalcium phosphate for the application of the injectable bone.[17] Copolymerization of propylene fumarate with ethylene glycol can be endowed with the elasticity of poly(propylene fumarate) for the cardiovascular stent. New materials for propylene fumarate polymers are continually investigated through copolymer synthesis, hybrid composites, and blends.

1.2.3.4 PEO and its derivatives

Polyethylene oxide (PEO) is one of the most important and widely used polymers in the biomedical applications due to its excellent biocompatibility.[18] It can be produced by anionic or cationic polymerization from ethylene oxide by initiators. PEO is used in the coating materials for the medical devices to prevent tissue and cell adhesion as well as preparation of biologically relevant conjugate and induction cell membrane fusion. PEO hydrogels can be fabricated by the cross-linking reaction such as gamma ray, electron beam irradiation, or chemical reaction. This hydrogel can be used in the application of drug delivery and tissue engineering. The hydroxyl in the glycol end group is very active, resulting in availability for chemical modifications. Attachment of bioactive molecules such as cytokines and peptides to PEO or poly(ethylene glycol) (PEG) allows to enhance the delivery efficiency of bioactive molecules (detailed explanation in section 1.3).

For the synthesis of biodegradable PEO, block copolymerization with PGA or PLA degradable units has been carried out to make biodegradable hydrogels. It can be polymerized as two- or three-block copolymers such as PEO-PLA, PEO-PLA-PEO, and PLA-PEO-PLA. For the biodegradable block, ε-caprolactone, δ-valerolactone, PLGA, etc., can be used. The characteristic of these series hydrogels shows temperature-sensitive phenomena. Sol state at room temperature changes to gel state at body temperature. Hence, biodegradable hydrogels were very useful for injectable cell-loading scaffolds.[19] After injection of the hybrid structure of chondrocyte cells and biodegradable hydrogels, hydrogels degraded *in vivo*, and then neocartilage tissue remains.

Also, copolymers of PEO and poly(propylene oxide) (PPO), such as PPO-PEO-PPO or PEO-PPO-PEO block copolymers, are the basis for Pluronics® and Tetronics®. Pluronics® forms a thermosensitive gel by the shrinking of hydrophobic segments of the copolymer PPO.[1] The physicochemical property of hydrogels can be varied with the composition and structure of the ratio of PPO and PEO. Some of them have been approved by the FDA and the Environmental Protection Agency (EPA) for the applications in food additives, pharmaceutical ingredients, and agricultural products. Although the polymer is not degraded by the body, the gels dissolve slowly, and then the polymer is eventually cleared. These polymers have been applied for the treatment of skin burns and protein delivery. These injectable hydrogels have advantages such as no surgical operation, easy pore size manipulation, and no need for the fabrication process of complex shapes.

1.2.3.5 Polyvinylalcohol

PVA is a hydrogel resulting in the containing of water content, especially similar to that of cartilage. It is relatively biocompatible, swelling to hold a large amount of water, is easily sterilized, and is easily fabricated and molded into desired shapes. It has reactive pendant alcohol groups that are available for modification such as chemical cross-linking, physical cross-linking, or incorporation of an acrylate group resulting in the improvement of mechanical properties. Changing the ratio of PVA and water, the molecular weight of PVA, and the quantity and duration of freeze/thaw cycles can control the physical properties of PVA hydrogels. PVA has been applied to cartilage regeneration due to similar mechanical properties, breast augmentation, diaphragm replacement, and bone replacement.[20] One of the significant drawbacks is it is not fully biodegradable because of the lack of labile bonds within the polymer backbone. So, it is recommended that low-molecular-weight PVA around 15,000 g/mole, which can be penetrated through the kidney, might be applied to the tissue engineering scaffolds.

1.2.3.6 Oxalate-based polyesters (polyoxalate)

Polyoxalate is one of absorbable polyesters that have oxalate linkages in their backbone. The oxalate-based polymers were first introduced by Shalaby and coworkers in the 1970s for suture-coating purposes.[21] Polyoxalate was reported to undergo ester hydrolysis to give oxalic acid and diol as by-products. The hydrophobicity was between PLA and PGA, and the hydrolytic stability decreased primarily with an increase in methylene fraction in the repeating unit. Copolyoxalate was also synthesized to increase the melting temperature by using cyclic and aromatic diols. The chemical composition influenced hydrolytic stability and tissue reaction. The poly(1,4-cyclohexylenedicarbinyl-*co*-hexamethylene oxalate) exhibited an increasing fast *in vivo* weight loss as the 1,4-cyclohexylenedicarbinyl content decreased and slight and mild tissue reactions during the first five days of implantation. Poy(oxalate-*co*-oxamide) exhibited higher hydrophilicity and a faster degradation profile compared with copolyoxalate. When formulated into films, poy(oxalate-*co*-oxamide) showed higher cell attachment and proliferation than PLGA.[22]

1.2.3.7 Polyphosphazene

Polyphosphazene consists of an inorganic backbone of alternating single and double bonds between phosphorous and nitrogen atoms, while most of the polymer comprises the carbon-carbon organic backbone.[20] It has side groups that can react with another functional group resulting in block or star polymers. Biological and physical properties can be controlled by the substitution of functional side groups. The wettability as hydrophilicity, hydrophobicity, and amphiphilicity of polyphosphazene might depend on the properties of side groups. It can be fabricated into films, membranes, and hydrogels for scaffolds applications by cross-linking or grafting modifications. Cytocompatibility of highly porous polyphosphazene scaffolds was observed with the possibility of skeletal tissue engineering. Also, the blend of polyphosphazene with PLGA has been investigated to modify and determine miscibility and degradability.

1.2.3.8 Biodegradable polyurethane

Polyurethane is one of the most widely used polymeric biomaterials in biomedical fields due to unique physical properties such as durability, elasticity, elastomer-like character, fatigue resistance, compliance, and tolerance. Moreover, the plentiful reactivity on the functional group of a polyurethane backbone can achieve the attachment of biologically active biomolecules and the adjustment of hydrophilicity/hydrophobicity.[23] Typical biodegradable polyurethane is composed of an amino acid–based hard segment, such as lysine diisocyanate, and a polyol soft segment, such as hydroxyl donor–like polyester and sugar. Hence, the degradation products of this nontoxic lysine diisocyanate–based urethane polymer are nontoxic lysine and the polyol. By the covalent bonding of various proteins, such as cytokines, growth factors, and peptides, introduced in the polymer backbone, the controlled release of the bioactive molecules can be achieved by the degradable manner of polyurethane scaffolds.

Mechanisms of degradation are hydrolysis, oxidation, and thermal and enzymatic manner. Both the chemistry and the composition of soft and hard segments play an important role in the degradability of polyurethane. Poly(urethane-urea) matrices with lysine diisocyanate as the hard segment and glucose, glycerol, or PEG as soft segments. Toxicity, induction of foreign-body reactions, and antibody formation were not observed in the *in vivo* experiment. The elucidation of long-term safety and biocompatibility for biodegradable polyurethane must be continuously conducted for the successful application of tissue engineering scaffold substrates.

1.2.3.9 Other synthetic polymers

Many synthetic polymers, either degradable or nondegradable, are newly launched and tested to mimic the natural tissue and wound-healing environment. Examples are poly(2-hydroxyethylmethacryrate) hydrogel, injectable poly(*N*-isopropylacryamide) hydrogel, and polyethylene for neocartilage, poly(iminocarbonates) and tyrosine-based poly(iminocarbonates) for bone and cornea, cross-linked collagen/PVA films and an injectable biphasic calcium phosphate/methylhydroxypropylcellulose composite for bone regeneration materials, a polyethylene oxide-*co*-polybutylene terephthalate for bone bonding, poly(*ortho*-ester) and its composites with ceramics for tissue-engineered bone, synthesized conducting polymer polypyrrole/hyaruronic acid composite films for the stimulation of nerve regeneration, and peptide-modified synthetic polymers for the stimulation of cell and tissue.[1,5,7]

It is very important for the design and synthesis of more biodegradable and biocompatible scaffold biomaterials to mimic the natural ECM in terms of bioactivity, mechanical properties, and structures. The more biocompatible biomaterials tend to elicit less of an immune response and to reduce inflammatory response at the implantation site, combined with scaffold-manufacturing methods.[13]

1.2.4 *Natural Polymers*

Many naturally occurring scaffolds can be observed as biomaterials for regenerative medicine and tissue engineering purposes. One of the typical examples is the ECM that is a very complex biomaterial and controls cell function. For the ECM of regenerative medicine and tissue engineering, natural and synthetic scaffolds are designed to mimic specific functions. Natural polymers are fibrins, collagens (gelatin), alginate, proteins, albumin, gluten, elastin, fibroin, hyarulonic acid, cellulose, starch, chitosan (chitin), sclerolucan, elsinan, pectin (pectinic acid), galactan, curdlan, gellan, levan, emulsan, dextran, pullulan, heparin, silk, chondroitin 6-sulfate, polyhydroxyalkanoates, etc. (Chemical structures are shown in Fig. 1.2C.) Much of the interest in these natural polymers comes from their biocompatibility, relatively abundance and commercial availability, and ease of processing.[16]

1.2.4.1 Fibrin

Fibrin plays a major role during wound healing, such as forming a hemostatic barrier to prevent bleeding and to support a natural scaffold for fibroblasts. The actual polymerization is triggered by the conversion of fibrinogen to fibrin monomers by thrombin, and gelation occurs quickly within 30~60 seconds. One advantage of using fibrin in this manner is the ability to completely fill the defect by gelling *in situ*. A fibrin sealant composed of fibrinogen and thrombin in addition to antifibronolytic agents has been already used in surgical application for sealing lung tears, cerebrospinal fluid leaks, and bleeding ulcer because of its natural role in wound healing. A fibrin sealant might be made from autologous blood as the patient's own blood or from recombinant proteins.[24] Fibrin gels can degrade either through hydrolytic or through proteolytic means. Fibrinogen is commercially available from several manufacturers, resulting in relatively low cost of fabrication of fibrin gels. More recently, many works have been done in the development fibrin as a potential tissue engineering scaffold matrix, especially cartilage formed from a fibrin/chondrocyte construct. Biochemical and mechanical analysis demonstrated cartilage-like properties. A fibrin/PLGA hybrid composite has been investigated to optimize the reduction of inflammatory reaction of PLGA for the application of tissue-engineered cartilage and intervetebral disc substitutes, as shown in chapter 16.[25]

1.2.4.2 Collagen

At least 22 types of collagen exist in the human body. Among these, collagen types I, II, and III are the most abundant and ubiquitous kind. The conformation of the collagen chain is triple helices that are packed or processed into microfibrils. Molecularly, the three repeating amino acid sequences, such as glycine, proline, and hydroxyproline, form protein chains, resulting in the intertwinement in a triple-helix arrangement. Type I collagen is the most abundant and is the major constituent of bone, skin, ligament, and tendon, whereas type II collagen is a major collagen in cartilage. Collagen can promote cell adhesion, as demonstrated by the Asp-Gly-Glu-Ala peptide in type I collagen that functions as a cell-binding domain. Due to the abundance and ready accessibility of these tissues, they have been frequently used as a source for the preparation of collagen.[1]

The purified collagen materials obtained from either molecular technology or fibrillar technology are subjected to additional processing to fabricate the materials into useful scaffold types for specific tissue-engineered organs. Collagen can be processed into several types such as membrane (film and sheet), porous material (sponge, felt, and fiber), gel, solution, filamentous material, tubular material (membrane and sponge), and composite matrix for the application of tissue repair, patches, bone and cartilage repair, nerve regeneration, and vascular and skin repair with/without cells.[26] Physicochemical properties of collagen can be improved by a variety of homogeneous and heterogeneous composites. Homogeneous composites can be formed between ions, peptides, proteins, and polysaccharides into a collagen matrix by means of ionic and covalent bonding, entrapment, entanglement, and coprecipitation. Heterogeneous composites like collagen/synthetic polymers, collagen/biological polymers, and collagen/ceramic hybrids (collagen/nano-hydroxyapatite and collagen/calcium phosphate) have been investigated to achieve distinct properties for tissue-engineered products.[20]

1.2.4.3 Alginate

Alginate originating from seaweed is composed of two repeating monosaccharides, L-guluronic acid and D-mannuronic acid. Repeating strands of these monosaccharides can form linear, water-soluble polysaccharides. Gelation occurs by interaction of divalent cations (e.g., Ca^{2+}, Mg^{2+}) with blocks of guluronic acid from different polysaccharide chains, as shown in Fig. 1.6. From this good gelation property, the encapsulation of calcium alginate beads impregnated with various pharmaceutics, cytokines, or cultured cells has been extensively investigated. Varying the preparation condition of gelation can control the structure and physicochemical properties. Calcium alginate scaffolds did not degrade by hydrolytic reaction, whereas they can be degraded by a chelating agent such as ethyleneaminetetraaceticacid (EDTA) or by enzymes. Also, the diffusion of calcium ions from an alginate gel can cause dissociation between alginate chains, resulting in a decrease of mechanical strength over time. One of the disadvantages of the alginate matrix is a potential

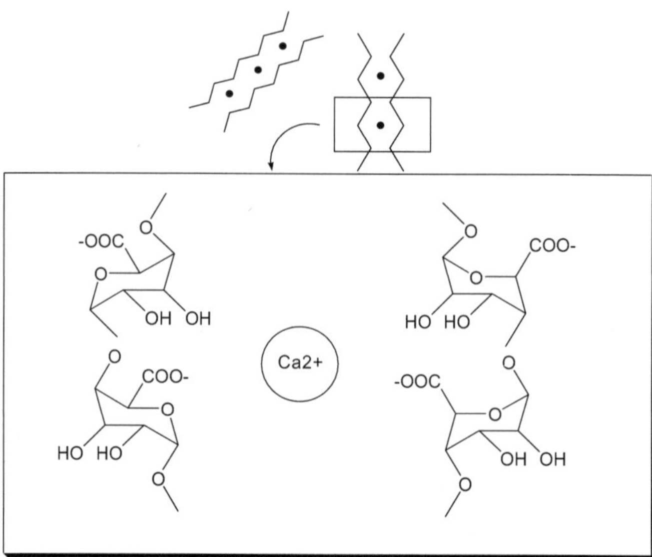

Figure 1.6. Schematic representation of the guluronate junction zone in alginate; egg-box model. The circles represent calcium ions.

immune response and the lack of complete degradation, since alginate is not native to the human body.[1,20]

Many researchers have studied the encapsulation of chondrocytes. Growth plate chondrocytes, fetal chondrocytes, and bone marrow–derived mesenchymal stem cells have been encapsulated in alginate.[20] In each system, the chondrocytes demonstrated a differentiated phenotype, producing ECM and retaining cell morphology of typical chondrocytes. Also, novel hybrid composites such as alginate/agarose (a thermosensitive polysaccharide), alginate/fibrin, alginate/collagen, and alginate/hyaruronic acid and different gelling agents (water, sucrose, sodium chloride, and calcium sulfate) were investigated to optimize the advantages of each component material for the tissue-engineered cartilage.[20–22] As a result, this hybrid material can offer the reason why the microenvironments of composite materials affect chondrogenesis. For the application of alginate to regenerative medicine and tissue engineering, the purification of alginate is extremely important.[27]

1.2.4.4 Small intestine submucosa

Porcine small intestine submucosa (SIS) is one of the important materials for natural ECM scaffolds.[28] Many researchers have described systematically that an acellular resorbable scaffold material derived from the SIS has been shown to be rapidly resorbed, to support early and abundant new blood vessel growth, and to serve as a template for the constructive remodeling of several body tissues, including musculoskeletal structures, skin, body wall, dura mater, urinary bladder, and blood vessels.[1] The SIS material consists of naturally occurring ECM that has been shown to be rich in components that support angiogenesis, such as fibronectin; glycosaminoglycans, including heparin; several collagens, including types I, III, IV, V, and VI; and angiogenic growth factors, such as basic fibroblast growth factor and vascular endothelial cell growth factor.[29] For these reasons, SIS scaffold has been successfully used to reconstruct for urinary bladder, vascular grafts, cartilage, and bone, alone or as a composite with synthetic polymers and inorganic biomaterials.

1.2.4.5 Silk

Silk is widely used in clinics as suture material. It is composed of a filament core protein such as fibroin and a gluelike coating such as sericin protein. Silk from the silkworm (*Bombyx mori*) and orb-weaving spiders (e.g., *Nephia clavipes*) has been explored to understand the fabrication mechanisms and to exploit the properties of these proteins for use as scaffolds biomaterials. Very recently, silk fibroin has been increasingly tested and used for the innovative biomaterials application because of relatively good biocompatibility, slow degradability, and remarkable mechanical properties of the scaffold materials.[30] Also, the ability to control the molecular structure and morphology through versatile processability and the techniques of surface modification have expanded the utility for the protein of silk fibroin for the scaffold materials for regenerative medicine and tissue engineering. Scaffolds with a variety of shapes, such as films, fibers, sponges, meshes, membranes, and yarns, have been shown to support stem cell adhesion, proliferation, and differentiation *in vitro* and promote tissue repair *in vivo*. In addition, 3D

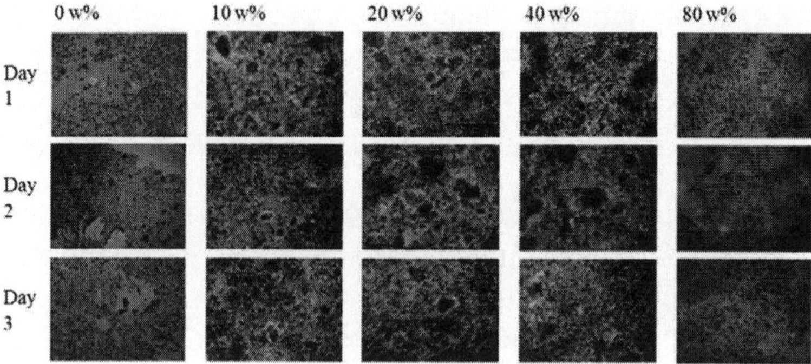

Figure 1.7. Immunocytochemical staining of RPE cells on PLGA, silk/PLGA film after 1, 2, and 3 days (magnification x100).[31]

silk fibroin scaffolds are promising for engineering a range of tissues: bone, ligament, cartilage, skin, nerve, etc. With the advances of genetic and biochemical engineering, the application of native silk proteins will expand exponentially in the area of regenerative medicine and tissue engineering. Figure 1.7 shows the regeneration of the retina using PLGA/silk hybrid scaffold–seeded retinal pigment epithelial (RPE) cells. From the immunocytochemical staining of RPE cells on a PLGA, silk/PLGA film after one, two, and three days with 100 magnification, silk plays an important role for the activation of the growth and proliferation of RPE cells.[31]

1.2.4.6 Hyaluronan

Hyaluronic acid, a natural glycosaminoglycans polymer, can be found abundantly within cartilaginous ECM. It has some disadvantages in the natural form, such as high water solubility, fast resorption, and fast tissue clearance times, resulting in no conduciveness for biomaterials. In order to overcome these undesirable characteristics, chemical modifications have been done to increase biocompatibility, tailor the degradation rate, control water solubility, and fit the mechanical property. To increase hydrophobicity, esterification was carried out to increase the hydrocarbon content of the added alcohol, resulting in tailored degradation rates since hydrophobicity directly influences hydration and the de-esterification reaction.[1,20] Another approach such as the condensation reaction between the carboxylic groups of unmodified hyaluronan molecules with the hydroxyl groups on the other hyalunonic acid has been performed to fabricate the sponge form. And then, bone marrow–derived mesenchymal progenitor cells were seeded to induce chondrogenesis and osteogenesis on this scaffold, resulting in successfully supported mesenchymal stem cell proliferation and differentiation for osteochondral application. Also, sulfation reaction onto hyaluronan gel has been investigated to create a variety of sulfated derivates ranging from one to four sulfate groups per disaccharide subunit. A cross-linking network, hydrogel, can be formed using diamines of individual hyaluronic acid chains together. Chondrocytes were seeded onto sulfated hyaluronic acid hydrogel and showed good cell compatibility for the tissue-engineered cartilage.

1.2.4.7 Chitosan

Chitosan, a polysaccharide derived from chitin, is composed of a simple glucosamine monomer and is similar in physicochemical properties to many glycosaminoglycans. It has been widely recognized that it is relatively biocompatible and biodegradable, does not evoke a strong immune response, and is relatively cheap due to its abundance and good reactivity with diverse methods of chemical processing. Chitin is typically extracted from arthropod shells by means of acid/alkali treatment to hydrolyze acetamido groups from N-acetylglucosamine, resulting in the production of chitosan. It has a molecular weight of 800,000~1,500,000 g/mole and dissolves easily compared with native chitin polymer.[1,5,7,20] For the application of the tissue-engineered cartilage, a 3D composite, such as chondroitin sulfate A/chitosan hydrogel scaffold, was prepared. This hydrogel supported the differentiated phenotype of seeded articular chondrocytes and type II collagen and proteoglycan production. Also, the organic/inorganic hybrid scaffold as the chitosan/tricalcium phosphate scaffold was fabricated for tissue-engineered bone. When osteoblastic cells collected from rat fetal calvary was seeded onto a chitosan/tricalcium phosphate scaffold, the cells proliferated in a multiplayer manner and deposited a mineralized matrix.

1.2.4.8 Agarose

Agarose is another type of marine-source polysaccharide purified extract from a sea creature such as agar or agar-bearing algae. One of the unique properties of agarose is the formation of a thermally reversible gel, which starts to set at concentrations in excess of 0.1% concentration at a temperature around 40°C and a gel-melting temperature of 90°C. Agarose gel has been widely used in electrophoresis for proteins and nucleic acids. Its good gelling behavior might be applicable to suitable injectable bone substitutes and a cell carrier matrix.[16] Allogenic chondrocyte–seeded agarose gels have been used as a model to repair osteochondral defects *in vivo*. The repaired tissues were scored histologically based on the intensity and extent of proteoglycan and type II collagen immunoassaying, structural features of the various cartilaginous zones, integration with host cartilage, and morphological features and arrangement of chondrocytic cells. The allogenic chondrocyte/agarose-grafted repairs had a higher semiquantitative score than control grafts. These results showed the good potential for the application of tissue engineering.[32] The more detailed studies such as the *in vivo* mechanical properties, biocompatibility and toxicity, and the balance degradation and synthesis kinetics of agarose-based tissue-engineered products must follow for the successful agarose application.

1.2.4.9 Acellular dermis

Acellular human skin removed of all cellular components may be one of the most significant ECMs for the scaffolds matrix. An acellular dermis may be seeded with fibroblasts and keratinocytes to fabricate a dermal-epidermal composite for the regeneration of skin. AlloDerm™ (LifeCell, Branchburgh, NJ, USA) is a typical commercialized product as a split-thickness acellular allograft prepared from human cadaver skin and cryopreserved for off-shelf use.[33] It has been successful in the treatment of burn patients due to a nonantigenic dermal scaffold that includes elastin, proteoglycan, and a basement membrane.

1.2.4.10 Polyhydroxyalkanoates

Polyhydroxyalkanoates are entirely natural and obtained from a microorganism *Alcaligen eutrophus*, a gram-negative bacteria. The physical properties of polyhydroxybutyrate (PHB) are similar to those of nondegradable polypropylene. Its copolymers with hydroxyvalerate—poly(hydroxybutylate-*co*-hydrovalerate) (PHBV) —have a modest range of mechanical properties and a correspondingly modest range of chemical compositions of each monomer and processing condition. Also, these polymers can be manufactured with many features, such as fibers, meshes, sponges, films, tubes, and matrices, through standard processing techniques due to good processability.

The family of polyhydroxyalkanoates shows no acute inflammation, abscess formation, or tissue necrosis in the tissue adjacent in the form of nonporous discs or cylinders.[1] In order to optimize the mechanical property of PHBV, an organic/inorganic hybrid composite such as PHBV/hydroxyapatite was developed for the tissue-engineered bone due to osteoconductive activity of hydroxyapaitite.[10] Also, Schwann cell–seeded PHB was applied for the regeneration of nerve in the shape of a conduit to guide and induce neonerve tissue at the nerve ends. Good nerve regeneration in PHB conduits in comparison with nerve grafts was observed. The shapes, mechanical strength, porosity, thickness, and degradation rate of PHB and its copolymers can be engineered.

1.2.4.11 Other natural polymers

Natural polymers except those discussed in earlier sections are proteins, albumin, gluten, elastin, fibroin, cellulose, starch, sclerolucan, elsinan, pectin (pectinic acid), galactan, curdlan, gellan, levan, emulsan, dextran, pullulan, heparin, chondroitin 6-sulfate, etc. Although they were not explained in this section, they are of interest due to being the most abundant biopolymers on earth and due to their unusual and useful functional properties. Typical properties are (i) biocompatibility and nontoxicity, (ii) easy processing as film and gel status, (iii) heat stability and thermal processability over a broad temperature range, and (iv) water solubility.[1,16] For the successful application for the regenerative medicine and tissue engineering scaffolds of these natural polymers, *in vivo* and *in vitro* experiments and physicochemical modification must be performed in the near future.

1.2.5 *Bioactive Molecules Release System for the Regenerative Medicine and Tissue Engineering*

Bioactive molecules as growth factors are polypeptides that transmit signals to modulate cellular activity and tissue development such as cell patterning, motility, proliferation, aggregation, and gene expression. As in the development of the tissue-engineered organs, regeneration of functional tissue requires maintenance of cell viability and differentiated function, encouragement of cell proliferation, modulation of the direction and speed of cell migration, and regulation of cellular adhesion. For example, transforming growth factor-β_1 (TGF-β_1) might be required to induce osteogenesis and chondrogenesis from bone marrow–derived mesenchymal stem cells. Also, brain-derived neurotrophic factor (BDNF) can be enhanced to regenerate spinal cord injury.

Also small molecules can control the differentiation of stem cells to specified cells. Hydroxybutylate or β-mercaptoethanol can be differentiated to neuronal cell from bone marrow–derived mesenchymal stem cells.

The easiest method for the delivery of the bioactive molecules is the injection near the site of cell differentiation and proliferation.[2,3] The most significant problem of the direct injection method of bioactive molecules is the relatively short half-life, the relatively high molecular weight and size, very low tissue penetration, and potential toxicity of systemic level.[34]

One promising way of the improvement technique of their efficacy is the locally controlled release of bioactive molecules for a desired release period by the impregnation into a scaffold. Through impregnation into a scaffold, protein structure and biological activity can be stabilized to a certain extent, resulting in prolonging the release time at the local site. The duration of bioactive molecules release from a scaffold can be controlled by the types of biomaterials used, the loading amount of cytokine, the formulation factors, and the fabrication process. The release mechanisms are largely divided into categories: (i) diffusion-controlled, (ii) degradation-controlled, and (iii) solvent-controlled release mechanism through the selection of biomaterials. The mechanism of biodegradable scaffolds materials was degradation controlled, whereas that of the nondegradable one was diffusion and/or solvent controlled. The desired release pattern such as constant, pulsatile, and time-programmed behaviors along the specific site and the type of injury can be achieved by the appropriate combination of these mechanisms. Also, the cytokine release system might be designed in a variation with geometries and configurations such as scaffold, tube, nose, microsphere, injectable forms, fiber, etc.[1,10] One of the serious problems during the fabrication of a cytokine-loaded scaffold is the denaturation and deactivation of cytokines, resulting in the loss of biological activity. Hence, an optimized method must be developed for a stabilized cytokine release scaffold.

Another available emerging technology is the tethering to the surface, that is, immobilization of protein on the surface of the scaffold matrix. Immobilization of insulin and transferrin to the poly(methylmetacrylate) films stimulates the growth of fibroblast cell compared to the same concentrations of soluble or physically adsorbed proteins. For the enhancement of cytokine activity, a PEO chain was applied as a short spacer between the surface of the scaffold and the cytokine. Tethered EGF, immobilized to the scaffold through the PEO chain, showed more improved DNA synthesis or cell rounding compared to the physically adsorbed EGF surface.[35]

Conjugation of cytokine with an inert carrier prolongs the short half-life of protein molecules. Inert carriers are albumin, gelatin, dextran, and PEG. PEGylation—that means PEG-conjugated cytokine—is most widely used for the release. It appears to decrease the rate of cytokine degradation, attenuate the immunological response, and reduce clearance by the kidneys.[1,35] Also, the PEGylated cytokine can be impregnated into scaffold materials by physical entrapment for sustained release. This conjugation method can be applied to the delivery of proteins and peptides. Immobilized arginin-glycine-aspartic acid (RGD) and tyrosin-leucine-glycine-serine-arginine (YIGSR), which are typical ECM proteins, onto the biomaterials can enhance cell viability, function, and recombinant products in cell.

Gene-activating scaffolds are being designed to deliver to targeted genes, resulting in the stimulation of specific cellular responses at the molecular level.[2,3] Modification of bioactive molecules with resorbable biomaterials systems obtains specific interactions with cell integrins, resulting in cell activation. These bioactive bioglasses and macroporous scaffolds also can be designed to activate genes that stimulate regeneration of living tissue.[9] Gene delivery would be accomplished by complexation with positively charged polymers, encapsulation, and gel by means of a scaffold structure. Methods of gene delivery for gene-activating scaffolds are almost the same as those of proteins, drugs, and peptides.

1.3 Scaffold Fabrication and Characterization

1.3.1 *Fabrication Methods of Scaffolds*

Engineered scaffolds may enhance the functionalities of cells and tissues to support the adhesion and growth of a large number of cells by providing a large surface area and pore structure within a 3D structure. Porosity needs provide enough space, permit cell suspension, and penetrate the 3D structure. Also, these porous structures provide to promote ECM production, to transport nutrients from nutrient media, and to excrete waste products.[1,5,7,9,10] From these points of view, an adequate pore size and a uniformly distributed and interconnected pore structure to allow for easy distribution of cells throughout the scaffold structure are very important. The scaffold structure is directly related to fabrication methods. Over 20 methods have been proposed.[36–38]

1.3.1.1 Electrospinning method

This technique is relatively old and normally engaged in polymer engineering to spin small-diameter fibers. Very recently, many researchers have applied this technique to scaffolds in the regenerative medicine area due to its potential as mimicking Mother Nature. The morphology and architecture, thickness, two- or 3D structure, surface micro- and nano-porosity, and orientation of the spun fiber can be controlled by types of polymers, solvents, devise settings, etc. Nano-electrospinning of PGA, PLA, PLGA, caprolactone copolymers, collagen, elastin, etc., has been extensively developed, as depicted in section 1.4.[5,38] For example, electrostatic processing can consistently produce PGA fiber diameters at or below 1 μm. By controlling the pickup of these fibers, the orientation and mechanical properties can be tailored to a specific need of the injured site. Also, collagen electrospinning was performed utilizing type I collagen dissolved in 1,1,1,3,3,3-hexafluoro-2-propanol with 0.083 g/mL concentration. The optimally electrospun type I collagen nonwoven fabric appeared with an average diameter of 100 ± 40 nm, resulting in biomimicking fibrous scaffolds.

Bioactive molecules-loaded electrospinning nanofibers could be also fabricated by the controlling of the manufacturing factor; thus the application of this ECM-mimicking scaffolds has been expanded as multifunctional regenerative medicine and tissue engineering.

1.3.1.2 PGA nonwoven sheet

The most common and commercialized one is the PGA nonwoven sheet with porosity around 97% and 1~5 mm thickness, which is tested almost of tissue-engineered organs. The fiber diameter is 13~14 μm, and it will be swelling to 20 μm after the immersion in a cell culture medium. The monofilament of PGA for the suture was fabricated into a nonwoven sheet by needle punch. In order to stabilize dimensionally and provide mechanical integrity, the fiber-bonding technology by heat and by PLGA or PLA solution spray-coating methods were developed.[39] Figure 1.4 shows the shape of nose and ear tissue-engineered products as hybrid reticular cartilage from rabbit chondrocytes and PGA nonwoven fibers. Recently, fibrous woven and knitted 3D scaffolds have been suggested.

1.3.1.3 Porogen-leaching methods

Porogen-leaching methods are combined with polymerization, solvent casting, gas foaming, or compression molding of natural and synthetic scaffolds biomaterials with leaching of pore-generating particles such as sodium chloride crystals, sodium tartrate, and sodium citrate sieved using a molecular sieve.[1,13,15,25,27,28,31,34] These methods have been broadly used to make 3D scaffolds in the whole area of the scaffolds of tissue engineering.

PLGA, PLA, collagen, poly(*ortho* ester), or SIS- and DBP-impregnated PLGA scaffolds were successfully fabricated with a biodegradable sponge structure by this method with above 93% porosity and desired pore size of 1,000 μm. Using the solvent-casting/particulate-leaching method, complex geometries such as tube, nose, and specific organ types, not so much as nano-composite hybrid scaffolds, could be fabricated by means of conventional polymer processing techniques like calendaring, extrusion, and injection. Complex geometry can be fabricated from porous film lamination.[39] The advantage of this method is easy control of porosity and geometry. However, disadvantages of this method are (i) the loss of water-soluble biomolecules or cytokines during leaching of the porogen, (ii) the possibility of porogen remaining as salt that can harmfully affect cell culture, (iii) the different geometry surface and cross section, and (iv) pore tortuosity.

1.3.1.4 Gas-foaming method

The gas-foaming method is a sudden expansion of CO_2 gas under high pressure, resulting in the formation of a sponge structure because of the nucleation and expansion in the dissolved CO_2 scaffold matrix. PLGA scaffolds with above 93% porosity and around 100 μm median pore size were developed by this method.[40] The significant advantage is no loss of bioactive molecules in the scaffold matrix due to no need for the leaching process and no residual organic solvent, whereas the disadvantage is the presence of skimming film layers on the scaffold surface, resulting in another process to further remove this skin layer.

1.3.1.5 Phase separation method

The phase separation method is divided into freeze-drying, freeze-thawing, freeze-immersion precipitation, and emulsion freeze-drying.[14] Phase separation by freeze-drying can be induced in a polymer solution with appropriate concentration by rapid freezing. Then the used solvent is removed by freeze-drying, resulting in a porous structure as a portion of the solvent. There are collagen scaffolds with pores between 50~150 μm, collagen-glycosaminoglycan blend scaffolds with an average pore size between 90~120 μm, and chitosan scaffolds with a pore size from 1~250 μm varied with freezing conditions.[41] Also, scaffold structures of synthetic polymers such as PLA or PLGA were successfully made with over 90% porosity and 15~250 μm size by this phase separation method. The freeze-thaw technique induces phase separation between a solvent and a hydrophilic monomer upon freezing, followed by the polymerization of the hydrophilic monomer by means of ultraviolet irradiation and removal of the solvent by thawing. This leads to the formation of a macroporous hydrogel. A similar method is freeze-immersion precipitation. A polymer solution is cooled, immersed in a nonsolvent, and then vaporized, leading to a porous scaffold structure. Also, the emulsion freeze-drying method is useful for the fabrication of a porous structure, as shown in chapter 33. Mixtures of a polymer solution and a nonsolvent were thoroughly sonicated, frozen quickly in liquid nitrogen at –198°C, and then freeze-dried, resulting in a sponge structure. The advantage of these techniques is the loading of hydrophilic or hydrophobic bioactive molecules, whereas the disadvantages are a relatively small pore size and difficulty in controlling the precise pore structure.[1,5,14]

1.3.1.6 Rapid prototyping

There are several rapid prototype methods for the scaffold application, such as solid free form (SFF, chapters 25 and 31), 3D printing™ (3DP, chapter 29), selective laser printing (SLS, chapter 45), selective laser ablation (SLA) and stereolithography (STL), and 3D fiber deposition (3DF) such as fused deposition modeling (FDM). These methods reveal to be the most promising to satisfy many of the general requirements for scaffold biomaterials. Rapid prototype skills with highly sophisticated computer-aided design (CAD)/computer-aided manufacturing (CAM) robotic units can fabricate a fine-tunable porosity, pore size and shape, and a completely interconnected pore network that might facilitate

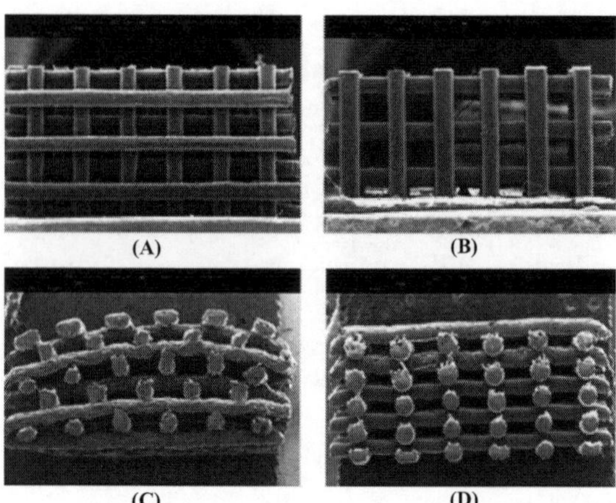

Figure 1.8. An SPCL scaffold with different geometry fabricated by the rapid prototype method for the application of the regeneration of the spinal cord.

better cell migration and nutrient perfusion. Figure 1.8 shows a starch-polycaprolactone (SPCL) scaffold with different geometry fabricated by the rapid prototype method for the application of the regeneration of the spinal cord. After seeding the olfactory ensheathing cells, cells in SPCL scaffolds proliferated continuously for 28 days and no adverse effects of SPCL were observed. It seems that the SPCL scaffolds with a highly porous microstructure favor nutrient supply, waste removal, and cell migration, facilitating cell proliferation.[42]

One of the first rapid prototyping devices to be developed for the tissue engineering application was 3DP. Briefly, the desired feature of a diseased organ form can be fabricated by depositing by a CAD/CAM-controlled manner a jet of solvent on top of a polymer powder bed. The solvent binds the powder, resulting in the formation of a pattern of fibers, built layer by layer. Similary, SLS consists of projecting a laser beam that can sinter the powder by local high temperature on a polymeric powder bed. Laser ablation is carried out to ablate the scaffold materials in a specific form using a laser beam.

These techniques are able to fabricate the periodic structures with well-defined, controlled, and completely interconnected porosity.[38]

3DF is an FDM technique as molten polymeric biomaterials, hydrogel, and paste biomaterials are extruded from a CAM-controlled robotic unit on a stage in the form of a fiber. These filaments are deposited to form a layer, and a 3D scaffold could be built with a layer-by-layer strategy following a CAD pattern. Different fiber architecture might be performed by altering the angle of deposition between subsequent layers, resulting in different pore shapes. Recently, in order to improve more delicate scaffolds, a multidispensing system was used for depositing different materials at the same time to produce constructs with different physicochemical properties. The drawback of this method might be the high temperatures involved during fabrication of molten polymers.

1.3.1.7 Injectable gel method

Injectable gel scaffolds also have been reported and explained in section 1.3. An injectable, gel-forming scaffold may provide several advantages such as it (i) can fill any shape of defects due to flowable materials, (ii) may load various types of bioactive molecules and cells by simple mixing, (iii) does not contain residual solvents that may be present in a performed scaffold, and (iv) does not require a surgical procedure for placement. Typical examples are thermosensitive gels such as Pluronics and PEG-PLGA-PEG triblock copolymers, pH-sensitive gels such as chitosan and its derivates, ionically cross-linked gels such as alginate, and fibrin gels, hyaluronan gels, etc., as already introduced in section 1.2. In the near future, multifunctional gels such as tissue-specific injectable scaffolds materials with very fast sol-gel transition and that can be fully degradable for a desired period will be present.

Also, newly hybridized fabrication techniques such as organic/inorganic and synthetic/natural at the nanosized level to biomimic are continuously developed for the application of tissue-engineered scaffolds.

1.3.2 *Physicochemical Characterization of Scaffolds*

For the successful achievement of 3D scaffolds, several characterization methods are needed. These can be divided into four categories: (i) morphology like porosity, pore size, and surface area, (ii) mechanical properties like compressive and tensile strength, (iii) bulk properties like degradation and its relevant mechanical properties, and (iv) surface properties like surface energy, chemistry, and charge.

Porosity is defined as the fraction of the total volume occupied by voids and appears in percentage. Most widely used methods for the measuring of porosity are mercury porosimetry, scanning electron microscopy, and confocal laser microscopy.

Mechanical properties are extremely important when designing tissue-engineered products. To determine the mechanical properties of a porous structure, conventional testing instruments can be used. Mechanical tests can be divided into (i) creep tests, (ii) stress-relaxation tests, (iii) stress-strain tests, and (iv) dynamic mechanical tests. These test methods are similar to those of conventional biomaterials everywhere.[1,5,7]

The rate of degradation of manufactured scaffolds is one of the most important factors for designing tissue-engineered products. Ideally, the scaffold constructs provide mechanical and biochemical supports until entire tissue regeneration occurs without any changes, and then completely biodegrade at a rate consistent with tissue generation. Immersion studies are commonly conducted to track the degradation of a biodegradable matrix. So, the changes of weight loss and molecular weight can be evaluated by chemical balance, scanning electron microscopy, and gel permeation chromatography. From these results, we can expect the mechanism of biodegradation.

It is generally recognized that the adhesion and proliferation of different types of cells on polymeric materials depend largely on surface characteristics such as wettability (hydrophilicity/hydrophobicity of surface free energy), chemistry, charge, roughness, and rigidity. Especially, 3D applications for tissue engineering are more important to cell migration, proliferation, DNA/RNA synthesis, and phenotype presentation on the scaffold materials. Surface chemistry and charge can be analyzed by electron-scanning chemical analysis and streaming potential, respectively. Also, wettability of the scaffold surface can be measured by contact angle with static and dynamic methods.

1.3.3 *Sterilization Method for Scaffolds*

Sterilizability of polymeric scaffold biomaterials is an important aspect of the properties because, especially, polymers have lower thermal and chemical stability than other materials such as ceramics and metals, and consequently, they are more difficult to sterilize using conventional techniques. Commonly used sterilization techniques are dry heat sterilization, autoclaving, radiation, and ethylene oxide gas (EOG) sterilization. Also, plasma glow discharge and electron beam sterilization have been recently proposed due to their conveniences[5].

In dry heat sterilization, the temperature varies between 160°C and 190°C. This is above the melting and softening temperatures of many linear polymers like PLGA, resulting in shrinking of the scaffold dimension. PLA scaffolds were sterilized by 129°C for 60 seconds, resulting in minimal change in tensile properties. One of the significant problems was the decrease of molecular weight, which might affect the degradation kinetics of the polymers. The only polymers that can safely be dry-sterilized are polytetrafluoroethylene (PTFE) and silicone rubber. However, ceramic and metallic scaffolds were safe in the range of this temperature. Steam sterilization (autoclaving) is performed under high steam pressure at relatively low temperature (125°C–130°C). In the case of the family of poly(α-hydroxy ester)s, the trace of water can deteriorate the PLGA backbone.

Chemical agents such as EOG and propylene oxide gases and phenolic and hypochloride solutions are widely used for sterilizing all biomaterials since they can be used at relatively low temperatures. Chemical agents sometimes cause polymer deterioration even when sterilization takes place at room temperature. However, the time of exposure is relatively short (overnight), and almost all of the scaffolds can be sterilized with this method. Especially, the cold EOG sterilization method is the most widely method in the condition of 35°C and 95% humidity. While the hot EOG method of 60°C and 95% humidity can cause shrinkage of PLGA scaffolds. One of the significant problems is the residual EOG caused harm on the surface and within the polymer. So, it is important that the scaffolds be subjected to adequate degassing or aeration subsequent to EOG sterilization so that the concentration of residual EOG is reduced to acceptable levels.

Radiation sterilization using isotopic ^{60}Co can also deteriorate polymers since at high dosage the polymer chains can be dissociated or cross-linked according to the characteristics of the chemical structures. At a 2.5 Mrad dose, the tensile strength and molecular weight of PLGA decreased. Also, there is a rapid decrease in the molecular weight of PGA nonwoven felt with increasing doses of radiation. Thus, it is very important to bear in mind that the properties and useful lifetime of the PLGA implanted can be significantly affected by irradiation. The physical properties continued to deteriorate with time following irradiation.

Sterilization methods might significantly affect the physicochemical properties of the scaffold matrix. The specific effects with various methods are determined by the kinds of scaffolds materials themselves, the scaffold preparation methods, and the sterilization factors. It is essential that a new standardization for sterilizing scaffold devices be designed and established.

1.4 Conclusions

Very recently, regenerative medicine and tissue engineering show tremendous potential as revolutionary research pushes. Also, many successful results have been reported in the potential for regenerating tissues and organs such as skin, bone, cartilage, nerve of peripheral and central nerves, tendon, muscle, cornea, bladder and urethra, and liver, as well as composite systems like a human phalanx and joint on the basis of "intelligent" scaffold biomaterials from polymers, ceramic, metal, composites, and their hybrids.[41]

As previously emphasized, scaffold materials must contain the site of cellular and molecular induction and adhesion and must allow for the migration and proliferation of cells through porosity. They should also maintain strength, flexibility, biostability, and biocompatibility to mimic a more natural, 3D environment. With the advances of more sophisticated CAD/CAM techniques, we can mimic the nature of the human body's structure. Also, the combination of the cells and redesigned bioactive scaffolds has attempted to expand to a tissue level of hierarchy.

From this point of view, novel scaffold biomaterials, novel scaffolds fabrication methods, novel characterization methods, and a translational clinical protocol must be developed.

Appendix I: Published Textbook in the Area of Regenerative Medicine and Tissue Engineering

(1) X. Peter and H. Jennifer, *Scaffolding in Tissue Engineering* (CRC PrILlc, New York, 2005).

(2) E. Regine, E. Dieter, and P. Ralf, *Cell and Tissue Reaction Engineering* (Springer Verlag, New York, 2008).

(3) A. Haverich and H. Graf, *Stem Cell Transplantation and Tissue Engineering* (Springer Verlag, New York, 2002).

(4) O. P. Bernhard and N. B. Sangeeta, *Tissue Engineering* (Prentice Hall, New Jersey, 2003).

(5) D. Shi, *Biomaterials and Tissue Engineering* (Springer Verlag, New York, 2004).

(6) L. Qin, *Advanced Bioimaging Technologies in Assessment of the Quality of Bone and Scaffold Materials* (Materials Research Society, San Francisco, 2007).

(7) L. Rui and J. S. Roman, *Biodegradable Systems in Tissue Engineering and Regenerative Medicine* (CRC PrILlc, New York, 2004).

(8) G. Khang, M. S. Kim, and H. B. Lee, *A Manual for Biomaterials/Scaffold Fabrication Technology* (World Scientific, Singapore, 2007).

(9) C. Laurencin and L. Nair, *Nanotechnology and Tissue Engineering* (CRC PrILlc, New York, 2008).

(10) K. Lewandrowski and D. Wise, *Tissue Engineering and Biodegradable Equivalents: Scientific and Clinical Applications* (Marcel Dekker, New York, 2001).

(11) Kumar and S. S. R. Challa, *Tissue, Cell and Organ Engineering* (John Wiley & Sons, New York, 2007).

(12) M. D. Michael, *Introduction to Biomedical Engineering* (Prentice Hall, New Jersey, 2010).

(13) H. Vasif, K. Lewandrowski, and J. Y. Michael, *Tissue Engineering and Novel Delivery Systems* (Marcel Dekker, New York, 2003).
(14) A. Anthony and R. P. Lanza, *Methods of Tissue Engineering* (Academic Press, London, 2001).
(15) C. Steven, *Advanced Biomaterials-Characterization, Tissue Engineering, and Complexity* (Materials Research Society, San Francisco, 2007).
(16) L. L. Hench and J. R. Jones, *Biomaterials, Artificial Organs and Tissue Engineering* (CRC PrILlc, New York, 2005).
(17) Bronzino and D. Joseph, *Tissue Engineering and Artificial Organs* (CRC PrILlc, New York, 2007).
(18) V. M. Goldberg and A. I. Caplan, *Orthopedic Tissue Engineering* (Taylor & Francis, London, 2007).
(19) N. Dib, D. A. Taylor, and E. B. Diethrich, *Stem Cell Therapy and Tissue Engineering for Cardiovascular Repair* (Springer Verlag, New York, 2007).
(20) G. Farshid, *Functional Tissue Engineering* (Springer Verlag, New York, 2003).
(21) P. H. Anthony and V. H. Paul, *Biopolymer Methods in Tissue Engineering* (Humana Press, New Jersey, 2004).
(22) A. K. Dillow and A. M. Lowman, *Biomimetic Materials and Design: Biointerfacial Strategies, Tissue Engineering, and Targeted Drug De* (Marcel Dekker, New York, 2002).
(23) H. Mori and H. Matsuda, *Cardiovascular Regeneration Therapies Using Tissue Engineering Approaches* (Springer Verlag, New York, 2005).
(24) J. W. Fluhr and P. Elsner, *Tissue Engineering in Dermatology* (Skarger, Basel, 2009).
(25) J. Denstedt, *Biomaterials and Tissue Engineering in Urology* (CRC Press, New York, 2009).
(26) J. J. Mao and A. Mikos, *Translational Approaches in Tissue Engineering and Regenerative Medicine* (Artech House, London, 2007).
(27) A. Anthony, *Foundations of Regenerative Medicine: Clinical and Therapeutic Applications* (Academic Press, New York, 2009).
(28) S. Matteon, *Strategies in Regenerative Medicine* (Springer Verlag, New York, 2008).

(29) T. Philippe, *Stem Cells and Regenerative Medicine* (Nova Science, New York, 2008).

(30) A. Julie, L. William, and Stanford, *Stem Cells in Regenerative Medicine: Methods and Protocols* (Humana Press, New York , 2008).

(31) L. Walter, *Stem Cells and Regenerative Medicine* (World Scientific, Singapore, 2007).

(32) K. Ursulal and F. Alimorad, *Orthopedic Regenerative Medicine: A Nutritional Guide* (Authorhouse, Bloomington, 2009).

(33) W. Carsten, *Polymers for Regenerative Medicine* (Springer Verlag, New York, 2007).

(34) A. Atala and R. Lanza, *Principles of Regenerative Medicine* (Academic Press, New York, 2007).

(35) Y. V. Ioannis, *Regenerative Medicine* (Springer Verlag, New York, 2005).

(36) D. Sheng, *Chemical and Functional Genomic Approaches to Stem Cell Biology and Regenerative Medicine* (John Wiley & Sons, Manhattan, 2007).

(37) M. Kusano and S. Shioda, *New Frontiers in Regenerative Medicine* (Springer Verlag, New York, 2006).

(38) K. Lee and D. L. Kaplan, *Tissue Engineering II* (Springer Verlag, New York, 2007).

(39) J. Morser and S. I. Nishikawa, *The Promises and Challenges of Regenerative Medicine* (Springer Verlag, New York, 2007).

(40) S. David, *Regenerative Biology and Medicine* (Academic Press, New York, 2007).

(41) S. Robin and S. Suzanne, *Stem Cell Medicine: The New Adult Stem Cell Regenerative Therapy for Cancer, Spinal Injuries, Multip* (Hatherleigh Press, London,2009).

(42) K. Suzanne, *Umbilical Cord Blood: A Future for Regenerative Medicine* (World Scientific Pub Co Inc, Singapore, 1990).

Appendix II: Periodically Published Journal in the Area of Regenerative Medicine and Tissue Engineering

(1) *Advanced Drug Delivery Reviews* (Elsevier Science BV, 60(2), Netherlands).

(2) *Journal of Biomedical Materials Research. Part A and B* (John Wiley & Sons, 84A(3), Unites States).
(3) *Advanced Materials* (Wiley-VCH, 21(32/33), Unites States).
(4) *Progress in Polymer Science* (Pergamon-Elsevier Science, 36(2), Unites States).
(5) *AIChE Journal* (John Wiley & Sons, 54(12), Unites States).
(6) *Biomaterials* (Elsevier SCI, 31(14), Netherlands), http://www.biomaterials.org/
(7) *The Journal of Surgical Research* (Academic Press Elsevier Science, 144(1), Unites states)
(8) *Biotechnology Advances* (Pergamon-Elsevier Science, 26(1), England).
(9) *Biopolymers* (John Wiley & Sons, 89(5), Unites States).
(10) *Tissue and Cell* (Churchill Livingstone, 26(5), Scotland).
(11) *Journal of Tissue Viability* (Elsevier Science, 19(2), England).
(12) *Materials Science and Engineering* (Elsevier Science SA, 59(1–6), Switzerland).
(13) *Chemical and Biophysical Research Communications* (Academic Press Elsevier Science, 345(2), Unites States).
(14) *Journal of Bioscience and Bioengineering* (Soc Bioscience Bioengineering Japan, 108(4), Japan).
(15) *Journal of Biomechanics* (Elsevier Science, 40(2), England).
(16) *European Journal of Cardio-Thoracic Surgery* (Elsevier Science Bv, 19(4), Netherlands).
(17) *Pharmacology & Therapeutics* (Wiley-Blackwell Publishing, 105(2), Unites States).
(18) *Acta Biomaterialia* (Elsevier Science Ltd, 7(4), England).
(19) *International Review of Cytology* (Elsevier Academic Press, 262, Unites States).
(20) *The Journal of Nutritional Biochemistry* (Elsevier Science, 7(8), Unites States).
(21) *Trends in Biotechnology* (Elsevier Science London, 20(8), Netherlands).
(22) *Journal of Plastic, Reconstructive & Aesthetic Surgery* (Elsevier Science, 62(4), England).
(23) *Journal of Controlled Release* (Elsevier Science Bv,64(1–3), Netherlands).

(24) *British Journal of Plastic Surgery* (Elsevier Science, 58(8), England).
(25) *Gastroenterology* (W B Saunders Co-Elsevier, 129(3), United States).
(26) *International Journal of Biochemistry* (Pergamon-Elsevier Science, 3(17), England).
(27) *International Journal of Oral and Maxillofacial Surgery* (Churchill Livingstone, 35(10), Scotland).
(28) *International Journal of Tissue Regenerations* (Hanrimwon, Korea, ijtr@catholic.ac.kr).

References

1. G. Khang, S. J. Lee, M. S. Kim, and H. B. Lee, Biomaterials: Tissue-engineering and scaffolds, in Ed. S. Webster, *Encyclopedia of Medical Devices and Instrumentation,* 2nd ed. (John & Wiley Press, New York, 2006), pp. 366–383.
2. S. Petit-Zeman, *Nature Biotech.*, **19**, 201, 2001.
3. L. G. Griffith and G. Naughton, *Science*, **295**, 1009, 2002.
4. P. Shah, A. Hillel, R. Silverman and J. Elisseeff, Cartilage tissue engineering, in *Eds.* A. Atala, R. Lanza, J. Thomson, and R. Nerem, *Principles of Regenerative Medicine* (Academic Press, Burlington, 2008), pp. 1176–1197.
5. G. Khang, M. S. Kim and H. B. Lee, in Eds. G. Khang, M. S. Kim, and H. B. Lee, *A Mannual for the Fabrication of Tissue Engineered Scaffolds* (World Scientific, Singapore, 2007).
6. M. E. Furth and A. Atala, Current and future perspectives of regenerative medicine, in Eds. A. Atala, R. Lanza, J. Thomson, and R. Nerem, *Principles of Regenerative Medicine* (Academic Press, Burlington, 2008), pp. 2–15.
7. H. B. Lee, G. Khang, and J. H. Lee, Polymeric biomaterials, in Eds. J. B. Park and J. D. Bronzino, *Biomaterials: Principles and Applications* (CRC Press, Boca Raton, FL, 2003).
8. W. G. Billotte, Ceramic biomaterials, in Eds. J. B. Park and J. D. Bronzino, *Biomaterials: Principles and* Applications (CRC Press, Boca Raton, FL, 2003).
9. L. L. Hench and J. M. Polak, *Science*, **295**, 1014, 2002.

10. F. R. A. Rose and R. O. C. Oreffo, *Biochem. Biophys. Res. Commun.*, **292**, 1, 2002.
11. J. C. Frician, R. Bareille and F. Rouais, *J. Dent. Res.*, **77**, 406, 1998.
12. G. Khang, C. S. Park, J. M. Rhee, S. J. Lee, Y. M. Lee, I. Lee, M. K. Choi, and H. B. Lee, *Macromol. Res.*, **9**,267, 2001.
13. S. J. Yoon, S. H. Kim, H. J. Ha, Y. K. Ko, J. W. So, M. S. Kim, Y. I. Yang, G. Khang, and H. B. Lee, *Tissue Eng.*, **14(4)**, 539, 2008.
14. G. Khang and H. B. Lee, Cell-synthetic surface interaction: Physicochemical surface modifications, in Eds. A. Atala and R. Lanza, *Principles of Tissue Engineering* (Academic Press, Orlando, FL, 2001).
15. H. L. Kim, S. J. Kim, H. Yoo, M. Hong, D. Lee, and G. Khang, *Intern. J. Tissue Regen.*, **1(2)**, 81 (2010).
16. W. H. Wong and D. J. Mooney, Synthesis of properties of biodegradable polymers used as synthetic matrices for tissue engineering, in Eds. A. Atala and D. J. Mooney, *Synthetic Biodegradable Polymer Scaffolds* (Birkhauser, Boston, MA, 1996).
17. L. J. Suggs, R. S. Krishna, C. A. Garcia, S. J. Peter, J. M. Anderson, and A. G. Mikos, *J. Biomed. Mater. Res.*, **42**, 312, 1998.
18. J. M. Harris, Ed., *Poly(ethylene glycol) Chemistry: Biotechnical and Biomedical Applications* (Plenum, New York, NY, 1997).
19. M. H. Kim, H. N. Hong, J. P. Hong, C. J. Park, S. W. Kwon, S. H. Kim, G. Khang, and M. Kim, *Biomaterials*, **31(6)**, 1213, 2010.
20. B. L. Seal, T. C. Otero, and A. Panitch, *Mater. Sci. Eng.*, **R34**, 147, 2001.
21. S. J. Holland and B. J. Tighe, *J. Cont. Rel.*, **4**, 155, 1986.
22. Y. Song, J. Kwon, B. Kim, Y. Jeon, G. Khang, and D. Lee, *J. Biomed. Mater. Res. A.*, 2011.
23. S. Agarwal, R. Gassner, N. P. Piesco, and S. R. Ganta, Biodegradable urethanes for biomedical applications, in Eds. K-U. Lewandrowski, D. L. Wise, D. J. Trantolo, J. D. Gresser, M. J. Yasemski, and D. E. Altobeli, *Tissue Engineering and Biodegradable Equivalents: Scientific and Clinical Applications* (Marcel Dekker, New York, NY, 2002).
24. C. J. Dunn and K. L. Goa, *Drugs*, **58**, 863, 1999.
25. M. Sha'ban, S. J. Yoon, Y. K. Ko, H. J. Ha, S. H. Kim, J. W. So, R. B. H. Idrus, and G. Khang, *J. Biomater. Sci. Polymer Ed.*, **19(9)**, 1219, 2008.
26. S.-T. Li, Chap. 6, Biologic biomaterials: Tissue-derived biomaterials (Collagen), in Eds. J. B. Park and J. D. Bronzino, *Biomaterials: Principles and Applications* (CRC Press, Boca Raton, FL, 2003).

27. H.-S. Park, D.-S. Ham, Y.-H. You, J. Shin, J.-W. Kim, J.-H. Jo, O. Y. Kim, G. Khang, and K.-H. Yoon, *Tissue Eng. Regen. Med.*, **7(5)**, 523, 2010.
28. S. J. Kim, E. H. Jo, O. Y. Kim, E. Y. Lee, J. E. Song, D. Lee, and G. Khang, *Intern. J. Tissue Regen.*, **1(2)**, 94, 2010.
29. S. F. Badylak, R. Record, K. Lindberg, J. Hodde, and K. Park, *J. Biomater. Sci., Polym. Ed.*, **9**, 863, 1998.
30. Y. Wang, H.-J. Kim, G. Vunjak-Novakovic, and D. L. Kaplan, *Biomaterials*, **27**, 6064, 2006.
31. E. H. Jo, S. J. Kim, S. J. Cho, G. Y. Lee, O. Y. Kim, E. Y. Lee, W. H. Cho, D. Lee, and G. Khang, *Polymer(Korea), in press*, 2011.
32. D. A. Lee, S. P. Frean, P. Lee, and D. L. Bader, *Biochem. Biophys. Res. Commun.*, **251**, 1998.
33. C.-J. Gustafson and G. Katz, *Burns*, 25, 331, 1999.
34. C. Kim, S. H. Kim, A Y. Oh, J. M. Rhee, and G. Khang, *Polymer(Korea)*, **32(6)**, 529 (2008).
35. S. Y. Lee, J. Lim, G. Khang, Y. Son, P.-H. Choung, S.-S. Kang, S. Y. Chun, H.-I. Shin, S.-Y. Kim, and E. K. Park, *Tissue Eng. Part A*, **15(9)**, 2491, 2009.
36. B. E. Chaignaud, R. Langer, and J. P. Vacanti, The history of tissue engineering using synthetic biodegradable polymer scaffolds and cells, in Eds. A. Atala and D. J. Mooney, *Synthetic Biodegradable Polymer Scaffolds* (Birkhauser, Boston, MA, 1996).
37. T. M. Freyman, I. V. Yannas, and L. J. Gibson, *Progress Mater. Sci.*, **46**, 273–282, 2001.
38. L. Moroni, J. R. de Wijin, and C. A. van Blitterswijk, *J. Biomater. Sci., Polym. Ed.*, **19(5)**, 543, 2008.
39. R. C. Thompson, M. C. Wake, M. J. Yasemski, and A. G. Mikos, *Adv. Polym. Sci.*, **122**, 245, 1995.
40. J. J. Barry, H. S. Gidden, C. A, Scotchford, and S. M. Howdle, *Biomaterials*, **25**, 3559, 2004.
41. C. J. Woolverton, J. A. Fulton, S. T, Lopina, and W. J. Landis, Mimicking the natural tissue environment, in Eds. K-U. Lewandrowski, D. L. Wise, D. J. Trantolo, J. D. Gresser, M. J. Yasemski, and D. E. Altobeli, *Tissue Engineering and Biodegradable Equivalents: Scientific and Clinical Applications* (Marcel Dekker, New York, NY, 2002).
42. D. Lee, Y. Song, C. Kim, H. Yoo, N. A. Silva, R. A. Sousa, C. M. Alves, A. J. Salgado, N. Sousa, R. L. Reis, and G Khang, *Intern. J. Tissue Regen.*, **1(1)**, 21, 2010.

Part II

CERAMIC AND METAL SCAFFOLD

Chapter 2

INNOVATIVE BIOINSPIRED SIC CERAMICS FROM VEGETABLE RESOURCES

M. López-Álvarez,[a] P. González,[a*] J. Serra, A. de Carlos,[b] S. Chiussi,[a] and B. León[a]

[a] *Department of Applied Physics, University of Vigo*
[b] *Department of Biochemistry, Genetics and Immunology, University of Vigo, Campus Lagoas-Marcosende, 36310 Vigo, Spain*
*pglez@uvigo.es

Bioinspired silicon carbide (SiC) ceramics is a very promising material for bone tissue engineering and regenerative medicine applications, due to its outstanding mechanical properties, interconnected hierarchic porosity, and biocompatible behavior. This innovative material is derived from vegetable resources, as woods, algae, and plants, following a ceramization process based on the molten-Si infiltration of carbon templates obtained by controlled pyrolysis of vegetable precursors. The final SiC ceramics retains the combined macro- and microporosity of the original vegetable structure, which resembles the natural hierarchical structure of bone tissue.

The *in vitro* biocompatibility of the SiC ceramics obtained from sapelli wood (*Entandrophragma cylindricum*) was evaluated by using the preosteoblastic cell line MC3T3-E1. Scanning electron microscopy (SEM) and confocal laser scanning microscopy (CLSM) demonstrate that cells seeded onto the SiC ceramics were able to attach, spread, and proliferate properly with the maintenance of

Handbook of Intelligent Scaffolds for Tissue Engineering and Regenerative Medicine
Edited by Gilson Khang

www.panstanford.com

the typical osteoblastic morphology throughout the time of culture. Alizarin Red staining and quantification revealed a higher and earlier level of differentiation on SiC ceramics than on tissue culture polystyrene (TCP) used as reference. These results demonstrate that the porous and hierarchical microstructure of SiC promotes the differentiation of osteoblastic cells.

2.1 Introduction

This chapter deals with the development of an innovative, bioinspired biomaterial based on SiC ceramics derived from vegetable resources, which mimics the interconnected hierarchic porosity of the bone structure, being a promising approach for the generation of porous ceramic scaffolds for bone tissue engineering and regenerative medicine.

In the last decade much effort has been made to engineer ideal scaffolds for bone tissue regeneration (high porosity, proper pore size, biocompatibility, biodegradability, osteoinductivity, etc.), but none of the current materials fulfills all demands. Thus, a broad range of solutions have been developed for each particular function, for instance, devices with high mechanical stability for large bone defects in load-bearing long bones and moldable or injectable materials for craniofacial surgery.[1–4]

The most intriguing aspects in scaffold design are due to the fact that bone is a highly hierarchical three-dimensional (3D) composite structure composed of an organic part (collagen, cells, and proteins) and an inorganic component formed by specific phases of calcium and phosphorous, especially needle- or plate-shaped crystals of carbonate-rich hydroxyapatite. Furthermore, the bone architecture depends on its location in the body and the local mechanical loads withstood at the site.[2,5]

Beyond its chemical composition, the design of an ideal matrix focused on regeneration of bone tissue should mimic the natural structure and architecture of bone. Scaffolds should have an internal structure intelligently designed, with a predetermined density, pore shape, and size, with appropriate interconnection pathways. High porosity levels are necessary to support migration and proliferation of osteoblasts and mesenchymal cells, bone tissue ingrowth, vascular invasion, nutrient delivery, and matrix deposition in empty spaces. In fact, the main critical factor affecting bone formation is the presence of a combined macro- and microporosity, since macropores (size $>100\ \mu m$) have a critical impact on osteogenic outcomes, promotion of vascularization, and mass transportation of nutrients and waste products,[4] while micropores (size around $10\ \mu m$) favor capillary formation. Currently, it is commonly accepted that 3D scaffolds should also contain nanoporosity to allow diffusion of molecules for nutrition and signaling.[2]

Pore interconnection also plays a key role in the overall biological system, since it provides the channel for cell distribution and migration, allowing efficient *in vivo* blood vessel formation. Furthermore, pore wall roughness contributes to increase the surface area, protein adsorption, and ion exchange.[1,3]

Following these considerations, efforts should be addressed to the development of synthetic biomaterials that tailor the remarkable biomechanical properties and hierarchical structure of bone tissue as an organized assembly of structural units at increasing size levels that provide optimum fluid transfer and self-healing.[6] Open structure–based scaffolds with appropriate pore size, interconnectivity, and total porosity should be modeled and developed in order to provide *in vivo* blood vessel invasion and neobone tissue ingrowth within the scaffolds.

The new generation of biomaterials for use as scaffolds for bone tissue engineering and regenerative medicine applications should mimic the smart structures present in nature. Computer-assisted design and rapid prototyping techniques may be used to generate intelligent scaffolds with defined architecture. Nevertheless, bioderived materials represent an exciting approach since they take advantage of the knowledge and perfection of materials developed by evolution over millions of years and they take inspiration from the most complex naturally organised chemical and biological structures.

Recently, in this context, bioinspired SiC ceramics obtained from vegetable structures, as different woods and plants, have been a matter of interest.[7–12] This innovative biomaterial is produced by molten Si infiltration of carbon templates obtained by controlled pyrolysis of vegetable precursors, and the final product is a ceramic material with open and interconnected porosity.[9,12,14,15] Moreover, the biodiversity of the natural grown vegetable structures offers a large variety of templates, with different density, morphology, pore shape, pore size, and interconnection, leading to bioinspired materials with optimized microstructure and tailorable properties, similar to those of the tissue to be regenerated.[16,17]

Besides the scaffold architecture, it is also relevant to notice that bone tissue engineering scaffolds should have enhanced osteoinductive functions and should promote the specific morphogenic process by controlling the chemistry of the scaffolds and, therefore, inducing cells to differentiate in a predetermined manner and to regenerate by themselves the desired tissue according to physiological pathways.[1,2] Consequently, as part of the integral design of scaffolds, the incorporation of bone growth factors or specific peptide sequences into the scaffolds should be considered. Moreover, the functionalization of the scaffold surface and the incorporation of bonelike chemical composition coatings, such as hydroxyapatite, substituted apatites, or silica-based glasses, can improve the bioactive and osteoconductive properties of the scaffolds.[3]

The experimental studies conducted to date on the development of SiC ceramics based on biologically derived structures indicate that this innovative material maintains the suitable structure-morphology for optimum ingrowth of bone tissue.[11] It has been demonstrated as well that the bioderived ceramics can be tailored simultaneously matching the mechanical properties of the host[15,16] and becoming bioactive by a suitable coating.[10,11,15,18] Moreover, the SiC ceramics is radiographically distinguishable from the new bone, and in addition the porous structure can be used as a delivery system for drugs and growth factors, which opens the door to a whole new generation of bioderived scaffolds for regenerative medicine applications.

2.2 Bioinspired SiC Ceramics

Bioinspired SiC ceramics can be produced from a large variety of natural cellulose templates. In fact, the fabrication of bio-SiC from soft and hardwoods, such as beech (*Fagus sylvatica*), eucalyptus (*Eucalyptus globulus*), sapelli (*Entandrophragma cylindricum*), oak (*Quercus robur*), etc., has been reported.[9,12–14]

Basically the bioceramization process consists of the infiltration of molten Si in carbonaceous scaffolds derived from vegetable precursors.[9,12,14,17] The whole fabrication process of bioinspired SiC ceramics can be described in three main steps:

(a) Precursors selection: In this first stage, the appropriate vegetable source is selected according to the desired properties of the final product, using a criteria of density, size, and interconnection of the pores. Depending on the nature of the vegetable, the natural samples are freeze-dried or dried with warm air in order to avoid the collapse of the vegetable architecture and to keep the original microstructure as it is.

(b) Pyrolyzation process: The natural vegetable samples are exposed to the pyrolysis process in a high-temperature furnace under inert conditions with a well-controlled ramp of 2°C/min up to 800°C, keeping the sample at this temperature for one hour. After cooling at the rate of 20°C/min, a carbonaceous scaffold that keeps the basic microstructural features of the starting natural material is obtained.[12,13,17] In this step, about 75$\pm$5% of the starting weight of the vegetable is lost, mainly in the form of water vapor and other volatiles products. Volume is reduced at the same time by about 60$\pm$5%. These reductions appear to be independent of the type of vegetable, which is very useful for precursor selection aimed at a specific final density.[13]

(c) Infiltration process: The final step is the molten Si infiltration of the carbonaceous scaffold in vacuum conditions. For this purpose, the carbon sample is introduced in a controlled heating furnace where the temperature is increased at 5–10°C/min until 1,550°C, exceeding the Si melting point (1,410°C), with 30 minutes of permanence at this last temperature. After the cooling of the sample with a well-controlled ramp of 20°C/min, the final SiC ceramics is obtained.[12,13,17] The Si infiltrates the carbon preform rapidly by capillarity. Through the connected porosity the molten Si reaches all pores and channels within a few seconds. The subsequent SiC formation reaction is spontaneous and exothermic. The heat generated during the reaction raises the liquid Si temperature and enhances the C solubility. The final product is then SiC with a certain amount of unreacted Si and a small percentage of unreacted C. If the channel diameter is comparable to the C wall thickness, the pores can be blocked by the precipitation of SiC, preventing the complete reaction of the C preform in those zones, where unreacted C would remain inside the sample. In the final step, most of the excess Si is removed from the ceramics by cycling it at high temperature over a carbon fiber cloth that soaks up the residual Si. The infiltration process does not significantly change the size and shape of the pieces (under 0.1%), so this step is oriented to final near-net-shape products.[12,13]

To illustrate the bioceramization process, Fig. 2.1 shows typical carbonaceous scaffolds, after vegetable pyrolyzation, and the

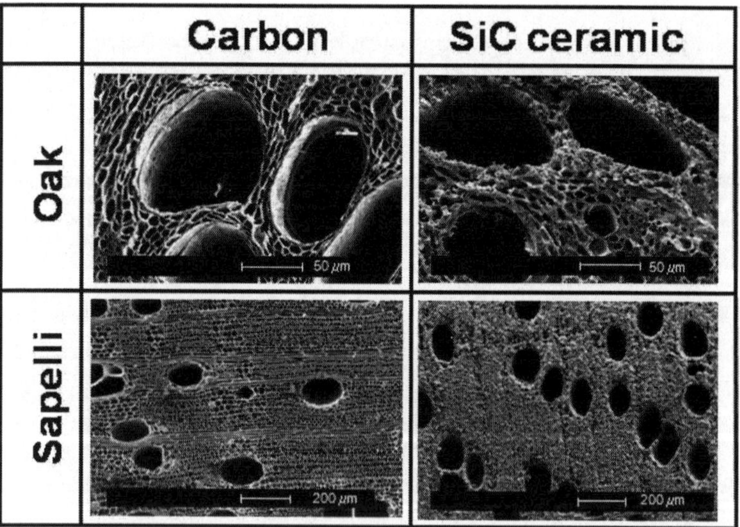

Figure 2.1. Final SiC ceramics and intermediate carbonaceous products obtained in the bioceramization process from oak (*Quercus robur*) and sapelli (*Entandrophragma cylindricum*). The ceramics preserves the original porosity and interconnectivity, and the natural microstructure is mimicked by faceted SiC grains.

final bioinspired SiC ceramics obtained from two different natural resources.

Oak (*Quercus robur*) and sapelli (*Entandrophragma cylindricum*) are angiosperms with bimodal pore distribution in the axial direction typical of hardwoods. Their hierarchic microstructure shows the existence of large-vessel cells, called tracheas (pore size between 50 and 200 μm in diameter), and smaller cells, called libriform "fibers" (size between 2 and 10 μm), with thick walls that are responsible for imparting mechanical resistance. SEM micrographs (Fig. 2.1) show how the cell structure of the vegetable precursor is reproduced along the process up to the microscopic level. The carbonaceous templates preserve the anatomical details of the original vegetable, like the wood vessels and the perforation plates that connect two vessel elements. The Si infiltration is very homogeneous inside the material, confirming the interconnectivity of pores. The ceramic microstructure is basically a well-connected SiC skeleton that replicates the wood cell wall structure, where the microporosity is partially filled with unreacted Si. More detailed analysis of the microstructure indicates that the SiC grains are present both in microcrystalline (μSiC) and in nanocrystalline (nSiC) form.[12,19]

Porosity measurements (Fig. 2.2) obtained by image processing from SEM micrographs of wood-derived SiC ceramics support the fact that a large variety of porous scaffolds can be produced by an adequate selection of the starting vegetable precursors. Thus, pine-based SiC offers a high amount of regular pores with a narrow distribution of pore areas centered at 300 μm^2, while hardwoods (beech and sapelli) provide SiC ceramics with higher pore sizes but lower frequency of pores.

The marine medium also offers great biodiversity, algae and plants, providing interesting precursors for ceramic scaffolds. Fig. 2.3 shows SEM micrographs of SiC ceramics produced from marine plant species, such as *Juncus maritimus* and *Zostera marina.* Both marine plants have a totally developed vascular system that ensures the porosity and interconnectivity required for a successful bioceramization process. The surface details and internal structure of the original plants are preserved. Thus, SEM micrographs reveal that regular macrochannels of 100 μm with aligned microchannels

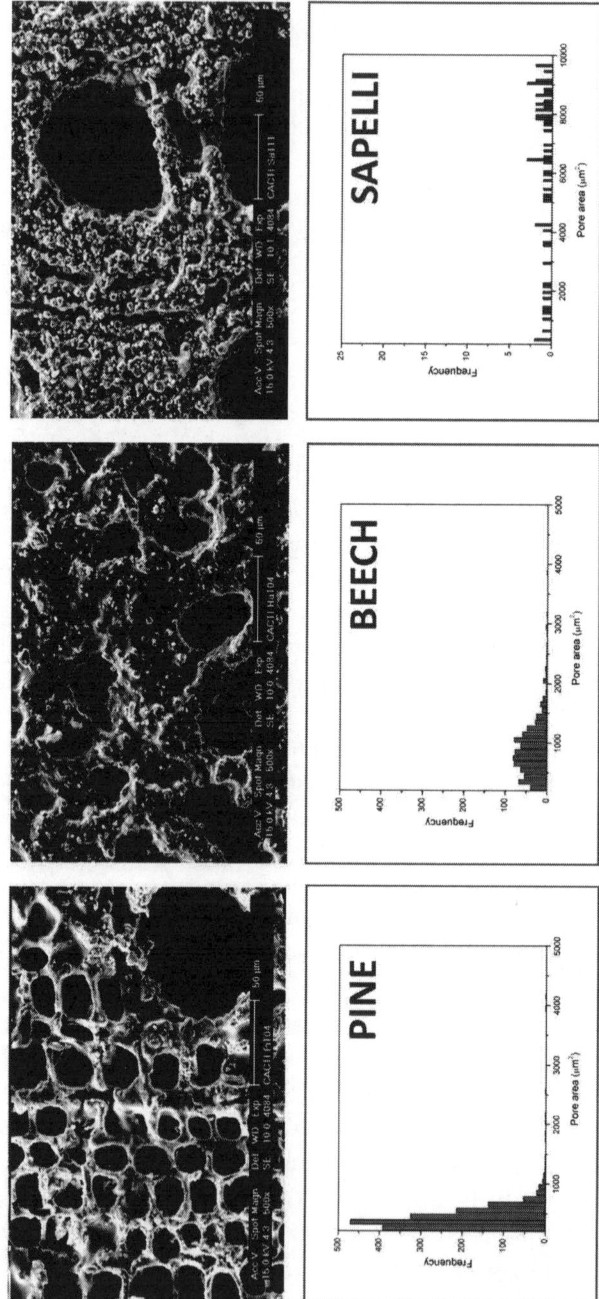

Figure 2.2. Porous size distribution of three bioinspired SiC ceramics derived from pine, beech, and sapelli.

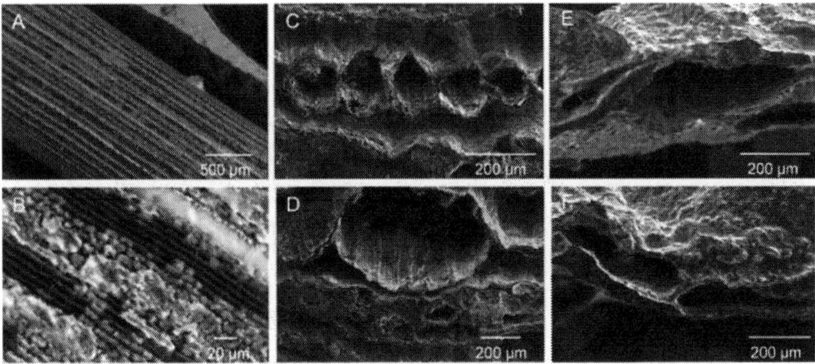

Figure 2.3. SEM micrographs of SiC ceramics obtained from the marine plants (a, b) *Juncus maritimus* and (c, d) *Zostera marina* and (e, f) the macroalgae *Laminaria ochroleuca.*

of 7 μm remain on the surface of *Juncus*-based SiC (Fig. 2.3a,b) and very large cavities are preserved inside the porous matrix of *Zostera*-based SiC (Fig. 2.3c,d). It has been also demonstrated that several marine macroalgae species can be used as vegetable precursors for the production of SiC bioinspired materials,[13] for instance, *Laminaria ochroleuca, Undaria pinnatifida, Saccorhiza polyschides*, and *Cystoseira baccata*. As an example, details of the microstructure of algae-derived SiC ceramics is shown in Fig. 2.3e,f.

Finally, it has been reported that mechanical tests[11,15,16] indicate the suitability of bioderived SiC ceramics for medical devices in terms of structural requirements, as results compare positively with the density, elastic modulus, compressive strength, and bending strength of a typical human cortical bone. For instance, the room-temperature compressive strength in the longitudinal (biological precursor growth) direction of sapelli-derived bio-SiC is from 1160 $\pm$ 100 (s.d.) MPa to 210 $\pm$ 20 (s.d.) MPa; in the radial direction, it is from 430 $\pm$ 50 (s.d.) MPa to 120 $\pm$ 10 (s.d.) MPa, depending on the amount of infiltrated residual Si. The fracture toughness reaches values between 2 and 3 MPa $(m)^{1/2}$, and the elastic modulus ranges from 25 to 230 GPa.[19,20] Therefore, bioinspired SiC ceramics can be tailored by an appropriate precursor selection, depending on the requirements of a particular type of bone in the body that should be repaired.

2.3 Biocompatibility Studies

The ability of bioinspired SiC ceramics to sustain cell attachment, spreading, proliferation, and differentiation was assessed by seeding MC3T3-E1 preosteoblastic cells on the materials up to 28 days. Cellular interaction with sapelli-based SiC ceramics was compared to cell growth on standard TCP, which represents the experimental "gold standard" for *in vitro* studies of osteoblasts.

The MC3T3-E1 cells are characterized by their capacity to differentiate into osteoblasts and osteocytes to form, finally, calcified bone tissue *in vitro*. The initial interaction with any surface involves protein adsorption to the surface, contact of rounded cells, attachment of cell to substrate, and spreading of cells. Cells then start to proliferate across the surface and pores to establish cell-to-cell contact points, and finally they synthesize the extracellular matrix. After the deposition of extracellular matrix, cell mineralization occurs with calcium phosphate deposits in the crystalline hydroxyapatite (HA) form.[21,22]

Cell proliferation was assessed and quantified with the Cell Proliferation Kit I (MTT) (Roche Applied Science) along the time of culture (Fig. 2.4a). MC3T3-E1 cells proliferated properly throughout the culture time on the tested material. The behavior of the cells on the SiC ceramics followed the same pattern as those grown on the TCP, with a progressive increase of the cellular density until the fourteenth day of incubation. An expected decrease of the cell proliferation on TCP on the twenty-first day was detected on both materials. At this time, cells had already covered the whole surface and began to express the osteoblastic genes. The increase of cell differentiation can have caused a decrease of proliferation. These results confirmed the noncytotoxicity of the sapelli-based SiC ceramics with a healthy behavior of the cells along the time of culture.

Cell adhesion and morphology was analyzed by SEM (Fig. 2.5). First, it is important to notice the microstructural properties of this bioinspired material (micrographies, Fig. 2.5a–c), where the long tubular cells and sap channels (tracheae), used by the angiosperm species to transport and store nutrients, can be clearly observed.[15] On the first day of incubation (Fig. 2.5a,d), cells appeared closely attached to the SiC ceramics with the flattened morphology typical of

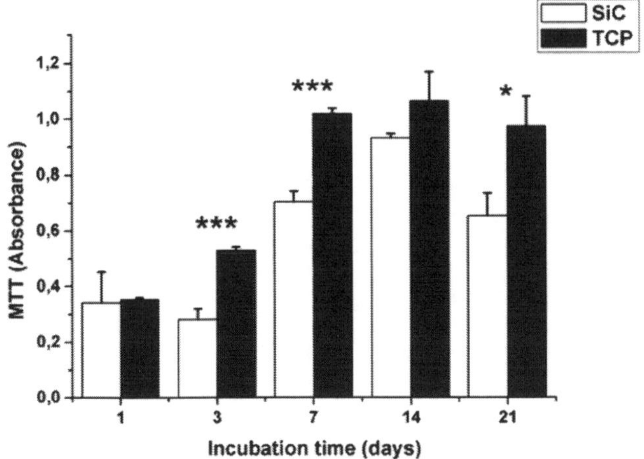

Figure 2.4. Cell proliferation results after incubation of MC3T3-E1 cells up to 21 days on SiC ceramics and TCP as control. (Statistical significant differences: $^{*}p < 0.05$, $^{**}p < 0.01$ and $^{***}p < 0.005$).

healthy osteoblasts. After three days of incubation (Fig. 2.5b,e), the cells, connected to each other by filopodia and lamellipodia, almost covered the whole surface, including some of the channels. At day 7 of incubation (Fig. 2.5c,f), several layers of cells covered the surface and the channels were almost completely filled. Micrographs showed a healthy appearance of the cells, which attached and spread

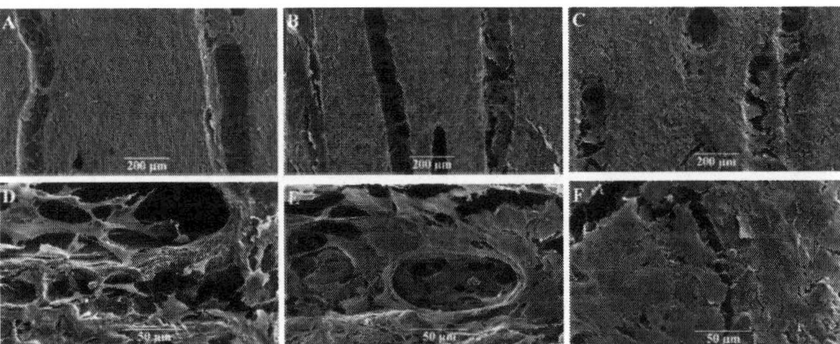

Figure 2.5. SEM images showing the attachment and spreading of the MC3T3-E1 cell line on bioinspired ceramics. (a, d) For 1 day; (b, e) for 3 days; and (c, f) for 7 days. Two magnifications are shown: (a, b, c) 100x and (d, e, f) 500x.

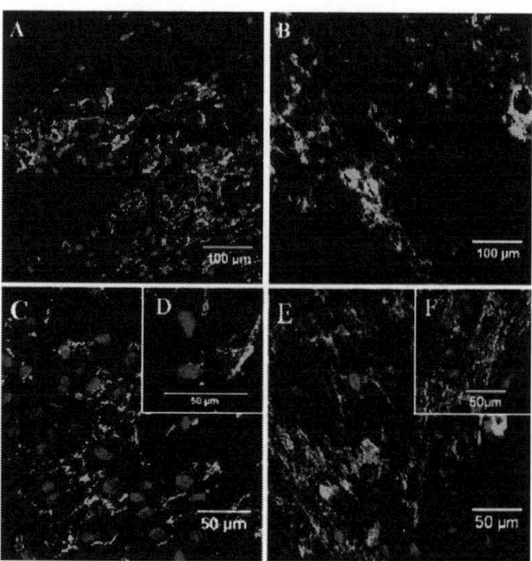

Figure 2.6. CLSM images of preosteoblastic cell line MC3T3-E1 growth on sapelli-based SiC ceramics along the time of culture. (a, c) For 1 day; (d) for 3 days; (b, e, f) for 7 days. Alexa Fluor 488 Phalloidin solution was the antibody used to visualize the cell cytoskeleton filamentous actin, or F-actin, (green) and propidium iodide solution to label cell nuclei (red). See also Color Insert.

properly, without any signs of cytotoxicity, maintaining the cell-to-cell and cell-to-material contact by filopodia and lamellipodia.

Besides the cell adhesion and morphology, the cytoskeleton organization was also analyzed by CLSM (Fig. 2.6). F-actin filaments and the progressive covering of the SiC ceramics surface was observed, proving the tested material to be cell friendly. Cytoskeleton organization was progressively higher throughout the time of culture, as shown by the increasing densification of actin filaments, observed after seven days of incubation (Fig. 2.6b–f). After one day of cell culture (Fig. 2.6a,c), images showed oval nuclei and some evidence of mitosis, which means that cells were already spreading over the SiC surface. At three days of cell culture (Fig. 2.6d), the nuclei of cells and F-actin fibers are clearly visible. After seven days of incubation, cells displayed an organization with numerous cell-to-cell contacts and presented a complex network of actin fibers (Fig. 2.6e,f).

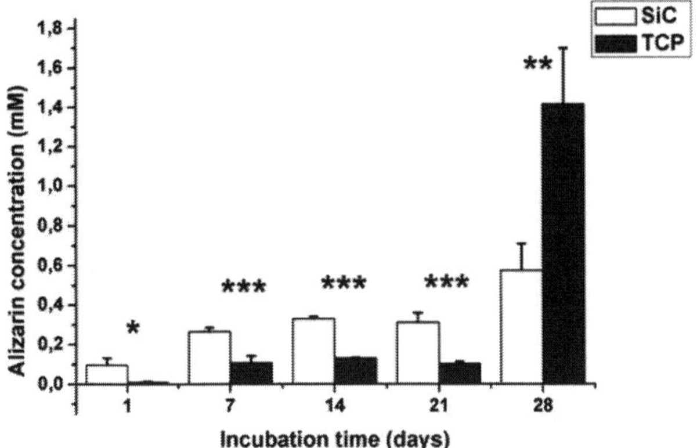

Figure 2.7. Quantification of calcium deposits, stained by Alizarin Red over the bioceramics and TCP, up to 28 days of cell culture. (Statistical significant differences: $^{*}p < 0.05$, $^{**}p < 0.01$, and $^{***}p < 0.005$).

Once the osteoblastic cells covered the whole surface and pores of the sapelli-based ceramics, they started to synthesize the extracellular matrix, followed by the deposition of calcium phosphate. These calcium deposits were determined by Alizarin Red staining and subsequent quantification (Fig. 2.7). Calcium deposits revealed a higher and earlier level of mineralization on SiC ceramics than on TCP during the first 21 days of incubation. Calcium deposits were detected inside the channels of the SiC ceramics even at 7 days of incubation and over and around the ceramics at 14, 21, and 28 days. These results confirmed that the surface microstructure of the SiC favored the earlier differentiation of the osteoblastic cells.

In addition to this assay, and to confirm the presence of the extracellular matrix and the mineralization process, SEM was used to evaluate the ceramics at 28 days of incubation (Fig. 2.8). The SEM micrographs show the whole surface of SiC covered by a thick layer of cells (Fig. 2.8a). When the surface was analyzed closely, a premineralized extracellular matrix was observed over the cell layer (Fig. 2.8b–e). Collagen fibers, oriented in all directions to form the trabecular bone, were also found (Fig. 2.8f).

It has been previously reported[23,24] that the surface microstructure of different materials has a great influence on the cellular acceptance of the materials. Structural features of the surface

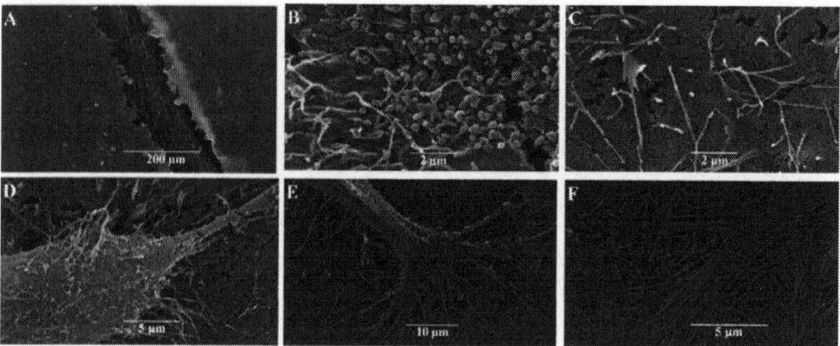

Figure 2.8. SEM images, at different magnifications, of MC3T3-E1 evaluation at 28 days on sapelli-based SiC ceramics.

modulate the expression of phenotypic markers and influence the way cells respond to regulatory factors. In fact, it has been shown that osteoblasts attach, spread, and proliferate more rapidly on smooth surfaces than on rough surfaces, while their differentiation is enhanced by rough morphologies.[23,24] It is also important to note that pores and pockets that resemble pores have an effect beyond that produced by growth factors and other signaling molecules. The protective space in excavations or inside pores was essential in several works to stimulate differentiation of precursors to osteoblasts.[25–27] The excellent differentiation levels of MC3T3-E1 on sapelli-based SiC ceramics could be explained by the existence of a vascular system remaining from the original wood, which has a microstructure with sheltering spaces favoring the mineralization of cells.

The bioinspired ceramics offers a great performance as a biomaterial, and the result is even better if we take into account that it has not been covered by a bioactive coating, where the initial periods for mineralization usually begin at 14–15 days of incubation.[23,28,29] However, deeper analysis of the calcium deposition nodules is required to prove the crystallinity of the calcium phosphate compounds formed.

2.4 Conclusions and Outlook

This chapter introduces the development of an innovative porous ceramic scaffold for bone tissue regeneration based on bioinspired

SiC obtained from vegetable resources. The new material preserves the interconnected hierarchic porosity of the starting natural precursors, such as wood, algae, or marine plants. The whole fabrication process, from precursor selection to the production of bioderived SiC ceramics, and *in vitro* cell proliferation and differentiation studies are presented.

Bioinspired SiC scaffolds mimic the original structure of the starting precursors. Therefore, considering the great biodiversity of nature, the microstructure of the scaffolds can be tuned by an appropriate selection of the starting vegetable resources. The *in vitro* biocompatibility studies demonstrate that this porous ceramic scaffold presents a good biological response. The preosteoblastic cell line MC3T3-E1 attached, spread, and proliferated properly on the sapelli-based SiC ceramics, following the same proliferation pattern as those grown on TCP controls. The unique interconnected hierarchical porosity and microstructure of this bioinspired ceramics favored an earlier mineralization of the osteoblastic cells. The initial steps in which the collagen fibers appear to start forming trabecular bone were found.

The versatility of the SiC fabrication process encourages further studies on the design of highly performing bioinspired materials from marine precursors (e.g., algae and marine plants), looking for adequate microstructures for highly porous tissue-engineered scaffolds. In this way, the replication of the desired bone architecture would favor the migration of osteogenic cells and new bone formation inside the porous structure of the material. Finally, the focus should be on advanced bioactive scaffolds enabling internal growth of tissue and site-specific delivery of bioactive signaling factors (temperature, pH, concentration, internal stimuli, etc). This approach should include issues such as release and retention of growth factors, surface functionalization and coating with bioactive ceramic materials.

Acknowledgments

This work was supported by the UE-Interreg IIIA (SP1.P151/03) Proteus project, the POCTEP 0330IBEROMARE1P project, and Xunta de Galicia (Projects 2006/12 and 2010/83). The authors acknowledge the helpful contributions of Dr. J. Martínez, Dr. A. R. de Arellano-López, Dr. F. M. Varela-Feria, Dr. E. Solla, Dr. J. P. Borrajo, and L. Rial and the technical assistance of Dr. J. Méndez and I. Pazos (CACTI, Universidade de Vigo) with SEM and CLSM analysis.

References

1. J. P. Fisher, A. G. Mikos, and J. D. Bronzino, Eds., *Tissue Engineering* (CRC Press, Boca Raton, 2007).
2. M. Santin, Ed., *Strategies in Regenerative Medicine* (Springer, NY, 2009).
3. B. D. Ratner, A. S. Hoffman, and F. J. Schoen, Eds., *Biomaterials Science: An Introduction to Materials in Medicine* (Elsevier Academic Press, London, 2004).
4. P. K. Chu and X. Liu, Eds., *Biomaterials Fabrication and Processing Handbook* (CRC Press, Boca Raton, 2008).
5. M. Vallet-Regí and D. Arcos, *Biomimetic Nanoceramics in Clinical Use* (RSC Publishing, Cambridge, 2008).
6. V. I. Sikavitsas, J. S. Temenoff, and A. G. Mikos, *Biomaterials,* **22**, 2581 (2001).
7. T. Ota, M. Takahashi, T. Hibi, M. Ozawa, and H. Suzuki, *J. Am. Ceram. Soc.,* **78**, 3409 (1995).
8. P. Greil, T. Lifka, and A. Kaindl, *J. Eur. Ceram. Soc.,* **18**, 1961 (1998).
9. J. Martínez-Fernández, F. M. Varela-Feria, and M. Singh, *Scripta Materialia,* **43**, 813 (2000).
10. P. González, J. Serra, S. Liste, S. Chiussi, B. León, M. Pérez-Amor, J. Martínez-Fernández, A. R. de Arellano-López, and F. M. Varela-Feria, *Biomaterials,* **24**, 4827 (2003).
11. P. González, J. P. Borrajo, J. Serra, S. Chiussi, B. León, J. Martínez-Fernández, F. M. Varela-Feria, A. R. de Arellano-López, A. de Carlos, F. M. Muñoz, M. López, and M. Singh, *J. Biomed. Mater. Res.,* **88A**, 807 (2009).
12. A. R. de Arellano-López, J. Martínez-Fernández, P. González, C. Domínguez, V. Fernández-Quero, and M. Singh, *J. Appl. Ceram. Technol.,* **1**, 95 (2004).
13. F. M. Varela-Feria, J. Martínez-Fernández, A. R. de Arellano-López, and M. Singh, *J. Eur. Ceram. Soc.,* **22**, 2719 (2002).

14. M. Singh, J. Martínez-Fernández, and A. R. de Arellano-López, *Curr. Opin. Sol. Stat. Mater. Sci.*, 7, 247 (2003).
15. P. González, J. Martínez-Fernández, A. R. de Arellano-López, and M. Singh, In Eds. B. Basu, D. S. Katti, and A. Kumar, *Advanced Biomaterials: Fundamentals, Processing and Applications* (Wiley, NJ, 2009), p. 347.
16. P. González, J. P. Borrajo, J. Serra, S. Liste, S. Chiussi, B. León, K. Semmelmann, A. de Carlos, F. M. Varela-Feria, J. Martínez-Fernández, and A. R. de Arellano-López, *Key Eng. Mater.*, **254**, 1029 (2004).
17. M. López-Álvarez, L. Rial, J. P. Borrajo, P. González, J. Serra, E. Solla, B. León, J. M. Sánchez, J. Martínez Fernández, A. R. de Arellano-López, and F. M. Varela-Feria, *Mater. Sci. Forum*, **587**, 67 (2008).
18. A. de Carlos, J. P. Borrajo, J. Serra, P. González, and B. León, *J. Mater. Sci: Mater. Med.*, **17**, 523 (2006).
19. V. S. Kaul, K. T. Faber, R. Sepúlveda, A. R. de Arellano-López, and J. Martínez-Fernández, *Mater. Sci. Eng.*, **428**, 225 (2006).
20. B. Kardashev, B. I. Smirnow, A. R. de Arellano-López, J. Martínez-Fernández, and F. M. Varela-Feria, *Mater. Sci. Eng.*, **442**, 444 (2006).
21. L. C. Baxter, V. Frauchiger, M. Textor, I. Ap Gwynn, R. G. Richards, P. Bongrand, and D. Brunette, *Eur. Cell. Mater.*, **4**, 1 (2002).
22. A. J. Salgado, M. E. Gomes, A. Chou, O. P. Coutinho, R. L. Reis, and D. W. Hutmacher, *Mater. Sci. Eng.*, ***C* 20**, 27 (2002).
23. C. C. Barrias, C. C. Ribeiro, M. Lamghari, C. S. B. Miranda, and M. A. Barbosa, *J. Biomed. Mater. Res. Part A*, **72**, 57 (2005).
24. L. Le Quehennec, M. A. Lopez-Heredia, B. Enkel, P. Weiss, Y. Amouriq, and P. Layrolle, *Acta. Biomater.*, **4**, 535 (2008).
25. P. Ducheyne and Q. Qiu, *Biomaterials*, **20**, 2287 (1999).
26. M. R. Urist, *Science*, **150**, 893 (1965).
27. A. J. García and P. Ducheyne, *J. Biomed. Mater. Res.*, **28**, 947 (1994).
28. M. E. Gomes, V. I. Sikavitsas, E. Behravesh, R. L. Reis, and A. G. Mikos, *J. Biomed. Mater. Res.*, **A 67**, 87 (2003).
29. C. K. Hee, M. A. Jonikas, and S. B. Nicoll, *Biomaterials*, **27**, 875 (2006).

Chapter 3

PRODUCTION OF THREE-DIMENSIONAL HIERARCHICAL NANO Ti-BASED METALS SCAFFOLDS FOR BONE TISSUE GRAFTS

Shuilin Wu,[a,b] Xiangmei Liu,[a,b] Paul K. Chu,[a*] Tao Hu,[a] Kelvin W. K. Yeung,[a] Jonathan C. Y. Chung,[a] and Zushun Xu[a,b]

[a]*Department of Physics & Material Science, City University of Hong Kong, Tat Chee Avenue, Kowloon, Hong Kong*

[b]*Ministry-of-Education Key Laboratory for the Green Preparation and Application of Functional Materials, School of Materials Science and Engineering, Hubei University, Wuhan 430062, China*

*paul.chu@cityu.edu.hk

It is necessary for ideal bone implant materials to possess a hierarchical structure. This chapter describes the fabrication of three-dimensional (3D), hierarchical, porous titanium (Ti)-based metal scaffolds that fully resemble the natural structure of bones from the macro to nano scale. The fabrication process is based on the chemical reaction between the spontaneous titanium oxide layer on the exposed surface and a concentrated sodium hydroxide (NaOH) solution. A nanoskeleton layer forms on the complex surface, and subsequently one-dimensional (1D) nano titanate belts and wires directly grow on this layer. This hierarchical structure resembles the lowest level of hierarchical organization of human bone tissues. The nanostructured surface exhibits

Handbook of Intelligent Scaffolds for Tissue Engineering and Regenerative Medicine
Edited by Gilson Khang

www.panstanford.com

superhydrophilicity and favors the cell adhesion and proliferation. In addition, the *in situ* growth of nanophase materials on the scaffolds may increase the bonding strength between the nanophase materials and the scaffolds, thus minimizing the risk induced by debris shed from the surface of the nano scaffolds. The remarkable simplicity of this hydrothermal surface nano-structuring process makes it widely accessible as an enabling technique in applications of metal-based nanophase materials to bone implants.

3.1 Introduction

A normal adult has 206 bones located in different positions in the body. These bones play important roles in the normal life of human beings because they provide the basic structural support and protect important organs of the body. In addition, various physical activities are carried out by bones with the help of muscles. However, accidents can lead to traumatic or nontraumatic destruction, and there are needs for bone implants. More than 500,000 bone grafting operations are performed annually in the United States.[1,2] The supply of autografting is limited despite good clinical applications, and allografting and xenografting can easily induce pathogen transfer or rejection reactions.[3] Moreover, synthetic matrices often integrate poorly with host tissues, resulting in possible infection or other adverse body responses due to their poor corrosion, wear, and other properties.[4–6] Therefore, it is necessary to develop biocompatible bone scaffolds for bone repair or reconstruction in order to overcome the limitations of traditional therapies.[4,7–9] Before a fully biomimetic bone scaffold can be produced, it is necessary to learn the basic structures of the human bone.

Figure 3.1 shows the basic structure of bones. The cross sections of the compact bone, showing cylindrical osteons with blood vessels running along Haversian canals (in the center of each osteon), are displayed in Fig. 3.1a.

The metabolic substances can be transported by the intercommunicating systems of canaliculi, lacunae, and Volkmann's canals, which are connected with marrow cavities.[12] The various

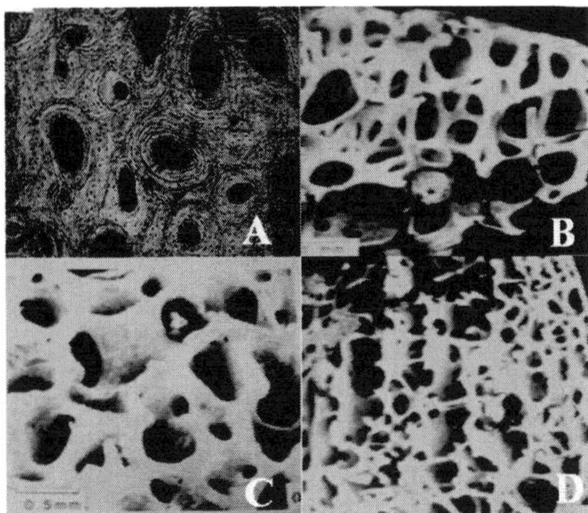

Figure 3.1. Images showing the structure of bones. (a) Optical micrograph of transverse cross section showing the microstructure of compact lamellar bone—human femora; (b) SEM image of a specimen taken from the femoral head, showing a low-density, open-cell, rodlike structure of cancellous bone; (c) SEM image of a specimen from the femoral head, showing a higher-density, roughly prismatic cell structure of cancellous bone; (d) SEM image of a specimen from the femoral condoyle, of intermediate density, showing a stress-oriented, parallel-plate structure of cancellous bone with rods normal to the plates.[10,11]

interconnecting systems are filled with body fluids, and their volume can be as high as 19%.[12] The cellular structure of cancellous bone is shown in Figs. 3.1b, 3.1c, and 3.1d.[11]

Cancellous bones are made up of an interconnected network of rods or plates. A network of rods produces low-density, open cells, whereas one of the plates gives high-density, virtually closed cells. In practice, the relative density of cancellous bones varies from 0.05 to 0.7, and technically speaking, bones with a relative density of less than 0.7 are classified as cancellous.[11]

Human bones are mainly composed of collagen (20 wt. %), hydroxyapatite mineral (69 wt. %), and water (9 wt. %).[12] Other organic materials, such as proteins, polysaccharides, and lipids, are also present in small quantities.[13] The diameter of the collagen microfibers varies from 100 to 2,000 nm, and these collagen fibers

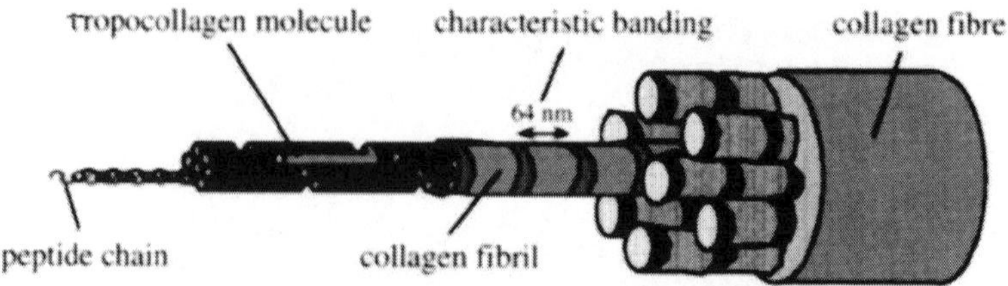

Figure 3.2. Hierarchical organization of collagen fiber.[2]

consist of carefully arranged arrays of tropocollagen molecules, which are long, rigid molecules (300 nm long and 1.5 nm wide) composed of three left-handed helices of peptides, as shown in Fig. 3.2.

Bones contain mostly type-I collagen, with some type-V collagen, and the molecules are organized into collagen fibrils formed by the assembly of tropocollagen molecules in a 3/4 staggered parallel array.[2] The typical hierarchical structure of human bones is illustrated in Fig. 3.3. Besides the hierarchical structure of collagen, the hydroxyapatite (HA) crystals present in the form of plates or needles are about 40–60 nm long, 20 nm wide, and 1.5–5 nm thick,[10–12] and the mineral phase present in the bone is not a discrete aggregation of

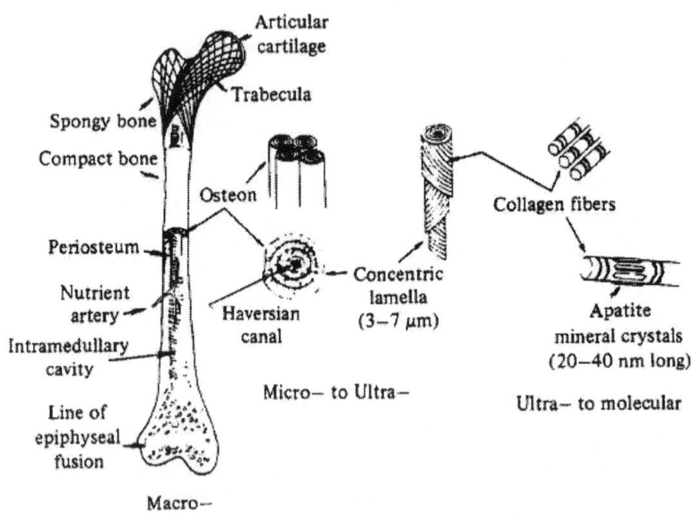

Figure 3.3. Hierarchical structures in a human long bone.[12,14]

the HA crystals but is made up of a continuous phase, evidenced by very good strength of the bone after complete removal of the organic phase.[12]

Due to this unique hierarchical structure of human bones with a macroporous structure and nano-scale organization, ideal bone implant materials must possess the hierarchical structure. This implies that the scaffolds should have not only sufficient 3D internal space and interconnective open channels that allow for mass transport, cell migration, attachment, and proliferation, as well as tissue ingrowth, but also 1D nanophase materials on the surface in order to match the lowest organization of human bones.

Because of excellent mechanical properties and biocompatibility, Ti-based biometals, such as Ti, Ti6Al4V, and nickel-titanium (NiTi), are used widely in various orthopedic applications. We have produced 3D porous NiTi and Ti scaffolds using capsule-free hot isostatic pressing (CF-HIP) powder metallurgy (PM), and the process is described here.

3.2 Fabrication and Characteristics of Macroporous Ti-Based Alloys

CF-HIP is a proven PM process to fabricate porous materials. With the use of a space holder, such as ammonia hydrogen carbonate (NH_4HCO_3), porous Ti-based metal scaffolds with adjustable porous structures, such as Ti and NiTi, have been produced successfully.[15,16] Fig. 3.4 shows the general fabrication process of Ti-based metal scaffolds using CF-HIP. In this process, metallic elemental powders including titanium are the starting materials with the ratios following the chemical compositions of the given Ti-based metals. They are mixed with a certain proportion of space holder powders (NH_4HCO_3) and pressed hydraulically under cold pressure to produce green compacts. These green compacts are preheated at 200°C to remove the space holders to acquire more initial internal pores, and these pores are filled with argon gas when the HIP unit is evacuated and backfilled with argon gas at high pressure. Afterward, these compacts are sintered at high temperatures in two steps, an initial shrinkage stage and a final expansion stage.

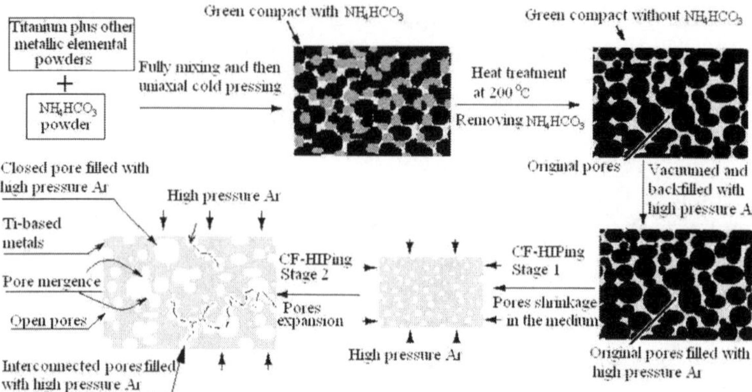

Figure 3.4. Fabrication process of macroporous Ti-based metals scaffold by CF-HIP.[15]

The typical surface morphology of the 3D macroporous Ti-based metal scaffolds is depicted in Fig. 3.5. It can be observed that most pores are interconnecting and their sizes are in the range of 50~800 μm.

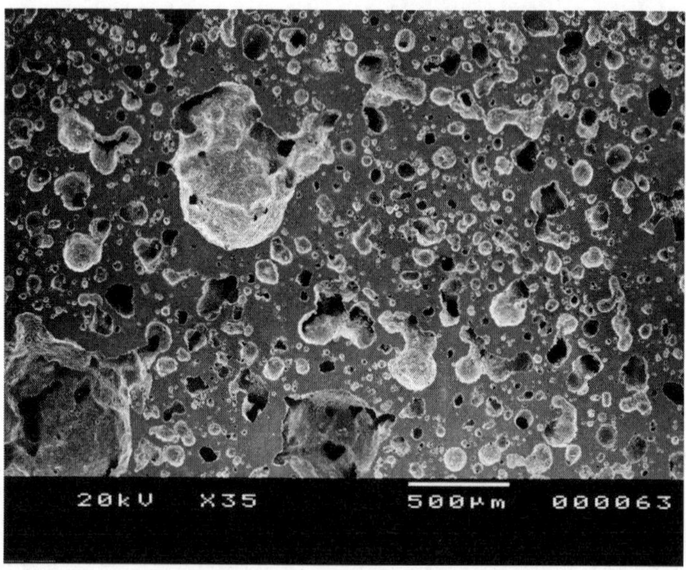

Figure 3.5. Typical porous structure of macroporous Ti-based metals by CF-HIP (NiTi).

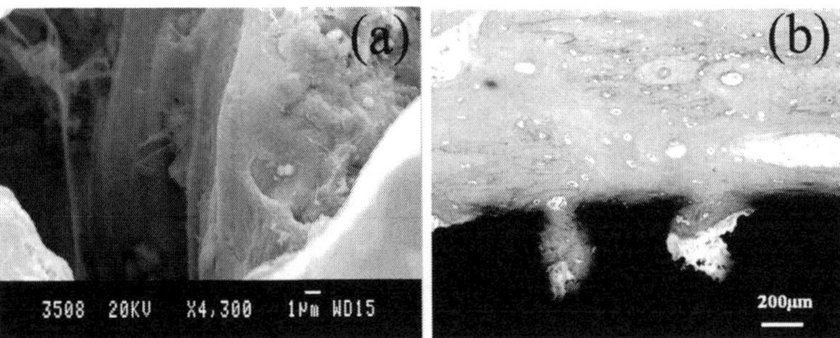

Figure 3.6. *In vitro* and *in vivo* tests of 3D macroporous Ti-based metal scaffolds produced by CF-HIP with NH_4HCO_3 as a space holder. (a) SEM image of cell in-growth on internal exposed surface of the NiTi scaffold; (b) Optical image of bone tissue (pink color) growth on the entire exposed surface of the Ti scaffold (black color). See also Color Insert.

The porosity can be controlled from 21% to 56% (in volume), and the open porosity can reach 70% using this method.[15,17]

Our previous investigations show that both 3D porous NiTi and Ti scaffolds produced by CF-HIP exhibit good mechanical properties and similar Young's modulus with human bone and that cells can attach smoothly and grow on the surface of the internal pores, as illustrated in Fig. 3.6a. After a short-term (3 months) *in vivo* test, the porous scaffolds also bode well for bone tissue in-growth, as shown in Fig. 3.6b.

3.3 Natural Growth and Characterization of 1D Nano Titanates

In view of the hierarchical organization of bones on the nano scale and the fact that nanophase materials significantly influence tissue acceptance and cell behavior[18–20] it is necessary to modify the surface of 3D macroporous Ti-based metal scaffolds to achieve a hierarchical structure on the nano scale to enhance its acceptability by nano HA crystals, tropocollagen, and other proteins from bone cells. Because of the 3D macroporous structure of Ti-based metal scaffolds, it is very difficult to treat the complex topographies using traditional line-of-sight techniques such as laser nitriding, physical vapor deposition (PVD), and even chemical vapor deposition (CVD). Some non-line-of-sight techniques can treat almost the entire surface of the scaffolds but only induce the formation of a bioactive layer without nanophase materials.[21,22] Although the fabrication of nanostructures by layer-by-layer processing[23] pH-induced self-assembly,[24] colloidal self-assembly,[25] electron beam lithography (EBL),[20] and interference lithography (IL)[26] have been proposed, few studies have reported the natural growth of bioactive nanophase materials directly on the surface of 3D macroporous scaffolds with complex topographies.

Our recent work reveals that a lower-temperature hydrothermal treatment can induce the formation of 1D nano titanates on the entire exposed areas of 3D porous Ti-based metals such as Ti and NiTi.[16] This fabrication process of 1D nanowires/nanobelts is schematically illustrated in Fig. 3.7. The 3D porous Ti-based metal plates fabricated by CF-HIP are put in a Teflon-lined autoclave in a concentrated NaOH aqueous solution. The autoclave is heated to 60°C~180°C for different time durations. The treated plates are washed in deionized water to remove the remaining alkaline solution and then dried at 60°C in an oven. During the heating process,

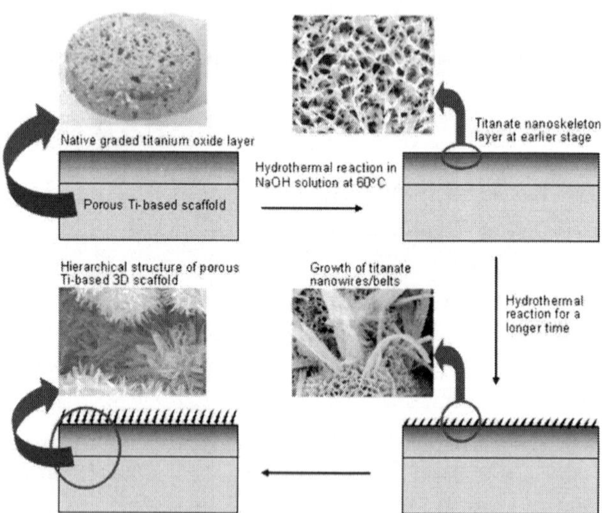

Figure 3.7. Fabrication process of 1D nanophase materials on Ti-based metals (NiTi & Ti).[16]

a titanate nanoskeleton layer forms on the exposed surface in the early stage via a reaction between native titanium oxide and NaOH.

As time elapses, some titanate nanobelts and nanowires nucleate and grow on the skeleton. Finally, a hierarchical Ti-based metal scaffold is formed. The typical surface morphology of this hierarchical structure is exhibited in Fig. 3.8. The formation of nano titanates on titanium plates and titanium dioxide powders has previously been reported, but there is still a dispute on the phase composition of these nano titanates. Peng and Chen[27] ascribe the formation of titanates to the direct reaction between Ti and the NaOH aqueous solution when the Ti plate is immersed in a concentrated alkaline solution. The reaction is shown in the following equation:[27]

$$6Ti + 2NaOH + 11H_2O \rightarrow Na_2Ti_6O_{13} + 12H_2(g) \qquad (3.1)$$

When the starting material is titanium dioxide, Sun *et al.* attribute the formation of nano titanates to the following hydrothermal reaction:[28]

$$4TiO_2 + 2NaOH \rightarrow Na_2Ti_4O_9 + H_2O \qquad (3.2)$$

Figure 3.8. Morphologies of a hydrothermally treated 3D porous NiTi scaffold. (a) Typical surface morphology of an exposed internal pore after treatment at 60°C for 240 hours; (b) Morphology of the wall of the exposed pore in (a); (c) Morphology of the bottom of the pore in (a); (d) General growth condition of nanowires/nanobelts on the exposed surface after treatment at 60°C for 240 hours.[16]

Using the same starting materials and the same process, Yang *et al.*[29] have obtained nano titanates composed of $Na_2Ti_2O_4(OH)_2$ using the following hydrothermal reaction:

$$2TiO_2 + 2NaOH \rightarrow Na_2Ti_2O_4(OH)_2 \quad (3.3)$$

Robert Armstrong proposes that hydrothermally synthesized nano titanates are sodium hydrogen titanates with a general formula of $Na_yH_{2-y}Ti_nO_{2n+1} \cdot xH_2O$.[30] Our results confirm that these 1D nano titanates produced by the reaction between the matrix scaffold and the concentrated alkaline solution are composed of a predominant phase of $Na_2Ti_2O_5{\cdot}H_2O$ and traces of other phases.[16] In the case of porous Ti-based scaffolds produced by PM, the hydrothermal reaction mechanism is more complex because the metallic Ti and surface native titanium dioxides may react with the concentrated alkaline solution. In addition, our investigation discloses that these 1D nanowires/nanobelts grown on the scaffold are unstable under high-energy electron impact.[16] As shown in the TEM picture in Fig. 3.9, the length of these 1D nanobelts on the surface can reach 10 μm.

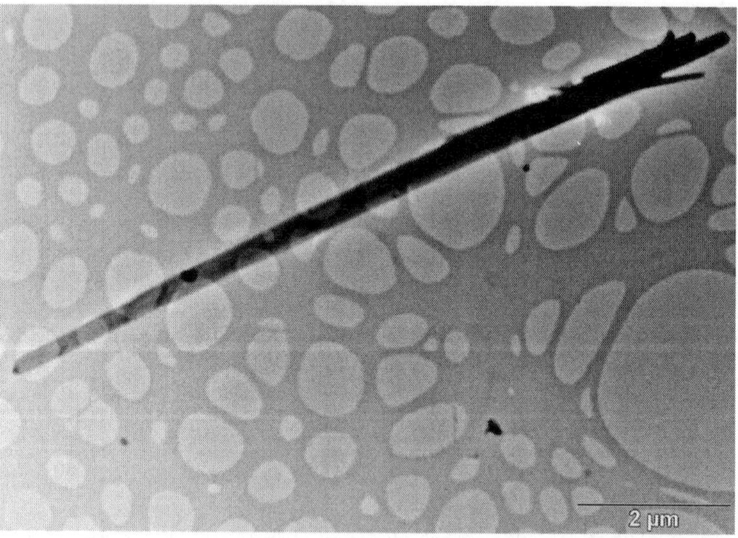

Figure 3.9. TEM image of single 1D nanobelt of titanate scratched from the hydrothermal-treated Ti-based metal scaffold.

From the perspective of surface modification of 3D porous scaffolds, the biggest advantage of this hydrothermal process is that the alkaline solution can reach the entire exposed surface, irrespective of the complex topography, due to its non-line-of-sight nature. In contrary, most other reported techniques can only be used to fabricate nano/micro porous structures on planar samples with simple topographies.[20,31–33] Furthermore, this hierarchical scaffold has no distinct interface between the substrate and the nanostructured layer, which is different from hybrid composites filled with a layer of organic nano matrix by self-assembly.[34] The natural growth directly from the substrate strengthens the bonding between the scaffolds and nanowires/nanobelts, favoring a smooth junction between bone tissues and scaffolds, benefiting the long-term fixation of the 3D scaffolds, which can possibly extend the lifetime of the implants. Our recent contact-angle measurements indicate that 1D surface nano titanates can significantly improve the hydrophilicity of both Ti and NiTi scaffolds.[16] On the other hand, other biodegradable polymer scaffolds, such as poly(L-lactic acid) (PLLA),[19,35] poly(DL-lactic-*co*-glycolic acid) (PLGA),[35] polystyrene (PS),[36] and poly(ethylene terephthalate) (PET),[37] have to undergo surface modification before *in vitro* or *in vivo* tests due to their hydrophobic nature. In practice, good hydrophilicity can improve the biocompatibility of biomaterials. Our cell culture tests show that these surface 1D titanates nanowires/nanobelts significantly enhance cell adhesion and proliferation (Fig. 3.10).

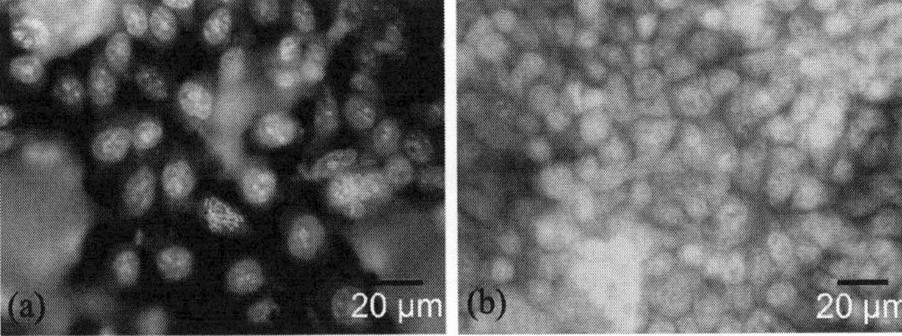

Figure 3.10. Microscopic view of cell growth on the surface of microporous NiTi with or without nanostructured titanate. (a) Untreated surface and (b) surface with nanostructured titanate.[16] See also Color Insert.

This is because the surface nanophase materials increase the active points where cells and proteins can accept and attach, besides the good hydrophilicity.

3.4 Conclusions and Outlook

This chapter focuses on the fabrication of a biomimetic Ti-based metal scaffold to highly resemble the hierarchical structure of human bones on both the macro and nano scales. A new PM method, CF-HIP, with a removable space holder, is utilized to fabricate 3D macroporous Ti-based metal scaffolds with adjustable porous structures. Cell culture results show that these metal scaffolds produced by CF-HIP have good cytocompatibility and are suitable for cell in-growth. Short-term *in vivo* implantation indicates that this macroporous structure of Ti-based metal scaffolds favors bone tissue in-growth. A facile hydrothermal process is effective in treating the entire exposed surface of the scaffolds due to its non-line-of-sight nature. Furthermore, the hydrothermal reaction between the Ti-based metal scaffolds and a concentrated alkaline solution can easily induce the formation of 1D titanate nanowires/nanobelts on the exposed surface, mimicking the hierarchical organization of human bones on the lowest level. These hydrophilic 1D titanate nanowires/nanobelts favor cell attachment and proliferation. Before the clinical application of these nano Ti-based metal scaffolds, two key problems must be solved. The first one is how to precisely control the growth direction, nano size, and nano shape. The second one is to understand the interfacial structure so as to control the bonding strength between the surface 1D nanophase materials and Ti-based metal scaffolds to avoid debris shed from the nanophase materials. It is also essential to carry out pertinent *in vivo* animal evaluations to investigate the effects of these nanophase materials on the growth of bone tissues.

Acknowledgments

This work was supported by the Hong Kong Research Grants Council (RGC) General Research Funds (GRF), Grant nos. City U

112306 and 112307; the National Natural Science Foundation of China Grant no. 50901032; the Ministry of Education Specialized Research Fund for the Doctoral Program of Universities Grant no. 20094208120003; the Hubei Provincial Natural Science Foundation Grant no. 2009CBD359; and the Hubei Provincial Middle-Young Research Fund Grant no Q20101010.

References

1. S. N. Parikh, *Orthopedics*, **1301** (2002).
2. K. A. Hing, *Phil. Trans. R. Soc. Lond.* **A362**, 2821 (2004).
3. R. Murugan and S. Ramakrishna, *Tissue Eng.*, **1845** (2007).
4. S. J. Hollister, *Nat. Mater.* **4**, 518 (2005).
5. A. P. Sclafani, J. R.Thomas, A. J. Cox, and M. H. Cooper, *Arch. Otolaryngol. Head Neck Surg.*, **328** (1997).
6. G. Chen, T. Ushida, and T. Tateishi, *Macromol. Biosci.*, **67** (2002).
7. F. T. Moutos, L. E. Freed, and F. Guilak, *Nat. Mater.* **162** (2007).
8. V. S. Lin, M. C. Lee, S. O'Neal, J. McKean, and K. L. P. Sung, *Tissue Eng.* **443** (1999).
9. J. L. Drury and D. J. Mooney, *Biomaterials*, **4337** (2003).
10. J. L. Katz, In *Symposia of the Society for Experimental Biology, Number XXXIV: The Mechanical Properties of Biological Materials* (Cambridge University Press, 1980), p. 99.
11. L. J. Gibson and M. F. Ashby, Eds., *Cellular Solids* (Cambridge Press, Cambridge, 1997).
12. J. B. Park, *Biomaterials Science and Engineering* (Plenum Press, NY, 1987).
13. S.L. Gunderson and R. C. Schiavone, In Ed. S. M. Lee, *International Encyclopedia of Composites* (VCH Publishers, NY, 1991), p. 324.
14. W. Suchanek and M. Yoshimura, *J. Mater. Res.*, **94** (1998).
15. S. L. Wu, C. Y. Chung, X. M. Liu, P. K. Chu, J. P. Y. Ho, C. L. Chu, Y. L. Chan, K. W. K. Yeung, W. W. Lu, K. M. C. Cheung, and K. D. K. Luk, *Acta Mater.*, **3437** (2007).
16. S. L. Wu, X. M. Liu, T. Hu, P. K. Chu, J. P. Y. Ho, Y. L. Chan, K. W. K. Yeung, C. L. Chu, and T. F. Hung, *Nano Lett.*, **3803** (2008).
17. S. L. Wu, X. M. Liu, Y. L. Chan, C. Y. Chung, P. K. Chu, C. L. Chu, K. O. Lam, K. W. K. Yeung, W. W. Lu, and K. D. K. Luk, *Surf. Coat. Technol.*, **2458** (2008).

18. A. S. Andersson, J. Brink, U. Lidberg, and D. S. Sutherland, *IEEE Trans. Nanobiosci.*, **49** (2003).
19. K. M. Woo, J. H. Jun, V. J. Chen, J. Y. Seo, J. H. Baek, H. M. Ryoo, G. S. Kim, M. J. Somerman, and P. X. Ma, *Biomaterials*, **335** (2007).
20. M. J. Dalby, N. Gadegaard, R. Tare, A. Andar, M. O. Riehle, P. Herzyk, C. D. W. Wilkinson, and R. O. C. Oreffo, *Nat. Mater.*, **997** (2007).
21. S. L. Wu, P. K. Chu, X. M. Liu, C. Y. Chung, J. P. Y. Ho, C. L. Chu, S. C. Tjong, K. W. K. Yeung, W. W. Lu, and K. M. C. Cheung, *J. Biomed. Mater. Res.*, **A79**, 139 (2006).
22. S. L. Wu, X. M. Liu, Y. L. Chan, J. P. Y. Ho, C. Y. Chung, P. K. Chu, C. L. Chu, K. W. K. Yeung, W. W. Lu, and K. M. C. Cheung, *J. Biomed. Mater. Res.*, **A81**, 948 (2007).
23. E. Jan and N. A. Kotov, *Nano Lett.*, **1123** (2007).
24. J. D. Hartgerink, E. Beniash, and S. I. Stupp, *Science*, **1684** (2001).
25. C. Huwiler, T. P. Kunzler, M. Textor, J. Voros, and N. D. Spencer, *Langmuir*, **5929** (2007).
26. J. H. Jang, C. K. Ullal, T. Gorishnyy, V. V. Tsukruk, and E. L. Thomas, *Nano Lett.*, **740** (2006).
27. X. Peng and A. Chen, *Adv. Funct. Mater.*, **1355** (2006).
28. X. Sun, X. Chen, and Y. Li, *Inorg. Chem.*, **4996** (2002).
29. J. Yang, Z. Jin, X. Wang, W. Li, J. Zhang, S. Zhang, X. Guo, and Z. Zhang, *Dalton Trans.*, **3898** (2003).
30. A. R. Armstrong, G. Armstrong, J. Canales, and P. G. Bruce, *Angew. Chem.-Int. Edit.*, **2286** (2004).
31. J. Y. Lim and H. J. Donahue, *Tissue Eng.*, **1879** (2007).
32. M. Karlsson and L. Tang, *J. Mater. Sci: Mater. Med.*, **1101** (2006).
33. W. Sun, J. E. Puzas, T. J.Sheu, X. Liu, and P. M. Fauchet, *Adv. Mater.*, **19** (921).
34. T. D. Sargeant, M. O. Guler, S. M Oppenheimer; A Mata, R. L. Satcher, D. C. Dunand, and S. I. Stupp, *Biomaterials*, **161** (2008).
35. A. G. Mikos, M. D. Lyman, L. E. Freed, and R. Langer, *Biomaterials*, **55** (1994).
36. J. A. Neff, K. D. Caldwell, and P. A. Tresco, *J. Biomed. Mater. Res.*, **511** (1998).
37. M.Yamamoto, K. Kato, and Y. Ikada, *J. Biomed. Mater. Res.*, **29** (1997).

Chapter 4

BIOCERAMIC SCAFFOLD—BONE TISSUE ENGINEERING

Willi Paul and Chandra P. Sharma*

Division of Biosurface Technology, Biomedical Technology Wing,
Sree Chitra Tirunal Institute for Medical Sciences & Technology,
Thiruvananthapuram 695012, India
*sharmacp@sctimst.ac.in

Natural bone consists of calcium phosphate with nanometer-sized needle-like crystals of approximately 5–20 nm width by 60 nm length. Synthetic calcium phosphates and bioglass are biocompatible and bioactive as they bond to bone and enhance bone tissue formation. This property is attributed to their similarity with the mineral phase of natural bone except its constituent particle size and is considered the most appropriate candidate for bone tissue engineering. Calcium phosphate ceramics have been used in dentistry and orthopedics for over 30 years because of these properties. Several studies indicated that incorporation of growth hormones into these ceramic matrices facilitated increased tissue regeneration. Nanophase calcium phosphates can mimic the dimensions of constituent components of natural tissues and can modulate enhanced osteoblast adhesion and resorption with long-term functionality of tissue engineered implants. This mini review discusses some of the recent developments in nanophase ceramic matrices utilized for bone tissue engineering.

Handbook of Intelligent Scaffolds for Tissue Engineering and Regenerative Medicine
Edited by Gilson Khang

www.panstanford.com

4.1 Introduction

Tissue engineering was defined by Langer and Vacanti[1] as an interdisciplinary field that applies the principles of engineering and life sciences toward the development of biological substitutes that restore, maintain, or improve tissue function. It can also be defined as the use of a combination of cells, engineering materials, and suitable biochemical factors to improve or replace biological functions in an effort to effect the advancement of medicine. The total market for the regeneration and repair of tissues and organs was estimated to be $25 billion worldwide[2] in 2001. As per a recent market forecast the worldwide market and emerging technologies for tissue engineering and regenerative medicine products will exceed $118 billion in 2013.[3] Although transplantation of organs has become an established and successful method of therapy, the severe scarcity of donor organs and immune rejection have become a major limitation and have stimulated the tissue engineering field. Two different approaches were being employed in assisted self-assembly of cells. In the first approach, donor cells and growth factors were seeded *in vitro* into a biodegradable scaffold. After the required period for cell growth and multiplication, the scaffold was surgically implanted into the body for the generation of healthy new tissue. In the second approach the scaffold was implanted into the damaged area along with the cells and the growth factors to stimulate tissue regeneration. Scaffolding architecture plays an important role in these three-dimensional (3D) matrix supports composed of cells and an extracellular matrix similar to natural tissue organization. Mimicking the natural tissue organization contributes significantly to the biological function of the tissue-engineered material. Enormous advances have been made in the field of materials science with the advent of nanotechnology. Bone tissue engineering is a specific area in nanotechnology where the development of nanostructured biomaterials may be able to replace hard and soft skeletal tissue and biocompatible materials for tissue genesis. Other related areas include creation of nanoporous biocapsules for cellular therapy. Tissue engineering stands to benefit most from nanotechnology because of the growing ability to fabricate complex nanostructured materials. This chapter reviews some of the important developments in the utilization of nanoceramic matrices in bone tissue engineering.

4.2 Bioceramics

The word "ceramic" can be traced back to the Greek term *keramos*, meaning "a potter" or "pottery." *Keramos*, in turn, is related to an older Sanskrit root meaning "to burn." Thus the early Greeks used the term to mean "burned stuff" or "burned earth" when referring to products obtained through the action of fire upon earthy materials.[4] Ceramics is defined in the *Encyclopedia Britannica* as "objects created from such naturally occurring raw materials as clay minerals and quartz sand, by shaping the material and then hardening it by firing at high temperatures to make the object stronger, harder, and less permeable to fluids." This broad classification includes structural clay products, whitewares, refractories, glasses, abrasives, cements, and advanced ceramics, which are further divided into more specific classes, from dinnerware to the National Aeronautics and Space Administration's (NASA's) reusable, lightweight ceramic tile for its space shuttle program. Bioceramics is a class of advanced ceramics that is defined as ceramic products or components employed in medical and dental applications, mainly as implants and replacements. It is biocompatible and can be inert, bioactive, and degradable in a physiological environment, which makes it an ideal biomaterial. However, it is brittle, with poor tensile strength, which makes it unsuitable for load-bearing applications. Materials that are classified as bioceramics include alumina, zirconia, calcium phosphates, silica-based glasses or glass ceramics, and pyrolytic carbons.

Calcium phosphates include tricalcium phosphates (TCPs), hydroxyapatite (HA), and tetracalcium phosphates. HA found in bone is a poorly crystalline, nonstochiometric apatite formed by nanosized needle-like crystals. Unlike tetracalcium phosphates and TCPs, HA does not break down under physiological conditions. In fact, it is thermodynamically stable at physiological pH and actively takes part in bone bonding, forming strong chemical bonds with surrounding bone. This property has been exploited for rapid bone repair after major trauma or surgery. HA can be termed bioactive and TCP as bioresorbable. Calcium phosphate–based ceramics has been utilized for treating bone defects as it is similar to natural bone and is highly biocompatible. However, its utility toward load-bearing applications is limited because of its poor mechanical strength. Calcium phosphate found in natural bone is in the form of nanometer-sized needle-like crystals of approximately 5–20 nm width by 60 nm length, with a poorly crystallized, nonstoichiometric apatite phase containing other trace ions, which gives strength to the bone. Therefore nanosized calcium phosphate is a good choice as a bone scaffold material.

4.3 Bone Tissue Engineering

The design of biomaterials with surface properties similar to physiological bone that demonstrates the basic criteria of osteoinduction and osteoconduction promotes the formation of new bone and improved orthopedics. Bone tissue engineering is comparatively a new approach in the repair of bony defects. The recent developments of tissue engineering in the field of orthopedic research make it possible to envisage the association of autologous cells and proteins that promote cell adhesion with osteoconductive material to create osteoinductive materials.[5] The nature of interaction between osteoblast cells and their substrate can influence the ability of these cells to produce an osteoid matrix around an implant, which, in turn, will determine the fate of the implant.[6] Materials used as a bone tissue engineering scaffold should be biocompatible and porous to be osteoinductive and osteoconductive and should have the mechanical property compatible with the natural bone tissue. They should also induce cell anchorage and remodel the extracellular matrix in order to integrate with the surrounding tissue.[7,8] The World Health Organization, along with 37 countries and the United Nations, has proclaimed the year 2000–2010 as the "Bone and Joint Decade." This global initiative is intended to improve the lives of people with musculoskeletal disorders, such as arthritis, and to advance understanding and treatment of musculoskeletal disorders through prevention, education, and research.

The chemical composition of the tissue engineered scaffold is crucial for the resorbability and osteoconductive properties, together with its internal porous structure for vascular growth. Porous bioceramics HA and TCP were demonstrated to have good osteoconductive properties resulting in a good functional recovery; however, the resorption was poor.[9] A novel polymer-nano HA porous scaffold was prepared without the use of organic solvents, which exhibited significantly higher cell growth, alkaline phosphatase activity, and mineralization *in vitro* compared with a scaffold prepared with organic solvents. Also, on implantation, the histological evaluation showed that bone formation was more extensive on the scaffold prepared without using organic solvents.[10] It has been proposed that the use of culture-expanded osteoprogenitor cells in conjunction with HA bioceramics significantly improves the repair of critical-size long-bone defects.[11] A macroporous HA scaffold coated with polylactic-co-glycolic acid (PLGA)-bioglass demonstrated the formation of an apatite layer on sample surfaces immersed in the simulated body fluid (SBF) for five days. This also had increased compressive strengths and seems to be a possible scaffold for bone tissue engineering.[12] HA ceramic matrices cultured with mesenchymal stem cells (MSCs) obtained from the respective patient's bone marrow cells were forced to differentiate into osteoblasts, and a bone matrix was formed with HA ceramic. This tissue-engineered HA was used to fill the patient's bone cavity after tumor curettage and exhibited an immediate healing potential and has been suggested as an alternative to autologous bone grafts.[13] The same has been found effective as a "bone graft" substitute for spinal fusion.[14] The three general criteria that have been considered for tissue regeneration are a) the ability of the cells to maintain their desired function without immunological rejection, b) a suitable 3D scaffold that helps in cell growth, and c) efficient delivery of growth factors facilitating tissue ingrowth.[1]

4.4 Research Perspective

Ceramics are widely used as bone substitutes for the past several years. HA and TCP are approved by the Food and Drug

Administration (FDA) as bone substitutes in humans. Ceramic-based bone graft substitutes include calcium phosphate, calcium sulfate, and bioglass used alone or in combination. Cell spreading is an essential function of a cell that has adhered to any surface and precedes the function of cell proliferation. Out of the bone and the ceramic material interactions that take place at the material surface, the interaction of osteoblasts is crucial in determining the tissue response at the biomaterial surface.[15] Attachment and spreading of specific bone-forming cells in cell culture have been utilized for predicting the behavior of the calcium phosphate materials *in vivo*.[16] The process of cell interaction on materials is highly dynamic and depends on various parameters influencing the cell responses. It is well known that the size and shape of the cell-spreading area, as well as the number, size, shape, and distribution of focal adhesion plaques are decisive for further migratory, proliferative, and differentiation behavior of anchorage-dependent cells.[17] Cells usually do not survive if the attached cells are round and are not spreading. If the cell material contact area is significantly high, the cells tend to skip the proliferation phase and enter sooner the differentiation program. If the adhesion is intermediate the cells are most active in migration and proliferation.

Needle-like ceramic nanoparticles based on calcium phosphate and zinc phosphate with an average particle size of less than 100 nm have been developed.[18] A nanophase matrix (ZnCaMgP) developed from these nanoparticles in the presence of magnesium ions exhibited excellent attachment and proliferation of osteoblast cells. This demonstrates its superior surface properties that are favorable for cell growth, cell matrix interaction, and tissue formation. For developing an ideal nanophase bone graft material, factors that are capable of triggering osteogenesis, that is, osteoinductive growth factors like bone morphogenic proteins, also need to be incorporated into the nanophase matrix. An important objective of bone tissue engineering is to develop improved scaffold materials or an arrangement to control osteoblast behavior, significantly affecting its response. Osteoblastic cells on HA exhibited unique attachment and subsequent behavior *in vitro*, which may explain why mineralized tissue formation is better on HA. Divalent cations, including Mg^{2+}, are known to be active in cell adhesion mechanisms.[19,20]

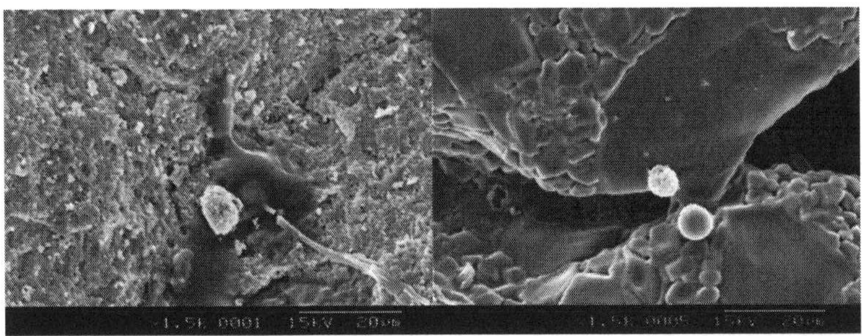

Figure 4.1. Osteoblast adhesion onto (a) an HA matrix and (b) a CaZnMgP matrix after 24 hours.

Minerals like zinc and magnesium are known to aid bone growth, calcification, and bone density.[21,22] Biphasic calcium phosphate ceramics containing zinc also promotes osteoblastic cell activity *in vitro*.[23] The present investigation demonstrates that the cells are spreading well on the matrix containing an optimum amount of calcium, zinc, and magnesium. It seems that the presence of calcium and magnesium encourages the spreading and adhesivity of osetoblasts cells onto bioceramic matrices. Cells are attached and spread completely on the ZnCaMgP matrix, and this matrix appears to be comparable to the HA matrix, as shown in Fig. 4.1. This makes the ZnCaMgP matrix a promising candidate as a bone tissue engineering scaffold similar to HA. Further, since the ceramic matrices are made from nanoparticles, it mimics the way nature itself lays down minerals.

Nanofibers up to a size range of 240 nm have been developed on the basis of the sol-gel and electrospinning technique, and bioceramic matrices have been developed for possible tissue engineering applications.[24] Attempts have been made to culture osteoblasts onto a nano-HA ceramic matrix. This *in vitro* cultured bone may show further bone-forming capability after *in vivo* implantation. This tissue engineering approach is being tried on patients with skeletal problems.[25] Adhesion of osteoblasts, synthesis of alkaline phosphatase, and deposition of calcium-containing mineral increased on PLGA scaffolds containing nanophase titania, emphasizing the utility of nanotechnology in orthopedic tissue engineering materials.[26]

Molecular and cellular interaction on calcium phosphate 3D scaffolds has been reviewed by several investigators.[27–31] They have superior properties for the stimulation of bone formation related to their specific interactions, such as ionic exchanges, superficial molecular rearrangement, and cellular activity, with extracellular fluids and cells. Bone tissue engineering approaches are successful in most of the clinical cases; however, the supply of oxygen and nutrients to the cells in the inner part of the implanted scaffolds remains a major concern, requiring additional investigations.[32] It has been concluded by one study that the extracellular Ca^{2+} and inorganic phosphate (Pi) significantly influence the growth and osteogenic differentiation of MSCs and the cellular effect of these ions needs to be considered when designing or constructing scaffolds for bone tissue engineering with calcium phosphate ceramic scaffolds.[33]

4.5 Basic Questions in Bone Tissue Engineering

As already discussed, the basic requirements for bone tissue engineering are a suitable matrix or scaffold, a requirement of specific cells that interact positively with the scaffold, and the introduction of proper growth factors at a proper concentration that can induce differentiation of cells into oesteoblasts—the most important property being the biocompatibility of the scaffold besides its ability to be utilized by clinicians, that is, how the scaffold can be sterilized, etc. The scaffold should facilitate adhesion and proliferation of cells, in addition to extracellular matrix deposition, allowing tissue ingrowth.[34] An ideal scaffold material should be a 3D porous structure with interconnected pores. It should be biodegradable but with a controlled degradation rate with adequate mechanical strength.[1] Only inorganic materials like calcium phosphates, HA, and glass ceramics, along with a natural coralline HA matrix, have been identified as potential bioceramic scaffold materials for applications in bone tissue engineering.[35] Although HA and bioglass matrices are considered potential scaffolds,[36,37] the biodegradability of these materials is relatively slow. Modifications of these materials need to be actively considered for improving their biodegradability. Composite materials developed by combining with biodegradable polymers, introducing macroporocity, etc., are some of the strategies to improve degradation, bone ingrowth, and osteoconductivity.[38] The pore size of the scaffold is an important parameter for good cellular distribution. It has been suggested that high porosity is beneficial in cell growth and distribution.[39] It has been suggested that a pore size of 200–400 microns as optimum for cell ingrowth, however, is cell specific.[40–43]

Bone tissue mainly consists of matrix-producing osteoblasts, tissue-resorbing osteoclasts, and osteocytes. All these cells are required for developing an engineered bone tissue. However, osteoblasts and osteocytes may be derived from stem cells that synthesize and regulate the mineralization. Human bone marrow-derived MSCs can be differentiated into the osteogenic lineage by culturing the cells in the presence of the osteogenic differentiation supplements dexamethasone, ascorbic acid, and β-glycerophosphate.[44,45] Osteogenic differentiation of mouse and human embryonic stem cells has been established *in vitro* by culturing the cells in a medium supplemented with ascorbic acid, β-glycerophosphate, dexamethasone,[46–48] bone morphogenetic protein 2 (BMP2), compactin,[49] or vitamin D_3.[50]

For favoring tissue ingrowth, growth factors are also usually incorporated into the scaffold materials. These include BMPs, basic fibroblast growth factor (bFGF), transforming growth factor-beta (TGF-β), vascular endothelial growth factor (VEGF), etc., that are used commonly because of their osteoinductive properties and vascularisation.[38] BMPs are the regulatory molecules that are involved in skeletal tissue formation during embryogenesis, growth, adulthood, and healing[51] and are known to be very important regulators of proliferation and osteogenic differentiation of MSCs.[52] BMP-2 and BMP-7 have been demonstrated to enhance bone regeneration in various experiments.[53–56]

4.6 Conclusion

Compared with natural human bone, the presently studied polymer scaffolds and ceramic scaffolds have insufficient elastic stiffness and compressive strength. The possible suggested alternative is a polymer-cermaic composite and nanocermaic scaffolds that should also have good resorbable properties.[42,43] Nanostructured materials or surface coatings will continue to improve the biocompatibility of a growing range of devices and scaffolds for tissue-engineered products, especially; the work on nanofibrillar networks produced by self-assembly is expected to converge with advances made in cell biology to provide functional scaffolds for tissue engineering applications in the following years to come. Therapeutic approaches for tissue-engineered repair of bone defects have attempted to mimic the natural process of bone repair by delivering a source of cells capable of differentiating into osteoblasts, inductive growth and differentiation factors, or bioresorbable scaffolding matrices to support cellular attachment, migration, and proliferation. However, an ideal engineered material from the point of view of a suitable 3D matrix, its technology for industrial production, a suitable cell source and method of seeding onto that matrix, and the ideal conditions for its proliferation with a precise combination of a suitable growth factor is yet to be standardized. A sufficient databank has been generated, and we could expect a positive response to these points in the coming few years. Nanostructured bioceramics seems to be an ideal matrix for bone tissue engineering since it satisfies all the requirements of a tissue-engineered matrix and can be developed into a successful tissue-engineered bone during this "bone and joint decade."

Acknowledgments

We are grateful to Prof. K. Mohandas, director, and Dr. G. S. Bhuvaneshwar, head, Biomedical Technology (BMT) wing of Sree Chitra Tirunal Institute for Medical Sciences & Technology (SCTIMST), for providing facilities for the completion of this chapter. This work has been partially funded by the Department of Science & Technology, New Delhi, under the Drugs and Pharmaceutical Research Programme, FADDS project #8013.

References

1. R. Langer and J. P. Vacanti, *Science,* **260**, 920 (1993).
2. *Tissue Engineering: Technologies, Markets, and Opportunities,* 3rd ed. (Drug and Market Development Publishing, MA, 2001).
3. *Worldwide Markets and Emerging Technologies for Tissue Engineering and Regenerative Medicine* (Life Science Intelligence, CA, 2009).
4. E. W. Washburn, H. Ries, and A. L. Day, Report of the Committee on Definition of the Term Ceramics, *J. Am. Ceram. Soc.*, **3**, 526 (1920).
5. K. Anselme, *Biomaterials,* **21**, 667 (2000).
6. D. A. Puleo and R. Bizios, *J. Biomed. Mater. Res.*, **26**, 291 (1992).
7. S. Scaglione, A. Braccini, D. Wendt, C. Jaquiery, F. Beltrame, R. Quarto, and I. Martin, *Biotechnol. Bioeng.*, **93**, 181 (2006).
8. H. Yoshikawa and A. Myoui, *J. Artificial Organs,* **8**, 131 (2005).
9. M. Mastrogiacomo, A. Muraglia, V. Komlev, F. Peyrin, F. Rustichelli, A. Crovace, and R. Cancedda, *Orthod. Craniofac. Res.*, **8**, 277 (2005).
10. S. S. Kim, M. S. Park, O. Jeon, C. C. Yong, and B. S. Kim, *Biomaterials,* **27**, 1399 (2006).
11. R. Cancedda, M. Mastrogiacomo, G. Bianchi, A. Derubeis, A. Muraglia, and R. Quarto, *Novartis Found. Symp.*, **249**, 133 (2003).
12. X. Huang and X. Miao, *J. Biomater. Appl.*, **21**, 351 (2007).
13. T. Morishita, K. Honoki, H. Ohgushi, N. Kotobuki, A. Matsushima, and Y. Takakura, *Artificial Organs,* **30**, 115 (2006).
14. K. K. Tan, G. H. Tan, B. S. Shamsul, K. H. Chua, M. H. Ng, B. H. Ruszymah, B. S. Aminuddin, and M. Y. Loqman, *Med. J. Malaysia,* **60**, C:53 (2005).
15. A. Hunter, C. W. Archer, D. S. Walker, and G. W. Blunn, *Biomaterials,* **16**, 287 (1995).
16. U. Meyer, A. Büchter, H. P. Wiesmann, U. Joos, and D. B. Jones, *Euro. Cells Mater.*, **9**, 39 (2005).
17. L. Bačáková, E. Filová, F. Rypáček, V. Švorčík, and V. Starý, *Physiol. Res.*, **53**, s35 (2004).
18. W. Paul and C. P. Sharma, *J. Mater. Sci. Mater. Med.*, **18**, 699 (2007).
19. J. H. Fitton, *Proceed. 5th Ann. Conf. Australian Soc. Biomater.*, **p.A 22**, (1995).
20. R. U. Hynes, *Cell,* **69**, 11 (1992).

21. A. Higashi, T. Nakamura, S. Nishiyama, M. Matsukura, S. Tomoeda, Y. Futagoshi, M. Shinohara, and I. Matsuda, *J. Am. Coll. Nutr.*, **12**, 61 (1993).
22. K. L. Tucker, M. T. Hannan, H. Chen, L. A. Cupples, P. W. F. Wilson, and D. P. Kiel, *Am. J. Clin. Nutr.*, **69**, 727 (1999).
23. Y. Sogo, A. Ito, K. Fukasawa, T. Sakurai, and N. Ichinose, *Mater. Sci. Tech.*, **20**, 1079 (2004).
24. H. W. Kim and H. E. Kim, *J. Biomed. Mater. Res. B. Appl. Biomater.*, **77**, 323 (2006).
25. H. Ohgushi, J. Miyake, and T. Tateishi, *Novartis Found. Symp.*, **249**, 118 (2003).
26. T. J. Webster and T. A. Smith, *J. Biomed. Mater. Res. A*, **74**, 677 (2005).
27. T. Guda, M. Appleford, S. Oh, and J. L. Ong, *Curr. Top. Med. Chem.*, **8**, 290 (2008).
28. F. Barrère, C. A. van Blitterswijk, K. de Groot, *Int. J. Nanomed.*, **1**, 317 (2006).
29. S. Srouji, T. Kizhner, and E. Livne, *Regen. Med.*, **1**, 519 (2006).
30. A. El-Ghannam, *Expert Rev. Med. Devi.*, **2**, 87 (2005).
31. H. Yoshikawa and A. Myoui, *J. Artificial Organs*, **8**, 131 (2005).
32. R. Cancedda, P. Giannoni, and M. Mastrogiacomo, *Biomaterials*, **28**, 4240 (2007).
33. Y. K. Liu, Q. Z. Lu, R. Pei, H. J. Ji, G. S. Zhou, X. L. Zhao, R. K. Tang, and M. Zhang, *Biomed. Mater.*, **4**, 25004 (2009).
34. A. J. Salgado, O. P. Coutinho, and R. L. Reis., *Macromot. Biosci.*, **4**, 743 (2004).
35. H. P. Wiesmann, U. Joos, and U. Meyer, *J. Oral Maxillofac. Surg.*, **33**, 523 (2004).
36. A. M. Ng, K. K. Tan, and M. Y. Phang, *J. Biomed. Mater. Res. A*, **85**, 301 (2008).
37. Q. Z. Chen, I. D. Thompson, and A. R. Boccaccini, *Biomaterials*, **27**, 2414 (2006).
38. W. J. Habraken, J. C. Wolke, and J. A. Jansen, *Adv. Drug Deliv. Rev.*, **59**, 234 (2007).
39. L. A. Solchaga, J. E. Dennis, V. M. Goldberg, and A. I. Caplan, *J. Orthop. Res.*, **17**, 205 (1999).
40. L. M. Pineda, M. Busing, R. P. Meinig, and S. Gogolewski., *J. Biomed. Mater. Res.*, **31**, 385 (1996).

41. E. Tsuruga, H. Takita, Y. Wakisaka, and Y. Kuboki, *J. Biochem.*, **121**, 317 (1997).
42. R. E. Holmes, *Plast. Reconstr. Surg.*, **63**, 626 (1979).
43. V. Karageorgiou and D. Kaplan, *Biomaterials*, **27**, 5474 (2005).
44. N. Jaiswal, S. E. Haynesworth, A. I. Caplan, and S. P. Bruder, *J. Cell. Biochem.*, **64**, 295, (1997).
45. S. K. Both, A. J. C. van der Muijsenberg, C. A. van Blitterswijk, J. de Boer, and J. D. de Bruijn, *Tissue Eng.*, **13**, 3 (2007).
46. L. D. K. Buttery, S. Bourne, J. D. Xynos, H. Wood, F. J. Hughes, S. P. F. Hughes, V. Episkopou, and J. M. Polak, *Tissue Eng.*, **7**, 89 (2001).
47. R. C. Bielby, A. R. Boccaccini, J. M. Polak, and L. D. K. Buttery, *Tissue Eng.*, **10**, 1518 (2004).
48. V. Sottile, A. Thomson, J. McWhir, *Clon. Stem Cells*, **5**, 149 (2003).
49. B. W. Phillips, N. Belmonte, C. Vernochet, G. Ailhaud, and C. Dani, *Biochem. Biophys. Res. Commun.*, **284**, 478 (2001).
50. N. I. zur Nieden, G. Kempka, H. J. Ahr, *Differen. Res. Biol. Divers.*, **71**, 18 (2003).
51. C. A. Kirker-Head, *Adv. Drug. Deliv. Rev.*, **43**, 65 (2000).
52. S. Bobis, D. Jarocha, and M. Majka, *Folia. Histochem. Cytobiol.*, **44**, 215 (2006).
53. E. Tsiridis, A. Bhalla, A. Zubier, N. Gurav, M. Heliotis, and D. Sanjukta, *Injury*, **6**, S25 (2006).
54. K. Na, S. Kim, B. K. Sun, D. G. Woo, H. N. Yang, and H. M. Chung, *Biomaterials*, **28**, 2631 (2007).
55. J. Kim, I. S. Kim, T. H. Cho, K. B. Lee, S. J. Hwang, and G. Tae, *Biomaterials*, **28**, 1830 (2007).
56. P. C. Bessa, M. Casal, and R. L. Reis, *J. Tissue Eng. Regen. Med.*, **2**, 81 (2008).

Part III

INTELLIGENT HYDROGEL

Chapter 5

INDUCTION OF SOFT-TISSUE REGENERATION USING HYDROGELS OPTIMIZED FOR INFLAMMATORY RESPONSE

Nicholas P. Rhodes* **and John A. Hunt**
UK Centre for Tissue Engineering, University of Liverpool, Duncan Building, Daulby Street, Liverpool L69 3GA, UK
*npr@liverpool.ac.uk

Conventional therapies for the reconstruction of soft tissues often do not result in a satisfactory outcome. Tissue engineering strategies are likely to give improved results, in terms of both aesthetic and functional characteristics. Acellular therapies, if possible, are favored over delivery of cell-based solutions, for clinical and commercial reasons. In this chapter we describe our experiences with scaffolds intended to support the growth, proliferation, and differentiation of preadipocyte and mesenchymal progenitor cells. These materials were assessed *in vivo* in rats for their ability to stimulate the infiltration of progenitor cells of adipocytic lineage, the host response during medium-term intramuscular implantation, and the resultant differentiation into mature adipocytes. The most promising biomaterial was found to be a dodecyl-amidated hyaluronan gel, a viscous matrix with

Handbook of Intelligent Scaffolds for Tissue Engineering and Regenerative Medicine
Edited by Gilson Khang

www.panstanford.com

excellent retention properties. Whilst macrophages were observed inside the gel after the first week, the inflammatory response was weak, with small levels of cytokines such as transforming growth factor-β (TGF-β). Mature adipocytes were observed after week 8, with significant numbers of cells at week 12 of varying dimensions, indicating *de novo* adipogenesis rather than recruitment of mature cells. It is concluded that the amidated hyaluronon gel is an excellent material for regenerating soft tissue, which actively drives adipogenesis.

5.1 Introduction

There is a major clinical need for the reconstruction of soft tissues, mainly to repair congenital defects (e.g., Romberg's disease or Poland syndrome),[1,2] burns, injury, or resection of tumors. Traditional therapeutic practices have generally been found not to function adequately. Transplantation of mature, autologous fat tissue or free-fat transplantation,[3,4] for example, has an unreliable outcome ranging from complete resorption and gross shrinkage of the graft to the formation of oily cysts. Common fillers of small defects, such as collagen, native hyaluronic acid, or synthetic polymers such as poly(methyl methacrylate) (PMMA) and silicone, generally have a short residence time in subcutaneous tissue and can cause a chronic host response.[5–8]

In contrast, cell-based tissue engineering and regenerative therapeutic strategies have been heralded as a novel and promising approach to the reconstruction of defective tissues, including soft tissue. The stromal-vascular fraction of adipose tissue contains large quantities of pluripotent mesenchymal progenitor cells[9,10] that have a high proliferative and differentiation potential, making them ideal candidates cells for this application. In addition, it has been reported that adipogenic progenitor cells have a lower oxygen consumption than mature adipocytes,[11] constructive for engineered tissues, which have an inadequate vascular supply.

However, whilst various published *in vitro* studies clearly demonstrate that preadipocytes within a range of polymers will differentiate into mature fat tissue, *in vivo* results have not been as promising. It is clearly uncertain how the host response to these polymers has interacted with the differentiation process of adipogenesis *in vivo*. Previous studies of ours have shown clearly that exposure to various inflammatory environments prevents normal preadipocyte differentiation.

This chapter will describe some of our experiences in defining polymers, which control the host inflammatory response, and the application of that technology in gels for soft-tissue regeneration.

5.2 Hyaluronan as a Base Polymer for Regenerative Therapies

Hyualuronic acid (or hyaluronan) occurs within the extracellular matrix compartment of most tissues and is well known for being actively angiogenic.[12] It is found within developing tissue and is implicated in tissue morphogenesis.[13–15] A major stumbling block is that naturally occurring hyaluronan is a viscous fluid with limited retention and without significant mechanical properties (Fig. 5.1).

By chemical modification of the free carboxyl groups on the glucuronic acid component of the hyaluronan chain, it has been clearly

Figure 5.1. Naturally occurring hyaluronan in its native state.

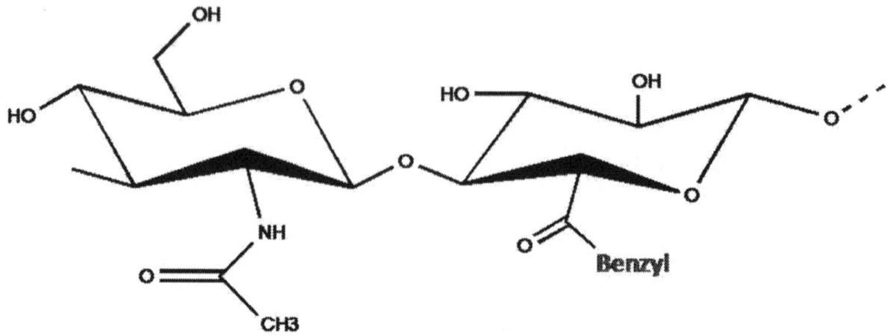

Figure 5.2. Chemical structure of esterified hyaluronan. Solubility of the biomaterial is controlled by the percentage of esterification.

demonstrated that the solubility of the hyaluronan molecule is drastically reduced such that it is solid even after extended periods in aqueous environments.[16] Furthermore, by varying the degree of carboxyl modification, the degradation rate can be tailored to meet the requirements of the implantation site and application.[17] It has been demonstrated in many studies that benzyl esterification of the carboxyl groups (Fig. 5.2) results in nontoxic degradation products,[18,19] including free hyaluronan, which is highly angiogenic.[12] Esterified hyaluronan-based materials that have been fashioned into nonwoven fleeces, membranes, and sponges have already been demonstrated as having functional efficacy when used as scaffolds in the reconstruction of skin,[20] cartilage,[21] and bone.[22,23] Moreover, it has been demonstrated that esterified hyaluronan sponges can support the differentiation of human adipocyte precursor cells *in vitro*[24]

5.2.1 *Experience with Esterified Hyaluronan*

Numerous *in vivo* studies carried out by us and by our collaborators have failed to demonstrate mature adipocyte differentiation within esterified hyaluronan sponges,[25–27] despite positive *in vitro* results.[24] In a study in humans, sponges were implanted for up to 16 weeks, both in acellular form and seeded with autologous preadipocytes.[26] In both, esterified hyaluronan was systematically replaced by a dense, fibrous collagenous extracellular matrix network. Although no calcification was observed, a population of inflammatory cells became well established throughout the

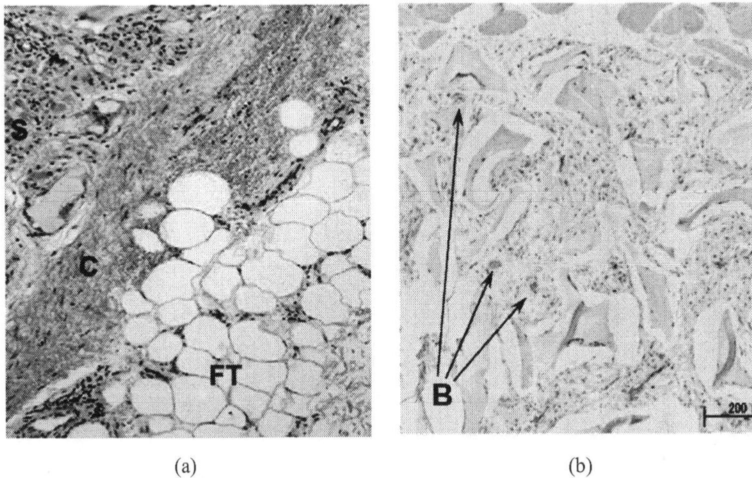

(a) (b)

Figure 5.3. Esterified hyaluronan implanted in (a) humans (16 weeks) and (b) rats (26 weeks); although blood vessels (B) were observed in limited numbers within the scaffolds, the cells within the scaffolds (S) were mostly macrophages. Fat cells (FT) were observed only outside the scaffolds and were part of the subcutaneous adipose layer. See also Color Insert.

scaffolds. Van Gieson staining demonstrated increased collagen deposition in seeded scaffolds compared to acellular specimens. Although seeded implants demonstrated the presence of vascular structures at the periphery of the scaffolds, few were observed deeper within the construct. Despite some distinctive differences between the seeded and acellular groups, no mature adipocytes were observed within either group (Fig. 5.3).

A similar effect was observed in pigs[27] and rats.[25] Macrophage infiltration was observed within the pores of the scaffold, and all mature adipocytes that were associated with the scaffold were observed only at the periphery. There were, however, no major differences between intramuscular and subcutaneous implantation in either species.

5.2.2 *Strategies for Controlling Inflammation*

Hyaluronan is a highly hydrated hydrogel and is therefore extremely hydrophilic. Benzyl esterification of the carboxylic acid moieties

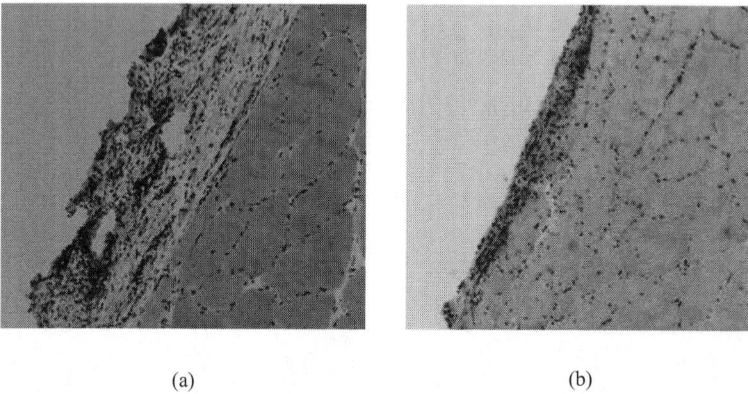

(a) (b)

Figure 5.4. (a) Highly and (b) moderately hydrophobic model polymers influencing the degree of inflammatory response.

confers a high degree of hydrophobicity into the scaffold. It was hypothesized that this factor alone would influence the inflammatory response to a large degree.

Rather than esterification, amidation of the carboxylic acid moieties of hyaluronan was judged to cause a reduction in hydrophobicity. By altering the carbon chain length of the attached amide group, the hydrophobic nature could be further enhanced or reduced.

The effect of hydrophobicity on the inflammatory response was tested in a rat model using a model polymer with varying siloxane content. It was demonstrated that moderate hydrophobicity lessened the degree of inflammation, whilst highly hydrophobic surfaces caused heavy macrophage infiltration (Fig. 5.4)

5.2.3 *Amidated Hyaluronan Biomaterials*

Various amidated hyaluronan derivatives, with varying carbon chain lengths, were evaluated, the two most promising being dodecyl (12C) and hexadecyl (16C) varieties. Initial investigation demonstrated that a reduced host inflammatory response was experienced with the dodecyl formulation, and so this was used in all further investigations (Fig. 5.5).

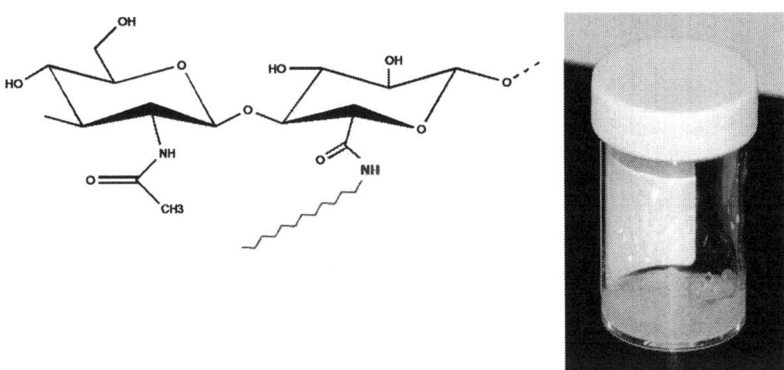

Figure 5.5. Dodecyl-amidated hyaluronan. The manufacturing process amidates approximately 6–7% of the hyaluronan carboxylic acid groups. The amidated polymer is a powder when dry and forms a viscous gel when hydrated, shown here at 30 mg/mL.

Amidated gels were injected in 200 μL aliquots through a 21-gauge needle directly into the dorsolumbar muscle of adult Wistar rats after creation of a small pocket using the injecting needle. Implantation time was up to 12 weeks. Gel implantation sites were explanted immediately following the sacrifice of the animals and fixed in a paraformaldehyde lysine phosphate (PLP) fixative and cold resin embedded in glycol methacrylate (GMA) resin. Samples were sectioned to 4 μm and analyzed using the following tinctural histological stains: van Gieson, von Kossa, and hematoxylin and eosin. Enzyme staining was performed for chloroacetate esterase. Immunohistochemistry was performed to specifically identify a number of cell types, subpopulations, and activation states using the following antibodies: ED1 (monocytes and immature macrophages), CD163, major histocompatibility complex I (MHC-I), MHC-II, CD54, α-β receptor (T-cells), γ-δ receptor (T-cell subset), CD2, CD4, natural killer cells, B-cells, vimentin, and TGF-β.

These experiments were repeated in a pig model to evaluate cross-species efficacy, with subcutaneous injection of amidated hyaluronan in the ear, in both acellular and autologous preadipocyte-seeded forms. Implantation was performed for up to six weeks, and explants analyzed by light and scanning electron microscopy following fixation in formalin and thin sectioning.

5.3 Results of Implantation of Amidated Gels

There was little-observed cell infiltration of the gels implanted intramuscularly in the rat after one week due to the high viscoelasticity of the gel, but cells were apparent after four weeks. Cells had infiltrated throughout the gel and into the center of the gel mass where the cells were actively producing an extracellular matrix. There was no appearance of gel degradation before 8 weeks, and this was still only minimal after 12 weeks. The presence of neovessels was observed in all samples beyond the two-week time period.

Immunohistochemical analysis of the implants demonstrated very low overall levels of inflammation before eight weeks that reached a peak at eight weeks and then declined thereafter. This was particularly so for the staining of the presence of immature and differentiating macrophages and the expression of MHC-I antigen.

The positively stained cells were within both the gel and at the host-gel interface. After 12 weeks, the positive staining was confined to the interface only, with a reduction in the inflammatory response within the gel samples. Overall, MHC-II expression was low for all implantation periods. There was discrete TGF-β expression, possibly associated with the vasculature, but this was at a low level.

Mature adipocytes were present within all 12-week implants (Fig. 5.6), the variable size indicating progression of fat droplet accumulation and therefore adipocyte development and maturation. Although the presence of a certain number of mature adipocytes can be put down to infiltration from the indigenous subdermal fat layer, many examples of gel without association with this layer containing adipocytes were observed, strongly suggesting *de novo* adipogenesis. Tissue integration with the host was excellent, with no fibrous layer evident at the interface between the gel and the indigenous tissue (Fig 5.7).

Neovessels were observed from early time periods, demonstrating an angiogenic response due to the presence of the gel. We propose that the inflammatory cell infiltration with low levels of activation-related expression (e.g., MHC-II) linked with increased angiogenesis was responsible for *in vivo* adipogenesis.

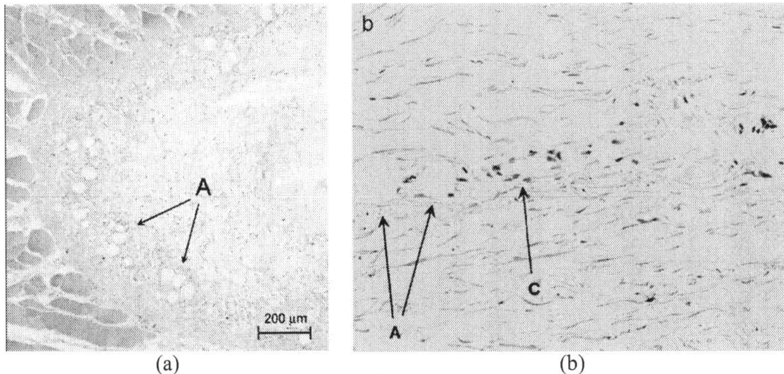

Figure 5.6. (a) Amidated hyaluronan gel implanted in a rat for 12 weeks; (b) Note the accumulation of mature adipocytes around neovessels, a common observation. A = adipocytes; C = capillary.

The observations in the pig model were broadly similar. Adipose tissue was observed around newly formed small vascular structures. There was minor inflammation but no substantial foreign body cell infiltration. Consistently, small adipocytes and disseminated capillaries were observed in the amidated hyaluronan samples (Fig. 5.7).

We believe that it is this combination of factors that was observed within the implanted gels, with low levels of lymphocyte infiltration

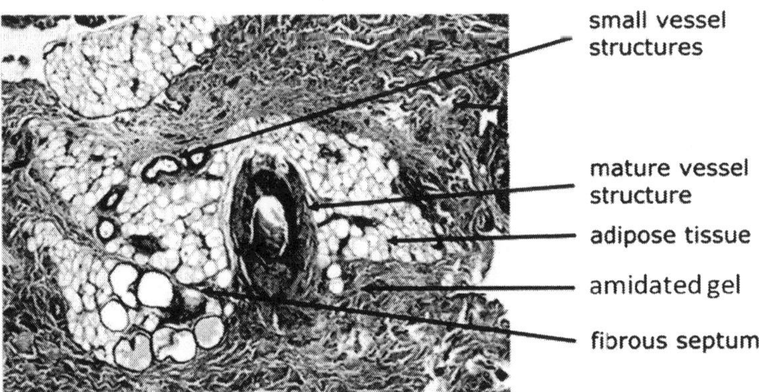

Figure 5.7. Amidated hyaluronan gel implanted in a pig for 6 weeks. Note the accumulation of mature adipocytes around neovessels, a common observation.

and activation, coupled with differentiating macrophages, which leads to the conditions for neoadipogenesis.

5.4 Conclusions and Outlook

This study demonstrated that the implantation of amidated hyaluronan gel in rats induces low overall levels of inflammation. Infiltration was predominantly due to macrophages.

De novo adipogenesis was indicated at 12 weeks, suggesting that the dodecyl-amidated hyaluronan gel could be implanted without cells to induce the regeneration of soft tissue. The results in this study may be interpreted as a demonstration of an inflammatory cytokine profile for the generation of neoadipose tissue. It is hypothesized that the inflammatory cytokine signaling was not intense but was great enough to induce the infiltration of adipocyte progenitor cells in the gel. As the host inflammatory response declined after four weeks, the conditions were correct for the differentiation of the infiltrating cells into adipocytes, possibly following proliferation under the action of TGF-β, influenced by the angiogenic action of the degrading gels.

Acknowledgments

This work was partly supported by the European Commission, under the Competitive and Sustainable Growth Programme. The contribution of Fidia Advanced Biopolymers for the construction and supply of amidated hyaluronan gels is gratefully acknowledged. We acknowledge the collaboration of the University of Dusseldorf, the University Hospital of Gent, and the University of Aachen in the human and pig experiments.

References

1. G. D. Finch and C. J. Dawe, *J. Pediatr. Orthop.*, **23**, 99 (2003).
2. A. Fokin and F. Robicsek, *Ann. Thorac. Surg.*, **74**, 2218 (2002).

3. E. Billings and J. W. May, *Plast. Reconstr. Surg.*, **83**, 368 (1989).
4. R. A. Ersek, *Plast. Reconstr. Surg.*, **87**, 219 (1991).
5. E. A. Balazs and J. L. Denlinger, *The Biology of Hyaluronan* (Wiley and Sons, NY, 1989).
6. R. Shafir, A. Amir, and E. Gue, *Plast. Reconstr. Surg.*, **106**, 1215 (2000).
7. T. F. Wilkie, *Plast. Reconstr. Surg.*, **60,** 179 (1977).
8. A. Mattie and F. V. Nicolle, *Aesthetic Plast. Surg.*, **14**, 227 (1990).
9. J. Smahel, *Br. J. Plast. Surg.*, **42,** 207 (1989).
10. P. A. Zuk, M. Zhu, P. Ashjian, D. A. De Ugarte, J. I. Huang, H. Mizuno, Z. C. Alfonso, J. K. Fraser, P. Benhaim, and M. H. Hedrick, *Mol. Biol. Cell.*, **13,** 4279 (2002).
11. von Heimburg, K. Hemmrich, S. Zachariah, S. H. Staiger, and N. Pallua, *Respir. Physiol. Neurobiol.*, **146**, 107 (2005).
12. R. Montesano, S. Kumar, L. Orci, and M. S. Pepper, *Lab. Invest.*, **75**, 249 (1996).
13. M. Culty, H. A. Nguyen, and C. B. Underhill, *J. Cell Biol.*, **116**, 1055 (1992).
14. T. C. Laurent and J. R. E. Fraser, *Feder. Am. Soc. Exp. Biol. J.*, **6**, 2397 (1992).
15. B. P. Toole, In Ed. E. D. Hay, *Cell Biology of the Extracellular Matrix* (Plenum, NY, 1981), p. 259.
16. S. Iannace, L. Ambrosio, L. Nicolais, A. Rastrelli, and A. Pastorello, *J. Mater. Sci.: Mater. Med.*, **3**, 59.
17. L. Benedetti, R. Cortivo, T. Berti, A. Berti, F. Pea, M. Mazzo, M. Moras, and G. Abatangelo, *Biomaterials,* **14**, 1154 (1993).
18. R. Cortivo, P. Brun, A. Rastrelli, and G. Abatangelo, *Biomaterials,* **12**, 727 (1991).
19. L. Hume, K. Kyyronen, L. Benedetti, V. Stella, and E. Topp, *Abstr. Pap. Am. Chem. Soc.*, **203**, 67 (1992).
20. V. Zacchi, C. Soranzo, R. Cortivo, M. Radice, P. Brun, and G. Abatangelo, *J. Biomed. Mater. Res.*, **40**, 187 (1998).
21. J. Aigner, J. Tegeler, P. Hutzler, D. Campoccia, A. Pavesio, C. Hammer, E. Kastenbauer, and A. Naumann, *J. Biomed. Mater. Res.*, **42**, 172 (1998).
22. R. Z. Gao, J. E. Dennis, L. A. Solchaga, V. M. Goldberg, and A. I. Caplan *Tissue Eng.*, **8**, 827 (2002).
23. L. A. Solchaga, J. S. Temenoff, J. Z. Gao, A. G. Mikos, Caplan, A. I., and Goldberg, V.M., *Osteoarth. Cartil.*, **13**, 297 (2005).
24. M. Halbleib, T. Skurk, C. de Luca, D. von Heimburg, and H. Hauner, *Biomaterials,* **24**, 3125 (2003).

25. N. P. Rhodes, C. Di Bartolo, and J. A. Hunt, *Biomacromolecules,* **8**, 2733 (2007).
26. F. B. Stillaert, C. Di Bartolo,J. A. Hunt, N. P. Rhodes, E. Tognana, S. Monstrey, and P. N. Blondeel, *Biomaterials,* **29**, 3953 (2008).
27. N. P. Rhodes, *Biomaterials,* **28**, 5131 (2007).
28. K. Hemmrich, K. Van de Sijpe, N. P. Rhodes, J. A. Hunt, C. Di Bartolo, N. Pallua, P. Blondeel, and D. von Heimburg, *J. Surg. Res.*, **144**, 82 (2008).

Chapter 6

ENZYMATICALLY TRIGGERED *IN SITU* GEL-FORMING BIOMATERIALS FOR REGENERATIVE MEDICINE

Yoon Ki Joung, Kyung Min Park, and Ki Dong Park*

Department of Molecular Science and Technology, Ajou University, San 5, Woncheon, Yeongtong, Suwon 443-749, South Korea

*kdp@ajou.ac.kr

Enzymatically triggered hydrogels formed *in situ* are emerging biomaterials that may have applications as injectable scaffolds for regenerative medicine. Although *in situ* formation is attractive for tissue regeneration, due to the injectable function enabling noninvasive therapy, there are many problems to be solved before they can be used clinically. Utilizing enzymes in hydrogelation might overcome the problems of hydrogels formed *in situ*. Recent examples of these studies include peroxidase, transglutaminase, phosphatase, and others. Enzyme-triggered hydrogels show predominant properties in rapid gelation and controlled mechanical strength, compared with existing hydrogels formed *in situ*. Therefore, the enzyme-triggered hydrogels have significant promise as noninvasive injectable scaffolds for tissue regeneration.

Handbook of Intelligent Scaffolds for Tissue Engineering and Regenerative Medicine
Edited by Gilson Khang

www.panstanford.com

6.1 Introduction

Hydrogels are used widely as a type of biomaterial attractive to matrices for regenerative medicine on account of their biocompatibility and tissuelike properties, such as viscoelasticity, diffusive transport, and low mechanical irritation to the surrounding tissues when implanted *in vivo*, as well as interstitial flow properties.[1,2] When injected or implanted in a defect site, the hydrogel temporarily plays the role of the extracellular matrix (ECM) in the early stages of tissue formation from cells.[3] As tissue formation progresses, the hydrogel matrix should be certainly replaced by naturally secreted ECMs and spaces for proliferation and the migration of cells.[4] Therefore, the tissue-specific nature and biodegradation of hydrogels are essential properties to be temporally and spatially controlled during the process of tissue formation, along with various bioactivities and mechanical properties. In addition to the hydrogel properties, *in situ* gelation that causes a sol–gel transition in time for practical use has attracted considerable interest for injectable scaffolds. Over the last two decades, many studies concentrated on hydrogels formed *in situ* for the purpose of developing injectable scaffolds.[5–8] *In situ* gelation can be obtained when certain molecules (mainly polymers) are assembled, cross-linked, or polymerized to form a polymeric network by the responses to external stimuli (i.e., temperature, pH, ionic strength, etc.) or mild chemical reactions. At the early stage of this field, many studies focused on stimuli (especially temperature)-responsive hydrogels due to their nontoxic condition and reversible transitions. However, these hydrogels exhibited many problems, such as poor solubility, high viscosity, slow phase transition time, pH changes during degradation, and weak mechanical strength, even though the bioactive effect is promising. Hydrogels formed *in situ* using mild chemical reactions, such as the Michael reaction, also have similar problems to stimuli-responsive hydrogels. In this chapter, enzyme-triggered hydrogels are suggested as an alternative option for such hydrogels formed *in situ*. First, current studies and problems of hydrogels formed *in situ* are introduced. Second, the basic knowledge and current studies of enzyme-triggered hydrogels are presented. Finally, the recent examples and future prospects of hydrogels formed *in situ* triggered by a variety of enzymes are introduced.

6.2 *In situ*–Formed Hydrogels as Injectable Scaffolds

Since a hydrogel was first applied to soft lenses in the 1960s, a large number of hydrogels have been exploited and investigated for pharmaceutical and biomedical applications. Among these, hydrogels formed *in situ* are advantageous in avoiding invasive surgical implantation, because a polymer solution is injected and then forms a hydrogel with the desired shape in the body. In other words, a solution of the gel precursor mixed with cells can potentially be injected through a needle to fill a defect many times larger than the injection needle itself.

To accomplish such an injectable function, the hydrogel should be obtained using a particular cross-linking method enabling rapid and nontoxic gelation under mild aqueous conditions. *In situ* cross-linking methods for injectable hydrogels have been developed by the self-assembly of molecules by stimuli (temperature, pH, ionic strength, etc.), mild reactions, such as Michael and Schiff-base reactions, or photo-cross-linking or photo-polymerization.[9–11] These methods have accompanied a lack of function in practical uses, such as hydrogel strength, solubility, solution viscosity, and toxicity. In recent years, an attractive approach using an enzyme-catalyzed reaction was suggested for the *in situ* formation of hydrogels. This approach enables hydrogels not only to be formed under mild conditions for short periods but also to be strengthened due to chemical cross-linking, which are major issues in the development of hydrogels *in situ*.

In addition to an injectable function, a hydrogel as a matrix for regenerative medicine requires more complicated functions as well as physical and biological properties. These functions and properties include controlled biodegradation, insoluble signaling molecules, soluble growth factors, and tissue-specific matrix stiffness, which we will not deal with in this chapter because they have been reviewed in many books.

6.3 Enzyme-Triggered Hydrogels

Several of the benefits that can be obtained using enzymes in hydrogel formation are summarized as follows:[12]

(i) Enzymes can form covalent bonds between substrates, yielding bonds generally stronger and more permanent than physical cross-links.
(ii) Enzymes often exhibit a high degree of substrate specificity, potentially allowing the sparing of pendant peptides or polysaccharide moieties meant for interactions with cell surface receptors during the cross-linking reaction.
(iii) Catalysis under mild reaction conditions with regard to temperature, pressure, and pH ensures the safety of encapsulated cells or biomolecules.
(iv) High enantio-, regio-, and chemo-selectivity, as well as the regulation of stereochemistry, provides the development of new reactions for functional compounds for pharmaceuticals.
(v) High efficiency of enzymatic reactions provides a much lower viscous solution than physical hydrogels, enabling easy handling and homogeneous loading of cells or biomolecules.
(vi) Enzymatic cross-linking offers the potential for kinetic control of gel formation and the formation of homogeneous gels *in situ* via simple control of the enzyme concentration because the amount of enzyme present is one of the key determinants of the overall cross-linking rate. This type of kinetic control of gel formation can greatly assist in the ability to deliver cell-based therapies in a noninvasive manner.

In recent years, the *in vitro* synthesis of polymers through enzymatic catalysis (*enzymatic polymerization*) has been studied extensively.[13–21] These studies show that many families of enzymes, such as oxidoreductases, hydrolases, and transferases, can be used in the polymerization of useful polymers. In 2002, Kaplan *et al.* reported a trial of hydrogel formation using enzymes (peroxidases) and poly(aspartic acid). Seven peroxidases were evaluated for cross-linking poly(aspartic acid)s, and the properties of the resulting hydrogels, such as gelation time, swelling, sol–gel ratio, and mechan-

Table 6.1. Substrates, gelation times, and mechanical strengths of major enzyme-triggered hydrogels.

Enzyme	Substrate	Gelation time	Mechanical strength	Reference
HRP	(HA)-TA	~10–400s	850–3, 000 Pa[b]	Kurisawa *et al.* Calabro *et al.*
	Chitosan-PA	~5–400s	0.5–2.5 N[a]	Feijen *et al.*
	Chitosan-GA/PA	~10–250s	>10^3 Pa[b]	Sakai *et al.*
	Dextran-TA	~10–250s	~10^4 Pa[b]	Feijen *et al.*
	Gelatin-TA	~5–150s	0.07–0.2 N[a]	Sakai *et al.*
	Alginate-TA	~10–250s	0.01–0.1 N/mm^2[c]	Sakai *et al.*
	Carboxymethyl-cellulose-TA	~5–500s	N.A.	Sakai *et al.*
	Tetronic-TA	~2–120s	200–10^4 Pa[b]	Park *et al.*
	HA (or gelatin)-TA /Tetronic-TA	~5–60s	500–10^4 Pa[b]	Park *et al.*
	Tetronic-RGD/TA	~2–180s	500–10^4 Pa[b]	Park *et al.*
TGase	PEG-peptides	15–50 min	N.A.	Griffith *et al.* Messersmith *et al.*
	soy protein isolate	20–250 min	10–400 Pa[b]	Zhang *et al.*
	gelatin	2–3 hr	~10^2 Pa[b]	Kobayashi *et al.* Payne *et al.*
	fibrinogen	<1 hr	~200–3000Pa[b]	Hubbell *et al.* Lutolf *et al.*
Phosphatase	Fmoc-peptides	3-30 min	~20–600Pa[b]	Xu *et al.* Ulijn *et al.*
	oligopeptides	2-30 min	200-4000 Pa[b]	Xu *et al.*

[a] Repulsion force by compression.
[b] Storage modulus.
[c] Stress by strain.
Abbreviations: HRP, horseradish peroxidase; HA, hyaluronic acid; TGase, transglutaminase; TA, tyramine; RGD, Arg-Gly-Asp; PA, 3-(p-hydroxyphenyl)propionic acid; GA, glycolic acid; Fmoc, *N*-(fluorenylmethyloxy-carbonyl); N.A., not available.

ical strength, were characterized.[22] Many types of enzymes have the potential for chemical cross-linking and polymerization. Table 6.1 lists the major enzyme-triggered hydrogels formed *in situ*.

6.3.1 *HRP-Catalyzed Systems*

Peroxidase catalysis is an oxidation of a donor to an oxidized donor through the action of hydrogen peroxide, liberating two water molecules.[23] Several studies have focused on an interesting enzyme of peroxidase species. Horseradish peroxidase (HRP) is a single-chain α-type hemoprotein that catalyzes the decomposition of

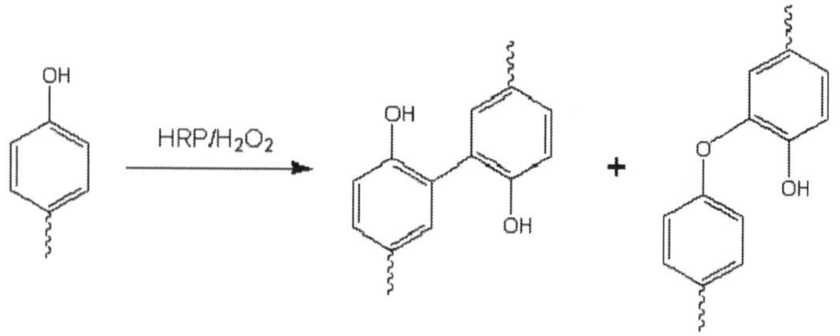

Figure 6.1. Cross-linking reaction of phenol substrates catalyzed by HRP.

hydrogen peroxide at the expense of aromatic proton donors.[24] HRP is an iron-containing porphyrin-type structure that catalyzes the coupling of a number of phenol and aniline derivatives using hydrogen peroxide as the oxidant. Fig. 6.1 shows the catalytic cycle of HRP for a phenol substrate.

Initial studies on the utilization of HRP concentrated on the polymerization of various types of phenol derivatives, even though they were not meant for biomedical applications. Most involved the polymerization of alkyl or aryl phenol derivatives.[25–28] Such studies were extended to the polymerization of tyrosine-containing peptides in buffer media.[29–33] Recently the HRP-catalyzed oxidation reached the polymerization of sugar-based phenols.[34] HRP has been applied to cross-linking of polymers, yielding hydrogels formed *in situ*. Kurisawa *et al.* prepared hyaluronic acid (HA) derivatives conjugated with tyramine (TA) to utilize a simple enzyme-catalyzed system in biocompatible hydrogels formed *in situ*.[35,36] They reported that the time when the solutions formed hydrogels was determined by the concentration of components, in which an HA derivative solution was mixed with HRP and hydrogen peroxide. In addition, the degradation of the formed hydrogels could be controlled by varying the concentration, meaning changes in the cross-linking density of the polymeric networks. They focused on making scaffolds for tissue regeneration, in which HA was selected as the main backbone of the network structure for this reason. The HA-based hydrogels were also developed by Calabro *et al.* in a similar manner for the purpose of scaffolds for cartilage regeneration.[37,38] Feijen's group applied dextran to the same system as Kurisawa's studies.[39] Dextran has been used in a range of hydrogels for drug delivery and tissue engineering, due to the good solubility in water, ease of chemical modification, and biocompatibility. They also demonstrated that the properties of the dextran hydrogels, such as gelation time, degradation time, swelling ratio, and elasticity, could be controlled by the concentrations of HRP, hydrogen peroxide, and polymers. This means that enzyme-triggered hydrogels have considerable potential as injectable biomaterials for tissue regeneration. In a subsequent study, researchers applied chitosan to a HRP-triggered system.[40] Chitosan was modified with glycolic acid and successively phloretic acid, followed by hydrogel characterization. The chitosan hydrogel was finally tested as a cellular matrix by chondrocyte encapsulation and culture, demonstrating cell viability for two weeks. Sakai *et al.* also suggested a novel derivative of chitosan, conjugated with 3-(*p*-hydroxyphenyl) propionic acid, which was soluble at neutral pH and *in situ* cross-linkable by an HRP reaction. The properties of chitosan hydrogels and their effect on cells were characterized.[41] The researchers prepared alginate derivatives for the HRP-triggering system.[42] Alginate is a natural polymer undergoing ionic gelation with divalent cations (e.g., calcium ion) and has a biocompatible nature. Alginate was modified with hydroxyl phenyl groups and characterized for hydrogel properties containing volume changes, mechanical properties, and gelation times. The alginate derivative is both ionically and enzymatically cross-linkable. Sakai's studies were extended to carboxymethylcellulose-based hydrogels. The hydrogel was characterized in the same manner as other hydrogels, and in particular, the cytotoxicity was tested using Crandell Rees feline kidney (CRFK) cells.[43] Gelatin was also applied to the HRP-triggerable system for hydrogels formed *in situ* by Sakai's group.[44] In their study, gelatin was conjugated with TA and the gelatin time and mechanical properties were characterized. The gelatin hydrogel was tested by a cell culture and injected subcutaneously into mice to test its feasibility as an injectable biomaterial for tissue engineering and drug delivery. The results showed that the coated and injected hydrogels are biocompatible and feasible for a cellular matrix. Recently, our group prepared a novel multiarm, amphiphilic copolymer derivative (Tetronic-TA) that was cross-linkable *in situ*

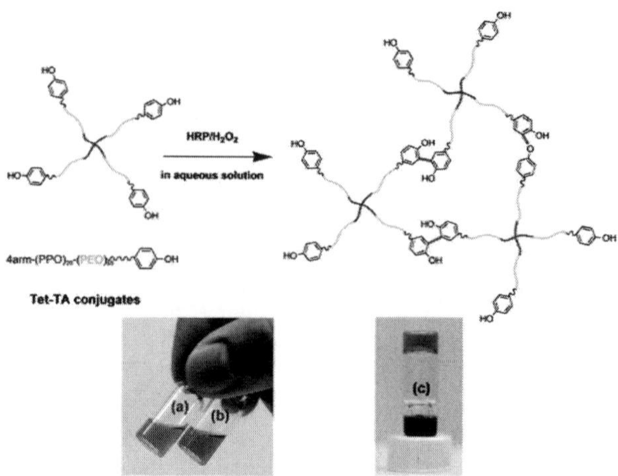

Figure 6.2. Schematic route of enzyme-triggered cross-linking of Tetronic–TA conjugate and images showing gelation. (a, b) Each solution contains HRP or H_2O_2; (c) inversed vial contains hydrogel. See also Color Insert.

by HRP (Fig. 6.2). The gelation time, degradation, and mechanical properties were characterized.[45] Unlike previous studies, conjugates of a synthetic four-arm amphiphilic copolymer (Tetronic) with TA were designed with various terminal links of the amphiphile to control the degradation of hydrogels. The utilization of synthetic polymers as a backbone chain resulted in strengthened mechanical properties as well as the simplified modulation of gelation and degradation. Our group also evaluated the effects of hydrogen peroxide and HRP on cytotoxicity under the condition of a cell-encapsulated culture for biomedical applications and revealed the nontoxic level of hydrogen peroxide and the nontoxic nature of HRP in cell culture (Fig. 6.3).

On the basis of this study, our efforts have been concentrated on the development of more tissue-compatible hydrogels in order to utilize injectable biomaterials for versatile biomedical applications. In further studies, phenol derivatives of natural polymers were cross-linked with Tetronic-TA to prepare hydrogels with improved biocompatibility (Fig. 6.4).[46] Combining HA or gelatin with the Tetronic-based hydrogel resulted in a significant improvement in biocompatibility. In addition, a route for the *in situ* conjugation of

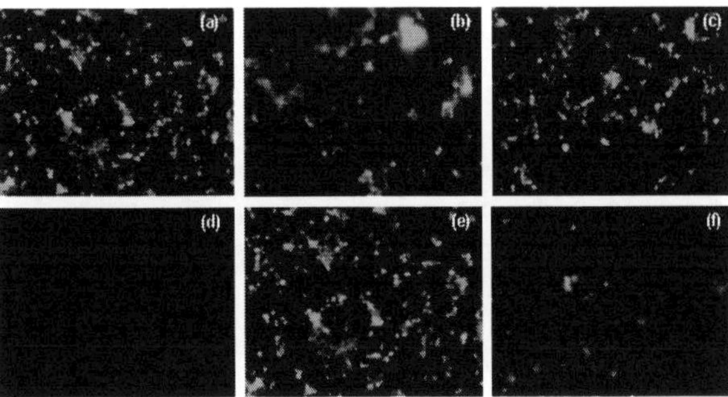

Figure 6.3. Fluorescence microscopic images of MC3T3-E1 cells in the Tetronic–TA hydrogel by live/dead assay. The hydrogel formed with 0.25 wt% H_2O_2 and (a) 0.6 mg/mL HRP; (b) 0.15 mg/mL HRP and 0.035 mg/mL HRP; (c) the hydrogel formed with 0.6 mg/mL HRP; (d) 1 wt% H_2O_2; (e) 0.25 wt% H_2O_2; (f) 0.063 wt% H_2O_2.

the Arg-Gly-Asp (RGD) peptide to the Tetronic hydrogel based on the HRP-triggering system was designed to examine cell adhesion on the hydrogels. The results showed that cell adhesion was enhanced in an RGD concentration–dependent manner. These studies are now progressing to obtain adjusted physical and biological properties as injectable biomaterials for a range of biomedical applications, including tissue regeneration.

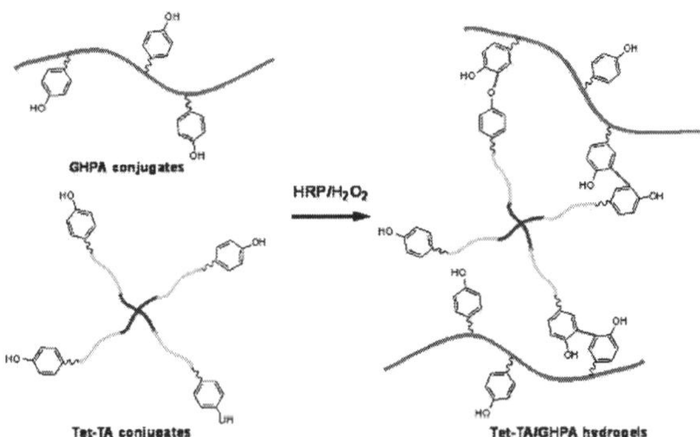

Figure 6.4. Scheme of Tetronic/gelatin hydrogel catalyzed by HRP and hydrogen peroxide. See also Color Insert.

6.3.2 *TGase-Catalyzed Systems*

Transglutaminase (TGase) is a family of enzymes whose function is to form an amide linkage between the γ-carboxamide group of certain peptidyl glutamine residues and primary amines, such as the ε-amino group of lysine (Fig. 6.5).[47–51] These calcium-dependent enzymes are ubiquitous in fluids and the ECM throughout the body, forming cross-links in the skin, liver, and blood clots. Therefore, the procedure can be adapted to human therapeutic use through the correct choice of a human-derived enzyme.

Griffith prepared enzymatically cross-linked hydrogels for the first time based on a poly(ethylene glycol) (PEG)-glutamine conjugate and synthetic peptides containing a lysine residue, catalyzed by TGase.[52] He confirmed that the ratio and total concentration of the macromers determine hydrogel formation and the gel properties, such as swelling ratio and entanglement ratio. He also examined the gelation kinetics of the PEG hydrogel.[53] Various types of PEG were used to predict the gelation kinetics, and the gelation time was controlled by varying the macromer and enzyme concentrations within ~10–60 minutes. Thompson prepared fibrinogen-based hydrogels *in situ,* using an interesting strategy that uses phototriggerable liposome and TGase. He utilized the photosensitive release of entrapped liposomal Ca^{2+} as an initiator of the TGase-mediated formation of fibrinogen hydrogels. The gelation arose from TGase-catalyzed cross-linking of predominately α and γ chains of fibrinogen. Gelatin was also applied to TGase-catalyzed systems by two groups. Payne *et al.* prepared gelatin hydrogels catalyzed by TGase and characterized the mechanical properties and cell entrapment, even though they targeted the microfluidic biosensor system.[54] Kobayashi *et al.*

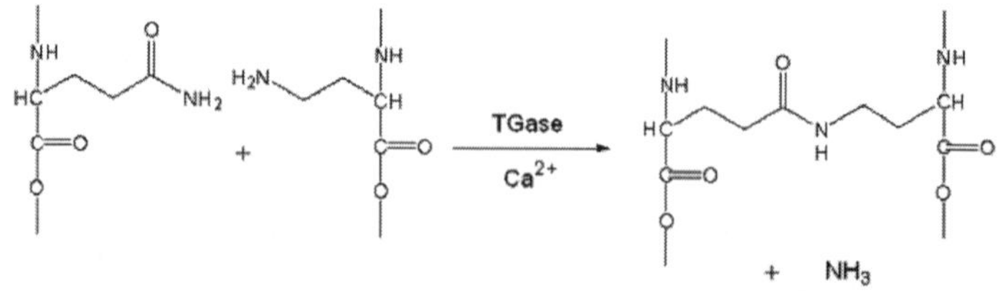

Figure 6.5. Scheme of the glutamine-lysine reaction catalyzed by calcium-dependent TGase.

prepared TGase-catalyzed gelatin hydrogels as potential scaffolds for tissue engineering.[55] They incorporated the RGD peptide, a cell adhesion factor, into the gelatin hydrogel and tested the proliferation of fibroblasts (NIH3T3) on a hydrogel matrix, resulting in significantly improved cell proliferation. Messersmith reported a rational design of short peptide substrates of TGase conjugated with a biocompatible polymer, PEG, with the intention of preparing an hydrogel *in situ* for many biomedical applications.[56] He also incorporated L-3,4-dihydroxylphenlylalanine (DOPA), an adhesive amino acid found in marine mussel proteins, into his hydrogel system. The TGase-triggered system also produced a cross-linked protein hydrogel. Zhang prepared a protein-based hydrogel by cross-linking a soy protein isolate (SPI) in the presence of microbial TGase (MTGase), in which the gelation time and mechanical strength were tunable.

Factor XIIIa, a member of TGase enzymes, is a coagulation factor and a catalyst for cross-linking fibrinogens for fibrin gels.[57] Hubbell designed exogenous bifunctional peptides into a fibrin gel, which is cross-linkable with thrombin-activated factor XIIIa.[58] The hydrogel was characterized as an enzyme-triggered proteolytic matrix. Hubbell and Lutolf prepared a factor XIIIa–triggerable fibrinlike hydrogel conjugated with a cell-adhesive peptide, RGD. The degradation of the hydrogel was tested by matrix metalloproteinase (MMP) derived from cells.[59]

6.3.3 *Other Enzyme-Catalyzed Systems*

In addition to HRP and TGase, other enzymes have also been exploited for *in situ* hydrogel formation. Xu *et al.* reported the first example of supramolecular hydrogels that respond to ligand-receptor interactions.[60] The researchers designed a compound, *N*-(flurenyl-9-methoxycarbonyl) (Fmoc)-D-Ala-D-Ala, as a receptor, targeting ligand-receptor binding with vancomycin (Van) as its ligand. The Fmoc-dipeptides are stereo-specific compounds, where Van shows no response to its enantiomer, and cause hydrogelation by π–π overlap of these amino acids. The microstructure was shown to be a nanofibrous network of stacked amino acids. They also reported another example of the enzymatic formation of supramolecular hydrogels.[61,62] In addition, they used an enzymatic reaction to convert the ionic group on an amino acid derivative into a neutral group, resulting in the formation of a small-molecular hydrogelator and supramolecular hydrogels. Gelation was processed by the dephosphorylation of the PO_4^{3-} of Fmoc tyrosine phosphate under basic conditions, which are mediated by alkaline phosphatase, one of the components of the kinase/phosphatase switches that regulate the protein activity. They prepared amphiphilic Fmoc–amino acid derivatives as hydrogelators to adjust the hydrophilic/hydrophobic balance of amino acids for hydrogelation. In a recent paper, Ulijn *et al.* reported a novel approach of enzyme-triggered hydrogelation using proteases, which are enzymes that normally hydrolyze peptide bonds in aqueous media, to perform the reverse reaction to produce an amphiphilic peptide hydrogelator that self-assembles to form a nanofibrous structure.[63] The researchers employed thermolysin from *Thermoprotelolyticus rokko* as a suitable enzyme to test this approach. Thermolysin has been used in reverse hydrolysis reactions and has a well-known preference for the hydrophobic/aromatic residues on the amine side of the peptide bond. The researchers prepared Fmoc-conjugated peptide amphiphiles that self-assemble into nanofibrous structures derived from π-stacking of the highly conjugated fluorenyl group, further stabilized by the formation of helical structures. They applied this system to a novel PEG-based hydrogel particle that was responsive to enzymes for controlled release.[64] A copolymer of PEG and acrylamide form a hydrogel particle incorporated with Fmoc-peptides with an enzyme-cleavable linker.

6.4 Conclusions and Outlook

The chapter provides information on enzymatically triggered *in situ* gel-forming biomaterials, which have attracted considerable attention among intelligent scaffolds for tissue regeneration. Hydrogels formed *in situ* have been developed steadily and are still under investigation because of their potential applications. However, many efforts to accomplish practical hydrogels as injectable biomaterials have ended in failure due to insufficiencies in safe cell encapsulation, solidity after hydrogelation, and signaling molecules to cells. Enzyme-triggered hydrogels have many advantages as injectable scaffolds, which can overcome the limitations of previous injectable biomaterials. Among them, HRP-catalyzed systems are probably more advantageous than other enzyme-triggered systems that show a slow gelation time of several tens to hundreds of minutes and relatively lower mechanical strength. The substrate for enzyme catalysis is also important for achieving successful scaffolding for tissue regeneration. Synthetic polymer derivatives do not retain their bioactivity, even though they are varied in chemical design. Natural biomolecules containing polysaccharides and proteins/peptides have disadvantages such as relatively poor physical and mechanical properties despite good biocompatibility and bioactivity. Therefore, a rational design of both advantages can be effective, for example, the joint of synthetic and natural polymers. To our knowledge, enzyme-triggered hydrogels have undoubtedly provided an advance that allows not only the delivery of cells safely into the body but also the immobilization of bioactive macromolecules. On the basis of these advances, the next challenges of injectable hydrogels for intelligent scaffolds will be to design and decorate microenvironments in a matrix to take the place of natural ECMs.

Acknowledgments

This work was supported by the Nano-Biotechnology Project (Regenomics) (B020214) and the Pioneer Research Program for converging technology (M10711060001-08M1106-00110) through the Korea Science and Engineering Foundation funded by the Ministry of Education, Science and Technology.

References

1. T. R. Hoare and D. S. Kohane, *Polymer*, **49**, 1993 (2008).
2. B. J. Baroli, *Pharm. Sci.*, **96**, 2197 (2007).
3. E. Martin, C. R. Simone, G. S. Ronald, S. M. Blanca, J. A. Hubbell, E. W. Franz, and P. L. Matthias, *Biomacromolecules*, **8**, 3000 (2007).

4. Y. K. Joung, J. H. Choi, K. M. Park, D. H. Go, J. W. Bae, and K. D. Park, *Biomed. Mater.*, **2**, 269 (2007).
5. K. M. Park, Y. K. Joung , K. D. Park, S. Y. Lee, and M. C. Lee, *Macromol. Res.*, **16**, 517 (2008).
6. X. Z. Shu, Y. C. Liu, F. S. Palumbo, Y. Lu, and G. D. Prestwich, *Biomaterials*, **25**, 1339 (2004).
7. A. B. Pratt, F. E. Weber, H. G. Schmoekel, R. Muller, and J. A. Hubbell, *Biotechnol. Bioeng.*, **86**, 27 (2004).
8. S. Sharifi, H. Mirzadeh, M. Imani, M. Atai, and F. Ziaee, *J. Biomed. Mater. Res. A*, **84A**, 545 (2008).
9. X. Z. Shu, Y. Liu, F. S. Palumbo, Y. Luo, and G. D. Prestwich, *Biomaterials*, **25**, 1339 (2004).
10. T. Ito, Y. Yeo, C. B. Highley, E. Bellas, C. A. Benitez, and D. S. Kohane, *Biomaterials*, **28**, 3418 (2007).
11. D. D. Hou, X. M. Tong, H. Q. Yu, A. Y. Zhang, and Z. G. Feng, *Biomed. Mater.*, **2**, S147 (2007).
12. S. Kobayashi, H. Uyama, and S. Kimura, *Chem. Rev.*, **101**, 3793 (2001).
13. S. Kobayashi, S. Shoda, and H. Uyama, *Adv. Polym. Sci.*, **121**, 1 (1995).
14. S. Kobayashi, S. Shoda, and H. Uyama, In Ed. J. S. Salamone, *The Polymeric Materials Encyclopedia* (CRC Press, Boca Raton, 1996), p. 2102.
15. S. Kobayashi, S. Shoda, and H. Uyama, In Ed. S. Kobayashi, *Catalysis in Precision Polymerization* (John Wiley & Sons, Chichester, 1997), chapter 8.
16. H. Ritter, In Ed. R. Arshady, *Desk Reference of Functional Polymers, Syntheses and Applications* (American Chemical Society, Washington, DC, 1997), p. 103.
17. R. A. Gross, D. L. Kaplan, G. Swift, Eds., *ACS Symposium Series* Vol. 684 (American Chemical Society, Washington, DC, 1998).
18. S. Kobayashi, H. Uyama, In Ed. A.-D. Schlu¨ter, *Materials Science and Technology-Synthesis of Polymers* (Wiley-VCH, Weinheim, 1998), chapter 16.
19. H. Joo, Y. J. Yoo, and J. S. Dordick, *Korean J. Chem. Eng.*, **15**, 362 (1998).
20. S. Kobayashi and H. Uyama, In Ed. J. I. Kroschwitz, *Encyclopedia of Polymer Science and Technology,* 3rd ed. (John Wiley & Sons, NY, 2002).
21. S. Kobayashi, H. Uyama, and M. Ohmae, *Bull. Chem. Soc. Jpn.*, **74**, 613 (2001).
22. S. J. Sofia, A. Singh, and D. L. Kaplan, *J. Macromol. Sci. A Pure Appl. Chem.*, **A39**, 1151 (2002).

23. M. P. J. Van Deurzen, F. Van Rantwijk, and R. A. Sheldon, *Tetrahedron*, **53**, 13183 (1997).
24. D. Buchanan and J. A. Nicell, *Biotechnol. Bioeng.*, **54**, 251 (1997).
25. H. Uyama, H. Kurioka, I. Kaneko, and S. Kobayashi, *Chem. Lett.*, 423 (1994).
26. A. Akkara, K. J. Senecal, and D. L. Kaplan, *J. Polym. Sci. Polym. Chem. Ed.*, **29**, 1561 (1991).
27. S. Alva, P. L. Nayak, J. Kumar, and S. K. Tripathy, *J. Macromol. Sci. Pure Appl. Chem.*, **A34**, 665 (1997).
28. J. S. Dordick, M. A. Marletta, and A. M. Klibanov, *Biotech. Bioeng.*, **30**, 31 (1987).
29. T. Michon, M. Chenu, N. Kellershon, M. Desmadril, and J. Gueguen, *Biochemistry*, **36**, 8504 (1997).
30. P. Wang and J. S. Dordick, *Macromolecules*, **31**, 941 (1998).
31. P. Wang, B. D. Martin, S. Parida, D. G. Rethwisch, and J. S. Dordick, *J. Am. Chem. Soc.*, **117**, 12885 (1995).
32. Kobayashi, H. Sumitomo, and Y. Ina, *Polym. J.*, **17**, 567 (1985).
33. T. Uchida, T. Serizawa, and M. Akashi, *Polym. J.*, **31**, 970 (1999).
34. S. Tawaki, Y. Uchida, Y. Maeda, and I. Ikeda, *Carbohyd. Polym.*, **59**, 71 (2005).
35. Kurisawa, J. E. Chung, Y. Y. Yang, S. J. Gao, and H. Uyama, *Chem. Commun.*, **34**, 4312 (2005).
36. F. Lee, J. E. Chung, and M. Kurisawa, *J. Control. Rel.*, **134**, 186 (2009).
37. A. Calabro *et al.*, *U.S. Patent No. 7*, **368**, 502 (2008).
38. Calabro *et al.*, *U.S. Patent No. 7*, **465** 766, (2008).
39. R. Jin, C. Hiemstra, Z. Zhong, and J. Feijen, *Biomaterials*, **28**, 2791 (2007).
40. R. Jin, L. S. M. Teixeira, P. J. Dijkstra, M. Karperien, C. A. van Blitterswijk, Z. Y. Zhong, and J. Feijen, *Biomaterials*, **X**, 1 (2009).
41. S. Sakai, Y. Yamada, T. Zenke, and K. Kawakami, *J. Mater. Chem.*, **19**, 230 (2009).
42. S. Sakai and K. Kawakami, *Acta Biomater.*, **3**, 495 (2007).
43. Y. Ogushi, S. Sakai, and K. Kawakami, *J. Biosci. Bioeng.*, **104**, 30 (2007).
44. S. Sakai, K. Hirose, K. Taguchi, Y. Ogushi, and K. Kawakami, *Biomaterials*, **X**, 1 (2009).
45. K. M. Park, Y. M. Shin, Y. K. Joung, H. S. Shin, and K. D. Park, *Biomacromolecules*, 11, 706–712 (2010).
46. K. D. Park, K. M. Park, and Y. K. Joung, *PCT Appl.*, (2009).

47. D. Aeschlimann and M. Paulsson, *Thromb. Hemo.*, **71**, 402 (1994).
48. D. D. Clarke, M. J. Mycek, A. Neidle, and H. Waelsch, *Arch. Biochem. Biophys.*, **79**, 338 (1959).
49. J. E. Folk and J. S. Finlayson, *Adv. Protein Chem.*, **31**, 1 (1977).
50. S. S. Greenberg, P. J. Birchbichler, and R. H. Rice, *FASEB J.*, **5**, 3071 (1991).
51. L. Lorand and S. M. Conrad, *Mol. Cell Biochem.*, **58**, 9 (1984).
52. J. J. Sperinde and L. G. Griffith, *Macromolecules*, **30**, 18, 5255, (1997).
53. J. J. Sperinde and L. G. Griffith, *Macromolecules*, **33**, 5476 (2000).
54. T. Chen, D. A. Small, M. K. McDermott, W. E. Bentley, and G. F. Payne, *Biomacromolecules*, **4**, 1558 (2003).
55. A. Ito, A. Mase, Y. Takizawa, M. Shinkai, H. Honda, K.-I Hata, M. Ueda, and T. Kobayashi, *J. Biosci. Bioeng.*, **95**, 196 (2003).
56. B.-H. Hu and P. B. Messersmith, *J. Am. Chem. Soc.*, **125**, 14298 (2003).
57. T. J. Sanborn, P. B. Messersmith, and A. E. Barron, *Biomaterials*, **23**, 2703 (2002).
58. J. C. Schense and J. A. Hubbell, *Bioconjugate Chem.*, **10**, 75 (1999).
59. Ehrbar, S. C. Rizzi, R. G. Schoenmakers, B. S. Miguel, J. A. Hubbell, F. E. Weber, and M. P. Lutolf, *Biomacromolecules*, **8**, 3000 (2007).
60. Y. Zhang, H. Gu, Z. Yang, and B. Xu, *J. Am. Chem. Soc.*, **125**, 13680 (2003).
61. Z. Yang, G. Liang, M. Ma, Y. Gao, and B. Xu, *Small*, **3**, 558 (2007).
62. Z. Yang, H. Gu, D. Fu, P. Gao, J. K. Lam, and B. Xu, *Adv. Mater.*, **16**, 1440 (2004).
63. S. Toledano, R. J. Williams, V. Jayawarna, and R. V. Ulijn, *J. Am. Chem. Soc.*, **128**, 1070 (2006).
64. D. Thornton, R. J. Mart, and R. V. Ulijn, *Adv. Mater.*, **19**, 1252 (2007).

Chapter 7

THERMO-SENSITIVE INJECTABLE SCAFFOLDS FOR REGENERATIVE MEDICINE

Moon Suk Kim,[a*] Jae Ho Kim,[a] Gilson Khang,[b] and Hai Bang Lee[a]

[a] *Department of Molecular Science and Technology, Ajou University, Suwon 443-759, Korea*
[b] *Department of Polymer/Nano Science and Technology, Chonbuk National University, 664-14, Duckjin, Jeonju 561-756, Korea*
*moonskim@ajou.ac.kr

The main goal of this chapter is to provide an introduction of currently available *in situ*–forming hydrogel scaffolds. *In situ*–forming hydrogel scaffolds are based on the idea that if the liquid-to-gel phase transition takes place at body temperature, certain materials can be prepared as a liquid and then form a *in situ* macroscopic gel by a syringe injection of their aqueous solutions at target sites. In tissue engineering, the gel can be applied as a scaffold with unique properties of easy handling and minimal invasiveness. This chapter summarizes *in situ*–forming hydrogel scaffolds, especially via electrostatic and hydrophobic physical interactions, and briefly explains their application as scaffolds in tissue engineering.

Handbook of Intelligent Scaffolds for Tissue Engineering and Regenerative Medicine
Edited by Gilson Khang

www.panstanford.com

7.1 Introduction

Hydrogels are three-dimensional, hydrophilic, polymeric networks capable of imbibing large amounts of water or biological fluids.[1] Among several hydrogels, the *in situ*-forming hydrogels are based on the idea that if a hydrogel undergoes a simple liquid-to-gel phase transition under physiological conditions, this biomaterial could be prepared as a liquid and then form the *in situ* macroscopic gel spontaneously at the site of injection (Fig. 7.1).[2]

During the last decade, research interest has shifted from *ex vivo*-fabricated hydrogel scaffolds to *in situ*-forming hydrogel scaffolds, largely due to minimal invasiveness. Specifically, a hydrogel solution at room temperature can easily incorporate various cells and/or growth factors by simple mixing and offer the advantage of avoiding surgical procedures to fill the tissue void, thus facilitating patient compliance and quality of life via comfort.[3]

The method of preparing *in situ*-forming hydrogels can fall into two main categories, self-assembly by reversible either electrostatic or hydrophobic interactions and nonreversible chemical reactions after introducing nonreversible covalent bonds. Among these *in situ*-forming hydrogels, we are focusing on self-assembling hydrogels, which can be formed by physical interactions in response to biological conditions.

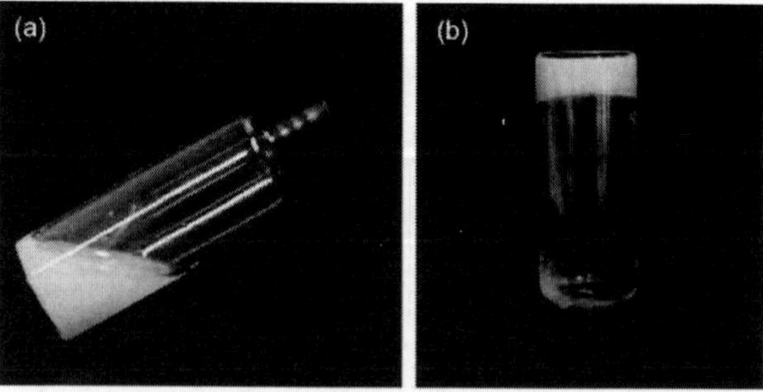

Figure 7.1. Images of hydrogel in (a) the liquid state at 25°C and (b) the gel state at 37°C. See also Color Insert.

In general, the most interesting biological stimulus is temperature.[2,3] Thus, both electrostatic and hydrophobic physical interactions by change of temperature under physiological conditions can be exploited in the design of *in situ*-forming hydrogels. This chapter provides an overview of creating *in situ*-forming hydrogels. Particular interest is given to *in situ*-forming hydrogel scaffolds that are applicable in scaffolds for regenerative medicine.

7.2 *In situ*–Forming Hydrogels Formed by Electrostatic Interactions

An electrostatic interaction can be induced by molecular interactions between electric charges such as cationic and anionic polyelectrolytes.[4,5] The magnitude of an electrostatic interaction between two electric charges is directly proportional to the magnitude of each of the charges. Thus, the intrinsic strength of electric charge is a very important point to create *in situ*-forming hydrogels. There are various anionic and cationic polymers with several properties. Recent studies have shown that a mixture of various anionic and cationic polymers is capable of forming a gel *in situ*.[4–6] The formation of a gel at certain temperatures can be attributed to the dominance of electrostatic association between electric charges such as cationic and anionic group inside the polymers.

Typical anionic polymers are alginate, sodium carboxymethylcellulose (CMC), hydroxypropylmethylcellulose (HPMC), tripolyphosphate, etc., as natural materials and polyacrylic acid derivatives as synthetic biomaterials.[6] And cationic polymers include polyethyleneimine (PEI), chitosan, polylysine, etc. Among several anionic and cationic polymers, introduction of *in situ*-forming hydrogels based on chitosan, alginate, and CMC via electrostatic interactions and their potential in *in situ*-forming hydrogel scaffolds are described later.

7.2.1 In situ–*Forming Chitosan Hydrogel Scaffolds*

Chitosan, an amino-polysaccharide obtained by an alkaline deacetylated derivative of chitin, is a natural and abundant polymer.[7] Its

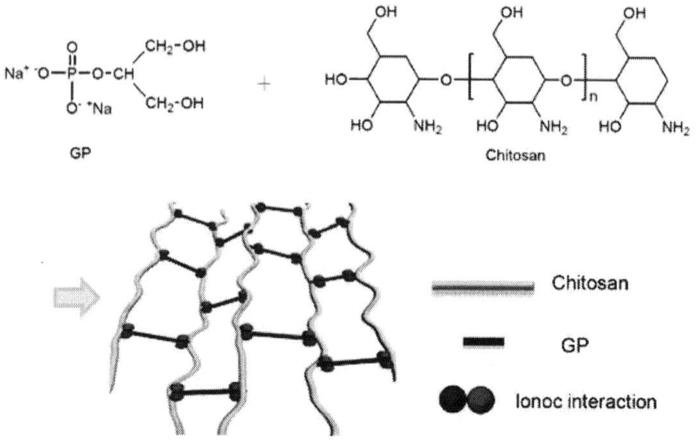

Figure 7.2. Schematic image of *in situ*-forming chitosan hydrogel in the presence of a GP disodium salt.

degradation occurs primarily by enzyme action and hydrolysis. The products of chitosan are saccharides and glucosamines, which are constituents of normal metabolites in mammals. Chitosan is also suggested to have minimal foreign body reaction, antithrombogenic properties, and stimulating effects on the immunosystem against viral and bacterial infections.[8] Thus, chitosan is used as an attractive candidate material for biomedical applications. It has a polycationic carbohydrate structure. It has been suggested that the cationic nature of chitosan provides a suitable point to act with anionic materials. Water-soluble charged materials may also form gels upon reaction with di- or trivalent counter ions. An electrostatic interaction can occur between polycationic chitosan and anionic glycerol phosphate (GP) disodium salt as one example (Fig. 7.2).[9,10]

The ammonium cations of the chitosan chains and the phosphate anions of GP induce electrostatic interactions, which are the main molecular forces involved in forming a gel with the desired mechanical properties.

A substantial change in the viscosity of GP-containing chitosan solutions is observed according to temperature changes (Fig. 7.3).

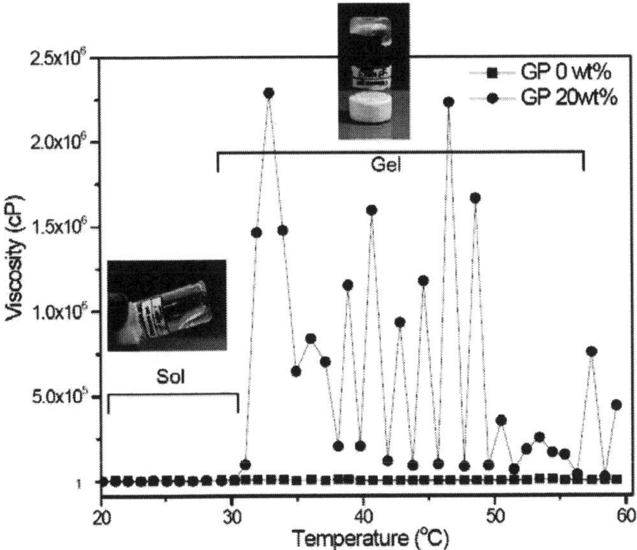

Figure 7.3. Viscosity vs. temperature curves for chitosan solutions without and with GP.

Besides GP, many anionic compounds such as polylactide (PLA), hyaluronic acid, thioglycolic acid, calcium phosphate cement, fucoidan, and heparin, can act as an electrostatic interaction point to make *in situ*-forming chitosan gel.[11–16]

When a chitosan solution in the presence of anionic compounds is injected into the body, it successfully forms a gel scaffold and, furthermore, maintains this structure for at least a few months. In addition, *in situ*-forming chitosan gel scaffolds act as suitable biocompatible substrates for the attachment and proliferation of stem cells (embryonic stem cells [ESCs], bone marrow stem cells [BMSCs], muscle-derived stem cells [MDSCs], or adipose-derived stem cells [ADSCs]). These characteristics suggest that *in situ*-forming chitosan gel scaffolds may be a promising alternative to *ex vivo*-fabricated scaffolds.[17–19]

With the eventual goal of applying *in situ*-forming chitosan gel scaffolds for injectable therapeutic tissue engineering, *in situ*-forming chitosan gel scaffolds can be utilized in tissue-engineered skin, cartilage, cardiac, bone, and nerve regeneration.[20]

The *in situ*–forming chitosan gel scaffolds described in these works offer numerous advantages as a noninvasive alternative for regenerative medicine.

7.2.2 In situ–*Forming Alginate Hydrogel Scaffolds*

Alginate, derived from seaweed, is an anionic linear polysaccharide with mannuronic and glucoronic acid as repeating units.[21] Alginates are a well-known example of ionically cross-linkable materials. Aqueous solutions of alginate form three-dimensional hydrogels when mixed with divalent or polyvalent cations.[22–24] Alginate gels were extensively studied for tissue engineering applications as a cell encapsulation material as well as an injectable three-dimensional matrix for *in vivo* cell delivery.[25]

Several reports demonstrated alginate gels as carriers of various cells for cartilage, bone, and nerve repair.[26–28] Gelation time and mechanical properties of *in situ*–forming alginate scaffolds have a dependence on the polymer concentration and the type of cation. Calcium cross-linking has been shown to yield gels with good mechanical properties. The use of alginate as *in situ*–forming alginate hydrogel scaffolds exhibits enhanced immunogenicity and poor bioresorbability, which may lead to adverse tissue interactions.

7.3 *In situ*–Forming Hydrogels Formed by Hydrophobic Interactions

By far the most studied type of physical interactions useful for *in situ*–forming hydrogels are hydrophobic interactions. Changes in physical interactions caused by environmentally induced swelling were first studied several decades ago.[29,30] Dusek *et al.* predicted that changes in external stimuli might result in phase transitions via hydrophobic interactions,[31] a theory since verified experimentally by several researchers.[32,33] The main advantage of these types of interactions is the ability to tailor their physical properties to respond to a particular physiological stimulus. Various *in situ*–forming hydrogels from typical biodegradable or bioabsorbable synthetic polymers reported so far are shown in Fig. 7.4.

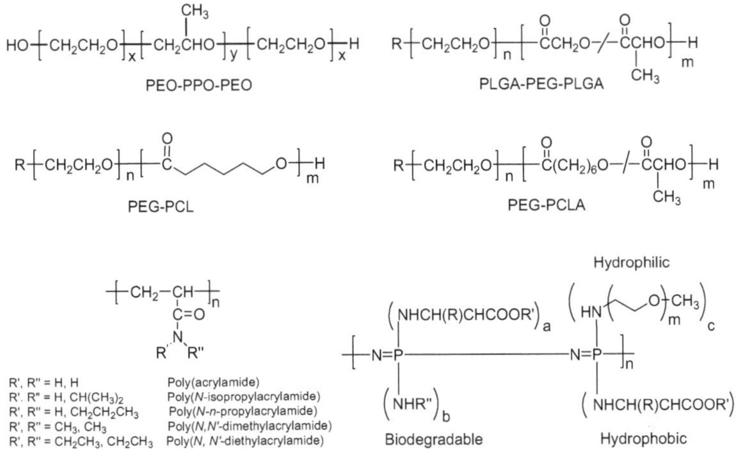

PEO-PPO-PEO

PLGA-PEG-PLGA

PEG-PCL

PEG-PCLA

R', R'' = H, H Poly(acrylamide)
R', R'' = H, $CH(CH_3)_2$ Poly(*N*-isopropylacrylamide)
R', R'' = H, $CH_2CH_2CH_3$ Poly(*N*-*n*-propylacrylamide)
R', R'' = CH_3, CH_3 Poly(*N*,*N*'-dimethylacrylamide)
R', R'' = CH_2CH_3, CH_2CH_3 Poly(*N*, *N*'-diethylacrylamide)

Hydrophilic

Biodegradable

Hydrophobic

Figure 7.4. Typical polymer structure of *in situ*-forming hydrogels via hydrophobic interactions.

Synthetic polymers used for tissue engineering must possess specific bulk and surface properties, easy handling at room temperature, amenability to sterilization, and structural integrity during *in vivo* application. For successful application in the regenerative medicine, further important polymer properties are adjustable degradation, which should correspond with the production rate of a new extracellular matrix by the incorporated cells to maintain a balance of assembly of new tissue and degradation of the scaffold material. Another crucial prerequisite for the practical *in vivo* application of scaffolds is an understanding of the potential for a host tissue response after implantation.

Poly(ethylene glycol) (PEG) is a nontoxic and biocompatible material that is widely used in biomedical applications.[34] This makes for a longer-acting medicinal effect and reduces toxicity, and it allows longer dosing intervals. on the basis of this reason, PEG may provide a useful starting point to make *in situ*-forming hydrogels. In addition, the PEG–polyester copolymers have a relatively easier-to-design molecular criterion for achieving optimal tissue engineering. Thus, these *in situ*-forming hydrogels from synthetic polymers, especially PEG– polypropylene glycol (PPG) and PEG–polyester block copolymers, will be highlighted in the next section.

7.3.1 *PEG–PPG Block Copolymers as* in situ*–Forming Hydrogel Scaffolds*

Triblock copolymers (poly[ethylene oxide]-b-poly[propylene oxide] -b-poly[ethylene oxide], PEO–PPO–PEO), known as Pluronics® (BASF) or Poloxamer (ICI), are initially commercialized nonionic surfactants. In the past few decades, Pluronics perhaps is the most intensively investigated thermo-response polymer for biomedical application.[35]

PEG and PPG blocks exhibit hydrophilic and hydrophobic properties, respectively. By adjusting the composition, aqueous solutions of commercial Pluronics or Poloxamer series can exhibit phase transitions from sol-to-gel at physiological temperature.[36,37] These block copolymer solutions became gel-faster at body temperature, but the gels formed are mechanically weak and have limited stability and short residence times, resulting in relatively rapid erosion of the gels at the injection site. In addition, block copolymers are nonbiodegradable polymers. To circumvent the drawbacks of common Pluronic gels, multiblock Pluronic copolymers via linking biodegradable carbonate, ester, disulfide, urea, or urethane bonds were designed and synthesized.[38–40]

Despite these disadvantages, Pluronics have been tried in tissue engineering as *in situ*–forming hydrogels. Matthew *et al.* investigated the effect of mammalian cell culture medium compared to pure water on the gelation properties of Pluronics. There is no difference in the thermodynamics of gel formation between cell culture medium and pure water.[41]

Pluronics represent a bio-inert environment attributed by the hydrophilicity and flexibility of PEO chains.[42] It was shown that a Pluronic gel can be a good substrate for hematopoietic stem cells, supporting their culture and preservation more than conventional tissue culture dishes.[43,44] Maheu *et al.* have studied the effect of Pluronics on osteoblast viability and phenotype maintenance *in vitro*. When supplemented with osteoactive factors, a Pluronic gel could be used as a cell carrier without altering osteoblast phenotype for bone tissue engineering.[45] A Pluronic gel was also reported as a scaffold for osteoblastic differentiation of rabbit mesenchymal stem cells.[46] To increase clinical demand for adipose tissue, a Pluronic gel has also been developed as a suitable scaffold for engineering adipose tissue constructs.[47] Use of a Pluronic gel as a suitable scaffold for adult-derived or somatic lung progenitor cells constructs resulted in the development of tissue with less inflammatory reaction compared to polyglycolic acid (PGA).[48]

7.3.2 *PEG–Other Degradable Polyesters as* in situ*–Forming Hydrogel Scaffolds*

Block copolymers contain both PEG as hydrophilic and polyester as hydrophobic segments investigated as a candidate for *in situ*–forming hydrogel scaffolds in order to endow good mechanical properties, biodegradability, long residence times, and no toxicity.

In 1997, Kim *et al.* reported the injectable gel system using biodegradable triblock copolymers, composed of poly(D,L-lactide-co-glycolide) (PLGA) and PEG.[49] After that, *in situ*-forming hydrogel scaffolds using various block copolymers composed of PEG and polyester blocks have been investigated extensively as well. The reported polyester blocks are poly(ε-caprolactone) poly(ε-caprolactone-*co*-D,L-lactic acid) poly(δ-valerolactone) (PVL), poly(trimethylene carbonate) (PTMC), poly(ε-carprolactone-*co*-trimethylene carbonate) (PCL-*co*-PTMC), poly(ε-carprolactone-*co*-1,4-dioxan-2-one) (PCL-*co*-PDO), poly(propylene fumarate) (PPF), poly-[(*R*)-3-hydroxybutyrate] (PHB), etc.[50–55]

The aqueous solutions of most block copolymers form a solution at ambient temperatures, due to the attribution to the dominance of hydrophilic association between the hydrophilic PEG segments and the surrounding water molecules. With an increase in temperature, however, the hydrophobic association between the polyester blocks causes the dehydration of hydrophobic segments and leads to the formation of hydrophobic interactions between hydrophobic domain states, which eventually results in a gel at body temperature (Fig. 7.5).

The sol–gel transition temperature was influenced by the copolymer composition and specifically by the length of PEG and polyester blocks. The gel formation occurring at body temperature makes this system an ideal candidate for *in situ*-forming hydrogel scaffolds for tissue engineering. We and some groups developed *in*

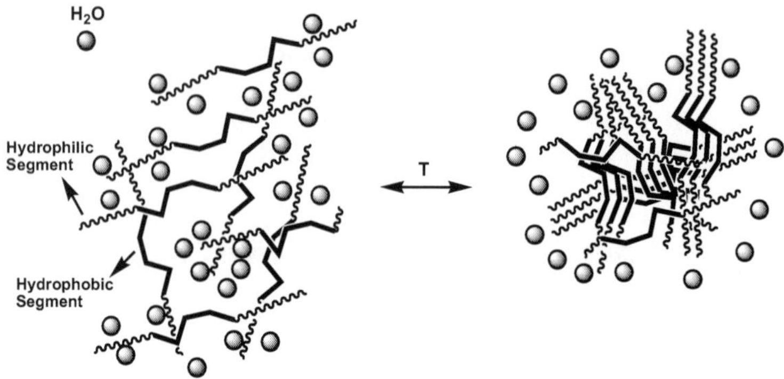

Figure 7.5. Schematic image of *in situ* gel formation via hydrophobic interactions.

situ-forming hydrogel scaffolds for autologous bone and articular cartilage tissue engineering using mesenchymal stem cells and chondrocytes.[54,56–58]

7.3.3 *Other Polymers as* in situ*–Forming Hydrogel Scaffolds*

One example is poly(organophosphazenes) as *in situ*-forming hydrogel scaffold. Poly(organophosphazenes) grafted with mPEG as hydrophilic and amino acid esters as hydrophobic segments were reported as a new class of biodegradable and *in situ*-forming hydrogels in 1999.[59] The polymers' showing their use as hepatocytes carriers holds promise to maintain good viability and liver-specific activity.[60]

Poly(N-isopropylacrylamide) (PNiPAAm) and its copolymers have also been popular in an attempt to yield hydrogels with thermal responsiveness and improved properties. PNiPAAm exhibits a sharp phase transition in an aqueous solution at body temperature.[61,62] Recent developments on PNiPAAm-based hydrogels include their use for cell encapsulation and delivery.[63] PNiPAAm as an *in situ*-forming hydrogel scaffold in combination with chondrogenic differentiation factors and growth factors (transforming growth factor β3) holds promise for orthopedic tissue engineering applications.[64] However, the limited issue of PNiPAAm that has to be addressed is nonbiodegradability. Future work using PNiPAAm as an *in situ*-forming gel scaffold is directed into further focusing on biodegradability to improve its potential in tissue engineering.

7.4 Conclusions and Outlook

With the rapid progress in regenerative medicine, the larger demand for scaffolds in tissue engineering is increasing. Research in the area of development of *in situ*-forming hydrogel scaffolds for tissue engineering has been well established in the past years. Typically, because aqueous solutions of *in situ*-forming hydrogel materials handled in this chapter are liquid at ambient temperature, they can easily incorporate various cells and/or biological factors by simple mixing. In addition, they offer the advantage of avoiding surgical procedures for application in regenerative medicine, because cells or biological factors are thus able to be *in situ* entrapped simply by injection of their aqueous solutions at target sites with minimal invasiveness. Recently, *in situ*-forming hydrogel scaffolds are focusing on biocompatibility, biodegradability, easy handling, and mechanical integrity during *in vivo* application and finally facilitating patient compliance and quality of life via comfort.

It is likely that *in situ*-forming hydrogel scaffolds can fulfill the requirements for specific applications in the future. Although much progress has been made in the fundamental research of *in situ*-forming hydrogel scaffolds, the various biomaterials will undoubtedly make *in situ*-forming hydrogel scaffolds an important topic in both chemistry and material sciences in the next decade.

References

1. K. Y. Lee and D. J. Mooney, *Chem. Rev.*, **101**, 1869 (2001).
2. L. Yu and J. Ding, *Chem. Soc. Rev.*, **37**, 14731 (2008).
3. L. Klouda and A. G. Mikos, *Eur. J. Pharm. Biopharm.*, **68,** 34 (2008).
4. T. Chakraborty, I. Chakraborty, and S. Ghosh, *Langmuir*, **22**, 9905 (2006).
5. L. Gärdlund, L. Wågberg, and R. Gernandt, *Colloids Surf. A Physicochem. Eng. Asp.*, **218**, 137 (2003).

6. B. D. Ratner, A. S. Hoffman, F. J. Schoen, and J. E. Lemons, *Biomaterials Science: An Introduction to Materials in Medicine*, 2nd ed. (Academic Press, NY, 2004).
7. D. L. Nettles, S. H. Elder, and J. A. Gilbert, *Tissue Eng.*, **8**, 1009 (2002).
8. N. Ishikawa, Y. Suzuki, M. Ohta, H. Cho, S. Suzuki, M. Dezawa, and C. Ide, *J. Biomed. Mater. Res. A*, **83**, 33 (2007).
9. A. Chenite, M. Buschmann, D. Wang, C. Chaput, and N. Kandani, *Carbohydrate Polym.*, **46**, 39 (2001).
10. N. Kashyap, B. Viswanad, G. Sharma, V. P. Ramarao, and M. N. V. Ravi Kumar, *Biomaterials*, **28**, 2051 (2007).
11. Y. Hong, Y. Gong, C. Gao, and J. Shen, *J. Biomed. Mater. Res A*, **85**, 628 (2008).
12. H. Tan, C. R. Chu, K. A. Payne, K. G. Marra, *Biomaterials*, **30**, 2499 (2009).
13. C. E. Kast, W. Frick, U. Losert, and A Bernkop-Schnürch, *Int. J. Pharm.*, **256**, 183 (2003).
14. H. H. Xu, M. D. Weir, and C. G. Simon, *Dent. Mater.*, **24**, 1212 (2008).
15. M. Fujita, M. Ishihara, M Simizu, K. Obara, T. Ishizuka, Y. Saito, H. Yura, Y. Morimoto, B. Takase, T. Matsui, M. Kikuchi, and T. Maehara, *Biomaterials*, **25**, 699 (2004).
16. S. Nakamura, M. Nambu, T. Ishizuka, H. Hattori, Y. Kanatani, B. Takase, S. Kishimoto, Y. Amano, H. Aoki, T. Kiyosawa, M. Ishihara, and T. Maehara, *J. Biomed. Mater. Res. A*, **85**, 619 (2008).
17. Y. Yin, F Ye, J Cui, F Zhang, X Li, and K Yao, *J. Biomed. Mater. Res. A*, **67**, 844 (2003).
18. K. S. Kim, H. H. Ahn, J. H. Lee, J. Y. Lee, B. Lee, H. B. Lee, and M. S. Kim, *Biomaterials*, **29**, 4420 (2008).
19. M. H. Cho, K. S. Kim, H. H. Ahn, Y. N. Shin, M. S. Kim, G. Khang, B. Lee, and H. B. Lee, *Tissue Eng.*, **14**, 1099 (2008).
20. E. Yilmaz, *Adv. Exp. Med. Biol.*, **553**, 59 (2004).
21. A. Gutowska, B. Jeong, and M. Jasionowski, *Anat. Rec.*, **263**, 342 (2001).
22. X. Li, T. Liu, K. Song, L. Yao, D. Ge, C. Bao, X. Ma, and Z. Cui, *Biotechnol. Prog.*, **22**, 1683 (2006).
23. B. J. Willenberg, T. Hamazaki, F. W. Meng, N. Terada, and C. Batich, *J. Biomed. Mater. Res. A*, **79**, 440 (2006).

24. E. J. Caterson, W. J. Li, L. J. Nesti, T. Albert, K. Danielson, and R. S. Tuan, *Ann. N. Y. Acad. Sci.*, **961**, 134 (2002).
25. A. Pielesz and M. K. Bak, *Int. J. Biol. Macromol.*, **43**, 438 (2008).
26. G. M. Williams, T. J. Klein, and R. L. Sah, *Acta Biomater.*, **1**, 625 (2005).
27. L. Wang, R. M. Shelton, P. R. Cooper, M. Lawson, J. T. Triffitt, and J. E. Barralet, *Biomaterials*, **24**, 3475 (2003).
28. S. Sakai, H. Masuhara, Y. Yamada, T. Ono., H. Ijima, and K. Kawakami, *J. Biosci. Bioeng.*, **100**, 127 (2005).
29. J. Kopecek, J. Vacík, and D. Lím, *J. Polym. Sci.*, **19**, 147 (1971).
30. Y. H. Bae, T. Okano, R. Hsu, and S. W. Kim, *Makromol. Chem. Rapid Commun.*, **8**, 481 (1987).
31. K. Dusek and D. Patterson, *J. Polym. Sci.*, **6**, 1209 (1968).
32. A. Suzuki and T. Tanaka, *Nature*, **346**, 345 (1990).
33. I. Ohmine and T. Tanaka, *J. Phys. Chem.*, **77**, 5725 (1982).
34. J. H. Lee, H. B. Lee, and J. D. Andrade, *Prog. Polym. Sci.*, **20**, 1043 (1995).
35. D. S. Hart and S. H. Gehrke, *J. Pharm. Sci. Mar.*, **96**, 484 (2007).
36. M. J. Song, D. S. Lee, J. H. Ahn, D. J. Kim, and S. C. Kim, *J. Polym. Sci Part A: Polym Chem.*, **42**, 772 (2004).
37. E. Ruel-Gariépy and J. C. Leroux, *Eur. J. Pharm. Biopharm.*, **58**, 409 (2004).
38. M. L. Adams, A. Lavasanifar, and G. S. Kwon, *J. Pharm. Sci.*, **92**, 1343 (2003).
39. P. N. Lan, S. Corneillie, E. Schacht, M. Davies, and A. Shard, *Biomaterials*, **17**, 2273 (1996).
40. J. H. Lee, Y. M. Ju, and D. M. Kim, *Biomaterials*, **21**, 683 (2000).
41. J. E. Matthew, Y. L. Nazario, S. C. Roberts, and S. R. Bhatia, *Biomaterials*, **23**, 4615 (2002).
42. W. Hu, J. J. Rathman, and J. J. Chalmers, *Biotechnol. Bioeng.*, **101**, 119 (2008).
43. A. Higuchi, K. Sugiyama, B. O. Yoon, M. Sakurai, M. Hara, M. Sumita, S. Sugawara, and T. Shirai, *Biomaterials*, **24**, 3235 (2003).
44. A. Higuchi, N. Aoki, T. Yamamoto, Y. Gomei, S. Egashira, Y. Matsuoka, T. Miyazaki, H. Fukushima, Y.M. Lee, S. Jyujyoji, and S.H. Natori, *Biomacromolecules*, **7**, 1083 (2006).

45. A. Higuchi, N. Aoki, T. Yamamoto, T. Miyazaki, H. Fukushima, T.M. Tak, S. Jyujyoji, S. Egashira, Y. Matsuoka, and S.H. Natori, *J. Biomed. Mater. Res. A*, **79**, 380 (2006).
46. J. M. Brunet-Maheu, J. C. Fernandes, C. A. de Lacerda, Q. Shi, M. Benderdour, P. Lavigne, *J. Biomater. Appl.*, doi:10.1177/0885328208096534 (2008).
47. J. W. Huang, W. J. Chen, S. K. Liao, C. Y. Yang, S. S. Lin, and C. C. Wu, *Chang Gung Med. J.*, **29**, 363 (2006).
48. A. V. Vashi, E. Keramidaris, K. M. Abberton, W. A. Morrison, J. L. Wilson, A. J. O'Connor, J. J. Cooper-White, and E. W. Thompson, *Biomaterials*, **29**, 573 (2008).
49. J. Cortiella, J. E. Nichols, K. Kojima, L. J. Bonassar, P. Dargon, A. K. Roy, M. P. Vacant, J. A. Niles, and C. A. Vacanti, *Tissue Eng.*, **12**, 1213 (2006).
50. B. Jeong, Y. H. Bae, D. S. Lee, and S. W. Kim, *Nature*, **388**, 860 (1997).
51. M. S. Kim, K. S. Seo, G. Khang, S. H. Cho, and H. B. Lee, *J. Polym. Sci., Part A: Polym. Chem.*, **42**, 5784 (2004).
52. S. W. Kim, H. J. Kim, K. E. Lee, S. S. Han, Y. S. Sohn, and B. Jeong, *Macromolecules*, **40**, 5519 (2007).
53. M. S. Kim, H. Hyun, G. Khang, and H. B. Lee, *Macromolecules*, **39**, 3099 (2006).
54. E. Behravesh, A. K. Shung, S. Jo, and A. G. Mikos, *Biomacromolecules*, **3**, 153 (2002).
55. J. P. Fisher, S. Jo, A. G. Mikos, and A. H. Reddi, *J. Biomed. Mater. Res. A*, **71**, 268 (2004).
56. X. J. Loh, S. H. Goh, and J. Li, *Biomaterials*, **28**, 4113 (2007).
57. H. K. Kim, W. S. Shim, S. E. Kim, K. H. Lee, E. Kang, J. H. Kim, K. Kim, I. C. Kwon, and D. S. Lee, *Tissue Eng. Part A*, **15**, 923 (2009).
58. M. S. Kim, S. K. Kim, S. H. Kim, H. Hyun, G. Khang, and H. B. Lee, *Tissue Eng.*, **12**, 2863 (2006).
59. H. A. Declercq, M. J. Cornelissen, T. L. Gorskiy, and E. H. Schacht, *J. Mater. Sci. Mater. Med.*, **17**, 113 (2006).
60. S. C. Song, S. B. Lee, J. I. Jin, and Y. S. Sohn, *Macromolecules*, **32**, 2188 (1999).
61. K. H. Park and S. C. Song, *J. Biosci. Bioeng.*, **101**, 238 (2006).
62. M. Nakayama, T. Okano, T. Miyazaki, F. Kohori, K. Sakai, and M. Yokoyama, *J. Control. Rel.*, **28**, 46 (2006).

63. Y. Pei, J. Chen, L. Yang, L. Shi, Q. Tao, B. Hui, and J. Li, *J. Biomater. Sci. Polym. Ed.*, **15**, 585 (2004).
64. H. Hatakeyama, A. Kikuchi, M. Yamato, and T. Okano, *Biomaterials*, **27**, 5069 (2006).
65. K. Na, J. H. Park, S. W. Kim, B. K. Sun, D. G. Woo, H. M. Chung, and K. H. Park, *Biomaterials*, **27**, 5951 (2006).

Chapter 8

PHOTOCURABLE HYDROGEL FOR TISSUE REGENERATION

Min Soo Bae and Il Keun Kwon*

Department of Maxillofacial Biomedical Engineering and
Institute of Oral Biology, School of Dentistry,
Kyung Hee University, 1 Hoegi-dong,
Dongdaemun-gu, Seoul 130-701, South Korea
*kwoni@khu.ac.kr

8.1 Introduction

An exciting and revolutionary strategy to treat patients who need new organs or tissues is the engineering of man-made organs or tissues (Fig. 8.1). Tissues or organs can be potentially engineered with a number of different strategies. Tissue-specific cells are isolated from a small tissue biopsy from the patient and harvested *in vitro*. The cells are subsequently mixed into three-dimensional polymer scaffolds that act like the natural extracellular matrices found in tissues. These scaffolds deliver the cells to the desired site in the patient's body, provide a space for new tissue formation, and potentially control the structure and function of the engineered tissue.[1,2] A variety of tissues are being engineered using this approach, including fabricated artery, bladder, skin, cartilage, bone, ligament, and

Handbook of Intelligent Scaffolds for Tissue Engineering and Regenerative Medicine
Edited by Gilson Khang

www.panstanford.com

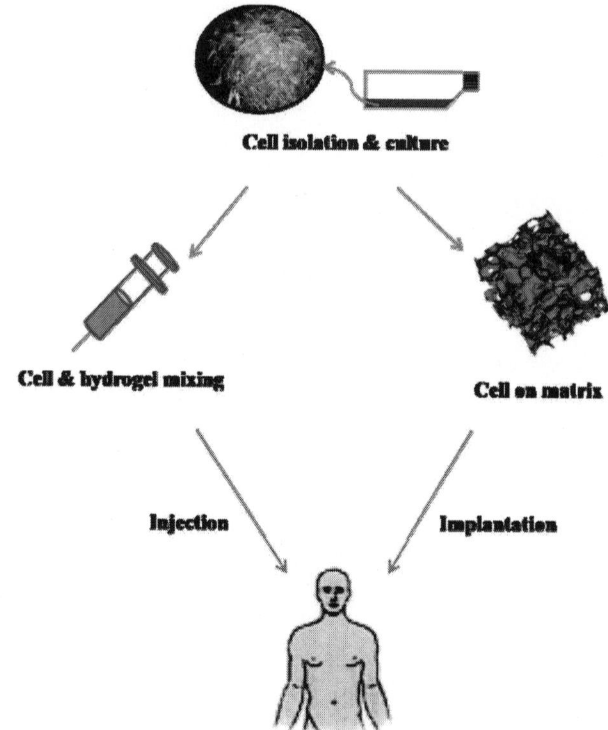

Figure 8.1. Schematic illustration of typical tissue engineering approaches. Cells are obtained from a small biopsy from a patient, expanded *in vitro*, and transplanted into the patient either by injection using a needle or other minimally invasive delivery approach or by implantation at the site following an incision (cut) by the surgeon to allow placement.

tendon. Several of these tissues are now at or near clinical stages.[3–8] In addition, various approaches have been introduced to form differentiated or undifferentiated cells (i.e., stem cells) into the desired cell phenotype.[9]

A critical element in virtually all tissue engineering approaches is the polymer scaffold. With good design the polymer mimics many roles of extracellular matrices found in tissues. Extracellular matrices, comprised of various amino acids and sugar-based macromolecules, bring cells together and control the tissue structure, regulate functions of the cells, and allow the diffusion of nutrients,

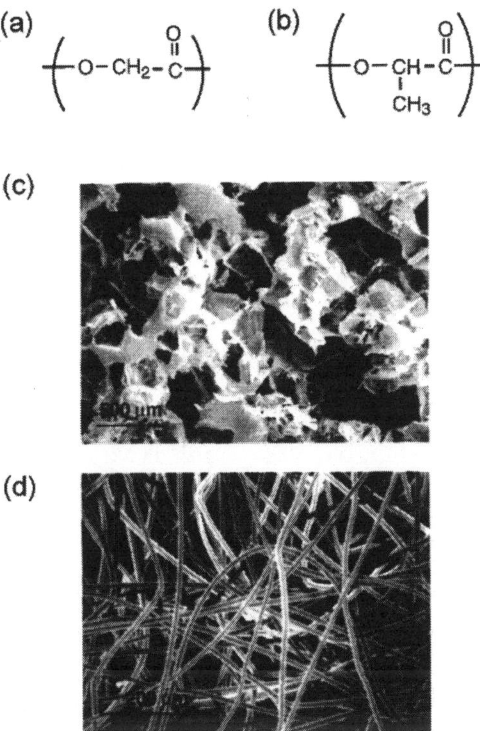

Figure 8.2. Chemical structure of (a) poly(glycolic acid), (b) poly(lactic acid), typical structures of (c) porous scaffolds of poly(lactide-*co*-glycolide), and (d) nonwoven fabrics of poly(glycolic acid). These latter materials have been widely used for tissue engineering applications. (Reprinted from Ref. 12; copyright 1998 John Wiley & Sons, Inc.).

metabolites, and growth factors.[10] To date, various types of polymers have been studied and utilized in tissue engineering.[11] Aliphatic polyesters, including poly(glycolic acid) (PGA), poly(lactic acid) (PLA), and the copolymer polylactic-co-glycolic acid (PLGA) of these materials, are the most widely used synthetic polymers (Fig. 8.2).[12,13] These polymers have a long history of use in medical applications and are considered safe in many situations by the Food and Drug Administration (FDA). However, the use of these types of polymer scaffolds requires the surgeon to make incisions (cuts) sufficiently large to enable placement of the polymer/cell constructs.

An exciting alternative approach to cell delivery for tissue engineering is the use of polymers (i.e., hydrogels) that can be injected into the body. This approach enables the clinician to transplant the cell and polymer combination in a minimally invasive manner. Hydrogels have structural similarity to the macromolecular-based components in the body and are considered biocompatible.[14] Hydrogels have found numerous applications in tissue engineering as well as in drug delivery. Tissue engineering is the most recent application of hydrogels, in which they are used as scaffolds to engineer new tissues (Fig. 8.1).[15]

8.2 Photopolymerization of Hydrogels

Visible or ultraviolet (UV) light can interact with light-sensitive compounds called photoinitiators to create free radicals, which can initiate polymerization to form cross-linked hydrogels.[16] The use of light to polymerize or cure materials *in vivo* has been practiced extensively in dentistry to form sealant and dental restorations *in situ*.[17,18] Photopolymerization has also been used in electronic materials, printing materials, optical materials, membranes, polymeric materials, and coatings and surface modifications.[16] Photopolymerization has several advantages over conventional polymerization techniques. These include spatial and temporal control over polymerization, fast curing rates (less than a second to a few minutes) at room or physiological temperatures, and minimal heat production.[19]

One major advantage of photopolymerization is that hydrogels can be created *in situ* from aqueous precursors using photopolymerization in a minimally invasive manner, for example, using laparoscopic devices,[20,21] catheters,[22,23] or a subcutaneous injection with transdermal illumination.[24] Fabrication of polymers *in situ* is attractive for a variety of biomedical applications because this allows one to form complex shapes that adhere and conform to tissue structures. Polymerization conditions for *in vivo* applications, however, are difficult since biological systems require a narrow range of acceptable temperatures and pH, as well as the absence of toxic materials, such as most monomers and organic solvents (Table 8.1). Some photopolymerization systems can overcome these limitations

Table 8.1. Predicted and observed order of errors for eigenvalues based on various combinations of stiffness and mass matrices.

	Predicted order of error in representing			Observed error for
Type	K	M	λ	λ
k2r-m2c	$O(h^2)$	$O(h^4)$	$O(h^2)$	$O(h^2)$
k2r-m2l	$O(h^2)$	$O(h^2)$	$O(h^2)$	$O(h^2)$
k3e-m3c	$O(h^2)$	$O(h^6)$	$O(h^2)$	$O(h^2)$
k3e-m3l	$O(h^2)$	$O(h^4)$	$O(h^2)$	$O(h^2)$
k3r-m3a	$O(h^4)$	$O(h^6)$	$O(h^3)$	$O(h^3)$
k3e-m3b	$O(h^5)$	$O(h^7)$	$O(h^4)$	$O(h^7)$
k3r-m3d	$O(h^7)$	$O(h^8)$	$O(h^6)$	$O(h^8)$

Note: The rapid conversion of $O(h^2)$ to $O(h^4)$ by ion–molecule reactions.
Source: Takagi (1996).

because the polymerization conditions are sufficiently mild (low light intensity, short irradiation time, physiological temperature, and low organic solvent levels) to be carried out in the presence of cells and tissues.

8.2.1 *Photoinitiators of Photocurable Hydrogel*

Being the simplest diatomic molecule, it is natural that H_2^+ and its isotopomers represent benchmark systems for dissociative recombination, in particular with respect to the comparison of experiment and theory. The experimental and theoretical studies of DR of H_2^+ and its isotopomers in their zeroth vibrational levels are described in some detail in two recent reviews,[2,25] and only a brief summary of the present situation will be given here (Fig. 8.3)[16,26,27].

8.3 Photopolymerizable Materials

Polymerization of monomers using visible or UV irradiation has been thoroughly investigated.[16,28] While such systems work well for many applications, they cannot be generally utilized in tissue engineering because most monomers are cytotoxic. As a result, photopolymerizable hydrogels for tissue engineering applications

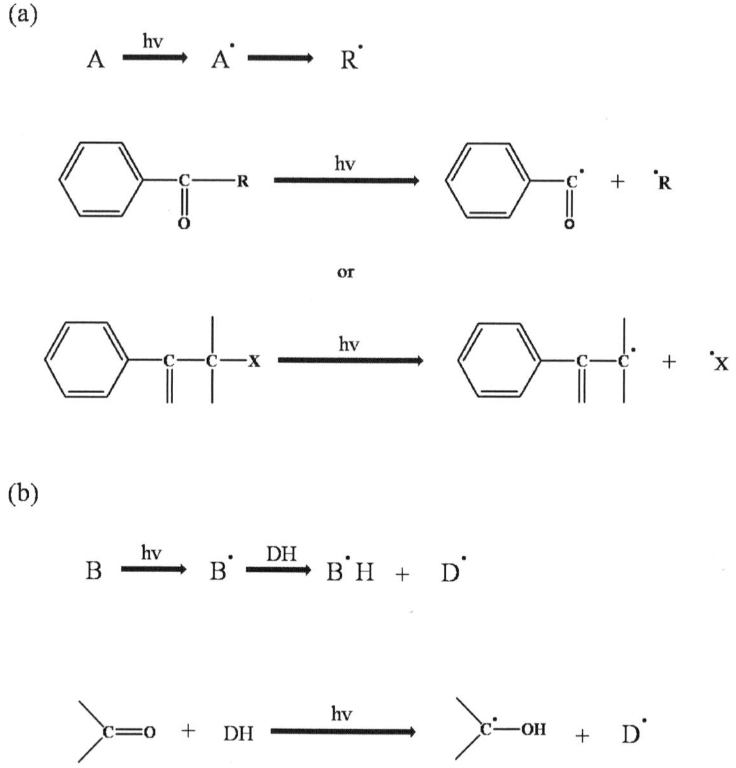

(a)

$$A \xrightarrow{h\nu} A^{\bullet} \longrightarrow R^{\bullet}$$

(b)

$$B \xrightarrow{h\nu} B^{\bullet} \xrightarrow{DH} B^{\bullet}H + D^{\bullet}$$

Figure 8.3. Photoinitiators that promote radical photopolymerization. (a) Radical photopolymerization by photocleavage. Upon exposure to light, these photoinitiators undergo cleavage to form radicals that promote the photopolymerization reaction. (b) Radical photopolymerization by hydrogen abstraction. Upon UV irradiation, these photoinitiators undergo hydrogen abstraction from the H-donor (DH) to generate radicals.

have generally been formed from macromolecular hydrogel precursors. These are water-soluble polymers with two or more reactive groups. Examples of photopolymerizable macromers include poly(ethylene glycol) (PEG) acrylate derivatives,[29] PEG methacrylate derivatives,[30,31] polyvinyl alcohol (PVA) derivatives,[25,32–33] and modified polysaccharides such as hyaluronic acid (HA) derivatives[34,35] and dextran methacrylate.[36,37]

8.3.1 *Photocurable Hydrogel from Natural Polymers*

8.3.1.1 Photo-cross-linkable collagen and gelatin

Gelatin, or denatured collagen, has several desirable properties for use as a biomaterial. These properties include its biological interaction capabilities and numerous side groups for modification.[38] Previously, chemical cross-linking was initiated using various bifunctional reagents such as glutaraldehyde, but these gels resulted in cases of local cytotoxicity and calcification.[38,39] Van Den Bulcke *et al.*[38] derivatized gelatin by reacting it with methacrylic anhydride to introduce photo-cross-linkable moieties. They noted that the degree of substitution of the gelatin and the storage conditions controlled the overall strength of the gel.[38] Brinkman *et al.*[40] also used methacrylic anhydride to incorporate an acrylate moiety onto the lysine and hydroxylysine residues of the collagen backbone (Fig. 8.4a). Photo-cross-linking of the macromer using visible light irradiation and in the presence of rat aortic smooth muscle cells was used to prepare gels.[40] A significant increase in the mechanical integrity of the construct upon cross-linking was observed, although it was still sufficiently less than the tensile strength required for some *in vivo* applications, such as vascular grafts. The researchers also showed that the triple helical conformation, as well as cell viability, was maintained during the cross-linking process.[40]

In addition to the reaction with methacrylic anhydride, collagen has been derivatized using the 1-ethyl-3-(3-dimethyl aminopropyl) carbodiimide/*N*-hydroxysuccinimide (EDC/NHS) conjugation method to add photosensitive cinnamate moieties.[39] This method produced photo-cross-linked gels with mechanical properties comparable with those cross-linked using glutaraldehyde.[39] Moreover, collagen and gelatin have been reacted with other reagents to facilitate light-induced chemical cross-linking. Gelatin has also been modified with styrene groups by reaction with 4-vinylbenzoic acid. This modified gelatin has also been reacted with PEG diacrylate and investigated as a tissue-adhesive glue for arterial repair.[41] The release rate of albumin from these gels is indirectly related to the gelatin concentration and degree of modification, and the adhesiveness of the gel was greater than that of fibrin glue.[42] Styrenederivatized gelatin has also been investigated for

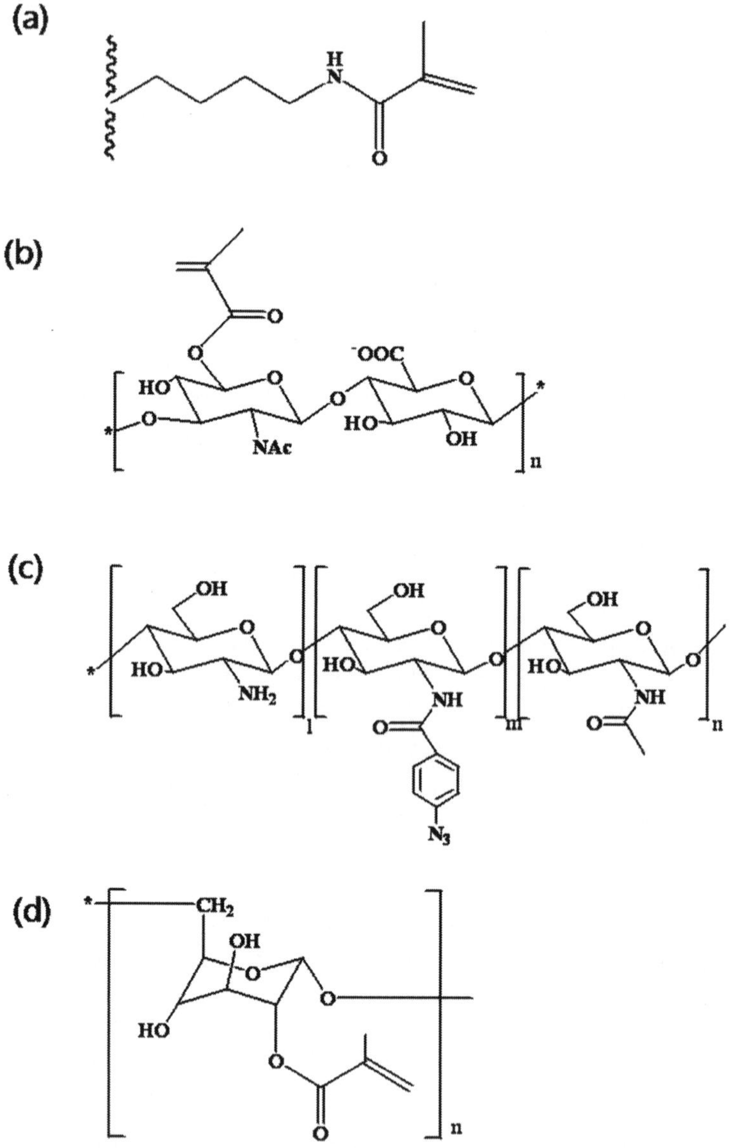

(a)

(b)

(c)

(d)

Figure 8.4. Examples of various photopolymerizable natural polymers, including (a) collagen methacrylamide, (b) methacrylate-hyaluronic acid, (c) AZ-CH-LA, and (d) methacrylate-dextran.

applications such as delivery vehicles,[43–45] nerve guides,[46,47] and cell carriers for chondrocyte transplantation.[48]

8.3.1.2 Photo-cross-linkable hyaluronic acid

HA is a naturally derived nonimmunogenic, nonadhesive glycosaminoglycan composed of D-glucuronic acid and *N*-acetyl-D-glucosamine and is found in many connective tissues.[49–57] HA undergoes enzymatic degradation[49,51–55,57,58] and plays a role in the promotion of cell motility and proliferation,[49,51,55,58] wound healing,[49,52,53,58] angiogenesis,[52–54] and the reduction of long-term inflammation.[49,52–54] It is attractive for biomaterial applications because it can be modified with various functional groups so that covalent cross-linking reactions are possible.[49–64] Smeds *et al.*[58] modified HA for photopolymerization with the addition of a methacrylate group by reaction with methacrylic anhydride (Fig. 8.4b).[57] They investigated the influence of the degree of methacrylation on bulk mechanical properties such as swelling, compression, and creep compliance. It was demonstrated that the hydrogels prepared from methacrylated HA are capable of swelling up to 14 times their dry weight and are stronger and more resilient than corresponding alginate gels.[56]

8.3.1.3 Photo-cross-linkable chitosan

Chitosan is a partially deacetylated form of chitin, a natural linear homopolymer of b-1,4-linked *N*-acetyl-D-glucosamine.[65–67] Useful advantages of chitosan include its hemostatic activity,[66] immunological activity, and ability to accelerate wound healing.[65,66] It has been modified by reaction with N,N,N',N'-tetramethylethylenediamine, EDC, and 4-o-b-D-galactopyranosyl-(1,4)-D-gluconic acid to introduce azide and lactose moieties for photo-cross-linking (Fig. 8.4c). This modified chitosan (AZ-CH-LA) has primarily been investigated as a tissue adhesive,[65,67,68] and it has been shown that AZ-CH-LA demonstrates higher binding strengths than fibrin glue and is effective at sealing punctures in thoracic aorta and lungs in rabbit models.[68] Azide-modified chitosan has also been investigated as a matrix for drug delivery[66,69] and as a cell-seeded material for micromolding[70] and photolithography.[71] Moreover, chitosan has also been modified by a reaction with 4-vinylbenzoic acid. This material has been used to create tubular structures upon irradiation with visible light.[72]

8.3.2 *Photocurable Hydrogel from Synthetic Polymers*

Anseth *et al.*[73] originally modified polyanhydrides with cross-linkable methacrylate groups to produce polymer networks that have controlled degradation and mechanics. There is an added potential of network formation directly in the body. The general structures for selected dimethacrylated anhydride monomers are shown in Fig. 8.5. In general, the core of the molecule consists of hydrophobic repeating units, such as sebacic acid, carboxyphenoxy propane, or carboxyphenoxy hexane. Other anhydride monomers, including methacrylated tricarballylic acid and methacrylated pyromellitylimidoalanine, have also been synthesized to impart greater cross-linking density and a biologically recognized component, respectively.[74]

Photo-cross-linkable polyanhydrides have also been extensively characterized with respect to reaction behavior and material properties. Even with mild initiation conditions (e.g., with 0.1 wt% initiator and 7 mW/cm^2 UV light, polymerization occurs within three

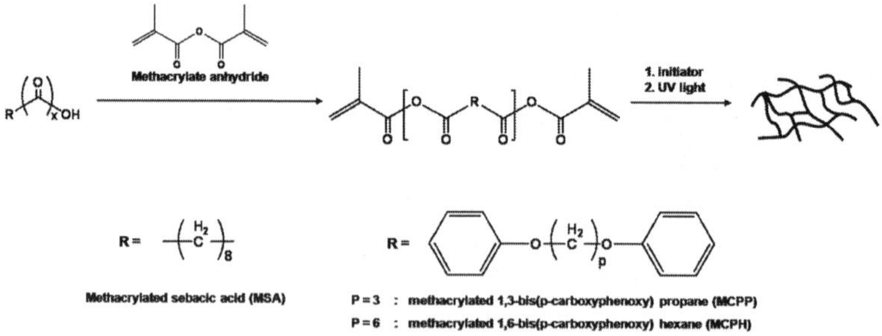

Figure 8.5. Photo-cross-linkable polyanhydrides, typically synthesized via the reaction of acids with methacrylate anhydride. Dimethacrylated anhydride monomers, methacrylated sebacic acid (MSA), methacrylated 1,3-*bis*(*p*-carboxyphenoxy) propane (MCPP), and methacrylated 1,6-*bis*(*p*-carboxyphenoxy) hexane (MCPH), as well as a general polymerization.

minutes), polymerization times are well within acceptable clinical time scales, allowing for network formation intraoperatively.[75] In general, densely cross-linked networks formed from multifunctional anhydride monomers degrade via a surface erosion mechanism, with mass loss only at the surface erosion zones on the exposed areas of the polymer. These erosion zones allow only minimal penetration of water into the polymer, so hydrolysis occurs exclusively near the polymer surface and is dictated by the polymer chemistry. For example, a disk (~1.7 mm thickness) composed entirely of methacrylated sebacic acid degrades in approximately three days, whereas a disk composed entirely of the more hydrophobic methacrylated carboxyphenoxy hexane takes more than one year to completely degrade.[76] Because erosion occurs only at the polymer surface, structural integrity is maintained for longer degradation periods than in polymers that degrade throughout their bulk. For example, more than 90% of the tensile modulus of poly(methacrylated sebacic acid) and poly(methacrylated carboxyphenoxy hexane) networks are maintained for up to 40% mass loss.[76] On the other hand in bulk eroding systems, polymer chains are cleaved relatively homogeneously throughout the network and mechanical properties can plummet even when little mass loss has occurred.[77,78]

8.3.2.1 Photo-cross-linkable poly(ethylene glycol)

PEG, a water-soluble polymer, has a long history of use in biomaterials. This is primarily because of the extreme hydrophilicity of PEG, which decreases the adsorption of proteins and can be used to alter the interaction between materials and tissues and cells. Additionally, the end groups on PEG are easily modified through a variety of synthetic reactions. For instance, the reaction of PEG with acryloyl chloride or methacryloyl chloride in the presence of triethylamine is a simple technique for adding photoreactive vinyl groups.[79] Photopolymerizable PEG hydrogels have been used for numerous applications, including as membranes for the encapsulation of islets of Langerhans,[79–81] as barriers to reduce intimal thickening after balloon angioplasty,[82] as matrices for chondrocyte encapsulation in cartilage regeneration,[83–87] for osteoblast encapsulation in bone regeneration,[88] for the encapsulation of mesenchymal stem cells,[89] and for the delivery of nitric oxide to reduce platelet adhesion and smooth muscle cell proliferation.[90] Additionally, Elbert and Hubbell[91] used diacrylated PEG macromers to fabricate hydrogels incorporating a variety of peptides by first reacting the macromers with a cysteine-containing peptide via a conjugate addition reaction and then by using subsequent photopolymerization into a network.

PEG hydrogel networks are formed from functionalized PEG molecules, such as PEG-di(meth)acrylate (Fig. 8.6). Polymerization is initiated by reactive centers, such as radicals, generated from thermal energy, redox reactions, or the photo-cleavage of initiator molecules. These free radicals propagate through unsaturated vinyl bonds on the PEG macromolecular monomers, and chain polymerization occurs. The propagation of free radicals through multiple carbon–carbon double bonds on the constituting PEG monomers results in covalently cross-linked, high-molecular-weight kinetic chains. The functionalities of the chain-growth polymerized hydrogels can be expanded through the copolymerization of other functional di(meth)acrylated macromers.

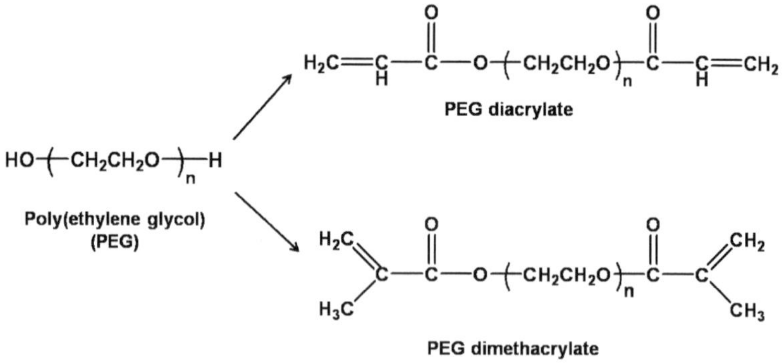

Figure 8.6. Chemical structures of a PEG macromer and its di(meth)acrylate derivatives that often solution polymerize to form hydrogel networks useful for biomaterial applications.

8.3.2.2 Transition in theoretical work

PVA has been used in a variety of biomedical applications, such as contact lenses, tendon repair, and drug delivery and as scaffolds for a wide variety of tissue engineering applications, including bone, cartilage, and heart valves.[92,93] Hydrogels prepared from this synthetic polymer are of particular interest because of its biocompatibility, high water content, tissue-like elasticity, and ease of fabrication and sterilization.[93–97] Perhaps the most attractive feature of using PVA is the abundance of pendant hydroxyl groups along the backbone, which allow for various chemical modifications such as the introduction of methacrylate groups or biological molecules like fibronectin.[92–101]

Previously, PVA hydrogels have been cross-linked using chemical means such as aldehydes[100] and physical mechanisms such as repeated freeze–thaw cycles to induce crystallinity.[96,100] Limitations to these methods include toxicity of the aldehydes and stability of physical cross-links. Muhlebach *et al.*[102] were the first to use photopolymerization to cross-link PVA by modification of the aforementioned pendant hydroxyl groups with acrylic acid and methacrylic acid for contact lens applications.[96,102] Several other groups have been successful in introducing various functional groups for photopolymerization by reaction with various reagents such as glycidyl acrylate,[99] aminobutyraldehyde diethyl acetal (Fig. 8.7),[92] and 2-isocyanatoethyl methacrylate.[97] A wide range of mechanical properties and degradation times are obtainable by varying the concentration of reactive groups and the number and type of degradable groups.[92,94,100] These hydrogels degrade via hydrolysis of ester bonds until reverse gelation occurs, when there is no longer an infinite network present but only branched, soluble chains.[95] Nuttelman *et al.*[96] engineered a hydrogel scaffold by preparing PLA end-capped with hydroxyethyl methacrylate, which was subsequently grafted onto PVA. This copolymer demonstrated control over degradation and greater cell adhesion based on the number of lactide repeat units per side chain.[96] Copolymers of PVA and PEG, as well as PVA and chondroitin sulfate, have been created and used to encapsulate chondrocytes for cartilage tissue engineering applications.[97,101]

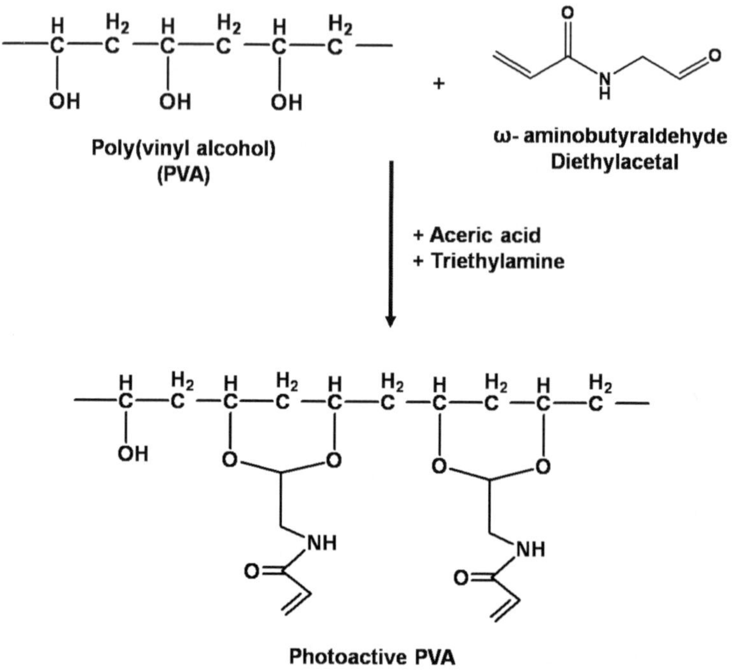

Figure 8.7. Chemical structure of photoactive PVA. Pendant hydroxyl groups were converted to photopolymerizable groups by reaction with aminobutyraldehyde diethyl acetal.

8.3.2.3 Photo-cross-linkable poly(hydroxyl esters)

Poly(hydroxyl esters), such as PLA, PGA, and PCL, are among the most thoroughly investigated synthetic biomaterials.[103,104–110] However, poly(a-hydroxyl esters) are highly crystalline polyesters that lack modifiable side groups, which could be used to facilitate better material–cell interactions.[105–110] Several investigators have combined degradable PLA and photopolymerization technology to manipulate the backbone with amino acids to allow for further functionalization.[109,110] PLA films have also been surface-grafted with polyacrylic acid (PAA) and poly(acrylamide) using photopolymerization in an attempt to generate a bioactive surface with rapid degradation.[108]

Various copolymers containing blocks of poly(hydroxyl esters) and other synthetic polymers have been prepared. For example, PCL-b-PEG-b-PCL networks have been synthesized, and they exhibit a greater compressive modulus and degradation rate than PCL alone.[111] Networks based on adipic acid, 4-hydroxycinnamic acid, and PCL diols with elastomeric properties have also been prepared.[112,113] PCL has also been combined with trimethylene carbonate, and the reaction behavior and bulk properties have been characterized.[114–122] Furthermore, star-PCL-b-PLA macromers have been synthesized.[113,114] These networks show dependence in physical properties and degradation on the length of the copolymer block.[123,124] Han and Hubbel[105,106] and Ju *et al.*[107] prepared lactide-based PEG networks emanating from glycerol centers. The degradability of these networks was controlled by the ratio of lactide and PEG.[105,107]

8.4 Summary and Outlook

In summary, a wide range of different precursor molecules are being developed that form networks via photoinitiated polymerizations for tissue engineering and molecule delivery applications. The diversity in polymer properties, ranging from highly cross-linked, hydrophobic networks to loosely cross-linked, swollen hydrophilic hydrogels, expands the applicability of these biomaterials. As more knowledge and understanding of biological events such as tissue healing and cell–material interactions becomes available, the materials used in these applications will become smarter and more complex. In particular, there is great interest in designing biological function into polymers, such as enzymatically degrading cross-links and adhesion sites, to better control the interactions between cells and materials. Also, as advances are made in the area of polymer synthesis, it is becoming easier to tailor polymer properties and structure to match desired properties. Novel materials may open up avenues for unique material behavior such as externally triggered properties (e.g., temperature induced), shape-memory polymers, and the patterning of structures three-dimensionally. The next decade is bound to produce previously unimaginable biomaterials that use photopolymerization technology and can help to overcome shortages of donor tissue available for transplantation to patients.

References

1. A. J. Putnam and D. J. Mooney, *Nat. Med.*, **2**, 824 (1996).
2. J. J. Marler, J. Upton, R. Langer, and J. P. Vacanti, *Adv. Drug Deliv. Rev.*, **33**, 165 (1998).
3. L. E. Niklason, J. Gao, W. M. Abbott, K. K. Hirschi, S. Houser, R. Marini, and R. Langer, *Science*, **284**, 489 (1999).
4. F. Oberpenning, J. Meng, J. J. Yoo, and A. Atala, *Nat. Biotechnol.*, **17**, 149 (1999).
5. B. Pomahac, T. Svensjo, F. Yao, H. Brown, and E. Eriksson, *Crit. Rev. Oral Biol. Med.*, **9**, 333 (1998).
6. P. X. Ma and R. Langer, *J. Biomed. Mater. Res.*, **44**, 217 (1999).
7. R. F. Service, *Science*, **289**, 1498 (2000).
8. V. S. Lin, M. C. Lee, S. O'Neal, J. McKean, and K. L. P. Sung, *Tissue Eng.*, **5**, 443 (1999).
9. C. A. Heath, *TIBTECH*, **16**, 17 (2000).
10. B. Alberts, D. Bray, J. Lewis, M. Raff, K. Roberts, and J. D. Watson, In *Molecular Biology of the Cell* (Garland Publishing, NY, 1994), p. 971.
11. B. S. Kim and D. J. Mooney, *TIBTECH*, **16**, 224 (1998).
12. L. D. Harris, B.S. Kim, and D. J. Mooney, *J. Biomed. Mater. Res.*, **42**, 396 (1998).
13. R. C. Thomson, A. G. Mikos, E. Beahm, J. C. Lemon, W. C. Satterfield, T. B. Aufdemorte, and M. J. Miller, *Biomaterials*, **20**, 2007 (1999).
14. M. S. Jhon and J. D. Andrade, *J. Biomed. Mater. Res.*, **7**, 509 (1973).
15. A. C. Jen, M. C. Wake, and A. G. Mikos, *Biotech. Bioeng.*, **50**, 357 (1996).
16. A. B. Scranton and C. N. Bowman, *Photopolymerization Fundamentals and Applications* (ACS Publishers, New Orleans, 1996).
17. A. Maffezzoli, A. Della Pietra, S. Rengo, L. Nicolais, and G. Valletta, *Biomaterials*, **15**, 1221 (1994).
18. M. G. Fleming, W. A. Maillet, *J. Can. Dent. Assoc.*, **65**, 447 (1999).
19. C. Decker, *J. Coat. Technol.*, **59**, 97 (1987).
20. J. L. Hill-West, S. M. Chowdhury, M. J. Slepian, and J. A. Hubbell, *Proc. Natl. Acad. Sci. U S A*, **91**, 5967 (1994).

21. J. L. West, J. A. Hubbell, *Proc. Natl. Acad. Sci. U S A*, **93**, 13188 (1996).
22. J. L. Hill-West, S. M. Chowdhury, R. C. Dunn, and J. A. Hubbell, *Fertil. Steril.*, **62**, 630 (1994).
23. J. L. Hill-West, S. M. Chowdhury, A. S. Sawhney, C. P. Pathak, R. C. Dunn, and J. A. Hubbell, *Obstet. Gynecol.*, **83**, 59 (1994).
24. J. Elisseeff, K. Anseth, D. Sims, W. McIntosh, M. Randolph, and R. Langer, *Proc. Natl. Acad. Sci. U S A*, **96**, 3104 (1999).
25. B. Muller, *US Patent. US, Ciba-Geigy Corporation*, **286**, 035 (1995).
26. K. T. Nguyen and J. L. West. *Biomaterials*, **23**(22), 4307 (2002).
27. J. Elisseeff, K. Anseth, D. Sims, W. McIntosh, M. Randolph, M. Yaremchuk, and R. Langer, *Plast. Reconstr. Surg.*, **104**(4), 1014 (1999).
28. C. Decker, *J. Coat. Technol.*, **59**, 97 (1987).
29. A. S. Sawhney, C. P. Pathak, J. J. van Rensburg, R. C. Dunn, and J. A. Hubbell, *J. Biomed. Mater. Res.*, **28**(7), 831 (1994).
30. J. Elisseeff, W. McIntosh, K. Anseth, S. Riley, P. Ragan, and R. Langer, *J. Biomed. Mater. Res.*, **51**, 164 (2000).
31. I. S. Kim, Y. I. Jeong, and S. H. Kim, *Int. J. Pharm.*, **205**, 109 (2000).
32. B. Muller, *US Patent. US, CIBA Vision Corporation* **678**, 437 (1996).
33. P. Martens and K. S. Anseth, *Polymer*, **41**, 7715 (2000).
34. T. Matsuda, M. J. Moghaddam, H. Miwa, K. Sakurai, and F. Iida, *ASAIO J.*, **M154**, 38 (1992).
35. P. Bulpitt and D. Aeschlimann, *J. Biomed. Mater. Res.*, **47**, 152 (1999).
36. S. H. Kim and C. C. Chu, *J. Biomed. Mater. Res.*, **49**, 517 (2000).
37. S. H. Kim and C.C. Chu, *J. Biomater. Appl.*, **15**, 23 (2000).
38. A.I. Van Den Bulcke, B. Bogdanov, N. De Rooze, E. H. Schacht, M. Cornelissen, and H. Berghmans, *Biomacromolecules*, **1**, 31 (2000).
39. C. M. Dong, X. Y. Wu, J. Caves, S. S. Rele, B. S. Thomas, and E. L. Chaikof, *Biomaterials*, **26**, 4041 (2005).
40. W.T. Brinkman, K. Nagapudi, B. S. Thomas, and E. L. Chaikof, *Biomacromolecules*, **4**, 890 (2003).
41. C. L. Li, T. Sajiki, Y. Nakayama, M. Fukui, and T. Matsuda, *J. Biomed. Mater. Res. B*, **66B**, 439 (2003).
42. H. Okino, Y. Nakayama, M. Tanaka, and T. Matsuda, *J. Biomed. Mater. Res.*, **59**, 233 (2002).
43. T. Manabe, H. Okino, M. Tanaka, and T. Matsuda, *Biomaterials*, **25**, 5867 (2004).
44. T. Masuda and M. Furue, *Tissue Eng.*, **10**, 523 (2004).

45. H. Okino, R. Maeyama, T. Manabe, T. Matsuda, and M. Tanaka, *Clin. Cancer Res.*, **9**, 5786 (2003).
46. E. Gamez, Y. Goto, K. Nagata, T. Iwaki, T. Sasaki, and T. Matsuda, *Cell Transplant*, **13**, 549 (2004).
47. E. Gamez, K. Ikezaki, M. Fukui, and T. Matsuda, *Cell Transplant*, **12**, 481 (2003).
48. A. Hoshikawa, Y. Nakayama, T. Matsuda, H. Oda, K. Nakamura, and K. Mabuchi, *Tissue Eng.*, **12**, 2333 (2006).
49. J. A. Burdick, C. Chung, X. Q. Jia, M. A. Randolph, and R. Langer, *Biomacromolecules*, **6**, 386 (2005).
50. X. Q. Jia, J. A. Burdick, J. Kobler, R. J. Clifton, J. J. Rosowski, S. M. Zeitels, and R. Langer, *Macromolecules*, **37**, 3239 (2004).
51. A. Khademhosseini, G. Eng, J. Yeh, J. Fukuda, J. Blumling, R. Langer, and J. A. Burdick, *J. Biomed. Mater. Res. A*, **79A**, 522 (2006).
52. J. B. Leach, K. A. Bivens, C. N. Collins, and C. E. Schmidt, *J. Biomed. Mater. Res. A*, **70A**, 74 (2004).
53. J. B. Leach, K. A. Bivens, C. W. Patrick, and C. E. Schmidt, *Biotechnol. Bioeng.*, **82**, 578 (2003).
54. J. B. Leach and C. E. Schmidt, *Biomaterials*, **26**, 125 (2005).
55. Y. D. Park, N. Tirelli, and J. A. Hubbell, *Biomaterials*, **24**, 893 (2003).
56. K. A. Smeds and M. W. Grinstaff, *J. Biomed. Mater. Res.*, **54**, 115 (2001).
57. K. A. Smeds, A. Pfister-Serres, D. L. Hatchell, and M. W. Grinstaff, *J. Macromol. Sci. Pure Appl. Chem.*, **36**, 981 (1999).
58. K. S. Masters, D. N. Shah, L. A. Leinwand, and K. S. Anseth, *Biomaterials*, **26**, 2517 (2005).
59. C. Chung, J. Mesa, G. J. Miller, M. A. Randolph, T. J. Gill, and J. A. Burdick, *Tissue Eng.*, **12**, 2665 (2006).
60. C. Chung, J. Mesa, M. A. Randolph, M. Yaremchuk, and J. A. Burdick, *J. Biomed. Mater. Res. A*, **77A**, 518 (2006).
61. A. Khademhosseini, J. Yeh, S. Jon, G. Eng, K. Y. Suh, J. A. Burdick, and R. Langer, *Lab Chip.*, **4**, 425 (2004).
62. K. S. Masters, D. N. Shah, G. Walker, L. A. Leinwand, and K. S. Anseth, *J. Biomed. Mater. Res. A*, **71A**, 172 (2004).
63. D. Miki, K. Dastgheib, T. Kim, A. Pfister-Serres, K. A. Smeds, M. Inoue, D. L. Hatchell, and M. W. Grinstaff, *Cornea*, **21**, 393 (2002).
64. D. L. Nettles, T. P. Vail, M. T. Morgan, M. W. Grinstaff, and L. A. Setton, *Ann. Biomed. Eng.*, **32**, 391 (2004).

65. M. Ishihara, K. Nakanishi, K. Ono, M. Sato, M. Kikuchi, Y. Saito, H. Yura, T. Matsui, H. Hattori, M. Uenoyama, and A. Kurita, *Biomaterials*, **23**, 833 (2002).

66. S. R. Jameela, S. Lakshmi, N. R. James, and A. Jayakrishnan, *J. Appl. Polym. Sci.*, **86**, 1873 (2002).

67. K. Ono, Y. Saito, H. Yura, K. Ishikawa, A. Kurita, T. Akaike, and M. Ishihara, *J. Biomed. Mater. Res.*, **49**, 289 (2000).

68. K. Ono, M. Ishihara, Y. Ozeki, H. Deguchi, M. Sato, Y. Saito, H. Yura, M. Sato, M. Kikuchi, A. Kurita, and T. Maehara, *Surgery*, **130**, 844 (2001).

69. Y. Yeo, J. A. Burdick, C. B. Highley, R. Marini, R. Langer, and D. S. Kohane, *J. Biomed. Mater. Res. A*, **78A**, 668 (2006).

70. J. Fukuda, A. Khademhosseini, Y. Yeo, X. Y. Yang, J. Yeh, G. Eng, J. Blumling, C. F. Wang, D. S. Kohane, and R. Langer, *Biomaterials*, **27**, 5259 (2006).

71. J. M. Karp, Y. Yeo, W. L. Geng, C. Cannizarro, K. Yan, D. S. Kohane, G. Vunjak-Novakovic, R. S. Langer, and M. Radisic, *Biomaterials*, **27**, 4755 (2006).

72. T. Matsuda and T. Magoshi, *Biomacromolecules*, **3**, 942 (2002).

73. K. S. Anseth, V. R. Shastri, and R. Langer, *Nat. Biotechnol.*, **17**, 156 (1999).

74. J. S. Young, K. D. Gonzales, and K. S. Anseth, *Biomaterials*, **21**, 1181 (2000).

75. D. S. Muggli, A. K. Burkoth, S. A. Keyser, H. R. Lee, and K. S. Anseth, *Macromolecules*, **31**, 4120 (1998).

76. D. S. Muggli, A. K. Burkoth, and K. S. Anseth, *J. Biomed. Mater. Res.*, **46**, 271 (1999).

77. F. Tsuji, *Polymer*, **43**, 1789 (2002).

78. M. Vert, J. Mauduit, and S. M. Li, *Biomaterials*, **15**, 1209 (1994).

79. G. M. Cruise, D. S. Scharp, and J. A. Hubbell, *Biomaterials*, **19**, 1287 (1998).

80. G. M. Cruise, O. D. Hegre, F. V. Lamberti, S. R. Hager, R. Hill, D. S. Scharp, and J. A. Hubbell, *Cell Transplant*, **8**, 293 (1999).

81. G. M. Cruise, O. D. Hegre, D. S. Scharp, and J. A. Hubbell, *Biotechnol. Bioeng.*, **57**, 655 (1998).

82. J. L. West and J. A. Hubbell, *Proc. Natl. Acad. Sci. U S A*, **93**, 13188 (1996).

83. S. J. Bryant and K. S. Anseth, *Biomaterials*, **22**, 619 (2001).

84. J. Elisseeff, K. Anseth, D. Sims, W. McIntosh, M. Randolph, and R. Langer, *Proc. Natl. Acad. Sci. U S A*, **96**, 3104 (1999).
85. J. Elisseeff, K. Anseth, D. Sims, W. McIntosh, M. Randolph, M. Yaremchuk, and R. Langer, *Plast. Reconstr. Surg.*, **104**, 1014 (1999).
86. J. Elisseeff, W. McIntosh, K. Anseth, S. Riley, P. Ragan, and R. Langer, *J. Biomed. Mater. Res.*, **51**, 164 (2000).
87. J. Elisseeff, W. McIntosh, K. Fu, T. Blunk, and R. Langer, *J. Orthopaed. Res.*, **19**, 1098 (2001).
88. J. A. Burdick and K. S. Anseth, *Biomaterials*, **23**, 4315 (2002).
89. C. R. Nuttelman, M. C. Tripodi, and K. S. Anseth, *J. Biomed. Mater. Res. A*, **68A**, 773 (2004).
90. K. S. Bohl and J. L. West, *Biomaterials*, **21**, 2273 (2000).
91. D. L. Elbert and J. A. Hubbell, *Abstr. Pap. Am. Chem. S*, U442219 (2000).
92. K. H. Schmedlen, K. S. Masters, and J. L. West, *Biomaterials*, **23**, 4325 (2002).
93. C. R. Nuttelman, D. J. Mortisen, S. M. Henry, and K. S. Anseth, *J. Biomed. Mater. Res.*, **57**, 217 (2001).
94. P. Martens,T. V. Holland, C. N. Bowman, and K. S. Anseth, *Abstr. Pap. Am. Chem.*, **S222**, U420 (2001).
95. P. J. Martens, C. N. Bowman, and K. S. Anseth, *Polymer*, **45**, 3377 (2004).
96. C. R. Nuttelman, S. M. Henry, and K. S. Anseth, *Biomaterials*, **23**, 3617 (2002).
97. S. J. Bryant, K. A. Davis-Arehart, N. Luo, R. K. Shoemaker, J. A. Arthur, and K. S. Anseth, *Macromolecules*, **37**, 6726 (2004).
98. K. S. Anseth, A. T. Metters, S. J. Bryant, P. J. Martens, J. H. Elisseeff, and C. N. Bowman, *J. Control. Rel.*, **78**, 199 (2002).
99. S. J. Bryant, P. Martens, J. Elisseeff, M. Randolph, R. Langer, and K. Anseth, Eds., *The Wiley Polymer Networks Group Review Series* (John Wiley & Sons, NY, 1999), p. 395.
100. P. Martens and K. S. Anseth, *Polymer*, **41**, 7715 (2000).
101. P. J. Martens, S. J. Bryant, and K. S. Anseth, *Biomacromolecules*, **4**, 283 (2003).
102. A. Muhlebach, B. Muller, C. Pharisa, M. Hofmann, B. Seiferling, and D. Guerry, *J. Polym. Sci. Pol. Chem.*, **35**, 3603 (1997).
103. A. S. Sawhney, C. P. Pathak, and J. A. Hubbell, *Macromolecules*, **26**, 581 (1993).
104. D. S. W. Benoit, A. R. Durney, and K. S. Anseth, *Tissue Eng.*, **12**, 1663 (2006).

105. D. K. Han and J. A. Hubbell, *Macromolecules*, **29**, 5233 (1996).
106. D. K. Han and J. A. Hubbell, *Macromolecules*, **30**, 6077 (1997).
107. Y. M. Ju, K. D. Ahn, J. M. Kim, J. A. Hubbell, and D. K. Han, *Polym. Bull.*, **50**, 107 (2003).
108. A. V. Janorkar, A. T. Metters, and D. E. Hirt, *Macromolecules*, **37**, 9151 (2004).
109. J. Elisseeff, K. Anseth, R. Langer, and J. S. Hrkach, *Macromolecules*, **30**, 2182 (1997).
110. G. John and M. Morita, *Macromolecules*, **32**, 1853 (1999).
111. H. Kweon, M. K. Yoo, I. K. Park, T. H. Kim, H. C. Lee, H. S. Lee, J. S. Oh, T. Akaike, and C. S. Cho, *Biomaterials*, **24**, 801 (2003).
112. M. Nagata and S. Hizakae, *Macromol. Biosci.*, **3**, 412 (2003).
113. M. Nagata and Y. Sato, *Polymer*, **45**, 87 (2004).
114. I. K. Kwon and T. Matsuda, *Biomaterials*, **26**, 1675 (2005).
115. T. Matsuda, I. K. Kwon, and S. Kidoaki, *Biomacromolecules*, **5**, 295 (2004).
116. T. Matsuda and M. Mizutani, *Macromolecules*, **33**, 791 (2000).
117. T. Matsuda and M. Mizutani, *J. Biomed. Mater. Res.*, **62**, 395 (2002).
118. T. Matsuda, M. Mizutani, and S. C. Arnold, *Macromolecules*, **33**, 795 (2000).
119. M. Mizutani, S. C. Arnold, and T. Matsuda, *Biomacromolecules*, **3**, 668 (2002).
120. M. Mizutani and T. Matsuda, *Biomacromolecules*, **3**, 249 (2002).
121. M. Mizutani and T. Matsuda, *J. Biomed. Mater. Res.*, **61**, 53 (2002).
122. M. Mizutani and T. Matsuda, *J. Biomed. Mater. Res.*, **62**, 387 (2002).
123. B. G. Amsden, G. Misra, F. Gu, and H. M. Younes, *Biomacromolecules*, **5**, 2479 (2004).
124. A. O. Helminen, H. Korhonen, and J. V. Seppala, *Macromol. Chem. Physic.*, **203**, 2630 (2002).

Chapter 9

HYALURONAN-BASED HYDROGEL SCAFFOLDS

Insup Noh
Department of Chemical Engineering,
Seoul National University of Science and Technology,
138 Gongnung-gil, Nowon-gu, Seoul 139-743
Republic of Korea
insup@seoultech.ac.kr

Hyaluronic acid (HA)-based hydrogels have many promising potential applications in tissue engineering and other medical devices such as implants and drug/cell carriers due to their outstanding biological and chemical properties. This chapter discusses the biology and chemistry of HA and its chemical derivertization, fabrication methods of HA hydrogels, and the potential applications of HA hydrogels as a scaffold for tissue engineering and as a carrier for cell delivery and pharmaceuticals.

9.1 Introduction

A hydrogel is a three-dimensional, hydrophilic, polymeric network containing large amounts of water or biological fluids.[1] Since the introduction of hydrogels as soft contact lenses,[2] their applications

Handbook of Intelligent Scaffolds for Tissue Engineering and Regenerative Medicine
Edited by Gilson Khang

www.panstanford.com

have increased tremendously, and nowadays they are favored in a wide range of implants, tissue engineering, pharmaceutical and cell therapeutics, and other biomedical applications.[3–5] Both natural and synthetic polymers have been employed for the fabrication of hydrogels, such as HA, chondroitin sulfate, chitosan, fibrin, heparan sulfate, alginate, dextran, poly(vinyl alcohol) (PVA), poly(ethylene oxide) (PEO), Pluronic gel, and polyvinylidone, depending on their purposes. These medical-grade polymers have been converted to hydrogels via cross-linking methods. Numerous methods of cross-linking of polymer chains for hydrogel formation have been developed by various methods of either chemical or physical cross-linkings depending on their target diseases and applications.[6,7]

Among the above medical grade polymers for hydrogel networks, HA has been chosen as a highly attractive natural polymer for hydrogel fabrication, and numerous research groups have worked on the development of synthesis methods of novel HA-based hydrogels so as to take advantages of its excellent biological and chemical properties. Different synthesis methods, such as self-cross-linkings, photopolymerization, and physical gelation, were developed according to the needs of hydrogels specific for application purposes and target sites such as skins, blood vessels, bone and cartilages.

Recently injectable, that is, *in situ*, HA hydrogels that turn into a macroscopic gel at the local site of injection,[8–10] depending on surrounding circumstances such as time, pH, enzymes, energy sources, temperatures, and others, have drawn interest. Their advantages in biomedical and pharmaceutical engineering applications include convenient and safe delivery of bioactive molecules including stem cells, patient comfort, minimal side effects, easy handling and cost reduction. *In situ* HA hydrogelation has been achieved by various methods, such as photopolymerization, normally irreversible covalent bonds either with or without cross linker molecules, or self-assembly by mechanisms of either reversible or irreversible chemical reactions. Self-cross-linking hydrogels in response to a certain external stimuli (e.g., temperature, pH, time lapse) and hydrogels that release their contents in response to biological stimulus or its degradation[11,12] have been recognized to be of most interest among these gelling systems. In the following sections, various strategies to create gelling methods are outlined, including introduction of chemical and biological characteristics of HA. Particular attention is given to its useful chemical methods that are applicable in tissue engineering as a scaffold and pharmaceuticals as a carrier.

9.2 Characteristics of Hyaluronic Acid in Biomedical Engineering

HA, also called hyaluronate or hyaluronan depending on its polyanion states, is a unique, non-sulfated glycosaminoglycan found in the extracellular matrix, ocular vitreous, and joint fluid in all animals and a naturally derived polymer composed of disaccharide repeat units of glucuronic acid and *N*-acetylglucosamine, linked by 1-3 bonds. It is obtained by either extraction from chicken roosts or bacterial fermentation. The fermentation of HA is a unique and highly controlled process. While most of the glycosaminoglycans are synthesized in a cell's Golgi network, it is expressed at the plasma membrane and immediately extruded out of a cell and into the extracellular matrix during fermenting process. The molecular weight of native HA normally ranges over several million Daltons depending on its fermenting methods. HA showed important and specific biological roles in maintenance of stem cell's pluripotency and differentiation, homeostasis, acceleration of wound healing, and anticancer, vascular disease, and antiarthritis therapy. Due to its unique biological roles, it has been employed as a polymeric network for both hydrogel and solid scaffolds for its applications in various tissue regenerations and pharmaceuticals, even though its biological mechanisms have not been completely elucidated yet. Three functional groups in HA—carboxylic acid, hydroxyl, and acetamide groups—have been employed as reaction sites for chemical modifications (Fig. 9.1). Fabrication of HA-based hydrogels has been obtained mainly through the chemistry of those functional groups.[13]

Biological characteristics of HA have been extensively studied, such as its bindings to specific receptors, cluster determinant 44 (CD44), intracellular adhesion molecule 1 (ICAM-1), and the receptor for hyaluronate-mediated motility (RHAMM) expressed by most cells, including those implicated in the pathology of fibroblasts,

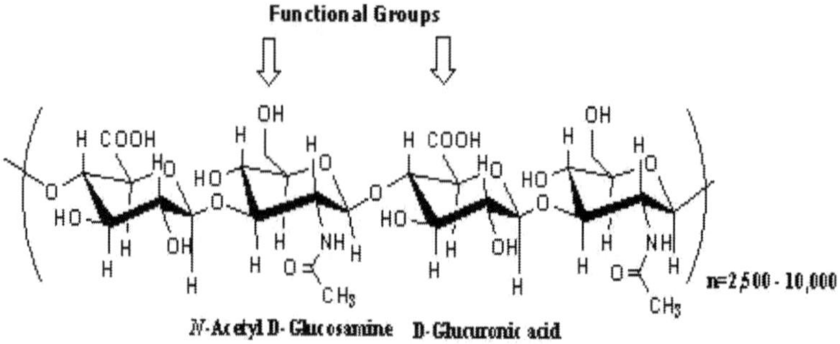

Figure 9.1. Chemical structure of hyaluronic acid and its functional groups; 6′-OH of *N*-acetyl D-glucose amine and –COOH of D-glucuronic acid.[16]

some endothelial cells and osteoarthritis, such as leukocytes, chondrocytes, and synoviocytes.[14] The wide distribution of these receptors on the membrane of many cell types accounts for its diverse and specific biological effects. As an example, the binding of HA to two hyalhedrisn such as RHAMM and CD44 has been reported to trigger intracellular signal events such as cytokine release and protein phosphorylation cascades, as well as stimulation of cell cycle proteins, e.g. in leukocytes. The receptor interaction of HA through CD44 stimulates transduction and other cell-signaling pathways that modulate cell functions, such as its adhesion, migration, proliferation, and endocytosis as well as HA degradation and uptake.[15] The interaction through RHAMM has been known to regulate cellular responses to growth factors and plays a role in cell migration in fibroblasts. HA degrades by the actions of reactive oxygen intermediates and three types of enzymes, hyaluronidase, β-D-glucuronidase, and β-N-acetyl-hexoaminidase, from macrophases, fibroblasts and endothelial cells. In addition to theseenzymatic degradation, HA has been known to contribute on pathological processes such as arthritis, wound-healing responses, vascular diseases and hemocompatibility by its concentration and molecular weight distribution, thus being employed as diagnostic markers for disease.

These chemical and biological properties of HA led to a recognition of its applications in tissue engineering scaffolds, leading to the necessity of the development methods for chemical modification and hydrogel fabrication. The following sessions discuss detailed chemistries of HA derivatives and their applications in hydrogel fabrication.

9.3 HA Derivatives

9.3.1 *Ester Derivatives*

As mentioned in the previous session, HA is a very flexible, viscous, biodegradable, and hydrophilic natural polymer. To form structural rigidity and to be less susceptible to enzymatic degradation, conversion of hydrophilic HA networks to more hydrophobic ones was performed by chemical grafting of hydrophobic hydrocarbons to the carboxylic acid in HA. The chemical modification was processed through an alkylation of HA via an alkyl halide, obtaining HA derivatives with different degrees of modifications of its available carboxyl groups. Depending on the degree and types of alkylations, corresponding degrees of rigidity and hydrophobicity of HA were achieved. As an example, Fidia Advanced Biopolymers (Italy) has yielded diverse commercial HA esters with hydrophobic domains (Hyaff-11 by grafting ethyl esters [Fig. 9.2] and Hyaff-11 by grafting benzyl esters), thus fabricating fiber, membrane, and sponge forms of HA. The Hyaff-11 microparticles demonstrated better results of bioactive molecules and cell delivery and at the same time have been employed as scaffolds for cartilage regeneration and mesenchymal stem cell cultures.

9.3.2 *Carbodiimide* ($R_1N{=}C{=}NR_2$)

Chemical reactions of HA via 1-ethyl-3-(3'-dimethylaminopropyl) carbodiimidehave been performed between its carboxylic acid and amine groups of the targeting bioactive macromolecules. Since the

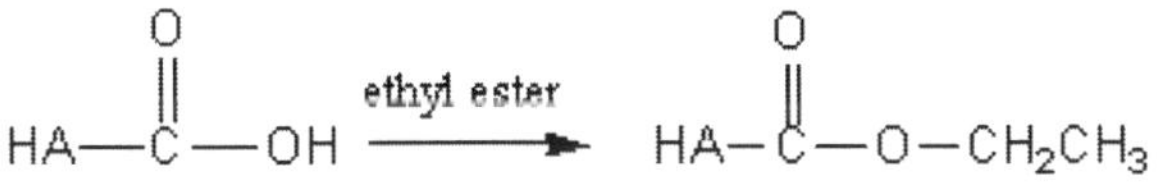

$$\text{HA}-\overset{\overset{\displaystyle O}{\|}}{C}-\text{OH} \xrightarrow{\text{ethyl ester}} \text{HA}-\overset{\overset{\displaystyle O}{\|}}{C}-O-CH_2CH_3$$

Figure 9.2. Example of formation of hydrophobic HA.[16]

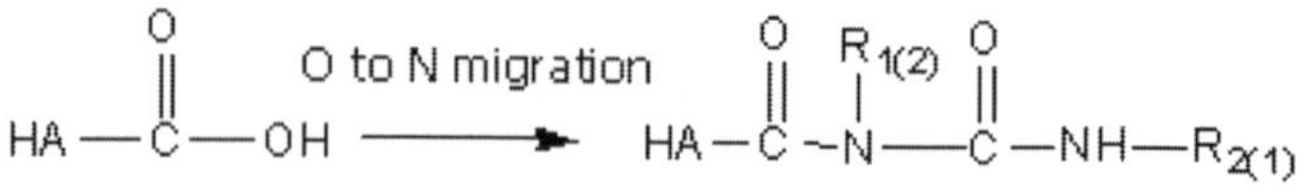

$$HA-\overset{O}{\overset{\|}{C}}-OH \xrightarrow{\text{O to N migration}} HA-\overset{O}{\overset{\|}{C}}-\underset{}{\overset{R_{1(2)}}{\overset{|}{N}}}-\overset{O}{\overset{\|}{C}}-NH-R_{2(1)}$$

Figure 9.3. Carbodiimide-mediated chemical reaction of HA.[17–21]

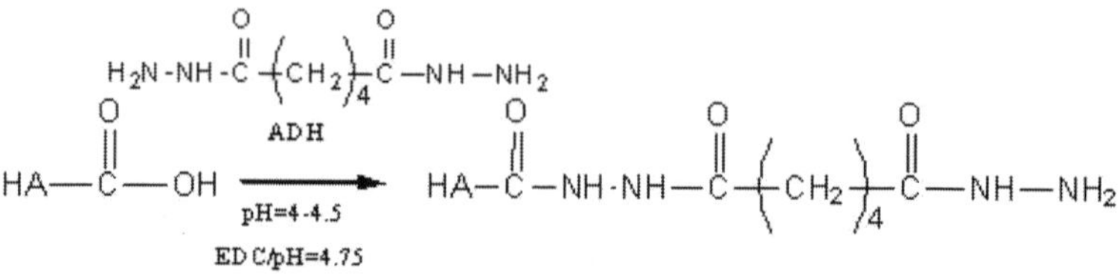

$$H_2N\text{-}NH\text{-}\overset{O}{\overset{\|}{C}}\text{-}(CH_2)_4\text{-}\overset{O}{\overset{\|}{C}}\text{-}NH\text{-}NH_2$$

$$HA-\overset{O}{\overset{\|}{C}}-OH \xrightarrow[\substack{pH=4\text{-}4.5 \\ EDC/pH=4.75}]{ADH} HA-\overset{O}{\overset{\|}{C}}-NH\cdot NH-\overset{O}{\overset{\|}{C}}-(CH_2)_4-\overset{O}{\overset{\|}{C}}-NH-NH_2$$

Figure 9.4. Functionalization of HA with primary amines.[24,25]

carbodiimide reaction is very sensitive to pH, frequently leading to the formation of unreactive intermediate acylurea from carboxylic acid, depending on the reaction conditions, very low coupling yields could be obtained (Fig. 9.3).[17–21]

To avoid this problem, amino dihydrazaide has been employed by the Prestwich group and others, forming functional amine end groups.[22] Further grafting of bioactive molecules was performed under mild basic conditions to its new primary amine end groups via the succinate–NHS ester reaction (Fig. 9.4).[23]

9.3.3 *Sulfydrylation (HA–SH)*

Functionalization of HA with thiols has been developed for coupling bioactive compounds (Fig. 9.5) and a subsequent self-coupling reaction with unsaturated groups via Michael-type addition reactions by the research groups of Hubbell[26–29] and Noh.[7,30–32] After coupling the amino dihydrazide to the carboxylic acid of HA, an HA derivative with disulfide groups is formed. The reduction of disulfide leads to sulfydryl end groups, which are important for further chemical reactions such as acrylates end groups.

As another example of an HA derivative with thiol end groups, the HA has been coupled under basic conditions with ethylene sulfide with some precipitation, probably due to ethylene sulfide oligomerization (Figs. 9.6 and 9.7).[34,35] HA with sulfydrylate was obtained

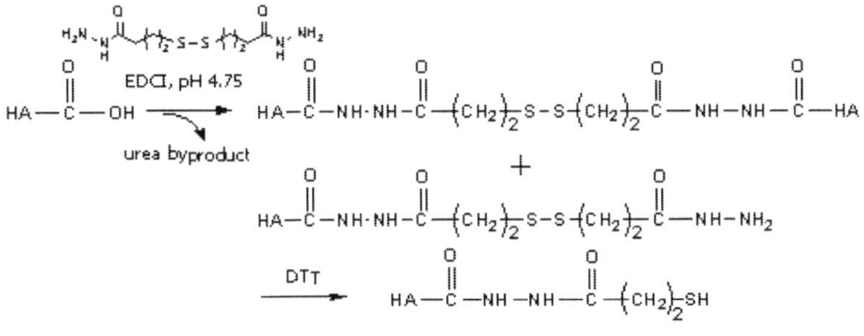

Figure 9.5. Functionalization of HA with primary amines and thiols via HA-COOH.[33]

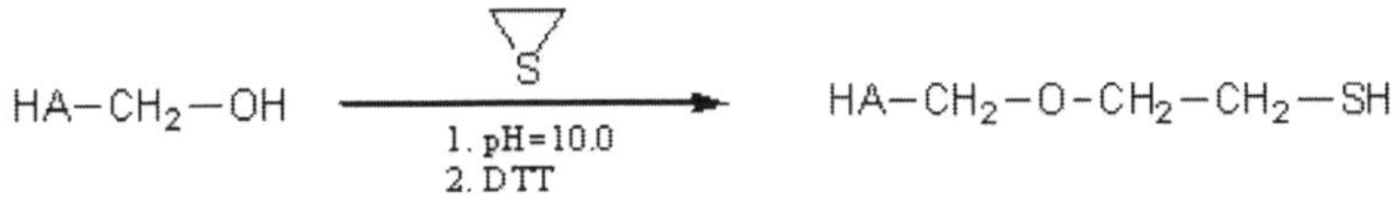

Figure 9.6. Synthesis of HA–SH.[34,35]

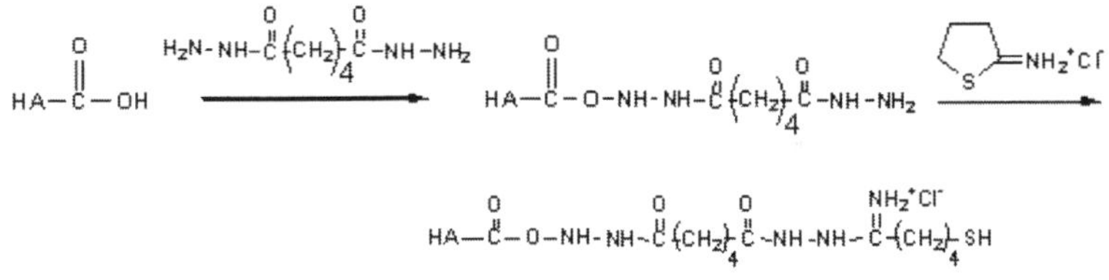

Figure 9.7. Functionalization of HA with thiols via HA-OH.[34,35]

by acidifying after processing of both filtering and reducing with reducting agents such as DTT. Further derivertization of fluroscein was performed, grafting either flurorescein, 4-(hydroxymercuri) benzoate, or iodoacetate to this thiol group.

9.3.4 *Sulfation*

Molecules with sulfate groups have been grafted to HA to mimic hemocompatibility of heparan sulfate (Fig. 9.8) by employing the mechanisms of electrostatic repulsion of anionic charges of the sulfate groups.[16] The anionic charges of the HA-heparan sulfates

$$HA-CH_2-OH \longrightarrow HA-CH_2-O-SO_3^-$$

Figure 9.8. Sulfation of HA.[16]

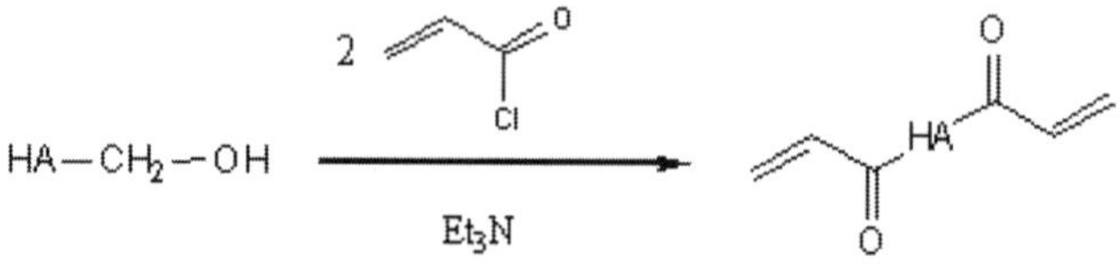

$$HA-CH_2-OH \xrightarrow[Et_3N]{2\ \text{acryloyl chloride}} \text{HA diacrylate}$$

Figure 9.9. Acrylation of HA with acylchlorides.[36]

reduced both platelet adhesion and thrombosis formation. Sulfation was obtained by chemically grafting sulfur trioxide pyridine to the hydroxyl group in HA, thus increasing anionic charges of HA.

9.3.5 *Acrylates*

Graftings of unsaturation domains to HA have been tried to functionalize for further chemical reactions through either photochemistry with an adequate initiator or Michael-type addition reactions without an initiator. Hydroxyl groups of HA have been employed to produce their reactions with acryloyl chlorides (Fig. 9.9).[36] These acrylate end groups have been employed for further sites of photopolymerization and Michael-type addition reactions (see the following session).

9.4 Fabrication of Hyaluronic Acid Hydrogels

HA has demonstrated many inherent biological and chemical advantages for biomaterials applications such as hydrogels and scaffolds for tissue engineering and carriers in drug and cell delivery system. Hydrogels have been created by employing HA-derivatives through many chemical methods such as bifunctional cross-linkers, enzymes, intra- and inter-molecular reactions, photochemical reactions, prefabrication reactions with different biocompatible polymers, etc. The following sessions describe detailed methods of chemical modification and hydrogelation of HA by using HA, HA derivatives, and cross-linkers.

9.4.1 *Hydrogel Formation by Direct Cross-Linking Methods*

9.4.1.1 Diepoxy cross-linking method

Diepoxy cross-linkers have been employed for fabrication of HA hydrogels via hydroxyl groups by using PEO diglycidyl ether, ethylene glycol diglycidylether, 1,4-butanediol diglycidyl ether, and polyglycerol polyglycidyl ether under basic conditions.[21,37] While diepoxy compounds demonstrated formation of ester linkages between carboxyl groups at acidic condition, they formed ether linkages between hydroxyl groups at basic condition. Further two-step experiments were performed on the basis of these results: an HA hydrogel was formed with diepoxyoctane via ester intra- and intermolecular linkages between HA at acid condition at a first step, and then via ether linkages at basic condition.[38] This method demonstrated a yield of hydrogel formation through a cross-linking method (Fig. 9.10).

9.4.1.2 Bifunctional amine cross-linkers

As mentioned earlier in the HA derivertization session, carbodiimide reactions are processed via the reactions between carboxylic acid and amines. Simple fabrication of HA hydrogels via this carbodiimide reaction is achieved by using solely carbodiimide to induce intra- and intermolecular cross-linking of HA with amine-derived HA (Fig. 9.11).[38]

To overcome no induction of similar reactions with HA in solution due to the formation of unreactive acylureas, Kuo *et al.*[39]

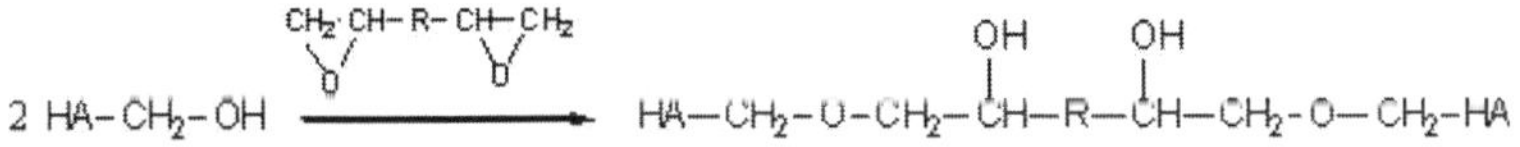

$$2\ \mathrm{HA{-}CH_2{-}OH} \xrightarrow{\mathrm{CH_2{-}CH{-}R{-}CH{-}CH_2}\ (\text{diepoxide})} \mathrm{HA{-}CH_2{-}O{-}CH_2{-}CH(OH){-}R{-}CH(OH){-}CH_2{-}O{-}CH_2{-}HA}$$

Figure 9.10. Hydrogel formation via chemical reaction of diepoxides with HA-OH.[16,21]

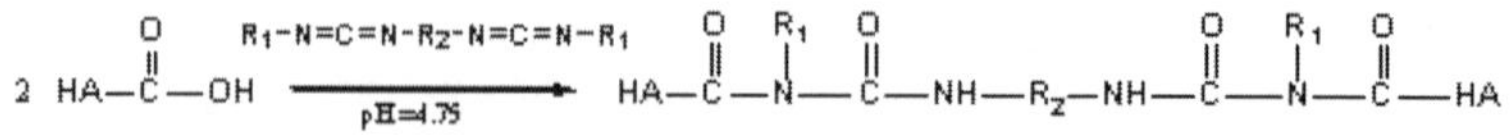

$$2\ \mathrm{HA{-}\overset{\displaystyle O}{\overset{\|}{C}}{-}OH} \xrightarrow[\mathrm{pH=4.75}]{\mathrm{R_1{-}N{=}C{=}N{-}R_2{-}N{=}C{=}N{-}R_1}} \mathrm{HA{-}\overset{\displaystyle O}{\overset{\|}{C}}{-}\overset{\displaystyle R_1}{\overset{|}{N}}{-}\overset{\displaystyle O}{\overset{\|}{C}}{-}NH{-}R_2{-}NH{-}\overset{\displaystyle O}{\overset{\|}{C}}{-}\overset{\displaystyle R_1}{\overset{|}{N}}{-}\overset{\displaystyle O}{\overset{\|}{C}}{-}HA}$$

Figure 9.11. Hydrogel formation via Chemical reactions of bifunctional amines to HA-COOH.[19–21]

developed a new *bis*-carbodiimide cross-linking technique. Several *bis*-carbodiimides were synthesized and covalently bound to HA through a mechanism similar to the acylurea formation, thus producing aromatic or aliphatic inter-molecular cross-linking of HA. Finally, several efforts to make reactions between HA's carboxyl groups and the amines of bifunctional cross-linkers have been attempted. Several research groups made use of reactions via carbodiimide and 1-hydroxybenzotriazole to couple activated amines (e.g., adipic dihydrazide [ADH]) to HA's carboxyl groups. Carbodiimide and *N*-hydroxysulfosuccinimide were used for coupling reactions between HA' carboxylic acid and primary amines (e.g., 1,4-diaminobutane dihydrochloride) under mild basic conditions.[18] Similarly, Vercryusse *et al.* and Oh *et al.* have employed a variety of polyvalent hydrazide cross-linkers (two to six hydrazides per cross-linker and the succinate–NHS ester reaction) for synthesis of HA hydrogels via carbodiimide-mediated reactions[40] and grafting of bioactive molecules as well as for testing of its molecular degradation (Figs. 9.12 and 9.13).[23]

Recently, HA hydrogels fabricated through ADH cross-linking have been prepared with some defined mesh sizes for tests of its adhesion and proliferation of the lung carcinoma cell line H460M[41] and the adhesion barrier with antiadhesion drugs such as the glucocorticoid receptor agonist budesonide within the abdominopelvic cavity following surgery or other injuries.[42] While the defect sites either with or without hydrogel addition induced a large amount of adhesion, the site with hydrogel containing budesonide

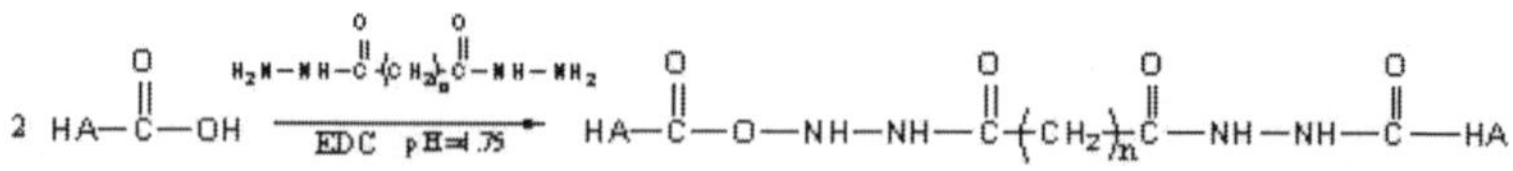

$$2\ \mathrm{HA{-}\overset{\displaystyle O}{\overset{\|}{C}}{-}OH} \xrightarrow[\mathrm{EDC\ \ pH=4.75}]{\mathrm{H_2N{-}NH{-}\overset{O}{\overset{\|}{C}}{-}(CH_2)_n{-}\overset{O}{\overset{\|}{C}}{-}NH{-}NH_2}} \mathrm{HA{-}\overset{\displaystyle O}{\overset{\|}{C}}{-}O{-}NH{-}NH{-}\overset{\displaystyle O}{\overset{\|}{C}}{-}(CH_2)_n{-}\overset{\displaystyle O}{\overset{\|}{C}}{-}NH{-}NH{-}\overset{\displaystyle O}{\overset{\|}{C}}{-}HA}$$

Figure 9.12. Hydrogel formation via chemical reactions of dihydrazide with HA.[17–21,24,25]

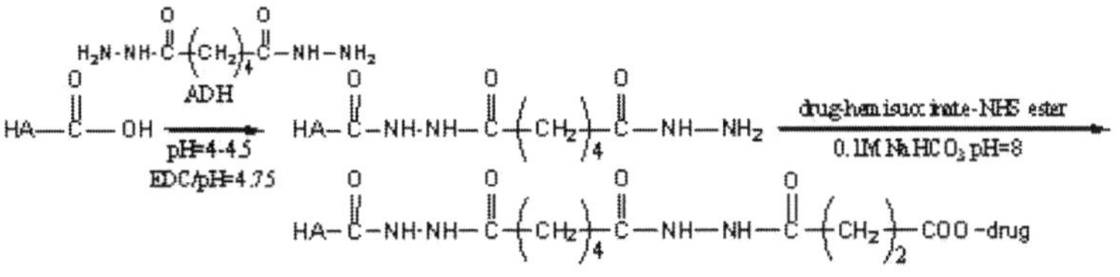

Figure 9.13. Hydrogel formation via chemical reactions of HA and ADH and subsequent coupling of bioactive drugs.[21,24,25]

demonstrated significant reduction of both adhesion and inflammation in animals, indicating the effective prevention of postsurgical adhesion by concurrent anti-adhesive drug delivery. This HA hydrogel was prepared by employing both HA–ADH and HA–aldehyde (Fig. 9.14) through conjugations of ADH and sodium periodate, respectively. This anti-inflammatory function of HA was also studied

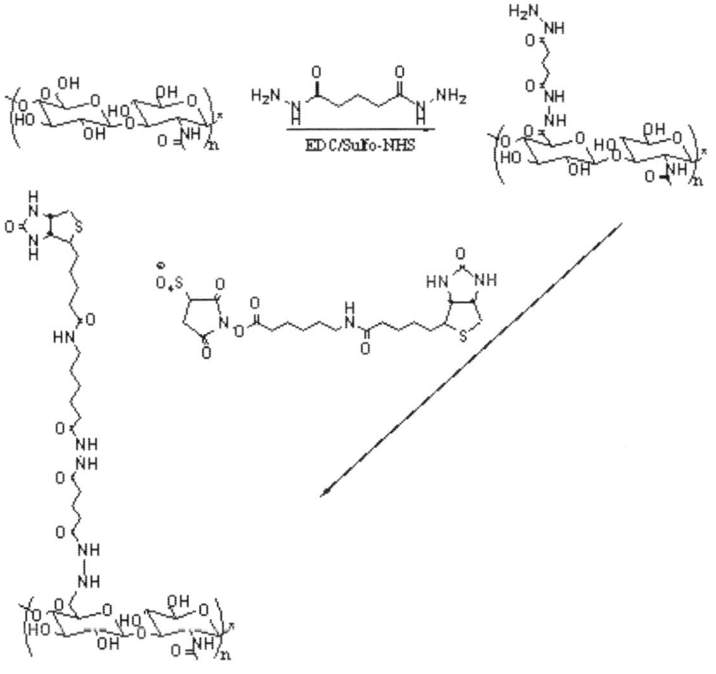

Figure 9.14. Introduction of biotin to the HA backbone.[18,44]

by grafting dexamethasone to HA–ADH via succinate chemistry.[43] Another example of an HA hydrogel synthesis via ADH was the grafting of HA–ADH with biotin to functionalize and pattern.[44]

9.4.1.3 Divinyl sulfone

HA hydrogels were fabricated by cross-linking HA with divinyl sulfone,[13,16,33,45] such as Biomatrix's Hylan-B gel and Genzyme's Hylaform gel.[13,45,23] Sulfate groups grafted to the HA network repulse surrounding substances by acting as anionic charges and mimic a hemocompatibility of heparan sulfate (Fig. 9.15), leading to the formation of a higher degree of cell repulsiveness. This sulfation was obtained by a reaction of divinyl sulfone in either sulfur trioxide pyridine or bis(2-ethylhexyl)sulfosuccinate with the hydroxyl group in HA. HA hydrogel formation, Prevelle, Mentor in USA, via divinyl sulfone was also achieved via sulphonyl-bis-ethyl cross-linking between HA's hydroxyl groups at basic condition. Different types of HA hydrogels, ranging from soft gels to solid forms, were synthesized, such as membranes, micro-particles and tubes, by controlling the reaction conditions such as divinyl sulfone in sodium bis(2-ethylhexyl)sulfosuccinate reverse micelle system under basic conditions for micro-particle synthesis. Compared with native HA, the above relatively soft Hylan-Bgel has an extended residence time *in vivo*, particularly in sites where low mechanical force was imposed on the implant. Further evaluation of sulfonylated HA hydrogels was performed by employing as a matrix for potential neural tissue engineering of the central nervous system trauma of chick dorsal root ganglia.[46] Furthermore, cross-linked thiolated HA supported neurite extensions for the entire duration of the culture period, whereas fibrin cultures exhibited collapsed and degenerating

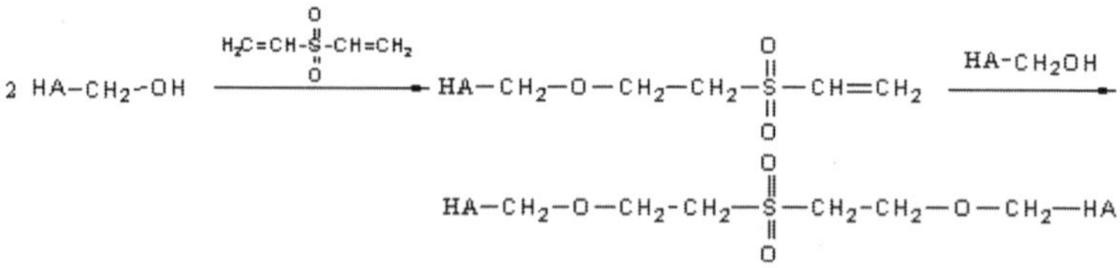

$$2\ \mathrm{HA{-}CH_2{-}OH} \xrightarrow{\mathrm{H_2C{=}CH{-}S(=O)_2{-}CH{=}CH_2}} \mathrm{HA{-}CH_2{-}O{-}CH_2{-}CH_2{-}S(=O)_2{-}CH{=}CH_2} \xrightarrow{\mathrm{HA{-}CH_2OH}}$$

$$\mathrm{HA{-}CH_2{-}O{-}CH_2{-}CH_2{-}S(=O)_2{-}CH_2{-}CH_2{-}O{-}CH_2{-}HA}$$

Figure 9.15. Cross-linking of HA via divinyl sulfone.[13,16,33]

extensions beyond 60 hours. These sulfonated hydrogels have been evaluated for stem cell delivery, scaffolds for scaffold for tissue engineering of cartilage and therapeutic applications. Hyalform and Prevelle have low stiffness (G'), which is a measurement of gel harness or a measurement of resistance to deformation. These hydrogels have been employed as dermal fillers.

9.4.1.4 *In situ* HA hydrogels

- Photo-cross-linking methods[47,48]

The previous session indicated that a wide variety of HA hydrogels can be synthesized by various cross-linking methods, but most of these prefabricated hydrogel fabrication techniques have sometimes difficulties to be applied under physiological conditions. To control gel formation conditions and take advantages of cell or bioactive moleculeencapsulation, the *in situ* photocross-linking method has been developed, a method that reacts only on exposure to the appropriate wavelength of light sources.[49] Since *in situ* hydrogels has been developed by using PEO flanked with oligo(hydroxy acids) and (metha)acrylates (Fig. 9.16),[26] HA hydrogel was developed by Schmidt *et al.*[47,50–53] The functional groups for photopolymerization come from the acrylated groups of HA, inducing rapid polymerization on irradiation with visible light in the presence of a suitable photoinitiator, for example, 2-hydroxy-1-[4-(2-hydroxyethoxy)phenyl]-2-methyl-1-propanone.[16]

Furthermore, photo polymerization demonstrated normally capability of the hydrogel proceeding under physiological conditions without detrimental side effects to bioactive molecules or encapsulated cells in the gel. To this end, HA has been further modified with diverse photo-cross-linkable groups, including cinnamoyl,

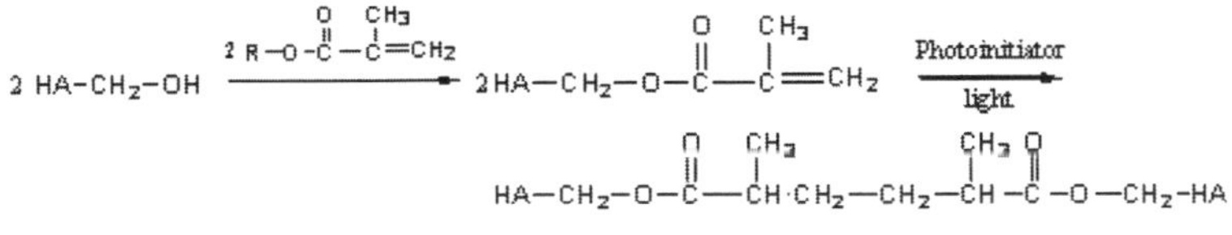

$$2\ \mathrm{HA{-}CH_2{-}OH} \xrightarrow{2\ \mathrm{R{-}O{-}C(=O){-}C(CH_3){=}CH_2}} 2\ \mathrm{HA{-}CH_2{-}O{-}C(=O){-}C(CH_3){=}CH_2} \xrightarrow[\text{light}]{\text{Photoinitiator}}$$

$$\mathrm{HA{-}CH_2{-}O{-}C(=O){-}CH(CH_3){\cdot}CH_2{-}CH_2{-}CH(CH_3){-}C(=O){-}O{-}CH_2{-}HA}$$

Figure 9.16. HA hydrogel by photopolymerization.[16]

and thymine[54]; glycidyl methacrylate via methacrylic anhydride[48] methacrylol group,[55] glycidyl methacrylate[56]; and acrylic acid[57] Additionally, sulfated HA has been modified with 4-azidoaniline, yielding an HA derivative suitable for patterning features as small as 100 mm,[58] which may be useful for directing cell adhesion or growth in a new hydrogel.

The degradation time of the hydrogels was controlled under physiological conditions from 1 day to 4 months, depending on the type of α-hydroxy acids, for example, glycolic or lactic acid as well as the mechanisms such as hydrolysis and enzyme types. This photopolymerization led to the formation of highly hydrated nonadhesive gels, due to their extreme hydrophilicity and cell-nonadhesiveness by employed HA itself and counterpart polymers such as PEO. The synthesized *in situ* HA hydrogels[59] were effective in the prevention of scar formation after pelvic surgery in animal models and the prevention of thrombosis and reduction of long-term intimal thickening by smooth muscle cells when applied as a mechanical barrier on severely injured arteries and abdominal surgery.[57] Smeds *et al.* reported on the use of two types of methacrylate-modified HA that, on photopolymerization, formed visco-elastic gels, possessing mechanical properties similar to those of, for example, nucleus pulposus and meniscus.[59]

HA hydrogels by photopolymerization have been applied to tissue engineering[47,60] and stem cell encapsulation by incorporating a spacing biocompatible macromolecules such as poly(ethylene glycol) and bioactive oligopeptide such as RGD[47,61] and its self-renewal.[62] As examples, application of HA hydrogels to corneal healing in New Zealand rabbit eyes demonstrated that the corneal wound was fully sealed and no vitreous fluid leaked from the eye after three times of 10 second exposures of an argon ion laser. These photopolymerized HA hydrogels demonstrated candidates for suture-less closure of wounds and tissue regeneration in complex shapes and sizes at not-easily-accessible sites, including corneal lacerations. Stem cells for osteogenic differentiation were encapsulated in the acrylated or methacrylated hyperbranched polyglycerol-based HA hydrogel with an initiator and UV light with different doses (Fig. 9.17). Degrees of cell damages were dependent on the choice of initiator and conditions of UV irradiation. The

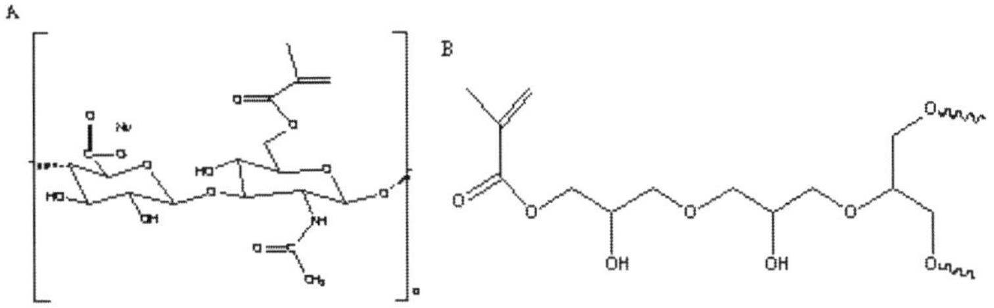

Figure 9.17. Cross-linkable methacrylated HA (A) and methacrylated hyperbranched polyglycerol, of which a fragment is depicted (B).[61]

results indicated that the cell viability and cell cycle progression of exposed mesenchymal stem cell monolayers were affected by initiators, probably due to radicals generated during UV irradiations, but osteogenic cell differentiation was not affected.

To control the mechanical properties of HA hydrogels, patterning of the gel swelling was further studied by employing methacrylyoyl-grafted HA and triethylamine as base catalyst and exposing the HA solution to UV light for 0.5–20 minutes at the first step and adding up sodium hydroxide as the second step.[52] Degradation of the HA hydrogel was further controlled by grafting polylactide as a spacing domain between the HA and the methacrylate functional group.[63] This HA-polylactide hydrogel was expected to have dual degradation sites such as the HA itself by hyaluronidase and ester linkage of HA-lactide by hydrolysis (Fig. 9.18).

- Enzyme and oligopeptide-based *in situ* hydrogels

An *in situ* HA hydrogel was also fabricated by utilizing characteristics of both oligopeptides and enzymes to mimick micro-environment of ECM in the body. A oligopeptide-mediated HA hydrogel was synthesized by using *hexa*-histidine (His) (Fig. 9.19).[64] After activation of HA (Fig. 9.19), *N*-(5-amino-1-carboxypentyl) iminodiacetic acid (NTA) was coupled to the activated HA side chains. To chelate Zn(II) ions, $ZnSO_4$ was added to the HA-NTA solution. α-helical peptides of 70 amino acid residues carrying a *hexa*-His peptide at both termini and the brain-derived neurotrophic factor (BDNF) carrying a His at the C-terminus were coordinated with the Zn(II) ions chelated to HA chains for both the cross-linking of HA and the tethering of BDNF.

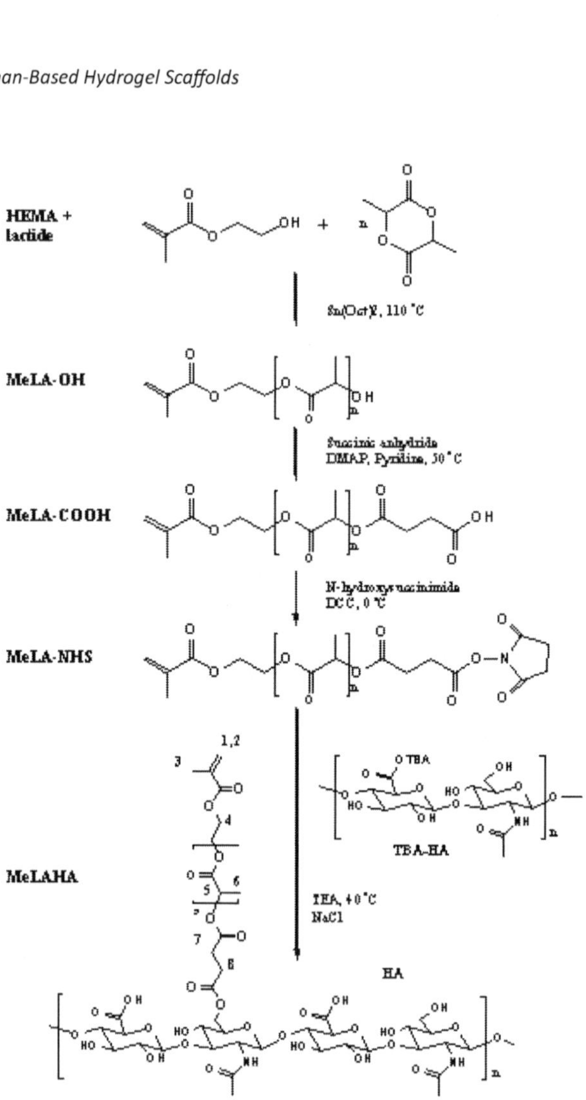

Figure 9.18. Synthesis of methacrylated lactide HA63 (HEMA = 2-hydroxyethyl methacrylate, Me = methacrylate, LA = lactic acid, DMAP = dimethylaminopyridine, TEA = triethylamine, DCC = dicyclohexylcarbodiimide, TBA = tetrabutylammonium)[52].

Mixing the Zn–HA solution and BDNF with His and a cross-linking peptide with His at both termini (CLP), solutions turned automatically into an HA hydrogel.

In situ HA hydrogels by enzymes were fabricated by employing HA–tyramine (HA–Tyr) conjugates in the presence of hydrogen

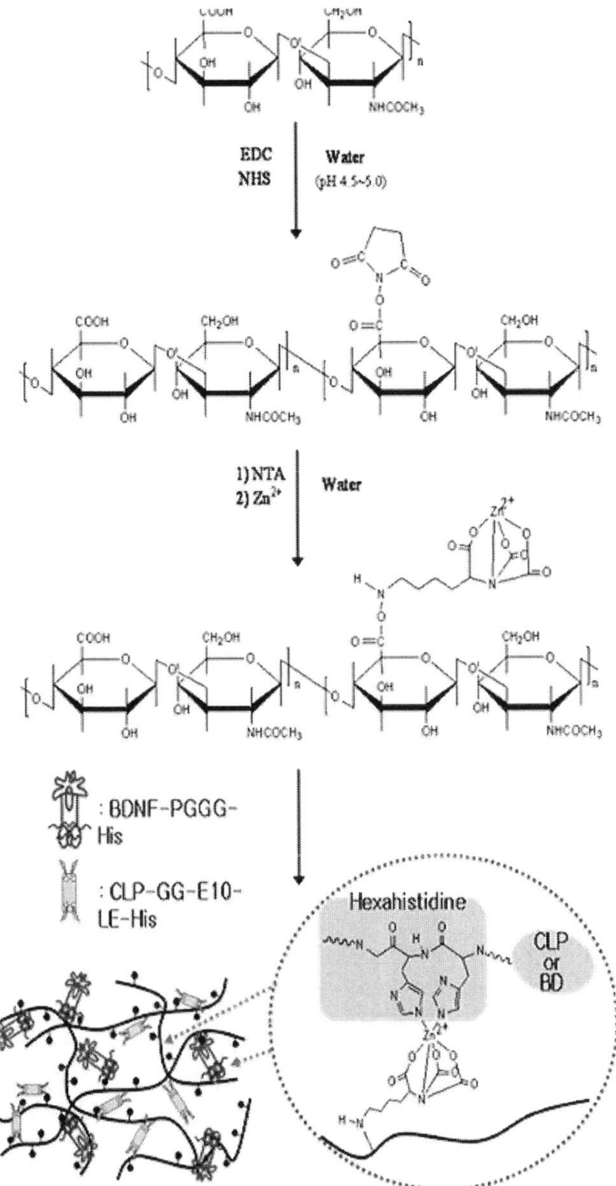

Figure 9.19. Derivatization of HA followed by cross-linking with CLP and incorporation of BDNF–His.[64] BDNF = brain-derived neurotrophic factor (11 amino acids), His = hexahistidine, E10 = Kelasve; (70 amino acids).[64]

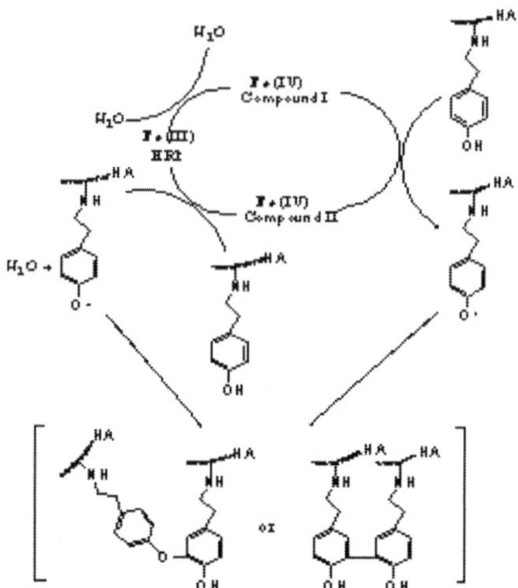

Figure 9.20. Cross-linking of HA-Tyr– conjugates by an enzymatic coupling reaction.[65]

peroxide (H_2O_2) and horseradish peroxidase (HRP), yielding enzymatically cross-linked, highly elastic, and biodegradable HA hydrogels (Fig. 9.20).[65] The HRP-mediated coupling reaction of phenol moieties in HA–Tyr conjugates occurred either via a carbon–carbon bond at the *ortho* positions and a carbon–oxygen bond between the carbon atom at the *ortho* position and the phenoxy oxygen. While the gelation time was dependent on the precursor concentrations of enzyme or H_2O_2/Tyr ratios, the gel was degraded over several months, due to slow hydrolysis of the newly formed urethane bonds. In related hydrogels where the Tyr units were linked to the HA via an ester-containing diglycolic group, degradation was, however, controlled more rapidly within several days. A double syringe equipped with a mixing chamber was normally utilized for the synthesis of HA hydrogels: one compartment contained the HA-Tyr solution, whereas the other one contained a mixture solution of H_2O_2 and HRP. This double syringe system facilitated the administration of the hydrogel system, leading to its applications in protein delivery and scaffold for tissue engineering. Furtehr grafting of dextran-tyramine to HA was turned into an injectable hydrogel for cartilage tissue engineering in the presence of HRP and H_2O_2.

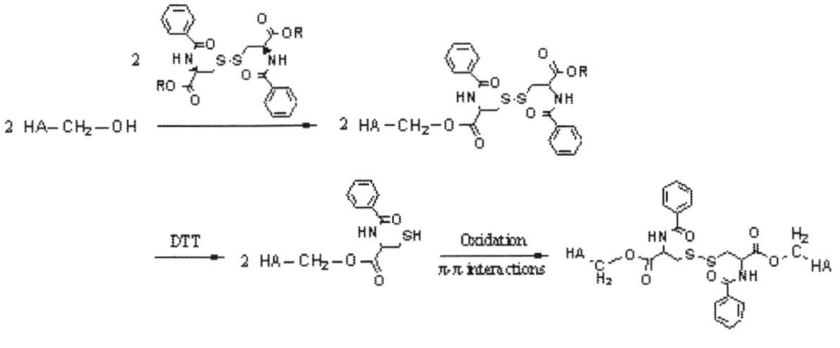

Figure 9.21. Activation of carboxylic acid of DBC by employing DCC/NHS and synthesis of HA–DBC and reduction of its disulfide bonds to obtain HA–BC-L-benzoyl-cysteine; DBC, *tetra*-butylammonium hydroxide (TBA–OH), 5,5-*di*-thio-bis(2-nitrobenzoic acid) (DTNB).[66]

- Oxidation-based *in situ* hydrogels

Another type of *in situ* HA hydrogel for tissue engineerings such as regeneration of intervertebral disk was obtained by oxidation-based auto-cross-linking of HA by using sodium periodate. Fabio Salvatore Palumbo *et al.* synthesized auto-cross-linking HA hydrogels having a fibril structure (Fig. 9.21).[66] The activation of the carboxylic groups of *N*,*N*-dibenzoyl-L-cystine (DBC) was performed with the activators NHS and dicyclohexylcarbodimide. The solution of DBC–NHS was reacted with HA–tetrabutyl ammonium (TBA) in the presence of a catalyst (see Fig. 9.21). After reduction of the -S-S- bridges to -SH groups with a reducing agent(DTT), HA–benzoylcysteine (HA–BC) was obtained and solutions of HA–BC were automatically turned into an HA gel via oxidation mechanims.[66,67]

- Michael-type reaction method

Fabrication of *in situ* HA hydrogels via the Michael-type reaction mechanism is well-known chemistry,[30] and it is carried out over time by employing the end-group reaction between unsaturations and thiol groups of HA (Fig. 9.22).[26–32] While the HA-NH_2 or HA-acrylate has been obtained by using ADH by the cross-linking with bis(sulfosuccinimidyl) suberate or methacrylated HA (HA–MA) or acrylated HA, the HA with thiol groups (HA-SH) were synthesized

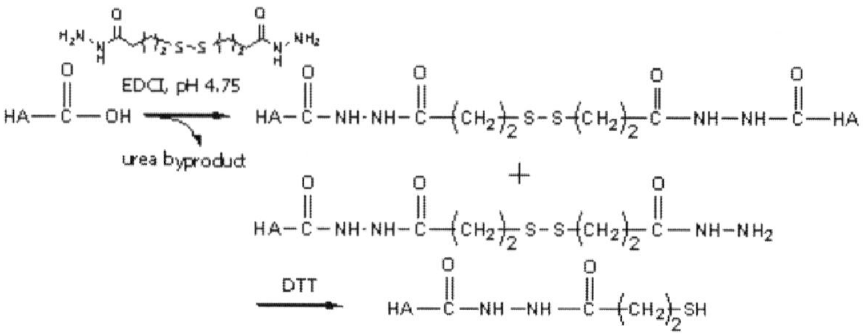

Figure 9.22. Cross-linking reaction. The reaction between HA–aldehyde (HAA) and PVA–hydrazide (PVAH) is highly selective and results in the formation of a network held together by stable hydrazone bonds.[75]

with sodium *tetra*-thionate, 3,3-*di*-thiobis(propanoic hydrazide), or *di*-vinyl sulfone with the hydroxyl groups of HA and then reduced with reducing agents such as DTT.[23,76] Detailed chemical mechanisms of Michael-type reactions have been reported elsewhere by the author.[77] The vinylsulfone groups react quite rapidly under mild conditions (pH 7, room temperature) with thiol groups, resulting in cross-linked hydrogels. The properties of the thiol-modified HA gel could be controlled by 1) the molecular weight of starting HA, 2) the degree of grafted of thiols and (metha)acrylates among other factors.[27]

9.4.1.5 HA–aldehyde hydrogels

By utilizing the reaction mechanisms that both formaldehyde and glutaraldehyde cross-link proteins via amine-grafted HA derivatives for tissue preservation and regeneration,[25,68] grafting of formaldehyde to low molecular weight HA was used to create a cross-linked HA hydrogel, Biomatrix's Hylan-A that is water soluble but more viscous and elastic than native HA.[21,43] The *in situ* reaction through Schiff's base formation of aldehyde groups of the oxidized low molecular weight HA with amino residues of collagen II in the counterpart polymer was used for the preparation of self-cross link hydrogels for intervertebral disc tissue engineering. The gelation time depended on the pH and ionic strength of the buffer, and was typically about several minutes in a phosphate buffer solution. The strength of the HA gels increased with the degrees of their oxidation and aldehydes.

The presence of grafted aldehyde groups was observed by a multiphoton confocal microscope on fluorescence staining. The aldehyde groups were used for covalent cross-linking with HA-ADH. The resulting cross-linked networks were highly pliable and had break strength of 500 Pa elastic modulus at 200%–300% strain. The mechanical characteristics of these doubly cross-linked hydrogel were similar to those of vocal fold lamina propria. *In vitro* cell proliferation results by the MTT assay showed that the proliferation of the fibroblasts. After loading a low-molecular-weight model drug, rhodamine into the HA gel, its release was monitored using UV-Vis spectroscopy. Rhodamine-loaded HA hydrogels maintained their ability to form doubly cross-linked networks when mixed with the HA–ADH solution. Most of the entrapped rhodamine was burst-liberated from the HA gel in the first several hours and completely released over 3 days. These novel hydrogels demonstrated potential as biodegradable, control-released hydrogels for vocal fold tissue regeneration.

Glutaraldehyde was used to cross-link amine-grafted HA to yield hydrogels with high resistance toward biodegradation. Tomihata and Ikade used acidic glutaraldehyde solutions to fabricate cross-linked HA films.[69] Hu *et al.* compared glutaraldehyde and carbodiimide-mediated reactions of HA and found that glutaraldehyde-based HA gelation yielded highly cross-linked HA materials suitable for use as 3-D scaffolds for tissue engineering.[70] Because of HA's proven properties of biocompatibility and biodegrdation, the hydrogels were evaluated on their further possible application in prevention of peritoneal adhesions, a serious consequence of abdominal surgery, which could lead to pain, bowel obstruction, or even infertility. Yoon Yeo *et al.*[71] also synthesized *in situ* HA hydrogels by employing HA derivatives such as HA–NH_2 and HA–aldehyde for the prevention of postoperative abdominal adhesions in a rabbit model.[18,71,72] They tested degree of HA degradation by hyaluronase by measuring the amount of released glucuronic acid from the cross-linked HA gels. The effects of HA degradation on the productions of tissue-type plasminogen activator and plasminogen activator inhibitor-1 were measured to develop a combined adhesion barrier and drug delivery system for prevention of postoperative peritoneal adhesions. Plasminogen activator and plasmin are related to cell migration, tissue adhesion and connective tissue remodeling according to degradation of implanted HA scaffolds after surgery. Particles of cross-linked HA hydrogels were also fabricated through aldehyde chemistry using a sodium bis(2-ethylhexyl) sulfosuccinate/*iso*-octane reverse micelle system for potential application in vocal fold regeneration and poly(lactic-*co*-glycolic acid) nanoparticle dispersion in aldehyde- and ADH -modified HA and then combined these two particles via a double-barreled syringe for an adhesion barrier combined with drug delivery.[73,74] These HA gel particles formed a flexible and durable hydrogel through their physical interactions. The HA particles showed prolonged *in vivo* residence time, temporal release of therapeutics, and matching viscoelasticity for use as a scaffold for vocal fold tissue engineering. As a hybrid type of HA hydrogels, HA–aldehyde was coupled with ADH-grafted PVA and the obtained HA–PVA hydrogel was tested for bone tissue engineering by adding bone morphogenetic protein-2 (BMP-2) (Fig. 9.23).[75]

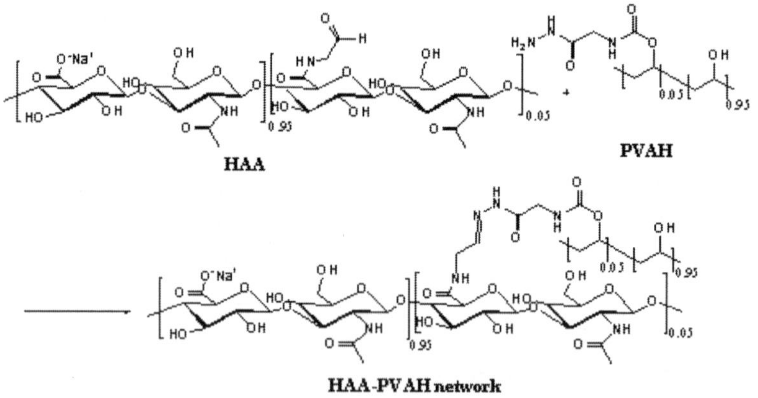

HAA

PVAH

HAA-PVAH network

Figure 9.23. Derivatives of HA–NH_2 and HA–SH for Michael-type addition reactions.[33]

9.4.1.6 Azaide

In situ chemical gelation of aqueous HA solution has been developed by grafting with either azaide or alkyne terminal functionality through copper-catalyzed azide-alkyne cyclo-addition. Azaide and alkyne-grafted HAs were obtained by reacting 11-azido-3,6,9-trioxaundecan-1-amine and proparlgylamine with separate HA via EDC/NHS chemistry (Fig. 9.24).[78] When these two types of HA derivatives were mixed together, they gave rise to a 1,3-dipolar cyclo-addition reaction, showing quick gelation in the presence of catalyst Cu(I). By carrying out the HA gelation process in aqueous solutions of benzidamine and doxorubicin, respectively, the swollen HA hydrogel acted as drug reservoirs.[78] The doxorubicin release was well-controllable depending on the gel's degree of cross-linking and degradation. Finally, formation of the self -cross-linking gels using aqueous suspensions of *Saccharomices cerevisiae* yeast cells allowed scaffold fabrication, inside which cells were homogeneously distributed and dhered well to the inner pores surfaces. The obtained HA gels have positive features in the choice of scaffolds for tissue engineering and drug delivery systems.

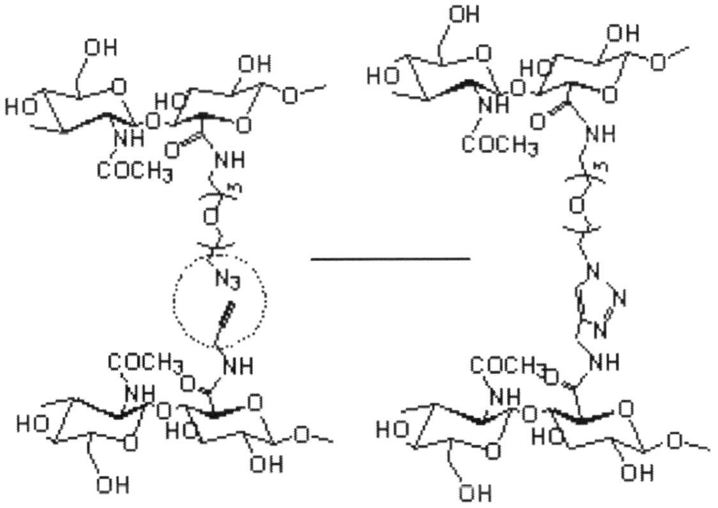

Figure 9.24. Formation of HA-based auto-cross-linking gels via azaide reactions.[78]

9.5 Hyaluronic Acid-Based Hybrid Hydrogels

9.5.1 *HA–Collagen/Oligopeptide Hydrogels*

Biodegradable polymer scaffolds serve a central role in tissue engineering by directing cellular processes based on their structural and biochemical properties. A natural extracellular matrix (ECM), among all the scaffolds, induce biological tissue engineering and remodeling such as wound dressing and bones in animal.[79] To mimic these biological and physic-chemical characteristics of the natural extracellular matrix, collagen, among other natural polymers, has been employed by many research groups, due to its abundance in the ECM and outstanding cellular behaviors, as well as other important biological signals. HA–collagen and HA–biotin/collagen hydrogels were synthesized by grafting HA with the PEO-diepoxide (Fig. 9.25) to provide support for cell adhesion. The HA–collagen hydrogels demonstrated a continuous exterior and an interconnected porous interior, with pore diameters ranging from 6 to 9 μm. HA and HA–collagen hydrogels degraded in the presence of hyaluronidase and collagenase, indicating that HA–collagen scaffolds were controllably degraded. Complete degradation of the hydrogels occurred within 14 days in hyaluronidase (100 U/mL) and 3 days in collagenase (66 U/mL).[44] Another fabrication of HA–collagen hydrogel was reported by photopolymerization by employing methacrylation of HA, thus obtaining a hydrogel with collagen interpenetrated.[50]

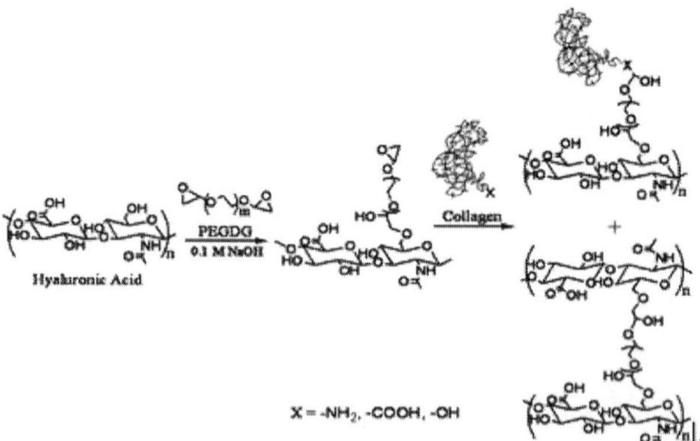

Figure 9.25. Synthesis of the HA and the HA–collagen hydrogel.[44]

Even though ideal scaffolds and hydrogels for tissue engineering have been suggested by directly utilizing the ECM, as described earlier, there were still unresolved problems such as denaturation and uncontrollability of natural proteins during their synthesis and applications. Thus the design of HA hybrid scaffolds focused on mimicking the protein's oligopeptide sequence such as cell adhesion domains. Many research groups focused on fabrications of cell-specific or cell-ubiquitous oligopeptide-based HA or hybrid hydrogels. HA-based hybrid scaffolds were fabricated by incorporating some specific oligopeptide domains from the ECM to induce mainly cell adhesiveness. ECM-mimicking oligopeptide-incorporated HA-based hydrogel for tissue engineering that induce cell adhesion and migration have been studied by Hubbell *et al.*[28,29] They used cell-specific adhesive oligopeptides as cross-linkers in hydrogel's network. The bioactive oligopeptide sequences were cell adhesive or cleavable by surrounding cells onto the gel network or enzymes such as matrix metallo-proteinase (MMP), respectively. The MMP is a family of enzymes that have many physiological roles, including mainly the breakdown of ECM networks during tissue remodeling and disease in animals. Therefore the integration of cell-adhesive and MMP-cleavable oligopeptide domains is a reasonable approach toward introducing ECM biological properties to artificial scaffolds for tissue engineering. Human fibroblasts have been known to invade the pores of the hydrogel through integrin-binding oligopeptide domains such as Arg-Gly-Asp-Ser-Pro. Furthermore, in the case of adding bone morphogenic protein-7 as a biological signal into the hydrogel for tissue engineering, it has been applied as a scaffold for bone tissue engineering with bioactive molecule delivery. As another example of the applications of cell-adhesive oligopeptides, immobilization of Arg-Gly-Asp was made to the HA hydrogel network and it demonstrated successful brain tissue engineering.[80] Another oligopeptide application to the hydrogel was performed by Noh *et al.* and Kim *et al.* by employing low-molecular-weight HA, which demonstrated different physical and biological properties from high-molecular-weight HA.[30–32] They demonstrated both chemical incorporation of the Arg-Gly-Asp peptide into the HA polymer network and physical incorporations of both mesenchymal stem cells and bone morphogenic protein-7 into the low molecular weight HA gel. The incorporated oligopeptide demonstrated induction of cell adhesion by mimicking microenvironment of ECM in the body, and thus better bone tissue regeneration was observed in the calvarial defect model in the rat, which is important in its application in tissue engineering.

9.5.2 *HA–Natural Polymer Hybrid Hydrogels*

An HA–natural polymer hydrogel was fabricated via diverse methods by employing, mainly, chitosan among others. These natural polymers could be chitosan, dextran, alginate, chondroitin sulfate, gelatin, collagen and etc. HA–chitosan hydrogels have been reported by many research groups because of their convenient chemistry and the biological advantages of each polymer.[81–85] Chitosan had biological properties in cartilage and bone regeneration. The simple hydrogel formation was obtained by a polyelectrolytic complex of anionic HA-aldehyde and cationic chitosan for the test of tissue regeneration of articular cartilage.[82] A biocompatible and biodegradable HA–chitosan hybrid hydrogel was also derived from water-soluble chitosan and oxidized HA on mixing, without a chemical cross-linking agent. The resulting gelation was attributed to the Schiff base reaction between the amino and aldehyde groups of chitosan and HA, respectively. The HA–chitosan complex demonstrated the *in vitro* encapsulation of bovine articular chondrocytes within the hybrid hydrogel matrix and support of cell survival and cell retention with chondrocytic morphology, thus indicating the possibility of its application in cartilage tissue engineering via cell encapsulation.

An injectableHA–chitosan hydrogel was also formed via both succinic activation of chitosan (N-succinyl-chitosan) and periodate activation of HA (HA–aldehyde) for its applications in cartilage tissue engineering via chondrocyte encapsulation.[83] Another type of HA–chitosan hydrogel was obtained via EDC/NHS coupling chemistry by further grafting of poly(*N*-*iso*propylacrylamide) with an initiator in order to examine their physicochemical characteristics, for example, thermorepsonsive gel formation, *in vitro* drug release and *in vivo* pharmacodynamics.[84] The sol–gel transition behavior was observed as one of the physical property tests, and both hydrophilic and lipophillic drugs, including nalbuphine, indomethacin, and the nalbuphine prodrug, were used as model drugs in an *in vitro* drug release experiment.

An HA–dextran hybrid hydrogel, among other HA-based hybrid hydrogels, was also fabricated to evaluate either its cytotoxicity or its drug delivery for its application as an injectable form for implantation in ferret vocal fold, even though other numerous studies have been done on dextran or HA only.[53,85]

9.5.3 *HA–Synthetic Polymer Hybrid Hydrogels*

HA–synthetic polymer hybrid hydrogels were fabricated by employing hydrophilic polymers such as PEO, PVA, Pluronic polymers, and others. Among those polymers, PEO was adapted in many experiments as a counterpart of HA, due to its excellent biocompatibility and well-known physicochemical characterizations in biomedical applications. An HA–PEO hydrogel, using thiol functionalization, was developed by Zheng Shu *et al.* for the design of HA-PEO hydrogels for tissue engineering and drug delivery.[33] Thiolated HA derivatives were chemically coupled to α, β–unsaturated esters and the amide of PEO. After induction of *in situ* HA-PEO hydrogel by mixing HA–SH and PEO–diacrylates in several minutes, it was applied as a scaffold for bone tissue engineering.[31] Low cytotoxicity of human tracheal scar fibroblasts was observed in the HA-SH precursor solution. Also *in vivo* chondrogenesis of mesenchymal stem cells in the HA–PEO photopolymerized hydrogel was studied with addition of transforming growth factor (TGF)-β3 in a nude mouse. Mesenchymal stem cells in the HA-PEO hydrogels containing TGF-β3 produced higher-quality cartilage on the basis of expression of cartilage-specific genes and production of proteoglycan and collagen II. When used independently, TGF-β3 and HA alone induced cartilage-specific gene expression and collagen II production with reduction of collagen I production, even though TGF-β3 was essential for proteoglycan production.[86] HA–PEO hydrogels synthesized by HA–ADH and methacrylated PEO were also applied to the regeneration of ischemic heart[84] and postoperative pericardial adhesion prevention.[87]

The promotion of tendon graft incorporation within the bone tunnel was also tested by HA–PEO hydrogels grafted with BMP-2 by the method of photopolymerization.[88] HA-BMP-2 complex was used to stimulate periosteal progenitor cells to direct fibro-cartilagenous attachment and new bone formation in an extra-articular tendon-bone healing model. PEO–diacrylate with HA-BMP-2 was turned into a HA–PEO hydrogel by photo-polymerization.

The HA-Pluoronic F127 hydrogel was also synthesized by Shu-Hui Hsu *et al.* for its applications in drug delivery.[89] Fabrication of the HA-PVA hydrogel, another example of an HA–synthetic polymer hybrid scaffold, was discussed in the previous HA–aldehyde session in detail.

9.6 Conclusions and Outlook

Recognition and discoveries of the physicochemical andbiological properties of HA and its applications in tissue engineering, cell delivery and pharmaceuticals has led to the development of many novel and promising synthesis strategies of HA-derivatives and HA-based hydrogels and scaffolds. The modified HA has been applied as biomaterials for orthopedics, anti-adhesion barrier, cardiovascular disease, ophthalmology, dermatology and wound healing, neural and glial, drug/cell delivery and various tissue engineering. This overview of the methods of HA chemical modification based on its chemistry and biology as well as the methods of hydrogel fabrication described earlier illustrates diverse HA hydrogel systems and its possible applications. While some approaches have been extensively evaluated on their biocompatibility and *in vivo* potential in tissue engineering and therapeutic applications, leading to their commercialization in visco-elastic hydrogels such as Synvisc® from Biomatrix, OrthoVisc® from OrthoBiotech and Hyalgan® from Fida for injections for osteoarthritis and in implants of artificial intraocular lens such as Healon® from Advanced Medical Optics and Opegan R® from Seikagaku, others are still under progress such as recently carriers for stem cell delivery. Whether some of the HA-based hydrogels will make it to the clinic will depend on the choice of right biocompatible HA-derivatives, and their *in vivo* performance in specific applications and target diseases. Novel HA-based hydrogels have to demonstrate rigorously proven biocompatibility, controlled degradation, specific interaction with cells and ECM components in the body and subsequent tissue regeneration behavior in biomedical applications. Since the HA-based hydrogels mentioned in the main text showed their clinical potentials and advantages, we expect that research about HA derivertization and subsequent HA-based hydrogel for biomedical and therapeutic applications will keep focusing on novel and clinicable methods of hydrogel fabrication.

Acknowledgments

This research was supported by the Pioneer Research Center Program through the National Research Foundation of Korea funded by the Ministry of Education, Science and Technology (2011-0001696).

References

1. N. A. Peppas, P. Bures, W. Leobandung, and H. Ichiwaka, *Euro. J. Pharm. Biopharm.*, **27** (2000).
2. Wichterle and D. Lim, *Nature*, **117** (1960).
3. A. S. Hoffman, *Adv. Drug Del. Rev.*, **3** (2002).
4. N. E. Fedorovich, J. Alblas, J. R. De Wijn, W. E. Hennink, A. J. Verbout, and W. J. A. Dhert, *Tissue Eng.*, **1905** (2007).
5. S. R. Van Tomme and W. E. Hennink, *Expert Rev. Med. Dev.*, **147** (2007).
6. W. E. Hennink and C. F. van Nostrum, *Adv. Drug Del. Rev.*, **13** (2002).
7. M. S. Kim, D. Y. Kim, S. Y. Jo, H. K. Kang, Y. D. Park, K. B. Lee, I. S. Kim, S. J. Hwang, and I Noh, *Biomater. Res.*, **83** (2008).
8. Hatefi and B. Amsden, *J. Control. Rel.*, **9** (2002).
9. B. Packhaeuser, J. Schnieders, C. G. Oster, and T. Kissel, *Euro. J. Pharm. Biopharm.*, **445** (2004).
10. J. D. Kretlow, L. Klouda, and A. G. Mikos, *Adv. Drug Del. Rev.*, **263** (2007).
11. S. Tanna, M. J. Taylor, T. S. Sahota, and K. Sawicka, *Biomaterials*, **1586** (2006).
12. N. Kashyap, B. Viswanad, G. Sharma, V. Bhardwaj, P. Ramarao, and M. N. V. Ravi Kumar, *Biomaterials*, **2051** (2007).

13. P. A. Band, The chemistry biology and medical applications of hyaluronan and its derivatives, **33** (1998).
14. P. Ghosh and D. Guidolin, *Semin. Arthritis Rheum.*, **10** (2002).
15. N. Volpi, *Current Med. Chem.*, **1799** (2006).
16. G. E. Wnek and G. L. Bowlin, *Biomaterials and Biomedical Engineering*, 2nd ed. (2008).
17. L. Benedetti, R. Cortivo, T. Berti, A. Berti, F. Pea, M. Mazzo, M. Moras, and G. Abatangelo, *Biomaterials*, **1154** (1993).
18. P. Bulpitt and D. Aeschlimann, *J. Biomed. Mater. Res.*, **152** (1999).
19. T. Pouyani, J. W. Kuo, G. S. Harbison, and G. D. Prestwich, *J. Am. Chem. Soc.*, **5972** (1992).
20. J. J. Young, K. M. Cheng, T. L. Tsou, H. W. Liu, and H. J. Wang, *J. Biomater. Sci. Polym. Edn.*, **767** (2004).
21. J. B. Leach and C. E. Schmidt, Hyaluronan in Encyclopedia of Biomaterials and Biomedical Engineering, 2, Edited by C. E. Wnek and C. L. Bowlin, Informa, **1421** (2008).
22. T. Pouyani and G. D. Prestwich, *Bioconjug. Chem.*, **339** (1994)
23. E. J. Oh, S. W. Kang, B. S. Kim, G. Jiang, I. H. Cho, and S. K. Hahn, *J. Biomed. Mater. Res.*, **685** (2008).
24. Y. Luo, M. R. Ziebell, and G. D. Prestwich, *Biomacromolecules*, **208** (2000).
25. Y. Luo, K. R. Kirker, and G. D. Prestwich, *J. Controlled Release*, **169** (2000).
26. A. S. Sawhney, C. P. Pathak, and J. A. Hubbell, *Macromolecules*, **581** (1993).
27. J. A. Burdick and G. D. Prestwich, *Adv. Healthcare Mater.*, **H41** (2011)
28. M. P. Lutolf and J. A. Hubbell, *Nature Biotechnol.*, **47** (2005).
29. M. P. Lutolf, J. L. Lauer-Fields, H. G. Schmoekel, A. T. Metters, F. E. Weber, G. B. Fields, and J. A. Hubbell, *Proc. Natl. Acad. Sci.*, **5413** (2003).
30. Noh, G. W. Kim, Y. J. Choi, M. S. Kim, Y. Park, K. B. Lee, I.S. Kim, S.J. Hwang, and G. Tae, *Biomed. Mater.*, **116** (2006).
31. J. J. Kim, I. S. Kim, Y. Park, G. Tae, K. B. Lee, S. J. Hwang, I. Noh, S. H. Lee, and K. Sun, *Biomaterials*, **1830** (2007)
32. Kim, Y. Park, G. Tae, K. B. Lee, S. J. Hwang, I. S. Kim, I. Noh, and K. Sun, *J. Biomed. Mater. Res., Part A*, **967** (2009)
33. X. Z. Shu, Y. Liu, F. Palumbo, and G. D. Prestwich, *Biomaterials*, **483** (2004).
34. M. A. Serbana, G. Yanga, and G. D. Prestwich, *Biomaterials*, **1388** (2008).

35. S. K. Hahn, J. K. Parkb, T. Tomimatsua, and T. Shimoboji, *Int. J. Biol. Macromol.*, **374** (2007).
36. V. B. Gerasimas, V. M. Chernoglazov, and A. A. Klesov, *Biokhimiia*, **1086** (1980).
37. N. Yui, T. Okano, and Y. Sakurai, *J. Control. Rel.*, **105** (1992).
38. Tomihata and Y. Ikada, *J. Biomed. Mater. Res.*, **243** (1997).
39. J. W. Kuo, D. A. Swann, and G. D. Prestwich, *Bioconjug. Chem.*, **232** (1991).
40. K. P. Vercruysse, D. M. Marecak, J. F. Marecek, and G. D. Prestwich, *Bioconjug. Chem.*, **686** (1997).
41. David, V. Dulong, D. L. Cerf, L. Cazin, M. Lamacz, and J. P. Vannier. *Acta Biomaterialia*, **256** (2008).
42. Y. Yeo, M. Adil, E. Bellas, A. Astashkina, N. Chaudhary, and D. S. Kohane, *J. Control. Rel.*, **178** (2007).
43. T. Ito, I. P. Fraser, Y. Yeo, C. B. Highley, E. Bellas, and D. S. Kohane, *Biomaterials*, **1778** (2007).
44. T. Segura, B. C. Anderson, P. H. Chung, R. E. Webber, K. R. Shull, and L. D. Shea, *Biomaterials*, **359** (2005).
45. E. A. Balazs, P. A. Bland, J. L. Denlinger, A. I. Goldman, N. E. Larsen, E. A. Leshchiner, A. Leshchiner, and B. Morales. *Blood Coagu. Fibrinol.*, **173** (1991).
46. E. M. Horn, M. Beaumont, X. Z. Shu, A. Harvey, G. D. Prestwich, K. M. Horn, A. R. Gibson, M. C. Preul, and A. Panitch. *J. Neurosurg. Spine*, **133** (2007).
47. J. B. Leach, K. A. Bivens, C. W. Patrick, Jr., and C. E. Schmidt. *Biotechnol. Bioeng.*, **578** (2003).
48. K. A. Smeds, A. Pfister-Serres, D. L. Hatchell, and M. W. Grinstaff, *J. Macromol. Sci., Part A, Pure Appl. Chem.*, **981** (1999).
49. K. S. Anseth and J. A. Burdick, *Mater. Res. Soc. Bull.*, **130** (2002).
50. S. Suri and C. E. Schmidt, *Acta Biomaterialia* (2009)
51. J. B. Leach, K. A. Bivens, C. W. Patrick, Jr., and C. E. Schmidt, *Biotechnol. Bioeng.*, **578** (2003).
52. S. A. Zawko, S. Suri, Q. Truong, and C. E. Schmidt, *Acta Biomaterialia*, **14** (2009).
53. S. A. Zawko, Q. Truong, C. E. Schmidt, *J. Biomed. Mater. Res.*, **87A**, 1044 (2008).
54. S. R. Van Tommc, C. Storm, and W. E. Hennink. *Int. J. Pharm.*, **1** (2008).
55. T. Matsuda, M. J. Moghaddam, H. Miwa, K. Sakurai, and F. Iida, *ASAIO J.*, **154** (1992).

56. T. Matsuda and T. Magoshi, *Biomacromolecules*, **942** (2002).
57. G. Chen, Y. Ito, Y. Imanishi, A. Magnani, S. Lamponi, and R. Barbucci, *Bioconjug. Chem.*, **730** (1997).
58. T. Sawada, K. Tsukada, K. Hasegawa, Y. Ohashi, Y. Udagawa, and V. Gome. *Human Reprod.*, **353** (2001).
59. K. A. Smeds and M. W. Grinstaff, *Biomed. Mater. Res.*, **115** (2001).
60. C. Chung, J. Mesa, G. J. Miller, M. A. Randolph, T. J. Gill, and J. A. Burdick, *Tissue Eng.*, **2665** (2006).
61. N. E. Fedorovich, M. H. Oudshoorn, D. van Geemen, W. E. Hennink, J. Alblas, and W. J. Dhert, *Biomaterials*, **344** (2009).
62. S. Gerecht, J. A. Burdick, L. S. Ferreira, S. A. Townsend, R. Langer, and G. Vunjak-Novakovic, *Proc. Natl. Acad. Sci.*, **11298** (2007).
63. S. Sahoo, C. Chung, S. Khetan, and J. A. Burdick, *Biomacromolecules*, **1088** (2008).
64. T. Nakaji-Hirabayashi, K. Kato, and H. Iwata, *Biomaterials* (2009).
65. F. Lee, J. E. Chung, and M. Kurisawa, *J. Control. Rel.*, **186** (2009).
66. F. S. Palumbo, G. Pitarresi, A. Albanese, F. Calascibetta, and G. Giammona, *Acta Biomaterialia* (2009)
67. T. W. Thannhasuer, Y. Konishi, and H. A. Scheraga, *Methods Enzymol.*, **115** (1987).
68. P. L. Lu, J. Y. Lai, D. H. Ma, and G. H. Hsiue, *J. Biomater. Sci., Polym. Edn.*, **1** (2008).
69. G. L. Ellman, *Arch. Biochem. Biophys.*, **443** (1958).
70. K. Tomihata and Y. Ikada, *J. Polym. Sci., Part A, Polym. Chem.*, **3553** (1997).
71. Y. Yeo, C. B. Highley, E. Bellas, T. Ito, R. Marini, R. Langer, and D. S. Kohane, *Biomaterials*, **4698** (2006).
72. X. Jia, G. Colombo, R. Padera, R. Langer, and D. S. Kohane, *Biomaterials*, **4797** (2004).
73. Sahiner, A. K. Jha, D. Nguyen, X. Jia, *J. Biomater. Sci, Polymer Edn.*, **223** (2008).
74. Y. Yeo, T. Ito, E. Bellas, C. B. Highley, R. Marini, D. S. Kohane, *Ann. Surg.*, **819** (2007).
75. K. Bergman, T. Engstrand, J. Hilborn, D. Ossipov, S. Piskounova, and T. Bowden, *J. Biomed. Mater. Res., Part A*, **23** (2008).
76. A. L. Brown, E. M. Srokowski, X. Z. Shu, G. D. Prestwich, K. A. Woodhouse, *Macromol. Biosci.*, **648** (2006).

77. G. W. Kim, Y. J. Choi, M. S. Kim, Y. Park, K. B. Lee, I. S. Kim, S. J. Hwang, I. Noh, *Curr. Appl. Phys.*, **28** (2007).
78. V. Crescenzi, L. Cornelio, C. Di Meo, S. Nardecchia, R. Lamanna, *Biomacromolecules*, **1844** (2007).
79. J. K. Kim, J. S. Lee, H. J. Jung, J. H. Cho, J. I. Heo, Y. H. Chang, *J. Nanosci. Nanotechnol.*, **3852** (2007).
80. F. Z. Cui, W. M. Tian, S. P. Hou, Q. Y. Xu, I. S. Lee, *J. Mater. Sci: Mater. Med.*, **1393** (2006).
81. W. Wang, *J. Mater. Sci.: Mater. Med.*, **1259** (2006).
82. S. R. Frenkel, G. Bradica, J. H. Brekke, S. M. Goldman, K. Ieska, P. Issack, M. R. Bong, H. Tian, J. Gokhale, R. D. Coutts, R. T. Kronengold, *Osteoarthr. Cartil.*, **798** (2005).
83. T. Huaping, R. C. Constance, A. P. Karin, G. M. Kacey, *Biomaterials*, **2499** (2009).
84. J. Y. Fang, J. P. Chen, Y. L. Leu, and J. W. Hu, *Euro. J. Pharm. Biopharm.*, **626** (2008).
85. J. Trudel, S. P. Massia, *Biomaterials*, 3299 (2002).
86. B. Sharma, C. G. Williams, M. Khan, P. Manson, J. H. Elisseeff, *Plast. Reconstr. Surg.*, **112** (2007).
87. S. J. Yoon, Y. H. Fang, C. H. Lim, B. S. Kim, H. S. Son, Y. Park, K. Sun, *J. Biomed. Mater. Res., Part B* (2009)
88. R. Connors, J. Muir, G. Reiss, P. Kouretas, M. Whitten, T. Sorenson, A. Albanil, G. Prestwich, and D. Bull, *J. Surg. Res.*, **237** (2007).
89. S. H. Hsu, Y. L. Leu, J. W. Hu, and J. Y. Fang, *Chem. Pharm. Bull.*, **453** (2009).

Part IV

ELECTROSPINNING NANOFIBER

Chapter 10

GUIDANCE OF CELL ADHESION, ALIGNMENT, INFILTRATION, AND DIFFERENTIATION ON ELECTROSPUN NANOFIBROUS SCAFFOLDS

Sang Jin Lee[*] and James J. Yoo

Wake Forest Institute for Regenerative Medicine,
Wake Forest University Health Sciences Medical Center Boulevard,
Winston-Salem, North Carolina 27157, USA
*sjlee@wfubmc.edu

The interactions between cells and the extracellular matrix (ECM) are fundamental processes that govern signaling for cell adhesion, alignment, migration, and differentiation. Electrospun scaffolds are a novel tool that provides a nanofibrous architecture that mimics ECM's three-dimensional environment. As such, scaffolds generated by electrospinning can be used to observe and understand cellular behaviors in a nano-scaled configuration. This chapter illustrates methods in which guidance of cell adhesion, alignment, infiltration, and differentiation on electrospun nanofibrous scaffolds can be measured. In addition, we provide evidence that these cellular behaviors can be controlled by varying electrospinning fabrication parameters.

Handbook of Intelligent Scaffolds for Tissue Engineering and Regenerative Medicine
Edited by Gilson Khang

www.panstanford.com

10.1 Introduction

Topographical cues generated by the ECM have significant effects upon cellular behaviors, including adhesion, proliferation, alignment, migration, and differentiation. Studies have shown that substratum topography has direct effects on the ability of cells to orient, migrate, and produce an organized cytoskeletal arrangement.[1] The topography of a native tissue matrix is a complex structure, comprising pores, fibers, ridges, and other nano-scaled features. A fundamental understanding of cell-substrate interactions is important for tissue engineering applications and the development of medical implant devices. Electrospinning technology has been widely employed to fabricate tissue-engineered scaffolds that mimic native ECM architecture. Thus, an understanding of the interactions between electrospun fibrous scaffolds and mammalian cells is crucial to the successful production of target tissues and organs. Although many factors contribute to the successful generation of functional tissues, which include biomaterial selection and composition, this chapter will focus on the structural and morphological effects of electrospun nanofibers on cells.

Electrospinning techniques permit fabrication of fiber diameters in the range of several micrometers to tens of nanometers. This technology offers many advantages, which include the production of nano-scaled fibers with specific spatial orientations, high aspect ratios, high surface areas, and controlled pore geometries. Some of these parameters can be optimized to enhance cellular growth and function *in vitro* and *in vivo*, leading to improved cell adhesion, cellular protein expression, and improved diffusion of oxygen and nutrients.[2] Electrospun scaffolds fabricated with collagen mimic natural ECM, consisting of randomly oriented collagen fibers with nano-scale architecture. This architecture is desirable as it facilitates cell adhesion and proliferation upon cell seeding. In addition, electrospinning offers the ability to control scaffold composition, structure, and mechanical properties.[3–5] In this chapter we present evidence that electrospun nanofibrous scaffolds can guide cell adhesion, alignment, infiltration, and differentiation and that this can be controlled by applying various fabrication parameters.

10.2 Electrospinning Technology

Electrospinning has been widely used as a fabrication method to generate nanofibers for various tissue engineering applications. The nano-scaled fiber structures generated by this method are designed to support cell adhesion and guide cellular behavior. This technology offers the ability to control scaffold composition, structure, and mechanical properties.

Electrospinning requires a high-voltage power supply, a syringe pump, a polymer solution to be spun, and a grounded collection surface. Solutions containing various materials are exposed to a high-voltage power supply at 5–25 kV potential between the solution tip and the grounded surface (Fig. 10.1). The solution is delivered with a syringe through a blunt-tip needle at a flow rate of 3–10 mL/hr using a syringe pump. Fibers can be collected onto a specially designed grounded mandrel at a distance of 5–30 cm from the needle tip. Controlling certain variables during electrospinning allows for controlling morphology, diameter, and alignment of fibers[6] (Table 10.1).

The fabricated electrospun nanofibers possess unique features and properties, including an extremely high surface-area-to-volume ratio, which allows for enhanced cellular interactions with scaffolds.

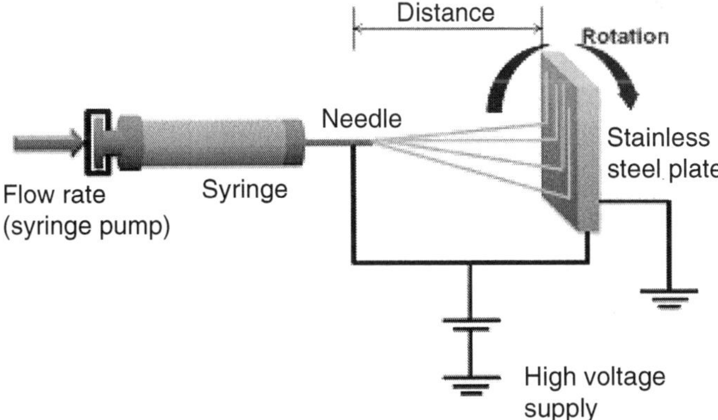

Figure 10.1. Electrospinning methodology. Fibers are collected on a grounded surface. The electrostatic field causes splaying of solution, and solutions of sufficient viscosity and surface tension form fibrous mats that adhere to the grounded surface.

Table 10.1. Fabrication parameters for controlling the configuration of electrospun fibers[6]

Fabrication parameters	Effect on fiber morphology
Solution concentration/viscosity	↓– bead formation, ↑ – larger fibers
Conductivity/solution charge density	↑– uniform bead-free fibers, smaller fibers
Surface tension	No conclusive link
Polymer molecular weight/viscosity	↑– reduced the number of beads
Dipole moment and dielectric constant	Successful spinning
Flow rate	↓– smaller fibers, ↑ – larger fibers
Field strength/voltage	↓– larger fibers, ↑ – smaller fibers
Distance between needle tip and collector	↓– smaller fibers, ↑ – larger fibers
Needle tip design and diameter	↑needle diameter – larger fibers
Collector geometry	Controlled fiber orientation, construct shape
Ambient parameters (temperature and humidity)	↑temperature – ↓ viscosity, ↑ humidity – appearance of circular pores on the fibers

In addition, three-dimensional electrospun nanofibrous scaffolds with high porosity and pore interconnectivity provide a favorable environment for cells using a variety of synthetic and naturally derived biomaterials. Thus, this technique creates a nano-scaled surface environment that supports cellular adhesion and proliferation.

10.3 Cellular Interactions with Electrospun Fibrous Scaffolds

The interactions of cells with biomaterials are critically important for the successful outcome of tissue engineering applications. Thus, the behavior of cells grown on a biomaterial surface, including adhesion to the material, development of appropriate cellular structures, cell growth, differentiation, and maintenance of proper cellular function, must be investigated in order to obtain insight into the biocompatibility of the biomaterial. It is well known that adhesion and proliferation of different types of cells on biomaterials depend mostly on surface characteristics, such as wettability (hydrophilicity/hydrophobicity), chemistry, charge, roughness, and rigidity. It is also known that cells are sensitive to the topography of the supporting surface of the biomaterials. Surface topography affects the formation of a fibrous capsule around implants, the inflammatory response at the tissue–implant interface, fibroblast attachment, angiogenesis, and many other cellular processes, such as cellular differentiation, DNA/RNA transcription, cell metabolism, ECM production, and phenotypic expression.[7–10] Studies of the interactions between surface topography and cells have encompassed a wide variety of cell types and substratum features, including roughness,[9,11] microgrooves,[12,13] ridges, micropores,[8] wells, and nodes, which can influence cellular behaviors. In the following sections, we will focus on the cellular behavior, including cell adhesion, alignment, infiltration, and differentiation, on electrospun nanofibrous scaffolds.

10.3.1 *Cell Adhesion*

The contact of cells on the substrate is mediated by cellular components such as integrins, adhesive proteins, and the actin cytoskeleton. Cell adhesion is a complex process involving physical interactions, chemical binding events, and biological signaling processes. It plays a central role in the regulation of cellular behaviors such as cell proliferation, differentiation during development, and the modulation of cell migration in the injured region, metastasis, and angiogenesis. Thus, cell adhesion has been one of the critical criteria for evaluating the integration of implanted biomaterials. The topography of the surface of a biomaterial can influence cell adhesion and proliferation, which further affects cellular functions. The electrospinning process provides the opportunity to fabricate tissue-engineered scaffolds with a nano-scaled topography and high porosity similar to natural ECM. These nanofibers provide a suitable material and environment for tissue engineering since they can be used to enhance cell adhesion and proliferation.

Focal adhesions of cells can be regulated by the topography of biomaterials. Focal adhesion lies at the convergence of integrin adhesion, signaling, and the actin cytoskeleton, and it is known to control signaling complexes and integrin function. For instance, the diameters of electrospun fibers can be controlled by varying the fabrication parameters (Table 10.1). This provides different topographical cues. Figure 10.2 shows the cytoskeletal organization and

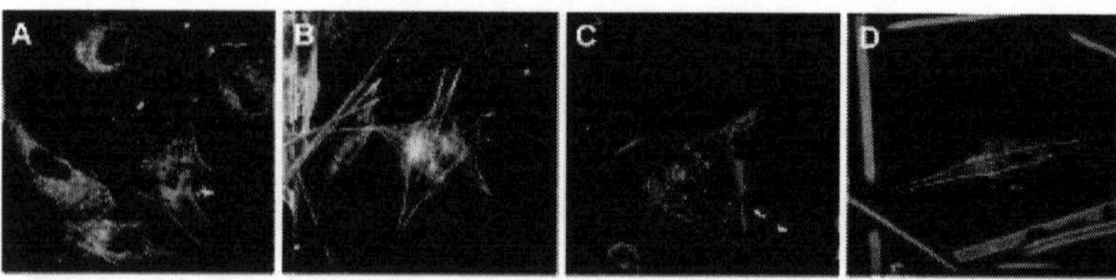

Figure 10.2. Cytoskeletal and focal adhesion of human aortic endothelial cells on electrospun nanofibers with various fiber diameters; (A) 500 nm, (B) 1 μm, (C) 2 μm, and (D) 5 μm ($\times$600 magnification). Red: F-actin, Green: vinculin (unpublished data).

focal adhesion of endothelial cells (EC) on electrospun polycaprolactone (PCL)/collagen scaffold with different fiber diameters. EC on nano-scaled fibers (0.5 μm, Fig. 10.2a) show a better-developed cytoskeletal organization and improved focal adhesion compared with other fiber diameters. Because the fiber diameters of electrospun nanofiber scaffolds are orders of magnitude smaller than the size of most cells, the cells are able to organize around the fibers or spread and attach to adsorbed proteins at multiple focal points.[14,15] Furthermore, Finne–Wistrand *et al.* have shown that the electrospun nanofibrous mats can enhance adhesion and proliferation of mesenchymal stem cells compared with a flat smooth surface.[16]

Several studies have demonstrated that electrospun fibrous scaffolds can enhance cellular responses, including cell adhesion and maintenance of cell phenotype maintenance.[17–20] Shih *et al.* demonstrated the adhesion, proliferation, motility, and differentiation of human mesenchymal stem cells on electrospun type I collagen nanofibers of different diameters.[21] These indicate that nano-scaled fibers can support initial cell adhesion, which then affects further cell proliferation and differentiation.

10.3.2 *Cell Alignment*

Nanofiber scaffolds are known to guide cellular attachment and orientation. As such, nanofibers fabricated by electrospinning provides a controlled environment that allows for enhanced cellular orientation that leads to accelerated tissue function. Skeletal muscle is a tissue that possesses unidirectional cellular orientation to function

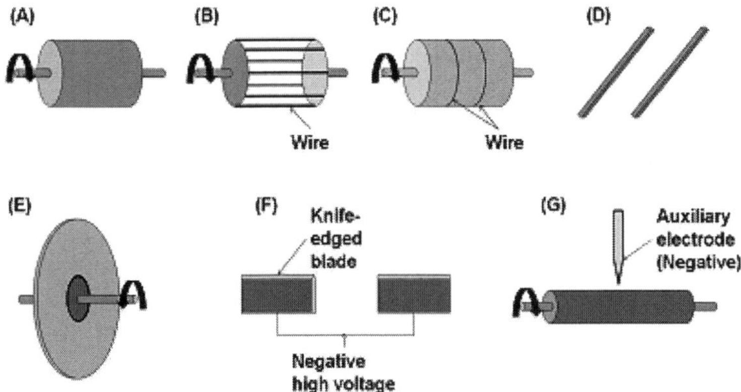

Figure 10.3. Schematic diagrams of various collectors for aligned electrospun fibers: (A) rotating drum,[22,23] (B) rotating wire drum, (C) rotating drum with wire,[24] (D) parallel electrodes,[25] (E) rotating disk,[26] (F) knife-edge electrode,[27] and (G) auxiliary electrode.[28]

properly. Therefore, the use of electrospun scaffolds with aligned nanofiber architecture would result in cellular alignment. A conventional collection of nanofibers on the surface of a rotating drum or mandrel can simply produce aligned fibers with high rotation velocities. To achieve fiber alignment, the collector in the electrospinning assembly must be correctly designed. Schematic diagrams of various collecting substrates used for spinning aligned nanofibrous assemblies are shown in Fig. 10.3. This type of geometric arrangement would be appropriate for use in scaffold-based tissue engineering, as the cells grow preferentially in the direction of the fiber orientation.

A simple way to achieve uniaxial alignment in the nanofiber deposition is to increase the rotational speed of the collector. Figure 10.4 shows that nanofibrous scaffolds with different fiber angles were produced by electrospinning at various rotation rates. Progressive alignment of fiber orientation was observed as the rotation rate increased. Histograms for orientation of electrospun scaffolds indicate a close correlation between fiber distribution and the wrapped normal distribution function.

Electrospun fiber orientation can influence cell proliferation, in addition to controlling cell orientation and tissue growth. It is

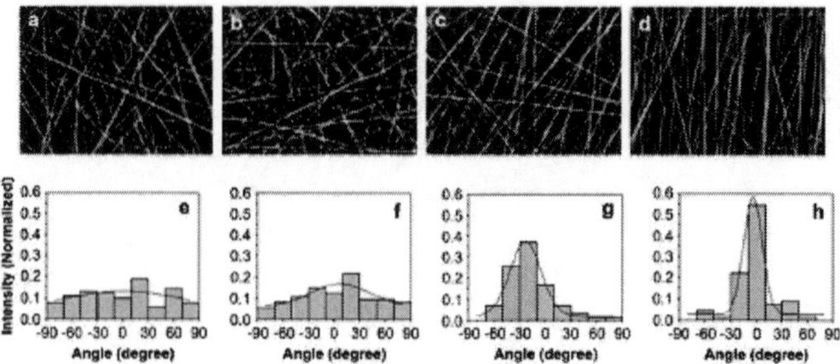

Figure 10.4. Fiber angles for the different rates of rotation: (a–d) SEM images of electrospun PCL/collagen nanofibers (×4.0K magnification) and (e–h) normalized histograms of fiber angle; (a,e) static, (b,f) 800 rpm, (c,g) 1,500 rpm, and (d,h) 2,350 rpm.[22]

known that musculoskeletal tissues such as skeletal muscle exhibit significant anisotropic mechanical properties and highly oriented cells underneath the ECM. An essential step in engineering functional skeletal muscle tissue is to mimic the structure of native tissue, which is comprised of highly oriented myofibers formed from fused mononucleated muscle cells. It is well known that the structure and organization of muscle fibers dictate tissue function. The ability to efficiently organize muscle cells to form aligned myotubes *in vitro* would greatly benefit efforts in skeletal muscle tissue engineering. There are many factors that can guide cellular growth and orientation, including uniaxial mechanical stimulation generated by a bioreactor. This stimulation facilitates muscle cell orientation and accelerates muscle tissue formation. It has been demonstrated that aligned nanofibers greatly influence muscle cell organization and enhance myotube formation. The myotubes formed on aligned nanofibrous scaffolds are highly organized and are significantly longer than myotubes formed on randomly oriented scaffolds.[22] On the aligned nanofiber scaffolds, the myotubes are highly organized when compared to those grown on randomly oriented nanofibers and culture dishes (Fig. 10.5). The diameter of the myotubes was not significantly different between the aligned and randomly oriented nanofiber scaffolds; however, the myotubes on the aligned nanofibrous scaffolds were more than twice the length of the myotubes on

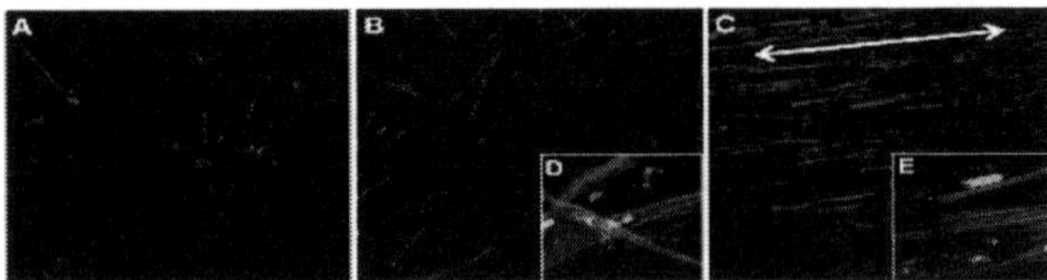

Figure 10.5. F-actin of human skeletal muscle cells (hSkMCs) on electrospun PCL/collagen nanofiber meshes: (A) culture dish, (B) randomly oriented, and (C) aligned electrospun nanofibrous scaffolds (×40 magnification). Laser confocal microscopy images of F-actin of hSkMCs seeded on the (D) randomly oriented and (E) aligned scaffolds (×600 magnification).[22] See also Color Insert.

the randomly oriented scaffolds, which supports the hypothesis that aligned nanofibers promote myotube formation.[22] These findings indicate that electrospun aligned nanofibrous scaffolds are able to guide and orient skeletal muscle cells into organized structures that closely mimic native tissues. The aligned nanofibrous scaffolds are able to promote skeletal muscle morphogenesis into parallel-oriented myotubes. The use of aligned nanofibrous scaffolds seeded with skeletal muscle cells may provide implantable functional muscle tissues for patients with large muscle defects.

In neural tissue engineering, guidance of axonal growth is important to facilitate axonal regeneration through injured environments in the central nervous system (CNS) and peripheral nervous system (PNS). The longitudinal distribution of axons in both the CNS and the PNS makes it essential to design scaffolds that can guide regenerating axons along their natural axis of growth. Experiments using dorsal root ganglia (DRG) neurons seeded onto electrospun fibers have demonstrated the ability of aligned fibers to facilitate directed neurite outgrowth in comparison to randomly oriented fibers.[29] Rat hippocampal progenitor cells cultured on grooved substrates demonstrate increased cell alignment and neuronal differentiation.[30,31] Further, aligned electrospun fibers have shown promise in fostering robust regeneration *in vivo* within a rat peripheral nerve injury model.[32] Therefore, manipulation of electrospinning parameters is crucial to produce highly aligned fibers that facilitate nerve regeneration through an injury site as quickly and as efficiently as possible.

10.3.3 *Cell Infiltration*

A technical concern with the use of electrospinning is the small pore size, which may limit cellular infiltration into the nanofibrous scaffold. For cell migration or infiltration to occur, the pore size of a scaffold should be greater than the size of a cell; a value of 10 μm has been suggested as the minimum pore size necessary for cellular infiltration. As an example, in the engineering of blood vessels using electrospun nanofibrous scaffolds, pore size does not present a major problem with respect to coating the lumen using EC. However, it may limit the ability of smooth muscle cells (SMC) to colonize the outer portion of the neovessel and remodel the ECM. To overcome this limitation, several approaches have been designed to generate larger pores within electrospun scaffolds. These include the use of a salt-leaching technique[33,34] and co-electrospinning with a water-soluble polymer as sacrificial fibers.[35] However, these approaches may compromise the mechanical properties and dimensional stability of the scaffold when the fibers dissolve and collapse. Microintegration is another approach developed by Stankus *et al.*[36] for producing a highly cellularized construct. However, this process may limit homogeneity, sterility, and cell survival, due to the long-term processing times required.

Controlled variables of electrospinning include the solution concentration, flow rate, electric field strength, distance between tip and collector, needle tip design, and collector composition and geometry. These parameters can determine the fiber morphology, diameter, and alignment. It is demonstrated that fiber diameter can be controlled using various parameters, as indicated.[37,38] The increase of fiber diameter of the electrospun fibrous scaffold increases the pore area of the scaffold. To address this technical challenge, studies have been directed toward evaluating cellular infiltration into electrospun fibrous scaffolds with various fiber diameters. To evaluate *in vitro* cell infiltration, scaffolds with four different fiber diameters (500 nm, 1 μm, 2 μm, and 5 μm) were used. NIH/3T3 fibroblasts were seeded onto electrospun fibrous scaffolds with different fiber diameters up to 21 days. Figure 10.6 shows that fiber diameters of the electrospun scaffolds can affect cellular infiltration with time *in vitro*. These results also indicate that fiber diameter is closely

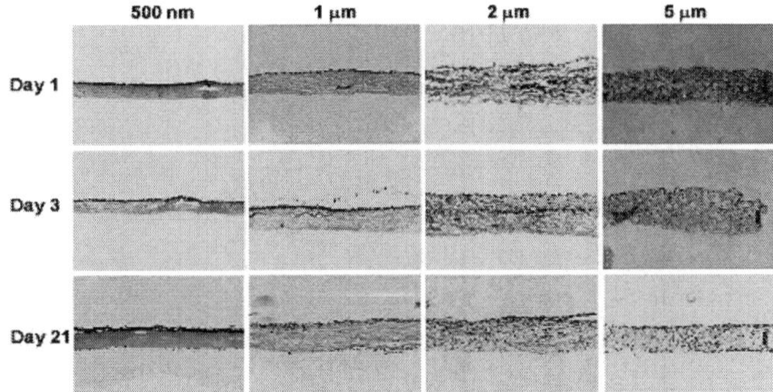

Figure 10.6. Cellular infiltration into the electrospun scaffolds with various fiber diameters. NIH/3T3 fibroblasts were seeded into electrospun PCL/collagen scaffolds up to 21 days. Nuclei stained by DAPI (×40 magnification) (unpublished data).

related to the pore area in electrospun scaffolds. Overall, obtaining adequate cellular infiltration into electrospun nanofibrous scaffolds while maintaining their mechanical properties and structural integrity is still a major challenge.

10.3.4 *Cellular Differentiation*

Cellular differentiation on electrospun fibers is closely related to cell-substrate interactions. Neural stem cell (NSC) proliferation can be promoted using fibroblast growth factor-2 (FGF-2).[39] In addition, NSC can be preferentially differentiated into neurons using retinoic acid and forskolin,[40] into astrocytes using leukemia inhibitor factor (LIF) and bone morphogenic protein (BMP),[41] and into oligodendrocytes with insulin-like growth factor (IGF) or platelet-derived growth factor (PDGF).[42,43] However, the effect of topographical cues on NSC proliferation and differentiation is poorly understood. It remains unclear how topographical features (specifically, nanofiber diameter and alignment) influence stem cell proliferation and differentiation, and this is partially due to a lack of reliable methods for producing fibers with well-defined diameters. Systematic characterization of nanotopographical regulation of cell behavior is important in understanding and eventually engineering on an artificial niche for *ex vivo* manipulation of stem cells. Christopherson *et al.* demonstrated that fiber diameter is an important parameter that affects adhesion, spreading, migration, proliferation, and lineage specification of the NSC under expansion and differentiation conditions.[30] As the fiber diameter increased, the NSC showed reduced migration, spreading, and proliferation in the presence of FGF-2 and a serum-free medium. Accumulated evidence suggests that electrospun nanofibrous scaffolds can partially mimic the topographical features of the natural ECM and influence cellular differentiation.

10.4 Summary

Electrospinning has evolved as a powerful tool in tissue engineering. This fabrication technology provides the ability to control biomaterial composition, fiber diameter, fiber alignment, geometry, and drug/protein incorporation into a scaffold. Nanofibers generated by electrospinning are able to support the adhesion and proliferation of a wide variety of cell types (Table 10.2); moreover, the cells are able to maintain their phenotypic and functional characteristics on nanofibrous scaffolds. Additionally, a growing body of evidence demonstrates that micro- to nano-scaled topography plays an important role in controlling the adhesion, proliferation, and survival of adult and embryonic stem cells in culture.[44] Therefore, electrospun nanofibrous scaffolds can serve as a tool for studying the topographical aspects of cellular interactions that would lead to improved tissue formation. Furthermore, these scaffolds can be functionalized by adding biochemical and mechanical cues to enhance cellular interactions for tissue engineering applications (Figure 10.7). However, specific interactions between cells and electrospun scaffolds functionalized through surface modifications and bioactive factor incorporation are still poorly understood. An understanding of the specific cues that enhance cell adhesion, proliferation, and guidance of cells seeded on a scaffold, as well as cues that could affect host cell infiltration, differentiation, and vascularization *in vitro* and *in vivo*, is crucial for the advancement of tissue engineering applications.

Table 10.2. Cellular interaction of various cell types with electrospun nanofibrous substrates.

Cell types	Materials	Fiber configuration	Outcomes	Reference
Mesenchyma l stem cells	PEUU	Fiber diameter and alignment	Ligament-like cell differentiation	38
	PLA	Fiber alignment	Cell alignment and *in vivo* differentiation	45
	PCL	Fiber alignment	Chondrogenic differentiation	46
	Collagen	Fiber diameter	Cell adhesion and differentiation	21
	PCL, collagen, PES	Randomly oriented fibers	Hepatocyte-like cell differentiation	47
	PLLDO	Randomly oriented fibers	Cell adhesion and proliferation	16
Osteoprogenitor cell (MC3T3-E1)	PLA	Fiber diameter	Cell spreading, proliferation, and differentiation	17
Embryonic stem cells	PCL	Fiber alignment	Neural cell differentiation	44
Skeletal muscle cell	PCL/collagen	Fiber alignment	Cell alignment and myotube formation	22
C2C12	PU	Fiber alignment, mechanical and electrical stimulation	Cell alignment and maturation	48
	PEU	Fiber alignment	Cell adhesion and alignment	49,50
DRG	PDO	Fiber alignment	Neurite outgrowth	51
	PLA	Fiber alignment	Neurite outgrowth	52
DRG, Schwann cells, OEC, fibroblasts	Collagen/PCL	Fiber alignment	Neurite outgrowth, cell adhesion and migration	29
Schwann cells	PCL	Fiber alignment	Cell maturation	32
Neural stem cells	PES	Fiber diameter	Cell differentiation and proliferation	30
Fibroblasts	PLGA	Fiber diameter and orientation	Cell adhesion and proliferation	38
	PMMA	Fiber alignment	Cell migration	53
NIH/3T3 fibroblasts	PGA/collagen	Fiber diameter and alignment	Cell adhesion	54
Vascular endothelial cells	PCL-PU	Fiber alignment	Cell attachment and proliferation	55
	PLCL	Fiber alignment	Phenotypic maintenance	56

Abbreviations: DRG, dorsal root ganglia; OEC, olfactory ensheathing cell; PEUU, poly(ester urethane) urea; PLA, polylactide; PCL, polycaprolactone; PLLDO, poly(L-lactide)-*co*-(1,5,-dioxepan-2-one); PU, polyurethane; PEU, polyesterurethane; PDO, polydioxanone; PES, polyethersulfone; PLGA, poly(lactide-*co*-glicolide); PMMA, poly(methyl methacrylate); PLCL, poly(lactide-*co*-caprolactone).

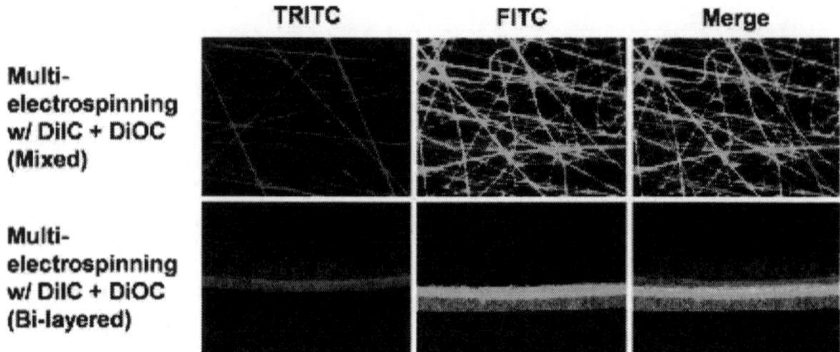

Figure 10.7. Fluorescent images of functionalized electrospun fibers incorporating different fluorescence dyes, DiIC (red) and DiOC (green). The multiple electrospinning technique showed mixed and layered fibrous structures with red- and green-labeled fibers simultaneously. Electrospinning technology shows potential as an adequate scaffolding system for tissue engineering applications (unpublished data). See also Color Insert.

Acknowledgments

We would like to thank Dr. Jennifer Olson for editorial assistance with this chapter.

References

1. R. G. Flemming, C. J. Murphy, G. A. Abrams, S. L. Goodman, and P. F. Nealey, *Biomaterials*, **20**(6), 573 (1999).
2. W. J. Li, C. T. Laurencin, E. J. Caterson, R. S. Tuan, and F. K. Ko, *J. Biomed. Mater. Res.*, **60**(4), 613 (2002).
3. S. J. Lee, S. H. Oh, J. Liu, S. Soker, A. Atala, and J. J. Yoo, *Biomaterials*, **29**(10), 1422 (2008).
4. S. J. Lee, J. J. Yoo, G. J. Lim, A. Atala, and J. Stitzel, *J. Biomed. Mater. Res. A*, **83**(4), 999 (2007).
5. J. Stitzel, J. Liu, S. J. Lee, M. Komura, J. Berry, S. Soker, G. Lim, M. Van Dyke, R. Czerw R, J. J. Yoo, and A. Atala, *Biomaterials*, **27**(7), 1088 (2006).
6. Q. P. Pham, U. Sharma, and A. G. Mikos, *Tissue Eng.*, **12**(5), 1197 (2006).

7. P. Clark, P. Connolly, A. S. Curtis, J. A. Dow, and C. D. Wilkinson, *Topograph. Control Cell Behav.: II. Multiple Grooved Substrata Dev.*, **108**(4), 635 (1990).
8. S. J. Lee, J. S. Choi, K. S. Park, G. Khang, Y. M. Lee, and H. B. Lee, *Biomaterials*, **25**(19), 4699 (2004).
9. J. Y. Martin, Z. Schwartz, T. W. Hummert, D. M. Schraub, J. Simpson, J. Lankford, Jr., D. D. Dean, D. L. Cochran, and B. D. Boyan, *J. Biomed. Mater. Res.*, **29**(3), 389 (1995).
10. R. Singhvi, A. Kumar, G. P. Lopez, G. N. Stephanopoulos, D. I. Wang, G. M. Whitesides, and D. E. Ingber, *Science*, **264**(5159), 696 (1994).
11. C. H. Lohmann, L. F. Bonewald, M. A. Sisk, V. L. Sylvia, D. L. Cochran, D. D. Dean, B. D. Boyan, and Z. Schwartz, *J. Bone. Miner. Res.*, **15**(6), 1169 (2000).
12. B. A. Dalton, X. F. Walboomers, M. Dziegielewski, M. D. Evans, S. Taylor, J. A. Jansen, and J. G. Steele, *J. Biomed. Mater. Res.*, **56**(2), 195 (2001).
13. J. H. Lee, S. J. Lee, G. Khang, and H. B. Lee, *J. Biomater. Sci. Polym. Ed.*, **10**(3), 283 (1999).
14. K. L. Elias, R. L. Price, and T. J. Webster, *Biomaterials*, **23**(15), 3279 (2002).
15. C. Xu, R. Inai, M. Kotaki, and S. Ramakrishna, *Tissue Eng.*, **10**(7–8), 1160 (2004).
16. A. Finne-Wistrand, A. C. Albertsson, O. H. Kwon, N. Kawazoe, G. Chen, I. K. Kang, H. Hasuda, J. Gong, and Y. Ito, *Macromol. Biosci.*, **8**(10), 951 (2008).
17. A. S. Badami, M. R. Kreke, M. S. Thompson, J. S. Riffle, and A. S. Goldstein, *Biomaterials*, **27**(4), 596 (2006).
18. K. N. Chua, W. S. Lim, P. Zhang, H. Lu, J. Wen, S. Ramakrishna, K. W. Leong, and H. Q. Mao, *Biomaterials*, **26**(15), 2537 (2005).
19. C. H. Lee, H. J. Shin, I. H. Cho, Y. M. Kang, I. A. Kim, K. D. Park, and J. W. Shin, *Biomaterials*, **26**(11), 1261 (2005).
20. E. Luong-Van, L. Grondahl, K. N. Chua, K. W. Leong, V. Nurcombe, and S. M. Cool, *Biomaterials*, **27**(9), 2042 (2006).
21. Y. R. Shih, C. N. Chen, S. W. Tsai, Y. J. Wang, and O. K. Lee, *Stem Cells*, **24**(11), 2391 (2006).
22. J. S. Choi, S. J. Lee, G. J. Christ, A. Atala, and J. J. Yoo, *Biomaterials*, **29**(19), 2899 (2008).
23. S. Y. Chew, J. Wen, E. K. Yim, and K. W. Leong, *Biomacromolecules*, **6**(4), 2017 (2005).
24. D. Zhang and J. Chang, *Nano Lett.*, **8**(10), 3283 (2008).

25. D. Li, Y. Wang, and Y. Xia, *Nano Lett.*, **3**(8), 1167 (2003).
26. S. Zhong, W. E. Teo, X. Zhu, R. W. Beuerman, S. Ramakrishna, and L. Y. Yung, *J. Biomed. Mater. Res. A*, **79**(3), 456 (2006).
27. W. E. Teo, and S. Ramakrishna, *Nanotechnology*, **16**, 1878 (2005).
28. L. S. Carnell, E. J. Siochi, N. M. Holloway, R. M. Stephens, C. Rhim, L. E. Niklason, and R. L. Clark, *Macromolecules*, **41**(14), 5345 (2008).
29. E. Schnell, K. Klinkhammer, S. Balzer, G. Brook, D. Klee, P. Dalton, and J. Mey, *Biomaterials*, **28**(19), 3012 (2007).
30. G. T. Christopherson, H. Song, and H. Q. Mao, *Biomaterials*, **30**(4), 556 (2009).
31. J. B. Recknor, D. S. Sakaguchi, and S. K. Mallapragada, *Biomaterials*, **27**(22), 4098 (2006).
32. S. Y. Chew, R. Mi, A. Hoke, and K. W. Leong, *Biomaterials*, **29**(6), 653 (2008).
33. T. G. Kim, H. J. Chung, and T. G. Park, *Acta Biomaterials*, **4**(6), 1611 (2008).
34. J. Nam, Y. Huang, S. Agarwal, and J. Lannutti, *Tissue Eng.*, **13**(9), 2249 (2007).
35. B. M. Baker, A. O. Gee, R. B. Metter, A. S. Nathan, R. A. Marklein, J. A. Burdick, and R. L. Mauck, *Biomaterials*, **29**(15), 2348 (2008).
36. J. J. Stankus, J. Guan, K. Fujimoto, and W. R. Wagner, *Biomaterials*, **27**(5), 735 (2006).
37. Q. P. Pham, U. Sharma, and A. G. Mikos, *Biomacromolecules*, **7**(10), 2796 (2006).
38. C. A. Bashur, L. A. Dahlgren, and A. S. Goldstein, *Biomaterials*, **27**(33), 5681 (2006).
39. F. H. Gage, P. W. Coates, T. D. Palmer, H. G. Kuhn, L. J. Fisher, J. O. Suhonen, D. A. Peterson, S. T. Suhr, and J. Ray, *Proc. Natl. Acad. Sci. U S A*, **92**(25), 11879 (1995).
40. X. Zhang, J. Cai, K. M. Klueber, Z. Guo, C. Lu, and W. I. Winstead, M. Qiu, and F. J. Roisen, *Stem Cells*, **24**(2), 434 (2006).
41. K. Nakashima, M. Yanagisawa, H. Arakawa, N. Kimura, T. Hisatsune, M. Kawabata, K. Miyazono, and T. Taga, *Science*, **284**(5413), 479 (1999).
42. J. Hsieh, J. B. Aimone, B. K. Kaspar, T. Kuwabara, K. Nakashima, and F. H. Gage, *J. Cell Biol.*, **164**(1), 111 (2004).
43. J. G. Hu, S. L. Fu, Y. X. Wang, Y. Li, X. Y. Jiang, X. F. Wang, M. S. Qiu, P. H. Lu, and X. M. Xu, *Neuroscience*, **151**(1), 138 (2008).

44. J. Xie, S. M. Willerth, X. Li, M. R. Macewan, A. Rader, S. E. Sakiyama-Elbert, and Y. Xia, *Biomaterials*, **30**(3), 354 (2009).
45. C. K. Hashi, Y. Zhu, G. Y. Yang, W. L. Young, B. S. Hsiao, K. Wang, *et al.*, *Proc. Natl. Acad. Sci. U S A*, **104**(29), 11915 (2007).
46. J. K. Wise, A. L. Yarin, C. M. Megaridis, and M. Cho, *Tissue Eng. Part A*, **15**(4), 913 (2009).
47. S. Kazemnejad, A. Allameh, M. Soleimani, A. Gharehbaghian, Y. Mohammadi, N. Amirizadeh, *et al.*, *J. Gastroenterol. Hepatol.*, **24**(2), 278 (2009).
48. I. C. Liao, J. B. Liu, N. Bursac, and K. W. Leong, *Cell Mol. Bioeng.*, **1**(2–3), 133 (2008).
49. S. A. Riboldi, M. Sampaolesi, P. Neuenschwander, G. Cossu, and S. Mantero, *Biomaterials*, **26**(22), 4606 (2005).
50. S. A. Riboldi, N. Sadr, L. Pigini, P. Neuenschwander, M. Simonet, P. Mognol, *et al.*, *J. Biomed. Mater. Res. A*, **84**(4), 1094 (2008).
51. W. N. Chow, D. G. Simpson, J. W. Bigbee, and R. J. Colello, *Neuron. Glia. Biol.*, **3**(2), 119 (2007).
52. H. B. Wang, M. E. Mullins, J. M. Cregg, A. Hurtado, M. Oudega, M. T. Trombley, *et al.*, *J. Neural. Eng.*, **6**(1), 016001 (2009).
53. Y. Liu, A. Franco, L. Huang, D. Gersappe, R. A. Clark, and M. H. Rafailovich, *Exp. Cell Res.*, **21** (2009).
54. F. Tian, H. Hosseinkhani, M. Hosseinkhani, A. Khademhosseini, Y. Yokoyama, G. G. Estrada, *et al.*, *J. Biomed. Mater. Res. A*, **84**(2), 291 (2008).
55. M. R. Williamson, R. Black, and C. Kielty, *Biomaterials*, **27**(19), 3608 (2006).
56. W. He, T. Yong, Z. W. Ma, R. Inai, W. E. Teo, and S. Ramakrishna, *Tissue Eng.*, **12**(9), 2457 (2006).

Chapter 11

FABRICATION OF TISSUE ENGINEERING SCAFFOLDS BY ELECTROSPINNING TECHNIQUES

Jiang Chang,[a,b,*] Wenguo Cui,[b] Yue Zhou,[b] and Lei Chen[a]
[a] *Biomaterials and Tissue Engineering Research Center, Shanghai Institute of Ceramics, Chinese Academy of Sciences, Shanghai, 200050, Peoples Republic of China*
[b] *School of Biomedical Engineering and Med-X Research Institute, Shanghai Jiao Tong University, 1954 Hua Shan Road, Shanghai 200030, People's Republic of China*
*jchang@mail.sic.ac.cn

Electrospinning is one of the popular and preferred techniques for the preparation of porous scaffolds which have great potential in tissue engineering applications because of the similarity of the structure of electrospun fibers to that of the extracellular matrix (ECM). Many specific designs of electrospinning setups have been developed to fabricate fibrous materials for different tissue engineering applications. By modifying the electrospinning parameters, fibers with different morphologies, diameters, and structures can be fabricated. Nanofibers can be bundled to improve their mechanical strength. Electrospun porous mats with different two-dimensional (2D) patterned architectures can be fabricated by designing electroconductive collectors with patterned structures. In addition, fibrous tubes with different macroscopic

Handbook of Intelligent Scaffolds for Tissue Engineering and Regenerative Medicine
Edited by Gilson Khang

www.panstanford.com

configurations and patterned architectures can also be achieved with well-tailored three-dimensional (3D) collectors. These nanofibrous scaffolds with specific macro- and microstructures providing biomimetic functionalization are currently being explored for skin, bone, cartilage, nerve, vascular, and controlled delivery of drugs. The development of tissue engineering techniques depends on the improvement of the design and fabrication techniques of nanostructured scaffolds as well as a profound understanding of the material structure-related biological processes.

11.1 Introduction

Tissue engineering is an interdisciplinary field that applies the principles of engineering and the life sciences to the development of biological substitutes that restore, maintain, or improve tissue/organ function.[1] Biomaterials play a pivotal role in tissue engineering by serving as matrices for cellular ingrowth, proliferation, and new tissue formation in 3D.[2] Nanofibrous scaffolds have a high surface-to-volume ratio, which is thought to enhance cell adhesion,[3] migration, proliferation, and differentiated function.[4] Therefore, several fabrication techniques, such as melt-blown,[5] phase separation,[6] self-assembly,[7] template synthesis,[8] and electrospinning,[9] have been employed to produce suitable polymeric nanofibrous materials for tissue engineering applications. Among them, electrospinning (Fig. 11.1) is one of the popular and preferred techniques to use because of the similarity of the structure of electrospun fibers to that of the ECM, and this technique is simple and cost effective and is able to produce continuous nanofibers of various materials, from polymers to ceramics. In addition, electrospinning seems to be the only method that can be further developed for large-scale production of continuous nanofibers for industrial applications.

Electrospinning is a process that was first conceived in the late nineteenth century by Lord Rayleigh. The first U.S. patent was issued in 1934 to Formhals.[10] In its simplest form, an electrospinning device essentially consists of a statitron to create an electric field between a grounded collector and a positively charged capillary

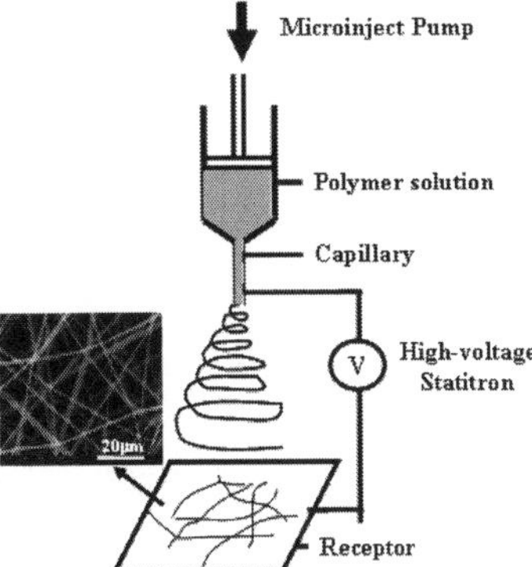

Figure 11.1. Schematic illustration of the basic setup of electrospinning.

filled with a polymer solution (Fig. 11.1). When the electrostatic charge becomes larger than the surface tension of the polymer solution at the capillary tip, a polymer jet is created. This fine polymer jet travels from the charged capillary to the grounded mandrel and allows for the production of continuous micro- to nano-scale polymer fibers, which can be collected in various orientations to create unique structures with different compositions and mechanical properties.[9]

Electrospun fibers have been considered for use as scaffolds for engineering tissues such as cartilages, bones, skin, blood vessels, the heart, nerves, tendons/ligaments, *etc.*[11] To prepare these different kinds of scaffolds, there is a wide variety of variables to be considered, including the choice of material, fiber orientation, porosity, surface modification, fibrous scaffold structure, *etc.* Natural and synthetic materials, as well as hybrid blends of the two, which can provide an optimal combination of mechanical and biomimetic properties, are all good candidates to be considered for fabricating electrospun fibrous scaffolds according to different application requirements. Fiber orientation (random, aligned, 2D and 3D micropatterned) and porosity/pore size of electrospun scaffolds can be controlled and optimized by adjusting the processing parameters, such as solution composition, the form of the collecting plate, *etc.* Due to the flexibility in controlling the structure of the scaffolds, which affects the properties of the materials, electrospun scaffolds of different structures have demonstrated superior potential in a number of different tissue engineering applications. This chapter reviews some recent developments related to the electrospinning technique, with a focus on the control of the structure and assembly of nanofibers.

11.2 Electrospun Nanofibers

Electrospun ultrafine fibers are generally collected as nonwoven and randomly arranged structures because of the bending instabilities associated with the electrically charged liquid jets of the electrospun solution. The morphology and diameter of elecrospun fibers are dependent on a number of processing parameters, including a) the intrinsic properties of the solution, such as the type of polymer, polymer concentration, viscosity, elasticity, electrical conductivity, and surface tension of the solvent; b) operational conditions, such as the feeding rate for the polymer solution, tip-to-collector distance, and the strength of the applied electric field; c) the design of the specific electrospinning setup; and d) the humidity and temperature of the surroundings.[12]

Chang *et al.* investigated the effect of electrospinning parameters on the diameter and morphology of bioactive glass (BG) nanofibers.[13] Figures 11.2a and 11.2b showed that the morphology of BG fibers changed from beaded fibers to smooth fibers with an increase in the poly(vinyl pyrrolidone) (PVP) concentration from 0.1

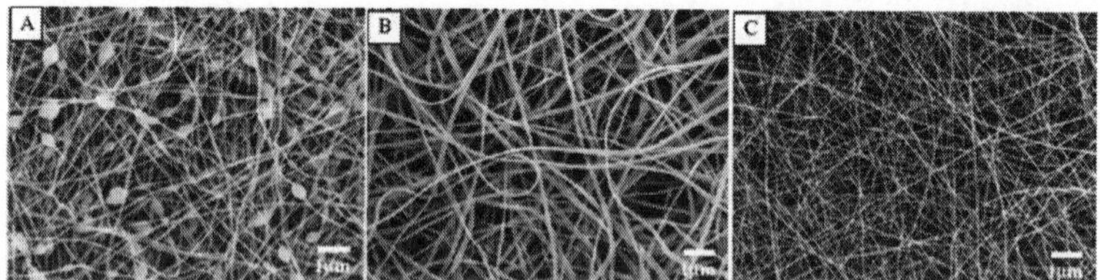

Figure 11.2. SEM images of bioactive glass nanofibers that were electrospun from an ethanol solution containing: (A) 0.1 g/mL PVP, 0.5 mL/mL BG sol, and no P123; (B) 0.2 g/mL PVP, 0.5 mL/mL BG sol, and no P123; (C) 0.2 g/mL PVP, 0.27 g/mL P123, and 0.33 mL/mL BG sol.

to 0.2 g/mL. Addition of pluronic P123 (EO_{20}–PO_{70}–EO_{20}) resulted in a decrease in the diameter of the BG fibers (Fig. 11.2c). Cui *et al.* designed an orthogonal experiment system to optimize the processing parameters of electrospining, which resulted in fibers with different morphologies and diameters by the controlling of different parameters.[14] Furthermore, the researchers found that the surface wettability of the fibrous mats could be influenced by the fiber diameters, bead size, and bead percentage.[15] Meanwhile, they also compared drug release properties of electrospun fibers with different diameters and found that fibers with larger diameters exhibited a longer period of nearly zero-order drug release compared with fibers with smaller diameters.[16]

Yarin *et al.* recently fabricated core-shell nanofibers by co-electrospining two different polymer solutions through a spinneret comprising two coaxial capillaries.[17] Yang *et al.* also fabricated core-shell structured, ultrafine electrospun fibers as carriers for therapeutic proteins by emulsion electrospinning. Both studies indicated that the core-shell structured electrospun fibers protected the structural integrity of the encapsulated protein during incubation in the medium.[18] Nanofibers with hollow interiors could also be fabricated using the co-electrospining technique for nanofluidics and hydrogen storage.[19]

By changing the compositions used for electrospinning, Chang *et al.* fabricated composite fibers of poly(butylenes succinate) (PBSU)/wollastonite/apatite via electrospinning and the biomimetic process (Fig. 11.3a). The fibrous structures of the composite scaffolds could be adjusted by the amount of wollastonite or the incubation time.[20] Cui *et al.* developed electrospun fibrous nanocomposites of hydroxyapatite (HA) and poly(DL-lactide) (PDLLA) through *in situ* growth (Fig. 11.3b).[21] In another study, gelatin was grafted on the PDLLA fibrous surface to induce HA deposition (Fig. 11.3c).[22] The nucleation and growth of HA crystals on a fibrous template could be controlled by induction conditions. These fibrous nanocomposites should have potential applications such as coatings on medical devices, scaffolds for tissue engineering, and fillers for fiber-enforced composites. In short, the morphology of electrospun fibers can be flexibly controlled by the electrospinning parameters to meet the requirements for different applications.

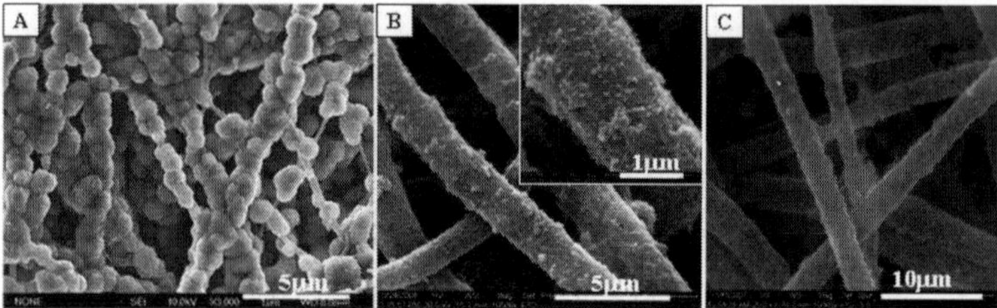

Figure 11.3. SEM micrographs of (a) the fiberous composite scaffolds of PBSU/wollastonite/apatite; (b) the PDLLA fiberous composite scaffolds coated with HA by *in situ* growth process; and (c) the PDLLA fiberous composite scaffolds coated with HA, which were induced by grafted gelatin on the surface of PDLLA fibers.

11.3 One-Dimensional Electrospun Fibrous Bundle

Electrospinning can get continuous fibers with diameters ranging from a nano- to a microscale. However, the mechanical strength of electrospun ultrafine fibers is usually significantly lower than that of textile fibers made from the same polymer, which limits their applications in many fields. Therefore, ultrafine electrospun fibers were bundled to improve their mechanical strength. Wang *et al.* used a novel self-bundling electrospinning method for generating macroscopically continuous, aligned electrospun fiber yarn to yield a higher-value tensile strength mat.[23] Placing conducting blades in line with a gap in between, Teo *et al.* collected an end-fixing fiber bundle consisting of highly aligned nanofibers.[24] In addition, it is also feasible to obtain a nanofiber bundle of micron size by dipping the electrospun fiber bundle in a liquid and drying it later. For instance, Bosworth *et al.* collected uniaxially aligned fiber bundles through spinning into a liquid reservoir, and these bundles could develop a temporary scaffold mimicking the hierarchical tendon structure.[25] Chang *et al.* found that when collectors with patterned topographical structures were used, the fibers deposited on the protrusions of the collectors were much denser than those aligned in parallel between the protrusions and the suspended fibers tended to adhere together to form fiber bundles on the protrusions template. Therefore, the fibers bundle could be collected between the protrusions, and alignment degree of the fibers could be controlled by the area of protrusions on the collectors.[26]

11.4 Two-Dimensional Electrospun Fibrous Membranes

Compared with the limited application of one-dimensional (1D) electrospun fibrous bundles, in tissue engineering, 2D electrospun fibrous membranes are much more widely used.

The prototype of 2D electrospun fibrous membranes was prepared using flat collecting plates and thus had a randomly distributed fibrous structure with relatively poor mechanical properties.[27,28] For more than a decade, researchers proved that specific topological architectures could promote favorable biological responses in tissue regeneration, such as enhanced protein adsorption, as well as enhanced cell attachment, proliferation, and rate of movement.[29] Since native ECMs found in tissues or organs have a defined architecture, which is significant for tissue function, efforts have been made to explore materials with ordered microstructures and patterns possessing specific functions. By using various electrospinning setups, such as rotating disk collectors, parallel electrodes, rotating wire drum collectors, *etc.*, scaffolds with aligned nanofibers or simple patterns can be fabricated.[27] Several studies have demonstrated the migration and proliferation of cells along the direction of electrospun fiber alignment.[30,31] In the peripheral nerve regeneration process, aligned nanofibrous scaffolds had the ability to direct axonal outgrowth and glial cell migration.[32] Xu *et al.* developed an aligned nanofibrous scaffold via electrospinning of a poly(L-lactide-co-ε-caprolactone) (PLCL,75:25) copolymer.[31] They found that the adhesion and proliferation of coronary artery smooth muscle cells were improved compared with plane polymer films, and they were able to migrate along the axis of the aligned fibers. Furthermore, proteins comprising the cytoskeleton of the smooth muscle cells were aligned parallel to the aligned fibers, demonstrating the cells' proclivity to organize along oriented fiber topography. All the results lead to the same conclusion that a nanofibrous scaffold containing aligned fibers is a favorable structure in oriented cell growth.

Considering the delicate structure and complexity of tissue, it is challenging to produce a 2D structure with accurately controlled fibrous structures or patterns. Generally, the 2D architecture of an electrospun mat is composed of two parts, the macroscopic organization of the nanofibers and the secondary structure of the individual fibers' arrangement. Although there has been a significant effort in developing systems to pattern or print substrates in a defined manner for tissue engineering applications, only a limited number of patterning approaches using electrospinning are available.[33] To design a mesoscopically ordered structure of matrices, Zhu *et al.* developed a slowly rotating frame cylinder fiber collector and obtained electrospun fibrous mats with porosity as high as 92.4% and average pore sizes of 132.7 μm.[34] They found that higher human dermal fibroblast viability, collagen deposition, and cell migration on the scaffold were achieved for fibrous mats with larger pore sizes and higher porosity.

A new method using electroconductive templates was employed to fabricate electrospun mats with controllable architectures and patterns, as shown in Fig. 11.4, and electrospun fibrous

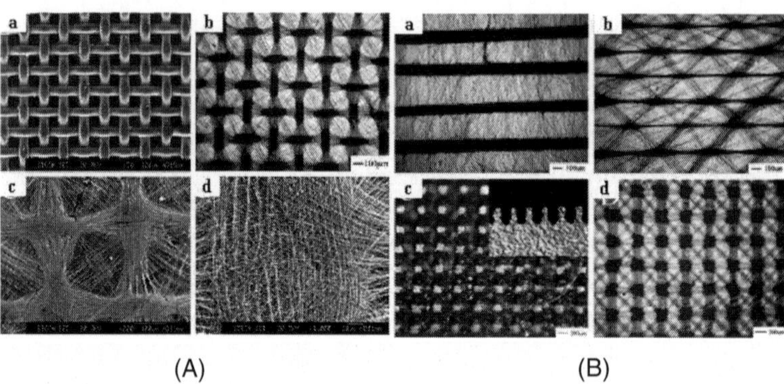

Figure 11.4. (A): (a) SEM image of a patterned collector; (b) optical image of an electrospun mat collected using the collector; c) SEM image of a typical unit in the electrospun mat collected using the collector; d) SEM image of the cross point in the electrospun mat. (B): Optical images of electrospun mats collected using collectors (a) without protrusions and (b) with protrusions; (c) optical image of a patterned collector composed of metal protrusions; d) optical image of electrospun mats collected using the collector in (c).

mats with different architectures and patterns could be collected through designing electroconductive templates with different microstructure.[26]

In this study, it was also demonstrated that protrusions on the template as well as the diameter of the electroconductive wires in the collectors were both important parameters, which might greatly affect the structures of the electrospun mats, and woven structures could be generated by a time-dependent control of the arrangement of the electroconductive protrusions on the collector. Recently, by using a stainless steel mesh as the template collector, Wang *et al.* successfully prepared electrospun nanofiber meshes with well-tailored architectures and patterns from biodegradable poly(ε-caprolactone) (PCL).[35] They found that electrospun PCL meshes gave a much higher cell proliferation rate compared with randomly oriented PCL mats using a mouse osteoblastic cell line (MC3T3-E1). The cells grew and elongated along the fiber orientation directions. The resultant cellular organization and distribution mimicked the topological structures of the PCL electrospun meshes. Therefore, electrospun nanofiber scaffolds with tailored architectures and patterns hold great potential for engineering functional tissues or organs, where an ordered cellular organization is essential.

11.5 Three-Dimensional Electrospun Fibrous Scaffolds

Electrospun fiber mats have been shown to provide an excellent environment for cell growth. However, many tissue engineering applications require specific assemblies with 3D architecture. From this point of view, electrospun fibrous tubes are of practical importance in vascular, neural, and tendinous tissue engineering.[27] A 3D conduit can be fabricated by depositing fibers over a rotating rod (<5 mm) as the collector. Utilizing this technique, Stitzel *et al.* fabricated a 12 cm long fibrous conduit by electrospinning a polymer solution containing a mixture of collagen type I, elastin, and poly(DL-lactide-*co*-glycolide) on a circular mandrel of diameter 4.75 mm for vascular grafts. Compliance tests showed that the conduit had a diameter change of 12%–14% within the physiologic pressure range, which is close to that of the native vessels.[36] Moreover, since the native artery is comprised of three distinct layers (intima, media, and adventitia), Boland *et al.* made an extremely daunting attempt to create a biomimicking vascular graft with a three-layered vascular construction. A collagen type I and elastin solution (w/w, 80/20) at a concentration of 0.083 g/mL in 1,1,1,3,3,3-hexafluoro-2-propanol (HFP) was electrospun onto a 4 mm internal diameter (ID) tubular mandrel and subsequently seeded with both fibroblasts and smooth muscle cells. Another tubular scaffold, this time of 2 mm ID, was created with a collagen type I and elastin solution (w/w, 30/70) and inserted into the 4 mm ID scaffold. The lumen was filled with a smooth muscle cell suspension and cultured for 3 days. Finally, a suspension of human umbilical vascular endothelial cells was injected into the lumen, and the entire construct was cultured for 2 additional days. Histological examination revealed that the scaffold was a three-layered construct with complete cellular infiltration.[37]

Considering some specific requirements of tubular scaffolds in tissue engineering applications, such as the variation in anatomic location and biological environment, it is also important to design and control microscopic and macroscopic 3D structures of tubes to create desired cellular responses.[38] A novel static method of fabricating 3D fibrous tubes composed of ultrafine electrospun fibers was recently achieved.[39] By using this unique technique, micro and macro single tubes with multiple micropatterns, multiple interconnected tubes, and many tubes with the same or different sizes, shapes, structures, and patterns can be prepared synchronously (Fig. 11.5). It is expected that electrospinning with the static collecting method using 3D collectors with specifically designed patterns and configurations has great potential for fabrication of fibrous tubes with controllable architectures and 3D configurations, which may be attractive in many biomedical and industrial applications.

These results indicate that electrospun tubes, without an additional biological coating or drug-loading treatment, are promising scaffolds for functional nervous regeneration. Electrospun tubes can be knitted in meshes and various frames, depending on the cytoarchitecture of the tissue to be regenerated. Moreover, these guidance conduits can be loaded with various fillers, such as collagen, fibrin, or self-assembling peptide gels, or with cytokines and seeded with

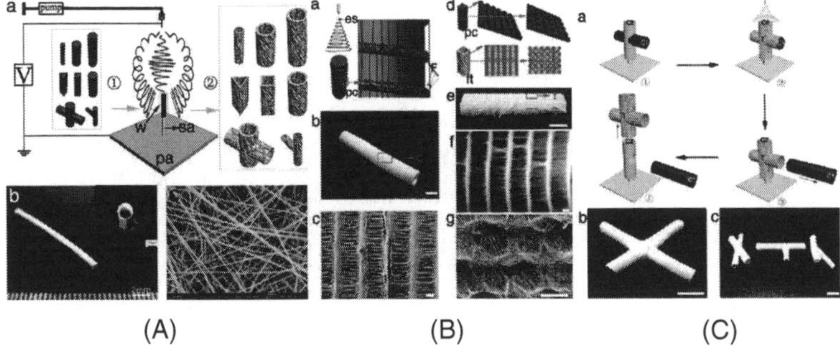

Figure 11.5. (A): (a) Schematic illustration of fabrication of fibrous tubes by the electrospinning technique using 3D columnar collectors; b) a fibrous tube with a diameter of 500 μm; c) SEM image of fiber assemblies of the tube in (b). (B): (a) Schematic illustration of collecting process using a cylindrical collector with equally spaced circular protrusions (es, electrospinning process; pc, patterned collector); (b) a fibrous tube with patterned architectures (scale bar = 5 mm); (c) magnified image of (b) (scale bar = 200μm); d) schematic illustration of collectors with two different patterns and relevant fibrous tube (ft, fibrous tube); (e) a fibrous tube with two different patterns (scale bar = 5 mm); (f,g) magnified images of two different patterns of (e) (scale bar = 200 μm). (C): (a) Schematic illustration of the process for fabrication of tubes with multiple interconnected tubular structures; (b) a crossing tube (scale bar = 5 mm); (c) tubes with various interconnected tubular structures (scale bar = 5 mm).

cells. The versatility of this technique gives room for further scaffold improvements.

11.6 Conclusions and Outlook

In this chapter, the technical development of fabricating electrospun fibers with different morphologies and configurations (1D, 2D, and 3D) for tissue engineering applications was reviewed. By modifying electrospinning parameters, fibers with different morphologies, diameters, and structures (core shell, hollow nanofibers, *etc.*) can be fabricated. Nanofibers can be bundled to improve their mechanical strength. Electrospun mats with different 2D patterned architectures can be fabricated by designing electroconductive collectors. With well-tailored 3D collectors, fibrous tubes with different macroscopic configurations and patterned architectures can also be achieved. These nanofibrous scaffolds with specific microstructures providing biomimetic functionalization are currently being explored for skin, bone, cartilage, nerve, vascular, and controlled delivery of drugs. The development of tissue engineering techniques depends on the improvement of the design and fabrication techniques of nanostructure scaffolds as well as a profound understanding of the material structure-related biological processes.

Acknowledgments

This work is supported by the National Basic Science Research Program of China (973 Program) (Grant No. 2005CB522704), the Natural Science Foundation of China (Grant No. 30730034 & 30600628), and the Science and Technology Commission of Shanghai Municipality (Grant No. 08JC1420800).

References

1. R. Langer and J. Vacanti, *Science*, **920** (1993).
2. L. L. Hench and J. M. Polak, *Science*, **1015** (2002).
3. R. G. Flemming, C. J. Murphy, G. A. Abrams, S. L. Goodman, and P. F. Nealey, *Biomaterials*, **573** (1999).
4. T. G. Kim and T. G. Park, *Tissue Eng.*, **221** (2006).
5. B. Gu, J. V. Badding, and A. Sen, *Polym. Prepr.*, **142** (2003).
6. Z. W. Ma, M. K. Ryujilnai, and S. Ramakrishna, *Tissue Eng.*, **101** (2005).
7. J. D. Hartgerink, E. Beniash, and S. I. Stupp, *Science*, **1684** (2001).
8. L. Feng, S. Li, H. Li, J. Zhai, Y. Song, L. Jiang, and D. Zhu, *Angew. Chem. Int. Edit.*, **1221** (2002).
9. C. P. Barnes, S. A. Sell, E. D. Boland, Da. G. Simpson, and G. L. Bowlin, *Adv. Drug. Del. Rev.*, **1413** (2007).
10. A. Formhals, *U.S. Patent*, **1975504** (1934).
11. Y. Z. Zhang, C. T. Lim, S. Ramakrishna, and Z. M. Huang, *J. Mater. Sci.: Mater. Med.*, **933** (2005).

12. D. Li and Y. N. Xia, *Adv. Mater.*, **1151** (2004).
13. W. Xia, D. M. Zhang, and J. Chang, *Nanotechnology*, **135601** (2007).
14. W. G. Cui, X. H. Li, S. B. Zhou, and J. Weng, *J. Appl. Polym. Sci.*, **3105** (2007).
15. W. G. Cui, X. H. Li, S. B. Zhou, and J. Weng, *Polym. Degrad. Stabil.*, **731** (2008).
16. W. G. Cui, X. H. Li, X. L. Zhu, G. Yu, S. B. Zhou, and J. Weng, *Biomacromolecules*, **1623** (2006).
17. Z. Sun, E. Zussman, A. L. Yarin, J. H. Wendorff, and A. Greiner, *Adv. Mater.*, **1929** (2003).
18. Y. Yang, X. H. Li, W. G. Cui, S. B. Zhou, R. Tan, C. Y. Wang, *J. Biomed. Mater. Res. A*, **374** (2008).
19. D. Li and Y. N. Xia, *Nano Lett.*, **933** (2004).
20. D. M. Zhang, J. Chang, and Y. Zeng, *J. Mater. Sci.: Mater. Med.*, **443** (2008).
21. W. G. Cui, X. H. Li, S. B. Zhou, and J. Weng, *J. Biomed. Mater. Res. A*, **831** (2007).
22. W. G. Cui, X. H. Li, J. G. Chen, S. B. Zhou, and J. Weng, *Cryst. Growth Des.*, **4576** (2008).
23. X. F. Wang, K. Zhang, M. F. Zhu, B. S. Hsiao, and B. Chu, *Macromol. Rapid Commun.*, **826** (2008).
24. W. E. Teo and S. Ramakrishn, *Nanotechnology*, **1878** (2005).
25. L. Bosworth, P. Clegg, and S. Downes, *Inter. J. Nano Biomater.*, **263** (2008).
26. D. M. Zhang and J. Chang, *Adv. Mater.*, **3664** (2007).
27. W. E. Teo1 and S. Ramakrishna, *Nanotechnology*, **R89** (2006).
28. Z. M. Huang, Y. Z. Zhang, M. Kotakic, and S. Ramakrishna, *Compos. Sci. Technol.*, **2223** (2003).
29. W. J. Li, R. Tuli, C. Okafor, A. Derfoul, K. G. Danielson, D. G. Hall, and R. S. Tuan, *Biomaterials*, **599** (2005).
30. C. H. Lee, H. J. Shin, I. H. Cho, and Y. M. Kang, *Biomaterials*, **1261** (2005).
31. C. Y. Xu, R. Inai, M. Kotaki, and S. Ramakrishna, *Biomaterials*, **877** (2004).
32. E. Schnell, K. Klinkhammer, S. Balzer, G. Brook, D. Klee, and P. Dalton, *Biomaterials*, **3012** (2007).
33. D. Sun, C. Chang, S. Li, and L. Lin, *Nano Lett.*, **839** (2006).
34. X. L. Zhu, W. G. Cui, X. H. Li, and Y. Jin, *Biomacromolecules*, **1795** (2008).

35. Y. Z. Wang, G. X. Wang, L. Chen, H. Li, T. Y. Yin, B. C. Wang, J. C. M. Lee, and Q. S. Yu, *Biofabrication*, **9** (2009).
36. J. Stitzel, J. Liu, S. J. Lee, M. Komura, J. Berry, S. Soker, G. Lim, M. V. Dyke, R. Czerw, J. J. Yoo, and A. Atala, *Biomaterials*, **1088** (2006).
37. E. D. Boland, J. A. Matthews, K. J. Pawlowski, D. G. Simpson, G. E. Wnek, and G. L. Bowlin, *Front. Biosci.*, **1422** (2004).
38. N. Gadegaard, M. J. Dalby, M. O. Riehle, A. S. G. Curtis, and S. Affrossman, *Adv. Mater.*, **1857** (2004).
39. D. M. Zhang and J. Chang, *Nano Lett.*, **3283** (2008).

Chapter 12

BIODEGRADABLE TUNABLE NANOFIBROUS MATRIX FOR REGENERATIVE MEDICINE

Shanta Raj Bhattarai,[a] Madhab Prasad Bajgai,[b] and Hak Yong Kim[c*]

[a]*Department of Pharmaceutical Sciences, Wayne State University, MI, USA*
Department of [b]*Bionanosystem and* [c]*Textile Engineering,*
Chonbuk National University, Jeonju, South Korea
*khy@chonbuk.ac.kr

Electrospinning is a novel method of producing and assembling nano-scale fibers into three-dimensional (3D) scaffolds that mimic the structure and biochemical environment of human body's tissues. Scaffolds can be fabricated from both natural and synthetic polymers and can even be manufactured into different sizes and shapes, including tubes, matrices, and coatings, among others. Fibers comprising scaffolds can be tuned or custom-engineered to a specific orientation (parallel/perpendicular) or thickness, including on the nanometer scale (a thousand times thinner than a human hair). What results from these capabilities is an implantable scaffold with the correct fiber diameter, orientation, and architecture, rendered virtually indistinguishable from native tissue and recognized as "self" by the body. However,

Handbook of Intelligent Scaffolds for Tissue Engineering and Regenerative Medicine
Edited by Gilson Khang

www.panstanford.com

electrospinning is not limited to polymers. Other elements of regenerative medicine, including cells, drugs, and bioactive factors, can be combined into a scaffold. Recently, many researchers have used electrospinning techniques to produce a variety of materials, including vascular grafts, nerve guides, tendon, and skin. In fact, electrospinning is an area of intense focus for academic biomaterials research, validating the importance of this technology platform. Moreover, electrospinning is a tunable manufacturing technique for the design of biomaterials with potentially reorganizable architecture for various cell and tissue growth.

12.1 Introduction

Nanofiber technology is one of the fastest-growing areas in the material science and engineering fields and requires interdisciplinary backgrounds, such as chemistry, material science, information technology, electronics, environmental technology, and biological technology. The last decade has seen significant progress in the production of nanofibers by electrospinning.[1–5] Nanofibers have drawn significant attention in many fields, including high-performance filter media, composites, textiles, catalysts, electronic and optical devices, drug delivery systems, and tissue engineering.[3,5] Due to the very small diameter of the polymer fibers obtained by electrospinning, the surface-area-to-volume ratio of the fibers is significantly high. The nonwoven matrices formed from nanofibers have very small pore sizes; however, the total porosity of the matrices will still be very high. This makes them excellent candidates for use in filtration and membrane application. Due to the high filtration efficiency, polymer nanofiber–based filtration devices are now commercially developed for industrial air filtration, as filters for gas turbine generators, and filters for heavy-duty engines. Recently it has been demonstrated that the filtration efficiency of these membranes can be increased by using charged fibers that would modify the electrostatic attraction of the particles.[2] Electrospinning has the potential to develop into the future fabrication process for developing protective clothing. Optimal protective clothing should satisfy properties such as it should be permeable to both air and water vapor, it should be reactive with toxic gases and chemicals, it should be insoluble in solvents, and it should be lightweight. As discussed before, due to the very small pore size of nanofiber matrices, they provide an impermeable barrier to toxic chemical agents. Also the high surface-area-to-volume ratio of nanofibers enables them to form lightweight fabrics with remarkable breathing properties.[4] Due to the simplicity and versatility of the electrospinning process, conductive fibers can be fabricated from polymers with unusual electrical, electronic, ionic, photoelectric, and piezoelectric properties. These fibers could be used in the fabrication of novel nanoelectronic devices, nanoelectronic machines, and sensors for high-technology applications.[6]

In electrospinning, polymer solutions are deposited as fibrous matrices, in which chain entanglements in a sufficiently high-polymer concentration in solution produce continuous fibers.[7] Although the process of electrospinning has been known for some time,[8] and attempts have been made previously to make vascular grafts using this technique,[9] the application of electrospun fibers as scaffolds for tissue engineering has recently investigated by many researchers.[3,5] The prototype electrospinning setup that has been employed in this study is illustrated schematically in Fig. 12.1, which consists of a syringe, a ground electrode (aluminum sheet on a rotating drum) 10–15 cm from the needle, and a high-voltage power supply. Jet initiation is achieved by charging the polymer solution, followed by injection through the capillary tip with a tip diameter of 0.5 mm. Because of its charge, the ejected solution is drawn toward the collector as a whipping jet.[10] During the jet's travel, the solvent

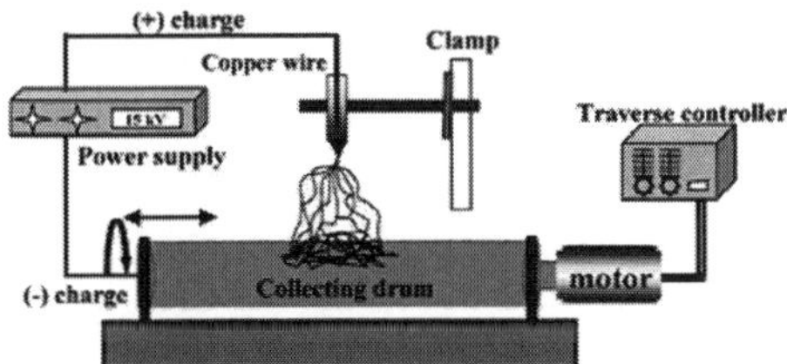

Figure 12.1. A schematic of electrospinning. Reprinted from Ref. 10 with permission from Elsevier.

gradually evaporates, leaving a continuous fiber that accumulates on the collector target.

Orientation can be controlled through rotation and translation of the collector drum while collecting the fibers. This process results in the production of a nonwoven fibrous matrix. These matrices can have a fiber diameter in the order of nanometers to microns. The structure, produced by the electrospinning process, is a nonwoven, 3D, porous, and nano-scale fiber-based matrix.

Another actively investigated area of the application of polymeric nanofibers is the biomedical field.[11] Both nondegradable and degradable polymers have been investigated for various applications. As early as 1977, it was proposed that nondegradable polymeric nanofibers can be used for developing prosthetic medical devices such as blood vessels and vascular grafts.[12] Recently it has been shown that electrospinning can be used to create porous, thin films with structural gradients and controlled morphology on prosthetic devices in order to facilitate the integration of the devices with the body and thereby enhance their biocompatibility. One of the major drivers of this progress is the potential use of nanofibrous structures as scaffolds for engineering tissues in regenerative medicine. Electrospun fibers are capable of emulating the nanofibrous architecture of the native extracellular matrix (ECM).[3] They can potentially provide *in vivo*–like nanomechanical and physicochemical signaling cues to the cells to establish apposite cell–scaffold interactions and promote functional changes between and within cells toward the synthesis of a genuine ECM over time. They are of particular interest in regenerative medicine and tissue engineering because they can be potentially tailored to mimic the natural ECM in terms of structure, chemical composition, and mechanical properties.[3,5] In this context, they serve as scaffolds to direct cellular behavior and function until host cells can repopulate and resynthesize a new natural matrix. The ECM molecular network surrounding the cells provides mechanical support and regulates cellular activities. The natural ECM in human tissue is mainly composed of proteoglycans (glycosaminoglycan [GAG]) and fibrous proteins, both with nano-scale structural dimensions. Studies have shown that scaffolds with nano-scale structures support cell adhesion and proliferation and function better than their microscale counterparts. In this context, this chapter presents a brief overview of the biomedical application of the electrospun nanofibrous scaffold.

12.2 Electrospun Nanofiber Matrices

Nanofibers are being explored for a variety of applications, and research in this area is rapidly expanding. A literature survey for nanofibers, based on a SciFinder Scholar search in the past nine years, is presented in Fig. 12.2, and it clearly demonstrates the rising interest in nanofibers. The potential medical application of polymer nanofiber matrices is in the area of tissue engineering. The role of biomaterials in tissue engineering is to act as a scaffold for cells to attach to and organize into tissue.[3,5] The ideal tissue engineering scaffold should mimic the ECM, the natural abode of cells. The structure and morphology of nanofiber matrices closely match the structure of ECM of natural tissue. Studies have shown that cells seeded on biodegradable polymeric nanofiber mats can attach, proliferate, and maintain their phenotype expression.[10,13] Development

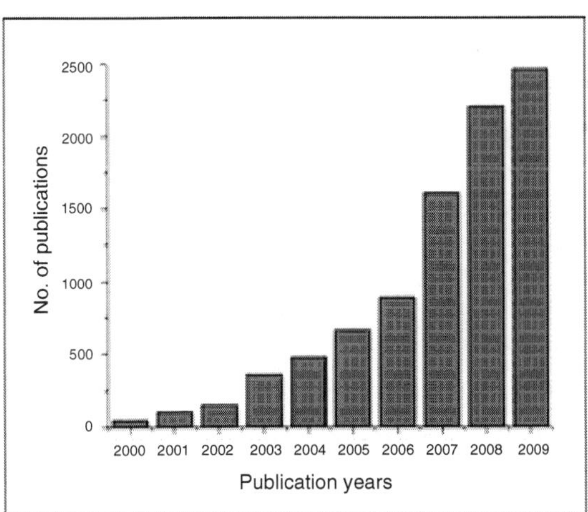

Figure 12.2. Research articles published in recent years using electrospun nanofibers. These numbers were based on the SciFinder Scholar search using the keywords "electrostatic spinning," "electrospinning," and "nanofibers."

of nanofiber-based 3D scaffolds using various biopolymers and synthetic biodegradable polymers has currently become an active area of research.

Polymeric nanofibers have also been investigated as a novel wound dressing and as hemostatic devices.[14] The high surface area of nanofiber matrices allows for oxygen permeability as well as prevents fluid accumulation at the wound site. On the other hand, the small pore size of the matrices efficiently prevents bacterial penetration, making them ideal candidates for wound dressings.Furthermore, flexibility of the electrospinning process allows for co-spinning polymers with drugs or proteins, thereby making the nonwoven nanofiber matrix a drug delivery matrix, which could enhance wound healing.

A number of synthetic polymer nanofibers with fiber diameters from a few tens to a few hundreds of nanometers have been fabricated for tissue engineering; these include polyglycolide (PGA), poly(L-lactic acid) (PLLA), and their copolymers poly(glycolide-*co*-lactide) (PLGA) and poly(ε-caprolactone) (PCL). Studies have demonstrated favorable biological responses of seeded cells, such as enhanced cell attachment and *in vitro* proliferation. It is important to have a 3D structure of a scaffold for cell attachment, growth, and migration. For tissue engineering, nanofibrous structures positively promote cell–matrix interactions.[10] Cells seeded on a nanofibrous scaffold should have an appropriate interaction with their environment, that is, the cells maintain a normal phenotypic shape, adhere onto the fibers, proliferate on the nanofibrous network, pack the structure, and finally integrate with the surrounding fibers to form a 3D cellular network (Fig. 12.3).

Recently, there has been a growing interest in the synthesis of natural polymer-based nanofibers because of their proven biocompatibility and their resorbable biodegradation products. Advantageous attributes of natural polymers include hydrophilicity, nontoxicity, and a less immune reaction, as well as enhanced cell adhesion and proliferation. Collagen, gelatin, hyaluronan, chitosan, and alginate are the most commonly used natural polymers in tissue engineering. In a few recent studies, collagen and chitosan have been successfully fabricated into nanofibers and have demonstrated good cellular compatibility.[13] The ability to generate nanofibrous

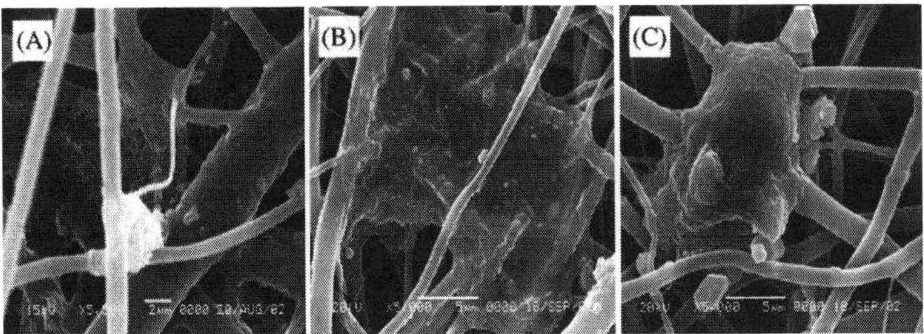

Figure 12.3. SEM micrographs of the interaction between cells and an electrospun nanofibrous structure after 5 days of culture. NIH 3T3 fibroblast cells adhering onto the fibers and proliferating on the nanofibrous network showed a packing structure after 5 days of culture. (A) Original magnification 5,500; (B) original magnification 5,000. The cells cross-linked the nanofibrous matrix and integrated with the surrounding fibers to form a 3D cellular network; (C) with original magnification 5,000. Reprinted from Ref. 10 with permission from Elsevier.

matrices from natural polymers, especially those derived from plants, may provide virtually unlimited resources for the development of tissue-compatible scaffolds for functional restoration of damaged or dysfunctional tissues. This restoration currently relies mainly on the autograft and allograft procedures—surgical procedures facing the challenges of limited resources, risk of infection, and viral transmission.

12.3 Electrospun Nanofiber Matrices as Tissue Regenerative Matrices

"Tissue regeneration," "tissue engineering," and "regenerative medicine" are related terms and sometimes used interchangeably. A new frontier in health care, regenerative medicine is an evolving therapeutic approach that utilizes living cells to repair or replace body tissue damaged by injury, disease, or the aging process. It is a multidisciplinary field involving biology, medicine, and engineering.[15] Regenerative therapies rely on the body's own natural ability to repair and regenerate and enable the body to heal by itself. Regenerative medicine also empowers scientists to grow living cells, tissues, and organs in the laboratory and to safely implant them into the human body for the purposes of healing. The nanofibrous matrix's platform technology develops "designer scaffolds" for the purposes of regenerative medicine, depending on the fabrication technique known as electrospinning. Biomedically, the electrospun nanofibrous matrix has been used in various desciplines like nanofiber skin grafts, blood vessel (vascular and cardiac) grafts, ligaments grafts, nerve grafts, skeletal muscle grafts, bone tissue grafts, articular cartilage tissue grafts, skin grafts, and biogenic molecule (drug, DNA, protein, and enzyme) delivery.

12.3.1 *Skin Grafts*

The skin is the largest tissue in the body. It is a modified epithelial tissue, with a keratinized layer of dead cells that provides a physical barrier to the outside world. Skin wounds normally heal by the formation of epithelialized scar tissue rather than by regeneration of full skin.[16] Of the two layers of the skin, epidermis and dermis, the epidermis has less capacity to heal; however, when large areas of the epidermis need to be replaced, normal regeneration is lacking. Further, the dermis has an enormous capacity to regenerate. The scar tissue that forms in the absence of the dermis lacks the elasticity, flexibility, and strength of the normal dermis.[16] Consequently, scar tissue limits movements, causes pain, and is cosmetically undesirable. Therefore, engineered skin tissue would be an excellent alternative, not only to close a wound, but also to stimulate the regeneration of the dermis. In the late '90s collaboration between clinical scientists and materials scientists at Sheffield in England made the first big improvement on this technique—the development of flexible synthetic surfaces on which keratinocytes could be easily cultured *in vitro*. The synthetic support medium allows for rapid culture, reducing waste, and makes the tissue very much easier to handle. The cultured keratinocytes plus the synthetic support form a flexible dressing that can be applied directly to the wound bed. Clinical studies have shown that cells migrate from the dressing to the wound and greatly accelerate healing rates, frequently resulting in complete remission for chronic ulcers that had resisted other treatments. Along with collagen, several other natural and synthetic

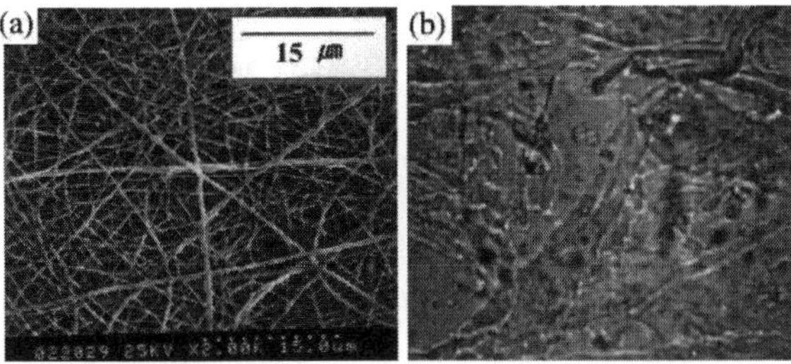

Figure 12.4. PU wound dressings: (a) representative SEM image of PU wound dressing prepared by electrospinning; (b) photograph of Tegaderm™ (×200). Reprinted from Ref. 14 with permission from John Wiley & Sons Inc.

polymers have been explored for skin tissue engineering; however, the use of these biomaterials as nanofibers has been very limited. Recently, researcher developed silk fibroin nanofiber matrices by electrospinning for skin tissue engineering.[17] Due to their high porosity and high surface-area-to-volume ratio, fibroin nanofibers coated with type I collagen were found to promote keratinocyte/fibroblast adhesion and spreading. Therefore, silk fibroin nanofibers show potential to be developed as a scaffold for skin tissue engineering. Khil *et al.*[14,18] studied polyurethane (PU) electrospun nanofiber matrices for the purpose of wound dressings (Figs. 12.4 and 12.5).

PU nanofiber matrices provided excellent oxygen permeability and controlled water evaporation. By virtue of these properties, the matrices allowed fluid from the wound to exude, while preventing dehydration of the wound. Further, the ultrafine porosity of the matrices disallowed invasion by exogenous microorganisms. These results indicated that PU nanofiber matrices showed potential to be developed as wound-dressing materials.

12.3.2 *Blood Vessel (Vascular and Cardiac) Grafts*

Vascular grafts are in large demand for coronary and peripheral bypass surgeries. Although synthetic grafts have been developed, replacement of vessels with purely synthetic polymeric conduits often leads to the failure of such grafts, especially in grafts less than

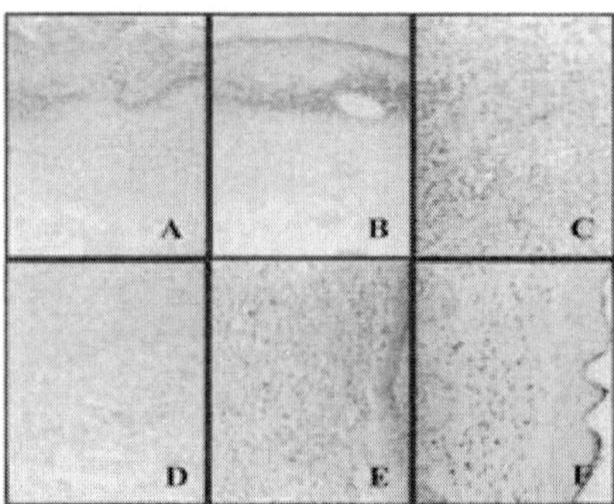

Figure 12.5. Histological findings of a wound (a) at 3rd day postwounding of the control group (HE stain 200); (b) at 3rd day postwounding, polyurethane-membrane group (HE stain 200); (c,d) at 6th day postwounding of the polyurethane-membrane-treated group (HE stain 200); (e) at 15th day postwounding of the control group (HE stain 200); (f) at 15th day postwounding of the polyurethane-membrane-treated group (HE stain 200). Reprinted from Ref. 14 with permission from John Wiley & Sons Inc.

6 mm in diameter or in areas of low blood flow, mainly due to the early formation of thrombosis. Moreover, commonly used materials lack growth potential, and long-term results have revealed several material-related failures, such as stenosis, thromboembolization, calcium deposition, and infection. Tissue engineering has become a promising approach for generating a biocompatible vessel graft with growth potential. Since the first success of constructing blood vessels with collagen and cultured vascular cells, there has been considerable progress in the area of vessel engineering. To date, tissue-engineered blood vessels (TEBVs) could be successfully constructed *in vitro* and be used to repair vascular defects in animal models. From the days of research on developing vascular grafts using materials that produce minimal interaction with the inflowing blood and adjacent tissues, researchers have come a long way in developing constructs at the nano scale that interact with cells and cause blood vessel formation.

Conventional electrospinning produces randomly oriented nanofibers; however, Mo *et al.* developed an aligned biodegradable PLLA–CL (75:25) nanofibrous scaffold using a rotating collector disc for collection of aligned electrospun nanofibers.[19] These aligned nanofibers were explored to fabricate tubular scaffolds that could be used for engineering blood vessels. Their results demonstrated that nano-sized fibers mimic the dimensions of the natural ECM, provide mechanical properties comparable to the human coronary artery, and form a well-defined architecture for smooth muscle cell (SMC) adhesion and proliferation.[20] Aligned fibers not only give structural integrity but also maintain vasoactivity as they provide the necessary mechanical strength needed to sustain high pressure of the human circulatory system.[21] Scientists studied the response of endothelial cells along with SMCs on the aligned nanofibers of PLLA-CL, and their results demonstrated that both the cell types showed enhanced adhesion and proliferation rates on the nanofibrous scaffold.[21] In addition, it was observed that the SMCs' cytoskeleton organization was along the direction of the nanofibers. These results suggested that aligned nanofibers may provide for a good scaffolding system for vascular tissue engineering.

PLA/PCL bilayered tubular scaffolds with a PCL inner layer of randomly oriented microfibers (1.5–6 μm) and a PLA outer layer composed of oriented fibers with diameters from 800 nm to 3 μm have been fabricated. The outside is composed of concentric layers of circumferentially oriented PLA fibers analogous to the tunica media layer of blood vessels. The concentric inner layers of randomly oriented PCL fibers represent the elastic lamina and tunica intima layers of a blood vessel. The bilayered PLA/PCL scaffolds have a maximum stress of 4.3 MPa and an elastic deformation up to 10% strain, which is strong enough and may provide the compliance necessary for substitute blood vessels. Cellular attachment, spreading, and proliferation are proven on these bilayered scaffolds using mouse fibroblasts and human venous myofibroblasts. The fibroblasts adhered and proliferated along the aligned fibrous network, revealing high cellular densities on the outer layer compared with the inner layer over 30 days of static culture. It is now established that there is a significant effect of nano-scale-textured surface roughness on cell response in terms of cell adhesion and proliferation. It is also known that cells attach and organize very well around fibers with diameters smaller than them.[21] Their study demonstrated enhanced spreading and proliferation of endothelial cells on modified polyethylene terephthalate (PET) nanofiber matrices, while preserving their phenotype. On the basis of this study, gelatin-modified PET nanofibers could be potential candidates for the engineering of vascular grafts. Researchers developed electrospun micro- and nanofibrous scaffolds from natural polymers, such as collagen and elastin, with the goal of developing constructs for vascular tissue engineering.[22] Their results demonstrated that electrospun collagen and elastin nanofibers were able to mimic the complex architecture required of vascular constructs and were able to provide good mechanical properties that are desired in the environment of the bloodstream. Their study indicated that micro- and nanofibrous scaffolds synthesized from natural polymers, such as collagen and elastin, could be useful in the engineering of artificial blood vessels.

12.3.3 *Ligament Grafts*

Ligaments have a unique combination of molecular, structural, and mechanical properties, but there is no single unique marker that can be used to distinguish between ligaments and other tissues. In addition, ligament properties vary with the anatomic location, age, and many factors associated with an injury or disease. On the structure–function level, the unique helical organization of collagen fiber bundles is thought to be essential for ligaments to perform their stabilizing function. Ligaments are bands of dense connective tissue responsible for joint movement and stability. Ligament ruptures result in abnormal joint kinematics and often irreversible damage of the surrounding tissue, leading to tissue degenerative diseases, which do not heal naturally and cannot be completely repaired by conventional clinical methods.[23] Recently, tissue engineering methods involving nanofibers have been successfully employed to meet this challenge.[23] In particular, aligned nanofibers enhanced cell response and hence were explored as scaffolds for ligament tissue engineering. Lee *et al.* studied the effects of PU nanofiber alignment and the direction of mechanical stimuli on the ECM generation of human ligament (anterior cruciate) fibroblasts (HLF).[24] Collagen and silk have been used clinically for decades (e.g., hemostats and surgical sutures) and more recently for tissue engineering.[25] Collagen type I gel provided a 3D environment for seeded ligament fibroblasts to proliferate and organize in response to mechanical stimulation.[26] However, the utility of collagen gels was limited by their inferior mechanical properties. Fibrous silk and collagen scaffolds were subsequently adopted in order to provide the necessary mechanical competence of the developing ligaments throughout the *in vitro* cultivation and following implantation *in vivo*. Conventional electrospinning produces randomly oriented nanofibers; however, the authors made use of a rotating target to achieve electrospun fiber alignment. The fibers were then seeded with HLF to study the influence of alignment on HLF behavior. Their results demonstrated that HLF were spindle shaped, oriented in the direction of nanofibers, and showed enhancement in the synthesis of ECM proteins (collagen) on aligned nanofibers compared with randomly oriented nanofibers. In addition, the authors also studied the effect of the direction of mechanical stimuli on the ECM produced by HLF. For this study the authors seeded HLF on PU nanofibers that were parallel, vertically aligned to the strain direction, and randomly oriented. The results demonstrated that HLF were more sensitive to strain in the longitudinal direction. Therefore, this study concluded that aligned nanofibrous scaffolds showed promise for use in ligament tissue engineering.[24]

12.3.4 *Nerve Grafts*

Nerve damage debilitates thousands of Americans each year and leads to problems such as paralysis. In addition to the physical and emotional outcomes, financial burdens amounting to nearly $7 billion in the United States alone are associated with nerve injury (American Paralysis Association, 1997). In fact, nearly 50,000 peripheral nerve repair procedures were performed in 1995 (National Center for Health Statistics, 1995). Nerve injuries can occur in the spinal cord in the central nervous system or, more commonly, in nerve bundles in the peripheral nervous system. Nerve injury can result from mechanical, thermal, chemical, or pathological mechanisms. In the nervous system, degeneration of neurons or glial cells or any unfavorable change in the ECM of neural tissue can lead to a wide variety of clinical disorders. Peripheral nerve injuries resulting from trauma, diseases, or tumor surgery require grafts to bridge the proximal and distalnerve ends. Autografts are the current gold standard; however, they are often limited in supply, require a prolonged surgical time, and can cause donor site morbidity. As an alternative, several conduit materials, both natural and synthetic, have been used as nerve grafts. Several biodegradable polymers were successfully electrospun into nerve grafts and tested for their efficacy to stimulate axonal regeneration through their entire length.[27,28] Neural tissue repair is a daunting challenge because almost all neural injuries lead to an irreversible loss of function. Neural tissue engineering aims to repair neural tissue by employing biological tools such as normal or genetically engineered cells and ECM equivalents, along with potent synthetic tools such as biomaterials for scaffold design and/or drug delivery systems.

Yang *et al.*[28] have studied the potential of PLLA-based electrospun nanofibrous scaffolds for the purpose of neural tissue engineering. Their study involved understanding the influence of nanofibrous scaffolds on neural stem cells (NSCs). Their results also indicated that randomly oriented nanofibers (150–350 nm) not only supported NSC adhesion but also promoted NSC differentiation.[28] Bhattarai *et al.*[29] have recently reported another study wherein they tried to understand the role of a natural-synthetic polymeric nanofiber comprised of well-blended chitosan and PCL in neural tissue engineering. The design combines technological advances in biocompatible polymers and nanotechnology to produce nanofibrous matrices with significantly improved mechanical and biological properties. Schwann cells on chitosan–PCL fibers exhibited the most significant spreading, as indicated by the large polar cell body, while Schwann cells on PCL nanofibers and chitosan–PCL film had smaller, spherical cell morphologies. Similarly, PC12 cells exhibited the most significant spreading on chitosan–PCL nanofibers with more extended neuritis compared with those on the control materials, which exhibited smaller, spherical cell bodies. As a model biomedical application, they constructed nerve guide conduits from this novel nano-featured material and demonstrated its excellent mechanical and biological properties *in vitro*.

12.3.5 *Skeletal Muscle Grafts*

Skeletal muscle is responsible for the control of voluntary movement and the maintenance of structural contours of the body. Muscle loss or deficiency is encountered in various pathological states (by disease or trauma), and attempts to correct them have been employed with limited success.[30] Mature skeletal muscle tissue is composed of multinucleated, postmitotic fibers that do not regenerate following injury. Locally, quiescent populations of myogenic progenitors exist, which will fuse with existing or damaged myotubes to form new ones. In major injuries where the muscle structure is irreversibly compromised, engineered muscle constructs may overcome problems of muscle transfers and provide a successful replacement device for muscle regeneration. Therefore, tissue engineering of skeletal muscle, although challenging, is an exciting alternative to surgical techniques for skeletal muscle regeneration. Skeletal muscle tissue engineering is a promising interdisciplinary specialty that aims at the reconstruction of skeletal muscle loss caused by traumatic injury, congenital defects, or tumor ablations. Due to the difficulty in procuring donor tissue, the possibilities for alternative treatment such as autologous grafting (e.g., muscle flaps) are limited. This process also presents consistent problems with donor site morbidity. Skeletal muscle tissue engineering tries to overcome this problem by generating new functional muscle tissue from autologous precursor cells (stem cells). Multiple stem cells from different sources can be utilized for restoration of differentiated skeletal muscle tissue using tissue engineering principles. Researchers have explored the use of electrospun microfibers made from degradable polyester urethane (PEU) as scaffolds for skeletal muscle tissue engineering.[31] On the basis of the above studies using primary human satellite cells (biopsy from a 38-year-old female), C2C12 (murin myoblast cell line), and L6 (rat myoblast cell line) indicated that the electrospun microfibers of PEU showed satisfactory mechanical properties and an encouraging cellular response in terms of adhesion and differentiation. On the basis of these studies, the electrospun microfibers of PEU show potential to be further explored as a scaffolding system for skeletal muscle tissue engineering.

12.3.6 *Bone Tissue Grafts*

Bone tissue occurs in different bones of the skeleton. Bone is a hard and rigid tissue. Like cartilage, bone consists of living cells with large amounts of ground substance or matrix. It is impregnated with organic salts, such as calcium carbonate (7%) and calcium phosphate (85%). Small amounts of sodium and magnesium are also present. In addition to this, the matrix contains numerous collagenous fibers and a large amount of water. Collagen fibers, together with bone cells, constitute the organic (living) matter in bone tissue. Bone serves several critical functions, including providing muscular support, producing blood cells and immune cells, and acting as a mineral reservoir to maintain electrolyte balance in the body. Bone may be lost after trauma, cancer, fractures, periodontitis, osteoporosis, and infectious disease, and presently there is no definitive method for regeneration.[32] Bone grafts are increasingly used; however, they are plagued by high failure rates of between 16% and 50%.[33] Autografts also present problems associated with a secondary surgery site, as well as a limited supply and morbidity of the donor site.[34] Similarly, metal implants have a high failure rate and often require a second surgical procedure.[35] The design of scaffolds for bone tissue engineering (BTE) is based on the physical properties of bone tissue, such as mechanical strength, pore size, porosity, hardness, and overall 3D architecture. For BTE, scaffolds with a pore size in the range of 100–350 μm and porosity greater than 90% are preferred for better cell/tissue ingrowth and hence enhanced bone regeneration.[36] To overcome the drawbacks of the current bone graft materials, BTE using bone marrow stem cells has been suggested as a promising technique for reconstructing bone defects. To understand the influence of mesenchymal stem cells (MSCs) on nanofibers, MSCs derived from the bone marrow of neonatal rats were seeded on a nanofibrous scaffold. The results indicated that the MSCs migrated inside the scaffold and produced an abundant ECM in the scaffold. In continuation to this study, Shin *et al.* tested PCL nanofibers, along with MSCs *in vivo* in a rat model. Their results demonstrated ECM formation throughout the scaffold, along with mineralization and type I collagen synthesis.[37] These studies demonstrated that PCL-based nanofibrous scaffolds are potential candidates for bone tissue engineering. In another study, hydroxyapatite (HA) has been used with β-tricalcium phosphate (β-TCP) to develop biodegradable nano-composite porous scaffolds.[38] β-TCP/HA scaffolds built from HA nanofibers with β-TCP as a matrix were used to fabricate porous scaffolds by a technique that integrated the gel-casting technique with the polymer sponge method.[38] *In vitro* results demonstrated that incorporation of HA nanofibers as a second component in β-TCP significantly increased the mechanical strength of the porous composite scaffolds. This study introduced nano-composites with HA nanofibers as a promising scaffolding system for load-bearing applications such as bone tissue engineering.

12.3.7 *Articular Cartilage Tissue Grafts*

Cartilage degeneration caused by congenital abnormalities or disease and trauma is of great clinical consequence, given the limited intrinsic healing potential of the tissue. Because of the lack of blood supply and subsequent wound-healing response, damage to cartilage alone, or chondral lesions, results in an incomplete attempt at repair by local chondrocytes. Full-thickness articular cartilage damage, or osteochondral lesions, allows for a normal inflammatory response but results in inferior fibrocartilage formation. To prevent progressive joint degeneration in diseases such as osteoarthritis, surgical intervention is often the only option. In spite of the success of total joint replacement, treatments for repair of cartilage damage are often less than satisfactory and rarely restore full function or return the tissue to its native normal state. The rapidly emerging field of tissue engineering holds great promise for the generation of functional tissue substitutes, including cartilage, by engineering tissue constructs *in vitro* for subsequent implantation *in vivo*. The basic principle is to utilize a biocompatible, structurally and mechanically sound scaffold that is seeded with an appropriate cell source and is loaded with bioactive molecules to promote cellular differentiation and/or maturation. Although recent progress has been made in engineering cartilage of various shapes and sizes for cosmetic purposes, the challenges of engineering a weight-bearing tissue, such as articular cartilage that consists of multiphasic cellular architecture, are significant.[39] There have been a number of successful approaches to tissue engineer cartilage, including the use of natural and synthetic biomaterial scaffolds, allogeneic and autologous sources of mature chondrocytes and chondroprogenitor cells, chondroinductive growth factors, such as the transforming growth factors-β (TGF-βs), and combinations thereof. PCL-based nanofibrous scaffolds by electrospinning showed better compatibility with fetal bovine chondrocytes (FBCs) and maintain chondrocytes in a mature, functional state.[40] Results further demonstrated that FBCs seeded on PCL nanofibers were able to maintain their chondrocytic phenotype by expressing cartilage-specific ECM genes such as aggrecan, collagen type II and IX, and cartilage oligomeric matrix protein.[40] FBCs exhibited a spindle or round shape on the nanofibrous scaffold in contrast to a flat, well-spread morphology, as seen when cultured on tissue culture polystyrene. Another interesting finding from this study was that a serum-free medium produced a more sulfated proteo-glycan-rich cartilaginous matrix compared with the same cultured in a monolayer on tissue culture polystyrene. These results demonstrated that the bioactivity of FBCs depends on the architecture of the scaffold and the composition of the culture medium. Hence, PCL nanofibers show potential to be further explored as scaffolds for cartilage tissue engineering.

12.3.8 *Drug, DNA, Protein, and Enzyme Delivery*

The utility of nanofibrous electrospun composite scaffolds has greatly expanded over the last decade so that they now serve as viable delivery vehicles for a host of different biomedical applications. The material properties of electrospun scaffolds are extremely advantageous for biogenic molecule delivery, in which site specificity and lower overall medicinal dosages lead to a potential industry-altering mechanism of delivering therapeutics. Different biogenic molecules used to predominantly treat infections and cancers can easily be incorporated and released at therapeutic dosages. Further, the inherent high porosity of these electrospun scaffolds allows for a more precisely controlled degradation, which is tunable by polymer composition and fiber morphology, leading to sustained release. Controlled delivery systems are used to improve the therapeutic efficacy and safety of drugs by delivering them to the site of action at a rate dictated by the need of the physiological environment. A wide variety of polymeric materials have been used as delivery matrices, and the choice of the delivery vehicle polymer is determined by the requirements of the specific application.[41]

With the availability of multiple gene delivery systems, the selection of the most appropriate gene delivery vehicle to meet the needs of a particular therapeutic application can be challenging. Viral- and plasmid-based delivery vehicles are currently used for the production of therapeutic proteins so as to elicit a desired biological response.[42] This would especially be useful in the field of tissue engineering, wherein it would be possible to cause the production of a desired protein (growth factor) that can enhance the process of tissue regeneration. Therefore, gene delivery systems have been explored for applications in the engineering of a variety of tissues. The most commonly used carrier-based systems for gene delivery are cationic liposomes and condensing agents such as poly(ethylenimine), and poly(l-lysine). More recently, biomaterial-based gene delivery systems have been explored and are proving to be a promising approach. Various biomaterials such as poly(ethylene glycol) (PEG), PLGA, and PLA–PEG copolymers are currently being investigated for gene and protein delivery.[43] Scaffolds for gene delivery need to provide structural stability and site-specific delivery of genes, along with protection of genes from the biological system until they are released. Further, the released DNA needs to retain its structural integrity until it is taken up by the desired cells.

PLGA and PLA–PEG block copolymer-based nanofibrous scaffolds have been used for plasmid DNA delivery.[44] Recently, α-chymotripsin was attached to electrospun polystyrene nanofibers (120 nm) as a catalytic system and examined for its catalytic efficiency in biotransformation.[45] This study indicated that the nanofibrous enzyme system had a higher hydrolytic activity (65%) than the immobilized enzyme and three times more nonaqueous activity than immobilized α-chymotrypsin in organic solvents. In a recent study, poly(vinyl alcohol) (PVA) electrospun nanofibers have been studied for protein delivery.[46] The coated and uncoated PVA nanofibrous scaffolds were then studied for release kinetics and bioactivity of the released proteins under physiological conditions. The results demonstrated that intact protein/enzyme was continuously released from both the nanofiber types and their bioactivity was preserved after release from the nanofibrous scaffolds. All the above studies demonstrated the potential of nanofibers as controlled delivery systems and hence demanded exploration at greater depth to enable this technology to benefit the patient.

12.4 Conclusions

Electrospinning offers a rapid and convenient way of producing scaffolds with nano-scale elements and has been utilized across a broad range of polymer systems and tissue engineering endeavors. Electrospinning allows the tissue engineer to specifically tailor materials to each specific application and cellular environment. However, it remains largely unknown what the best fiber diameter or interfiber distance is to optimize cell function. This is confounded by the fact that 3D electrospun nanofibrous environments are quite complex, and by changing a dimension such as fiber diameter or alignment, you automatically change the interfiber distance, which can influence cell migration. The underlying mechanisms for enhanced cellular response to nanostructures are only now beginning to be realized by using highly regular and reproducible nanostructured surfaces. However, transferring lessons learned from these model systems over to highly complex 3D electrospun scaffolds is the next step for the advancement of the electrospinning technique for tissue engineering. The need for improving the biomechanical properties of electrospun scaffolds is paramount and is a major obstacle currently tissue engineers face. As fiber diameter and surface functionalization have been shown to effect cell differentiation, gradient fiber scaffolds (through the depth of the scaffolds) could readily be produced using electrospinning to regenerate more complex tissue structures. These remarkable properties, combined with specific surface chemistry, provide cells with a 3D *in vivo* environment. Efforts are being made to improve mechanical properties of electrospun nanofibers to match the native tissue by a combination of several polymers, fiber diameter, and fiber orientation. In addition, higher porosity and wide pore diameters resulting from a combination of micronanofibers showed encouraging cell infiltration into the 3D construct. All of these properties need to be considered while constructing nanofiber-based grafts for tissue regeneration. However, to move forward from our current position, there is a clear need for further research into the effects of 3D nanofiber architecture, functionalization, and interfacial properties on cell behavior. In general, electrospinning utilizes several fluorinated and toxic organic solvents to dissolve polymers. Such toxic solvents might affect the structural conformation of several biopolymers and proteins and result in an undesired cellular response. A critical need exists to replace these toxic organic solvents with aqueous-based or less toxic solvents during electrospinning. Also more efforts need to be made to improve the efficiency of nanofiber production, packing, shipping, and handling. Such efforts can further improve the efficacy of nanofibers and can lead to the development of commercially viable nanofiber technology for a variety of biomedical applications. Mimicking the architecture of the ECM is one of the major challenges of tissue engineering. Among all the approaches used to prepare the ECM synthetically, the approach using nanofibers has shown the most promising results. Nanofibers can be formed using one of three prevailing techniques: electrospinning, self-assembly, and phase separation. Electrospinning is the most widely studied technique and has also shown the most promising results.

Nanofibers, irrespective of their method of synthesis, have provided for scaffolds with high surface area and enhanced porosity. These properties have been demonstrated to have a significant effect on cell adhesion, proliferation, and differentiation. Hence nanofibrous matrices are currently being explored as scaffolds for musculoskeletal tissue engineering (including bone, cartilage, ligament, and skeletal muscle), skin tissue engineering, neural tissue engineering, vascular tissue engineering, and controlled delivery of drugs, proteins, and DNA. The results of all recently published studies clearly indicate that nanofiber-based scaffolds show excellent potential to be developed for a variety of tissue engineering applications.

References

1. D. H. Reneker and I. Chun, *Nanotechnology*, **216** (1996).
2. P. P. Tsai, H. Schreuder-Gibson, and P. Gibson, *J. Electrostat.*, **333** (2002).
3. D. R. Nisbet, J. S. Forsythe, W. Shen, D. I. Finkelstein, and M. K. Horne, *J. Biomater. Appl.*, doi: 10. 1177/0885328208099086 (2008).
4. D. Smith and D. H. Reneker, *PCT/US*, 00/27737 (2001).
5. R. Vasita and D. S. Katti. *Int. J. Nanomed.*, **15** (2006).
6. A. G. MacDiarmid, W. E. Jones, I. D. Noris, J. Gao, A. T. Johnson, N. J. Pinto, J. Hone, B. Han, F. K. Ko, H. Okuzaki, and M. Llaguno, *Synth. Met.*, **27** (2001).
7. D. H. Reneker, A. L. Yarin, H. Fong, and S. Koombhongse, *J. Appl. Phys.*, **4531** (2000).
8. A. Formhals, *US Patent No.1*, **975**, 504 (1934).
9. G. E. Martin and I. Cockshott, *US Patent No. 4*, **043**, 331 (1977).
10. S. R. Bhattarai, N. Bhattarai, H. K. Yi, P. H. Hwang, D. I. Cha, and H. Y. Kim, *Biomaterials*, 2595 (2004).
11. Z. M. Huang, Y. Z. Zhang, M. Kotaki, and S. Ramakrishna, *Compos. Sci. Technol.*, **2223** (2003).
12. G. E. Martin, I. D. Cockshott, and F. J. T. Fildes, *US Patent*, **4044404** (1977).
13. N. Bhattarai, D. Edmondson, and O. Veiseh, *Biomaterials*, **6176** (2005).
14. M. S. Khil, D. I. Cha, H. Y. Kim, I. S. Kim, and N. Bhattarai, *J. Biomed. Mater. Res. B Appl. Biomater.*, **675** (2003).
15. R. Langer and J. Vacanti, *Tissue Eng. Sci.*, **260920** (1993).
16. R. A. F. Clark, A. J. Singer, R. P. Lanza, R. Langer, and J. Vacanti, Eds., *Principles of Tissue Engineering* (Academic Press: San Diego, 2nd ed., 2000), p. 855.
17. B. M. Min, G. Lee, S.H. Kim, Y. S. Nam, T. S. Lee, and W. H. Park, *Biomaterials*, **1289** (2004).
18. M. S. Khil, S. R. Bhattarai, H. Y. Kim, S. Z. Kim, and K. H. Lee, *J. Biomed. Mater. Res. B Appl. Biomater.*, **117** (2005).
19. X. Mo and H. J. Weber, *Macromol. Symp.*, **413** (2004).

20. X. Mo, C. Y. Xu, M. Kotaki, and S. Ramakrishna, *Biomaterials*, **1883** (2004).
21. C. Y. Xu, R. Inai, M. Kotaki, and S. Ramakrishna, *Biomaterials*, **877** (2004).
22. E. D. Boland, J. A. Matthews, K .J. Pawlowski, D. G Simpson, G. E. Wnek, and G. L Bowlin *Front. Biosci.*, **1422** (2004).
23. V. S. Lin, M. C. Lee, S. O Neal, J. McKean, and K. L. Sung, *Tissue Eng.*, **443** (1999).
24. C. H. Lee, H. J. Shin, I. H. Cho, Y. M. Kang, I. A. Kim, K. D. Park, and J. W. Shin, *Biomaterials*, **1261** (2005).
25. N. Minoura, M. Tsukada, and M. Nagura, *Biomaterials*, **430** (1990).
26. R. G. Young, D. L. Butler, W. Weber, A. I. Caplan, S. L. Gordon, and D. J. Fink, *Orthop. Res.*, **406** (1998).
27. S. Patel, K. Kurpinski, R. Quigley, H. Gao, B. S. Hsiao, M. Poo, and S. Li, *Nano Lett.*, **2122** (2007).
28. F. Yang, R. Murugan, S. Wang, and S. Ramakrishna, *Biomaterials*, **2603** (2005).
29. N. Bhattarai, Z. Li, J. Gunn, M. Leung, A. Cooper, D. Edmondson, O. Veiseh, M. H. Chen, Y. Zhang, R. G. Ellenbogen, and M. Zhang, *Adv. Mater.*, 2009.
30. C. A. DiEdwardo, P. Petrosko, and T. O. Acarturk, *Clin. Plast. Surg.*, **647** (1999).
31. S. A. Riboldi, M. Sampaolesi, P. Neuenschwander. G. Cossu, and S. Mantero, *Biomaterials*, **4606** (2005).
32. S. Kimakhe, S. Bohic, C. Larrose, A. Reynaud, P. Pilet, B. Giumell, D. Heymann, and G. Daculsi, *J. Biomed. Mat. Res.*, **18** (1999).
33. S. Stevenson, S. E. Emery, and V. M. Goldberg, *Clin. Ortho. Rel. Res.*, **324**, 66 (1996).
34. V. M. Goldberg and S. Stevenson, *Orthopedics*, **809** (1994).
35. S. E. Graves, D. Davidson, L. Ingerson, P. Ryan, E. C. Griffith, B. F. McDermott, H. J. McElroy, and N. L. Pratt, the Australian Orthopaedic Association National Joint Replacement Registry, *Med. J. Aus.*, **S31** (2004).
36. S. P. Bruder and A. I. Caplan, Bone regeneration through cellular engineering, in *Principles of Tissue Engineering*, Eds. R. P. Lanza, R. Langer, and J. Vacanti (Academic Press: San Diego, 2nd ed., 2000), p. 683.
37. Y. M. Shin, M. M. Hohman, M. P. Brenner, and G. C. Rutledge, *Polymer*, **9955** (2001).
38. Y. Zhang, C. T. Lim, S. Ramakrishna, and Z. M. Huang, *J. Mater. Sci. Mater. Med.*, **933** (2005).

39. S. H. Kamil, K. Kojimam, M. P. Vacanti, L. J. Bonassar, C. A. Vacanti, and R.D. Eavey, *Laryngoscope*, **90** (2003).
40. W. J. Li, K. G. Danielson, P. G. Alexander, and R. S. Tuan, Scaffolds, *Biomed. Mater. Res. A*, **1105** (2003).
41. J. Heller and A. S. Hoffman, Drug delivery system, in *Biomaterial science, an Introduction to Materials in Medicine*, Eds. B. D. Ratner, A. S. Hoffman, and F. J. Schoen, (Elsevier Academic Press: San Diego, 2nd ed., 2004), p. 629.
42. L. G. Fradkin, J. D. Ropp, and J. F. Warner, Gene-based therapeutics, in *Principles of Tissue Engineering*, Eds. R. P. Lanza, R. Langer, and J. Vacanti (Academic Press: San Diego, 2nd ed., 2000), p. 383.
43. A. M. Funhoff, S. Monge, R. Teeuwen, G. A. Kining, N. M. E. Schuurmans-Nieuwenbroek, D. J. A. Crommelin, D. M. Haddleton, W. E. Hennink, and C. F. Nostrum, *J. Controlled Release*, **711** (2005).
44. X. Fang and D. H. Reneker, *J. Macromo. Sci. Phys. B*, **169** (1997).
45. H. Jia, G. Zhu, B. Vugrinovich, W. Kataphinan, D. H. Reneker, and P. Wang, *Biotechnol. Prog.*, **1027** (2002).
46. J. Zeng, A. Aigner, F. Czubayko, T. Kissel, J. H. Wandorff, and A. Greiner, *Biomacromolecules*, **1484** (2005).

Chapter 13

PHBV/PROTEINS COMPOSITE NANOFIBROUS SCAFFOLDS FOR TISSUE ENGINEERING

K. M. Kamruzzaman Selim, Zhi-Cai Xing, and Inn-Kyu Kang*

Department of Polymer Science and Engineering, Kyungpook National University, Daegu 702-701, Republic of Korea

*ikkang@knu.ac.kr

In this chapter, the basic concept of the electrospinning technique, especially its uses regarding the fabrication of poly(3-hydroxybutyrate-co-3-hydroxyvalerate) (PHBV)/proteins nanofibrous scaffolds, is described. After interaction of an as-prepared PHBV/proteins composite with cells, the morphological and biological phenomena are comprehensively reviewed on the basis of selective research studies. The polymer nanocomposites exhibited higher cell adhesion, proliferation, and cell viability comparison with nonconjugated nanofibers. Thus the prepared nanocomposites have a potential for use in regard to tissue engineering, wound dressing, and other biomedical applications.

Handbook of Intelligent Scaffolds for Tissue Engineering and Regenerative Medicine
Edited by Gilson Khang

www.panstanford.com

13.1 Introduction

Electrospinning is the most elegant method for nanofiber production, which was introduced in the early 1930s.[1,2] Recently, much attention has been paid to the electrospinning process as a unique technique, because it can produce polymer nanofibers having a diameter in a range of several micrometers down to tens of nanometers.[3] In addition, electrospun nanofibrous scaffolds possess an extremely high surface-to-volume ratio, tunable porosity, an ability to conform over a wide variety of sizes and shapes, superior mechanical properties, and ease of controlling composition to achieve desired properties and functionality.[4] All these advantages of electrospun nanofibrous scaffolds have encouraged the researchers more interested in the past several years to rigorously investigate different composite materials to find their applications in various fields, such as filtration,[5] optical and chemical sensors,[6] electrode materials,[7] and biological/ biomedical scaffolds.[8,9] The electrospinning method has been actively explored recently in regard to biomedical fields. Natural polymers (i.e., collagen, gelatin, keratin, silk, fibrinogen, elastin, *etc.*) and synthetic polymers—that is, poly(lactic acid) (PLA), poly(glycolic acid) (PGA), poly(lactide-*co*-glycolide) (PLGA), poly(ε-caprolactone) (PCL), PHBV, *etc.*—are being co-electrospun to form nanocomposites in order to be used as wound dressings,[10] drug delivery systems,[11] and vascular graft and tissue engineering applications.[8,12]

To use nanocomposites in biomedical fields, some properties of composite materials, such as hydrophilicity, mechanical modulus and strength, biodegradability, biocompatibility, and specific cell interactions, need to be considered.[4] By selecting a combination of appropriate components and by adjusting the component ratio, properties of electrospun nanocomposites can be tailored to have desired new functions. On the basis of polymer physics, polymer blending, followed by electrospinning, constitutes one of two effective means to combine different polymers in yielding new materials properties.[4] For example, mixtures of collagen with elastin[13] as well as mixtures of chitosan with poly(ethylene oxide) (PEO) or poly(ethylene glycol) (PEG)[14] have been electrospun to fabricate nanocomposites for biomedical applications.

In this chapter, the basic concept regarding the electrospinning technique, especially its uses for the preparation of PHBV/gelatin and PHBV/collagen (PHBV–Col) scaffolds, is described. After the interaction of the prepared nanocomposites with cells, the morphological and biological phenomena of polymer composites have been comprehensively reviewed on the basis of selective research studies.

13.2 Electrospinning Technique

Formhals (1934)[15] patented a process whereby an experimental setup protocol was outlined for the production of polymer filaments by using electrostatic force. The process is referred to as electrospinning when it is used to spin fibers. In other words, electrospinning is a process that forms nanofibers through an electrically charged jet of polymer solution or polymer melt. To perform electrospinning, the polymer must be in liquid form, either as a molten polymer or as a polymer solution. The liquid polymer solution is passed through the electrospinning system to form nanofibers. A basic electrospinning system usually consists of three major components: a high-voltage power supply, a spinneret (e.g., a pipette tip/syringe), and a grounded collecting unit (Fig. 13.1).

When a charged polymer solution is fed through the spinneret under an external electric field, a suspended conical droplet is formed, whereby the surface tension of the droplet is in equilibrium

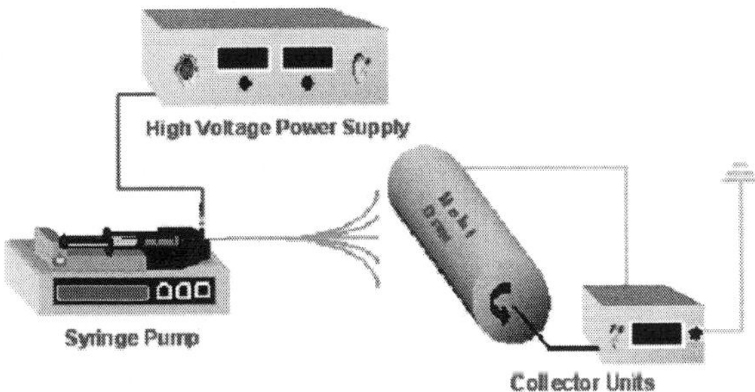

Figure 13.1. Schematic illustration of the electrospinning apparatus (adapted from Ref. 16).

with the electric field. When the applied electric field is strong enough to overcome the surface tension, a tiny jet is ejected from the surface of the droplet and drawn toward the collecting plate. During the jet propagation toward the collecting plate, the solvent in the jet stream gradually evaporates. The resulting product consists of a nonwoven fibrous scaffold with a large surface-area-to-volume ratio and a small pore size (in microns).[4] The electrospun fiber properties and morphology depend on many parameters, such as solution parameters (viscosity, surface tension, conductivity, etc.), processing conditions (voltage, feed rate, temperature, etc.), and ambient parameters (humidity, type of atmosphere, pressure, etc.).

13.3 Nanocomposite Preparation

Polymer nanofibers are chemically inert and do not have specific functions. So biomolecules or biocompitable materials need to be added to the surface of electrospun nanofibers for use in biomedical or other purposes. Biomolecules can be added to nanofibers through surface modification/coating by means of several methods, such as blending or physical coating, graft copolymerization via irradiation (γ-rays, electron beams, UV rays, etc.), plasma treatment, ozone or H_2O_2 oxidation, Ce^{4+} oxidation, etc. However, the blending method is the most effective in making PHBV/proteins composites that are useful for biomedical purposes.

13.3.1 *Principle of the Blending Method*

Polymers are first dissolved with a suitable solvent. Biomolecules, such as gelatin, keratin, and collagen, or biocompatable inorganic materials, such as hydroxyapatite, silver, etc., are then mixed homogeneously with the polymer solution. The resultant solution is then subjected to electrospinning in order to prepare nanocomposites. Compared with synthetic polymers (i.e., PLA, PGA, PLGA, PCL, PHBV, etc.), natural biopolymers (i.e., collagen, gelatin, keratin, silk, fibrinogen, elastin, etc.) have good biocompatibility. However, the processability of natural biopolymers is, in general, fairly poor.

Moreover, protein nanofibres easily lose their fibrous forms in an aqueous medium. Accordingly, they need to be cross-linked in order to retain proper structural stability. The stability of proteins in an aqueous medium can be achieved by blending them with a hydrophobic polymer, such as PHBV, PLA, PLGA, PCL, or PLGA.

13.4 Typical PHBV/Protein Nanocomposite Preparation

Several examples of PHBV/proteins nanocomposites prepared by the blending method have been discussed here briefly on the basis of selected published research studies.

13.4.1 *Electrospun PHBV/Collagen Composite Nanofibrous Scaffolds*

PHBV, a copolymer of microbial polyester, is one of the most promising materials for tissue engineering. It is a biodegradable, biocompatible, nontoxic, and thermoplastic polyester that is produced by bacteria. PHBV is being developed and commercialized as an ideal substitute for nonbiodegradable polymeric materials in regard to commodity applications, due to its biodegradability and easy processability. On the other hand, collagen is the main protein of connective tissue in animals, as well as the most abundant protein found in mammals, making up about 25% to 35% of the whole body's protein content. Collagen fibers are a major component of the extracellular matrix (ECM) that supports most tissues and gives the cells a structure from the outside.

Meng *et al.*[16] prepared PHBV–Col composite nanofibrous scaffolds by electrospinning. To prepare a PHBV–Col nanocomposite, PHBV and type I collagen were dissolved in 1,1,1,3,3,3-hexafluoro-2-isopropanol (HFIP). HFIP was chosen as a solvent because both PHBV and collagen are soluble in it. The mixed polymer solution was then delivered to a 20-gauge metal needle connected to a high-voltage power supply. Upon the application of high voltage, a jet of fluid was ejected from the needle. As the jet accelerated toward

Table 13.1. Electrospinning conditions for various amounts of polymer solution

	Concentration (wt%)	Voltage (kV)	Distance* (cm)	Flow rate (mL/h)
PHBV	2	7	15	1.5
PHBV-Col	2	12	22	1.0
Collagen	3	20	15	1.5

*Distance is from the spinneret to the collector.

a grounded collector, the solvent evaporated and a charge polymer fiber was deposited on the collector in the form of a nanofiber web. Nanofibers with various proportions of PHBV and collagen (7:3, 5:5, and 3:7) were prepared. Nanocomposite mats with more than 50% collagen were fragile. Therefore, 7:3 PHBV–Col was used in this study (Table 13.1).

The resulting fibers varied from 300 to 600 nm in diameter. The images demonstrate continuous fiber morphology, and the composite fibers do not contain beads (Fig. 13.2).

To study the surface morphology of PHBV and PHBV–Col nanofibers, the surface was imaged by a tapping mode using an atomic force microscope (AFM) (Fig. 13.3). The collagen nanofiber was so fragile that AFM observation was not possible. The PHBV nanofiber surface (Fig. 13.3a) showed a relatively homogeneous image, while PHBV–Col (Fig. 13.3b) showed a heterogeneous image. The presence of collagen in PHBV was considered to have induced such heterogeneity.

Figure 13.2. SEM micrographs of nanofibrous scaffolds; (a) PHBV, (b) PHBV–Col, and (c) collagen (adapted from Ref. 16).

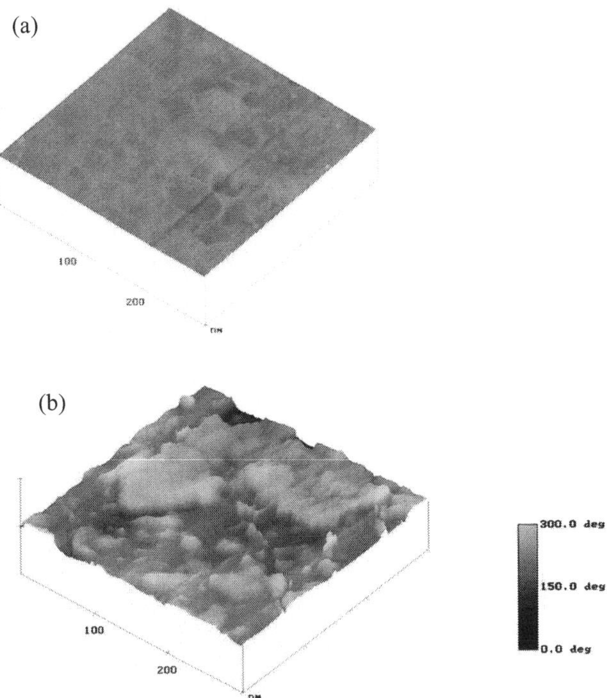

Figure 13.3. AFM phase images of nanofibrous scaffold surface; (a) PHBV and (b) PHBV-Col (adapted from Ref. 16). See also Color Insert.

13.4.2 *Electrospun PHBV/Gelatin Composite Nanofibrous Scaffolds*

Among natural biopolymers, gelatin can be obtained by denaturing collagen and has almost an identical composition and biological properties as those of the parent collagen. Much attention has been focused on the use of gelatin as a tissue engineering material, due to its low cost. To prepare a PHBV/gelatin composite scaffold, a transparent polymer solution was prepared by dissolving gelatin and PHBV in 2,2,2-trifluoroethanol (TFE) at a ratio of 50/50 with sufficient stirring at room temperature. To examine the effect of the gelatin content on fiber morphology, a 6 wt% polymer solution was prepared, using different ratios of PHBV and gelatin. To examine the effect of the polymer concentration on fiber morphology, a

Figure 13.4. SEM micrographs of electrospun fibers using a PHBV/gelatin solution at TFE 6 wt%; (a) 30/70, (b) 50/50, and (c) 70/30 (adapted from Ref. 17).

PHBV/gelatin (50/50) blend was dissolved, using TFE to prepare several solutions with concentrations ranging from 2% to 8 wt%. The PHBV/gelatin nanocomposite scaffold was then obtained by co-electrospinning PHBV and gelatin solutions.[17] As shown in the SEM photographs of Fig. 13.4, a highly uniform and the finest nanofiber could be obtained through different mixing ratios. To consider the mechanical properties and biocompatibility of the nanofibrous scaffold, a 50/50 mixing ratio of dope solution was fixed for the study. Figure 13.5 shows PHBV/gelatin nanofibers depending on the

Figure 13.5. SEM micrographs of electrospun fibers of a PHBV/ gelatin (50/50) mat using TFE solutions as a function of concentration; (a) 2 wt%, (b) 4 wt%, (c) 6 wt%, and (d) 8 wt% (adapted from Ref. 17).

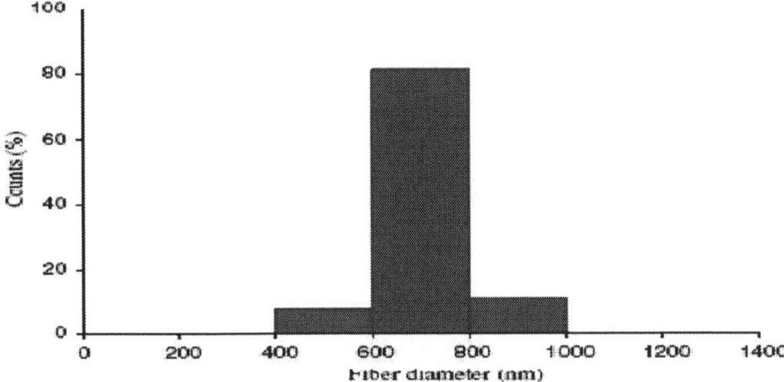

Figure 13.6. Diameter distribution of PHBV nanofibers electrospun from a 6 wt% PHBV–TFE solution (adapted from Ref. 17).

concentration of 2–8 wt%, at a 50/50 mixing ratio. At a concentration of 4 wt% and 6 wt%, respectively, continuous nanofibers without beads could be obtained.

The image analysis of PHBV/gelatin nanofibers (6 wt%, 50/50) revealed that their diameters ranged from 400 to 1000 nm, as shown in Fig. 13.6.

13.5 Interaction of As-Prepared Nanocomposites with Cells and Results Obtained Thereby

When as-prepared nanofibrous composites interact with cells, the nanocomposites show improved biocompatibility. They exhibit higher cell adhesion, proliferation, and viability comparison with the control nanofibers. Results obtained by the interaction of biomolecules containing nanocomposites with cells are discussed briefly in this section on the basis of selected research studies.

13.5.1 *PHBV–Col Nanocomposites*

Cell culture experiments indicated that the PHBV–Col nanofibrous scaffold accelerated the adhesion and growth of NIH3T3 cells more effectively than the PHBV nanofibrous scaffold, thus making the former a good candidate for tissue engineering.[16] Figure 13.7 shows the adhesion of NIH3T3 fibroblasts on PHBV and PHBV–Col nanofibrous

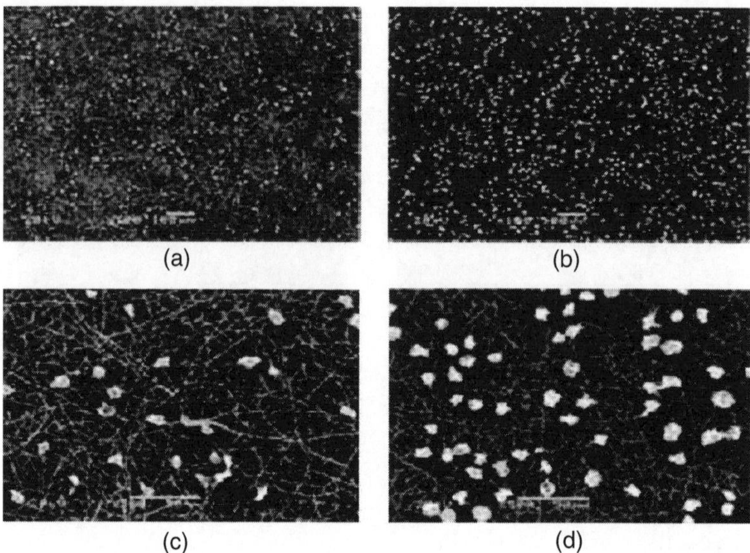

Figure 13.7. SEM images of NIH3T3 cells cultured for 4 h on PHBV (a,b) and PHBV–Col (c,d) nanofibrous scaffolds at an original magnification of 100× (a,c) and 500× (b,d) (adapted from Ref. 16).

scaffolds after a 4-hour culture. The cells adhered well on the surfaces of both nanofibrous scaffolds. More cells were observed on the surface of the PHBV–Col nanofibrous scaffold than that of PHBV.[47] Figure 13.8 shows the proliferation of cells on the nanofibrous scaffolds. Cell proliferation on the PHBV nanofibrous scaffold was significantly accelerated by the incorporation of type I collagen (PHBV–Col) ($P < 0.01$).[16]

13.5.2 *PHBV/Gelatin Nanocomposites*

Cell culture experiments showed that NIH3T3 cells had very favorable interactions with the PHBV/gelatin composite scaffold comparison with the PHBV film and the PHBV nanofibrous scaffold. Significantly, cellular infiltration into the PHBV/gelatin composite fibrous scaffold was demonstrated. It is concluded that co-electrospinning the ECM with synthetic polymers, such as PHBV, has many potential applications in tissue engineering.[4,8] As shown in Fig. 13.9, NIH3T3 fibroblasts were highly dispersed on to the PHBV

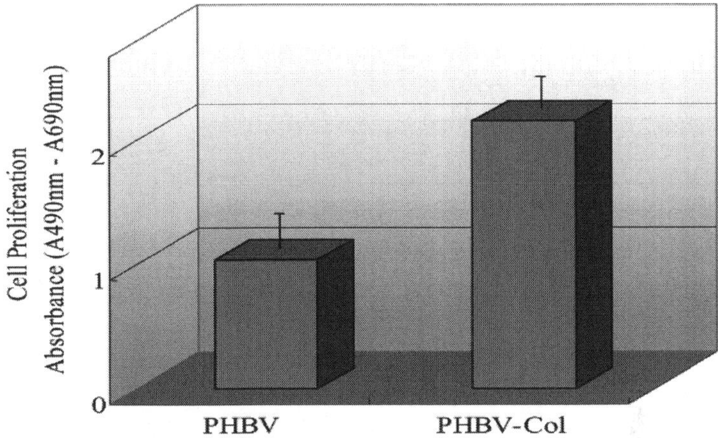

Figure 13.8. Proliferation of NIH3T3 cells cultured for 68 h. Data is expressed as mean ± SD ($n = 5$) for the specific absorbance (adapted from Ref. 16).

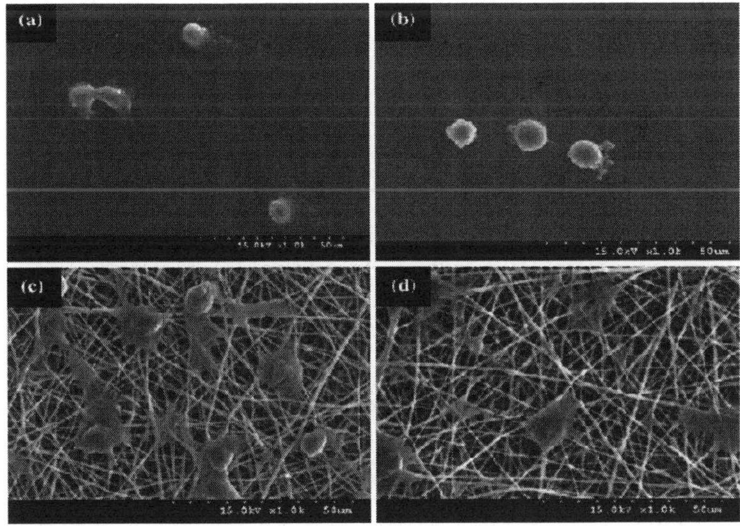

Figure 13.9. Adhesion of fibroblasts on a (a) culture dish, (b) PHBV film, (c) PHBV, and (d) PHBV/gelatin nanofibrous scaffolds for 4 h incubation (adapted from Ref. 17).

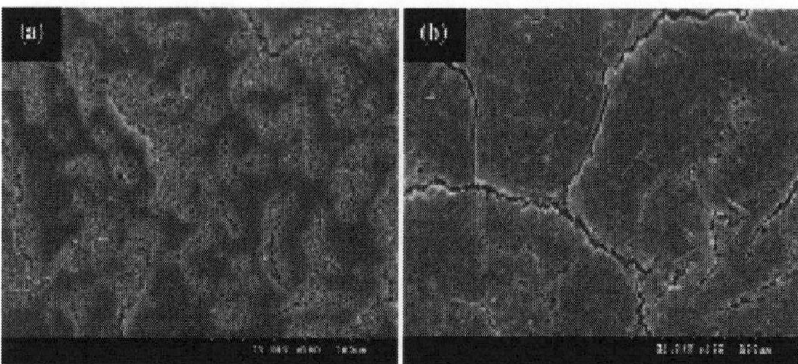

Figure 13.10. Morphology of fibroblasts on (a) PHBV and (b) PHBV/gelatin nanofibrous scaffolds for 1 d incubation (adapted from Ref. 17).

(Fig. 13.9c) and PHBV/gelatin scaffolds (Fig. 13.9d) comparison with the culture dish (Fig. 13.9a) and the PHBV film (Fig. 13.9b) after 4 hours of incubation.[4,8] NIH3T3 fibroblasts, cultured on PHBV and PHBV/gelatin nanofibers for 1 day, are shown in Fig. 13.10. As the results show, cells that adhered to the PHBV/gelatin scaffold (Fig. 13.10b) more quickly formed a monolayer compared with the PHBV nanofibrous scaffold (Fig. 13.10a), showing good tissue compatibility of the PHBV/gelatin nanofibrous scaffold.[17]

Figure 13.11 shows cross-sectional images of the PHBV/gelatin scaffold after fibroblast culture. After three days of incubation, a monolayer morphology was found. After six days of incubation,

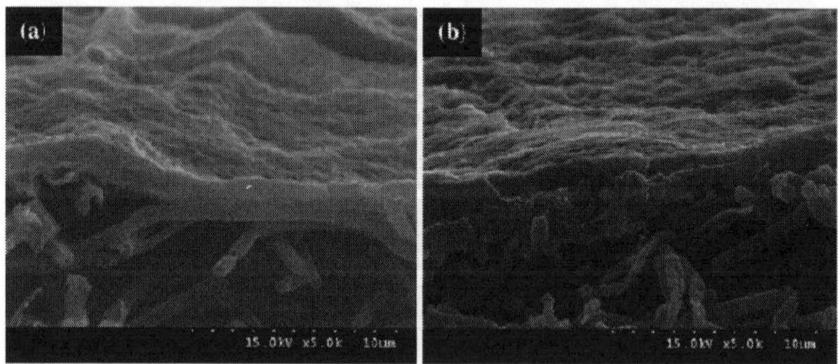

Figure 13.11. Cross-sectional images of a PHBV/gelatin nanofibrous scaffold after fibroblasts cultured after (a) 3 days and (b) 6 days incubation (adapted from Ref. 17).

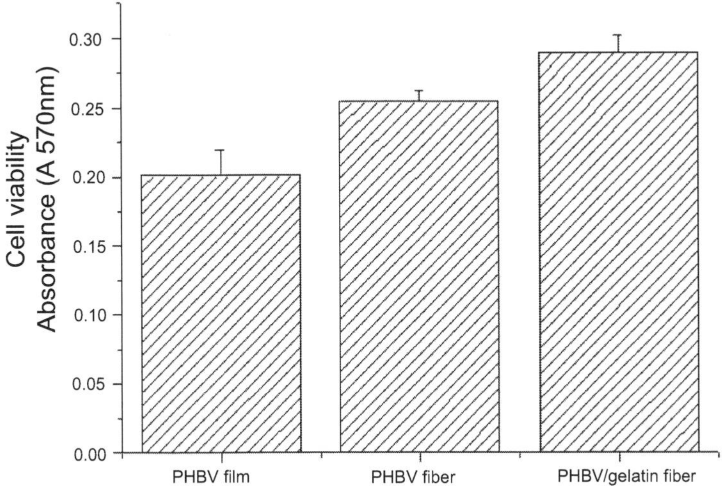

Figure 13.12. Cell viability formosan absorbance expressed as a measure of cell viability from NIH3T3 cells cultured onto nanofibrous scaffolds for 5 d. Data is expressed as mean ± SD ($n = 5$) of the specific absorbance (adapted from Ref. 17).

a thicker layer was formed, indicating that the composite scaffold allowed for cellular penetration or infiltration inside the PHBV/gelatin composite fibrous structure.[17]

After seven days of incubation, compared with the PHBV film, the cell viability of fibroblasts cultured on PHBV and PHBV/gelatin nanofibrous scaffolds was significantly enhanced ($p < 0.01$), as shown in Fig. 13.12.[17]

13.6 Conclusions

In this chapter, discussion has been mainly focused on the use of the electrospinning technique for the immobilization of natural biopolymers, such as collagen and gelatin, with a synthetic polymer, such as PHBV, to prepare nanocomposites on the basis of selected research studies. On the basis of the research results, it has been demonstrated that PHBV/proteins nanocomposites show higher cell adhesion, proliferation, and cell viability comparison with those of nonconjugated control nanofibers. The prepared nanocomposites could be used in tissue engineering applications.

Acknowledgment

This work was supported by grant No. RTI04-01-01 from the Regional Technology Innovation Program of the Ministry of Knowledge Economy (MKE).

References

1. J. Doshi and D. H. Reneker, *J. Electrostat.*, **35**, 151 (1995).
2. S. Ramakrishna, K. Fujihara, W. E. Teo, T. C. Lim, and Z. Ma, *An Introduction to Electrospinning and Nanofibers* (World Scientific Publishing: Singapore, 2005).
3. B. M. Min, G. Lee, S. H. Kim, Y. S. Nam, T. S. Lee, and W. H. Park, *Biomaterials*, **25**, 1289 (2004).
4. D. Liang, B. S. Hsiao, and B. Chu, *Adv. Drug Del. Rev.*, **59**, 1392 (2007).
5. H. Schreuder-Gibson, P. Gibson, L. Wadsworth, S. Hemphill, and J. Vontorcik, *Adv. Filtr. Separ. Technol.*, **15**, 525 (2002).
6. X. Wang, Y. G. Kim, C. Drew, B. C. Ku, J. Kumar, and L. A. Samuelson, *Nano Lett*, **4**, 331 (2004).
7. C. Kim, S. H. Park, W. J. Lee, and K. S. Yang, *Electrochimica Acta*, **50**, 877 (2004).
8. M. S. Khil, S. R. Bhattarai, H. Y. Kim, S. Z. Kim, and K. H. Lee, *J. Biomed. Mater. Res. B, Appl. Biomaterials*, **72B**, 117 (2005).
9. F. Yang, R. Murugan, S. Wang, and S. Ramakrishna, *Biomaterials*, **26**, 2603 (2005).
10. J. S. Choi, K.W. Leong, and H. S. Yoo, *Biomaterials*, **29**, 587 (2008).
11. C. Chen, G. Lv, C. Pan, M. Song, C. Wu, D. Guo, X. Wang, B. Chen, and Z. Gu, *Biomed. Mater.*, **2**, L1 (2007).
12. Y. Ito, H. Hasuda, M. Kamitakahara, C. Ohtsuki, M. Tanihara, I. K. Kang, and O. H. Kwon, *J. Bio. Sci. Bio. Eng.*, **100**, 43 (2005).
13. E. D. Boland, J. A. Matthews, K. J. Pawlowski, D. G. Simpson, G. E. Wnek, and G. L. Bowlin, *Front. Biosci.*, **9**, 1422 (2004).

14. B. Duan, C. Dong, X. Yuan, and K. Yao, *J. Biomater. Sci. Polym. Ed.*, **15**, 797 (2004).
15. A. Formhals, A process and apparatus for preparing artificial threads. US Patent 26, **975**, 504 (1934).
16. W. Meng, S. Y. Kim, J. Yuan, J. C. Kim, O. H. Kwon, N. Kawazoe, G. Chen, Y. Ito, and I. K. Kang, *J. Biomater. Sci. Polymer Ed.*, **18**, 81 (2007).
17. W. Meng, Z. C. Xing, K. H. Jung, S. Y. Kim, J. Yuan, I. K. Kang, S. C. Yoon, and H. I. Shin, *J. Mater. Sci.: Mater. Med.*, **19**, 2799 (2008).

Chapter 14

NANOFIBROUS SCAFFOLDING FOR BONE TISSUE ENGINEERING

Hae-Won Kim
Department of Nanobiomedical Science & WCU Research Center, Graduate School Dankook University, South Korea
Department of Biomaterials Science, School of Dentistry & Institute of Tissue Regeneration Engineering (ITREN), Dankook University, South Korea
kimhw@dku.edu

Nanofibrous matrices made of medical-graded materials have shown great potential for the regeneration of a variety of tissues, including bone. The extremely high surface area with tunable fiber diameters and compositions is attractive for alluring initial cellular events and stimulating development into specific tissue types. This chapter introduces recent studies on nanofiber technology for bone tissue engineering, with particular emphasis on the electrospinning approach. Generation of a bone-mimicking composition and functionalization of matrices with bioactive factors are some of the main focuses.

14.1 Introduction

Among the various types of scaffolds with three-dimensional (3D) morphology to facilitate tissue development, the nanofibrous structure has recently gained great attention in the tissue engineering

Handbook of Intelligent Scaffolds for Tissue Engineering and Regenerative Medicine
Edited by Gilson Khang

www.panstanford.com

area.[1,2] Tissue engineering is to deliver the concept of mimicking the native tissue structure involving cells and the extracellular matrix (ECM) under *ex vivo* conditions with the convergence of technological development and biological knowledge. This is possible, in most part, by engineering the ECM component, which can trigger cells to develop a tissue-mimicking composition and biological hierarchy. The nanofibrous structure, with its morphological feature, is thus regarded to mimic the tissue ECM, such as collagen and elastin fibrous proteins, those most abundant in native tissue.

Some methodologies are currently being used to create the nanofibrous structure, which include phase separation, self-assembly, and electrospinning.[3] Among these, electrospinning is the most widely used technique.[4–8] This is because not only is the processing simple and the setup cost-effective, but also the composition and fiber diameter (tens of nanometers to a few micrometers) can be easily tuned to the applications. Many studies have reported the potential of nanofibrous matrices in guiding tissue cells to adhere and spread and in stimulating the secretion of appropriate ECM molecules targeted to blood vessel, skin, cartilage, muscle, tooth, and bone. For bone tissue regeneration, there are growing interests in producing nanofibrous materials with compositions and morphologies appropriate for bone cell differentiation and bone-mimicking ECM production, which finally replaces the native function of bone.

This chapter will introduce some recent studies on nanofibrous materials for bone tissue regeneration, with particular emphasis on the electrospinning method. The main category will be focused on the generation of a bone-mimicking composition and functionalization of matrices with bioactive factors.

14.2 Nanofiber Production Tools and Electrospinning

To generate materials into a nanofibrous form, some methodologies have been introduced, which include phase separation, self-assembly, and electrospinning. Compared with the former two methods, electrospinning has been the most popularly used tool to

Table 14.1. Merits and limitations of methods of generating nanofibrous materials for tissue regeneration.

Tools	Merits	Limitations
Phase separation	Easy process Controllable size and shape Porosity control	Limited composition Low yield
Self-assembly	Small fiber diameter (<10 nm) Control of functional groups	Low yield Short fibers Limited composition
Electrospinning	Easy process Cost effective Broad spectrum of materials Fiber aligning possible	Difficulty in forming 3D shape and pore structure

generate nanofibers for tissue regeneration. Some merits and limitations of each tool are summarized in Table 14.1.

14.2.1 *Phase Separation*

The phase separation method utilizes the immiscibility of phases that comprises the material and removal-off part. When a material is dissolved in a solvent or co-solvents and undergoes cooling below the freezing point(s) of the solvent(s) used, phases are separated due to thermodynamic immiscibility. The separated phases form networks with various structures, depending on processing variables. Ma *et al.* have reported the production of biodegradable polymers with a nanofibrous structure by using the phase separation method.[9] A detailed procedure consists of the steps as follows: (i) polymer dissolution, (ii) gelation, (iii) solvent extraction, and (iv) freezing and freeze drying. They further reported the nanofibrous-structured surface improved the cell adhesion and expression of phenotypes that are specific to bone.

14.2.2 *Self-Assembly*

Generally, self-assembly is the method of constructing larger-scale materials using smaller molecular building blocks. Hartgerink *et al.* introduced the method of creating self-assembled nanofibers using peptide amphiphiles targeting for bone.[10] Small molecules named peptide amphiphiles are arranged in a concentric manner such that bonds can form among the arranged molecules, which further extend to the plane's normal direction, resulting in nanofiber morphology. Some biological functional moieties can be introduced in the peptides design, which can ultimately play important biological roles, such as bone mineralization.

14.2.3 *Electrospinning*

Electrospinning is considered a relatively simple and easy process to generate nanofibers with cost-effectiveness and mass producibility. It uses an electric field to spin material dissolved in a solution. When a high electric field is applied, surface charges collect at the tip of a needle and overcome the surface tension of the solution at the tip, becoming a jet and gathering onto a metal collector (Fig. 14.1). During the electrospinning, the solvent evaporates and fibrous material is obtained. The obtained fiber sizes range from tens of nanometers to a few micrometers, depending on the material type and processing variables. Generally the obtained fibrous network is random but can be aligned by modifying processing setups, such as collector parts. Moreover, morphological control, such as the creation of nanopores and the core shell structure, is possible by using porogens and designing a nozzle properly. A range of materials, mainly polymers, have been electrospun into nanofibers for medical uses. With its versatility and easy setup, research on electrospinning is growing rapidly in the tissue regeneration area.

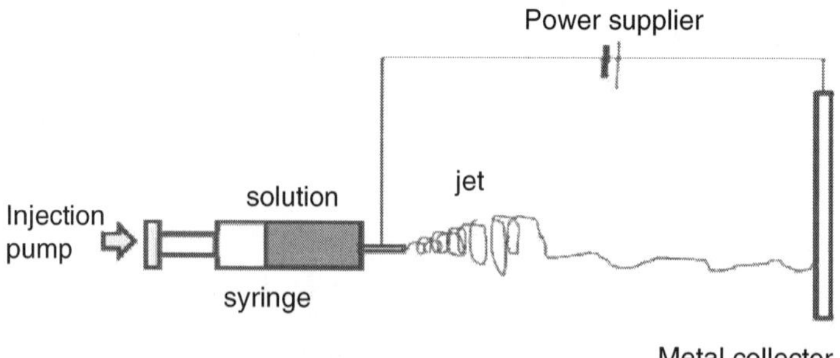

Figure 14.1. Schematic illustration of the electrospinning process to generate nanofibers.

14.3 Nanofibrous Materials for Bone Tissue Engineering

14.3.1 *Biodegradable Polymers*

As a matrix for bone cell culturing and regeneration of hard tissues, a range of biopolymers with synthetic origin have been used, which include poly(α-hydroxyl acid), such as poly(lactic acid) (PLA), poly(glycolic acid) (PGA), and poly(ε-caprolactone) (PCL), and poly(hydroxyalkenoate), such as poly(hydroxybutyrate) (PHB) and poly(hydroxybutyrate-co-hydroxyvalerate) (PHBV). Moreover, natural polymers such as collagen, gelatin, silk, and chitosan are also considered attractive material choices for bone regeneration.

For the electrospinning of synthetic polymers, organic solvents such as chloroform, dichloromethane, and tetrahydrofuran have been used, whilst a water-based solvent system is generally required to dissolve natural polymers. Otherwise, fluorinated solvents, such as hexafluoropropanol and trifluoroethanol, can be introduced to dissolve collagen and gelatin, as well as their mixtures with synthetic polymers.

PCL has long been studied as a degradable nanofiber matrix for bone regeneration.[11] Rat bone marrow stem cells (rBMSCs) were demonstrated to develop a bone matrix after 4 weeks of culturing on PCL nanofibers, such as secretion of type I collagen and calcium mineralization. PLA nanofibers also showed good responses to bone cells (MC3T3-E1).[12] Shin *et al.* investigated that PCL nanofibers well supported the growth of cells and bonelike tissue formation *in vivo* when the cell–nanofiber construct was implanted in the rat for 4 weeks, suggesting the potential of PCL nanofibers for bone tissue engineering.[13] PHB and PHBV electrtospun fibers were shown to have higher osteoblastic cell growth level than their equivalent flat films, while maintaining osteoblastic phenotypes.[14]

Although synthetic polymers have shown the potential as a cell and tissue matrix, their hydrophobic nature needs to be overcome to find extended uses in the tissue regeneration area. One simple approach is to blend the synthetic polymers with natural polymers, as the natural components will increase hydrophilicity and initial cellular events.[15,16] PCL was blended with gelatin in a co-solvent and electrospun into a nanofibrous matrix without the formation of beads. Compared with pure PCL, wherein cell growth was limited, the blended nanofiber showed better penetration of cells within the nanofiber matrix.[15] PLA was mixed with gelatin at various ratios (1:3, 1:1, and 3:1) and electrospun into nanofibers.[16] The blended nanofiber improved the water affinity of the matrix with respect to the PLA nanofiber. Osteoblastic cells (MC3T3-E1) were shown to have better proliferation behavior and expression of genes related with bone on the PLA–gelatin than on pure PLA. Along with biological functions, mechanical improvement was also noticed in the blending of collagen with PCL.[17] The collagen-added PCL presented a higher elongation rate than PCL, while preserving tensile strength, suggesting the role of collagen in the mechanical function. When implanted in a subcutaneous region, tissue cells penetrated actively into the pores of the collagen–PCL nanofiber, whilst only limited progress was noticed in the PCL nanofiber.[17]

Along with synthetic polymers, natural polymers, including collagen, gelatin, chitosan, and silk, have been sources for the bone regenerative nanofiber matrix. Among these, collagen is the one that has gained the most considerable attention.[18–20] Collagen (type I), as the major organic component of tissue, constitutes the main bone ECM with a nanofibrillar structure and performs mechanical and biological functions. Collagen nanofibers have been electospun with various diameters, which were well in the regime of native collagen fibrils. Electrospun collagen (type I) has shown to provide an active substrate condition for BMSCs to adhere and migrate.[19] However, organic solvents such as fluoroalcohols are generally used to dissolve collagen for electrospinning. In this course, the biological activity of collagen can be degraded. As demonstrated by Jeugolis *et al.*,[20] the use of fluoroalcohols as a solvent is detrimental to preserving native collagen characteristics, due to the degradation of the structure and possible biological functionality. Furthermore, crosslinking is generally needed after electrospinning in order to provide the nanofiber matrix with appropriate structural stability and chemical integrity to prevent premature degradation in a fluid. Using a water-based solvent, gelatin was successfully produced by introducing ethyl acetate as a co-dissolving solvent. Ethyl acetate significantly reduced the surface tension of water and, consequently, favored the jet spinning into the nanofiber form.[21]

The chitosan nanofiber has also been studied as a bone regenerative matrix. As a degradable and nontoxic polysaccharide biopolymer, chitosan has been developed into nanofibers by electrospinning. In particular, a chitosan solution should be prepared in acidic solvents. Shin *et al.* produced the chitosan nanofiber for use as a guided bone regeneration membrane in dentistry. The electrospun membrane showed good osteoblast responses, such as bone-associated gene expressions, and formed new bone in a rabbit calvarium defect after 4 weeks of implantation.[22] The silk fibroin nanofiber has also been developed as a bone cell matrix.[23] Silk fibroin is known to be cell compatible and biodegradable and has minimal inflammatory reaction.[23] Silk nanofibers electrospun with sizes of $\sim$500 nm to 1 μm were shown to guide the adhesion and growth of BMSCs[23] and osteoblasts.[24]

14.3.2 *Bioactive Inorganics*

Electrospinning of inorganics with various compositions has been well documented in many areas. For the tissue regeneration area, biocompatible compositions should first be considered. As one criterion required for bone regenerative materials, the "bone bioactivity" is considered, which generally means the ability to form a direct bond with native bone tissue through an interface containing a bone mineral–like phase. At the bone-bioactive interface, a series of biological reactions involved in the bone formation are significantly favored and stimulated. Currently available bone-bioactive inorganics include calcium phosphates and bioactive glasses/glass ceramics, and these are the fascinating choices of materials for electrospinning nanofibers. Although there have been significant levels of work on inorganic nanofibers, very little has been documented on inorganics with bone-bioactive compositions.

One challenging work has recently been performed on the generation of bioactive glass nanofibers by electrospinning.[25] Silica-based sol-gel glass with a bioactive composition ($70SiO_2 \cdot 25CaP \cdot 5P_2O_5$) was successfully produced with the help of a poly(vinyl butyral) binder in an ethanol-water-based solvent. Following heat treatment at 700°C, a well-developed nanofibrous web maintaining the initial morphology was produced. The fiber size ranged from tens to hundreds of nanometers as the sol concentration was varied (Fig. 14.2). When rBMSCs were cultured on the bioactive glass nanofiber, they favorably migrated over and secreted an osteoblastic differentiation phenotype. The differentiation level was significantly higher on the nanofiber form of bioactive glass than either on the dense sintered bioactive glass or on the nanofiber form of the

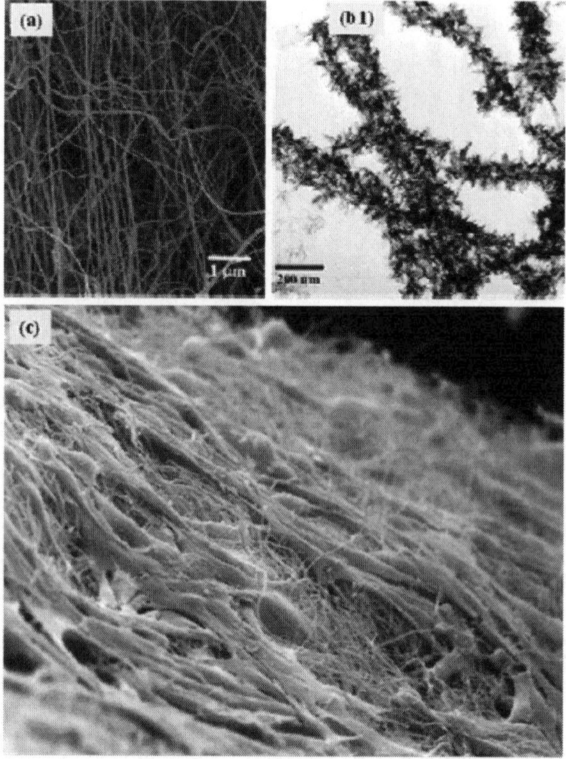

Figure 14.2. Bioactive inorganic nanofibers obtained by electrospinning: (a) SEM image of bioactive glass nanofibers produced by electrospinning and heat treatment; (b) TEM image of acellular bone bioactivity of nanofibers, showing the formation of bone mineral–like apatite on the surface after soaking in SBF for 3 days (adapted with permission from Ref. 25; copyright 2006 Wiley-VCH Verlag GmbH & Co.); (c) Rat BMSCs grown on bioactive glass nanofibers for 7 days showing cells were viable with active cytoplasmic extensions in contact with the underlying nanofibrous substrate.

PCL polymer, highlighting the merits of bioactive glass nanofibers in terms of both morphology and composition. Similar processing steps, involving the use of a sol-gel solution, mixing with polymer binder, and thermal treatment, have also been applied to produce bioactive inorganic nanofibers with different compositions, such as hydroxyapatite,[26–28] fluorohydroxyapatite,[26] and silica.[29]

Although bioactive inorganics may be a good source for stimulating bone cell differentiation, matrix formation and mineralization, poor mechanical strength, namely, brittleness, has restricted their use as matrices for bone regeneration. Some recent works have utilized inorganic nanofibers as a bioactive component for nanocomposites with polymers.[30,31] Nanofibrous bioactive glass added to a collagen solution was produced to a nanocomposite, wherein collagen fibrils and inorganic nanocomponents were organized on the nano scale (Fig. 14.3). Synthetic polymers, such as PLA and PCL,

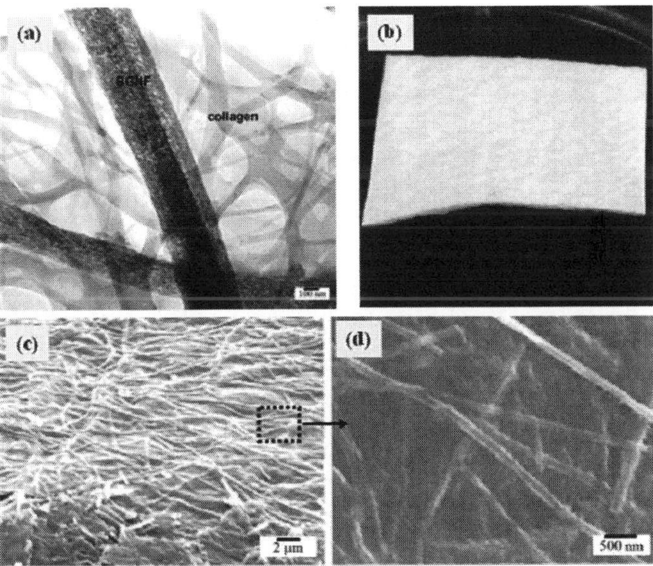

Figure 14.3. An electrospun bioactive glass nanofiber was used as an inorganic nanofiller for the production of nanocomposites with collagen: (a) TEM image showing the glass nanofiber (BGNF) and collagen fibers well organized. (b) Composite produced into a thin membrane. (c) SEM image of the nanocomposite showing the microstructure (adapted with permission from Ref. 30; copyright 2007 Wiley Interscience Co.).

were also improved in terms of bioactivity and osteoblast responses with the incorporation of small amounts of the bioactive nanocomponent. One chief merit of the nanocomponent is that complex-shaped composites, such as 3D scaffolds, are possibly produced without disintegration of the morphology, which consequently leads to improvement in the bioactivity and mechanical properties.

14.3.3 *Composite Nanofibers*

Bone is a kind of nanocomposite, consisting of a collageneous fiber network and embedded calcium phosphate (mostly hydroxyapatite) nanocrystallites. Thus combining the polymeric phase with bioactive inorganics is considered an appropriate way to better mimic the composition and structure of the bone matrix.[32] For successful electrospinning of polymer-inorganic composites, the properties of the solution should be carefully tailored and considered.

Recent studies have focused on the preparation of composite nanofibers by electrospinning. One elegant approach was made in the gelatin–hydroxyapatite composite.[33] Hydroxyapatite crystals were first prepared in the presence of gelatin by the precipitation of calcium and phosphorus precursors, resulting in a nanocomposite precipitate. Using transmission electron microscopy, bone mineral–like nanocrystallites were observed to be well dispersed and organized, which was facilitated by the nucleation role of gelatin amino acids (Fig. 14.4). After dissolving the nanocomposite precipitate in a solvent, it was possible to electrospin nanofibers containing 20% and 40% hydroxyapatite with fiber sizes of a few hundreds of nanometers. Compared with the gelatin nanofiber, the nanocomposite nanofibers showed better osteoblastic differentiation behavior, while preserving mechanical integrity. Using a similar approach, a hydroxyapatite–collagen nanofiber, which was to mimic the bone ECM, was also produced.[34] Moreover, a chitosan–hydroxyapatite nanofiber was developed as a functional bone matrix.[35]

Bioactive inorganics were also added to some synthetic polymers during the electrospinning.[36,37] In this case, however, the innate hydrophobicity of polymers hampered the involvement of

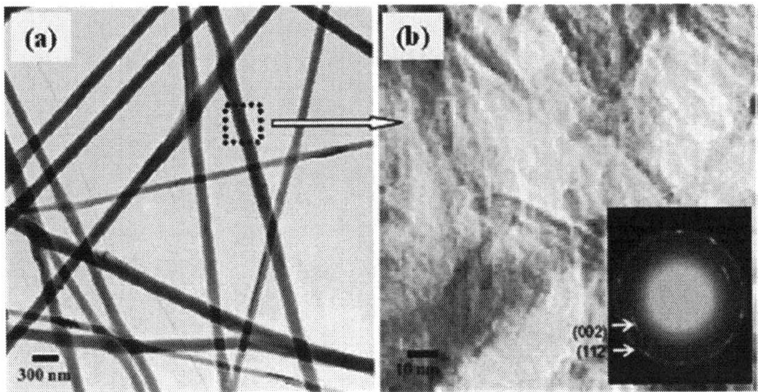

Figure 14.4. Bone mimetic gelatin–20% apatite nanofiber. TEM investigation of the organization of hydroxyapatite in the gelatin matrix (a hydroxyapatite crystalline pattern was revealed in the inset) (adapted with permission from Ref. 33; 2005 Wiley-VCH Verlag GmbH & Co.).

hydrophilic inorganic components significantly. The common solvents for dissolving synthetic polymers were also not effective for the dispersion of inorganic nanoparticles. Using ultrafine-sized $CaCO_3$ particles ($\sim$40 nm in size), PCL–$CaCO_3$ composite fibers could be produced.[36] The composite fibers showed favorable osteoblastic adhesion and growth and were suggested as a guided bone regeneration membrane. Our recent work has used a surfactant (fatty acid) to homogenize the hydroxyapatite nanoparticles that were obtained by a sol-gel method within a PLA–chloroform solution.[37] The amphiphilic nature of the surfactant was suggested to act as a mediator at the interface of the hydroxyapatite nanocrystallites and oil phase (PLA–chloroform). Fibers of a few micrometers in size were produced without the formation of beads, and the hydroxyapatite nanocrystallites were observed to be dispersed well within the PLA matrix. The composite fiber showed to enhance the growth and differentiation of osteoblastic cells with respect to pure PLA fibers. Conclusively, the nanofibers made from polymer-inorganic composites are believed to stimulate the biological events of bone-associated cells, such as cell adhesion and osteogenic differentiation. To realize this, homogenous dispersion of inorganics within polymeric solution is considered of special importance, which leads to obtaining a bead-free nanofibrous matrix.

14.4 Functionalization of Nanofibers for Bone Tissue Engineering

14.4.1 *Surface Modifications*

To enhance the biological responses of nanofibers as a matrix for bone tissue regeneration, the nanofiber surface is often tailored with a bone-mimetic composition or biological therapeutics. Given that the cells recognize initially the subsurface of a material, functionalizing the surface to trigger initial cell adhesion and to stimulate synthesis of bone ECMs and mineralization is considered a promising approach.

Among surface functionalization methods, the modification of the polymeric nanofiber surface with a bone mineral–like phase, mainly hydroxyapatite, has recently been exploited.[38–40] The PCL nanofiber matrix was activated with an alkaline solution, presoaked in calcium and phosphorus solutions and then immersed in a body-simulating medium.[38] A hydroxyapatite mineral phase was induced on the PCL nanofiber surface within a few days, and it covered the surface almost completely after 7 days (Fig. 14.5). The

Figure 14.5. Hydroxyapatite mineralized on the surface of a PCL nanofiber for bone regeneration use (adapted with permission from Ref. 38; copyright 2009 Wiley Interscience Co.).

apatite-mineralized PCL showed active growth of cells and upregulated osteoblast differentiation with respect to an untreated PCL nanofiber. A series of bone-related genes, including Runx2, collagen I, alkaline phosphatase, and osteocalcin, were stimulated on the mineralized PCL nanofiber. Moreover, periodontal ligament fibroblasts were also triggered to secrete bone cell phenotypes.[41] On the basis of these studies, the mineralized surface is believed to have the potential of stimulating bone cell activity. The surface of PLA nanofibers was also tailored with a mineral phase, wherein calcium was incorporated to speed up the mineral induction.[42]

The mineralized surface may find future applications in bone regeneration with the coupling of biological molecules or drugs to elicit therapeutic effects. Our recent study developed a biofunctional nanofiber surface, wherein a bioactive glass nanofiber was used as a substrate to induce a hydroxyapatite mineral phase in the coupling of fibronectin to target enhanced initial cellular events.[43] Fibronectin, as the major cell-adhesive protein, was effectively incorporated within the hydroxyapatite mineral, and the system demonstrated significant improvement in the initial osteoblast adhesion and spreading. Hydroxyapatite, as the major bone mineral phase, is known to have strong affinity to a series of proteins and biomolecules that are specific to bone, such as osteocalcin, osteopontin, and bone sialoprotein. Therefore, the functionalization of the nanofiber surface with hydroxyapatite and its subsequent coupling with bone-specific proteins are considered promising to enhance and target specific responses of cell and tissue.[44,45]

Along with the apatite mineral phase, biomolecules such as proteins and growth factors were introduced on the surface of the polymer nanofibers. Conjugating methods based on physical adsorption or chemical linkage have been introduced to endow the surface with biofunctionality. In this case, the preservation of stability of the biomolecules should be considered. Recent studies have shown the use of RGD (Arg-Gly-Asp) peptides in modifying the polymer nanofiber, wherein the polymer surface was linked with peptides through covalent bonds.[46,47] Results showed better cell responses at the initial stages, including cell adhesion, spreading, and growth, demonstrating the role of a cell-adhesive ligand that was tailored on the surface. Although most reported the use of peptides or proteins that dominate initial cellular events, future works are expected to follow on utilizing bone-specific biomolecules, such as growth factors and bone morphogenetic proteins. Depending on the types of biomolecules, the coupling method should be selected appropriately to maintain the biological activity of the molecules.

14.4.2 *Incorporation of Biomolecules*

To secure biological stability and therapeutic effects, biomolecules are often introduced within the nanofiber matrix. In this case, however, electrospinning conditions, such as solvent type, should be carefully considered in order not to degrade the therapeutic agents. A range of biomolecules have been incorporated within polymer nanofibers, including antibiotics, proteins, and genes.[48,49] Li *et al.* used silk and its blending with polyethylene oxide to incorporate bone morphogetic protein (BMP-2).[48] The mesenchymal stem cells were shown to have higher osteogenic differentiation and calcification on the BMP–nanofiber matrix. Nie *et al.* reported a gene delivery system of nanofibers, wherein BMP-2 plasmid DNA was incorporated into PLGA–hydroxyapatite composite nanofibers.[49] Instead of mixing genes directly with the nanofiber solution, DNA was preloaded within chitosan and then electrospun with the composite solution. Tansfection efficiency was enhanced on the nanofiber containing DNA–chitosan with respect to the control nanofiber group.

To entrap biomolecules within the nanofiber, a core-shell structure nanofiber has often been suggested. A dual-syringe apparatus is designed to separate the injection lines of two solutions either

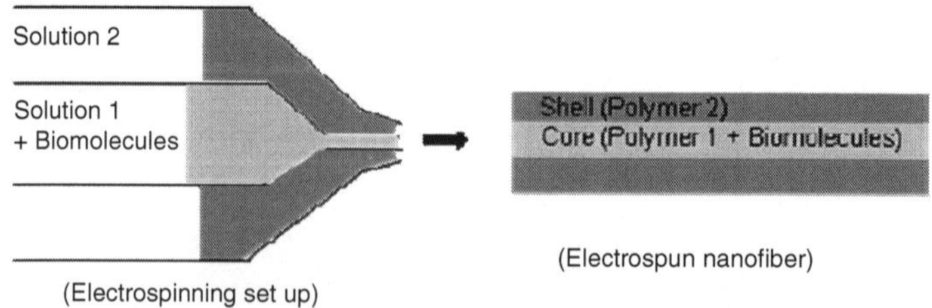

Figure 14.6. Electrospinning nozzle setup for preparing a core-shell nanofiber structure. Biomolecules incorporated within core polymer 1 will be released through the outer shell of polymer 2.

containing or not containing biomolecules (Fig. 14.6).[50,51] While the core solution secures drug-loading efficiency and stability, the shell polymer layer controls the release profile of the biomolecules. To gain optimal delivery potential, the selection of materials and solutions for the core and shell parts should be considered.

14.5 Concluding Remarks

Nanofibrous-structured materials hold great potential for the regeneration of tissues, including bone. Electrospinning has become a facile tool for creating nanofibers from a variety of compositions, such as degradable polymers and bioactive inorganics. In terms of mechanical and biological aspects, the composite approach is considered an appropriate way of producing a matrix that mimics the native bone ECM. Moreover, surface functionalization with bone mineral or bioactive molecules improves biological reactions at the interface, stimulating cell adhesion, growth, and matrix formation. Bone-targeting proteins and genes are currently introduced within the nanofiber matrix to elicit therapeutic effects that ultimately control cellular events and bone formation. With new ideas on electrospinning and the development of compositions qualified for an artificial bone ECM, bone tissue engineering using nanofibrous scaffolds may be realized in the near future.

Acknowledgments

This work was supported by the Priority Research Centers Program (grant#: 2009-0093829) and WCU (World Class University) program (grant#: R31-10069) through the National Research Foundation (NRF) funded by the Ministry of Education, Science and Technology.

References

1. M. M. Stevens, *Materials Today*, **11**, 18 (2008).
2. D. W. Hutmacher, *Biomaterials*, **21**, 2529 (2000).

3. L. A. Smith and P. X. Ma, *Coll. Surf. B. Biointerfaces*, **39**, 125 (2004).
4. T. J. Sill and H. A. Recum, *Biomaterials*, **29**, 1989 (2008).
5. C. P. Barnes, S. A. Sell, E. D. Boland, D. G. Simpson, and G. L. Bowlin, *Adv. Drug Del. Rev.*, **59**, 1413 (2007).
6. D. Liang, B. S. Hsiao, and B. Chu, *Adv. Drug Del. Rev.*, **59**, 1392 (2007).
7. Q. P. Pham, U. Sharma, and A.G. Mikos, *Tissue Eng.*, **12**, 1197 (2006).
8. Z. Ma, M. Kotaki, R. Inai, and S. Ramakrishna, *Tissue Eng.*, **11**, 101 (2005).
9. P. X. Ma and R. Zhang, *J. Biomed. Mater. Res.*, **46**, 60 (1999).
10. J. D. Hartgerink, E. Beniash, and S. I. Stupp, *Science*, **294**, 1684 (2001).
11. H. Yoshimoto, Y. M. Shin, H. Terai, and J. P.Vacanti, *Biomaterials*, **24**, 2077 (2003).
12. A. S. Badami, M. R. Kreke, M. S. Thompson, J. S. Riffle, and A. S. Goldstein, *Biomaterials*, **27**, 596 (2006).
13. M. Shin, H. Yoshimoto, and J. P. Vacanti, *Tissie Eng.*, **10**, 33 (2004).
14. K. Sombatmankhong, N. Sanchavanakit, P. Pavasant, and P. Supaphol, *Polymer*, **48**, 1419 (2007).
15. Y. Zhang, H. Ouyang, C. T. Lim, S. Ramakrishna, and Z. M. Huang, *J. Biomed. Mater. Res. Part B: Appl. Biomater.*, **72**, 156 (2004).
16. H. W. Kim, H. S. Yu, and H. H. Lee, *J. Biomed. Mater. Res. Part A*, **87A**, 25 (2007).
17. J. J. Lee, H. S. Yu, S. J. Hong, I. Jeong, J. H. Jang, and H. W. Kim, *J. Mater. Sci. Mater. Med.*, (2009).
18. J. A. Matthews, G. E. Wnek, D. G. Simpson, and G. L. Bowlin, *Biomacromolecules*, **3**, 232 (2002).
19. Y. R. V. Shin, C. N. Chen, S. W. Tsai, Y. J. Wang, and O. K. Lee, *Stem Cells*, **24**, 2391 (2006).
20. D. I. Zeugolis, S. T. Khew, E. S. Y. Yew, A. K. Ekaputra, Y. W. Tong, L. L. Yung, D. W. Hutmacher, C. Sheppard, and M. Raghunath, *Biomaterials*, **29**, 2293 (2008).
21. J. H. Song, H. E. Kim, and H. W. Kim, *J. Mater. Sci. Mater. Med.*, **19**, 95 (2008).
22. S. Y. Shin, H. N. Park, K. H. Kim, M. H. Lee, Y. S. Choi, Y. J. Park, Y. M. Lee, I. C. Rhyu, S. B. Han, S. J. Lee, and C. P. Chung, *J. Periodontol.*, **76** ,1778 (2005).
23. H. J. Jin, J. S. Chen, V. Karageorgiou, G. H. Altman, and D. L. Kaplan, *Biomaterials*, **25**, 1039 (2004).

24. C. Meechaisue, P. Wutticharoenmongkol, R. Waraput, T. Huangjing, N. Ketbumrung, P. Pavasant, and P. Supaphol, *Biomed. Mater.*, **2**, 181 (2007).
25. H. W. Kim, H. E. Kim, and J. C. Knowles, *Adv. Funct. Mater.*, **16**, 1529 (2006).
26. H. W. Kim and H. E. Kim, *J. Biomed. Mater. Res. Part B: Appl. Biomat.*, **22B**, 323 (2005).
27. Y. Wu, L. L. Hench, J. Du, K. L. Choy, and J. Guo, *J. Amer. Ceram. Soc.*, **87**, 1988 (2004).
28. X. Dai and S. Shivkumar, *J. Amer. Ceram. Soc.*, **90**, 1412 (2007).
29. S. Sakai, Y. Yamada, T. Yamaguchi, and K. Kawakami, *Biotechnol. J.*, **1**, 958 (2006).
30. H. W. Kim, J. H. Song, and H. E. Kim, *J. Biomed. Mater. Res. Part A*, **79A**, 698 (2006).
31. H. W. Kim, H. H. Lee, and G. S. Chun, *J. Biomed. Mater. Res. Part A*, **85A**, 651 (2008).
32. M. J. Olszta, X. Cheng, S. S. Jee, R. Kumar, Y. Y. Kim, M. J. Kaufman, E. P. Douglas, and L. B. Gower, *Mater. Sci. Eng. R.*, **58**, 77 (2007).
33. H. W. Kim, J. H. Song, and H. E. Kim, *Adv. Funct. Mater.*, **15**, 1988 (2005).
34. J. H. Song, H. E. Kim, and H. W. Kim, *J. Mater. Sci.: Mater. Med.*, **19**, 2925 (2008).
35. Y. Zhang, J. R. Venugopal, A. El-Turki, S. Ramakrishna, B. Su, and C. T. Lim, *Biomaterials*, **29**, 4314 (2008).
36. K. Fujihara, M. Kotaki, and S. Ramakrishna, *Biomaterials*, **26**, 4139 (2005).
37. H. W. Kim, H. H. Lee, and J. C. Knowles, *J. Biomed. Mater. Res. Part A*, **79**, 643 (2006).
38. H. S. Yu, J. H. Jang, T. I. Kim, H. H. Lee, and H. W. Kim, *J. Biomed. Mater. Res. Part A*, **88**, 747 (2009).
39. S. H. Park, T. I. Kim, Y. Ku, C. P. Ching, S. B. Han, J. H. Yu, S. P. Lee, H. W. Kim, and H. H. Lee, *J. Ceram. Soc. Jap.*, **116**, 31 (2008).
40. J. Chen, B. Chu, and B. S. Hsiao, *J. Biomed. Mater. Res. Part A*, **79**, 307 (2006).
41. S. H. Park, T. I. Kim, Y. Ku, C. P. Chung, S. B. Han, J. H. Yu, S. P. Lee, H. W. Kim and H. H. Lee, *J. Ceram. Soc. Jap.*, **116**, 31 (2008).
42. W. Cui, X. Li, S. Zhou, and J. Weng, *J. Biomed. Mater. Res. Part A*, **82A**, 831 (2007).

43. H. W. Kim, H. H. Lee, and J. C. Knowles, *J. Nanosc. Nanotech.*, **8**, 3013 (2008).
44. Q. Q. Hoang, F. Sicheri, A. J. Howard, and D. S. Yang, *Nature*, **425**, 977 (2003).
45. D. Wang, S. C. Miller, P. Kopečková, and J. Kopeček, *Adv. Drug Del. Rev.*, **57**, 1049 (2005).
46. T. G. Kim and T. G. Park, *Tissue Eng.*, **12**, 221 (2006).
47. J. F. Alvarez-barreto, M. C. Shreve, P. L. Deanqelis, and V. I. Sikavitsas, *Tissue Eng.*, **13**, 1205, (2007).
48. C. Li, C. Vepari, H. J. Jin, H. J. Kim, and D. L. Kaplan, *Biomaterials*, **27**, 3115 (2006).
49. H. Nie and C. H. Wang, *J. Control Rel.*, **120**, 111 (2007).
50. Z. C. Sun, E. Zussman, A. L. Yarin and J. H. Wendorff, *Adv. Mater.*, **15**, 1929 (2003).
51. H. Jiang, Y. Hu, P. Zhao, Y. Li, and K. Zhu, *J. Biomed. Mater. Res. Part B: App. Biomater.*, **79B**, 50 (2006).

Chapter 15

STRATEGIES TO ENGINEER ELECTROSPUN SCAFFOLD ARCHITECTURE AND FUNCTION

Aaron S. Goldstein,[a*] Christopher A. Bashur,[a] and Joel Berry[b]

[a] *Department of Chemical Engineering, Virginia Polytechnic Institute and State University, Blacksburg, VA 24061, USA*

[b] *Department of Biomedical Engineering, University of Alabama at Birmingham, Birmingham, AL 35294, USA*

*aaron.goldstein@vt.edu

15.1 Overview

With its initial application for the processing of biomaterials a decade ago, electrospinning has rapidly become a promising technology for the generation of scaffolds for tissue engineering applications, including bone,[1] ligament,[2] blood vessel,[3] peripheral nerve,[4] skin,[5] cartilage,[6] muscle,[7] heart, and heart valve.[8] The discovery of electrospinning as a phenomenon dates back at least 100 years,[9] but its rediscovery a decade ago coincided with the emergence of nanotechnology and tissue engineering. Specifically, the diameters of electrospun fibers (typically 0.1 to 5 μm)—which are comparable in size to large collagen fibrils (50–300 nm) and can guide cell

Handbook of Intelligent Scaffolds for Tissue Engineering and Regenerative Medicine
Edited by Gilson Khang

www.panstanford.com

attachment, spreading, and function[10]—have motivated efforts to apply this technology to cell- and tissue-contacting medical devices.

Despite this early promise, electrospun materials have not yet been clinically approved for medical applications. Ironically, their limitations as medical devices may be linked to their advantages as nanomaterials. Because they are comprised of submicron fibers, electrospun meshes are extremely soft, their tensile strength is dictated by the strength of fiber cross-links, the pores in meshes are on the order of a few microns, and the rate of mesh formation is on the order of several microns (in thickness) per minute. Consequently, electrospinning—as a technology—must be married to other technologies to achieve scaffold architectures that are tailored toward specific tissue applications.

This chapter examines some of the strategies that have been described in the literature to tailor the architecture of electrospun scaffolds—at the nanometer-, micron-, and millimeter-length scales—for specific applications. At the length scale of tens of nanometers to a few microns, fiber features create surface topographies that affect the attachment, orientation, and function of anchorage-dependent mammalian cells. At the length scale of microns, the porosity of the electrospun scaffolds affects the ability for cells to migrate, deposit an extracellular matrix, and self-assemble into tissue-like structures. Finally, at the millimeter-length scale, the electrospun scaffold must exhibit the shape—and ideally the mechanical properties—of the tissue it is designed to regenerate.

The organization of this review is built around key features of electrospun meshes at the nanometer-, micron-, and millimeter-length scales and highlights strategies that have been employed for tissue engineering applications. It concludes with considerations of the mechanical properties of electrospun structures and how they depend on the scaffold architecture.

15.2 Design of Fiber Topography to Affect Cell Function

Viability, proliferation, and phenotypic behavior of anchorage-dependent cells depend heavily on the micron-scale physical features of the biomaterial surface, which may be tuned through control of the electrospinning conditions. Specific features that are considered in this section are 1) the degree of fiber alignment, 2) the diameter of individual fibers, and 3) the roughness of fibers.

15.2.1 *Effect of Electrospun Fiber Alignment on Cell Morphology*

Simple electrospinning onto a stationary target produces a random distribution of fiber orientations, while the deposition onto a rotating drum,[11] rotating disc,[12] or between two grounded surfaces[13] can result in moderate to very high degrees of fiber alignment. Although seemingly similar, mammalian cells respond differently to these surfaces. For example, the projected cell area of osteoblasts[1] and fibroblasts[11] is diminished on random electrospun meshes. In contrast, when the fibers are aligned these cells selectively spread parallel to the direction of fiber alignment (Fig. 15.1). Similarly, embryonic hippocampal neurons and dorsal root ganglia (DRG) have been reported to extend axons[14] and neurites,[15] respectively, along fibers. Further, when these fibers are aligned a higher percentage of embryonic neurons exhibit polarization, while the lengths of DRG neurons is increased.

Two theories can explain these two effects of fibers on cell morphology. The first is related to the phenomenon of contact

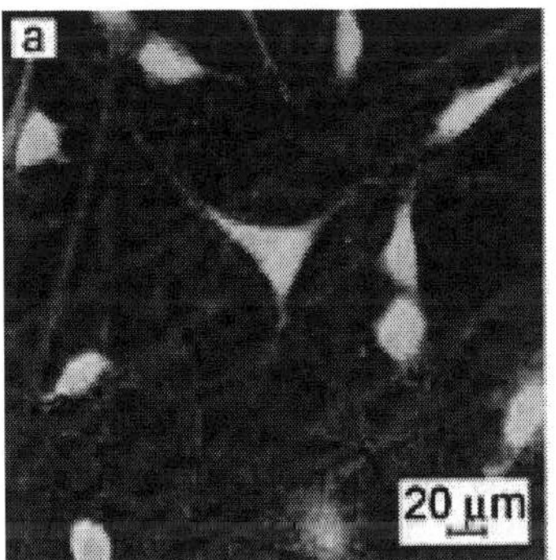

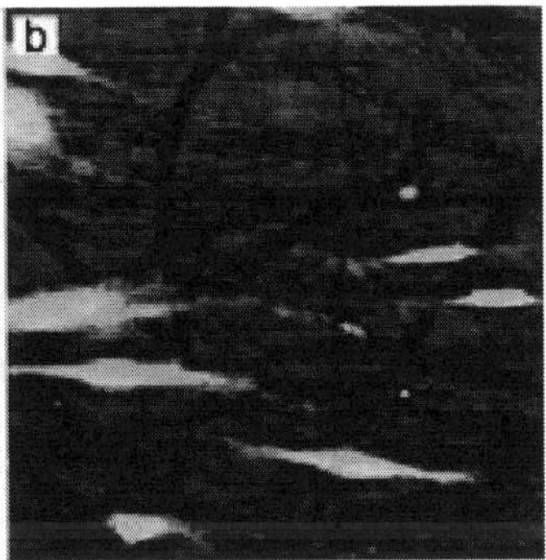

Figure 15.1. Cells on (a) random and (b) horizontally aligned fibers. The fiber alignment guides cell alignment.

guidance.[10] Because cells establish contacts along the elevated features of fiber substrates,[1] the diminished spreading on randomly electrospun meshes may be a consequence of the spatially limited adhesive features. In contrast, when fibers are aligned they guide adherent cells to extend lamellipodia outward along parallel fibers, resulting in cell alignment parallel to fiber alignment. The second explanation is related to the mechanical properties of the meshes. Because spreading of mammalian cells is reduced on less rigid materials,[16,17] and electrospun substrates can be more than an order of magnitude more compliant than the bulk polymer,[18] the projected cell area on random electrospun meshes should be less than that on cast films of the same polymer. However, when the fibers are aligned the mechanical properties of the electrospun meshes are anisotropic and stiffer in the direction of fiber alignment.[18–20] Consequently, cells may preferentially align parallel to the more rigid axis. A similar effect has been observed when cells are seeded into anisotropic gels that are uniaxially confined: the cells will align parallel to the confined (more rigid) axis.[21]

15.2.2 *Effect of Fiber Diameter on Cell Morphology*

Cell behavior can also be influenced by the diameter of electrospun fibers, and this can be varied by adjusting a number of electrospinning parameters, including the concentration of the polymer solution, the voltage potential, and needle diameter (see Subbiah *et al.*[22] for a review). The effect of fiber diameter on cell morphology appears to vary with cell type. For example, both the projected area and the aspect ratio of adherent fibroblasts are increased with increases in fiber diameter.[23] In contrast, neural stem cells have been shown to extend significantly longer neurites on aligned meshes of 300 nm fibers than on aligned meshes of 1.5 μm.[24] The effect of fiber diameter on cell adhesion and proliferation is ambiguous. Some data has suggested that larger fibers support modestly better cell adhesion and proliferation,[1] while other data suggests the opposite.[11,25] (The latter, however, may be a consequence of residual solvent in the fibers.)

15.2.3 *Effect of Fiber Roughness*

The incorporation of roughness/texture into the surface of electrospun fibers may be an attractive means to effect biological responses. For example, osteoblasts are very sensitive to nanotopography, and model studies on planar surfaces have shown that roughness affects cell shape,[26] cell attachment,[27,28] and expression of the mRNA encoding the proteins alkaline phosphatase, type I collagen, and the osteogenic transcription factor RUNX2.[29] One topographical feature that can be readily introduced into the surface of electrospun fibers is nanopores. These nanopores, on the order of 50–500 nm, can be introduced and systematically varied in size by adjusting electrospinning conditions, such as solvent volatility, solution concentration, and relative humidity.[30,31] Roughness can also be incorporated into the fiber surface by the addition of small particulates (e.g., ceramics, carbon nanotubes). Recently, Gupta *et al.*[32] compared two approaches for incorporating hydroxyapatite (HA) nanoparticles into electrospun fibers: combining them into an electrospinning solution (of poly[L-lactic acid-co-caprolactone] and gelatin) and electrospraying the HA from a separate syringe. They found that the latter approach increased the strain-to-failure of the meshes and enhanced alkaline phosphatase activity of adherent osteoblasts. Mei *et al.* examined the effect of combining HA and multiwall carbon nanotubes (MWNTs) into electrospun scaffolds.[33] In particular, they reported that HA and MWNTs together resulted in better cell proliferation of periodontal ligament cells but poorer proliferation of gingival epithelial cells. This result suggests that the combination of HA and MWNTs may be attractive for guided regeneration of the periodontium. Finally, roughness can be incorporated into the fiber surface by the controlled growth of a surface coating on the electrospun fibers. Examples of this include the growth of HA by immersion of electrospun meshes in a calcium- and phosphate-rich solution[34,35] and the growth of polypyrrole by immersion of meshes in a mixture of pyrrole, ferric chloride, and toluene sulphate.[36] In these cases the primary goal was to modify the surface chemistry. Nevertheless, the roughness of the resultant surfaces—which can affect cell response—was closely related to the growth rate.

15.3 Creation of Larger Pores to Facilitate Cell Entry into Scaffolds

One shortcoming of electrospun meshes is that cells seeded on their surfaces do not readily migrate into the interior.[37,38] This limitation has been attributed to the very small gaps between fibers through which the cells must pass, and while it may be mitigated by increasing the diameter of the electrospun fibers,[38] this can negate intrinsic advantages of using nanofibers over microfibers (e.g., topographic features to guide spreading, cell alignment, and expression of phenotypic markers). Therefore, several alternative strategies have been tested to enlarge the spaces between electrospun fibers in order to better permit cell migration into electrospun meshes, including 1) co-electrospinning, 2) combining electrospun fibers with larger extruded fibers, and 3) incorporating a porogen into scaffolds.

15.3.1 *Co-Electrospinning of a Sacrificial Polymer*

Co-electrospinning of two polymers from separate spinnerets has been examined by many groups as a means to increase cell penetration. Typically, one polymer is water soluble (e.g., gelatin, polyethylene oxide [PEO]) and is intended to dissolve away to reveal larger spaces between remaining fibers. Results with this approach have been modest. Baker *et al.*[37] using PEO to form sacrificial fibers found good cell infiltration into 0.8 mm thick scaffolds when the PEO content was greater than 50 wt% but showed that the resultant scaffolds could collapse in cell culture. In contrast, Ekaputra *et al.*[38] reported only modest improvements in cell penetration with co-electrospinning of PEO or gelatin sacrificial fibers. Indeed, they observed better cell penetration when they neglected the sacrificial polymer and simply increased the diameter of their fibers from 0.5 to 1.3 μm.

15.3.2 *Incorporation of Extruded Fibers*

The combination of large fibers with electrospun nanofibers also has been examined as a means to increase pore size and cell penetration into electrospun scaffolds. Pham *et al.*[39] combined electrospun polycaprolactone (PCL) nanofibers with electrospun PCL microfibers to form composite scaffolds. In this study the diameter of the microfibers could be tuned between 2 and 10 microns, while the nanofibers were 600 nm in diameter. The authors showed histological cross sections of the scaffolds and quantitatively characterized cell distribution. Unfortunately, they found that cells did not spontaneously migrate into their scaffolds, and the authors had to employ medium perfusion with a bioreactor to achieve a uniform cell distribution. The use of larger microfibers has also been considered. Marins *et al.*[40] constructed scaffolds with alternating layers of 400–1,400 nm electrospun nanofibers and 300-micron melt-extruded microfibers for bone tissue applications. Although they did not explicitly examine cell distribution within their resultant scaffolds, they reported increases in both cell number and alkaline phosphatase activity with the incorporation of nanofibers. Finally, the laboratories of Drs. Gatenholm (Chalmers University, Sweden[41]) and Freeman (Virginia Tech, USA, Fig. 15.2) have explored the possibility of constructing three-dimensional biomaterial scaffolds in which nanofibers are

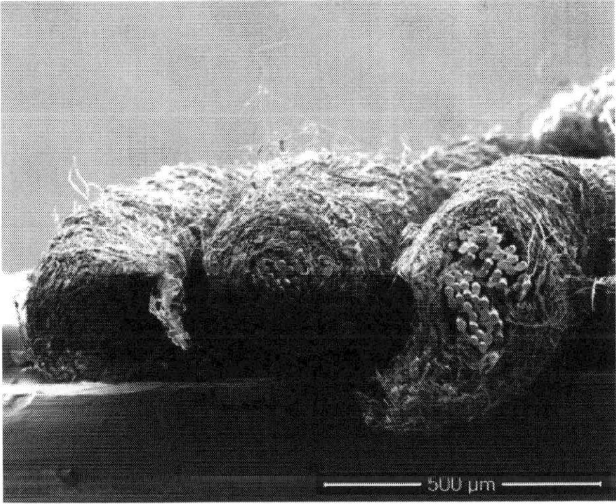

Figure 15.2. Thick layers of nanofibers deposited around bundles of microfibers (courtesy of J. W. Freeman).

electrospun around extruded microfibers. When the electrospun fibers are deposited loosely to facilitate cell migration[41], the microfibers serve to increase the mechanical properties of the scaffold. In contrast, when the fibers are deposited densely (e.g., Fig. 15.2), the microfibers may be subsequently removed to improve cell penetration into the scaffolds.

15.3.3 *Incorporation of Porogens into Electospun Meshes*

Researchers have endeavored to create pores by the incorporation of soluble porogens into the electrospun scaffolds. For example, the laboratories of Dr. Tae Gwan Park (KAIST, South Korea[42]) and Dr. Joseph Freeman (Virginia Tech, USA) have tried to incorporate salt particles into the developing electrospun scaffolds. This approach appears to increase cell density.[42] Recently Leong *et al.*[43] examined the effect of simultaneously growing ice crystals on the madrel while electrospinning. By varying the humidity they could vary the size of ice crystals and consequently the size of macropores. Both their *in vivo* and *in vitro* data indicate cell infiltration throughout 50–150-micron-thick scaffolds. However, the use of porogens to create macropores decreases in the mechanical properties of the resultant scaffolds.[43]

15.3.4 *Chemotaxis and other Considerations Regarding Cell Infiltration into Electrospun Meshes*

It is worth mentioning here that increasing the size of pores and channels in electrospun meshes may not necessarily translate into spontaneous cell infiltration. Chemical cues are often required to induce cells to migrate (i.e., chemotaxis), and in the absence of such stimuli cells may preferentially accumulate at the outer surfaces of biomaterials. It has been demonstrated previously that cells seeded uniformly in foam scaffolds will—under static culture conditions—depopulate the scaffold interior and accumulate at the scaffold surface, but will remain uniformly distributed when culture medium is perfused through the scaffold.[44,45] These findings suggest that cells prefer to grow on the external surfaces of biomaterial scaffolds, presumably because of the better availability of oxygen and nutrients. Consequently, implicit in any effective strategy to increase cell infiltration into electrospun meshes is a method to keep them there. Perfusion culture may be one such method, as it has been shown to maintain uniform cell distributions in macroporous scaffolds[44,45] and improve cell distribution in electrospun scaffolds.[39]

15.3.5 *Electrospraying or Electrospinning of Cells*

As an alternative to porogens and other methods to increase the pore size of electrospun meshes and facilitate migration of mammalian cells into electrospun scaffolds, methods have been developed to incorporate live cells into scaffolds during the electrospinning process. Stankus *et al.* have used both pressurized air to spray suspensions of cells[46] and voltage potentials of 8.5 to 10 kV to electrospray suspensions of cells (in a conventional culture medium)[46,47] during the electrospinning process. Despite the toxicity of the solvent used to electrospin their polymer (hexafluoroisopropanol in this case), Stankus *et al.* showed good distributions of cells throughout the scaffolds as well as evidence of cell proliferation within the scaffolds. Recently, van Aalst *et al.*[48] tested an alternative approach in which cells were electrospun from a poly(vinyl alcohol) (PVA) solution in such a way that they were embedded within electrospun PVA filaments. The cells were subsequently released by dissolution of the water-soluble PVA.

15.4 Creation of Three-Dimensional Architectures for Tissue-Specific Applications

Electrospun scaffolds must be processed into an appropriate three-dimensional form for each tissue engineering applications (i.e. bone, tendon/ligament, and cardiovascular). These three-dimensional structures can be intrinsically complex, but relatively simple architectures have been described that are suitable for cardiovascular, peripheral nerve, and connective tissue applications.

15.4.1 *Processing Techniques for Tube- and Cord-Shaped Structures*

The production of tube-shaped scaffolds has been achieved by two different electrospinning techniques. The first involves electrospinning fibers around a cylindrical mandrel that is subsequently removed. This simple technique—which has been used to prepare scaffolds for vascular[49] and peripheral nerve applications[50]—involves minimal handling and does not require any subsequent processing. However, the electrospinning process may be time intensive, as Li *et al.*[6] have reported that the achievement of meshes of 1 mm in thickness can take several hours. Further, the limited ability for cells to penetration into the resultant mesh (as discussed in the previous section) can also be problematic.

The second technique to produce a tube-shaped structure involves electrospinning a thin flat mesh and then rolling it into a tube.[51,52] This technique involves handling of the electrospun mesh but has the potential to achieve thick electrospun meshes without requiring several hours of electrospinning. Further, Hashi *et al.*[52] used this technique and seeded cells onto the electrospun mesh prior to rolling in order to obtain a tubular vascular graft consisting of multiple layers of electrospun fibers with cells distributed throughout. In principle, this same technique could also be used to produce three-dimensional grafts for connective tissues (e.g., tendon, ligament).

15.4.2 *Variations on the Tube Structure for Blood Vessel and Annulus Fibrosis Applications*

In addition to simple tubes, researchers have sought to create more complex architectures that better reflect the structure of blood vessels. For example, Zhang and Chang describe the electrospinning of fibers onto molds to create structures such as bifurcating blood vessels.[53] In addition, they showed that if patterns of raised domains are placed on the grounded target, then the deposition of fibers is not purely random but includes preferential attachment at discrete points. In addition, efforts have been made to create scaffolds for blood vessels with distinct luminal and adventitial layers by electrospinning different polymers for the separate layers.[54,55] Finally, Li *et al.*[19] showed that scaffolds could be constructed where the direction of fiber orientation systematically rotates by 90°. This micron-scale orientation of fibers mimics the alternating orientation of collagen fibrils in the discrete lamallae that comprise the annulus fibrosus.

15.5 Composite Scaffolds and the Spatial Heterogeneity

Composite scaffolds—consisting of two or more ingredients—are frequently attractive over homogeneous scaffolds because the combination of materials can lead to superior mechanical and biological properties. In addition, this heterogeneity can be controlled at the nano-, micro-, and millimeter-length scales.

The simplest approach for introducing heterogeneity into electrospun scaffolds is to combine various ingredients in a single spinneret and to electrospin this mixture to form composite filaments. For example synthetic and natural polymers have been blended together to improve the mechanical properties,[56] bioactivity,[57,58] and electrospinnability.[59] Various particles and small molecules have also been incorporated into electrospun fibers, including ceramics,[32] carbon nanotubes,[33] pharmaceutics (e.g., rifampin, paclitaxel, and doxorubicin[60,61]), and biologically active proteins.[51]

Additionally, heterogeneity can be incorporated by electrospinning two different solutions from separate spinnerets. If the two spinnerets are placed in close proximity with opposite polarities, the two polymers can attract one another to form an interpenetrating mixture with heterogeneity at the micron-length scale (Fig. 15.3). When the spinnerets are offset[37] or have the same polarity, the polymers deposit around different centers and produce a mesh with a spatial gradient of properties on the order of millimeters. Finally, if one spinneret is placed within the second, then heterogeneous fibers with distinct core and shell chemistries can be fabricated.[62,63] This coaxial electrospinning approach may also be a means to control the biological and mechanical properties of electrospun meshes[63] or to tune the drug release rate.[64]

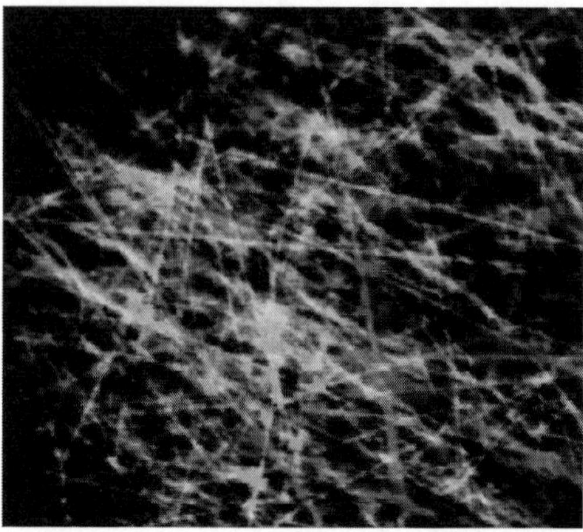

Figure 15.3. Poly(lactic-co-glycolic acid) co-electrospun from two spinnerets. The fluorescent dyes DiI (red) and DiO (green) were added to the separate solutions to distinguish the fibers. See also Color Insert.

Gradients on electrospun materials can also be achieved by modifying the electrospun filaments in a spatially graded manner. For example, Li *et al.*[65] grew a mineral gradient on PCL fibers. This resulted in spatial gradients in scaffold stiffness and osteoblast cell density on the resultant scaffolds. Separately, Valminkinathan *et al.*[66] created a concentration gradient of laminin covalently bound to electrospun PCL/poly(ethylene glycol) diamine fibers by combining a magnetic field and bioconjugate chemistries. They were then able to verify gradients in both the protein concentration and the density of adherent Schwann cells.

15.6 Mechanics of Scaffold Deformation and Failure Under Strain

Knowledge of the material properties of the nanofibers comprising electrospun tissue scaffolds will enable the understanding of mechanotransduction of cells seeded onto and within the scaffolds. The overall mechanical properties of any structure built from fibers depend on three distinct qualities: i) the architecture of the structure, ii) the physical and mechanical properties of individual fibers,

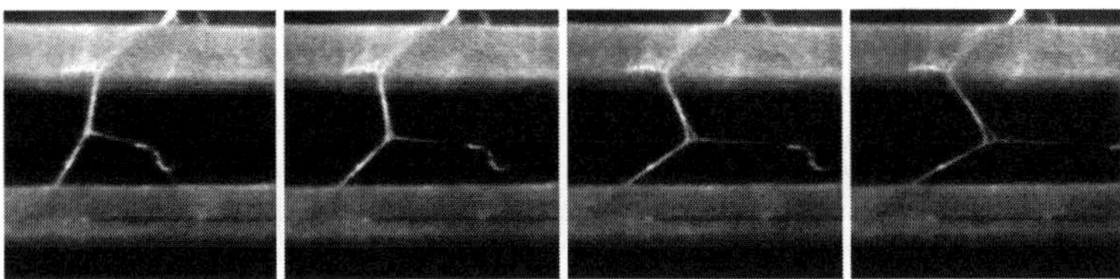

Figure 15.4. AFM/Optical microscope image series of a mechanical test of a junction of electrospun fibrinogen fibers. The junction is "unzipping" under the applied load (courtesy of Martin Guthold).

and iii) the density and strength of junctions between fibers. All three need to be known so that a structure with predictable mechanical properties can be designed. However, direct measurement of the mechanical properties of nanoscopic fibers is difficult. Nevertheless, this knowledge is needed to construct and test mechanical models of fiber networks and thus permit rational design of scaffold structures.

The recently developed technique of combined atomic force microscopy and fluorescence microscopy (AFM/FM) allows determination of the mechanical properties of individual nanoscopic fibers in buffer or ambient conditions[67] and has been applied to the characterization of native fibrin fibers, electrospun fibrinogen fibers, and electrospun collagen fibers (Fig. 15.4). This technique is ideally suited to investigate numerous natural and synthetic electrospun fibers as it can apply forces from 10^{-2} nN to 10^4 nN to fibers of a few to several hundred nanometers in radius and report elastic moduli from 10^6 Pa to 10^9 Pa. Carlisle *et al.* used this approach to extract a host of mechanical properties of the fibers, including extensibility, elastic limit, breaking strength of fibers, breaking strength of junctions, elastic (Young's modulus) and viscous component of the stretch modulus (stiffness), stress relaxation times, speed-dependent (frequency-dependent) stress–strain behavior, creep behavior, energy loss and energy storage per stretch cycle, toughness (absorbed energy), and strain-hardening/strain-softening behavior.

The nanolevel and microlevel strains in an electrospun matrix are key to understanding how tissue-level physiologic forces influence cell/matrix interaction. However, given the difficulty of directly measuring the stress–strain behavior of a bulk matrix comprised of nanoscopic electrospun fibers, other methods must be employed. The computational method known as finite element analysis (FEA) is an attractive approach to developing a detailed understanding of the nanolevel and microlevel strains generated by tissue-level physiologic forces. Stylianopoulos and Barocas[68] developed a structural model that uses volume-averaging theory to study the mechanical behavior of fibrous collagen networks. The model represented the network microstructure as a three-dimensional, fibrillar network and accounted for the three dimensionality and heterogeneity of the network, the interaction among the collagen fibers, and the realignment of the fibers upon the application of stress. The model successfully predicted mechanical features of fibrous collagen networks like the Poisson effect, the effect of network heterogeneity on the macroscopic deformation field, and the realignment of the collagen fibers. Their model did not account for the presence of cells or interstitial fluid in a tissue equivalent, which largely contribute to tissue viscoelasticity.

Stella *et al.*[69] combined computation and experimentation to study deformation mechanics of cell-seeded electrospun scaffolds. They electrospun polymer fiber scaffolds while simultaneously electrospraying viable vascular smooth muscle cells. The scaffolds were subjected to controlled biaxial stretch with three-dimensional cellular deformations and local fiber microarchitecture simultaneously quantified. They demonstrated that the local fiber geometry followed an affine behavior so that it could be predicted by macro-scaffold deformations. However, they found that local cellular deformations depended nonlinearly on changes in fiber microarchitecture and ceased at large strains where the scaffold fibers completely straightened.

15.7 Conclusions

Rationally designed electrospun meshes are intrinsically complex structures that exhibit controlled properties at the nanometer-, micron-, and millimeter-length scales. In this chapter strategies to control the meshes at these three length scales, including aligning the fibers, incorporating surface textures and porogens, and electrospinning around mandrels and molds, and, concurrently, to improve the mesh properties (e.g., chemical, mechanical, biologic) were described. In particular co-electrospinning and coaxial electrospinning techniques can be used to achieve heterogeneities on the nanometer-, micron-, and millimeter-length scales. Finally, this chapter concluded with consideration of the mechanical properties of electrospun meshes: how the mechanical properties of the individual fibers might be measured, and how the bulk properties of meshes might be predicted using computational models.

References

1. A. S. Badami, M. R. Kreke, M. S. Thompson, J. S. Riffle, and A. S. Goldstein, *Biomaterials*, **27,** 596–606 (2006).
2. C. H. Lee, H. J. Shin, I. H. Cho, Y. M. Kang, I. A. Kim, K. D. Park, and J. W. Shin, *Biomaterials*, **26**, 1261–1270 (2005).
3. B. W. Tillman, A. K. Yazdani, S. J. Lee, R. L. Geary, A. Atala, and J. J. Yoo, *Biomaterials*, **30**, 583–588 (2009).
4. J. D. Foley, E. W. Grunwald, P. F. Nealey, and C. J. Murphy, *Biomaterials*, **26**, 3639–3644 (2005).
5. S. G. Kumbar, S. P. Nukavarapu, R. James, L. S. Nair, and C. T. Laurencin, *Biomaterials*, **29**, 4100–4107 (2008).
6. W. J. Li, R. Tuli, C. Okafor, A. Derfoul, K. G. Danielson, D. J. Hall, and R. S. Tuan, *Biomaterials*, **26**, 599–609 (2005).
7. I. Liao, J. B. Liu, N. Bursac, and K. W. Leong, *Cell Mol. Bioeng.*, **1,** 133–145 (2008).
8. I. Vesely, *Circul. Res.*, **97**, 743–755 (2005).
9. J. Zeleny, *Phys Rev*, **10**, 1–8 (1917).
10. R. Singhvi, G. Stephanopoulos, and D. I. C. Wang, *Biotechnol. Bioeng.*, **43**, 764–771 (1994).
11. C. A. Bashur, R. D. Shaffer, L. A. Dahlgren, S. A. Guelcher, and A. S. Goldstein, *Tissue Eng. Part A*, (2009).
12. T. Courtney, M. S. Sacks, J. Stankus, J. Guan, and W. R. Wagner, *Biomaterials*, **27**, 3631–3638 (2006).
13. D. Li, Y. Wang, and Y. Xia, *Nano Lett*, **3**, 1167–1171 (2003).

14. J. Y. Lee, C. A. Bashur, N. Gomez, A. S. Goldstein, and C. E. Schmidt, *J. Biomed. Mater. Res. A*, **92**, 1398–1406 (2010).
15. J. Xie, M. R. Macewan, X. Li, S. E. Sakiyama-Elbert, and Y. Xia, *ACS Nano*, **3**, 1151–1159 (2009).
16. T. W. Thomas and P. A. DiMilla, *Med Biol Eng Comput*, **38**, 360–370 (2000).
17. J. B. Leach, X. Q. Brown, J. G. Jacot, P. A. DiMilla, and J. Y. Wong, *J. Neural. Eng.*, **4**, 26–34 (2007).
18. T. Stylianopoulos, C. A. Bashur, A. S. Goldstein, S. A. Guelcher, and V. H. Boracas, *J. Mech. Beh. Biomed. Mater.*, **1**, 326–335 (2008).
19. W. J. Li, R. L. Mauck, J. A. Cooper, X. Yuan, and R. S. Tuan, *J. Biomech.*, **40**, 1686–1693 (2007).
20. N. L. Nerurkar, D. M. Elliot, and R. L. Mauck, *J. Orth Res*, **25**, 1018–1028 (2007).
21. D. Karamichos, N. Lakshman, and W. M. Petroll, *Invest. Ophthalmol. Vis. Sci.*, **48**, 5030–5037 (2007).
22. T. Subbiah, G. S. Bhat, R. W. Tock, S. Parameswaran, and S. S. Ramkumar, *J Appl. Polym. Sci.*, **96**, 557–569 (2005).
23. C. A. Bashur, L. A. Dahlgren, and A. S. Goldstein, *Biomaterials*, **27**, 5681–5688 (2006).
24. Y. Yang, R. Murugan, S. Wang, and S. Ramakrishna, *Biomaterials*, **26,** 2603–2610 (2005).
25. I. K. Kwon, S. Kidoaki, and T. Matsuda, *Biomaterials*, **26,** 3929–3939 (2005).
26. J. Y. Lim, J. C. Hansen, C. A. Siedlecki, J. Runt, and H. J. Donahue, *J. R. Soc. Interface*, **2**, 97–108 (2005).
27. T. J. Webster, R. W. Sieel, and R. Bizios, *Biomaterials*, **20,** 1221–1227 (1999).
28. O. Zinger, G. Zhao, Z. Schwartz, J. Simpson, M. Wieland, D. Landolt, and B. Boyan, *Biomaterials*, **26**, 1837–1847 (2005).
29. M. J. Kim, C. W. Kim, Y. J. Lim, and S. J. Heo, *J. Biomed. Mater. Res. A*, **15**, 1023–1032 (2006).
30. S. Megelski, J. S. Stephens, D. B. Chase, and J. F. Rabolt, *Macromolecules*, **35**, 8456–8466 (2002).
31. C. L. Casper, J. S. Stephens, N. G. Tassi, D. B. Chase, and J. F. Rabolt, *Macromolecules*, **37**, 573–578 (2004).
32. D. Gupta, J. Venugopal, S. Mitra, V. R. Giri Dev, and S. Ramakrishna, *Biomaterials*, **30**, 2085–2094 (2009).

33. F. Mei, J. Zhong, X. Yang, X. Ouyang, S. Zhang, X. Hu, Q. Ma, J. Lu, S. Ryu, and X. Deng, *Biomacromolecules*, **8**, 3729–3735 (2007).
34. B. Mavis, T. T. Demirtaş, M. Gümüşderelioğlu, G. Gündüz, and U. Colak, *Acta Biomater.*, **5**, 3098–3111 (2009).
35. X. Li, J. Xie, X. Yuan, and Y. Xia, *Langmuir*, **24**, 14145–14150 (2008).
36. J. Y. Lee, C. A. Bashur, A. S. Goldstein, and C. E. Schmidt, *Biomaterials*, **30**, 4325–4335 (2009).
37. B. M. Baker, A. O. Gee, R. B. Metter, A. S. Nathan, R. A. Marklein, J. A. Burdick, and R. L. Mauck, *Biomaterials*, **29**, 2348–2358 (2008).
38. A. K. Ekaputra, G. D. Prestwich, S. M. Cool, and D. W. Hutmacher, *Biomacromolecules*, **9**, 2097–2103 (2008).
39. Q. P. Pham, U. Sharma, and A. G. Mikos, *Biomacromolecules*, **7**, 2796–2805 (2006).
40. A. Martins, S. Chung, A. J. Pedro, R. A. Sousa, A. P. Marques, R. L. Reis, and N. M. Neves, *J. Tissue Eng. Regen. Med.*, **3**, 37–42 (2009).
41. A. Thorvaldsson, H. Stenhamre, P. Gatenholm, and P. Walkenström, *Biomacromolecules*, **9**, 1044–1049 (2008).
42. T. G. Kim, H. J. Chung, and T. G. Park, *Acta Biomater.*, **4**, 1611–1619 (2008).
43. M. F. Leong, M. Z. Rasheed, T. C. Lim, and K. S. Chian, *J. Biomed. Mater. Res. A*, **91**, 231–240 (2009).
44. A. S. Goldstein, T. M. Juarez, C. D. Helmke, M. C. Gustin, and A. G. Mikos, *Biomaterials*, **22**(11), 1279–1288 (2001).
45. M. E. Gomes, C. M. Bossano, C. M. Johnston, R. L. Reis, and A. G. Mikos, *Tissue Eng.*, **12**(1), 177–188 (2006).
46. J. J. Stankus, J. Guan, K. Fujimoto, and W. R. Wagner, *Biomaterials*, **27**, 735–744 (2006).
47. J. J. Stankus, L. Soletti, K. Fujimoto, Y. Hong, D. A. Vorp, and W. R. Wagner, *Biomaterials*, **28**, 2738–2746 (2007).
48. J. A. van Aalst, C. R. Reed, L. Han, T. Andrady, M. Hromadka, S. Bernacki, K. Kolappa, J. B. Collins, and E. G. Loboa, *Ann. Plas. Surg.*, **60**, 577–583 (2008).
49. S. Kidoaki, I. K. Kwon, and T. Matsuda, *Biomaterials*, **26**, 37–46 (2005).
50. T. B. Bini, S. J. Gao, T. C. Tan, S. Wang, A. Lim, L. B. Hai, and S. Ramakrishna, *Nanotechnology*, **15**, 1459–1464 (2004).
51. S. Y. Chew, R. Mi, A. Hoke, and K. W. Leong, *Adv. Funct. Mater.*, **17**, 1288–1296 (2007).

52. C. K. Hashi, Y. Zhu, G. Y. Yang, W. L. Young, B. S. Hsiao, K. Wang, B. Chu, and S. Li, *Proc. Nat. Acad. Sci.*, **104**, 11915–11920 (2007).
53. D. Zhang and J. Chang, *Nano Lett.*, **8**, 3283–3287 (2008).
54. C. M. Vaz, S. van Tuijl, C. V. Bouten, and F. P. Baaijens, *Acta Biomater.*, **1**, 575–582 (2005).
55. V. Thomas, X. Zhang, S. A. Catledge, and Y. K. Vohra, *Biomed. Mater.*, **2**, 224–232 (2007).
56. J. Stitzel, J. Liu, S. J. Lee, M. Komura, J. Berry, S. Soker, G. Lim, M. Van Dyke, R. Czerw, and J. J. Yoo, *Biomaterials*, **27**, 1088–1094 (2006).
57. N. G. Rim, J. H. Lee, S. I. Jeong, B. K. Lee, C. H. Kim, and H. Shin, *Macromol. Biosci.*, **9**, 795–804 (2009).
58. M. P. Prabhakaran, J. R. Venugopal, and S. Ramakrishna, *Biomaterials*, **30**, 4996–5003 (2009).
59. H. J. Jin, J. Chen, V. Karageorgiou, G. H. Altman, and D. L. Kaplan, *Biomaterials*, **25**, 1039–1047 (2004).
60. J. Zeng, X. Xu, X. Chen, Q. Liang, X. Bian, L. Yang, and X. Jing, *J. Contr. Rel.*, **92**, 227–231 (2003).
61. J. Zeng, L. Yang, Q. Liang, X. Zhang, H. Guan, X. Xu, X. Chen, and X. Jing, *J. Contr. Rel.*, **105**, 43–51 (2005).
62. P. Zhao, H. Jiang, H. Pan, K. Zhu, and W. Chen, *J. Biomed. Mater. Res. A*, **83**, 372–382 (2007).
63. L. Wu, H. Li, S. Li, X. Li, X. Yuan, X. Li, and Y. Zhang., *J. Biomed. Mater. Res. A*, **92**, 563-574 (2010).
64. H. Jiang, Y. Hu, P. Zhao, Y. Li, and K. Zhu, *J. Biomed. Mater. Res. B Appl. Biomater.*, **79**, 50–57 (2006).
65. X. Li, J. Xie, J. Lipner, X. Yuan, S. Thomopoulos, and Y. Xia, *Nano Lett.*, **9**, 2763–2768 (2009).
66. C. M. Valmikinathan, J. Wang, S. Smiriglio, N. G. Golwala, and X. Yu, *Combinat. Chem. High Throughput Screen*, **12**, 656–663 (2009).
67. C. R. Carlisle, C. Coulais, M. Namboothiry, D. L. Carroll, R. R. Hantgan, and M. Guthold, *Biomaterials*, **30**, 1205–1213 (2009).
68. T. Stylianopoulos and V. H. Barocas, *Comput. Meth. Appl. Mech. Eng.*, **196**, 2981–2990 (2007).
69. J. A. Stella, J. Liao, Y. Hong, W. D. Merryman, W. R. Wagner, and M. S. Sacks, *Biomaterials*, **29**, 3228–3236 (2008).

Part V

NOVEL BIOMATERIALS FOR SCAFFOLD

Chapter 16

SYNTHETIC/NATURAL HYBRID SCAFFOLD FOR TISSUE REGENERATION

Gilson Khang,[a*] Soon Hee Kim,[a] Su Hyun Jung,[a] and Yun Sun Yang[b]

[a]*Dept of BIN Fusion Tech. and Dept of PolymerNano Sci. & Tech., Chonbuk Nat'l Univ., 664-14 Dukjin, Jeonju, 561-756 Korea*
[b]*Medipost Biomedical Research Institute, Medipost Co., Ltd., 1571-17, Seocho3-dong, Seocho-Gu, Seoul, 137-874 Korea*
*gskhang@chonbuk.ac.kr

For the scaffold materials, the family of poly(α-hydroxy acid)s, such as polyglycolide (PGA) and polylactide (PLA) and its copolymers like poly(lactide-*co*-glycolide) (PLGA), is extensively used or tested in the area of tissue-engineered organs as a bioerodible material due to good biocompatibility, controllable biodegradabilitiy, and relatively good processability. However, it is more desirable to endow the PLA, PGA, and PLGA scaffold with new functionality for tissue-engineered bioorgans. This chapter introduces the focus of synthetic/natural hybrid biomaterials as PLGA/fibrin in order to approach to a more natural three-dimensional (3D) environment and support biological signals for tissue growth and reorganization of chondrocyte and intervertebral disc *in vitro* and *in vivo*. Also, the reduction of inflammatory reaction of PLGA through the hybridization of demineralized bone particles (DBPs) and small intestine submucosa (SIS) has been reviewed for the supporting of the information of the design of tissue-engineered scaffolds.

Handbook of Intelligent Scaffolds for Tissue Engineering and Regenerative Medicine
Edited by Gilson Khang

www.panstanford.com

16.1 Introduction

Recently, numerous biodegradable polymeric biomaterials have been employed for devices for orthopedic surgery, scaffolds for tissue engineering, and vehicles for the drug delivery system. Implanted biomaterials and drug delivery vehicles have been reported to induce sequential events of immunologic reactions in response to injury caused by implantation procedures and result in acute inflammation marked by a dense infiltration of inflammation-mediating cells at the materials-tissue interface.[1–5] Prolonged irritations provoked by implanted biomaterials advance acute inflammation into a chronic adverse tissue response characterized by the accumulation of dense fibrotic tissue encapsulating the implants.[3]

PLGA is a member of a group of poly(α-hydroxy acid) that is among the few synthetic polymers approved for human clinical use by the Food and Drug Administration (FDA). Consequently, it has been extensively used and tested for scaffold material as a bioerodible material due to good biocompatibility, relatively good mechanical property, lower toxicity, and controllable biodegradability. It has been clinically utilized for three decades as sutures, bone plates, screws, and drug delivery vehicles, and its safety has been proved in many medical applications.[1] PLGA degrades by non-specific hydrolytic scission of its ester bonds into their original monomers, lactic acid and glycolic acid. During these processes, there is very minimal systemic toxicity; however, in some cases, the acidic degradation products can decrease the pH in the surrounding tissue, which results in a local inflammatory reaction and potentially poor tissue development, as shown in Fig. 16.1.[6] Also, its poor mechanical strength, small pore size, and hydrophobic surface properties for cell seeding have limited its usage.

Currently, biomaterials are endowed with biocompatibility through three different methods: coating with hydrophilic molecules, modifying surface characteristics using physiochemical methods, and impregnating bioactive substances. Previous reports showed that application of a mineral layer or localized delivery of an anti-inflammatory agent such as a corticosteroid with cytokine could effectively suppress inflammation and fibrosis of the implant.[7] Although the methods of such studies are experimentally

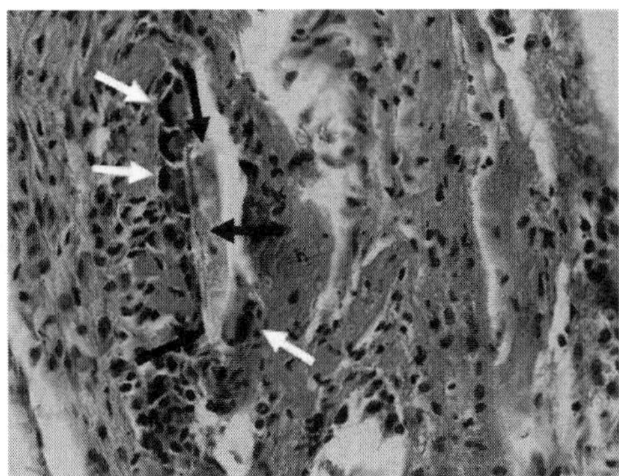

Figure 16.1. Foreign body granuloma. Small PLGA debris (black arrows) broken off from the PLGA film, surrounded by macrophages and multinucleated giant cells (white arrows). These induced macrophages and multinucleated giant cells were remaining over 2 months.

available, it is usually complicated to prepare the implants, and adverse effects of a specific growth factor have not been clearly defined. In addition, the mechanisms by which PLGA induces local inflammatory responses have not been discussed sufficiently.[8]

In our laboratory, the natural/synthetic hybrid scaffolds have been investigated during the last 15 years, such as SIS,[9,10] DBPs,[8,11] DBP gel,[12] fibrin,[13] keratin,[14] hyaluronic acid,[15] collagen gel,[4] silk,[4] and a 2-methacryloyloxyethyl phosphorylcholine (MPC) polymer (PMEH)[16] with PLGA to reduce cellular inflammatory response. In this chapter, we introduced (1) fibrin/PLGA hybrid scaffolds for the regeneration of disc and cartilage and (2) DBP/PLGA and SIS/PLGA hybrid scaffolds in terms of scaffold design for the reduction of host response and the augmentation of tissue formation.

16.2 Fibrin/PLGA Hybrid Scaffolds

16.2.1 *Fibrin/PLGA Hybrid Scaffolds for Cartilage Regeneration* in vivo *and* in vitro

Articular cartilage has a limited potential to repair; consequently, damaged articular cartilage will further degenerate and eventually turn into osteoarthritis. Autologous chondrocytes implantation was first published by Brittberg *et al.*[17] Recently, articular cartilage repair has been given much intention in orthopedic tissue engineering. Usually scaffolds are designed as a highly porous 3D structure to allow cells to accommodate and grow inside, as well as organize cells into a 3D tissue. Many trials have successfully cultured articular chondrocytes,[18] formed neocartilage tissue,[19] and transplanted autologous neocartilage to the defect, so biocompatible scaffolds that afford the proliferation of cartilage and accumulate the matrix have been widely investigated.[20]

As we discussed earlier, numerous attempts have been made for successful tissue reconstruction using PLGA-based scaffolds either by PLGA itself or in combination with natural polymers, such as collagen,[21] and extracellular matrices (ECMs) scaffolds, that is, SIS[9,10] as well as DBPs[8,11,12] in our previous studies. The incorporation of bioactive molecules in PLGA is believed to mediate cell behavior, for example, proliferation, differentiation, and function. To minimize cells lost during the *in vitro* seeding procedure, we used fibrin to immobilize cells and to provide homogenous cell distribution in PLGA scaffolds. Fibrin has been widely used for cartilage reconstruction purposes.[13,22,23] We hypothesized that fibrin would be an ideal cell carrier/transplantation matrix to enhance *in vitro* chondrogenesis of rabbit articular chondrocytes morphologically, histologically, biochemically, and phenotypically similar to the normal hyaline cartilage.

Articular cartilage was aseptically isolated from the femoral condyles and patellae of six-week-old New Zealand white rabbits, and isolated chondrocytes were cultured in a mixture of equal volume of F12/DMEM. PLGA (mole ratio 50:50, molecular weight 33,000 g/mole, Resomer RG 503 H) was purchased from Boehringer Ingelheim Pharma GmbH (Ingelheim, Germany). Microporous 3D PLGA scaffolds (0.2% w/v) were fabricated by the solvent-casting/salt-leaching technique using sodium chloride as a porogen. Each sample was assigned into two experimental groups: cultured chondrocytes were seeded into (1) PLGA scaffolds with fibrin (fibrin/PLGA) and (2) PLGA scaffolds without fibrin. One million cells per scaffold were incorporated and resuspended with the (1) commercially available fibrin glue kit from Greenplast®

(Green Cross P. D. Co., Yongin, Korea) and (2) culture medium. PLGA scaffolds were pretreated with calcium chloride solution for the PLGA/fibrin group prior to cell seeding. The chondrocytes-fibrin admixture was seeded into PLGA scaffolds and was allowed to polymerize within five minutes. The chondrocyte suspension in the culture medium was seeded directly into PLGA scaffolds and incubated for five minutes before adding the culture medium. All constructs were cultured for three weeks *in vitro* and then implanted at the dorsum of athymic nude mice. Resulted *in vitro* and *in vivo* constructs were harvested at one, two, and four weeks postimplantation.

The morphology of cells and the distribution of cartilaginous ECM in the fibrin/PLGA and PLGA were examined via histological staining (Fig. 16.2). Before implantation, hematoxylin and eosin (H&E) staining showed that cells and ECM filled the space of the fibrin/PLGA. A significant number of round morphological chondrocytes clusters and cartilaginous ECM formation were observed in the fibrin/PLGA hybrid construct than in the PLGA construct. The *in vitro* fibrin/PLGA constructs exhibited superior histoarchitectural characteristics of cartilage-like tissue compared with the control. The closely packed cartilage-isolated cells were homogeneously distributed in the ECM and exhibited rounded morphology with lacunae embedded in the basophilic ground substance. The pericellular and interterritorial matrix was strongly stained by the characteristic red of Safranin O, to indicate the presence of a proteoglycan-rich matrix in the construct. Cartilaginous ECM deposition was further demonstrated by positive Alcian blue staining to confirm the presence of accumulated glycosaminoglycans (GAGs).

The formation of cartilaginous tissue in fibrin/PLGA constructs was remarkably evident at each time point of one, two, and four weeks after *in vivo* implantation. In the fibrin/PLGA construct, an increase in the implantation period resulted in a lower cells-to-matrix ratio, similar to that of native hyaline cartilage. Cartilage-isolated cells with lacunae were sparsely distributed within the homogenous ECM in concert with the presence of a specific histochemical property of the accumulated proteoglycan-rich matrix and GAG. Whereas, in PLGA, the presence of chondrocytes in their natural round morphology increased with an increase in the

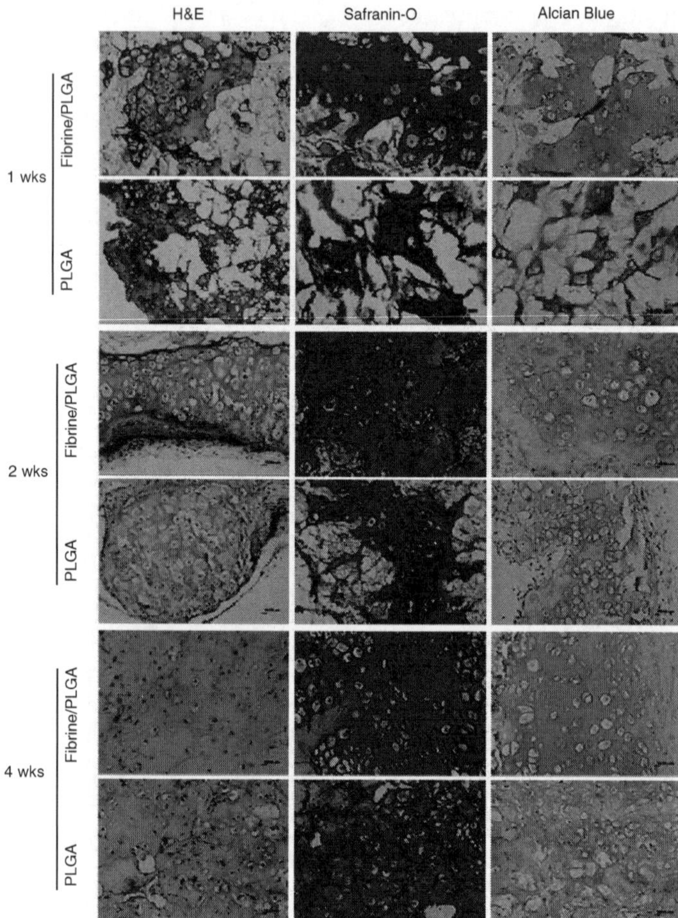

Figure 16.2. The morphology of cells and the distribution of cartilaginous ECM in the fibrin/PLGA hybrid scaffold and PLGA were examined via H&E staining, safranin O staining, and Alcian blue staining. See also Color Insert.

implantation period and resulted in a higher cells-to-matrix ratio. This phenomenon indicated the progress of cartilaginous tissue formation in the PLGA group. The difference between the fibrin/PLGA hybrid construct and the PLGA group was clearly visible in terms of the overall cartilaginous tissue formation, cells organization, and ECM distribution in the specimens. At the end of the *in vivo* experiment, the fibrin/PLGA constructs exhibited good-quality cartilage-like tissue compared with the PLGA group.

We analyzed collagen type II and type I immunolocalization of the fibrin/PLGA hybrid construct. Collagen type II was detected in all specimens. As shown in Fig. 16.3, collagen type II exhibited strong

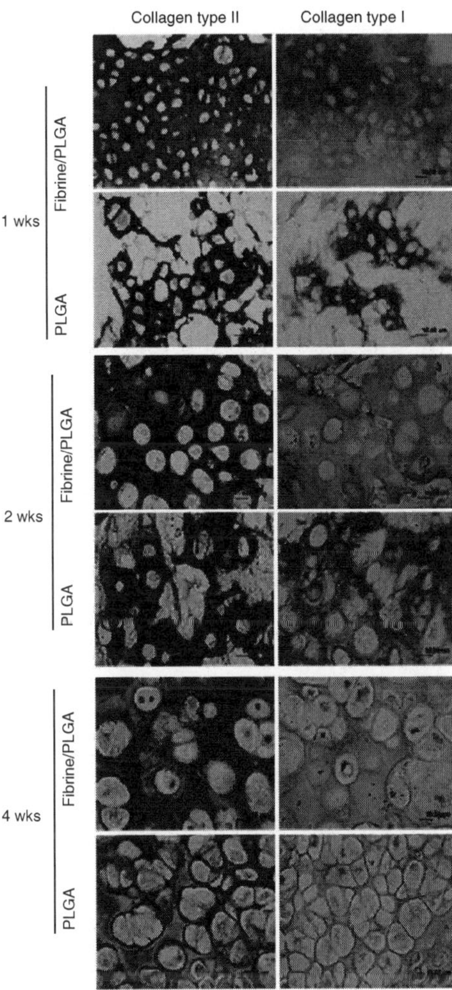

Figure 16.3. Collagen type II was detected in all *in vitro* specimens and at each time point of 1, 2, and 4 weeks postimplantation. See also Color Insert.

immunopositivity at the specific region of the *in vitro* fibrin/PLGA hybrid construct, mainly localized surrounding the pericellular and interterritorial matrix. After *in vivo* implantation, the fibrin/PLGA hybrid construct showed more homogeneous distribution of collagen type II than in the PLGA scaffold at each time point. Analysis of collagen type I in fibrin/PLGA hybrid constructs showed moderate positive immunoreactivity throughout the ECM of specimens at each time point of one and two weeks postimplantation. Interestingly, after four weeks' implantation, both fibrin/PLGA hybrid constructs and the PLGA group showed no collagen type I expression; however, in some cases, both groups showed very weak collagen type I expression.

The fibrin/PLGA hybrid scaffold promotes chondrogenesis of rabbit articular chondrocytes proven by means of morphology, histology, immunohistochemistry, chondrogenic gene expression, and sGAG production. This study suggests that the fibrin/PLGA hybrid scaffold may serve as a potential cell delivery vehicle and a structural basis for *in vitro* development of tissue-engineered articular cartilage constructs. Future studies using appropriate animal model for an autologous *in vivo* system are necessary to further validate the feasibility of fibrin/PLGA hybrid constructs for articular cartilage restoration.

16.2.2 *Fibrin/PLGA Hybrid Scaffolds for IVD* in vitro

The intervertebral disc (IVD) is the cornerstone of the joint complex comprising the spinal motion segment. The IVD functions to permit limited motion and flexibility, while maintaining segmental stability by absorbing and distributing external loads. The structure of the normal IVD includes the nucleus pulposus (NP) composed primarily of proteoglycans and collagen type II with a capacity to absorb and distribute load and the outer annulus fibrosus (AF) with well-organized layers consisting of collagen type II and collagen type I serving to stabilize the motion segment. The structure and function of the IVD may be altered by processes including normal physiological aging, mechanical factors, that is, trauma and repetitive stress, segmental instability of the spine, and inflammatory and biochemical factors. Patients of degenerative disc disease form the majority of patients with back pain of spinal origin. The IVD undergoes extrinsic morphological changes during their lifetime. In fact, most of the adult population will have a degenerative disc by the sixth decade of life. The current treatment options range from nonsteroidal anti-inflammatory drugs (NSAIDS) to invasive procedures, including spinal fusion and arthroplasty. Unfortunately these treatments are not solving the root of the problem, which is the degeneration of the IVD itself. As tissue engineering and regenerative research become more advanced, there is a trend toward reversing the etiology of the disease, and much effort has been put into regeneration of the IVD.[24]

Six-week-old New Zealand white rabbits were euthanized, and lumbar discs were obtained using an osteotome by means of an anterior approach in an *en bloc* fashion from L1–L2 to L7–S1 IVD. The AF and NP tissue was aseptically dissected from the lumbar discs. AF and NP cells were cultured separately in a mixture of equal volume of F12/DMEM supplemented with 10% fetal bovine serum (FBS) and (4-(2-hydroxyethyl)-1-piperazineethanesulfonic acid) HEPES buffer 1M. PLGA and fibrin/PLGA hybrid scaffolds were used—the same ones as in section 16.3.2.

The total sGAG production was normalized by the dried weight of each sample and represented as relative sGAG content in percentage (%). After one, two, and three weeks of *in vitro* culture, AF cells cultured in fibrin/PLGA exhibited 0.290±0.009, 0.341±0.004, and 0.443±0.014 relative sGAG content, respectively, while PLGA exhibited 0.255±0.025, 0.319±0.45, and 0.360±0.007 relative sGAG content, respectively, after one, two, and three weeks. Apparently, sGAG production was higher in fibrin/PLGA than in PLGA, and the differences between sGAG production magnitudes were significantly distinguished by week 2 (1.07-fold, p = 0.04) and week 3 (1.23-fold, p = 0.0003), as shown in Fig. 16.4a. As shown in Fig. 16.4b, after one, two, and three weeks of *in vitro* culture, AF cells cultured in fibrin/PLGA exhibited 0.169±0.021, 0.276±0.007, and 0.277±0.018 relative sGAG content, respectively, while PLGA exhibited 0.109±0.017, 0.208±0.0207, and 0.230±0.005 relative sGAG content, respectively, after one, two, and three weeks. The sGAG production in fibrin/PLGA was superior to PLGA throughout the *in vitro*

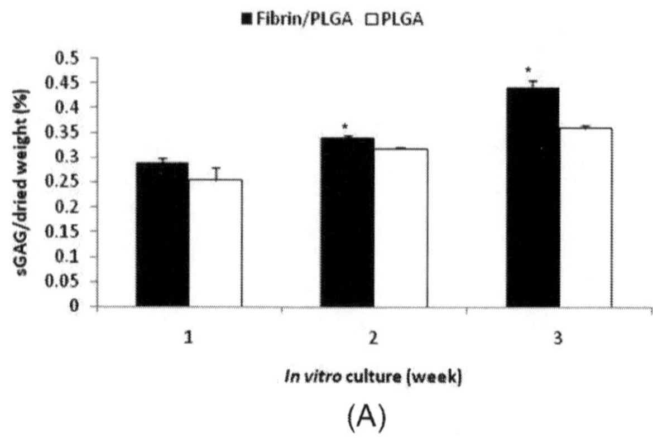

(A)

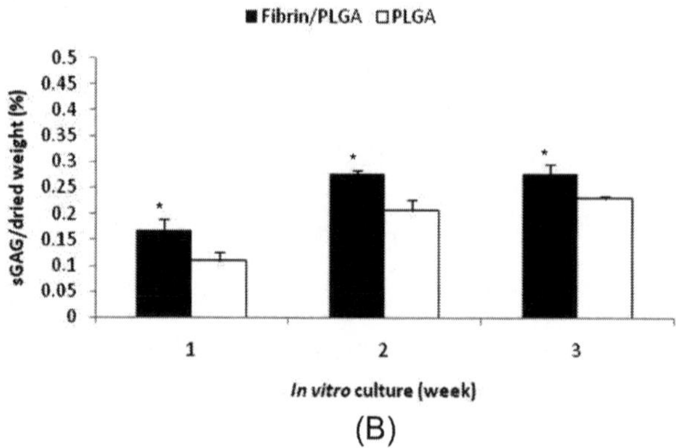

(B)

Figure 16.4. Total sGAG production was normalized by the dried weight of each sample and represented as relative sGAG content in percentage (%). The sGAG production was higher in fibrin/PLGA seeded with AF cells (a) and NP cells (b) than in the PLGA group.

culture. The magnitudes were significantly higher by 1.55-fold, 1.33-fold, and 1.2-fold after one week ($p = 0.008$), two weeks ($p = 0.009$), and three weeks ($p = 0.038$), respectively.

Approximately 1×10^5 cells per scaffold were cultured in the fibrin/PLGA and PLGA scaffolds. Seeded cells adhered onto the scaffolds, proliferated, and produced matrices filling in the void spaces of the scaffolds. No sign of cartilaginous tissue formation in fibrin/PLGA and PLGA was observed after one week of *in vitro* 3D

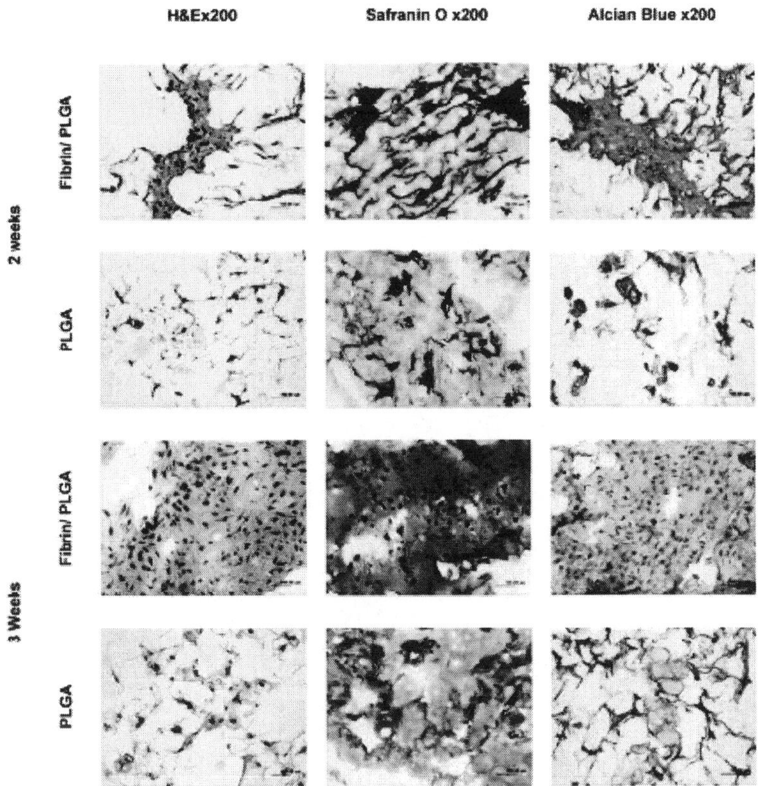

Figure 16.5. Safranin O indicating the presence of the proteoglycan-rich matrix corroborated with positive and Alcian blue staining, confirming GAG accumulation of the fibrin/PLGA and PLGA groups seeded with AF cells at 2 weeks' culture. See also Color Insert.

culture of AF and NP cells. At two weeks' culture, minimal cartilaginous tissue formation was observed in the fibrin/PLGA seeded with AF cells with several cells clusters filling up the void spaces of the scaffold (Fig. 16.5).

The newly synthesized ECM was strongly stained by the characteristic red of Safranin O, indicating the presence of the proteoglycan-rich matrix corroborated with positive Alcian blue staining confirming GAG accumulation. The formation of cartilaginous tissue in fibrin/PLGA was remarkably evident by three weeks of *in vitro* culture. The closely packed cells were homogeneously

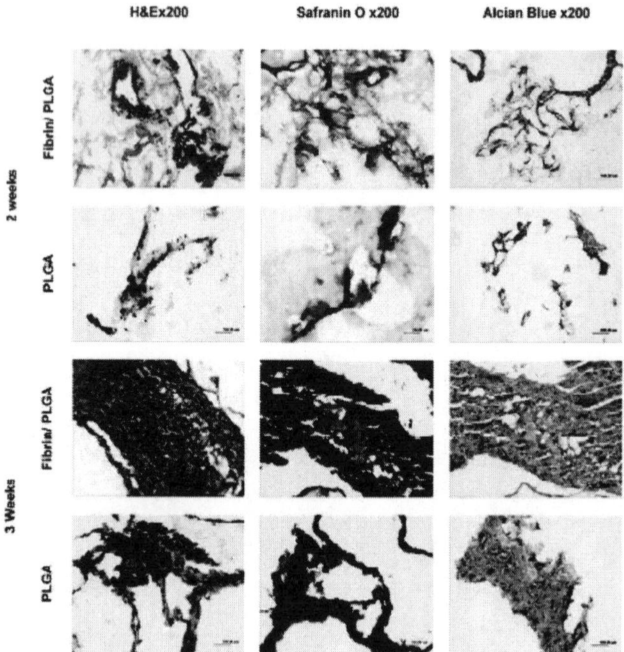

Figure 16.6. At 2 weeks, cartilaginous tissue formation was more prominent in fibrin/PLGA seeded with NP cells compared with PLGA. Clearly, the formation of cartilaginous tissue was evident by the third week of *in vitro* culture in fibrin/PLGA. The presence of an accumulated proteoglycan-rich matrix and GAG at the core region was significant and was intensely stained at 2 weeks and greatest at 3 weeks. Conversely, NP cells cultured in PLGA without fibrin demonstrated slower progress of cartilaginous tissue formation *in vitro*. The differences between the fibrin/PLGA and PLGA groups were distinguishable in terms of overall cartilaginous tissue formation, cell organization, and ECM distribution in all specimens. See also Color Insert.

distributed in the basophilic ECM in concert with the presence of specific histochemical property of a proteoglycan-rich matrix and GAG. Fibrin/PLGA developed a predominantly superior histoarchitecture compared with the PLGA group. In PLGA, a cluster of a few rounded cells was spotted throughout the specimen. For the NP specimen, as shown in Fig. 16.6, cartilaginous tissue formation was also more prominent in fibrin/PLGA than in PLGA.

Clearly, the formation of cartilaginous tissue was evident by the third week of *in vitro* culture in the fibrin/PLGA. The differences between the fibrin/PLGA and PLGA groups both in AF and in NP cases were distinguishable in terms of overall cartilaginous tissue formation, cells organization, and ECM distribution in all specimens. The presence of an accumulated proteoglycan-rich matrix and GAG at the core region was significant and was intensely stained at 2 weeks and greatest at 3 weeks. Conversely, AF and NP cells cultured in PLGA without fibrin demonstrated slower progress of cartilaginous tissue formation *in vitro*.

Fibrin supports higher cell proliferation, maintains phenotypic expression, and promotes greater sGAG production and cartilaginous tissue formation of AF and NP cells cultured in PLGA. This study suggests that a fibrin/PLGA hybrid scaffold may serve as a potential cell delivery vehicle and a structural basis for *in vitro* tissue-engineered IVD. This *in vitro* study revealed promising results; hence, future studies utilizing the *in vivo* system are necessary to further validate the development of tissue-engineered IVD using a fibrin and PLGA composite.

16.3 The Effect of DBPs on the Reduction of Inflammatory Reaction of the PLGA/DBP Hybrid Scaffold

DBPs have long been recognized as powerful inducers of new bone growth. Many authors reported that this osteoinductive property is mainly due to bone morphogenetic proteins (BMPs).[8,11,12] After the demineralization process using an acid solution, such as HCl, an acid-insoluble matrix of collagen and growth factors, including BMP, is left behind. In the bone defect site as well as in nonskeletal areas, DBPs induce osteogenesis without a fibrous reaction.[3,5] In a more recent study, we demonstrated that DBPs enhanced hydrophilicity of PLGA scaffolds with an increase of content and reduced the adverse cellular response associated with inflammation.[6] For this, we hypothesized that the inflammatory response of cells neighboring the PLGA implant may occur and can be reduced by impregnating DBPs into PLGA. We focused our attention on the early stages of the inflammatory reaction and used histological and molecular analyses to assess how cells and tissue responded to DBP-PLGA hybrid materials *in vivo* and *in vitro*. We evaluated the effect of five different ratios of DBP/PLGA hybrid materials on the cellular inflammatory response and tissue reaction induced by PLGA.[6]

16.3.1 *Cell Viability*

In order to evaluate the influence of DBP content in PLGA materials on cells, we analyzed the viability of mouse fibroblasts on PLGA and the five different ratios of DBP/PLGA scaffolds during the *in vitro* culture and found that DBPs enhanced initial attachment of fibroblasts on the scaffolds. At Day 1, the number of vital cells was significantly higher in the culture of the scaffold containing DBPs than that of the scaffold without DBPs (Fig. 16.7). Particularly, 20% and 40% DBP/PLGA scaffolds maintained the number of viable cells compared with PLGA scaffolds through three days. The range of DBP contents of PLGA scaffolds showed no adverse effects on fibroblast cell attachment, proliferation, and viability compared with PLGA scaffolds. No significant differences were found between the PLGA

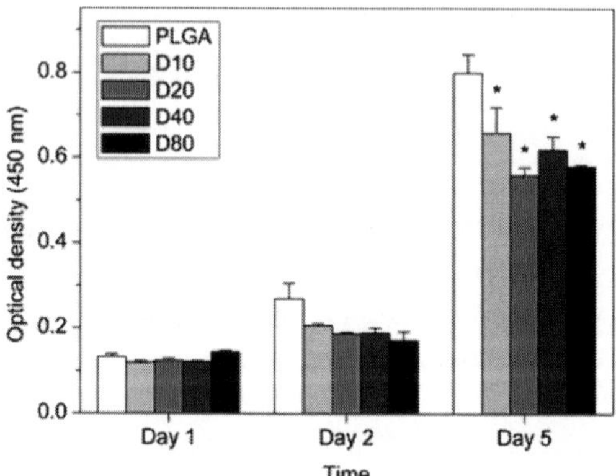

Figure 16.7. The number of viable fibroblasts on PLGA and DBP/PLGA scaffolds at 1, 2, and 3 days as determined by the MTT colorimetric assay. *,§,¶ correspond to $p < 0.05$ in comparison with PLGA scaffolds for each day. *Abbreviation*: MTT, (3-(4,5-Dimethylthiazol-2-yl)-2,5-diphenyltetrazolium bromide.

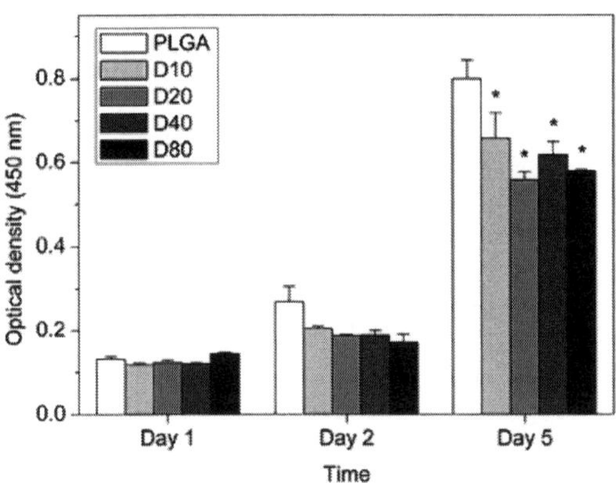

Figure 16.8. The number of viable HL-60 cells on PLGA and DBP/PLGA scaffolds at Days 1, 2, and 5, as determined by the MTT colorimetric assay. * corresponds to $p < 0.05$ in comparison with PLGA scaffolds for each day.

and the 10% DBP/PLGA scaffold at Days 2 and 3. Next we evaluated the viability of HL-60 cells on samples and did not find statistical significances between PLGA and DBP hybrid scaffolds at Days 1 and 2. By culture Day 5, proliferation of HL-60 cells with PLGA increased slightly more than with DBP/PLGA scaffolds (Fig. 16.8).

16.3.2 *Inflammatory Cytokine Expression*

To elucidate the cellular responses associated with inflammation on sample films, we measured the level of mRNA expression of tumor necrosis factor-α (TNF-α) and IL-1β from HL-60 cell in 48 hours after culture with PLGA or DBP/PLGA films, as shown in Fig. 16.9. TNF-α mRNA in HL-60 highly expressed following PLGA films compared with DBP/PLGA films; it was significantly lower following DBP/PLGA

films than PLGA films, with increases in contents of DBPs: 10%, 20%, 40%, and 80% of DBPs ($p < 0.005$, $p < 0.0001$, $p < 0.00005$, and $p < 0.00005$, respectively) (Fig. 16.10). The intensity of TNF-α expression of PLGA films was significantly 10 times higher or more than that of 40% DBP/PLGA films. HL-60 cells with 80% DBP/PLGA films rarely expressed TNF-α mRNA. Similarly, IL-1β mRNA expression decreased markedly with 40% and 80% DBP/PLGA films

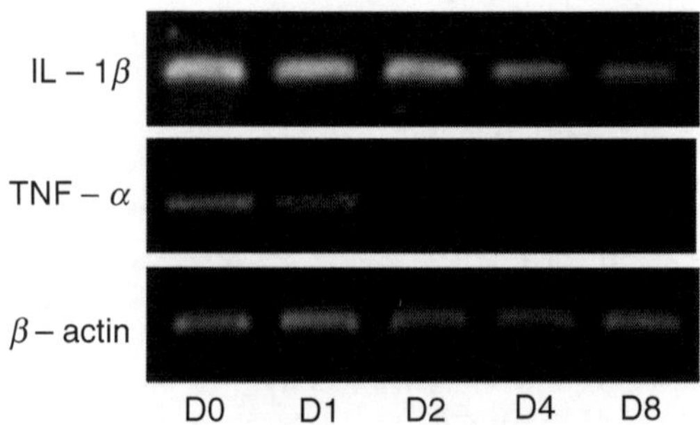

Figure 16.9. The number of viable HL-60 cells on PLGA and DBP/PLGA scaffolds at Days 1, 2, and 5, as determined by the MTT colorimetric assay. * corresponds to $p < 0.05$ in comparison with PLGA scaffolds for each day.

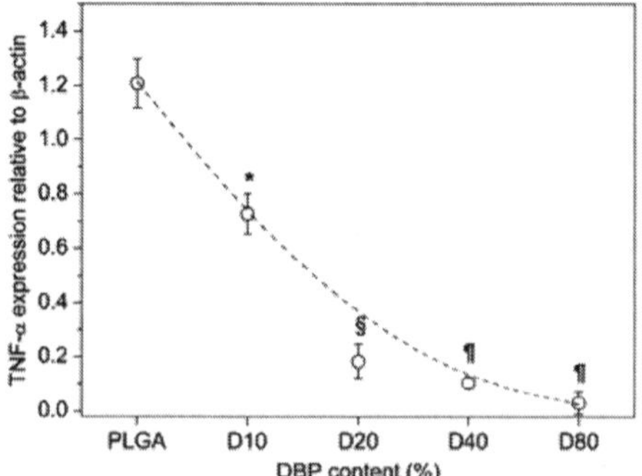

Figure 16.10. The expression of TNF-α mRNA on HL-60 cells fell continuously with an increase in the content of DBPs in PLGA. There is a significant decrease of TNF-α expression on HL-60 cells in each scaffold as DBP content increases from 0% to 80%.

compared with PLGA films. No significant differences of IL-1β mRNA expression were observed between PLGA, 10% DBP/PLGA, and 20% DBP/PLGA (Fig. 16.11).

16.3.3 In vivo *Tissue Response*

To further characterize the *in vivo* inflammatory response surrounding the implants, histological examination was performed at Day 5 after implantation. Remarkable inflammation was observed in

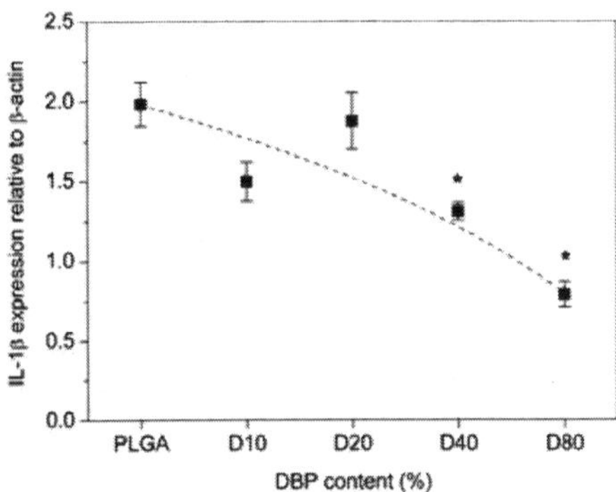

Figure 16.11. The expression of IL-1β mRNA on HL-60 cells fell continuously with an increase in the content of DBPs in PLGA. IL-1β mRNA expression decreased markedly on 40% and 80% DBP/PLGA films. * corresponds to $p < 0.05$ in comparison with PLGA films.

the tissue surrounding the PLGA films; however, this inflammatory reaction was progressively diminished with an increase in the content of DBPs in PLGA films. We observed numerous recruited neutrophils infiltrates with a large number of multinucleated giant cells (MNGCs) in adjacent tissue after PLGA film implantation. However, this inflammatory cellular response decreased as the

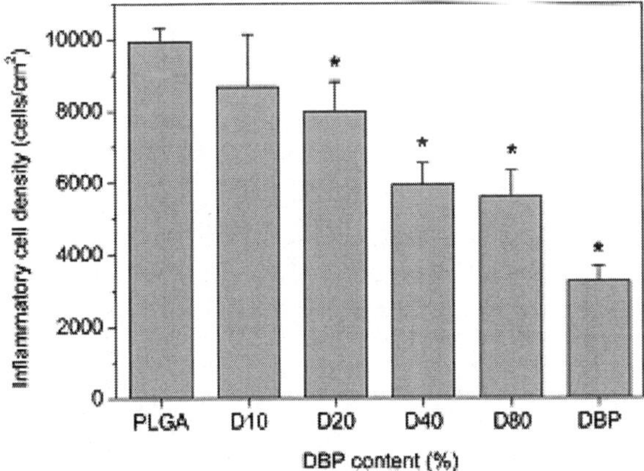

Figure 16.12. The inflammatory response to PLGA films, DBP hybrid PLGA films, and DBP. The density of inflammatory cells. *corresponds to $p < 0.05$ in comparison with PLGA films.

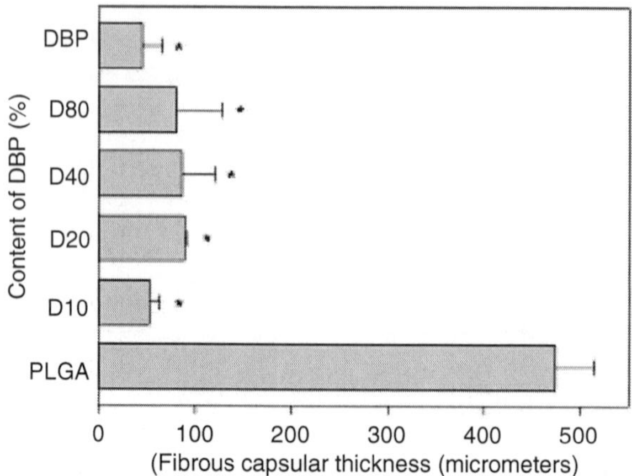

Figure 16.13. The inflammatory response to PLGA films, DBP hybrid PLGA films, and DBP. Fibrous wall thickness elicited by PLGA films was nearly five times that measured for DBP-impregnated PLGA films. *corresponds to $p < 0.00001$ in comparison with PLGA films.

content of DBPs continuously increased in PLGA films. The density of inflammatory cells following PLGA film implantation was approximately two times higher than that following 40% DBP/PLGA films. The DBP film had fewer inflammatory cells relative to the PLGA film (Fig. 16.12). PLGA, DBP/PLGA, and DBP films had a noticeable difference of fibrotic band encapsulation. The fibrotic thickness was significantly decreased in the DBP/PLGA hybrid and DBP film. The PLGA film had a five times or more broad fibrotic band than the other samples (Fig. 16.13). In 40% or 80% DBP/PLGA hybrid films, macrophages or foreign body giant cells were rarely observed in immediate contact with the DBP fragment surface, and a thin collagenous fibrous band surrounded the samples. The DBP film seldom recruited MNGCs compared with PLGA or DBP/PLGA hybrid films (Fig. 16.14).

These results indicate that the DBP hybrid PLGA film elicits decreased tissue reactivity that can result from biocompatibility of DBPs. TNF-α plays a role in the inflammatory response that activates leukocytes, enhances adherence of neutrophils and monocytes, promotes the migration of inflammatory cells into the intercellular space, and triggers local production of other pro-inflammatory

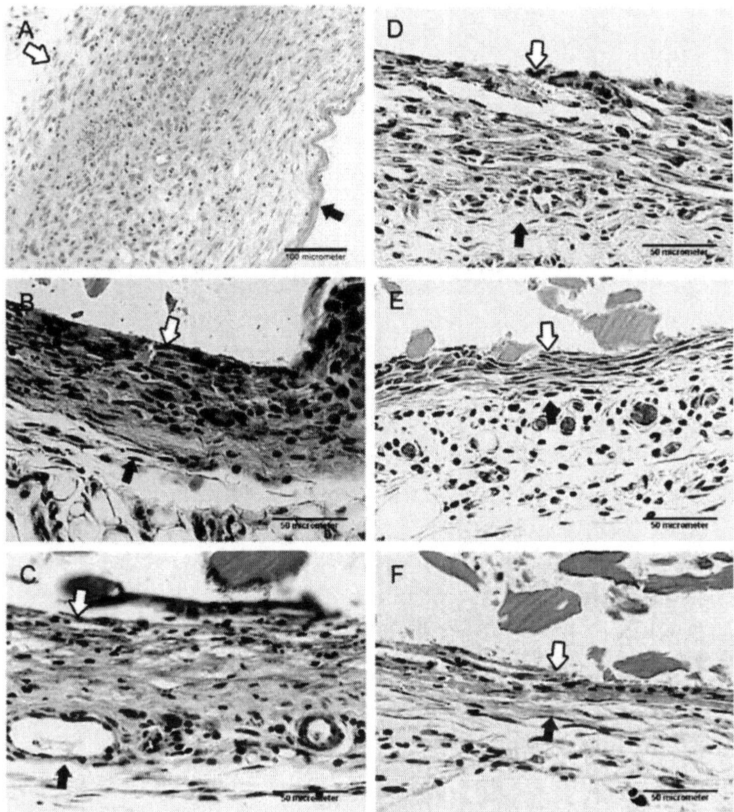

Figure 16.14. Photomicrographs of H&E stain sections of the directly bordering tissue after PLGA, DBP hybrid PLGA films, and DBP. (a) The tissue implanted with PLGA, bar length = 100 μm (x100); (b–f) the tissue implanted with 10% DBP/PLGA, 20% DBP/PLGA, 40% DBP/PLGA, 80% DBP/PLGA, and DBP, respectively, bar length = 50 μm (x400). Note that the number of inflammatory cells and fibrous band thickness in vicinity to the tissue-implanted samples decreased as the DBP content in PLGA films increased. Polymer-tissue interface surfaces are indicated by a white arrow. The fibrous wall thickness is represented by black and white arrows. See also Color Insert.

cytokines such as IL-1β. TNF-α and IL-1β are potent stimulators of fibroblast growth.[25] Therefore, temporospatial expression of these pro-inflammatory mediators allows fine-tuning of the inflammatory response.[6] DBPs have been reported as biocompatible inducers of bone formation, which is mainly responsible for bone morphogenetic proteins and several growth factors, including IGF and TGF-β1 in DBPs.[26]

In this study, we observed that DBP hybrid PLGA materials did not significantly affect the viability of HL-60 cells during 3D *in vitro* culture; however, their inflammatory response to DBP hybrid polymeric materials was obviously reduced. The probable cause for these results is that TNF-α expression may be strongly suppressed by bioactive substances released from DBPs; this pattern was enhanced with an increase in the content of DBPs in PLGA films, which can result in suppression of the IL-1β expression that reduces proliferation and fibrous capsular formation of fibroblasts.

DBP/PLGA materials may have more mechanical stability compared with PLGA materials without DBPs, which can led to decreased production of implant debris and reduction of tissue response. Whether this reduction of foreign body reaction occurs because the bioactive molecules released from DBPs induced suppression of local inflammation or DBP-impregnated PLGA materials provided a favorable surface or rigid structural support to cell remains to be determined.

In this study, we have shown that by impregnating DBPs in the PLGA materials, the inflammatory reaction could be effectively reduced *in vivo* and *in vitro*. This result suggests that hybridization of natural materials such as DBPs is suitable for control of an adverse tissue reaction of polymeric materials shown *in vivo*.

16.4 The Effect of SIS on the Host Tissue Response to PLGA/SIS Hybrid Scaffolds

SIS consists of more than 90% types I and III collagen, plus a wide variety of cytokines, including basic fibroblast growth factor (bFGF), transforming growth factor-β (TGF-β), epidermal growth factor (EGF), vascular endothelial growth factor (VEGF), and IGF-1, as well as GAGs, fibronectins, chondroitin sulfates, heparins, heparin sulfates, and hyaluronic acids.[27] These constituents are expected to facilitate the function of SIS as a tissue engineering scaffold by supporting cell attachment, proliferation,

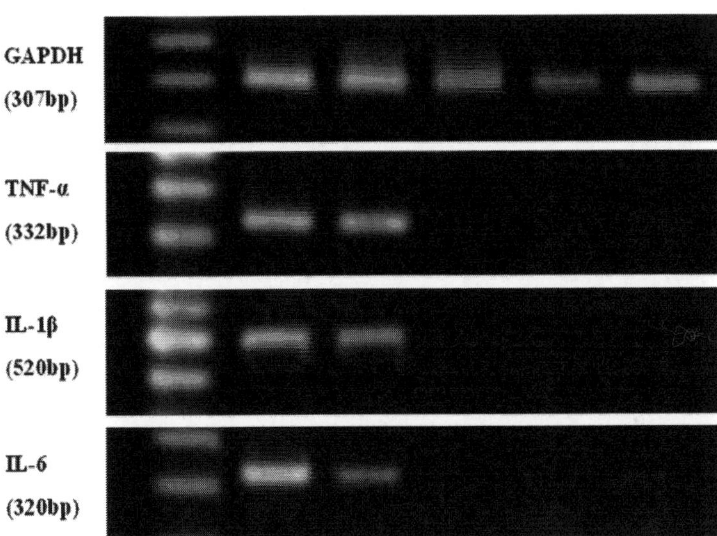

Figure 16.15. Representative gene expression band as IL-6, IL-1β, TNF-α, and GAPDH. These bands were normalized and showed in Fig. 16.16. *Abbreviation*: GADPH, Glyceraldehyde 3-phosphate dehydrogenase.

differentiation, and migration. We tried to compare the host tissue response to representative synthetic and natural biomaterials by assessing inflammation at the implanted area. We prepared and characterized PGLA/SIS films with five different ratios of SIS as 0%, 10%, 20%, 40%, and 80% to PLGA, performed subcutaneous implantation of these scaffolds into rats, and then compared the host tissue response by reverse transcriptase-polymerase chain reaction (RT-PCR) analysis of TNF-α, IL-1β, and IL-6 mRNA expression, as shown in Fig. 16.15, and then normalized (Fig. 16.16).

As shown in Fig. 16.16, TNF-α mRNA *in vivo* was highly expressed following PLGA films compared with SIS/PLGA films; it was significantly lower following SIS/PLGA films than PLGA films, with increases in contents of SIS: 10%, 20%, 40%, and 80% of SIS ($p < 0.05$). The intensity of TNF-α expression of PLGA films was significantly 10 times higher or more than that of 40% SIS/PLGA

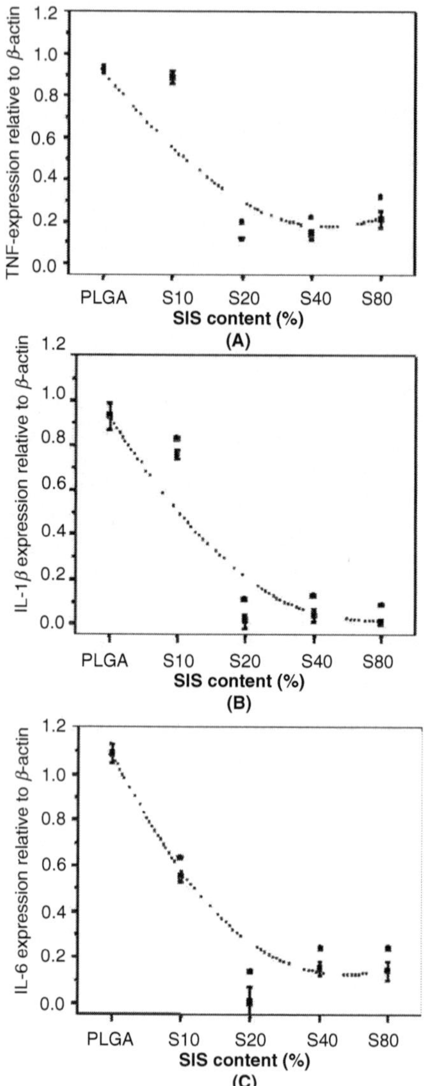

Figure 16.16. The expression of TNF-α IL-1β and IL-6 mRNA *in vivo* fell continuously with an increase in the content of SIS in PLGA. (a) TNF-α expression significantly decreased in each scaffold as SIS content increased from 0% to 80%; (b) IL-1β mRNA expression decreased markedly on 40% and 80% SIS/PLGA films; (c) IL-6 mRNA expression decreased markedly on 40% and 80% SIS/PLGA films.* corresponds to $p < 0.05$ in comparison with PLGA films.

films. Similarly, IL-1β mRNA expression decreased markedly with 40% and 80% SIS/PLGA films compared with PLGA films.

TNF-α plays a role of inflammatory response that activates leukocytes, enhances adherence of neutrophils and monocytes, promotes the migration of inflammatory cells into the intercellular space, and triggers local production of other pro-inflammatory cytokines such as IL-1β.[25] TNF-α, IL-6, and IL-1β, which are potent stimulators of fibroblast growth.[28] Therefore, temporospatial expression of these pro-inflammatory mediators allows fine-tuning of the inflammatory response.[29] SIS has been reported as a biocompatible inducer of cell growth, which is mainly responsible for several growth factors, as we discussed earlier.[27] In this study, we observed that their inflammatory response to SIS hybrid polymeric materials was obviously reduced. The probable cause for these results is that TNF-α expression may be strongly suppressed by bioactive substances released from SIS; this pattern was enhanced with an increase in the content of SIS in PLGA films, which can result in suppression of the IL-1β and IL-6 expression, which reduces proliferation and fibrous capsular formation of fibroblasts (data not shown).

16.5 Conclusions and Outlook

Tissue engineering, including regenerative medicine, shows tremendous potential as a revolutionary research push. Also, many successful results have reported the potential for regenerating tissues and organs such as skin, bone, cartilage, nerves of peripheral and central nervous systems, tendon, muscle, cornea, bladder and urethra, and liver, as well as composite systems like the human phalanx and joints, on the basis of scaffold biomaterials from polymers, ceramic, metal, composites, and their hybrids. The prerequisite physicochemical properties of scaffolds are (i) to support and deliver for cells, (ii) to induce, differentiate, and conduit tissue growth, (iii) to target the cell adhesion substrate, (iv) to stimulate cellular response, (v) to act as a wound-healing barrier, (vi) to be biocompatible and biodegradable, (vii) to ensure relatively easy processability and malleability into desired shapes, (viii) to be highly porous with large surface/volume, (ix) to have mechanical strength and dimensional stability, (x) to have sterilizability, (xi) to not induce inflammatory reaction and fibrotic capsule, and so on.[1–5]

From this point of view, the design and control over precise biochemical signals must be needed by the combination of a scaffold matrix and bioactive molecules, including genes, peptide molecules, and cytokines. Moreover, the combination of the cells and redesigned bioactive scaffolds has attempted to expand to a tissue level of hierarchy. In order to achieve this goal, novel hybrid scaffold biomaterials, novel scaffold fabrication methods, and novel characterization methods must be developed.

Acknowledgments

This work was supported by World Class University (R31-20029), SCRC(SC4110), and the MuscloBioorgan Center from KMOHW.

References

1. G. Khang, S. H. Kim, M. S. Kim, and H. B. Lee, "Hybrid, composite, and complex biomaterials for scaffolds," In Eds. A. Atala, R. Lanza, J. A. Thomson, and R. M. Nerem, *Principles of Regenerative Medicine* (Elsevier, San Diego, 2008), pp 636–655.
2. G. Khang, M. S. Kim, and H. B. Lee, In Eds. G. Khang, M. S. Kim, and H. B. Lee, *A Mannual for the Fabrication of Tissue Engineered Scaffolds* (World Scientific Publishing Co., Singapore, 2007).
3. G. Khang, S. H. Kim, J. M. Rhee, M. Sha'ban, and R. B. H. Idrus, "Synthetic/natural hybrid scaffold for cartilage and disc regeneration, In Ed. T. Tateishi, *Biomaterials in Asia* (World Scientific Publishing Co., Singapore, 2008).
4. G. Khang, S. J. Lee, M. S. Kim, and H. B. Lee, "Biomaterials: tissue-engineering and scaffolds," In Ed. S. Webster, *Encyclopedia of Medical Devices and Instrumentation,* 2nd ed. (John & Wiley Press, NY, 2006), pp 366–383.
5. G. Khang, D-W. Kim, and M. S. Kim, *Regenerative Medicine Series Book, Vol III. Tissue Engineering* (Hanrimwon Pub., Seoul, 2008).
6. S. J. Yoon, S. H. Kim, H. J. Ha, Y. K. Ko, J. W. So, M. S. Kim, Y. I. Yang, G. Khang, and H. B. Lee, *Tissue Eng.,* **14**(4), 539 (2008).

7. D. Lickorish, L. Guan, and J. E. Davies, *Biomaterials*, **28**(8), 1495 (2007).
8. J. Klompmaker, H. W. Jansen, R. P. Veth, J. H. de Groot, A. J. Nijenhuis, and A. J. Pennings, *Biomaterials*, **12**(9), 810 (1991).
9. J. Y. Lim, S. H. Kim, J. H. Choi, J. M. Rhee, H-S. Shin, and G. Khang, *Tissue Eng. Regen. Med.*, **5**(3), 437 (2008).
10. M. S. Kim, H. H. Ahn, Y. N. Shin, M. H. Cho, G. Khang, and H. B. Lee, *Biomaterials*, **28**, 5137 (2007).
11. S. J. Yoon, K. S. Park, B. S. Choi, G. Khang, M. S. Kim, J. M. Rhee, and H. B. Lee, *Key Eng. Mater.*, **342**, 161 (2007).
12. Y. K. Ko, S. H. Kim, J. S. Jeong, J. S. Park, J. Y. Lim, M. S. Kim, H. B. Lee, and G. Khang, *Polymer(Korea)*, **31**(6), 505 (2007).
13. M. Sha'ban, S. J. Yoon, Y. K. Ko, H. J. Ha, S. H. Kim, J. W. So, R. B. H. Idrus, and G. Khang, *J. Biomater. Sci., Polymer Edn.*, **19**(9), 1219 (2008).
14. Y. Oh, S. H. Kim, S. H. Jung, H. H. Hong, N. R. Jeon, S. J. Lee, M. van Dyke, J. J. Yoo, J. M. Rhee, and G. Khang, *Tissue Eng. Regen. Med.*, **5**(4), 855 (2008).
15. S. H. Jung, J. W. Jang, S, H. Kim, H. H. Hong, A. Y. Oh, J. M. Rhee, Y. S. Kang, and G. Khang, *Tissue Eng. Regen. Med.*, **5**(4), 643 (2008).
16. Y. Iwasaki, S-I. Sawada, N. Nakabayashi, G. Khang, H. B. Lee, and K. Ishihara, *Biomaterials*, **23**, 3897 (2002).
17. M. Brittberg, A. Lindahl, A. Nilsson, C. Ohlsson, O. Isaksson, and L. Peterson, *N. England J. Med.*, **331**(14), 889 (1994).
18. S. Munirah, B. S. Aminuddin, O. C. Samsuddin, K. H. Chua, N. H. Fuzina, and B. H. I. Ruszymah, *Tissue Eng. Regen. Med.*, **2**, 347 (2005).
19. S. Munirah, B. S. Aminuddin, K. H. Chua, O. C. Samsudin, A. H. M. Y. Badrul, B. Azmi, N. H. Fuzina, and B. H. I. Ruszymah, *J. Biosciences*, **17**(1), 9 (2006).
20. D. Eyrich, F. Brandl, B. Appel, H. Wiese, G. Maier, M. Wenzel, R. Staudenmaier, A. Goepferich, and T. Blunk, *Biomaterials*, **28**, 55 (2007).
21. G. Chen, T. Sato, T. Ushida, N. Ochiai, and T. Tateishi, *Tissue Eng.*, **10**, 323 (2004).
22. S. Munirah, S. H. Kim, B. H. I. Ruszymah, and G. Khang, *Euro. Cells Mater.*, **15**(2), 41 (2008).
23. S. Munirah, S. H. Kim, B. H. I. Ruszymah, and G. Khang, *J. Orthop. Surg. Res.*, **3**, 17 (2008).
24. Y. K. Ko, S. H. Kim, H. J. Ha, S. J. Yoon, J. M. Rhee, M. S. Kim, H. B. Lee, and G. Khang, *Key Eng. Mater.*, **342–343**, 173 (2007).
25. K. J. Tracey and A. Cerami, *Ann. Rev. Med.*, **45**, 491, 1994.

26. Blum, J. Moseley, L. Miller, K. Richelsoph, and W. Haggard, *Orthopedics*, **27**, 161 (2004).
27. S. H. Kim, K. S. Park, B. S. Choi, H. J. Ha, J. M. Rhee, M. S. Kim, Y. S. Yang, H. B. Lee, and G. Khang, *Adv. Exp. Med. Biol.*, **585**, 167 (2006).
28. K. Eid, S. Zelicof, B. P. Perona, C. B. Sledge, and J. Glowacki, *J. Orthop. Res.*, **19**(5), 962 (2001).
29. C. Gerad, *Nat. Immunol.*, **6**(4), 366 (2005).

Chapter 17

ARTIFICIAL BINDING GROWTH FACTORS

Takashi Kitajima and Yoshihiro Ito
Nano Medical Engineering Laboratory, RIKEN Advanced Science Institute 2-1 Hirosawa, Wako-shi, Saitama 351-0198 Japan
t-kitajima@riken.jp; y-ito@riken.jp

Growth factors act on cells via diffusible and nondiffusible mechanisms. For the design of growth factors for tissue engineering, regenerative medicine, and cell culture systems, it is important to consider their mechanisms of action at the cellular and molecular levels. Here, we summarized and discussed the designs and applications of gene-engineered binding growth factors.

17.1 Introduction

Growth factors play important roles in wound healing and tissue regeneration.[1] They are naturally occurring proteins capable of stimulating cellular growth, proliferation, and cellular differentiation. They are important for the regulation of a variety of cellular processes and typically act as signaling molecules between cells. They often promote cell differentiation and maturation, which vary among growth factors. For example, bone morphogenic proteins (BMPs) stimulate bone cell differentiation, while fibroblast growth

Handbook of Intelligent Scaffolds for Tissue Engineering and Regenerative Medicine
Edited by Gilson Khang

www.panstanford.com

factors (FGFs) and vascular endothelial growth factors (VEGFs) stimulate blood vessel differentiation (angiogenesis). The hepatocyte growth factor (HGF) also has angiogenic activity. In the field of tissue engineering, growth factors are now recognized as a fundamental requirement, in addition to cells and materials (either synthetic substrates or naturally occurring matrices).

The prolonged retention and activity of growth factors on cells or surrounding environment (i.e., the extracellular matrix [ECM] or artificial implant scaffold) is considered to be advantageous in regenerative medicine applications, until the repair process is initiated or even completed. However, most growth factors act in a diffusible manner and are generally unstable in a tissue environment, when exogenously applied; for example, the half-life of intravenously administered bFGF is very short (1–2 min).[2] Thus, many attempts have been made to improve the performance of growth factors (i.e., their active period and stability). One approach is the controlled release of growth factors via the development or modification of the carrier materials that are used for the delivery of these molecules.[3,4] The other approach is the modification of growth factors for immobilization on, or for high-affinity binding to, cells or scaffold biomaterials. This chapter focuses on the latter approach, especially on the engineering of growth factors using recombinant means to confer binding affinity.

17.2 Diffusible and Nondiffusible Actions of Growth Factors

Figure 17.1 summarizes the modes of action of growth factors in cells. Diffusible interactions (e.g., endocrine, paracrine, autocrine, and intracrine) and nondiffusible interactions (e.g., juxtacrine and matricrine) are known to occur. It is also known that each growth factor does not necessarily have a single mechanism of action at the cellular level.

Most growth factors act in a diffusible manner, interact with the cognate receptor on the cell membrane, and form a complex. This interaction induces the phosphorylation of the receptor and triggers

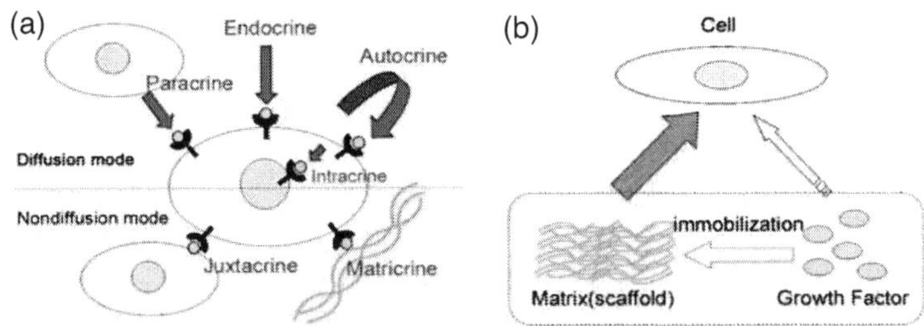

Figure 17.1. Modes of action of growth factors (a) and the concept of growth factor immobilization (b).

signal transduction in the cell. These complexes are then internalized, partially decomposed by lysosomes, and partially recycled to the cell membrane. Thus, internalization of the receptor/growth factor complexes reduces their number at the cell membrane, which leads to the desensitization of cells (down-regulation) and reduction of excessive response and overstimulation. In contrast, some growth factors are known to act in a nondiffusible manner by being present at the cell surface (juxtacrine) or by associating with specific substances, such as the ECM (matricrine). The nondiffusible mechanism was elucidated by the discovery of cell membrane–bound growth factors in the 1990s,[5,6] which include heparin-binding-epidermal growth factor (HB-EGF), transforming growth factor-β (TGF-β), tumor necrosis factor-α (TNF-α), colony stimulating factor 1 (CSF-1), and the c-kit ligand. These growth factors are hardly internalized but exhibit long-term activity without down-regulation.

The immobilization or high-affinity binding approach is considered as the means of choice to mimic the nondiffusible action of growth factors (Fig. 17.1b). Chemically immobilized growth factors were devised and investigated in the 1990s by our group, and later by other researchers, as reviewed recently by Ito.[7,8] These growth factors were covalently cross-linked to artificial materials and exhibited long-lasting activity, which suggests that internalization and down-regulation were blocked or retarded. The procedure of cross-linking is simple and applicable to various growth factors, without a need for prior modification of the growth factors themselves, although cross-linked molecules are randomly oriented on the surface. However, a possibility remains of damage or conformational changes after chemical treatment.

17.3 Gene-Engineered Binding Growth Factors

Gene engineering of growth factors aimed at conferring high binding affinity (substantial immobilization) is another means of immobilization. There are many reports on interactions between biological molecules. For example, the matrix protein fibronectin associates with collagen at high affinity via a polypeptide sequence termed collagen-binding domain (CBD). Fibronectin also comprises a binding domain for fibrin, cells, and heparin.[9] Binding growth factors were generated on the basis of the knowledge on this type of interaction, as the major design of these molecules is the fusion of growth factors with polypeptide sequences that have an affinity for target proteins or materials. Figure 17.2 shows examples of designs that we and others have employed. The resulting proteins exhibit collagen-binding, fibrin-binding, or cell-binding affinity.

Although this methodology requires steps for elaborate production of recombinant proteins, once the binding growth factors are generated, one can simply add them to the target materials for

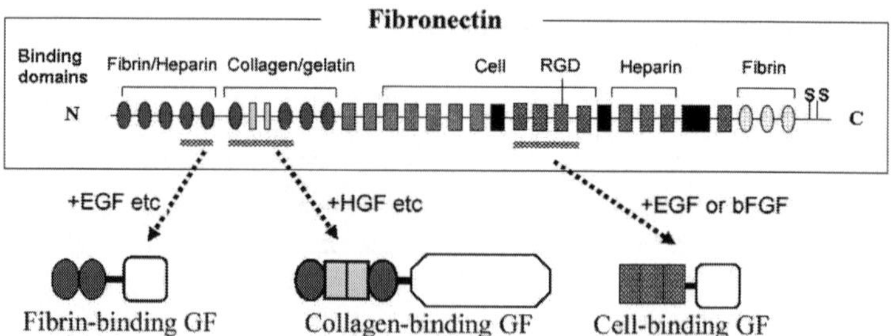

Figure 17.2. Construction of binding growth factors by fusion with fibronectin domains. DNA fragments encoding binding domains and growth factors were ligated and incorporated into expression vectors, and recombinant fusion proteins were produced using *E. coli* or eukaryotic expression systems. *Abbreviations*: GF, growth factors; RGD, cell-binding site sequence Arg-Gly-Asp.

binding, without harmful treatment. In addition, it is possible to achieve uniform orientation of molecules on the targets.

17.3.1 *Collagen-Binding Growth Factors*

Collagen is the major component of the ECM and is present in most tissues. It has been used as a material for scaffold preparation because of its biocompatibility. To date, four kinds of polypeptide sequences from the CBD of different proteins have been employed to generate the collagen-binding growth factor: von Willebrand Factor (vWF, 10 amino acids), bacterial collagenase ($\sim$24 kDa), human collagenase (eight amino acids), and fibronectin ($\sim$40 kDa or 27 kDa).

Nimni *et al.* produced a fusion protein of CBD with vWF and TGFβ1[13,14] or other growth factors. Collagen-binding TGFβ1 (rhTGFβ1-F2) had a lower biological activity compared with its native counterpart, both in the soluble state and in the collagen-bound state.[13] Yet, when combined with a collagen gel, rhTGFβ1-F2 stimulated the migration and proliferation of mesenchymal stem cells in the gel, as well as their differentiation to cartilage cells.[14] The osteogenic activity of this fusion protein was lower than that of native BMP2, without affinity to collagen. Other fusion proteins of EGF (EGF-CBD),[15] basic (b) FGF (rhbFGF-F2),[16] and BMP3 (rhBMP3-C)[17] all showed binding affinity to collagen and an activity similar to that of the original molecule: growth stimulatory activity on NIH3T3 cells (for EGF) and endothelial cells (for bFGF) and growth inhibitory activity on bone cells (for BMP3). It was not clearly demonstrated that these biological activities were exerted when they were combined with collagen.

Nishi *et al.*[18] produced other fusion proteins using the collagen-binding motif from bacterial collagenase fused to EGF or bFGF. Both were retained for 5–10 days when subcutaneously injected into rats (5–7 days), which suggests their tight binding to collagen or other ECM components. The EGF fusion protein exhibited only 1/10 of the original EGF activity, while the bFGF fusion protein showed an activity comparable with that of the native molecule. The CBD used may affect the activity of the fusion protein or its binding to the receptor, depending on the growth factors being fused. Brewster *et al.*[19] employed the same CBD to generate collagen-binding bFGF

(R136K-CBD). R136K is an engineered bFGF that has thrombin resistance. These authors used surface plasmon resonance (SPR) analysis to show that the binding affinity of the fusion protein was fourfold higher than that of native bFGF and R136K. R136K-CBD stimulated endothelial cell angiogenesis in fibrin gel. However, the advantage of the collagen-binding affinity of the fusion protein during angiogenesis was not clearly shown via comparison with unfused R136K.

A short polypeptide sequence of the human collagenase CBD was also employed to construct collagen-binding fusion proteins of platelet-derived growth factor BB (PDGF-BB),[20] bFGF,[21] BMP2,[22] and nerve growth factor (NGF).[23] Zhao *et al.*[21] compared a short polypeptide sequence of vWF with that of human collagenase as the fusion partner for bFGF and found that the latter provides a higher binding affinity to collagen. Considering that the same growth factors fused with different partner polypeptides were synthesized (as shown in Table 17.1), the comparison of the properties of these factors, as reported by Zhao *et al.*,[21] is important.

Fusion proteins of the CBD of fibronectin and EGF,[24] VEGF,[25] or HGF[26] were also produced. This fusion design (Fig. 17.2) was also applied to BMP4 (Kitajima *et al.*, unpublished). These fusion proteins exhibit binding affinity to major collagen types (I–V), which is not observed for the native growth factors. Moreover, they showed native biological activities in the collagen-bound state. These fusion proteins can be stored for longer periods than native growth factors. For example, collagen-binding EGF (FNCBD-EGF) bound to collagen-coated plates or diluted in culture medium showed no loss of EGF activity after one month of storage at 4°C, while native EGF lost its activity in the same conditions (Fig. 17.3). The stability of collagen-binding HGF (CBD-HGF) in culture medium at 37°C was also confirmed[26]: CBD-HGF continued to stimulate the proliferation of endothelial cells without supplementation, both in the soluble and in the collagen-bound state, for longer periods than native HGF. The fused CBD moiety may protect the growth factor moieties from degradation or denaturation.

When injected into muscle tissues, FNCBD-EGF bound to collagen distributes along muscle fibers, while EGF is not retained in the tissues.[27] It was also shown that FNCBD-EGF binds to type IV collagen (basement membrane) exposed on the inner surface of blood

Table 17.1. The gene-engineered binding growth factors reported to date, many of which are matrix-binding growth factors. It is well known that most growth factors have affinity to matrix proteins, and the ECM is considered as the reservoir of growth factors.[10,11] However, this affinity is not usually strong enough for prolonged retention of the growth factors at the ECM.[12] Therefore, artificial improvement of the matrix-binding affinity of growth factors has been tried.

Binding target	Growth factor	Fused polypeptide	Origin	Ref.
Collagen Gelatin	bFGF, TGFβ1, BMP3, EGF	CBD polypeptide[a] (10 amino acids)	vWF	14–17, 21
	EGF, bFGF	CBD (20 KDa)	bacterial collagenase	18, 19
	PDGF-BB, BMP2, bFGF, NGF	CBD polypeptide[b] (eight amino acids)	human collagenase	20–23
	EGF, VEGF, HGF	CBD (40 or 27 KDa)	Fibronectin	24–26
	EGF	Collagen (full size)	Collagen	29
Fibrin Fibrinogen	VEGF, BMP2, NGF, KGF, ephrin B2	FXIIIa substrate sequence[c] (eight or 12 amino acids)	α2-plasmin inhibitor	30–35
	EGF	FBD (11 Kda)	Fibronectin	37
Cell (integrin)	bFGF, EGF	Cell-binding domain (30 Kda)	Fibronectin	38, 39
	EGF	RGD	Fibronectin	40, 41
Cell (integrin and IGF-I R)	IGF-I	vitronectin (full size)	Vitronectin	42
Ti	BMP2	Ti-binding peptide motif[d] (six amino acids)	artificial (peptide library)	43
Solid surface	EGF	Fc region	immunoglobulin	44
Cellulose	SCF (extracellular domain)	cellulose-binding domain	bacterial xylanase	45

Abbreviations: FBD, fibrin-binding domain; IGF-IR, insulin-like growth factor-I receptor; IGF-I, insulin-like growth factor; KGF, keratinocyte growth factor; SCF, stem cell factor; Ti, titanium.

[a]Peptide sequence: WREPSFMALS; [b]Peptide sequence: TKKTLRT; [c]Peptide sequence: NQEQVSPL or LNQEQVSPRKK; [d]Peptide sequence: RKLPDA.

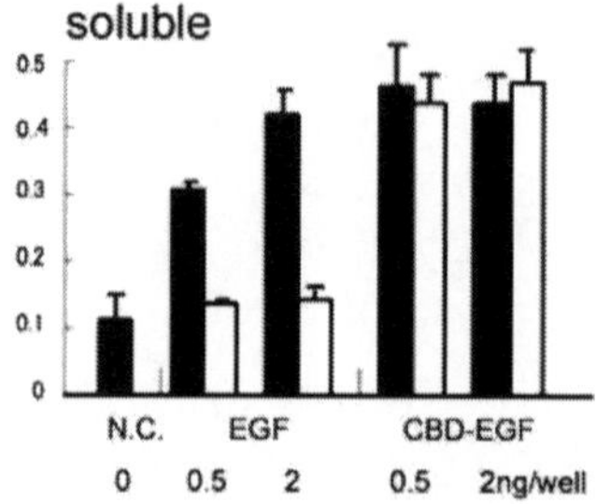

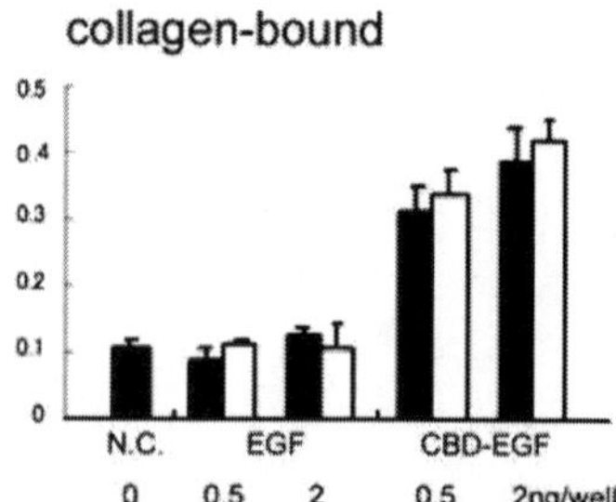

Figure 17.3. Stability of the growth stimulation activity of collagen-binding EGF. Fresh proteins or proteins stored at 4°C for 30 days in culture medium (left graph) or bound to collagen-coated wells (right graph) were applied to culture of human dermal fibroblasts (48-well plate). Filled column, day 0; open column, day 30. Cell growth activity was evaluated at seven days of culture using a WST1 colorimetric assay (absorbance at 450–650 nm; mean ± SD).

vessels after balloon-catheter injury. The application of CBD-HGF to the injured area enhances reendothelialization.[28]

Growth factors directly fused to collagen or other matrix proteins may represent the ultimate binding protein. Hayashi *et al.*[29] produced a fusion protein of EGF and type III collagen that adopts a triple helical structure, similarly to native collagen. This fusion protein can be incorporated into collagen fibrils to allow the preparation of collagen material that includes EGF.

17.3.2 *Fibrin-Binding Growth Factors*

Fibrin is an abundant protein present on the surface of injured tissues. In addition, it has been studied as a material for tissue engineering scaffolds. Therefore, growth factors with fibrin-binding affinity are considered useful for tissue regeneration. Sakiyama-Elbert *et al.*[30] produced a fusion protein between NGF and a short polypeptide of the α2-plasmin inhibitor (α2-pI); the activated coagulation factor XIII (FXIIIa) mediated the cross-linking of this sequence to fibrin. In addition, a plasmin cleavage sequence was inserted between the α2-pI sequence and the growth factor. The fusion protein (TG-P-NGF) was incorporated into fibrin clots by FXIIIa and was then released by plasmin when the tissues started to heal. TG-P-NGF was designed to be released "when cells demand." The activity of the fusion protein is approximately 40% that of native NGF, when assayed by neurite extension in PC12 cells. Unlike collagen-binding growth factors, this fusion protein binds to fibrin covalently. This may relate to the reduced activity observed in an *in vitro* assay.

The same or similar construct designs were applied to BMP2 (TG-pI-BMP-2),[31] KGF (P-KGF),[32] VEGF (α2-PI1-8-VEGF121),[33,34] and ephrinB2.[35] The TG-pI-BMP-2-mediated elevation of alkaline phosphatase activity in C3H10T1/2 cells was comparable with that observed for native BMP2, while its retention in a fibrin gel was higher than that of the native protein (60% vs. 25% of added amount). The KGF fusion protein (P-KGF) binds to fibrin gel and promotes the growth of epithelial cells after release from the gel. VEGF121 is a VEGF isoform that lacks the fibrin-binding sequence found in the major isoform (VEGF165).[36] Fusion with the α2-pI sequence enabled the incorporation of VEGF121 into a fibrin gel. In addition, the fusion protein (α2-PI1-8-VEGF121) stimulated the growth of endothelial cells on the gel in a dose-dependent manner.

Another fibrin-binding growth factor was reported recently[37]: a chimeric protein of the fibronectin fibrin-binding domain (FBD) and EGF. This fusion protein (FBD-EGF) binds to fibrin in the absence of the cross-linking enzyme (FXIIIa) and may be suitable for the preparation of fibrin-based scaffolds with EGF activity. The wound-healing potential of FBD-EGF was examined using an *in vitro* culture model of keratinocyte sheets; repair of the injured sheet was enhanced by fibrin-bound FBD-EGF but not by EGF alone.

17.3.3 *Cell-Binding Fusion Proteins*

Cell-binding growth factors were developed in the early 1990s by employing the fibronectin cell-binding domain (~30 kDa, including the RGD sequence). The EGF fusion protein (C-EGF)[38] promotes the proliferation of rat kidney fibroblasts. The bFGF fusion protein (FN-FGF)[39] stimulates the growth of endothelial cells (HUVEC); however, its efficacy as compared with native bFGF was not clearly shown in the same conditions. There are not many reports of attempts at constructing cell-binding growth factors. This can probably be explained by the fact that these fusion proteins would not be retained on the culture plates for long periods; thus, their efficacy is not persistent. An improved cell-binding EGF (containing an RGD sequence) was generated recently via the incorporation of hydrophobic linker sequences[40] and collagen-binding sequences.[41]

Van Lonkhyzen *et al.*[42] produced a chimeric protein of IGF-I and vitronectin. This protein (VN:IGF-I) associates with IGF-I-binding protein (IGF-BP). The complex of VN:IGF-I and IGF-BP binds to both the IGF-I receptor and to integrin (thus binds to cells) and stimulates receptor-mediated and integrin-mediated pathways. As a result, the VN:IGF-I/IGF-BP complex exhibits a higher cell proliferation activity compared with the IGF-I/IGF-BP complex.

17.3.4 *Other Binding Growth Factors*

Growth factors were also engineered to bind to artificial materials. BMP2 was fused with a polypeptide sequence that has affinity to titanium[43]; the peptide sequences were selected with a gene evolution approach using a phage display system. The fusion of this sequence allows the reversible binding of BMP2 to titanium, with retention of its biological activity.

Ogiwara *et al.*[44] produced an EGF fusion protein coupled with the Fc region of immunoglobulin G (IgG). The Fc region is often used as a fusion tag to allow extracellular domains of large membrane proteins or receptor molecules to adhere nonspecifically to solid surfaces. These authors used the hydrophobicity of the Fc region as a binding domain for hydrophobic artificial substrates, although this type of fusion protein is not considered as a binding growth factor.

Dohney *et al.*[45] fused the cellulose-binding domain of bacterial xylanase with the extracellular domain of the stem cell factor (SCF). The fusion construct binds to cellulose tightly and stimulates the proliferation of SCF-dependent murine and human cells.

17.4 Application of Engineered Binding Growth Factors

The binding growth factors were designed to continue to stimulate signaling for a long time, like chemically immobilized growth factors, as reviewed by Ito.[7,8] However, one report showed a very slow release and internalization of a bound growth factor.[46] One of the advantages of binding growth factors is the combination of natural tissues or artificial scaffolds by simple addition. Therefore, binding growth factors have been studied regarding their efficacy in wound healing, tissue reconstruction, and substitution.

17.4.1 *Skin Wound Repair*

The wound-healing effect of engineered EGF combined with matrix components was tested. Collagen-binding EGF (FNCBD-EGF) mixed with collagen gel stays in the wound, stimulates cell proliferation, and enhances the formation of granulation tissue in intractable skin wounds of diabetic mice (db/db).[27] Implantation of a collagen sponge combined with FNCBD-EGF induces epithelialization above or underneath the sponge, which results in wound closure.[24] Native EGF does not exhibit these wound-healing effects. Collagen-binding bFGF (rhbFGF-F2) dissolved in collagen solution accelerates wound healing, both in normal and in diabetic rats.[14]

Collagen-binding PDGF-BB combined with a collagen membrane accelerates the healing of rabbit dermal wounds (ischemic ulcers), with concomitant capillary formation.[47] Its effect, compared with native PDGF-BB, apparently improves after two weeks (but not after one week) of implantation. Hall *et al.*[15] showed that collagen-binding EGF (EGF-CBD) instillated into an injured colon without a carrier promotes the repair of the inflamed luminal surface.

Fibrin-binding KGF (P-KGF) in combination with fibrin showed a potential for wound healing of human skin tissues grafted on athymic mice,[32] although its efficacy was not great compared with that of native KGF/fibrin.

17.4.2 *Repair of Cardiovascular Tissues*

The angiogenic factors VEGF, bFGF, and HGF have been studied for their therapeutic potential to restore blood supply in ischemic tissues. The effects of gene-engineered constructs of these factors were also studied.

17.4.2.1 Materials that induce angiogenesis

Fibrin-binding VEGF121 (α2PI1-8-VEGF121) mixed with fibrin exhibits efficacious angiogenic properties, as assessed using a chorioallantoic membrane (CAM) assay[34]: the blood vessel density in the presence of fibrin-binding VEGF121 was approx 1.5 times higher than that induced by native VEGF121. In addition, α2PI1-8-VEGF121 induces the formation of nonleaky vessels on a Teflon chamber coated with fibrin and subcutaneously implanted.

Implantable scaffolds combined with collagen-binding growth factors have also been studied. Vessel migration into collagen scaffolds is enhanced by CBD-bFGF,[21] although the collagen-binding affinity does not vary much between native bFGF and CBD-bFGF. The difference in the angiogenic effect between native and engineered bFGFs was apparent at seven days but not at four days, after implantation. Collagen-binding HGF (CBD-HGF)[26] also enhances blood vessel migration into subcutaneously implanted collagen sponges that are combined with the fusion protein.

17.4.2.2 Artificial blood vessel and heart valve

Extended polytetrafluoroethylene (ePTFE) is used as the material for small-caliber artificial blood vessels. Figure 17.4a shows our preliminary design of artificial blood vessels with HGF activity; an ePTFE sheet was treated serially with an amphiphilic polymer and collagen and was then combined with CBD-HGF. Endothelial cells of the human coronary artery grew only on the surface, where CBD-HGF is present (Fig. 17.4b).

The effect of the application of CBD-HGF on endothelialization of implant materials was further studied. Porcine heart valves that were decellularized (i.e., mainly composed of collagen) were combined with CBD-HGF.[48] Endothelialization of the valves after implantation into a dog pulmonary artery was enhanced by CBD-HGF. Earlier coverage of valve surfaces by endothelial cells may prevent surface thrombosis and contribute to the durability of the implants.

A sheet of collagen preparation (decellularized porcine urinary bladder matrix) was combined with CBD-HGF and implanted into

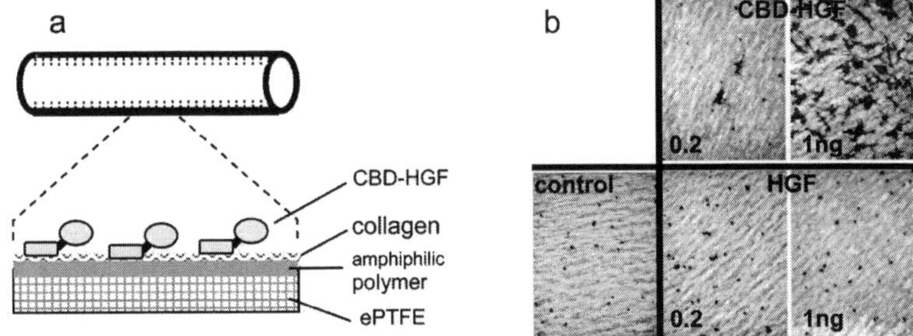

Figure 17.4. A design of an artificial blood vessel (a) and growth of coronary artery endothelial cells on vessels incubated with CBD-HGF or HGF (b). Cells were stained with an anti-CD31 antibody.

the right ventricular walls of a mechanically injured porcine heart.[49] An increase in contractility and electrical activity was noticed 60 days after surgery, which was accompanied with angiogenesis and the appearance of cardiomyocyte-like cells in the injured region.

17.4.3 *Nerve Regeneration*

Fibrin-binding NGF (TG-P-NGF) mixed with fibrin gel enhances neurite extension from chick embryo dorsal root ganglia (organ culture) by 50% relative to native NGF and by 350% relative to the fibrin gel alone, in spite of the reduced activity of the fusion construct.[30]

Collagen-binding NGF (CBD-NGF)[23] combined with a collagen membrane and implanted subcutaneously enhances nerve growth into the membrane; the CBD-NGF-mediated nerve fiber density is approximately twice that observed for native NGF. It also enhances healing of rabbit dermal ischemic ulcers.

17.4.4 *Bone Regeneration*

A collagen sponge impregnated with collagen-binding BMP3 (rh-BMP3-C) failed to induce ectopic bone formation in rats. This may coincide with a report that denies the osteoinductive activity of BMP3.[50] However, the authors showed that porous ceramics coated with collagen and combined with the recombinant protein induce cranial repair.

Chen *et al.*[22,51] subcutaneously implanted demineralized bone matrix (DBM) combined with collagen-binding BMP2 (BMP2-h) and showed ectopic bone formation in rats. They also showed that BMP2-h-loaded DBM was effective in the repair of rabbit mandible bone defects, with an efficiency that was 10%–20% higher than that of native BMP-2/DBM.

The examination of the effects of fibrin-binding BMP2 (TG-pI-BMP2) in a rat craniotomy defect model[31] revealed that the healing area was 76% larger than that observed for native BMP2. In addition to the craniofacial bones, this study was extended to the repair of long bones, as assessed by bone bridging.

17.5 Concluding Remarks

The major advantage of the gene-engineered binding growth factors is that the protein is able to bind to biological substances and artificial materials that can hardly retain native growth factors. In addition, the dose can be reduced. In some designs, the stability of the resultant molecules can be prolonged. These advantages are crucial to the production of scaffold materials equipped with additional functions.

Various binding growth factors were produced and examined over the past 15 years. Although fusion proteins with insufficient binding properties or with reduced biological activities have been reported in the early stages of these studies, recently designed proteins exhibit good properties. It now seems possible that any growth factor can be engineered to have desired binding properties (at least for matrix binding), without substantial loss of native activity. In addition, there has been an increase in the number of reports showing the efficacy of the engineered proteins during tissue repair, although there remain unsolved issues. Thus, the binding growth factors are considered as promising materials for regenerative medicine.

References

1. S. Werner and R. Grose, *Physiol. Rev.,* **83**, 835 (2003).
2. G. F. Whalen, Y. Shing, and J. Folkman, *Growth Factors,* **1**, 157 (1989).

3. M. J. Whitaker, R.A. Quirk, S. M. Howdle, and K. M. Shakesheff, *Pharm. Pharmacol.*, **53**, 1427 (2001).
4. R. Vasita and D. S. Katti, *Expert. Rev. Med. Devices.*, **3**, 29 (2006).
5. J. Massague, A. Pandiella, *Ann. Rev. Biochem.*, **62**, 515 (1993).
6. R. Iwamoto and E. Mekada, *Cytokine Growth Factor Rev.*, **11**, 335 (2000).
7. Y. Ito, *Soft Matter*, **4**, 46 (2008).
8. Y. Ito, In Eds. D. Puleo and R. Bizios, *Biological Interactions on Materials Surfaces: Understanding and Controlling Protein, Cell and Tissue Responses* (Springer, 2009), p. 173.
9. K. M. Yamada, In Ed. D. F. Mosher, *Biology of Extracellular Matrix: Series A. Fibronectin* (Academic Press, San Diego, 1989), p. 47.
10. J. Taipale and J. Keski-Oja, *FASEB J.*, **11**, 51 (1997).
11. M. E. Nimni, *Biomaterials*, **18**, 1201 (1997).
12. D. Moscatelli, *J. Biol. Chem.*, **267**, 25803 (1992).
13. T. L. Tuan, D. T. Cheung, L. T. Wu, A. Yee, S. Gabriel, B. Han, L. Morton, M. E. Nimni, and F. L. Hall, *Connect. Tissue Res.*, **34**, 1 (1996).
14. J. A. Andrades, B. Han, J. Becerra, N. Sorgente, F. L. Hall, and M. E. Nimni, *Exp. Cell. Res.*, **250**, 485 (1999).
15. F. L. Hall, A. Kaiser, L. Liu, Z. H. Chen, J. Hu, M. E. Nimni, R. W. Beart, Jr., and E. M. Gordon, *Int. J. Mol. Med.*, **6**, 635 (2000).
16. J. A. Andrades, J. A. Santamaria, L. T. Wu, F. L. Hall, M. E. Nimni, and J. Becerra, *Protoplasma*, **218**, 95 (2001).
17. B. Han, N. Perelman, B. Tang, F. Hall, E. C. Shors, and M. E. Nimni, *J. Orth. Res.*, **20**, 747 (2006).
18. N. Nishi, O. Matsushita, K. Yuube, H. Miyanaka, A. Okabe, and F. Wada, *Proc. Natl. Acad. Sci. U S A*, **95**, 7018 (1998).
19. L. P. Brewster, C. Washington, E. M. Brey, A. Gassman, A. Subramanian, J. Calceterra, W. Wolf, C. L. Hall, W. H. Velander, W. H. Burgess, and H. P. Greisler, *Biomaterials*, **29**, 327 (2008).
20. H. Lin, B. Chen, W. Sun, W. Zhao, Y. Zhao, and J. Dai, *Biomaterials*, **27**, 5708 (2006).
21. W. Zhao, B. Chen, X. Li, H. Lin, W. Sun, Y. Zhao, B. Wang, Y. Zhao, Q. Han, and J. Dai, *J. Biomed. Mater. Res. A*, **82**, 630 (2007).
22. B. Chen, H. Lin, Y. Zhao, B. Wang, Y. Zhao, Y. Liu, Z. Liu, and J. Dai, *J. Biomed. Mater. Res. A*, **80**, 428 (2007).
23. W. Sun, H. Lin, B. Chen, W. Zhao, Y. Zhao, and J. Dai, *J. Biomed. Mater. Res. A*, **83**, 1054 (2007).

24. T. Ishikawa, H. Terai, and T. Kitajima, *J. Biochem.*, **129**, 627 (2001).
25. T. Ishikawa, M. Eguchi, M. Wada, Y. Iwami, K. Tono, H. Iwaguro, H. Masuda, T. Tamaki, and T. Asahara, *Arterioscler. Thromb. Vasc. Biol.*, **26**, 1998 (2006).
26. T. Kitajima, H. Terai, and Y. Ito, *Biomaterials,* **28**, 1989 (2007).
27. T. Ishikawa, H. Terai, K. Yamamoto,K. Harada, and T. Kitajima, *Artif. Organs.*, **27**, 146 (2003).
28. N. Ohkawara, H. Ueda, S. Shinozaki, T. Kitajima, Y. Ito, H. Asaoka, A. Kawakami, E. Kaneko, and K. Shimokado, *J. Ater. Thromb.*, **14**, 185 (2007).
29. M. Hayashi, M. Tomita, and K. Yoshizato, *Biochim. Biophys. Acta,* **1528**, 187 (2001).
30. S. Sakiyama-Elbert, A. Panitch, and J. A. HubbellL *FASEB J.*, **15**, 1300 (2001).
31. H. G. Shmoekel, F. E. Weber, J. C. Schense, K. W. Graetz, P. Schwalder, and J. A. Hubbell, *Biotechnol. Bioeng.*, **89**, 253 (2005).
32. D. J. Geer, D. D. Swartz, and S. T. Andreadis, *Am. J. Pathol.*, **167**, 1575 (2005).
33. A. H. Zisch, U. Schenk, J. C. Schense, S. E. Sakiyama-Elbert, and J. A. Hubbell, *J. Control. Rel.*, **72**, 101 (2001).
34. M. Ehrbar, V. G. Djonov, C. Schnell, S. A. Tschanz, G. Martiny-Barton, U. Schenk, J. Wood, P. H. Burri, J. A. Hubbell, and A. H. Zisch, *Circ. Res.*, **94**, 1124 (2004).
35. M. Ehrbar, S. M. Zeisberger, G. P. Raeber, J. A. Hubbell, C. Schnell, and A. H. Zisch, *Biomaterials,* **29**, 1720 (2008).
36. A. Sahni and C. W. Francis, *Blood*, **96**, 3772 (2000).
37. T. Kitajima, M. Sakuragi, H. Hasuda, T. Ozu, and Y. Ito, *Acta Biomater.*, **5**, 2623 (2009).
38. Y. Kawase, Y. Ohdate, T. Shimojo, Y. Taguchi, F. Kimizuka, and I. Kato, *FEBS Lett.*, **298**, 126, erratum in *FEBS Lett.*, **301**, 124 (1992).
39. H. Hashi, M. Hatai, F. Kimizuka, I. Kato, and Y. Yaoi, *Cell Struct. Funct.*, **19**, 37 (1994).
40. I. Elloumi, R. Kobayashi, H. Funabashi, M. Mie, and E. Kobatake, *Biomaterials,* **27**, 3451 (2006).
41. E. Hannachi Imen, M. Nakamura, M. Mie, and E. Kobatake, *J. Biotechnol.* **139**, 19 (2009).
42. D. R. Van Lonkhuyzen, B. G. Hollier, G. K. Shooter, D. I. Leavesley, and Z. Upton, *Growth Factors,* **25**, 295 (2007).

43. K. Kashiwagi, T. Tsuji, and K. Shiba, *Biomaterials,* **30**, 1166 (2009).
44. K. Ogiwara, M. Nagaoka, C. S. Cho, and T. Akaike, *Biotechnol. Lett.,* **27**, 1633 (2005).
45. J. G. Doheny, E. J. Jervis, M. M. Guarna, R. K. Humphries, R. A. Warren, and D. G. Kilburn *Biochem. J.,* **339**, 429 (1999).
46. E. J. Jervis, M. M. Guarna, J. G. Doheny, C. A. Haynes, and D. G. Kilburn, *Biotechnol. Bioeng.,* **91**, 314 (2005).
47. W. Sun, H. Lin, H. Xie, B. Chen, W. Zhao, Q. Han, Y. Zhao, Z. Xiao, and J. Dai, *Growth Factors,* **25**, 309 (2007).
48. T. Ota, Y. Sawa, S. Iwai, T. Kitajima, Y. Ueda, C. Coppin, H. Matsuda, and Y. Okita, *Ann. Thorac. Surg.,* **80**, 1794 (2005).
49. T. Ota, T. W. Gilbert, D. Schwarzman, C. F. McTieman, T. Kitajima, Y. Ito, Y. Sawa, S. F. Badylak, and M. Zenati, *J. Thorac. Cardiovasc. Surg.,* **136**, 1309 (2008).
50. M. E. Bahamonde, and K. M. Lyons, *J. Bone Joint Surg. Am.,* **83-A**, S56 (2001).
51. B. Chen, H. Lin, J. Wang, Y. Zhao, B. Wang, W. Zhao, W. Sun, and J. Dai, *Biomaterials,* **28**, 1027 (2007).

Chapter 18

POROUS POLY(LACTIC-*CO*-GLYCOLIC ACID) MICROSPHERE AS CELL CULTURE SUBSTRATE AND CELL TRANSPLANTATION VEHICLE

Byung-Soo Kim
Department of Bioengineering, Hanyang University, 17 Haegdang-dong, Seongdong-gu, Seoul, Korea
bskim@hanyang.ac.kr

Porous poly(lactic-*co*-glycolic acid) (PLGA) microspheres can be utilized both as a cell culture substrate to expand cells *ex vivo* and as a cell transplantation vehicle for tissue engineering or cell therapy. *Ex vivo* cell expansion on porous PLGA microspheres and subsequent transplantation of the cell-microsphere constructs removes the trypsinization process used to harvest *ex vivo*-expanded cells for transplantation. This would be advantageous since trypsinization interrupts interactions between cultured cells and their extracellular matrices, facilitating apoptosis and consequently limiting the therapeutic efficacy of the transplanted cells. Stem cells can be cultured on macroporous PLGA microspheres in stirred suspension bioreactors, expanded, and differentiated into a specific lineage. The apoptotic activity of cells cultured on

Handbook of Intelligent Scaffolds for Tissue Engineering and Regenerative Medicine
Edited by Gilson Khang

www.panstanford.com

microspheres could be significantly lower than that of trypsinized cells. Cells cultured on macroporous PLGA microspheres could survive much better than trypsinized adipose-derived stem cells (ASCs) upon transplantation. Importantly, the implantation of cells cultured on macroporous PLGA microspheres could result in much more extensive tissue formation than the implantation of cells cultured on plates and trypsinized. *Ex vivo* cell expansion and transplantation using this system would improve the therapeutic efficacy of cells over the current methods used for tissue engineering.

18.1 Introduction

Cell therapy and tissue engineering often require *in vitro* cell culture for cell expansion or lineage-specific differentiation of stem cells. However, the current cell culture methods have several limitations. Trypsinization interrupts the interactions between cultured cells and extracellular matrices secreted from the cultured cells. This may result in high apoptotic activity and reduced therapeutic efficacy of the cells upon transplantation.[1] The large surface area of the culture substrate is required to produce clinically relevant numbers of transplantable cells, but the surface area of conventional, two-dimensional (2D) culture plates constrains *ex vivo* cell expansion.

Biodegradable macroporous microcarriers combined with three-dimensional (3D), stirred suspension bioreactors could be an alternative culture system to avoid the limitations of current *ex vivo* cell expansion methods for cell therapy and tissue engineering. A microcarrier is a microsphere with a diameter of a few hundred micrometers, provides a large surface area for cell attachment and growth, and makes feasible 3D suspension cultures in a stirred suspension bioreactor.[2] Microcarrier culture overcomes the limitation of cell growth area in the 2D culture plate and thus would be suitable for clinical-scale production of transplantable cells. Moreover, macroporous microcarriers further increase the surface area for cell attachment and growth and protect the cultured cells from shear stress produced in a stirred bioreactor. A macroporous microcarrier that is biodegradable and made of clinically approved biomaterials can also be used as a cell transplantation vehicle, following cell culture for tissue engineering or cell transplantation therapy, avoiding the necessity of trypsinization for the harvest of *ex vivo*–expanded cells.

This chapter describes a macroporous microcarrier made of PLGA, a biomaterial approved for clinical application and used as both an adipocyte culture substrate and a cell transplantation vehicle for adipose tissue engineering. Adipose tissue engineering has been proposed as a potential treatment for the augmentation of soft tissue loss due to cancer, trauma, or burns.[3] It has been shown that ASCs seeded onto scaffolds and implanted into animals regenerated adipose tissues.[4–6] Nonporous PLGA microspheres have been used for adipose tissue engineering.[7–9] In this chapter, macroporous PLGA microspheres were used as a cell culture substrate to expand human ASCs and as a cell transplantation vehicle for adipose tissue engineering. ASCs were cultured on macroporous PLGA microspheres in stirred suspension bioreactors to expand to a large number of cells and differentiate into an adipogenic lineage. Implantation of ASCs cultured on macroporous PLGA microspheres could result in better adipose tissue formation than implantation of ASCs cultured on plates, trypsinized, and subsequently mixed with macroporous PLGA microspheres.

18.2 Fabrication of Macroporous PLGA Microspheres

Macroporous microcarriers have been used for the culture and subsequent transplantation of bone marrow stromal cells for bone regeneration.[1] However, the microcarrier material used in the study was not suitable for clinical application. PLGA is a biocompatible, synthetic polymer approved by the U.S. Food and Drug Administration. It does not contain biological by-products or pathogens that are recognized as antigens or allergens. In this chapter, macroporous microspheres made of PLGA were used for adipose tissue engineering.

Macroporous microspheres were fabricated from 75:25 PLGA (molecular weight±80,000 Da, Birmingham Polymers, Birmingham, AL) using a previously described water/oil/water double-emulsion method.[10] The fabricated PLGA microspheres had open pores

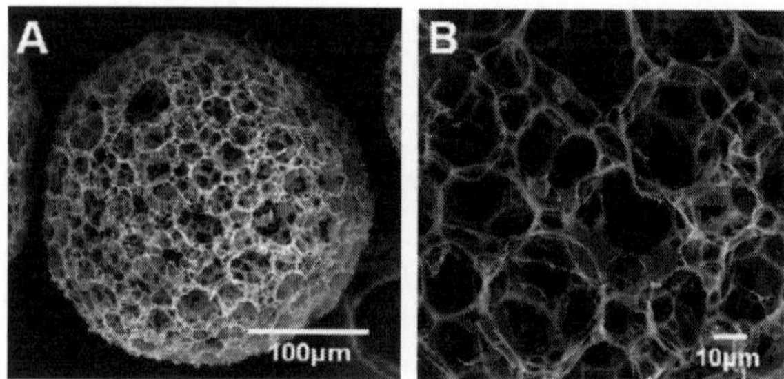

Figure 18.1. Scanning electron micrographs of (a) gross morphology and (b) cross section of open macroporous PLGA microspheres.

(Fig. 18.1a) on the microsphere surface and interconnected structures between the inner pores (Fig. 18.1b). The pore morphology in the cross section was similar to that on the surface, indicating that the porous structure was produced homogeneously throughout the microspheres. The average diameter of the microspheres was 372±43 mm. The average pore size on the microsphere surface was 36.1±6.9 mm, which is sufficient for cell infiltration and seeding within the microspheres.

18.3 Macroporous PLGA Microsphere as an ASC Culture Substrate

A microcarrier provides a large surface area for cell attachment and growth. Microcarrier culture combined with 3D suspension bioreactors overcomes the limitation of cell growth area in the 2D culture plate and thus would be suitable for clinical-scale production of transplantable cells. A macroporous microsphere with well-interconnected pores provides a larger surface area for cell adhesion and growth than a nonporous microsphere, and a larger number of cells can be expanded in a 3D suspension bioreactor. Furthermore, open macroporous microspheres facilitate the transport of nutrients and oxygen through the interconnected pores for cell growth and differentiation within the microspheres. The characteristics of open macroporous microspheres are desirable for high cell-seeding and cell-culture density.

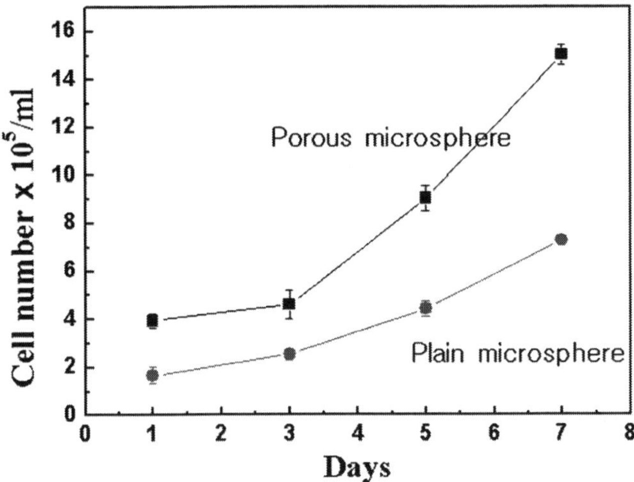

Figure 18.2. The number of human ASCs growing on macroporous PLGA microspheres (square) or nonporous PLGA microspheres (circle) in spinner flask culture. The cell inoculum density was 5×10^5 cells/mL.

Human ASCs were isolated from adipose tissue obtained by liposuction from informed and consenting patients using an enzymatic digestive process. The human ASCs were cultured in macroporous PLGA microspheres with a growth medium in spinner flasks for one week. One day after culture, 78.8±6.0% of the inoculated ASCs adhered to porous microspheres, while only 36.8±8.0% of the inoculated ASCs adhered to nonporous microspheres (Fig. 18.2), as determined according to DNA content measurement. ASCs cultured on macroporous PLGA microspheres in stirred suspension bioreactors expanded to a large number of cells. The number of ASCs cultured on the macroporous microspheres and nonporous microspheres increased 3.8- and 3.7-fold, respectively, over seven days (Fig. 18.2). Scanning electron microscopic examination of macroporous PLGA microspheres seeded with human ASCs and cultured in spinner flasks for seven days revealed that ASCs adhered well on the microsphere surface (Fig. 18.3a). Confocal microscopic examination revealed that cells were growing on pores throughout the macroporous microspheres (Fig. 18.3b).

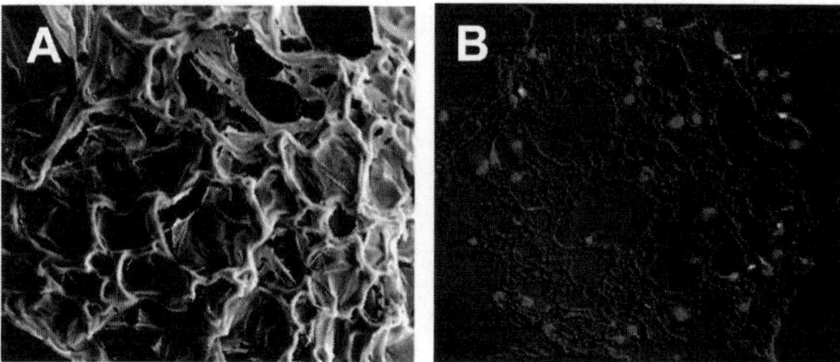

Figure 18.3. Human ASCs growing on macroporous PLGA microspheres on day 7 in spinner flask culture. (a) Scanning electron microscopic image and (b) confocal micrograph of DAPI staining of human ASCs cultured on macroporous PLGA microspheres. *Abbreviation*: DAPI, 4'-6-diamidino-2-phenylindole. See also Color Insert.

ASCs cultured on macroporous PLGA microspheres have the capacity to differentiate into mature adipocytes. ASCs were cultured on macroporous PLGA microspheres with a growth medium in spinner flasks for 7 days and were then cultured with the adipogenic differentiation medium for 10 days. The cells differentiated into an adipogenic lineage, as indicated by the cells staining positively for oil red O (Fig. 18.4).

ASCs cultured on macroporous PLGA microspheres had lower apoptotic activity than trypsinized ASCs. ASCs were cultured on plates or macroporous microspheres with growth medium for 7 days and subsequently with adipogenic differentiation medium for

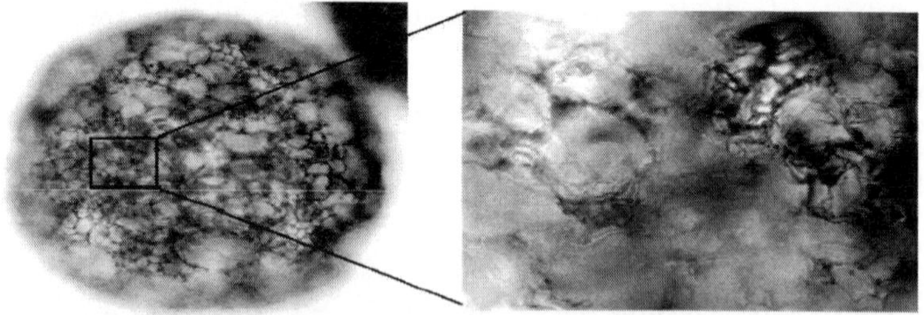

Figure 18.4. Human ASCs cultured on a macroporous PLGA microsphere and adipogenically differentiated for 10 days; oil red O staining. See also Color Insert.

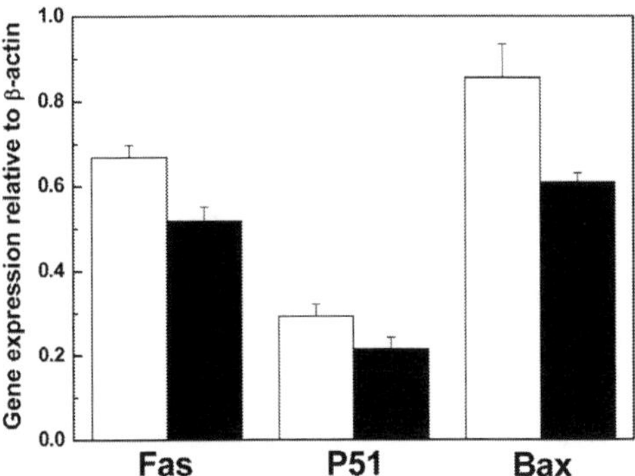

Figure 18.5. RT-PCR analysis for Fas, P51, and Bax expression of human ASCs cultured on plates, trypsinized, and subsequently cultured on macroporous PLGA microspheres for 1 day (white bars) and human ASCs cultured on macroporous PLGA microspheres.

10 days. Reverse transcriptase-polymerase chain reaction (RT-PCR) analysis showed that the ASCs cultured on macroporous PLGA microspheres exhibited significantly lower mRNA expression of Fas, P51, and Bax, which are proapoptotic genes, than those cultured on plates, trypsinized, and subsequently cultured on macroporous PLGA microspheres for one day (Fig. 18.5).

18.4 Macroporous PLGA Microsphere as an ASC Transplantation Vehicle

ASCs were cultured on macroporous PLGA microspheres with a growth medium for 7 days and then with the adipogenic differentiation medium for 10 days in spinner flasks and implanted into the subcutaneous space of mice. As a control, ASCs were cultured with the adipogenic differentiation medium on culture plates for 10 days, trypsinized, mixed with macroporous PLGA microspheres, and immediately injected into the subcutaneous space of athymic mice. Six weeks after implantation, ASCs cultured on microspheres and transplanted with the microspheres without trypsinization (group 1) had better *in vivo* adipogenic regenerative efficacy than ASCs cultured on plates, trypsinized, and subsequently mixed with microspheres (group 2). Oil red O staining of the implants retrieved at six weeks revealed that group 1 exhibited much more extensive *in vivo* adipogenesis than group 2 (Fig. 18.5). This could be due to the significantly lower apoptotic activity of ASCs cultured on macroporous PLGA microspheres (Fig. 18.5).

The macroporous PLGA microspheres have many advantages as cell transplantation vehicles. A microsphere is an injectable cell transplantation vehicle and can be easily injected into the body through a needle with no obstructions in the needle. A microsphere allows irregularly shaped defects to be filled easily and cells to be implanted using minimally invasive surgical procedures to regenerate tissues. A macroporous PLGA microsphere could be an appropriate injectable scaffold for adipose tissue engineering. Various types of injectable scaffolds have been used for adipose tissue engineering. Transplantation of ASCs cultured on nonporous PLGA microspheres resulted in adipose tissue formation.[7,8] However, nonporous PLGA microspheres do not provide a large surface area for cell adhesion and growth and thus may not be feasible for transplantation of cells at high density. ASC transplantation using fibrin, an injectable scaffold, resulted in good neo-adipose formation.[4,5] However, the volumes of the newly formed adipose tissues dramatically decreased

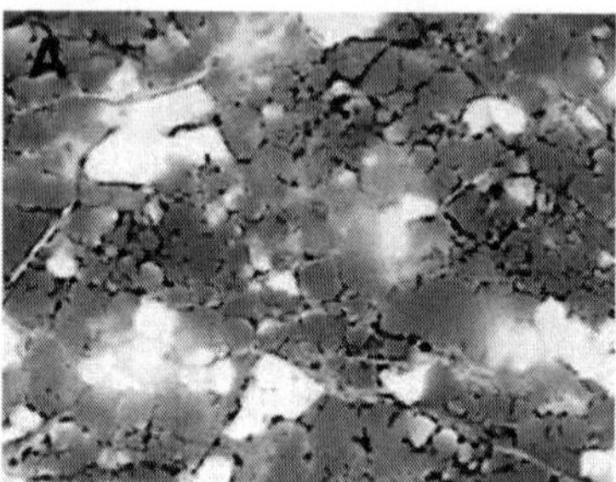

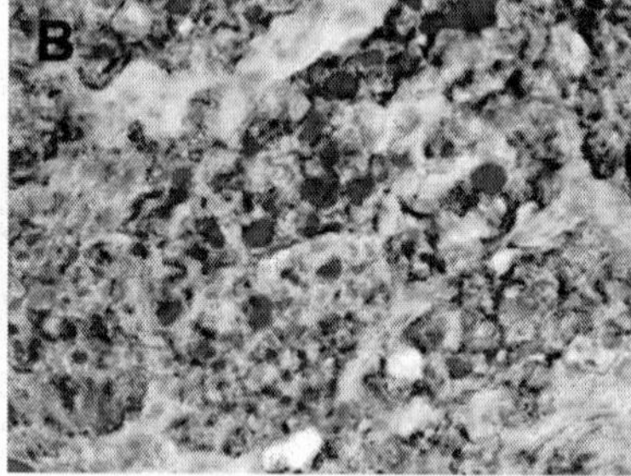

Figure 18.6. Histological analyses of the newly formed tissues 6 weeks after implantation into the subcutaneous dorsum of athymic mice; oil red O staining. (a) Implantation of ASCs cultured on macroporous PLGA microspheres in a spinner flask with the adipogenic differentiation medium for 10 days and (b) implantation of human ASCs cultured on plates with the adipogenic differentiation medium for 10 days, trypsinized, and subsequently mixed with macroporous PLGA microspheres. See also Color Insert.

to 30% to 50% of their initial volumes at four to six weeks.[4,5] In contrast, the use of macroporous PLGA microspheres resulted in the maintenance of newly formed tissue volume at 71% of the original implant volume at six weeks.

Another advantage of a macroporous PLGA microsphere over a nonporous PLGA microsphere is a less acidic environment within degrading PLGA implants. The acidic environment within degrading PLGA implants could negatively affect the viability of transplanted or host cells near and within the implant *in vivo*.[11] Since macroporous PLGA microspheres have large void volumes within them, the PLGA mass of macroporous PLGA microspheres would be much smaller than that of nonporous PLGA microspheres. The bulk density of macroporous PLGA microspheres is only ca. 2% of that of macroporous PLGA microspheres. In addition, the highly porous and interconnected pore structure of macroporous PLGA microspheres permits diffusion of degraded PLGA products throughout the device after implantation, resulting in a less acidic environment than with nonporous PLGA microspheres.

18.5 Concluding Remarks

Macroporous PLGA microspheres are suitable as both a cell culture substrate and a cell transplantation vehicle. After high-density cell expansion on macroporous PLGA microspheres, the cell-microsphere constructs can be transplanted for therapeutic applications. This process provide advantages over current methods, in which cells are cultured on 2D culture plates, harvested using trypsinization, and transplanted with or without a cell transplantation vehicle. *Ex vivo* cell expansion using macroporous microspheres in a 3D stirred suspension bioreactor has advantages for clinical-scale production of transplantable cells over cell expansion using conventional, 2D culture plates, which have a limited surface area for cell growth. The use of macroporous PLGA microspheres as a cell transplantation vehicle, following cell culture, avoids the trypsinization that is necessary for the harvest of expanded cells for transplantation and that may cause reduced cell viability and high apoptotic activity and limit the therapeutic efficacy of the transplanted cells. Apoptotic activity of ASCs cultured on macroporous PLGA microspheres was lower than that of trypsinized ASCs. The implantation of ASCs cultured on macroporous PLGA microspheres resulted in much more extensive adipose tissue formation than the implantation of ASCs cultured on plates, trypsinized, and subsequently mixed with macroporous PLGA microspheres. This method could be applied for adipose tissue augmentation or reconstruction in plastic and reconstructive surgery.

References

1. Y. Yang, F. M. V. Rossi, and E. E. Putnins, *Biomaterials,* **28**, 3110 (2007).
2. A. Werner, S. Duvar, J. Muthing, H. Buntemeyer, H. Lunsdorf, M. Strauss, and J. Lehmann, *Biotechnol. Bioeng.,* **68**, 59 (2000).
3. C. W. Patrick Jr., *Semin. Surg. Oncol.,* **19**, 302 (2000).
4. N. Torio-Padron, N. Baerlecken, A. Momeni, G. B. Stark, and J. Borges, *Aesthetic Plast. Surg.,* **31**, 285 (2007).
5. S. W. Cho, I. Kim, S. H. Kim, J. W. Rhie, C. Y. Choi, and B. S. Kim, *Biochem. Biophys. Res. Commun.,* **345**, 588 (2006).
6. S. W. Cho, S. S. Kim, J. W. Rhie, H. M. Cho, C. Y. Choi, and B. S. Kim, *Biomaterials,* **26**, 3577 (2005).
7. Y. S. Choi, S. N. Park, and H. Suh, *Biomaterials,* **26**, 5855 (2005).
8. Y. S. Choi, S. M. Cha, Y. Y. Lee, S. W. Kwon, C. J. Park, and M. Kim, *Biochem. Biophys. Res. Commun.,* **345**, 631 (2006).
9. C. W. Patrick Jr., B. Zheng, C. Johnston, and G. P. Reece, *Tissue Eng.,* **8**, 283 (2002).
10. T. K. Kim, J. J. Yoon, D. S. Lee, and T. G. Park, *Biomaterials,* **27**, 152 (2006).
11. H. J. Sung, C. Meredith, C. Johnson, and Z. S. Galis, *Biomaterials,* **25**, 5735 (2004).

Chapter 19

SUPPRESSION OF INFLAMMATORY REACTIONS ON MPC POLYMER SURFACES

Yasuhiko Iwasaki[a] **and Kazuhiko Ishihara**[b]

[a] *Department of Chemistry and Materials Engineering, Faculty of Chemistry, Materials and Bioengineering, Kansai University, 3-3-35 Yamate-cho; Suita-shi, Osaka 564-8680; Japan*

[b] *Department of Materials Engineering, Department of Bioengineering School of Engineering The University of Tokyo, 7-3-1 Hongo; Bunkyo-ku, Tokyo 113-8656; Japan*

yasu.bmt@kansai-u.ac.jp; ishihara@mpc.t.u-tokyo.ac.jp

The clarification of interactions between cells and biomaterials is indispensable for developing implantable devices and matrices. This chapter summarizes the suppression of inflammatory responses of cells and tissues in contact with phospholipid polymer surfaces with artificial cell membrane structures.

19.1 Introduction

Biomaterials implanted into the body induce a response that is different from the normal healing process.[1] The biological response

Handbook of Intelligent Scaffolds for Tissue Engineering and Regenerative Medicine
Edited by Gilson Khang

www.panstanford.com

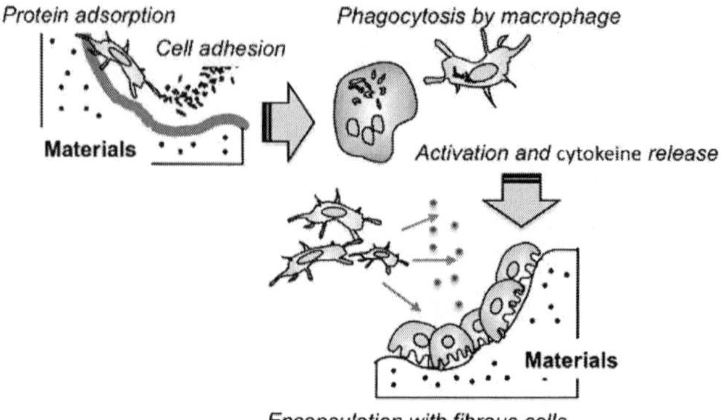

Figure 19.1. Inflammatory reactions observed on material surfaces.

to materials has been previously reviewed in detail.[2] Briefly, protein adsorption occurs immediately on materials after implantation. This phenomenon is not observed in the normal healing process. Many different proteins adsorb onto surfaces with native and denatured structures. Then, various types of cells, such as monocytes, leukocytes, and platelets, adhere to the surfaces and are activated to cause subsequent proinflammatory processes. Chronic inflammatory reactions at the biomaterial interface ensue, and the frustrated macrophages fuse together to form multinucleated foreign body giant cells that often persist for the lifetime of the implant. The end stage of the foreign body reactions involves walling off of the device by an avascular, collagenous fibrous tissue (Fig. 19.1).

Protein adsorption on biomaterials is then the trigger event of an unregulated foreign body reaction. In addition, secretion of cytokines from adherent cells provides important signals for inflammatory processes.

In this chapter, the suppression of inflammatory reactions with cells and tissues in contact with 2-methacryloyloxyethyl phosphorylcholine (MPC) polymers is summarized.

Cytocompatibility of the MPC polymers is discussed from the viewpoint of secretion of inflammatory cytokine mRNA from contacting cells. In addition, inhibition of the uptake of MPC polymer particles with phagocytes is described. The foreign body reaction of MPC polymers when implanted in subcutaneous tissue is also explained. After reading this chapter, the high potency of MPC polymers to regulate cell/material interfaces should be clarified.

19.2 Molecular Design and Fundamental Property of MPC Polymers

The cell membrane surface is considered the best surface for smooth interaction with blood components such as proteins and cells. The structure of the cell membrane, which is well known as the fluid-mosaic model, was proposed by Singer and Nicolson.[3] According to this model, amphiphilic phospholipids are arranged in a bilayer structure and proteins are located in or upon it. The distribution of these components is asymmetric. The phospholipids always flow dynamically, and the cell membrane maintains its strength by the supporting proteins. In all cells for which lipid compositional asymmetry has been described, negatively charged phospholipids such as phosphatidylserine are predominantly found on the inner, cytoplasmic side of the membrane, whereas the neutral, zwitterionic phosphorylcholine lipids such as phosphatidylcholines are located in the outer leaflet. The phosphatidylcholine surface provides an inert surface for the biological reactions of proteins and glycoproteins to occur smoothly on the membrane. This fact provides very significant information for the development of novel blood compatible polymers.

A new concept was proposed for making blood-compatible polymer materials that have good stability, processability, and applicability using a methacrylate monomer with a phosphorylcholine group, MPC (Fig. 19.2).[4] The synthesis of MPC was difficult. However, in 1987, Ishihara developed an improved synthetic route and purification method for MPC and a sufficient amount of MPC with excellent purity could be obtained.[5] Now, a Japanese chemical company produces MPC on an industrial scale and provides it worldwide. Thus, it became possible to prepare MPC polymers with various other methacrylates or styrene derivatives, and their blood compatibility was carefully evaluated.[6]

The introduction of other monomer units could change the properties of the MPC polymer. MPC can be polymerized with other vinyl

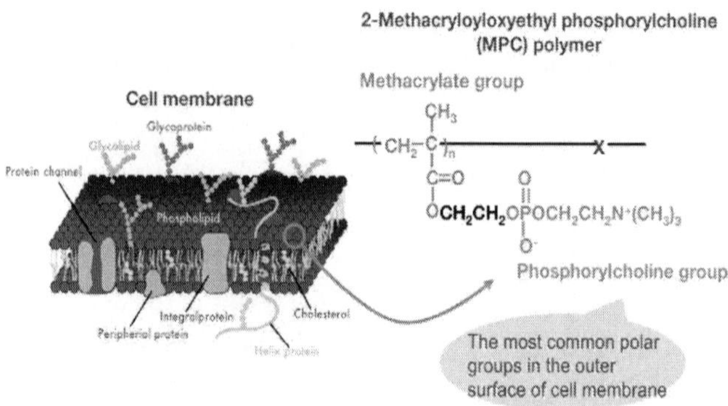

Figure 19.2. Chemical structure of MPC.

compounds by conventional radical polymerization. This means it is very easy to design MPC polymer structures to adapt to substrates that are modified with the MPC polymer. Controlling the monomer ratio determines the MPC unit composition in the polymer. MPC polymers were also prepared with other polymer architectures such as block-type copolymers and graft-type copolymers by a conventional radical polymerization technique.[7–8] Recent advancements in radical polymerization have provided well-defined polymers. That is, living radical polymerization of MPC could be reported, and small-molecular-weight-dispersion polymers were obtained.[9–10] In addition, living polymerization could facilitate the making of block-type polymers. Surface-initiated living radical polymerization has been proposed to prepare a brush-type polymer graft layer. These functional MPC polymers are useful for surface modification.

One of the MPC polymers used in implantable medical devices is poly(MPC-co-n-butyl methacrylate [BMA]) (PMB). We have investigated the blood compatibility of PMB, with special focus given to each blood component, for example, cells, plasma protein, phospholipids, and water.[11–18]

Figure 19.3 shows experimental columns containing polymer-coated poly(methyl methacrylate) beads and micrographs of the bead surface after contact with human blood without any anticoagulant. On the poly(BMA), which did not have MPC units, many adherent blood cells were observed and clots were formed.

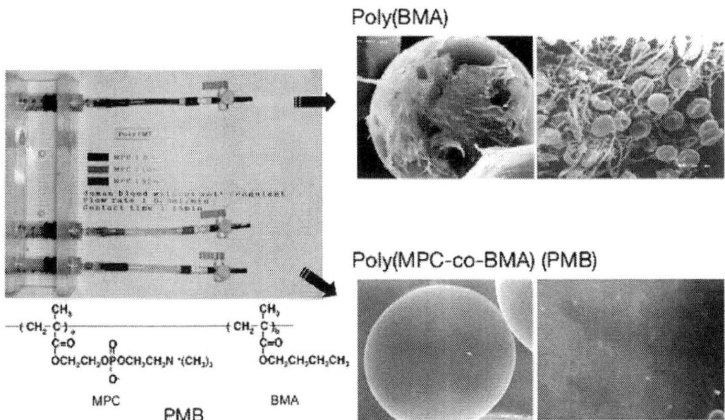

Figure 19.3. Microsphere columns after passage of human whole blood for 15 min and SEM images of polymer-coated beads packed in the columns. *Abbreviation*: SEM, scanning electron microscopy.

In contrast, PMB with a 30 mol% MPC unit (PMB30) effectively suppressed cell adhesion. While the adherent platelets may sometimes easily detach from the surface, they are strongly activated by contact with the polymer surface and induce embolization. Therefore, it is necessary to evaluate the activation of platelets barely contacting the polymer surface and those adhering to the surface to understand true blood compatibility. The activation of platelets after contact with PMB30 was evaluated by measuring the concentration of cytoplasmic calcium ions in the platelets.[11] The results clearly indicated that the activation of the platelets that contacted the PMB30 surface was less than that of those on glass and other polymer substrates. That is, PMB is one of an excellent group of antithrombogenic polymers for the suppression of cell adhesion and activation.

19.3 Secretion of HSP mRNA from Adherent Cells on MPC Copolymers

The macrophage is considered an important cell in the initial cellular response against implantable biomedical devices. Macrophages mediate inflammation by the secretion of inflammatory mediators, such as coagulation factors, lysosomal enzymes, cytokines, and growth factors. The proinflammatory cytokines, interleukin-1β (IL-1β), IL-6, and tumor necrosis factor-alpha (TNF-α) are multifunctional soluble mediators that affect various types of cells. The secretion level of these proinflammatory cytokines is different with the activation state.[19] IL-1β mediates the activation and proliferation of lymphocytes, fibroblasts, and endothelial cells.[20] Recently, numerous studies of the interaction between macrophages and materials have been conducted to understand the biocompatibility of materials.[21–27] In these reports, cytokine production from macrophages was evaluated using the enzyme-linked immunosorbent assay (ELISA).[22–24]

We have hypothesized that the MPC polymers reduce not only protein/material interactions but also cell/material interactions, including the inflammatory response of cells.

As shown in Fig. 19.4, the amount of proteins adsorbed on PMB from a cell culture medium containing fetal plasma was lower than

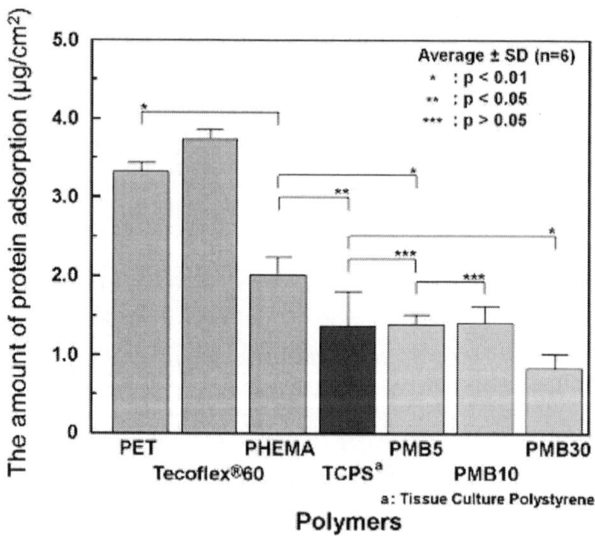

Figure 19.4. Amount of proteins adsorbed on polymer surfaces from cell culture medium. Mean values of six measurements and standard deviations are indicated.

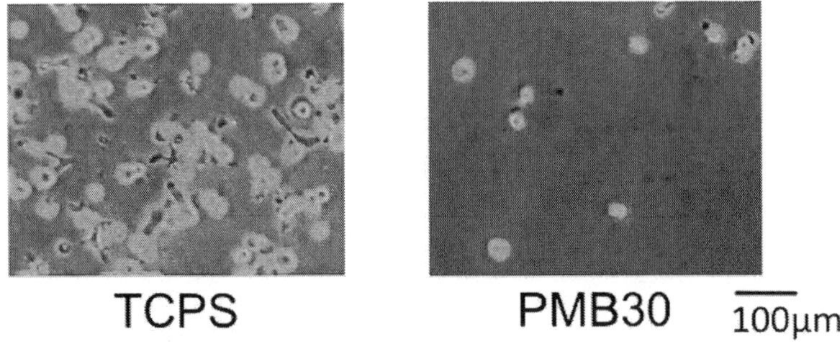

Figure 19.5. Phase-contrast micrographs of differentiated HL-60 cells adhered on polymer surfaces.

that on conventional biomedical polymers such as poly(ethylene telephatalate) (PET), segmented polyurethane, and Tecoflex® 60, even after coating with hydrophilic poly(2-hydroxyethyl methacrylate) (PHEMA).

However, there was no significant difference between the amount of protein adsorbed on tissue culture polystyrene (TCPS) and that on PMB5 (with 5 mol% MPC units) and PMB10 (with 10 mol% MPC units). On the basis of the increment of the MPC unit content to PMB30 (with 30 mol% MPC units), the amount of adsorbed protein was effectively decreased. This result suggests that the MPC units in the PMB play an important role in reducing protein adsorption.[11,13,15,17] These phenomena may be explained with the water structure on MPC polymer surfaces. That is, on MPC polymer surfaces, the free water fraction is larger than that on other hydrophilic polymers. This hypothesis remains, and much experimental data can support this hypothesis. When the free water fraction in the polymer is increased, the amount of adsorbed bovine serum albumin or bovine plasma fibrinogen is decreased. The effect of water-soluble polymers on the water structure has also been discussed by Kitano *et al.*[28] They explained that the water structure in a poly(MPC) aqueous solution is similar to that of natural water. Thus, the hydrophobic interaction between protein and polymer surfaces is too weak to adsorb proteins.

Figure 19.5 shows phase contract micrographs of differentiated HL-60 cells adhered on TCPS and PMB.

On TCPS, the number of adherent cells was clearly greater than that on PMB30. The cells that adhered on the PMBs were round in shape, but those on TCPS had a spreading morphology.

The expression of IL-1β mRNA in the cells cultured on the MPC polymers was then evaluated using reverse transcriptase-polymerase chain reaction (RT-PCR) analysis.[29] HL-60 cells were selected because they can differentiate into macrophage-like cells. In addition, we assessed the cell adhesion and protein adsorption in order to consider the relationship to the expression of IL-1β mRNA. The procedure was based on that reported by Kishida *et al.*[24,25]

The products of the PCR reaction were confirmed to be those of the gene transcripts by detection of an 821-bp band (β-actin mRNA) and a 566-bp band (IL-1β mRNA). The expression of the IL-1β mRNA was determined as the relative value to that obtained for β-actin. In general, amplified β-actin mRNA was used as an internal standard for the semi-quantification.[30,31]

RT-PCR analysis is a highly sensitive method and can detect various mRNAs from a small number of cells in contact with polymeric materials. As shown in Fig. 19.6, adherent HL-60 cells on the PMBs showed a lower expression of IL-1β mRNA than did those on other reference polymers. Kishida *et al.* reported the expression of IL-1β mRNA from HL-60 cells that adhered on various polymer surfaces.[32]

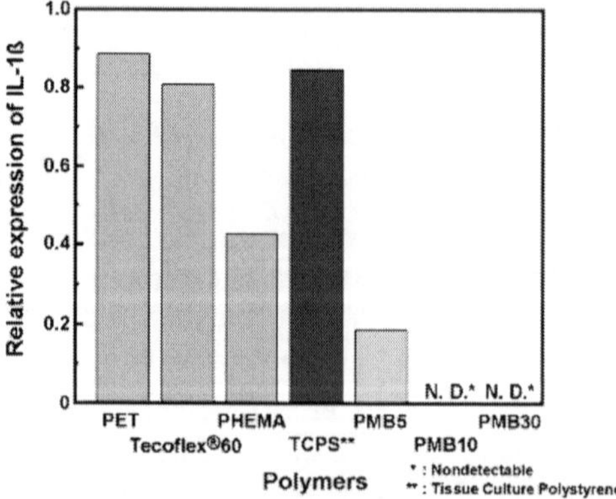

Figure 19.6. Expression of IL-1β mRNA transcripts in HL-60 cells on polymer surfaces as a standard of β-actin by RT-PCR after 24 h incubation. *Abbreviation*: N.D., not detected.

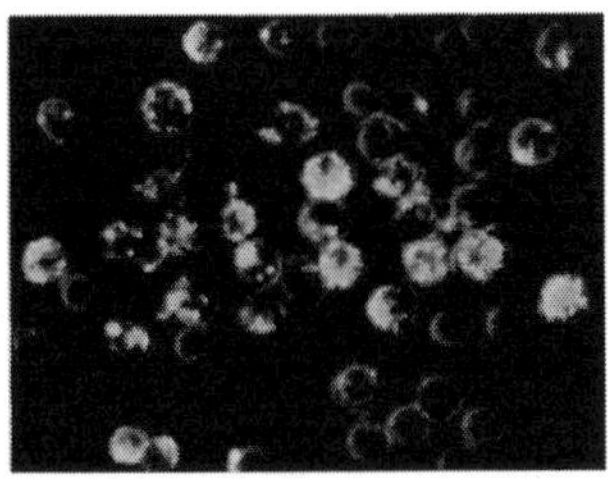
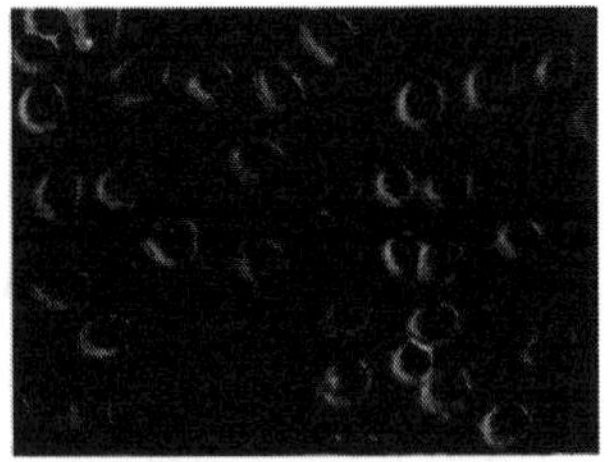

Figure 19.7. Phagocytosis of fluorescence-labeled particles by cultured mouse macrophages.

They described that the expression of IL-1β mRNA was affected by surface wettability and protein adsorption. In this study, the amount of protein on the TCPS and PHEMA was the same as that on PMB5 and PMB10. However, the expression of IL-1β mRNA on TCPS and PHEMA was high compared with that on PMB. A reduction in the secretion of cytokine and heat-shock protein (HSP) mRNA from cells in contact with aliphatic polyesters[33] or polyurethanes[34] blended with MPC was also identified.

Cellular uptake by macrophages with the phagocytosis mechanism is another cell response event. When the polymer particles are applied to the macrophages, the cells induce internalization of the particles. This is caused by molecular recognition of the ligands on the cell to the adsorbed proteins on the nanoparticles. In fact, the polystyrene nanoparticles (fluorescence labeled) can be accepted by the cellular uptake, as shown in Fig. 19.7. On the other hand, the surface of the nanoparticles is modified with the PMB30 by coating, and cellular uptake of the nanoparticles is effectively suppressed. This is caused by the excellent biocompatibility of PMB30.

19.4 Reduction of *in vivo* Host Responses to MPC Polymer Hydrogels

Hydrogels are insoluble, cross-linked polymer networks, which can absorb significant amounts of water.[35,36] They are also as flexible as soft tissue, a characteristic that minimizes their potential for irritating surrounding tissue. Some recent trends in hydrogel research for tissue engineering include macromolecular drug delivery and cell entrapment.[37,38] We have synthesized a novel biodegradable hydrogel consisting of MPC units with polyphosphates as macro-cross-linkers.[39] Polyphosphates appear interesting for biological and pharmaceutical applications because of their biocompatibility and structural similarities to naturally occurring nucleic and echoic acids. Recently, polyphosphates have been proposed for use in the field of tissue engineering as scaffolds and as gene carriers.[40–42]

We have previously investigated the *in vitro* biocompatibility of biodegradable poly(MPC) hydrogel (PCPG).[39] Poly(2-i-propyl-2-oxo-1,3,2-dioxaphospholane [IPP]-co-2-[2-oxo-1,3,2-dioxaphosphoroyloxyethyl methacrylate] [OPEMA]) (PIOP) was synthesized by the method previously described.[43] The MPC was polymerized with PIOP as a macro-cross-linker to form a hydrogel. Instead of MPC, PEGMA was used to make a reference hydrogel (PEPG), as shown in Fig. 19.8. Table 19.1 lists the degrees of hydration and the water fraction of the hydrogels cross-linked with the PIOP. The Heq value changed depending on the concentration of the PIOP. In this study, hydrogels that had 1.0 mol% PIOP as a cross-linker were used for the biocompatibility test.

Figure 19.9a shows optical micrographs of the subcutaneous tissue in contact with the hydrogels for seven days. Many inflammatory cells were observed around the PEPG. In contrast, such cells did not accumulate in the region of the tissue in contact with

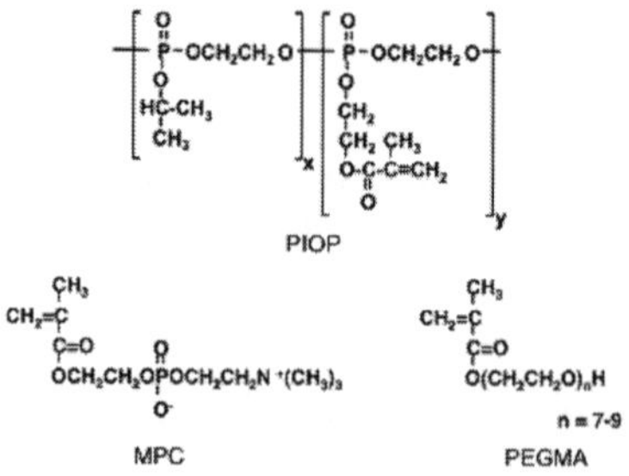

Figure 19.8. Chemical structures of macro-cross-linker (PIOP) and water-soluble monomers.

Table 19.1. Characterization of polyphosphate hydrogels.

Abbreviation	MPC/OPEMA unit in PIOP[a] (mol%)	Heq[b] (wt%)
PCPG 1.0	99.0/1.0	90.2
PEPG 1.0	99.0/1.0	83.3

[a] [Monomer] = 1.0 mol/L, [AIBN] = 5 mmol/L.
[b] Hydration degree (Heq) = (weight of water in polymer membrane/weight of polymer membrane saturated with water) × 100.

the PCPG. Mikos *et al.* reported that implantation of a biomaterial initiates a sequence of events similar to a foreign body reaction starting with an acute inflammatory response and, in some cases, leading to a chronic inflammatory response and/or development of granulation tissue, development of a foreign body reaction, and development of a fibrous capsule.[44] During the early stages, inflammatory cells such as neutrophils and monocytes accumulate on biomaterials implanted in soft tissue. On the PEPG, an acute inflammatory response occurred.

Figure 19.9b shows optical micrographs of the subcutaneous tissue in contact with the hydrogels for 28 days. A layer of fibrous cells

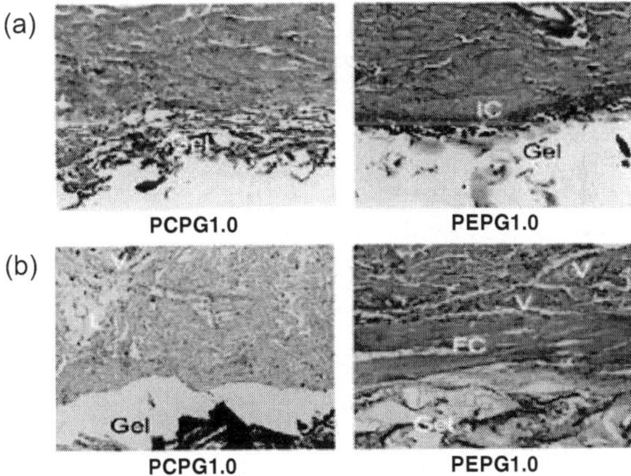

Figure 19.9. Histological micrographs of subcutaneous tissue in contact with hydrogel for (a) 7 days and (b) 28 days. The smashed gels are artifacts from histological preparations. *Abbreviations*: IC, inflammatory cell; L, lipid tissue; V, blood vessel; FC, fibrous capsule.

was observed on the PEPG, that is, a fibrous capsule had developed. In contrast, no fibrous layer was observed around the PCPG. These micrographs indicate that the PCPG did not induce an inflammatory response in the host and that there was a high degree of *in vivo* biocompatibility.

PEGMA has short ethylene glycol (EG) chains in the side chain. Although poly(ethylene oxide) (PEG) is believed to be a promising polymer for reducing nonspecific interaction with biomolecules,[45] the molecular weight and density of the chain might influence the degree of biocompatibility. In addition, PEG is susceptible to autoxidation in the presence of dioxygen and transition metal ions,[46,47] and the terminal hydroxyl group of PEG can be oxidized by alcohol dehydrogenase to an aldehyde. There is some concern that this aldehyde might react with proteins *in vivo* and with other molecules having amine groups. The aldehyde can be oxidized further by aldehyde dehydrogenase.[48,49]

In addition, it has been reported that PEGlation is not sufficient to reduce capsulation on a Teflon plate.[50] In this study, we observed the same phenomenon in a hydrogel having the EG chain.

Interactions with protein strongly relate to the biocompatibility of a matrix because the interaction may work as a trigger for host responses, especially foreign body reactions. We have clarified that PCPG does not have an adverse effect on the enzymatic activity of trypsin as a probe protein.[39]

After implantation for 28 days, PEPG and PCPG were allowed to remain in the subcutaneous tissue. The *in vivo* degradation of the hydrogel is now being investigated. Moreover, no cytotoxity was observed in the substances that degraded from the PCPG39. Very recently, three-dimensional (3D) cultivation of osteoblast cells has been performed in the PCPG.[51]

19.5 Newly Extracellular Matrices Generated from MPC Polymers

Recent cell engineering has progressed toward regenerated medicine and cells. Soon, cells with high functions should be obtained and treated by extensively developed nano- and biotechnology. Polymer matrices for cell cultures are very important for realizing these goals. Aliphatic polyesters, such as poly(lactic acid), poly(glycolic acid), and their copolymers have been commonly used as matrices for cell cultures.[52–54] These conventional biodegradable polymers are usually highly hydrophobic. The cell culture medium then difficultly penetrated the matrices.[55] The nutrients also did not penetrate these matrices. As a result, the seeded cells did not grow in the inner portion of the matrices. To solve this problem, MPC copolymers having enantiomeric polyesters on the side chains were synthesized.[56] Polymeric scaffolds were generated using the stereocomplex formations of the side chains. The wettability of the scaffolds was greatly improved in comparison with that of the scaffolds generated from conventional polyesters; cellular invasion into the MPC copolymer scaffolds was easily observed.

Spontaneous gelation of polymers can provide the desired size and shape for seeded and encapsulated cells. In addition, it is useful that the cross-linking network can be dissociated by the addition of chemical or physical stimulation. On the basis of these requirements, recoverable hydrogels have been prepared by mixing two kinds of aqueous polymer solutions. First, cross-linking of the polymer chains under physiological conditions is necessary for this purpose in order to avoid any reduced activity of the entrapped cells and biomolecules. The cytocompatibility and nutrient permeability of the hydrogel are important factors for the entrapped cells. Konno *et al.* prepared a new reversible hydrogel system composed of the MPC polymers, which can encapsulate the cells and proteins by mild treatment with a high degree of practicability and activity.[57] As the cross-linking mechanism between the MPC polymers in an aqueous medium, the specific reaction between boronic acid and multivalent alcoholic compounds was used. The boronic acid in a tetrahedral anionic structure produces stable complexes with the alcohol compounds, including polyvinyl alcohol (PVA), glucose, etc.[58–60] The hydrogel is reversibly dissociated by the addition of low-molecular-weight compounds such as D-glucose. They propose a new cell maintenance system called the *cell container* on the basis of this hydrogel. This system will be useful for maintaining the cells with a high activity and enabling their specific functions after being released from the hydrogel accompanied by dissociation of the

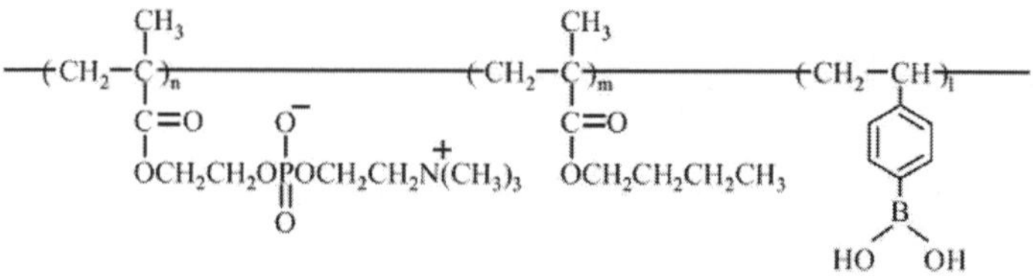

Figure 19.10. Chemical structure of PMBV.

hydrogel. To achieve these objectives, a cytocompatible water-soluble MPC polymer containing p-vinylphenylboronic acid (VPBA) units, that is, poly(MPC-co-BMA-co-VPBA) (PMBV) was synthesized (Fig. 19.10).

The water-soluble PMBV was used to make a hydrogel with PVA. When the aqueous polymer solutions contained PMBV and PVA, a hydrogel was formed within a short period after gentle shaking at room temperature. The hydrogel made from PMBV and PVA is also reversibly dissociated by the addition of glucose into the system.

To understand the behavior and state of the cells in the PMBV/PVA hydrogel as a means of evaluating the performance of

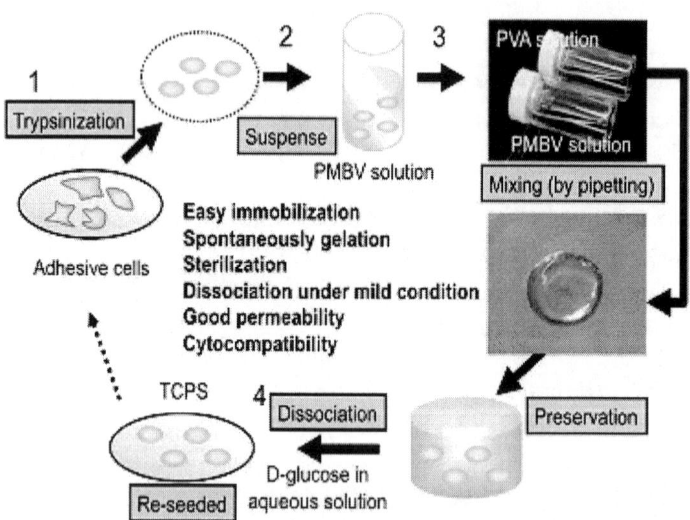

Figure 19.11. Immobilization and recovering cells in reversible-forming MPC polymer hydrogel.

the PMBV/PVA hydrogel as the cell container, cells were immobilized and cultured in the PMBV/PVA hydrogel. A schematic diagram of the cell culture procedure is shown in Fig. 19.11. The cells were suspended in the 5 wt% PMBV medium solution and mixed with the 5 wt% PVA aqueous solution for immobilization. The gelation occurred after gentle shaking for 10 seconds at room temperature. The PMBV was dissolved using DMEM containing 10% D-glucose.

Figure 19.12 shows phase-contrast microscope images of the various cells in the PMBV/PVA hydrogel. The cell morphology was spherical in shape, and the cells did not spread out in the PMBV/PVA hydrogel during the culture period.

The cell morphology correlates with the cellular activities and functions: strong cell adhesion and spreading often favor proliferation, while a round cell shape is required for cell-specific functions.[61,62] It is noteworthy that the cells did not aggregate with each other in the hydrogel. Therefore, the cell culture in the gel is useful for investigating individual cell-specific functions. In addition, number of cells did not increase significantly in the hydrogel.

After a six-day culture without changing the fresh medium, the PMBV/PVA hydrogel was dissociated by the addition of an excess amount of D-glucose and the cells that were recovered from the

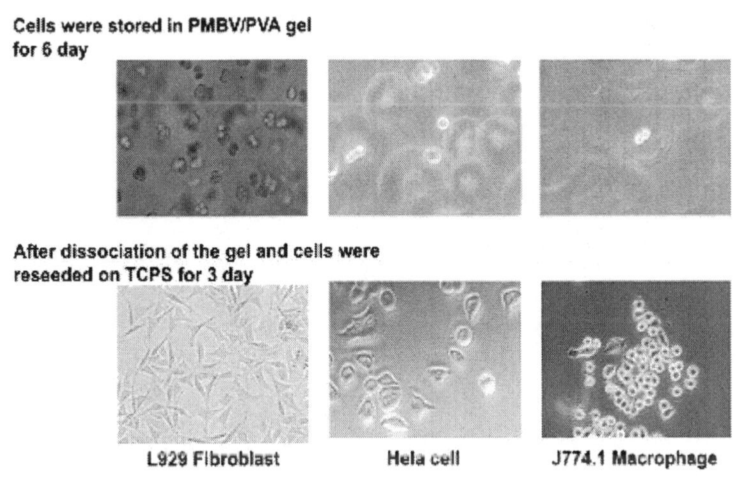

Figure 19.12. Phase-contrast microscope images of cells in the PMBV gel and after dissociation of the PMBV gel.

hydrogel were re-seeded on the TCPS. Figure 19.11 also shows phase-contrast microscope images of the cells recovered and cultured on the TCPS. Every cell quickly adhered and proliferated after seeding on the TCPS. It is considered that the viability of the cells was maintained even though they were in the PMBV/PVA hydrogel for six days. When the cells were seeded in/on conventional 3D matrices made from PLA or PGA, the cells barely penetrated into the matrices because of the hydrophobicity of these polymers, even if the materials have a pore structure. In addition, these 3D matrices were passively degraded by hydrolysis with time under physiological conditions. In contrast, the PMBV/PVA hydrogel could homogeneously disperse the cells inside the hydrogel and recover them by the addition of natural chemical compounds such as D-glucose in a shorter period. The cells encapsulated in the PMBV/PVA hydrogel maintain their morphology when they are in suspension with the cell culture medium. When recovered from the hydrogel, the cells demonstrated good adhesion to the solid substitute and proliferation on the surface. The cell container provides a mild and suitable environment for cells just like a cell suspension.[63] Treatment of the cells is made much easier by solidification with the PMBV/PVA hydrogel. The cells could also be recovered from the hydrogel at any time.

19.6 Conclusion

It was demonstrated that the MPC polymer did not induce inflammatory responses of cultivated cells and subcutaneous tissue. This phenomenon may be caused by the surface property of MPC polymer that suppresses biofouling. The hydrogels containing MPC units with high water content could be prepared easily and are useful in maintaining protein and cell functions. MPC polymers will be robust materials for advanced medical and cell/tissue engineering.

Acknowledgments

The authors would like to thank Dr. Tomohiro Konno, the University of Tokyo, and Dr. Shin-ichi Sawada, Tokyo Medical and Dental University, for their research efforts and valuable comments.

References

1. B. D. Ratner and S. J. Bryant, *Annu. Rev. Biomed. Eng.*, **6**, 41 (2004).
2. J. M. Anderson, *Annu. Rev. Mater. Res.*, **31**, 81 (2001).
3. S. J. Singer and G. L. Nicolson, *Science*, **175**, 723 (1972).
4. N. Nakabayashi and Y. Iwasaki, *Biomed. Mater. Eng.*, **14**, 345 (2004).
5. K. Ishihara, T. Ueda, and N. Nakabayashi, *Polym. J.*, **22**, 355 (1990).
6. T. Ueda, H. Oshida, K. Kurita, K. Ishihara, and N. Nakabayashi, *Polym. J.*, **24**, 1259 (1992).
7. K. Ishihara, T. Tsuji, T. Kurosaki, and N. Nakabayashi, *J. Biomed. Mater. Res.*, **28**, 225 (1994).
8. K. Sugiyama, K. Shiraishi, K. Okada, and O. Matsuo, *Polym. J.*, **31**, 883 (1999).
9. S. Yusa, K. Fukuda, T. Yamamoto, K. Ishihara, and Y. Morishima, *Biomacromolecules*, **6**, 663 (2005).
10. Y. I. Ma, E. J. Lobb, N. C. Billingham, S. P. Armes, A. L. Lweis, A. W. Lloyd, and J. Salvage, *Macromolecules*, **35**, 9306 (2002).
11. K. Ishihara, N. P. Ziats, B. P. Tierney, N. Nakabayashi, and J. M. Anderson, *J. Biomed. Mater. Res.*, **25**, 1397 (1991).
12. K. Ishihara, R. Aragaki, T. Ueda, A. Watanabe, and N. Nakabayashi, *J. Biomed. Mater. Res.*, **24**, 1069 (1990).
13. K. Ishihara, H. Oshida, Y. Endo, T. Ueda, A. Watanabe, and N. Nakabayashi, *J. Biomed. Mater. Res.*, **26**, 1543 (1992).
14. K. Ishihara, H. Oshida, Y. Endo, A. Watanabe, T. Ueda, and N. Nakabayashi, *J. Biomed. Mater. Res.*, **27**, 1309 (1993).
15. T. Ueda, A. Watanabe, K. Ishihara, and N. Nakabayashi, *J. Biomater. Sci. Polym. Ed.*, **3**, 185 (1991).
16. Y. Iwasaki, A. Mikami, K. Kurita, K. Ishihara, and N. Nakabayashi, *J. Biomed. Mater. Res.*, **36**, 508 (1997).
17. K. Ishihara, H. Nomura, T. Mihara, K. Kurita, Y. Iwasaki, and N. Nakabayashi, *J. Biomed. Mater. Res.*, **39**, 323 (1997).
18. Y. Iwasaki and K. Ishihara, *Anal. Bioanal. Chem.*, **381**, 534 (2005).
19. J. M. Anderson, A. Rodriguez, and D. T. Chang, *Semin. Immunol.*, **20**, 86 (2008).
20. C. A. Dinarello, *Blood*, **7**, 1627 (1991).
21. Y. Tabata and Y. Ikada, *Biomaterials*, **9**, 356 (1988).
22. T. L. Bonfield, E. Colton, R. E. Marchant, and J. M. Anderson, *J. Biomed. Mater. Res.*, **26**, 837 (1992).

23. T. L. Bonfield and J. M. Anderson, *J. Biomed. Mater. Res.*, **27**, 1195 (1993).
24. M. A. Cardona, R. L. Simmons, and S. S. Kaplan, *J. Biomed. Mater. Res.*, **26**, 851 (1992).
25. C. R. Jenny and J. M. Anderson, *J. Biomed. Mater. Res.*, **46**, 11 (1999).
26. C. Gretzer and P. Thomsen, *Biomaterials*, **21**, 1047 (2000).
27. J. B. Matthews, A. A. Besong, T. R. Green, M. H. Stone, B. M. Wroblewski, J. Fisher, and E. Ingham, *J. Biomed. Mater. Res.*, **52**, 296 (2000).
28. H. Kitano, K. Sudo, K. Ichikawa, M. Ide, and K. Ishihara, *J. Phys. Chem. B*, **104**, 11425 (2000).
29. S. Sawada, S. Sakaki, Y. Iwasaki, N. Nakabayashi, and K. Ishihara, *J. Biomed. Mater. Res.*, **64**, 411 (2003).
30. T. Ishizuka, S. Sawada, K. Sugama, and A. Kurita, *Clin. Exp. Immunol.*, **120**, 71 (2000).
31. T. Kinoshita, J. Imamura, H. Nagai, and K. Shimotohno, *Anal. Biochem.*, **206**, 231 (1992).
32. Kishida, S. Kato, K. Ohmura, K. Sugimura, and M. Akashi, *Biomaterials*, **17**, 1301 (1996).
33. Y. Iwasaki, S. Sawada, K. Ishihara, G. Khang, and H. B. Lee, *Biomaterials*, **23**, 3897 (2002).
34. S. Sawada, Y. Iwasaki, N. Nakabayashi, and K. Ishihara, *J. Biomed. Mater. Res.*, **79**, 476 (2006).
35. N. Peppas, Ed., *Hydrogels in Medicine and Pharmacy I-III* (CRC, Boca Raton, 1987).
36. D. Derossi, K. Kajiwara, Y. Osada, and A. Yamaguchi, Eds., *Polymer Gels-Fundamentals and Biomedical Applications* (Plenum, NY, 1989).
37. J. L. Drury and D. J. Mooney, *Biomaterials*, **24**, 4337 (2003).
38. B. Jeong, S. W. Kim, and Y. H. Bae, *Adv. Drug. Deliv. Rev.*, **17**, 37 (2002).
39. Y. Iwasaki, C. Nakagawa, M. Ohtomi, K. Ishihara, and K. Akiyoshi, *Biomacromolecules*, **5**, 1110 (2004).
40. C. Wan, H. Q. Mao, S. Wang, K. W. Leong, L. K. Ong, and H. Yu, *Biomaterials*, **22**, 1147 (2001).
41. J. Wang, P. C. Zhang, H. F. Lu, N. Ma, S. Wang, H. Q. Mao, and K. W. Leong, *J. Control. Rel.*, **83**, 157 (2002).
42. S. W. Huang, J. Wang, P. C. Zhang, H. Q. Mao, R. X. Zhuo, and K. W. Leong, *Biomacromolecules*, **5**, 306 (2004).
43. Y. Iwasaki, S. Komatsu, T. Narita, K. Akiyoshi, and K. Ishihara, *Macromol. Biosci.*, **3**, 238 (2003).

44. G. Mikos, L. V. McIntire, J. M. Anderson, and J. E. Babensee, *Adv. Drug. Deliv. Rev.*, **33**, 111 (1998).

45. J. M. Harris, Ed., *Poly(ethylene glycol) Chemistry: Biotechnical and Biomedical Applications*, (Plenum, NY, 1992).

46. R. Hamburger, E. Azaz, and M. Donbrow, *Pharm. Acta Helv.*, **50**, 10 (1975).

47. C. Crouzet, C. Decker, and J. Marchal, *Makromol. Chem.*, **177**, 145 (1976).

48. D. A. Herold, K. Keil, and D. E. Bruns, *Biochem. Pharmacol.*, **38**, 73 (1989).

49. T. Talarico, A. Swank, and C. Privalle, *Biochem. Biophys. Res. Commun.*, **250**, 354 (1998).

50. M. Shen, L. Martinson, M. S. Wagner, D. G. Castner, B. D. Ratner, and T. A. Horbett, *J. Biomater. Sci. Polym. Ed.*, **13**, 367 (2002).

51. C. Wachiralarpphaithoon, Y. Iwasaki, and K. Akiyoshi, *Biomaterials*, **28**, 984 (2007).

52. D. A. Barrera, E. Zylstra, P. Lansbury, and R. Langer, *Macromolecules*, **28**, 425 (1995).

53. G. Vunjak-Novakovic, B. Obradovic, I. Martin, P. M. Bursac, R. Langer, and L. E. Freed, *Biotechnol. Prog.*, **14**, 193 (1998).

54. S. L. Ishaug-Riley, G. M. Crane-Kruger, M. J. Yaszemski, and A. G. Mikos, *Biomaterials*, **19**, 1405 (1998).

55. L. Wald, A. G. Sarakinos, M. D. Lyman, A. G. Mikos, J. P. Vacanti, and R. Langer R. *Biomaterials*, **14**, 270 (1993).

56. J. Watanabe and K. Ishihara K., *Artif. Organs*, **27**, 242 (2003).

57. T. Konno and K. Ishihara, *Biomaterials*, **28**, 1770 (2007).

58. S. Kitano, Y. Koyama, K. Kataoka, T. Okano, and Y. Sakurai, *J. Control. Rel.*, **19**, 161 (1992).

59. T. D. James, K. R. A. S. Sandanayake, and S. Shinkai, *Angew. Chem., Int. Ed.*, **33,** 2207 (1994).

60. T. Miyata, T. Uragami, and K. Nakamae, *Adv. Drug Deliv. Rev.*, **54**, 79 (2002).

61. D. Mooney, L. Hansen, J. Vacanti, R. Langer, S. Farmer, and D. Ingber, *J. Cell Physiol.*, **151**, 497 (1992).

62. C. S. Chen, M. Mrksich, S. Huang, G. M. Whitesides, and D. E. Ingber, *Biotechnol. Prog.*, **14**, 356 (1998).

63. T. Konno and K. Ishihara, In Eds. A. Mahapatra and A. S. Kulshrestha, *Polymers for Biomedical Applications* (ACS Symposium Ser. 977) (American Chemical Society, Washington D. C., 2008), pp. 336–345.

Chapter 20

EXTRACELLULAR MATRIX–BASED SCAFFOLDS FROM SCRATCH

Willeke F. Daamen,[a*] Kaeuis A. Faraj,[a,c] Martin J. W. Koens,[a] Gerwen Lammers,[a] Katrien M. Brouwer,[a] Peter J. E. Uijtdewilligen,[a] Suzan T. M. Nillesen,[a] Luc A. Roelofs,[b] Jody E. Nuininga,[b] Paul J. Geutjes,[b] Wouter F. J. Feitz,[b] and Toin H. van Kuppevelt[a]

[a]*Radboud University Nijmegen Medical Centre, Department of Biochemistry (280), P.O. Box 9101, 6500 HB Nijmegen, The Netherlands*
[b]*Radboud University Nijmegen Medical Centre, Department of Urology (659), P.O. Box 9101, 6500 HB Nijmegen, The Netherlands*
[c]*aaP Bioimplant - EMCM BV, Middenkampweg 17, 6545 CH Nijmegen, The Netherlands*
*w.daamen@ncmls.ru.nl

In this chapter, the preparation of extracellular matrix (ECM)-based scaffolds from scratch is described. For applications in soft tissue engineering and regenerative medicine, scaffolds based on ECM proteins like collagen and elastin can be prepared with both a known architecture and a defined composition. In combination with appropriate modulating molecules, like growth factors, this may yield scaffolds containing the correct cues and signals to trigger regeneration of specific tissues *in vivo*. The construction of innovative scaffolds with a defined molecular composition, morphology and shape will be described in this chapter. The

Handbook of Intelligent Scaffolds for Tissue Engineering and Regenerative Medicine
Edited by Gilson Khang

www.panstanford.com

off-the-shelf availability may make such scaffolds especially beneficial for clinical use. The effect of growth factors bound to the scaffolds will be illustrated in a rat subcutaneous model.

20.1 Introduction

Cells signal to the ECM as much as the ECM signals to cells. All cells are surrounded by ECM molecules, which provide specific cues to the cells with regard to cellular processes, like proliferation, differentiation, migration, and cell-matrix interactions.[1] Since cells behave in a different way on various surfaces, one can use the surrounding matrix to guide cells in a certain direction.

Our aim is to engineer molecularly defined "smart" scaffolds which signal to the surrounding cells which tissue/organ to form. If one wants to study the effect of single ECM components on cell behavior and tissue remodeling, it is important to use highly purified components. Therefore, we have put considerable efforts into the purification of insoluble (fibrillar) type I collagen, insoluble elastin fibers and solubilized elastin.[2–4] From these basic components, tailor-made biomaterials can be produced.[5]

Next to the molecular makeup of scaffolds, the three-dimensional architecture is of major importance.[6] Cells in a porous bioscaffold typically synthesize their new ECM molecules aligning the presented scaffold,[7,8] and this offers the opportunity to direct the orientation of newly formed ECM. In such a way, the basket-weave collagen meshwork of normal skin may be mimicked in case of skin trauma, thus preventing the formation of densely aligned bundles present in scars,[9] thereby reducing scarring.

In this chapter, we will show the fabrication of collagen-/elastin-based scaffolds with a defined structure (dense flat films, porous scaffolds with various pore/lamellae structures, and tubular scaffolds), as well as scaffolds with a defined molecular composition, especially related to the incorporation of glycosaminoglycans and growth factors. An *in vivo* study using such acellular scaffolds will be presented, viz., the application of scaffolds with heparin and growth factors to boost angiogenesis in a rat subcutaneous model.

20.2 Scaffolds with a Specific Three-Dimensional Structure

20.2.1 *Flat Films*

For certain applications, for example, provision of structural support for epithelial cells (e.g., keratinocytes, urothelial cells), a two-dimensional structure is sufficient. Epithelia are generally only one or a few cell layers thick, and the tightly packed cells are resting on a basement membrane, containing mostly type IV collagen, heparan sulfate proteoglycan, and laminins.[10]

Compact collagen films (Fig. 20.1a) can be prepared by merely air-drying a homogenized and deaerated collagen suspension prepared in diluted acetic acid. The concentration of the collagen suspension will increase when the solvent evaporates. After evaporation a transparent film layer of the collagen remains, which appears rather dense by scanning electron microscopy (SEM) (Fig. 20.1c).

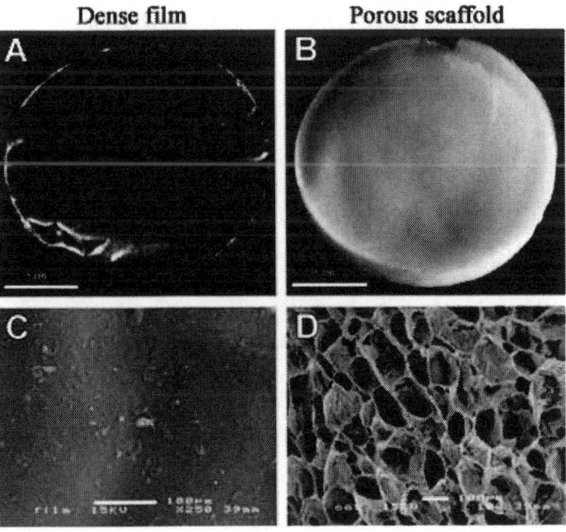

Figure 20.1. Macroscopic (a, b) and scanning electron microscopic images (c, d) of dense films (a, c) and porous scaffolds (b, d). Bar is 1 cm in (a) and (b) and 100 μm in (c) and (d).

For application in skin, for example, films are preferred over highly porous scaffolds, since cultured keratinocytes may infiltrate these scaffolds, instead of forming multiple cell layers on top.[6]

20.2.2 *Porous Scaffolds*

For tissues where ingrowth of cells is of importance, porous scaffolds are preferable (Fig. 20.1b, d). Flat, porous collagen or collagen-elastin scaffolds can be prepared by freezing and subsequently lyophilizing a suspension of collagen and elastin.[11] More specifically, a collagen(-elastin) suspension in diluted acetic acid can be applied to a mold, frozen, and lyophilized, resulting in porous scaffolds. The pore structure observed after freeze-drying resembles the ice crystal network formed during the freezing process. By varying the freezing temperature, and hence the velocity of freezing, the pore size of the scaffolds can be varied to some degree.[6,12] With faster freezing, smaller ice crystals are formed, leading to a scaffold with a smaller average pore diameter. In Fig. 20.2, scanning electron micrographs of scaffolds are shown which were frozen at three different temperatures. The largest and most honeycomb-like pores were formed when freezing at –20°C.

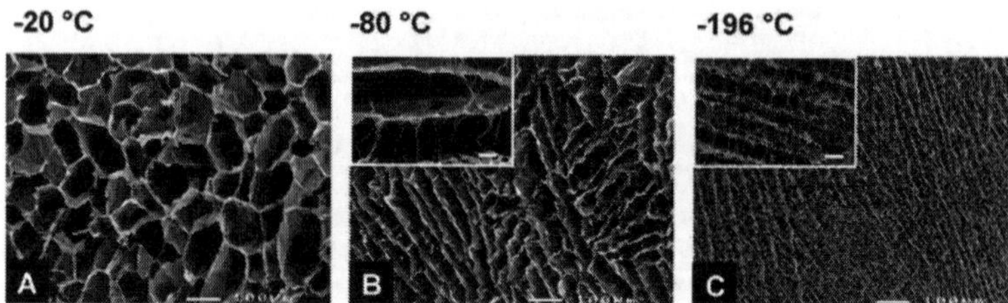

Figure 20.2. Scanning electron micrographs of porous collagen scaffolds prepared by freezing and lyophilization of an acidic collagen suspension. Pore size can be influenced by the freezing temperature. Using relatively high freezing temperature (–20°C) resulted in larger ice crystals compared with those obtained with lower freezing temperatures (–80 or –196°C). The insets in (b) and (c) show lamellae with interconnecting struts. Bars are 100 μm; bars in inserts are 20 μm. (Figure reproduced from Faraj *et al.* 2007 with permission).[6]

Porous collagen-based scaffolds were used in different *in vivo* animal models, including bladder,[13] urethra,[14] blood vessels,[15] oral mucosa,[16,17] and several *in utero* reconstructions.[18–21]

20.2.2.1 Unidirectional scaffolds

When ice crystal growth is controlled, scaffolds with more defined morphologies can be prepared using the freezing and lyophilizing technique. For instance, we can prepare scaffolds with unidirectional lamellae (Fig. 20.3) using a temperature gradient which forms between liquid nitrogen (−196°C) and ambient temperature (about 20°C). Finger-shaped ice crystals will start growing from the site of liquid nitrogen towards the site of higher temperature. After lyophilization, it results in a scaffold with unidirectional lamellar structures.[22] The individual lamellae are connected by thin struts of approximately 1–2 μm.

The principle is equivalent to the one described by Schoof *et al.*[23] Punches of these scaffolds with unidirectional lamellae have been used to guide Schwann cells *in vitro*, which may ultimately contribute to functional repair of peripheral nerve lesions.[24,25]

20.2.3 *Tubular Porous Scaffolds*

Tubular scaffolds are the materials of choice when replacing tubular structures in the body. Small-diameter tubes are necessary for

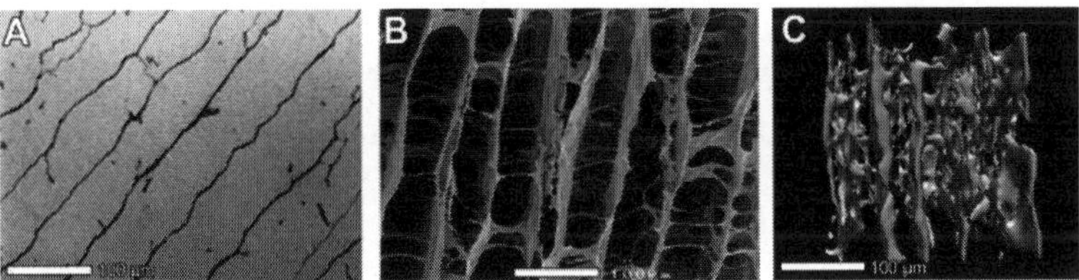

Figure 20.3. Scaffolds with unidirectional lamellae (connected by thin struts), prepared using a temperature gradient between −196°C and room temperature. Unidirectional scaffolds were visualized by (a) light microscopy (toluidine blue-stained epon sections), (b) SEM, and (c) micro-computed tomography after heavy metal staining. Bars are 100 μm. (Figure reproduced from Faraj *et al.* 2009 with permission).[22]

replacement of, for example, urethra and blood vessels ($\sim$6 mm[26] and $\leq$6 mm[27] in humans, respectively), whereas larger-diameter tubes are required to repair defects in for instance the trachea and esophagus (inner diameters of 1.2 cm[26] and 1.5–2 cm,[28] respectively).

Tubular porous scaffolds can be prepared by pouring the collagen(-elastin) suspension in a tubular mold, followed by the insertion of a cylindrical mandrel and subsequent freezing and lyophilization. By adjusting the inner diameter of the mold and the diameter of the mandrel, the wall thickness and inner diameter of the tubular scaffold can be modified. Again, by adjusting the freezing temperature, the pore size can be varied.

Figure 20.4 shows an example of a small-diameter collagen tube (inner diameter 3 mm).

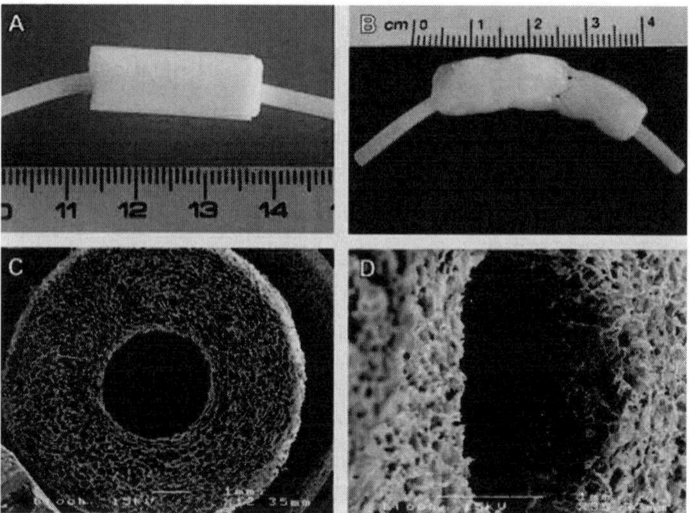

Figure 20.4. Macroscopic (A, B) and scanning electron microscopic (C, D) images of porous collagen tubes with a 3 mm inner diameter prepared by freezing and lyophilization. A) Macroscopic image of a dry collagen tube; B) two collagen tubes sutured to each other under wetted conditions showing its suturability. C) SEM image of the cross-section of a collagen tube and D) SEM image of the lumen of the tube. Bars are in cm in A, B and are 1 mm in C, D.

One application of such tubes may be urethra reconstruction, which may, for instance, be necessary in males with a congenital anomaly, such as hypospadias, which results in a mal-location of the urethra opening. Previously, we performed animal experiments with flat scaffolds in onlay urethroplasties.[14] In this experiment, the collagen scaffolds performed equally well to a Cook SIS 1 layer graft without noticeable voiding dysfunction or stone formation in the course of the experiment (up to nine months). While the urothelial cells fully populated the scaffold in one month, only few smooth muscle cells were found in the scaffold, even after nine months. More recently, we used the new collagen tubes to regenerate a critical defect of the rabbit urethra. With a tubular scaffold, additional suturing of flat scaffolds into a collagen tube is not necessary. This construct is now being optimized by the addition of components (e.g., growth factors) that specifically stimulate smooth muscle cell proliferation.[29]

20.3 Scaffolds with Defined Molecular Composition

20.3.1 *Purification of Scaffold Components*

Complex scaffolds with a predefined composition can be prepared when purified individual components are at hand. Only in this way, one can systematically study the effect of single components on tissue response *in vivo*. We have purified a number of basic scaffold components. Insoluble type I collagen is purified from bovine Achilles tendon using extractions with diluted acetic acid, NaCl, urea, acetone, and demineralized water.[2] Insoluble elastin fibers are isolated from equine ligamentum nuchae using extractions with NaCl, organic solvents, CNBr in formic acid, urea with diluted 2-mercaptoethanol, and a trypsin digestion.[3,30] Purity and fibril/fiber structure is analyzed by SDS-PAGE, amino acid analysis, transmission electron microscopy (TEM), and SEM. The isolated collagen fibrils still possess their characteristic striated banding pattern.[6] Purified elastin fibers are intact and free of microfibrils.[3,30] Glycosaminoglycans (heparan sulfate from bovine kidney and chondroitin sulfate from bovine trachea) can be purified using papain digestion, alkaline borohydride treatment, diethylaminoethyl cellulose (DEAE) ion-exchange chromatography, and glycosidase digestion.[31] Isolated glycosaminoglycans are characterized by agarose gel electrophoresis[32] and quantified using a modified Farndale assay.[33] Growth factors are produced in milligram quantities using recombinant DNA technology and purified using immobilized heparin affinity chromatography, for example, vascular endothelial growth factor (VEGF)[34] and fibroblast growth factor 2 (FGF2).[35]

20.3.2 *Cross-Linking and Covalent Attachment of Glycosaminoglycans*

Cross-linking of scaffolds will make them more resistant to degradation and will add to their mechanical stability. We generally use the nontoxic 1-ethyl-3-(3-dimethyl aminopropyl) carbodiimide (EDC) in combination with *N*-hydroxysuccinimide (NHS) because it yields zero-length cross-links, introducing an amide bond. The strength of the scaffold and the rate of biodegradation can be varied by the cross-linking conditions.[2]

Glycosaminoglycans can be bound to lyophilized scaffolds in the cross-linking step, thereby covalently attaching them to the scaffold.[31] In the human body, glycosaminoglycans are present in the ECM, where these linear polysaccharide chains are important for hydration of the tissue and binding and modulation of growth factors and cytokines.[36] When binding glycosaminoglycans, such as chondroitin sulfate, to collagen scaffolds, the water-binding capacity increases over 50%.[11]

20.3.3 *Binding of Growth Factors*

Binding of growth factors is often modulated by glycosaminoglycans, and we therefore use the intrinsic properties of glycosaminoglycans to bind, for example, heparin-binding growth factors to scaffolds. A wide variety of heparin-binding growth factors is known, including FGFs, VEGF, heparin-binding epidermal growth factor (hbEGF), placental growth factor (PlGF), hepatocyte growth factor (HGF), transforming growth factor-beta (TGF-β), and platelet-derived growth factor (PDGF).[37] *In vivo*, growth factors are stored in the ECM. One of the functions of glycosaminoglycans is to stabilize growth factors and to protect them from proteolytic degradation by their interactions, thereby providing a depot of growth factors.[38] When growth factors are bound to scaffolds via glycosaminoglycans, a sustained release is found.[35]

To bind growth factors to the scaffold, a scaffold containing glycosaminoglycans is incubated in a growth factor solution in phosphate buffered saline (PBS) at ambient temperature, followed by washings with PBS to remove unbound growth factors. The binding and release characteristics of the growth factors can be determined by sodium dodecyl sulfate polyacrylamide gel electrophoresis (SDS-PAGE), Western blotting, or autoradiography in case of the use of radiolabeled growth factors. The localization can be visualized with immunostainings. An example of the *in vivo* effect of growth factors bound to acellular collagen-glycosaminoglycan scaffolds is given next.

20.4 Acellular Scaffolds for *in vivo* Tissue Regeneration

In order to systematically study the effect of growth factors bound to collagen scaffolds via the glycosaminoglycan heparin, we prepared five different porous scaffolds composed of 1) cross-linked collagen only (COL), 2) cross-linked collagen with covalently incorporated heparin (COLH), 3) collagen, heparin, and bound FGF2 (1.6 ± 0.2 μg/mg scaffold, COLHF), 4) collagen, heparin, and bound VEGF (1.0 ± 0.3 μg/mg scaffold, COLHV), and 5) collagen, heparin, bound FGF2, *and* VEGF (1.0 ± 0.2 and 0.7 ± 0.3 μg/mg scaffold, COLHFV). Fully characterized scaffolds were subcutaneously implanted in rats and analyzed after 3, 7, and 21 days.[39]

Angiogenesis and numbers of hypoxic cells were analyzed. From seven days on, angiogenesis could be observed, both macroscopically (Fig. 20.5a) and by type IV collagen immunostaining (Fig. 20.5b). Most blood vessels were observed in the scaffold with both growth factors (COLHFV).

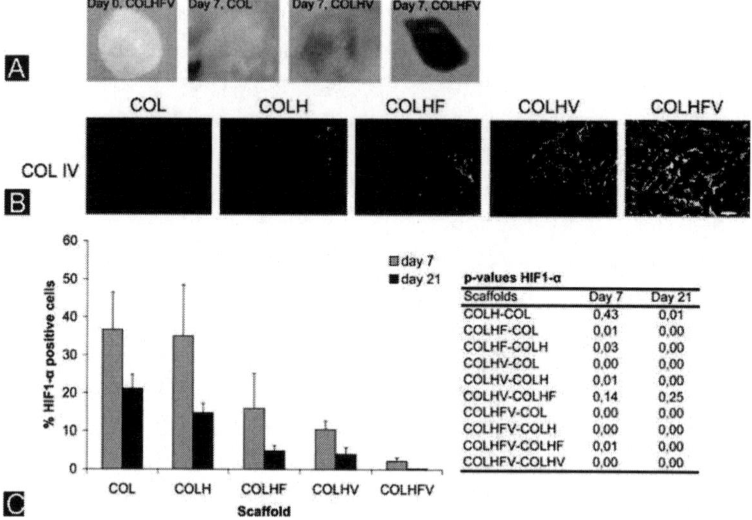

p-values HIF1-α

Scaffolds	Day 7	Day 21
COLH-COL	0,43	0,01
COLHF-COL	0,01	0,00
COLHF-COLH	0,03	0,00
COLHV-COL	0,00	0,00
COLHV-COLH	0,01	0,00
COLHV-COLHF	0,14	0,25
COLHFV-COL	0,00	0,00
COLHFV-COLH	0,00	0,00
COLHFV-COLHF	0,01	0,00
COLHFV-COLHV	0,00	0,00

Figure 20.5. Evaluation of the effect of heparin/growth factors using five different Col scaffolds.

(a) Macroscopical evaluation of a scaffold before implantation (day 0) and after implantation for 7 days. Note the red appearance of the Col scaffold containing heparin, FGF2, and VEGF (COLHFV). (B) Immunohistochemical evaluation of scaffolds 7 days after implantation and stained for type IV Col to indicate blood vessels. Note that the Col scaffold containing heparin, FGF2, and VEGF (COLHFV) shows the most and largest blood vessels. (c) Quantification by HIF1-α staining of the number of hypoxic cells present in scaffolds. Note that both at day 7 and day 21, the Col scaffold containing heparin, FGF2, and VEGF (COLHFV) had significantly fewer hypoxic cells than all other scaffolds. Bar is 50 μm. *Abbreviations*: Col, collagen; H, heparin; F, FGF-2; V, VEGF. HIF1-α, hypoxia-inducible factor 1, alpha subunit. (Figure reproduced from Nillesen *et al.* 2007 with permission).[39]

To study whether these blood vessels were capable of supplying the cells with oxygen, an immunostaining for hypoxia-inducible factor 1α (HIF1-α, part of a heterodimeric transcription factor) was performed, which stains hypoxic cells. The scaffolds with both growth factors (COLHFV) possessed significantly lower numbers of hypoxic cells on both days 7 and 21 compared with all other scaffolds (Fig. 20.5c). Already from day 7 on, almost all cells were provided with enough oxygen, whereas in the plain collagen scaffold

(COL), at 21 days still ~20% of cells were hypoxic. This clearly demonstrates the important effect growth factors have in acellular scaffolds in tissue engineering.

20.5 Future Outlook

The construction of scaffolds with both a defined morphology and a defined composition allows the preparation of the ECMs for tissue engineering from scratch. These artificial ECMs can be designed such that they harbor and provide the correct signals for the cells to populate the scaffolds and to regenerate the desired tissue, as seen in our *in vivo* experiment. When specific effector molecules (growth factors/morphogens) important in embryonic development of a particular organ are incorporated in the scaffolds, organogenesis may even be recapitulated in the adult situation.

Acknowledgments

Research described in this chapter was funded by the Dutch Program for Tissue Engineering (DPTE6735 and DPTE6739), the FP6-project EuroSTEC (contract LSHB-CT-2006-037409), ZonMW 920-03-304, and ZonMW 920-03-456.

References

1. W. P. Daley, S. B. Peters, and M. Larsen, *J. Cell Sci.*, **121**, 255 (2008).
2. J. S. Pieper, A. Oosterhof, P. J. Dijkstra, J. H. Veerkamp, and T. H. van Kuppevelt, *Biomaterials*, **20**, 847 (1999).
3. W. F. Daamen, T. Hafmans, J. H. Veerkamp, and T. H. van Kuppevelt, *Tissue Eng.*, **11**, 1168 (2005).
4. W. F. Daamen, S. T. Nillesen, R. G. Wismans, D. P. Reinhardt, T. Hafmans, J. H. Veerkamp, and T. H. van Kuppevelt, *Tissue Eng. Part A*, **14**, 349 (2008).
5. P. J. Geutjes, W. F. Daamen, P. Buma, W. F. Feitz, K. A. Faraj, and T. H. van Kuppevelt, *Adv. Exp. Med. Biol.*, **585**, 279 (2006).

6. K. A. Faraj, T. H. van Kuppevelt, and W. F. Daamen, *Tissue Eng.*, **13**, 2387 (2007).
7. W. F. Daamen, S. T. Nillesen, T. Hafmans, J. H. Veerkamp, M. J. van Luyn, and T. H. van Kuppevelt, *Biomaterials*, **26**, 81 (2005).
8. J. S. Pieper, P. B. van Wachem, M. J. van Luyn, L. A. Brouwer, T. Hafmans, J. H. Veerkamp, and T. H. van Kuppevelt, *Biomaterials*, **21**, 1689 (2000).
9. P. Martin, *Science*, **276**, 75 (1997).
10. M. Paulsson, *Crit. Rev. Biochem. Mol. Biol.*, **27**, 93 (1992).
11. W. F. Daamen, H. Th. B, van Moerkerk, T. Hafmans, L. Buttafoco, A. A. Poot J. H. Veerkamp, and T. H. van Kuppevelt, *Biomaterials*, **24**, 4001 (2003).
12. F. J. O'Brien, B. A. Harley, I. V. Yannas, and L. Gibson, *Biomaterials*, **25**, 1077 (2004).
13. J. E. Nuininga JE, H. van Moerkerk, A. Hanssen, C. A. Hulsbergen, J. Oosterwijk-Wakka, E. Oosterwijk E, R. P. de Gier, J. A. Schalken, T. H. van Kuppevelt, and W. F. Feitz, *Biomaterials*, **25**, 1657 (2004).
14. J. E. Nuininga JE, H. van Moerkerk, A. Hanssen, C. A. Hulsbergen, J. Oosterwijk-Wakka, E. Oosterwijk E, R. P. de Gier, J. A. Schalken, T. H. van Kuppevelt, and W. F. Feitz, *Eur. Urol.*, **44**, 266 (2003).
15. L. Buttafoco L., P. Engbers-Buijtenhuijs, A. A. Poot, P. J. Dijkstra, W. F. Daamen, T. H. van Kuppevelt, I. Vermes, and J. Feijen, *J. Biomed. Mater. Res. B Appl. Biomater.*, **77**, 357 (2006).
16. R. G. Jansen, A. M. Kuijpers-Jagtman, T. H. van Kuppevelt, and J. W. von den Hoff, *J. Craniofac. Surg.*, **19**, 599 (2008).
17. R. G. Jansen, T. H. van Kuppevelt, W. F. Daamen, A. M. Kuijpers-Jagtman, and J. W. von den Hoff, *Arch. Oral Biol*, **53**, 376 (2008).
18. A. J. Eggink, L. A. Roelofs, W. F. Feitz, R. M. Wijnen, R. A. Mullaart, J. A. Grotenhuis, T. H. van Kuppevelt, M. M. Lammens, A. J. Crevels, A. Hanssen, and P. P. van den Berg, *Fetal Diagn. Ther.*, **20**, 335 (2005).
19. A. J. Eggink, L. A. Roelofs, M. M. Lammens, W. F. Feitz, R. M. Wijnen, R. A. Mullaart, H. T. van Moerkerk, T. H. van Kuppevelt, A. J. Crevels, A. Hanssen, F. K. Lotgering, and P. P. van den Berg. *Fetal Diagn. Ther.*, **21**, 210 (2006).
20. L. A. Roelofs, A. J. Eggink, C. A. Hulsbergen-van de Kaa, R. M. Wijnen, T. H. van Kuppevelt TH, H. T. van Moerkerk, A. J. Crevels, A. Hanssen, F. K. Lotgering, P. P. van den Berg, and W. F. Feitz. *Fetal Diagn. Ther.*, **24**, 7 (2008).

21. L. A. Roelofs, A. J. Eggink, C. A. Hulsbergen-van de Kaa, P. P. van den Berg, T. H. van Kuppevelt, H. T. van Moerkerk, A. J. Crevels, F. K. Lotgering, W. F. Feitz, and R. M. Wijnen, *Tissue Eng. Part A*, **14**, 2033 (2008).
22. K. A. Faraj, V. M. J. I. Cuijpers, R. Wismans, X. F. Walboomers, J. A. Jansen, T. H. van Kuppevelt, and W. F. Daamen, *Tissue Eng. Part C*, **15**, 493 (2009).
23. H. Schoof, J. Apel, I. Heschel, and G. Rau, *J Biomed. Mater Res.*, **58**, 352 (2001).
24. A. Bozkurt, G. A. Brook, S. Moellers, F. Lassner, B. Sellhaus, J. Weis, M. Woeltje, J. Tank, C. Beckmann, P. Fuchs, L. O. Damink, F. Schugner, I. Heschel, and N. Pallua, *Tissue Eng.*, **13**, 2971 (2007).
25. A. Bozkurt, R. Deumens, C. Beckmann, L. Olde Damink, F. Schugner, I. Heschel, B. Sellhaus, J. Weis, W. Jahnen-Dechent, G. A. Brook, and N. Pallua, *Biomaterials*, **30**, 169 (2009).
26. S. Standring, N. R. Borley, P. Collins, A. R. Crossman, and H. Gray, *Gray's Anatomy: The Anatomical Basis of Clinical Practice*, 40th ed. (Elsevier Churchill Livingstone, 2008).
27. K. H. Yow, J. Ingram, S. A. Korossis, E. Ingham, and S. Homer-Vanniasinkam, *Br. J. Surg.*, **93**, 652 (2006).
28. H. J. Lin, K. T. Chen, N. P. Foo, C. C. Hsu, and H. R. Guo, *Br. J. Anaesth.*, **99**, 740 (2007).
29. J. E. Nuininga, M. J. Koens, D. M. Tiemessen, E. Oosterwijk, W. F. Daamen, P. J. Geutjes T. H. van Kuppevelt, and W. F. Feitz, *Tissue Eng. Part A*, **16**, 3319 (2010).
30. W. F. Daamen, T. Hafmans, J. H. Veerkamp, and T. H. van Kuppevelt, *Biomaterials*, **22**, 1997 (2001).
31. J. S. Pieper, T. Hafmans, J. H. Veerkamp, and T. H. van Kuppevelt, *Biomaterials*, **21**, 581 (2000).
32. C. H. van de Lest, E. M. Versteeg, J. H. Veerkamp, and T. H. van Kuppevelt, *Anal. Biochem.*, **221**, 356 (1994).
33. C. H. van de Lest, E. M. Versteeg, J. H. Veerkamp, and T. H. van Kuppevelt, *Biochim. Biophys. Acta*, **1201**, 305 (1994).
34. P. J. Geutjes, S. T. Nillesen, G. Lammers, W. F. Daamen, and T. H. van Kuppevelt, *Protein Expr. Purif.*, **69**, 76 (2010).
35. J. S. Pieper, T. Hafmans, P. B. van Wachem, M. J. van Luyn, L. A. Brouwer, J. H. Veerkamp, and T. H. van Kuppevelt, *J. Biomed. Mater. Res.*, **62**, 185 (2002).

36. R. Raman, V. Sasisekharan, and R. Sasisekharan, *Chem. Biol.*, **12**, 267 (2005).
37. M. Bernfield, M. Gotte, P. W. Park, O. Reizes, M. L. Fitzgerald, J. Lincecum, and M. Zako, *Annu. Rev. Biochem.*, **68**, 729 (1999).
38. S. Tumova, A. Woods, and J. R. Couchman, *Int. J. Biochem. Cell Biol.*, **32**, 269–288 (2000).
39. S. T. M. Nillesen, P. J. Geutjes, R. Wismans, J. Schalkwijk, W. F. Daamen, and T. H. van Kuppevelt, *Biomaterials*, **28**, 1123–1131 (2007).

Chapter 21

DESIGN OF BIOMIMETIC SCAFFOLDS FOR LIVER TISSUE ENGINEERING

Chong-Su Cho,[a] Hu-Lin Jiang,[a] Takashi Hoshiba,[b] and Toshihiro Akaike[c*]

[a]*Department of Agricultural Biotechnology, Seoul National University, Seoul 151-921, Korea*
[b]*Biomaterials Center, National Institute for Materials Science, Tsukuba 305-0044, Japan*
[c]*Department of Biomolecular Engineering, Tokyo Institute of Technology, Yokohama 226-8502, Japan*
*takaike@bio.titech.ac.jp

Recently, the development of biomaterials for liver tissue engineering has focused on the design of biomimetic scaffolds that can serve as a temporary mechanical support sufficient to withstand *in vivo* forces and maintain a potential space for tissue development through molecular recognition. Molecular recognition of scaffolds by hepatocytes has been achieved by chemical modification of biomaterials with galactose moiety as a specific ligand that can incur specific interactions with asialoglycoprotein receptors (ASGPR) on the hepatocytes. The biomimetic scaffolds mimic many roles of the extracellular matrix (ECM) in liver tissues, such as attachment, proliferation, migration, differentiation, and survival of hepatocytes. This chapter discusses the mechanism of

Handbook of Intelligent Scaffolds for Tissue Engineering and Regenerative Medicine
Edited by Gilson Khang

www.panstanford.com

specific interaction between galactose moieties in the artificial ECMs and ASGPR on the hepatocytes. Also, we explain the methods to design biomimetic scaffolds on the surface or the bulk of materials for liver tissue engineering. Furthermore, the criteria to design biomimetic scaffolds such as topology of the scaffolds, the coculture system, and the cell source are covered.

21.1 Introduction

Designs of biomimetic materials are very important for controlling cell-biomaterial interactions because biomimetic materials are materials that mimic a biological environment to elicit a desired cellular response, facilitating the fulfillment of their task.[1] The main task of the biomimetic material is not only to control specific interaction with a cell or tissue but also to provide a scaffold structure for tissue formation.[2] Specific ligands were incorporated into the biomimetic materials for enhancing cell adhesion[3] or for impacting cell proliferation or differentiation.[4,5]

Tissue engineering is an emerging interdisciplinary field in life sciences and aims to regenerate new biological tissue for replacing diseased tissues by using cells.[6] Tissue engineering generally requires a three-dimensional (3D) biomimetic scaffold for tissue regeneration because cell proliferation and differentiation, resulting in tissue regeneration, would be difficult if such a biomimetic scaffold that functions as a cell matrix is not provided.[7] An artificial ECM designed for tissue engineering mimics the functions of the ECM molecules naturally found in tissues.[8]

The artificial ECM should be considered including biomimetic scaffolds that serve as a temporary mechanical support sufficient to withstand *in vivo* forces and maintain a potential space for tissue development.[9] The tissue-specific function of seeded cells in the artificial ECM is strongly dependent on the specific cell surface receptors used by the cells to interact with the ECM, on interactions with surrounding cells, and on the presence of inductive growth factors.[10]

The liver plays an important role in many physiological functions such as metabolism, storage, synthesis, and release of vitamins, carbohydrates, proteins, lipids, and cyclic tetrapyrroles.[11] Also, the liver detoxifies and inactivates endogeneous and exogeneous substances such as toxins and metals.[12] Therefore, the design of biomimetic scaffolds for liver tissue engineering is critically important because hepatocyte growth and functions are affected by ECM features, hepatocytes are anchorage-dependent cells, and the hepatocytes rapidly lose liver-specific functions and viability when removed from the architectural liver.[13]

This review chapter summarizes the recent developments on the design of biomimetic scaffolds for liver tissue engineering. Also, the mechanism of the specific interaction between galactose moieties in the artificial ECMs and ASGPR on the hepatocytes is explained.

We discuss the methods to design biomimetic scaffolds on the surface or the bulk of materials for liver tissue engineering. The criteria to design biomimetic scaffolds such as topology of the scaffolds, the coculture system, and the cell source are covered.

21.2 Specific Interaction between Galactose Residue and ASGPR

Pricer *et al.*[14] first identified the ASGPR on the hepatocytes because circulating asialoglycoproteins (ASGPs) bind to and are degraded by hepatocytes. The hepatocytes have cell surface receptors that recognize and bind to molecules having exposed galactose, *N*-acetyl galactosamine, or glucose residues.[15] Wall *et al.*[16] reported that internalization occurred via coated pits and the tracers began to accumulate in a complex arrangement of larger smooth-surfaced vesicles. The ASGPR are integral membrane glycoproteins localized on the sinusoidal face of hepatocytes. The hepatic plasma membrane receptors mediate the specific binding and uptake of partially deglycosylated glycoproteins such as receptor-mediated endocytosis.[17]

21.3 Bulk Modification of Biomaterials

With bulk modification of biomaterials, galactose moieties are incorporated into the biomaterials, and the resulting recognition sites are present not only on the surface but also in the bulk of the

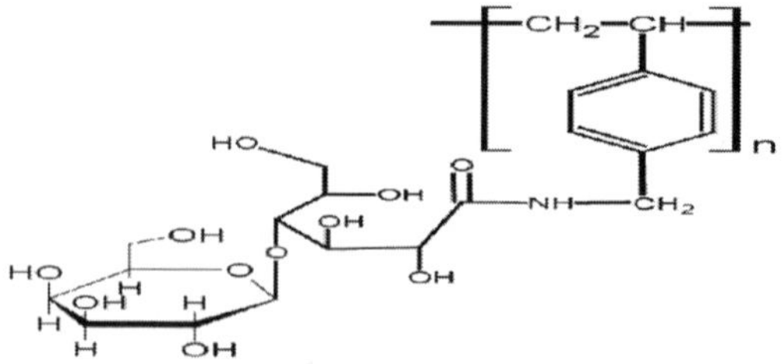

Figure 21.1. Chemical structure of PVLA (from Ref. 23).

biomaterials. Especially, bulk modification of biomaterials is beneficial to tissue engineering applications where injectable biomimetic materials are necessary to match the complex shape of native tissue at defect sites.[18]

Weigel *et al.*[19] firstly immobilized galactose groups to the poly(acryl amide) among the synthetic polymers. Rat hepatocytes were bound to a specific sugar in a Ca^{2+}-dependent manner,[20] and the cell binding to these surfaces was specifically inhibited by asialoorosomucoid.[21] Weisz *et al.*[22] also coupled galactose groups to the poly(acryl amide). Akaike *et al.* reported that galactose-carrying polystyrene (PS), as shown in Fig. 21.1, is an excellent artificial ECM to guide hepatocyte adhesion through the unique ASGPR-galactose interaction,[23] and the round morphology of hepatocytes adhered on the polymer was found to trigger the formation of spheroids of the hepatocytes in the presence of epidermal growth factor (EGF), leading to enhanced cell functions.[24]

Lopina *et al.*[25] immobilized galactose or glucose to the poly(ethylene oxide) hydrogels. The results indicated that hepatocytes adhered to the gels bearing galactose but not glucose. The amino galactose was conjugated with poly(styrene-co-maleic acid) (SMA) through amide bond linkage by Donati *et al.*[26] The galactose-coupled SMA showed about fivefold adhesion of HepG2 than SMA itself.

Natural polymers have been used as an artificial ECM because they have excellent physiological properties such as cell adhesion, mechanical properties similar to natural tissues, and biocompatibility, although they have several disadvantages such as risk of viral infection, antigenicity, instability, deterioration, and limited versatility in designing an ECM with specific properties.[7] Park *et al.*[27]

covalently coupled galactose moieties to chitosan, as shown in Fig. 21.2, to enhance galactose-specific recognition between galactose in the galactosylated chitosan (GC) and ASGPR of hepatocytes. The hepatocytes adhered to the surface at high concentration of GC showed round shapes and exhibited spheroid formation after 24 hours in the presence of EGF. Xyloglucan (XG) derived from tamarind seeds is composed of glucose units in the main chain and xylose and galactose units in the side chain. Seo *et al.*[28] studied specific interactions between galactose moieties in the XG and ASGPR on the hepatocytes adhered to the XG-coated PS surface. The hepatocyte adhesion to the XG-coated surface was dependent on the presence of Ca^{2+} ions, whereas that to the XG-coated surface could not be induced by Mg^{+2} ions, suggesting specific interactions between XG and hepatocytes. Gotoh *et al.*[29] introduced galactose moieties to silk fibroin (SF) created by the silkworm, *Bombyx mori*. The

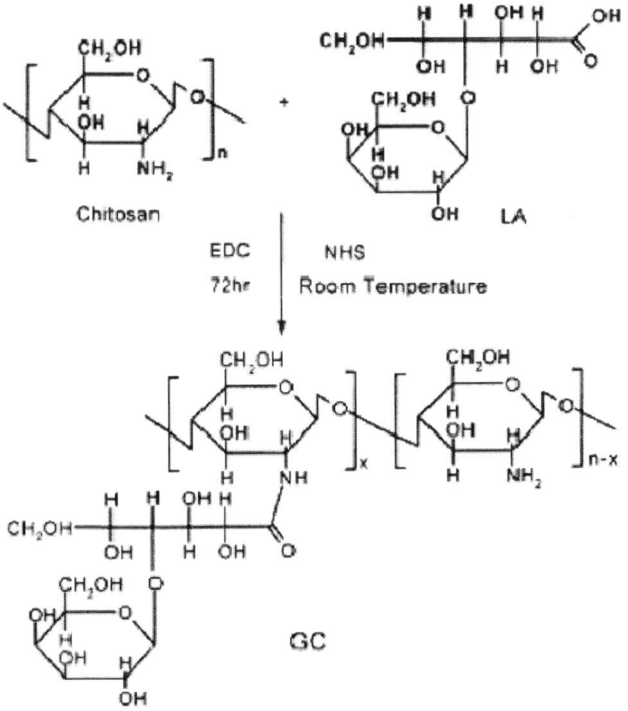

Figure 21.2. Synthesis scheme of GC (from Ref. 27).

attachment of hepatocytes on the galactosylated SF-coated dishes showed about an eightfold increase compared with that on uncoated dishes due to the specific interaction between galactose and ASGPR on the hepatocytes.

21.4 Surface Modification of Biomaterials

The surface modification of biomaterials with specific ligands is a simple way to make biomimetic scaffolds, although bulk properties dictate the mechanical properties of the ECM. Yoon *et al.*[30] introduced a galactose group into biodegradable poly(D,L-lactic-co-glycolic acid) (PLGA) films. The attachment as well as the viability of hepatocytes to the galactose-modified PLGA increased with increasing galactose concentrations on the surface, and the albumin secretion rate from the hepatocytes was enhanced. Ying *et al.*[31] immobilized galactose ligands on acrylic acid graft–copolymerized poly(ethylene terephthalate) (PET) film by plasma pretreatment. The hepatocytes cultured on the galactosylated surface exhibited good attachment and promoted spheroid formation of the attached cells. Also, the albumin and urea synthesis of hepatocytes cultured on the galactosylated surface was higher than that on the collagen-modified PET substrates. Kang *et al.*[32] similarly grafted lactose-carrying styrene to the PS dishes using plasma glow discharge. Yin *et al.*[33] used a silica surface to couple galactose ligands and to get the actual contact mechanics and adhesion strength of hepatocytes during two-dimensional cell spreading. Higashiyama *et al.*[34] coupled mixed ligands composed of fructose and galactose onto the PS culture plate. The suppression of apoptosis and necrosis was observed in hepatocyte spheroids cultured on the modified surfaces.

21.5 Criteria to Design Biomimetic Scaffolds for Liver Tissue Engineering

21.5.1 *Topology of the ECM*

Topology of the ECM is an important modulator to affect cell morphology, function, and physiological responsiveness because

3D ECMs induce more effectively hepatocyte functions than two-dimensional ones.[35] Especially, the design of 3D scaffolds to form hepatocyte spheroids is very important to enhance liver-specific functions in liver tissue engineering, because the structure of the spheroid resembles the tight cell-cell contact in native liver and tight junctions, which mimic the ultrastructure of the native lobule,[36] although the cells in large spheroids experience mass transfer limitations in the absence of a vascular network.

Encapsulation of hepatocytes in a biocompatible polymer allows incorporation of ECM features into the system and permits identification of encapsulated hepatocytes and analyses of their survival and function. Also, protection of encapsulated hepatocytes from immune surveillance *in vivo* is an important advantage of this system.[37] Yang *et al.*[38] encapsulated hepatocytes in galactosylated alginate capsules (GAC), as shown in Fig. 21.3. The results showed that higher cell viability and more spheroid formation of hepatocytes were obtained in the GAC than in the alginate capsules (AC).

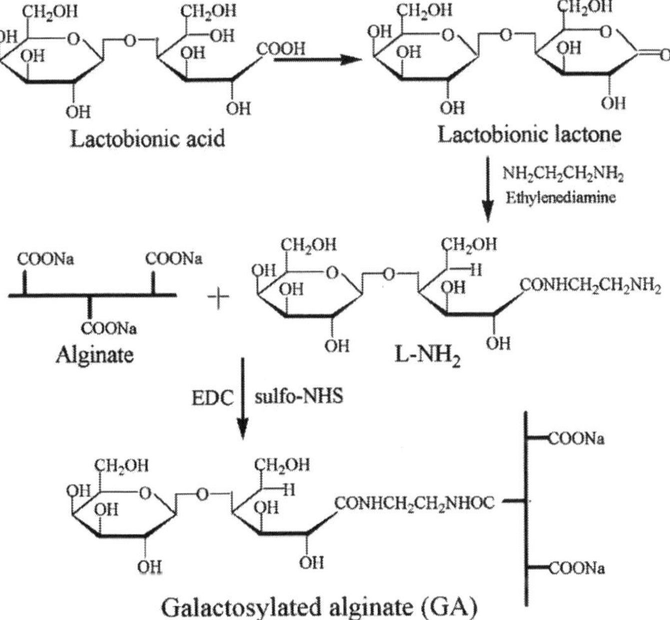

Figure 21.3. Reaction scheme of galactosylated alginate (from Ref. 38).

Moreover, liver functions of the hepatocytes such as albumin secretion and urea synthesis in the GAC were improved compared with those in the AC. During preparation of AC, galactosylated chitosan[39] or XG[40] as ECMs were entrapped together. Encapsulation of hepatocytes in the AC with ECMs exhibited a higher metabolic function in albumin secretion compared with those entrapped in AC themselves due to the enhanced multicellular spheroid formation of hepatocytes by the ECMs.

Hydrogels are 3D, hydrophilic, polymeric networks capable of imbibing large amounts of water.[41] Therefore, the soft and hydrophilic nature of hydrogels makes them suitable as scaffolds in tissue engineering after freeze-drying. Alginate, a block copolymer of β-D-mannuronic acid and α-L-guluronic acid, easily gels in the presence of divalent cations. Chung *et al.*[42] prepared an alginate/GC scaffold by mixing GC into an alginate sponge for hepatocyte attachment. The introduction of GC into an alginate sponge enhanced hepatocyte adhesion and mechanical properties of the sponge and better retained the differentiated cell functions. Also, the pore sizes of the sponge were greatly by the freezing temperature. Seo *et al.*[43] introduced heparin into an alginate/GC sponge to modulate the binding and signaling of various growth factors secreted by fibroblasts. The results indicated that the level of albumin secretion in alginate/GC/heparin sponges was markedly elevated compared with that in alginate/GC ones in the presence of hepatocyte growth factor. Hong *et al.*[44] coupled galactose moieties with gelatin for hepatocyte attachment. The galactosylated gelatin sponge exhibited higher hepatic-specific metabolic functions, such as the secretion of serum albumin and the synthesis of urea, than those cultured on a collagen type 1-coated monolayer. Chen *et al.*[45] introduced galactose moieties into a vegetable sponge for the culture of hepatocytes in a packed-bed bioreactor. The immobilized cells in the galactosylated sponge maintained high albumin and urea production rates during long-term perfusion culture.

The nanofibers are continuous and potentially providing for integrated manufacturing of 3D nanofibrous scaffolds with high porosity and high interconnectivity.[46] Highly porous nanofiber scaffolds can be prepared by electrospinning. Chua *et al.*[47] reported that a galactosylated poly(epsilon-caprolactone-co-ethyl ethylene phosphate) nanofiber scaffold promoted the mechanical stability of hepatocyte spheroids with smaller bulk and an integrated spheroid-nanofiber construct. Recently, Feng *et al.*[47] prepared nanofibrous GC scaffolds by electrospinning. The results showed that hepatocytes cultured on a GC nanofibrous scaffold formed stably immobilized 3D flat aggregates and exhibited superior cell bioactivity with higher levels of liver-specific functions than 3D spheroid aggregates formed on GC films.

21.5.2 *Coculture*

It is already reported that coculture of hepatocytes with other types of cells as the feeder cells can restore the liver function and regulate cell growth, migration, and/or differentiation,[29] although the exact mechanisms have not been elucidated yet. This may involve the cell-cell direct interactions between hepatocytes and feeder cells and the soluble factors from the feeder cells because the cell-cell contact interaction is important in the maintenance of hepatocytes.[50] Most commonly used feeder cells are epithelial- and fibroblast-like cells.[51] Human hepatocytes cocultured with human biliary epithelial cells restored the synthetic and metabolic liver function, ammonia detoxification, cytochrome P450 function, and protein expression and secretion.[52] Coculture of hepatocytes with fibroblasts improved the hepatocytes function and life span.[53]

Seo *et al.*[54] reported that the stabilized liver-specific functions such as albumin synthesis, ammonia elimination, and cytochrome P450 activity of hepatocytes cocultured with hepatocytes in the alginate/GC scaffold and 3T3 fibroblasts were enhanced compared with those monocultured with hepatocytes, as shown in Fig. 21.4.

Coculture of rat hepatocytes with sinusoidal cells gave cultures to survive longer with greater retention of the ultrastructural markers distinctive for all of these cell types.[55] The rat hepatocytes cocultured with 3T3-J2 murine fibroblasts retained their synthetic and detoxification functions on a long-term basis.[56] The rat hepatocytes viability cocultured with bone marrow cells were maintained as long as 21 days, while the hepatocytes culture alone could only be maintained for 7–10 days.[57] Also, the liver functions of the hepatocytes were increased under the coculture with mouse embryonic

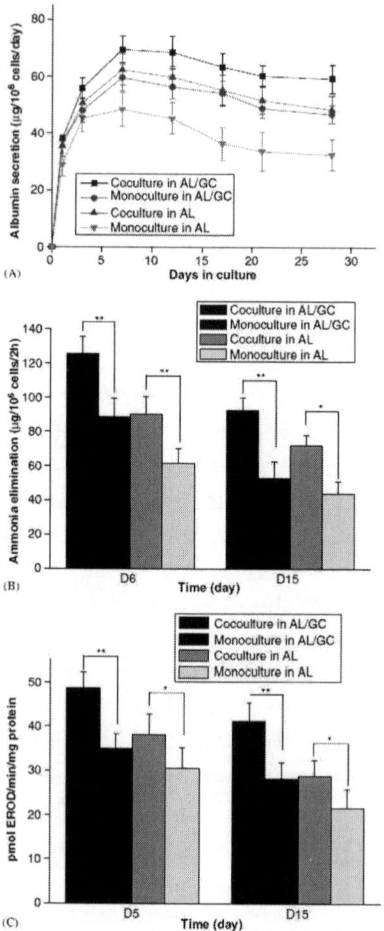

Figure 21.4. Comparison of liver-specific functions such as albumin secretion (a), ammonia elimination (b), and EROD activities (c) in AL and AL/GC sponges for monoculture and coculture of hepatocytes with NIH3T3. (a) Albumin secretion rates for both systems were measured over 28 days in culture. ELISA specific for mouse albumin was used to quantify protein in a 24-h conditioned medium. (b) Ammonia elimination was compared between both systems on days 5 and 16 of culture. (c) EROD deethylase activity by cytochrome P4501A1 was compared between both systems on days 6 and 16 of culture. EROD activity was measured using the highly fluorescent resorufin in both systems. All data is represented as mean $\pm$ SD$(n-1)$ (where $n = 6$). Statistics (two-tailed paired Student's t-test) compared monoculture with coculture; $*P < 0.05$ and $**Po < 0.01$. (From Ref. 54).

fiblasts,[58] stellate cells,[59] fetal hepatocytes,[60] and pancreatic islet cells.[61]

21.5.3 *Cell Sources*

Cell sources are very important for liver tissue engineering because it is necessary to choose cells that demonstrate the particular phenotype of interest.[62] The various cell sources are primary human hepatocytes, stem cells, immortalized human hepatocytes, the human hepatoma cell line, and primary xenogenic hepatocytes. The various cell sources for liver tissue engineering are discussed.

Human hepatocytes seem to be the natural choice, and they are very biocompatible to host for liver tissue engineering, although there are scarce, owing to a competing demand of liver transplantation, and there is the logistic problem of obtaining good-quality hepatocytes and cryopreservation of the cells.[63] However, liver-specific functions of the daughter cells after replication of adult human hepatocytes *in vitro* is generally unstable.

Xenotransplantation utilized hepatocytes of animals. The primary animal hepatocytes engraft efficiently and, being fully differentiated, can produce immediate metabolic support. The rodent hepatocytes have an apparently unlimited proliferative potential, although they have constitutive telomerase activity.[64] Primary porcine hepatocytes have been widely used because they exhibit P450 function and synthesis of urea, albumin, and other proteins and are activated by growth factors, similar to human hepatocytes, although they have several limitations such as the risk of xenozoonisis[65] and immune rejection by the patient.[62]

A new, promising way is the discovery of liver stem cells, because stem cells are an unique source of self-renewing cells within the human body and there are some stem cells present within[66] as well as outside the liver,[67] which can differentiate into fully mature hepatocytes. Hepatic stem cells were identified to exist in adult rodent and human liver by coexpression of stem cell and hepatocytic lineage markers. Hepatic oval cells are thought to share with hematopoietic stem cells.[68] Liver stem cells are also described as bipolar because they coexpress biliary and hepatocytic markers and they are capable of differentiating into both hepatocytes and bile duct epithelial cells.[69]

The bone marrow contains stem cells capable of migrating throughout the body and differentiating into various cell types, including liver cells.[63] Almost all liver cell types such as Kupffer cells and epithelial cells lining the bile ducts may originate from the bone marrow. In the bone marrow, there are certain populations of stem cell sources such as hematopoietic stem cells, marrow stromal stem cells, and multipotent adult progenitor cells.

Fetal hepatocytes are highly proliferative, which may facilitate engraftment and expansion of transplanted cell population.[70] The transplantation of fetal rat liver cells in hepatectomized rats results in the formation of 500–1,000 cell clusters within two to six months, showing 9 or 10 cell divisions, whereas adult hepatocytes transplanted under similar conditions undergo only 1 or 2 cell divisions, although ethical concerns need to be addressed.[71]

The establishment of an immortal hepatocyte cell line that can be continuously grown in unlimited quantities while retaining differentiated hepatocytes is highly desirable because present cell culture conditions have not allowed extensive proliferation of primary hepatocytes.[72] A number of proliferating human cell lines were immortalized by virtue of cultural conditions without the use of oncogenes or carcinogens, although considerable limitations inhibit their clinical utility due to the metabolic activity.[73] Also, cell immortalization can be generated by transferring specific oncogenes into these hepatocytes. Among them, the simian virus (SV)40T antigen was used for enhancing proliferation in hepatocytes.[74] Kobayashi *et al.*[75] used a reversible immortalization system with the SV40T antigen and Cre-Lox recombination targeting human hepatocytes in their final differentiated state.

Genetically engineered hepatocytes can already correct specific congential enzyme deficiencies; *ex vivo* gene therapy to correct human familial hypercholesterolemia using a retroviral vector has met with success.[76] An *ex vivo* approach to gene therapy for familial hypercholesterolemia (FH) has been developed in which the recipient is transplanted with autologous hepatocytes that are genetically corrected with recombinant retroviruses carrying the low-density lipoprotein (LDL) receptor.[77]

21.6 Summary

In this chapter, we have presented a summary of the published works on the design of biomimetic scaffolds for liver tissue engineering. The galactose-carrying synthetic and natural polymers among the artificial ECMs through bulk or surface modification of biomaterials can guide hepatocyte behaviors such as adhesion, morphology, proliferation, migration, differentiation, and survival through receptor-mediated interaction between ASGPR on hepatocytes and galactose moieties in the biomimetic scaffolds. Also, the topology of the ECM, coculture, and the cell source affected hepatocyte behaviors. The precise design of galactose-carrying biomimetic scaffolds as artificial ECMs will contribute to liver tissue engineering as a new system of bioartificial liver-assist devices or regeneration of liver tissue substitutes, although the hepatocellular-signaling pathway in the biomimetic scaffolds should be performed in the future.

Acknowledgments

We would like to acknowledge Miss Ji-Hye Seo for typewriting this chapter.

References

1. J. Tessman, A. Mikos, and A. Göpferich, *Biomaterials*, **24**, 4475 (2003).
2. S. Drotleff, U. Lungwitz, M. Breunig, A. Dennis, T. Blunk, J. Tessman, and A. Göpferich, *Eur. J. Pharm. Biopharm.*, **58**, 385 (2004).
3. E. Ruoslahti and M. D. Pierschbacher, *Science*, **238**, 491 (1987).
4. B. K. Mann, R. H. Schmedlen, and J. L. West, *Biomaterials*, **22**, 439 (2001).
5. P. R. Kuhl and L. G. Griffith-Cima, *Nature Med.*, **2**, 1022 (1996).
6. C. W. Patrick, A. G. Mikos, and L. V. McIntire, Eds., *Frontiers in Tissue Engineering* (Pergamon: Oxford).
7. F. Rosso, G. Marino, A. Giordano, M. Barbarisi, D. Parmeggiani, and A. Barbarisi, *J. Cell. Physiol.*, **203**, 465 (2005).
8. L. Fouser, L. Iruela-Arispe, P. Bornstein, and E. H. Sage, *J. Biol. Chem.*, **266**, 18345 (1991).

9. A. J. Putnam and D. J. Mooney, *Nature Med.*, **2**, 824 (1996).
10. P. A. Parsons-Winger and W. M. Saltzman, *Biotechnol. Prog.*, **9**, 600 (1993).
11. D. Zakim and T. D. Boyer, Eds., *Hepatology. A Textbook of Liver Disease (Vol.1)* (W. B. Saunders, Philadelphia, 1990).
12. M. W. Davis and J. P. Vacanti, *Biomaterials*, **17**, 365 (1996).
13. C. Chan, F. Berthiaume, B. D. Nath, A. W. Tilles, M. Toner, and M. L. Yarmuch, *Liver Transpl.*, **10**, 1331 (2004).
14. W. E. Pricer and G. Ashwell, *J. Biol. Chem.*, **246**, 4825 (1971).
15. E. Neufeld and G. Ashwell, Carbohydrate recognizing systems for receptor-mediated pinocytosis, in Ed. W. Lennarz, *Biochemistry of Glycoprotein and Proteoglycans* (Plenum Press, NY, 1979), p. 241.
16. D. A. Wall, G. Wilson, and A. L. Hubbard, *Cell*, **21**, 79 (1980).
17. G. Ashwell and J. Harford, *Annu. Rev. Biochem.*, **51**, 531 (1982).
18. H. Shin, S. Jo, and A. G. Mikos, *Biomaterials*, **24**, 4353 (2003).
19. P. H. Weigel, E. Schmell, Y. C. Lee, and S. Roseman, *J. Biol. Chem.*, **253**, 330 (1978).
20. P. H. Weigel, R. L. Schnaar, M. S. Kuhlenschmidt, E. Schmell, R. T. Lee, and Y. C. Lee *J. Biol. Chem.*, **254**, 10830 (1979).
21. P. H. Weigel, *J. Cell Biol.*, **87**, 855 (1980).
22. O. A. Weisz and R. L. Schmaar, *J. Cell Biol.*, **115**, 485 (1991).
23. K. Kobayashi, A. Kobayashi, and T. Akaike, Culturing hepatocytes on lactose-carrying polystyrene layer via asialoglycoprotein receptor-mediated interactions, in Eds. Y. C. Lee, and R. T. Lee, *Methods in Enzymology: Neoglycoconjugates Part B Biomedical Applications*, Vol. 247 (Academic Press, San Diego, CA, 1994), p. 409.
24. S. Tobe, Y. Takei, and K. Kobayashi, T. Akaike, *Biochem. Biophys. Res. Commun.*, **184**, 225 (1992).
25. S. T. Lopina, G. Wu, E. W. Merrill, and L. Griffith-Cima, *Biomaterials*, **17**, 559 (1996).
26. I. Donati, A. Gamini, A. Vetere, C. Compa, and S. Paoletti, *Biomacromolecules*, **3**, 805 (2002).
27. I. K. Park, J. Yang, H. J. Jeong, H. S. Bom, I. Harada, T. Akaike, S. I. Kim, and C. S. Cho, *Biomaterials*, **24**, 2331 (2003).
28. S. J. Seo, I. K. Park, M. K. Yoo, M. Shirakawa, T. Akaike, and C. S. Cho, *J. Biomater. Sci. Polym. Ed.*, **15**, 1375 (2004).
29. Y. Gotoh, S. Niimi, T. Hayakawa, and T. Miyashita, *Biomaterials*, **25**, 1131 (2004).

30. J. J. Yoon, Y. S. Nam, J. H. Kim, and T. G. Park, *Biotechnol. Bioeng.*, **78**, 1 (2002).
31. L. Ying, C. Yin, R. X. Zhuo, K. W. Leong, H. Q. Mao, E. T. Kang, and K. G. Neoh, *Biomacromolecules*, **4**, 157 (2003).
32. I. K. Kang, D. W. Lee, S. K. Lee, and T. Akaike, *J. Mater. Sci. Mater. Med.*, **14**, 611 (2003).
33. C. Yin, K. Liao, H. Q. Mao, K. W. Leong, R. X. Zhuo, and V. Chan, *Biomaterials*, **24**, 837 (2003).
34. S. Higashiyama, M. Noda, M. Kawase, and K. Yagi, *J. Biomed. Mater. Res. A*, **64**, 475 (2003).
35. F. Berthiaume, P. V. Moghe, M. Toner, and M. L. Yarmush, *FASEB J.*, **10**, 1471 (1996).
36. L. K. Hansen, C. C. Hsiao, J. R. Friend, F. J. Wu, G. A. Bridge, R. P. Remmel, F. B. Cerra, and W. S. Hu, *Tissue Eng.*, **4**, 65 (1998).
37. Z. Cai, Z. Shi, M. Sherman, and A. M. Sun, *Hepatology*, **10**, 855 (1989).
38. J. Yang, M. Goto, H. Ise, C. S. Cho, and T. Akaike, *Biomaterials*, **23**, 471 (2002).
39. X. L. Guo, K. S. Yang, J. Y. Hyun, W. S. Kim, D. H. Lee, K. E. Min, L. S. Park, K. H. Seo, Y. I. Kim, C. S. Cho, and I. K. Kang, *J. Biomater. Sci. Polym. Ed.*, **14**, 551 (2003).
40. S. J. Seo, T. Akaike, Y. J. Choi, M. Shirakawa, I. K. Kang, and C. S. Cho, *Biomaterials*, **26**, 3607 (2005).
41. N. A. Peppas, *Curr. Opin. Colloid Interface*, **2**, 531 (1997).
42. T. W. Chung, J. Yang, T. Akaike, K. Y. Cho, J. W. Nah, S. I. Kim, and C. S. Cho, *Biomaterials*, **23**, 2827 (2002).
43. S. J. Seo, Y. J. Choi, T. Akaike, A. Higuchi, and C. S. Cho, *Tissue Eng.*, **12**, 33 (2006).
44. S. R. Hong, Y. M. Lee, and T. Akaike, *J. Biomed. Mater. Res. A*, **67**, 733 (2003).
45. J. Chen and C. Lin, *J. Biosci. Bioeng.*, **102**, 41 (2006).
46. Y. Dzenis, *Science*, **304**, 1917 (2004).
47. K. N. Chua, W. S. Lim, P. Zhang, H. Lu, J. Wen, S. Ramakrishna, K. W. Leong, and H. Q. Mao, *Biomaterials*, **26**, 2537 (2005).
48. Z. Q. Feng, X. Chu, N. P. Huang, T. Wang, Y. Wang, X. Shi, Y. Ding, and Z. Z. Gu, *Biomaterials*, **30**, 2753 (2009).
49. L. Z. Chang and T. M. Chang, Coencapsulation of hepatocytes and bone marrow cells: *in vitro* and *in vivo* studies (2006).

50. S. N. Bhatia, U. J. Balis, M. L. Yamush, and M. Toner, *FASEB J.*, **13**, 1883 (1999).
51. B. Clement, C. Guguen-Guillouzo, J. P. Campion, D. Glaise, M. Bourel, and A. Guillouzo, *Hepatology*, **4**, 373 (1984).
52. L. M. Reid and D. M. Jefferson, *Hepatology*, **4**, 548 (1984).
53. S. Kaihara, S. Kim, B. S. Kim, D. J. Mooney, K. Tanaka, and J. P. Vacanti, *J. Pediatr. Surg.*, **35**, 1287 (2000).
54. S. J. Seo, I. Y. Kim, Y. J. Choi, T. Akaike, and C. S. Cho, *Biomaterials*, **27**, 1487 (2006).
55. K. Sugimachi, M. N. Sosef, J. M. Baust, A. Fowler, R. G. Tompkins, and M. Toner, *Cell Transplant*, **13**, 187 (2004).
56. M. K. Auth, R. E. Joplin, M. Okamoto, Y. Ishida, P. McMaster, J. M. Neuberger, R. A. Blaheta, T. Voit, and A. J. Strain, *Hepatology*, **33**, 519 (2001).
57. Z. Liu and T. M. Chang, *Artif. Cells Blood Substit. Immobil. Biotechnol.*, **28**, 365 (2000).
58. W. Kuri-Harcuch and T. Mendoza-Figueroa, *Differentiation*, **41**, 148 (1989).
59. L. Ricalton-Banks, C. Liew, R. Bhandari, J. Fry, and K. Skakesheff, *Tissue Eng.*, **9**, 401 (2003).
60. D. Van Poll, C. Sokmensuer, N. Ahmad, A. W. Tilles, F. Berthiaume, M. Toner, and M. L. Yarmush, *Tissue Eng.*, **12**, 2965 (2006).
61. K. W. Lee, S. K. Lee, J. W. Joh, S. J. Kim, B. B. Lee, K. W. Kim, and K. U. Lee, *Tissue Eng.*, **10**, 965 (2004).
62. K. M. Kulig and J. P. Vacanti, *Transpl. Immunol.*, **12**, 303 (2004).
63. M. Gewartowska and W. L. Olszewski, *Ann. Transplant*, **12**, 27 (2007).
64. K. Overturf *et al.*, *Am. J. Pathol.*, **151**, 1273 (1997).
65. J. W. Allen and S. N. Bhatia, *Tissue Eng.*, **8**, 725 (2002).
66. S. S. Thorgeirsson, *FASEB J.*, **10**, 1249 (1996).
67. N. D. Theise, M. Nimmakayalu, R. Gardner., P. B. Illei, G. Morgan, L. Teperman, O. Henegariu, and D. S. Krause, *Hepatology*, **32**, 11 (2000).
68. Y. Zhang, X. F. Bai, and C. X. Huang, *World J. Gastroenterol*, **9**, 201 (2003).
69. A. C. Piscaglia, C. Di Campli, G. Gasbarrini, and A. Gasbarrini, *Dig. Liver Dis.*, **35**, 507 (2003).
70. K. J. Allen and H. E. Soriano, *J. Lab. Clin. Med.*, **138**, 298 (2001).

71. J. S. Sandhu, P. M. Petkov, M. D. Dabeva, and D. A. Shafritz, *Am. J. Pathol.*, **159**, 1323 (2001).
72. K. Ohashi, F. Park, and M. A. Kay, *J. Mol. Med.*, **79**, 617 (2001).
73. P. Ferenci, *Clin. Liver Pis.*, **2**, 31 (1998).
74. H. Malhi and S. Gupta, *J. Hepatobil. Pancreat. Surg.*, **8**, 40 (2001).
75. N. Kobayashi, T. Okitsu, S. Nakaji, and N. Tanaka, *J. Artif. Organs*, **6**, 236 (2003).
76. L. A. Lee, *J. Clin. Invest.*, **108**, 367 (2001).
77. M. Grossman *et al.*, *Nat. Genet.*, **6**, 335 (1994).

Chapter 22

HYBRID POROUS SCAFFOLDS OF BIODEGRADABLE SYNTHETIC POLYMERS AND COLLAGEN FOR TISSUE ENGINEERING

Guoping Chen,* Naoki Kawazoe, and Tetsuya Tateishi

Biomaterials Center, National Institute for Materials Science, 1-1 Namiki, Tsukuba, Ibaraki 305-0044, Japan

*guoping.chen@nims.go.jp

Porous scaffolds play an important role in tissue engineering and regenerative medicine as a temporary support to control cell adhesion, proliferation, and differentiation and to guide the formation of new tissues and organs into desired shapes and complicated structures. The scaffolds must meet certain requirements for *in vitro* and *in vivo* applications. A number of porous scaffolds have been developed from biodegradable synthetic polymers and naturally derived polymers. Scaffolds prepared from synthetic polymers and naturally derived polymers have their respective advantages and drawbacks. Hybridization of the biodegradable synthetic polymers with naturally derived polymers has been employed to develop hybrid porous scaffolds in order to combine their advantages and avoid their drawbacks. The biodegradable

Handbook of Intelligent Scaffolds for Tissue Engineering and Regenerative Medicine
Edited by Gilson Khang

www.panstanford.com

synthetic polymers serve as mechanical skeletons to provide hybrid scaffolds with high mechanical strength to retain their designed shapes and sizes, while the naturally derived polymers contribute to good cell interaction and hydrophilicity to facilitate cell seeding and cell attachment. Hybrid porous scaffolds have been used for tissue engineering of a number of types of tissues. Recent developments of hybrid scaffolds and their application to tissue engineering are summarized.

22.1 Introduction

Tissue engineering and regenerative medicine have been widely used to treat defective and malfunctioning tissues and organs.[1] Some tissue-engineered products have been used for clinical applications, and some have been commercialized. A few methods have been developed in the approach for tissue engineering and regenerative medicine. They include cell therapy, scaffold-free, and scaffold-aided methods.[2–4] Cell therapy by directly injecting or applying somatic or stem cells to cleaned foci can be performed in low-invasive methods; however, the diffusion of cells from the foci has been a challenging issue. Reconstruction of tissues and organs by scaffold-free methods such as pellet culture and cell sheet engineering can avoid any undesirable effects from materials. Regeneration of large tissues and organs with complicated shapes by the scaffold-free method remains a challenge. The scaffold-aided method provides a scaffold to accommodate the seeded cells, to support cell adhesion and proliferation, and to guide the formation of tissue in the scaffold. The cells can be held together by the scaffold in large tissues and organs with complicated structures of specific shapes. Bioactive molecules can also be incorporated in the scaffold and their release controlled temporally and spatially to trigger a cascade of signaling events to direct cell functions. Scaffolds can also be designed to mimic the structures of extracellular matrices to provide cells with biomimetic *in vivo* microenvironments. However, the biomaterials used for scaffolds must be biocompatible and biodegradable to decrease or remove any side effects from the scaffold materials. The scaffold-aided method has been used for the tissue engineering of many tissues and organs such as skin, cartilage, bone, trachea, bladder, blood vessels, nerves, etc.

The scaffolds used for tissue engineering and regenerative medicine should permit cell adhesion, promote cell proliferation and differentiation, and be biocompatible, biodegradable, mechanically strong, and capable of being formed into desired shapes.[5–7] A number of scaffolds have been prepared from biodegradable synthetic polymers such as poly(glycolic acid) (PGA), poly(L-lactic acid) (PLLA), their copolymer of poly(lactic-*co*-glycolic acid) (PLGA), and biodegradable, naturally derived polymers such as collagen and hyaluronic acid.[8–10] Synthetic and naturally derived polymers have their respective advantages and disadvantages. Biodegradable synthetic polymers can easily be formed into designed shapes with relatively high mechanical strength. However, the scaffold surface of these polymers is relatively hydrophobic, which is not good for cell seeding. In contrast, naturally derived polymers have specific cell interaction peptides, and their scaffolds have hydrophilic surfaces, which are beneficial for cell seeding and cell attachment. However, naturally derived polymers are too weak mechanically. The two kinds of polymers have been hybridized to combine the advantages of both polymers and overcome their disadvantages. This chapter summarizes recent development of such hybrid scaffolds and their applications to tissue engineering and regenerative medicine.

22.2 PLGA-Collagen Hybrid

Hybrid porous scaffolds of biodegradable polymers such as PGA, PLLA, and PLGA and collagen have been prepared by introducing collagen microsponges in the openings of skeletons of biodegradable synthetic polymers (Fig. 22.1).[11] A PLGA-collagen hybrid sponge is prepared by hybridization by introducing collagen microsponges into the pores of a PLGA sponge (Fig. 22.2).[12] Observation of the cross section of the hybrid sponge with a scanning electron microscope (SEM) shows that collagen microsponges were formed in the pores of the PLGA sponge. SEM–electron probe microanalysis of elemental nitrogen indicates that the pore surfaces of the PLGA sponges are coated with collagen and microsponges of collagen had formed

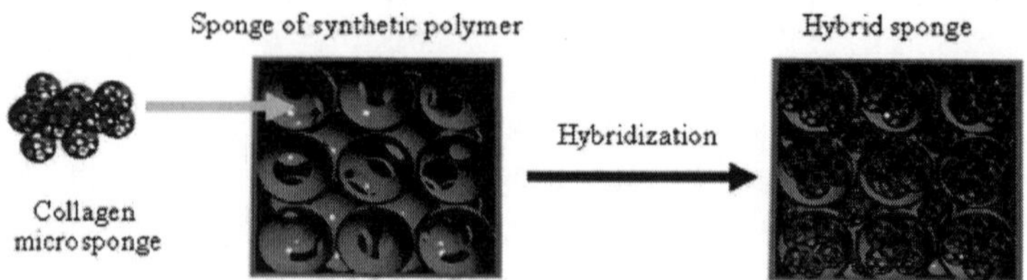

Figure 22.1. Preparation of a hybrid sponge of biodegradable synthetic polymers and collagen.

in the pores. All the surfaces of the PLGA sponges were covered with a thin layer of collagen. When a hybrid sponge is used for cell culture, the cells will contact the collagen instead of directly contacting the synthetic polymers.

The mechanical properties of PLGA-collagen were compared with those of PLGA sponges and collagen sponges. The ultimate tensile strength, modulus of elasticity, and static stiffness of the PLGA-collagen hybrid sponge are higher than those of the PLGA and collagen sponges in both the dry and wet states. The PLGA-collagen sponge has higher mechanical properties in the dry state than in the wet state. Serving as a mechanical skeleton, the PLGA sponge contributes greatly to the mechanical properties of the hybrid sponge. A strong mechanical skeleton provides the hybrid sponge with high mechanical strength to prevent any shrinkage and deformation during *in vitro* cell culture and *in vivo* implantation. The mechanical

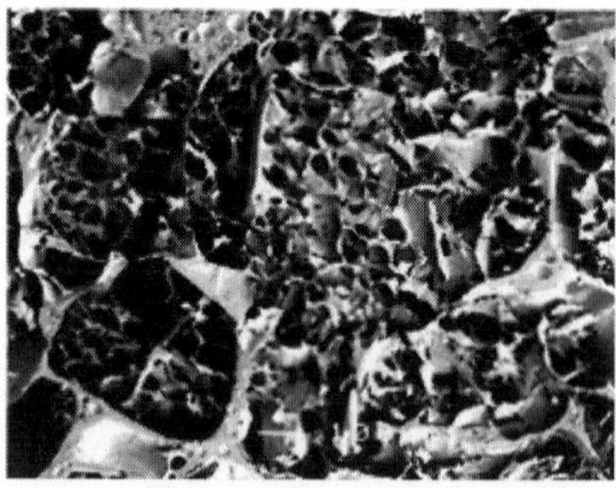

Figure 22.2. SEM photomicrograph of a PLGA-collagen hybrid sponge.

skeleton can also be formed into any desired shape to define the shape and structure of tissue-engineered tissues and organs.

The collagen microsponges in the pores and coated collagen layer on the surface can promote interaction with cells and thus facilitate cell adhesion and distribution. The introduction of collagen into the synthetic polymer sponges can also improve the wettability of the synthetic polymer sponges. The surface properties of biodegradable synthetic polymer sponges are hydrophobic. This property inhibits infiltration of the cell suspension solution into the porous scaffolds and, therefore, hinders smooth cell seeding. Pretreatment such as prewetting with ethanol, hydrolysis with an alkaline aqueous solution, and plasma treatment is required to change the surface property from hydrophobic to hydrophilic to facilitate cell seeding into the hydrophobic scaffolds. However, the water wettability of the PLGA-collagen hybrid sponge is increased. Improvement of wettability facilitates cell seeding and cell infiltration into the inner parts of the scaffolds.

The PLGA-collagen hybrid sponge has been used for the culture of bovine articular chondrocytes.[13] A PLGA sponge was used as a control. The chondrocytes were seeded into the PLGA sponge and PLGA-collagen hybrid sponge and cultured *in vitro* in a serum medium in a 5% CO_2 atmosphere at 37°C. Cell seeding into the PLGA sponge requires prewetting of the sponge with an aqueous ethanol solution. However, cell seeding in the PLGA-collagen hybrid sponge becomes much easier due to its improved wettability. Simply dropping the cell suspension solution onto the PLGA-collagen hybrid sponge causes the cell solution to penetrate the hybrid sponge. The cell-seeding efficiency of the PLGA-collagen sponge was higher than that of the PLGA sponge. More cells adhered to the PLGA-collagen hybrid sponge than to the PLGA sponge because the collagen promoted the cell adhesion in the hybrid sponge. The chondrocytes proliferated in the hybrid sponge during culture. The chondrocytes and their extracellular matrices gradually occupied the space in the hybrid sponge. Biochemical analysis demonstrated that the DNA and proteoglycan content in the hybrid scaffold increased with culture time. Most of the chondrocytes after four weeks' culture, and almost all the cells after six weeks, maintained their phenotypically rounded morphology. Immunohistological staining of type II collagen showed it was present in the extracellular matrix after four weeks and increased with time. The proliferated chondrocytes and the secreted extracellular matrices constructed new articular cartilage-like tissue. The scaffold slowly degraded, losing almost 36.9% of its mass after 10 weeks.

In vivo subcutaneous implantation in nude mice showed that a more homogeneous articular cartilage-like tissue was formed in the hybrid sponge than in the PLGA sponge. New tissues that formed in the hybrid sponges maintained the same shape as that of the hybrid sponge. The PLGA sponge also maintained its original shape after implantation. However, the control collagen sponge collapsed when implanted in the nude mice. Suppression of the surrounding tissues and contraction of the new tissue resulted in the collapse of the collagen sponge. The mechanically strong PLGA sponge and PLGA-collagen hybrid sponges protected themselves from deformation and collapse during *in vivo* implantation.

From the above results, it can be concluded that the PLGA-collagen hybrid sponge shows high mechanical properties and good cell interaction and thus promotes new tissue formation when used for tissue engineering. The PLGA-collagen hybrid sponge combines the advantages of both synthetic PLGA and naturally derived collagen. The hybridization method can also be used to prepare hybrid scaffolds of other biodegradable synthetic polymers and other naturally derived polymers.

22.3 PLGA-Collagen Hybrid Mesh

A mesh-type hybrid scaffold of biodegradable synthetic polymers and collagen has been developed by introducing weblike collagen microsponges in the openings of a mesh of biodegradable synthetic polymers (Fig. 22.3).[14] A PLGA-collagen hybrid mesh is one example of such a kind of mesh-type hybrid scaffold (Fig. 22.4). The formation of collagen microsponges in the openings of the PLGA mesh is obtained by introducing an aqueous collagen solution into the openings, freeze-drying, and cross-linking. After these procedures, weblike collagen microsponges form in the openings of the synthetic PLGA mesh. SEM observation showed that collagen microsponges

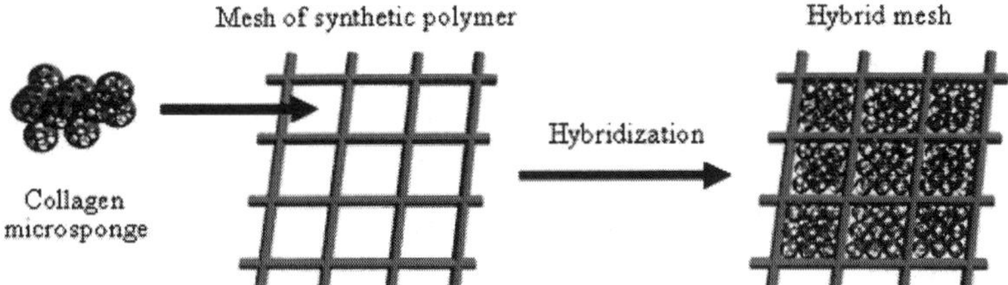

Figure 22.3. Preparation of a hybrid mesh of biodegradable synthetic polymers and collagen.

in each opening of the mesh were connected with each other by collagen fibers climbing over the PLGA fiber bundles of the polymer mesh. The unique structure is somewhat like spiderwebs formed in the networks of a net. The weblike collagen microsponges are expected to entrap cells, and the PLGA knitted mesh is expected to contribute high mechanical strength.

The mechanical properties of the PLGA-collagen hybrid mesh were evaluated by a static tensile test. The moduli of elasticity of the hybrid mesh, PLGA mesh, and collagen sponge were 35.4±1.4, 35.2±1.0, and 0.020±0.001 MPa, respectively. The hybrid mesh showed a slightly higher tensile strength than the PLGA knitted mesh and a much higher strength than the collagen sponge. The polymer mesh served as a skeleton and reinforced the hybrid mesh.

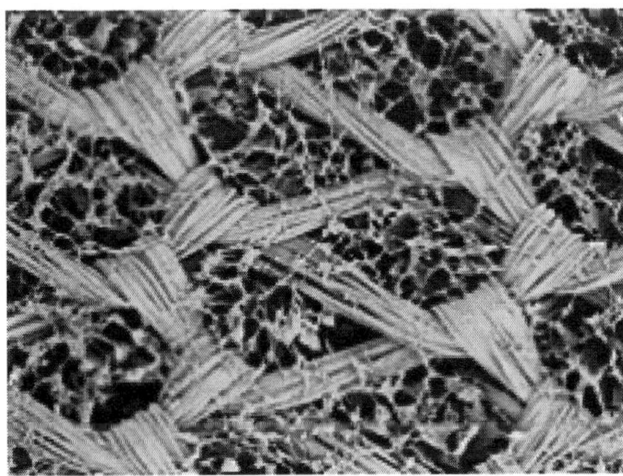

Figure 22.4. SEM photomicrograph of a PLGA-collagen hybrid mesh.

The PLGA-collagen hybrid mesh has been used for culturing fibroblasts, tenocytes, chondrocytes, and osteoblasts. The weblike collagen microsponges in the PLGA-collagen facilitated cell adhesion and distribution in the hybrid mesh.

The PLGA-collagen hybrid mesh has also been used for the tissue engineering of cartilage.[15] Chondrocytes isolated from bovine articular cartilage were seeded in the hybrid mesh by dropping a cell suspension in a culture medium onto the hybrid mesh. The chondrocytes adhered to the weblike hybrid mesh and showed a uniform distribution on the mesh. The cells continued to proliferate and regenerated a cartilaginous matrix, filling the voids in the hybrid mesh. The weblike collagen microsponges that formed in the openings of the knitted mesh not only prevented the seeded cells from passing through the hybrid mesh but also increased the specific surface area to provide sufficient surface for a spatially even cell distribution. After being cultured *in vitro* for one day, the cell/mesh sheets were used in single-sheet form to regenerate thin cartilage implants or in laminated form to yield thick cartilage implants. The thickness of the implant could be manipulated by varying the number of laminated mesh sheets. The cell/mesh sheets could also be rolled up in the shape of a cylinder, in which case the thickness of the implant was adjusted by the height and the diameter of the roll by the number of sheets used. As cells were seeded within each sheet of the hybrid mesh, the density and distribution of the seeded cells in the overall laminated or rolled form were as high and uniform as those in the single-sheet form. A round, disk-shaped single sheet, five-sheet, and 8 mm high roll implants after 4, 8, and 12 weeks' implantation retained their original shapes and appeared pearly white. The good mechanical properties of the hybrid mesh and spatial homogeneous cell distribution help the tissue-engineered cartilage implants retain the same shapes as the original scaffolds. In the cases of the laminated and rolled implants, the implanted sheets became integrated with each other. Histological examination showed that the chondrocytes had a round morphology. Alcian blue staining indicated that glycosaminoglycans (GAGs) were abundant and homogeneously distributed throughout the implants (Fig. 22.5). Toluidine blue staining demonstrated the typical metachromasia of articular cartilage. Type II collagen was detected by immunohistological staining. Northern

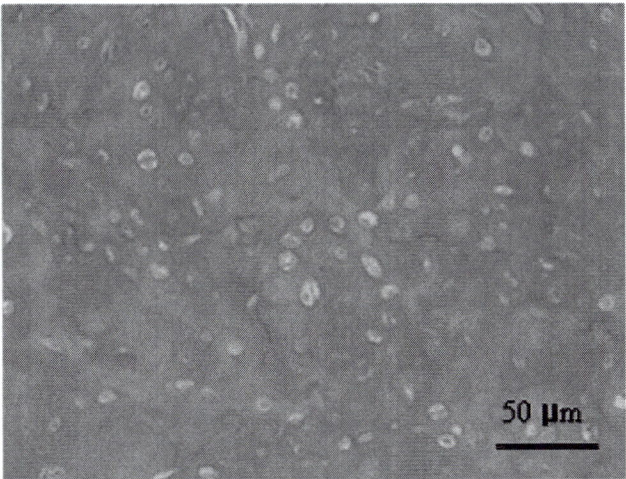

Figure 22.5. Alcian blue staining of a five-sheet implant after 12 weeks' implantation.

blot analysis showed that the genes encoding the articular cartilaginous matrices of type II collagen and aggrecan were expressed by chondrocytes and their expression was upregulated, while that for type I collagen was downregulated when the chondrocytes were cultured in the hybrid mesh. The dynamic complex modulus, structural stiffness, and phase lag approached those of native bovine articular cartilage. These results indicate that articular cartilage-like tissue was regenerated when chondrocytes were cultured in the PLGA-collagen hybrid mesh. The thickness and size of the engineered cartilage implants can be adjusted by using a single layer of mesh, laminating or rolling the hybrid mesh, and by controlling the size of the hybrid mesh. The similarity of the results for the five-sheet and roll implants to those for the single-sheet implant suggests that an increase in the implant thickness did not compromise cell viability, uniformity, or function. Use of the PLGA-collagen hybrid mesh as a single sheet or in the laminated state might be an effective method for the tissue engineering of cartilage of various thicknesses.

The PLGA-collagen hybrid mesh has been used for the culture of osteoblasts for the tissue engineering of bone.[16] The PLGA skeleton facilitates the formation of the hybrid mesh into desired shapes, and the collagen microsponges promote cell adhesion and uniform cell distribution throughout the scaffold. Osteoblasts established from rat marrow stroma adhered to the PLGA-collagen hybrid mesh and generated the desired-shaped bone. The PLGA-collagen hybrid mesh shows promise for use as a tool for custom-shaped bone regeneration in basic research on osteogenesis and for the development of therapeutic applications.

The PLGA-collagen hybrid mesh has also been used for osteochondral tissue engineering.[17] Canine articular chondrocytes were cultured in the hybrid mesh in Dulbecco's Modified Eagle's Medium (DMEM) containing 10% fetal bovine serum (FBS) and laminated to construct a chondral layer. Canine bone marrow–derived mesenchymal stem cells (MSCs) were cultured in the PLGA-collagen hybrid mesh in an osteogenic induction medium to construct an osteo-layer. The osteo- and chondral layers were sutured together and implanted subcutaneously in nude mice. The original round, disc shape of the osteochondral constructs was preserved during the implantation. The osteo- and chondral layers appeared red and glistening white, respectively. Histological examination of the implant specimens indicated that stromal cells and chondrocytes were evenly distributed throughout the scaffold. The laminated meshes were bound together, and the two layers had a distinct interface between them. The cells showed a round morphology in the chondral layer and a spindle-like morphology in the osteo-layer. In the chondral layer, spherical chondrocytes were surrounded by an abundant cartilaginous extracellular matrix. The round morphology and positive staining by safranin O and toluidine blue, together with the expression of genes encoding type II collagen and aggrecan, suggested the formation of neocartilage in the chondral layer. Expressions of genes encoding type I collagen and osteocalcin were detected in the osteochondral implant. The hybrid mesh supported the adhesion and proliferation of the stromal cells and the chondrocytes and promoted the formation of osteochondral-like tissue. Use of the PLGA-collagen hybrid mesh and lamination would be a useful strategy for osteochondral tissue engineering.

The PLGA-collagen hybrid mesh has been compared with the PLGA knitted mesh for the culture of human skin fibroblasts for the tissue engineering of dermal tissue.[18] The efficiency of cell seeding was much higher, and the cells grew more quickly in the hybrid mesh than in the PLGA mesh. The fibroblasts in the PLGA mesh grew from the peripheral PLGA fibers toward the centers of the openings, while those in the hybrid mesh also grew from the collagen microsponges in the openings of the mesh, resulting in more homogenous growth. The proliferated cells and secreted extracellular matrices were more uniformly distributed in the hybrid mesh than in the PLGA mesh. Histological staining of *in vitro*-cultured fibroblast/mesh implants indicated that the fibroblasts were distributed throughout the hybrid mesh and formed a uniform layer of dermal tissue having almost the same thickness as that of the hybrid mesh. However, the tissue formed in the PLGA mesh was thick adjacent to the PLGA fibers and thin in the centers of the openings. The results indicate that the weblike collagen microsponges formed in the openings of the PLGA knitted mesh increased the efficiency of cell seeding, improved cell distribution, and therefore facilitated rapid formation of dermal tissue having a uniform thickness.

Application of the PLGA-collagen hybrid mesh for the regeneration of vessel tissue has been attempted by using the hybrid mesh to patch the canine pulmonary artery trunk.[19] Hybrid meshes with or without autologous vessel cellularization were implanted. After two and six months' implantation, histological and biochemical examination showed that there was no thrombus formation in either group and the hybrid mesh was almost completely absorbed in both groups. Formation of an endothelial cell monolayer, a parallel alignment of smooth muscle cells, and a reconstructed vessel wall with elastin and collagen fibers was confirmed. At six months, the cellular and extracellular components in the hybrid mesh had increased to levels similar to those in native tissue. The hybrid mesh might be useful for *in situ* cellularization and the regeneration of autologous tissue in cardiovascular surgery.

22.4 PLLA-Collagen Hybrid Braid

A hybrid scaffold of a PLLA braid and a collagen sponge has been fabricated for ligament regeneration by forming collagen microsponges in the interstices of a PLLA braid (Fig. 22.6).[20] The

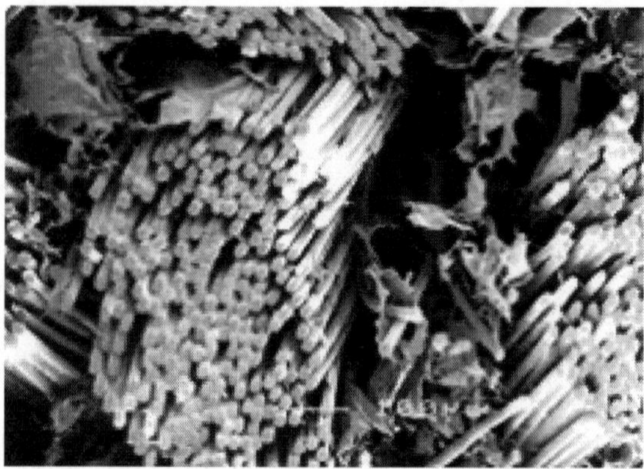

Figure 22.6. SEM photomicrograph of a PLLA-collagen hybrid braid.

PLLA braid provides high mechanical properties, while collagen microsponges provide good cell interaction with the hybrid braid. Cellular responses of ligament cells to the PLLA and PLLA-collagen hybrid braids were evaluated both *in vitro* and *in vivo*. More cells and greater homogeneous cell distribution were observed *in vitro* in the hybrid braid than in the PLLA braid. Greater fibroblast immigration and neoangiogenesis were detected in the hybrid braid than in the PLLA braid when implanted on a ruptured medial collateral ligament (MCL). Hybridization with collagen facilitated cell seeding and spatial cell distribution and promoted cell immigration and neoangiogenesis.

22.5 Biphasic Hybrid Porous Scaffold

A biphasic scaffold composed of an upper layer of a collagen sponge and a lower layer of a PLGA-collagen hybrid sponge has been reported for tissue engineering of osteochondral tissue.[21] At first, a biodegradable PLGA sponge cylinder was prepared by the porogen leaching method using NaCl particulates. Then, a collagen/PLGA-collagen biphasic sponge cylinder was prepared by introducing a collagen sponge into the pores of the PLGA sponge and forming a collagen sponge on one side of the PLGA sponge (Fig. 22.7). One layer of the biphasic scaffold was a highly porous collagen sponge. The

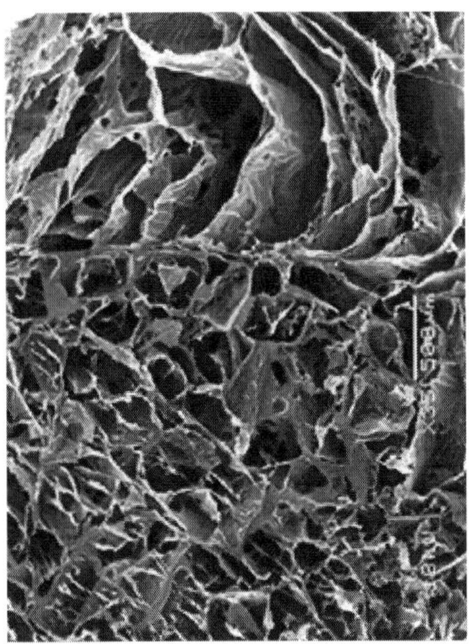

Figure 22.7. SEM photomicrograph of a collagen/PLGA-collagen biphasic sponge.

other layer was a hybrid sponge with a collagen sponge formed in the pores of the PLGA sponge. The collagen sponges in the two layers were connected.

The collagen/PLGA-collagen biphasic hybrid scaffold has been used for osteochondral tissue engineering. Canine bone marrow–derived MSCs were isolated from bone marrow aspirate of a one-year-old beagle and subcultured in DMEM containing 10% FBS. The subcultured MSCs were seeded into the collagen/PLGA-collagen biphasic sponge scaffold, cultured *in vitro* for one week, and transplanted into the knee of the same beagle. The implants were harvested after three months. Gross appearance showed that the defect treated with the cells/scaffold presented a smoother surface and better integration with the surrounding tissue than did a scaffold without cells. Histological examination of these specimens using hematoxylin and eosin and safranin O/fast green staining indicated that cartilage-like and underlying bonelike tissues were regenerated after four months' implantation. The cartilage-like tissue was stained intensively with safranin O and well integrated with the surrounding tissue. However, the defect implanted with the scaffold without the cells showed no evidence of hyaline cartilage regeneration.

22.6 Leak-Proof Hybrid Scaffolds

When seeded in porous scaffolds, cells do not always stay and remain in the scaffolds. They might pass through the porous structure of the scaffolds and leak from the open periphery, resulting in a loss of cells that affects cell distribution and, finally, the quality of the engineered tissue. To prevent cell leakage from porous scaffolds and increase cell-seeding efficiency, a novel type of hybrid scaffold has been developed that has the capacity to prevent cell leakage during cell seeding. The leak-proof hybrid scaffolds were prepared by covering all the surfaces of a collagen sponge except the top surface with a porous structure containing small pores to prevent cell leakage during cell seeding (Fig. 22.8).

To test this approach, a collagen sponge was first wrapped with a nondegradable nylon mesh.[20] The leak-proof collagen sponge was prepared in a mold by forming a collagen sponge in a polyethylene cylinder of which the bottom and inside walls were covered with a nylon membrane. SEM observation showed that a

Figure 22.8. Illustration of a leak-proof hybrid scaffold.

collagen sponge formed inside the mold and that it was cross-linked with the nylon membrane by forming collagen fibers that passed through the interstices of the nylon membrane. The climbing collagen fibers prevented the collagen sponge from detaching from the mold. Without the nylon membrane, the collagen sponge that formed inside the polyethylene mold shrank and detached from the mold. A polystyrene cylinder served as a support to maintain the shape of the collagen sponge. The nylon mesh helped in attaching the collagen sponge to the mold and preventing the collagen sponge from shrinking. Human bone marrow–derived MSCs were seeded in the collagen sponge and cultured in a proliferation medium for one week. Cell-seeding efficiency was $95.0 \pm 2.7\%$. The nylon membrane prevented leakage of the seeded cells. The MSCs adhered to the collagen sponge and were homogeneously distributed throughout the sponge. During culture, the cells proliferated and secreted extracellular matrices that filled the voids in the sponge. When this leak-proof scaffold was used for *in vivo* implantation, the collagen sponge in the center of the nylon mesh mold must be drilled out and then implanted. A cup-shaped PLLA sponge has been used to replace the nylon mesh mold to construct a leak-proof biodegradable hybrid scaffold.

The leak-proof PLLA-collagen hybrid scaffold was prepared by introducing a collagen sponge into the center of a cup-shaped skeleton of biodegradable synthetic polymers. The PLLA sponge cup was prepared using the porogen leaching method using a mold. The PLLA sponge cup was then filled with an aqueous collagen solution by introducing the collagen solution into the pores of the cup and the central space. The collagen solution–filled PLLA sponge cup was freeze-dried and cross-linked to form a cylinder-type PLLA-collagen hybrid sponge. The PLLA-collagen hybrid sponge had the same shape as the PLLA sponge cup. A collagen sponge was formed in the center of the PLLA sponge cup, and collagen microsponges were formed in the pores of the wall and bottom of the PLLA sponge cup. The hybrid sponge was composed of two parts, the central collagen sponge and the surrounding PLLA-collagen sponge cup. The central collagen sponge was connected with the collagen microsponges in the pores of the wall and bottom of the PLLA sponge cup by collagen fibers that passed through the interstices of the PLLA sponge.

The interconnection prevented the shrinkage of the central collagen sponge. The outer layer of the PLLA-collagen hybrid sponge should prevent cell leakage during cell seeding. At the same time, the high mechanical strength of the PLLA cup served as a mechanical skeleton and reinforced the hybrid sponge.

22.7 Conclusions

Several types of functional hybrid scaffolds have been prepared from biodegradable synthetic polymers and collagen. The initially developed hybrid porous scaffolds, such as the PLGA-collagen hybrid sponge, the PLGA-collagen hybrid mesh, and the PLLA-collagen hybrid braid, were developed by forming collagen microsponges in the pores, openings, or interstices of the PLGA sponge, the PLGA mesh, and the PLLA braid, respectively. The hybrid scaffolds combine the advantages of the two kinds of biodegradable polymers. The skeleton of the synthetic polymers defines the overall shape and size of the scaffolds and provides high mechanical properties, while the incorporated collagen microsponge contributes good cell interaction and facilitates cell seeding and distribution. Recently, a biphasic hybrid scaffold has been developed by forming a collagen sponge on one side of a PLGA-collagen hybrid mesh. Even more recently, leak-proof hybrid scaffolds have been developed by covering all the surfaces of the collagen sponge except the cell seeding surface with a skeleton having a structure composed of small pores. The leak-proof scaffolds not only combine the advantage of the biodegradable synthetic polymers and collagen, they also have the capacity to prevent cell escape from the porous scaffolds during cell seeding. These hybrid porous scaffolds have been demonstrated as being useful for tissue engineering of a variety of tissues such as skin, ligament, bladder, vascular tissue, cartilage, bone, and osteochondral tissue. The methods of hybridization introduced in the chapter are primarily used for the hybridization of biodegradable synthetic polymers and collagen. The methods can also be used to hybridize biodegradable synthetic polymers with other kinds of naturally derived polymers. More kinds of naturally derived polymers can be mixed and used for hybridization with biodegradable synthetic polymers. Bioactive molecules such as cell adhesion proteins, cell growth factors, and cytokines can also be supplemented in the naturally derived polymers for introduction in hybrid porous structures. The application of hybrid scaffolds will be further widened and used in the future for the tissue engineering of complicated tissues and organs such as liver, pancreas, and kidney.

Acknowledgments

This work was supported in part by the Ministry of Education, Culture, Sports, Science and Technology of Japan and in part by the New Energy and Industrial Technology Development Organization of Japan.

References

1. E. S. Place, N. D. Evans, and M. M. Stevens, *Nat. Mater.*, **8**, 457 (2009).
2. L. C. Amado, A. P. Amado, A. P. Saliaris, K. H. Schuleri, M. St John, J. S. Xie, S. Cattaneo, D. J. Durand, T. Fitton, J. Q. Kuang, G. Stewart, S. Lehrke, W. W. Baumgartner, B. J. Martin, A. W. Heldman, and J. M. Hare, *Proc. Natl. Acad. Sci. U S A*, **102**, 11474 (2005).
3. S. Kitahara, K. Nakagawa, R. L. Sah, Y. Wada, T. Ogawa, H. Moriya, and K. Masuda, *Tissue Eng. Part A*, **14**, 1905 (2008).
4. P. Langer, J. P. Vacanti, *Science*, **260**, 920 (1993).
5. G. Chen, T. Ushida, and T. Tateishi, *Macromol. Biosci.*, **2**, 67 (2002).
6. S. Hollister, *Nat. Mater.*, **4**, 518 (2005).
7. T. M. Franklin, L. E. Freed, and G. Farshid, *Nat. Mater.*, **6**,162 (2007).
8. V. Guarino, F. Causa, P. Taddei, M. di Foggia, G. Ciapetti, D. Martini, C. Fagnano, N. Baldini, and L. Ambrosio, *Biomaterials*, **29**, 3662 (2008).
9. J. I. Dawson, D. A. Wahl, S. A. Lanham, J. M. Kanczler, J. T. Czernuszka, and R. O. C. Oreffo, *Biomaterials*, **29**, 3105 (2008).
10. E. Gentleman, A. N. Lay, D. A. Dickerson, E. A. Nauman, G. A. Livesay, and K. C. Dee, *Biomaterials*, **24**, 3805 (2003).
11. G. Chen, T. Ushida, and T. Tatsuya, *Adv. Mater.*, **12**, 455 (2000).
12. G. Chen, T. Ushida, and T. Tatsuya, *J. Biomed. Mater. Res.*, **51**, 273 (2000).
13. G. Chen, T. Sato, T. Ushida, N. Ochiai, and T. Tatsuya, *Tissue Eng.*, **10**, 323 (2004).

14. G. Chen, T. Ushida, and T. Tatsuya, *J. Chem. Soc. Chem. Comm.*, **16**, 1505 (2000).
15. G. Chen, T. Sato, T. Ushida, R. Hirochika, R. Shirasaki, N. Ochiai, and T. Tetsuya, *J. Biomed. Mater. Res.*, **67A**, 1170 (2003).
16. K. Tsuchiya, T. Mori, G. Chen, T. Ushida, T. Tateishi, T. Matsuno, M. Sakamoto, and A. Umezawa, *Cell Tissue Res.*, **316**, 141 (2004)
17. G. Chen, T. Ushida, and T. Tatsuya, *Mat. Sci. Eeg. C-Bio S*, **26**, 124 (2006).
18. G. Chen, T. Sato, H. Ohgushi, T. Ushida, T. Tatsuya, and J. Tanaka, *Biomaterials*, **26**, 2559 (2005).
19. S. Iwai, Y. Sawa, H. Ichikawa, S. Taketani, E. Uchimura, G. Chen, M. Hara, J. Miyake, and H. Matsuda, *J. Thorac. Cardiovasc. Surg.*, **128**, 472 (2004).
20. A. Ide, M. Sakane, G. Chen, H. Shimojo, T. Ushida, T. Tateishi, Y. Wadano, and Y. Miyanaga, *Mat. Sci. Eeg. C-Bio S*, **17**, 95 (2001).
21. G. Chen, T. Sato, J. Tanaka,, and T. Tatsuya, *Mat. Sci. Eeg. C-Bio S*, **26**, 118 (2006).
22. G. Chen, D. Akahane, N. Kawazoe, K. Yamamoto, and T. Tateishi, *Mat. Sci. Eeg. C-Bio S*, **28**, 195 (2008).

Chapter 23

CHITIN AND CHITOSAN FOR TISSUE ENGINEERING APPLICATION

Sang Jun Park and Chun-Ho Kim*

Laboratory of Tissue Engineering, Korea Institute of Radiological & Medical Sciences 215-4, Gongneung, Nowon, Seoul 139-706, Korea

*chkim@kcch.re.kr

Many researchers have developed various artificial tissue, including skin, cartilage, bone, liver, and nerve, in tissue engineering. Chitosan, a polysaccharide abundant in nature, has great potential as a scaffold and cell delivery vehicle by its biodegradability, intrinsic antibacterial nature, minimal foreign body reaction, and nontoxicity. The scaffolds' size and shape, pore morphology, and mechanical property can be controlled for cells and application sites in the body. Chitosan has the ability to be prepared in various geometries and forms, such as porous sponges, beads, and thermo-gelling hydrogels, suitable for cell ingrowths and phenotype. In this chapter, the application and fabrication of porous chitosan sponges, beads, and thermo-gelling hydrogels are introduced.

23.1 Introduction

The goal of tissue engineering and regenerative medicine is the restoration or replacement of lost or damaged tissue or organs of the human body with transplantation of engineered tissue.[1] The

Handbook of Intelligent Scaffolds for Tissue Engineering and Regenerative Medicine
Edited by Gilson Khang

www.panstanford.com

engineered tissue consists of three basic elements, which are structural scaffolds, a source of cells, and a culture condition, including a cell signal. A three-dimensional scaffold takes roles of an extracellular matrix (ECM) analog, which functions as a necessary template for host infiltration and a physical support to guide the differentiation and proliferation of cells into the targeted functional tissue or organ.[2,3] The scaffolds require biocompatibility and biodegradability for being used in the body, an open-pore structure connected to pores, a surface for attachment, proliferation, and differentiation of cells, and also sufficient mechanical properties for support of cells and treatment.[4] The materials of the scaffolds are natural polymers such as collagen, hyaluronic acid, chitosan, alginate, etc., and these polymers can be sufficient for the above-mentioned scaffold conditions. Natural polymers as materials of scaffolds have been studied by many researchers because of their rich existence in nature. Some have been used in clinical trials.[5]

In recent years, considerable attention has been given to chitosan-based materials and their applications in the field of tissue engineering.[2,6–8] It is well known that chitosan is the second-most abundant polysaccharide on earth, after cellulose. Chitosan can be obtained by deacetylation of chitin, which is isolated from exoskeletons of crustaceans and also from cell walls of fungi or insects.[9] Chitosan has great potential as a biomolecule by the presence of reactive functional groups, gel-forming ability, high adsorption capacity, biodegradability, intrinsic antibacterial nature, minimal foreign body reaction, and nontoxicity.[10] Chitosan is also able to be molded in various geometries and forms, such as porous structures, suitable for cell ingrowths and phenotype.

In this chapter, the application and fabrication of porous chitosan sponges, beads, and thermo-gelling hydrogel scaffolds are introduced.

23.2 Chitin and Chitosan

Chitosan (poly[β-(1$\rightarrow$4)-2-amino-2-deoxy-D-glucose]) is formed through the N-deacetylation of chitin (poly[β-(1$\rightarrow$4)-2-acetamido-

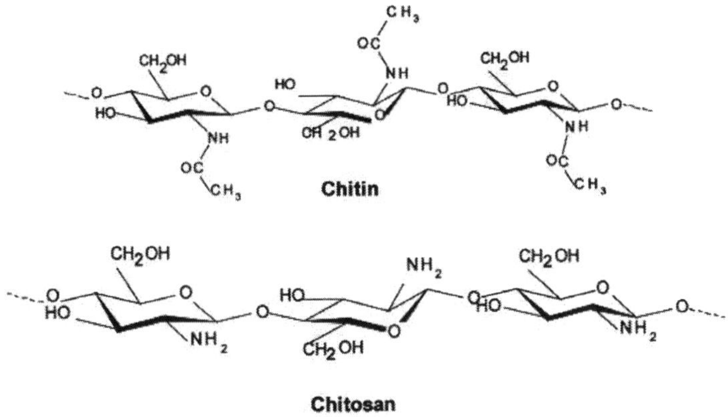

Chitin

Chitosan

Figure 23.1. Chemical structure of chitin and chitosan.

2-deoxy-D-glucose]), a copolymer of *N*-acetyl-glucosamine and *N*-glucosamine units linked by β-(1$\rightarrow$4) glycosidic bonds, in the presence of a hot alkali.[11] Figure 23.1 shows the chemical structure of chitin and chitosan. This deacetylation reaction process changes the molecular weight and degree of deacetylation of chitosan. The molecular weight of chitin is 1.03×10^6 to 2.5×10^6 and that of chitosan is 1×10^5 to 5×10^5.[12] Depending on the source and procedure, the molecular weight of chitosan may range from 300 kDa to over 1,000 kDa with a degree of deacetylation from around 30% to 95%.[1,2] Chitosan is a partially deacetylated derivative of chitin, and when the number of *N*-acetyl-glucosamine units is higher than 50%, the polymer is termed "chitin" generally. Conversely, when the number of *N*-glucosamine units is higher than 50%, the term "chitosan" is used.[13] However, because chitin and chitosan are not insoluble or soluble exactly in a dilute aqueous acidic solution according to the criteria of 50% *N*-acetyl-glucosamine or *N*-glucosamine units, it can be defined that chitin and chitosan are insoluble and soluble in a dilute aqueous acidic solution, respectively. Chitosan consisting of *N*-acetyl-glucosamine and *N*-glucosamine in the ratio of approximately 1:1 when prepared under homogeneous reaction conditions is soluble in water as well as in a dilute aqueous acidic solution. However, chitosan with the same extent under heterogeneous conditions is insoluble even in dilute aqueous acidic solutions.[11] It is from results of X-ray diffraction patterns of the two chitosans that the former chitosan is a random copolymer of *N*-acetyl-glucosamine and *N*-glucosamine units and the latter chitosan is a block copolymer.[14] The solubility of chitosan depends on the distribution of free amino and *N*-acetyl groups. The free amino groups are ionized in a dilute aqueous acidic solution of around pH 6 or less, and then the chitosan becomes soluble.[2] The solubility of chitosan can be varied depending on its molecular weight and chemical structure.

Several studies have examined the host tissue response to chitosan-based implants, because the chemical structure of chitosan is very similar to that of glycosaminoglycans (GAG) existing in living tissue.[15] In the results of the studies, chitosan is nontoxic, biodegradable, and biocompatible with living tissues[13] and degrades by lysozyme, the primary enzyme in body fluid, through hydrolysis of acetylated residues *in vivo*. The degradation rate of chitosan is inversely related to the degree of crystallinity and depends on the degree of deacetylation.[16,17] Chitosan also has a free amine group of a cationic charge and is responsible for electrostatic interactions with anionic GAG, proteoglycans, and other negatively charged molecules.[2] Chitosan can be combined with various materials such as alginate, collagen, hydroxyapatite, hyaluronic acid, polymethylmethacrylate, calcium phosphate, poly-L-lactic acid (PLLA), poly(D,L-lactic-co-glycolic acid) (PLGA), and growth factors for potential applications.[1,18–24]

Chitosan has a wide range of applications since it can be formulated into a variety of forms, including powders, gels, and films, and can also provide controlled release of growth factors and ECM components.[20–22] In the recent studies, chitosan has good characteristics for the attachment, proliferation, and viability of mesenchymal stem cells (MSCs).[23,24] On the other hand, a hemostasis bandage from chitosan was developed and commercialized by HemCon Medical Technologies, Inc., and was approved by the Food and Drug Administration (FDA). Chitosan attracts red blood cells, which have a negative charge, because chitosan has a positive charge. The red blood cells create a seal over the wound as they are drawn into the bandage, forming a very tight, coherent seal.[25] With these promising features, chitosan is considered a very interesting biomaterial to create various tissue analogs, including skin, cartilage, bone, liver, and nerve.

23.3 Chitosan Sponge Scaffolds

The chitosan scaffold for tissue engineering has to have a three-dimensional, open pore structure, and its pore size must be able to adjust. Control of scaffold pore size is critical for controlling cellular colonization rates and organization within an engineered tissue.[2] Pore morphology can also be expected to significantly affect scaffold degradation kinetics and the mechanical properties of the developing tissue.[2] Porous chitosan scaffolds are easily prepared by controlled freezing and lyophilization of chitosan solutions and gels that is a thermally induced phase separation (TIPS) process. Madihally and Kim prepared the porous sponge scaffolds by the TIPS process and reported that mean pore diameters of the scaffolds could be controlled by varying the freezing conditions in the TIPS process.[2,6]

Particularly, Lim *et al.* developed chitosan sponge scaffolds having interconnective micropores between macropores in the scaffold architecture by modifying the TIPS process, adding a nonsolvent for chitosan.[27] This method is detailed below.

23.3.1 *Preparation of Chitosan Sponge Scaffolds*

Chitosan sponge scaffolds were prepared by the TIPS and the modified TIPS process. Chitosan (86% deacetylated) was dissolved in 1% acetic acid aqueous solution to give 2% (w/v) chitosan solution. Ten milliliters each of 0% and 20% (v/v) n-butanol/1% acetic acid solution were dropped into 10 mL of the 2% (w/v) chitosan solution. The n-butanol was used as a nonsolvent for chitosan in the modified TIPS process. Final volumes of n-butanol were 0% and 10% in 1% chitosan solution. Each solution was homogenized to obtain the n-butanol/chitosan mixture. The mixed n-butanol/chitosan solution was poured onto flat-bottomed molds, then quickly frozen at −70°C for 24 hours, and lyophilized at −70°C for 24 hours. The porous chitosan sponge scaffolds were washed serially with 100%, 70%, and 50% alcohol for one hour each and then washed with deionized

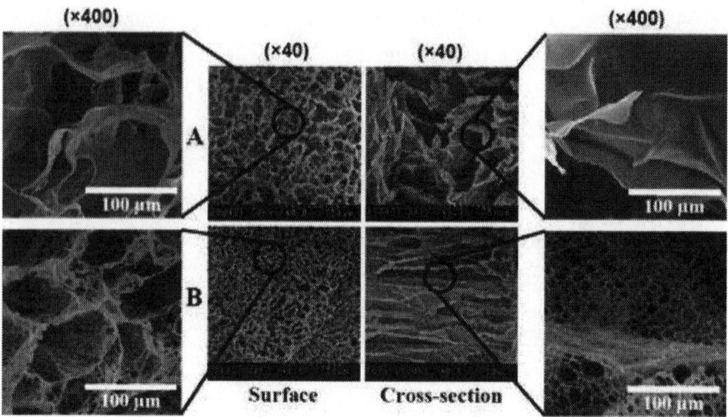

Figure 23.2. SEM micrographs of chitosan sponge scaffolds. (a) The surface and cross section of the CS chitosan sponge scaffold prepared by the TIPS process and (b) the μ-CS chitosan sponge scaffold prepared by the modified TIPS process.

water until the filtrate reached neutral pH. The resulting chitosan sponge scaffolds were freeze-dried again under the same conditions as described earlier. The chitosan sponge scaffolds were codenamed CS and μ-CS according to fabrication by the TIPS and the modified TIPS process, respectively.

Figure 23.2 shows the morphologies of the porous chitosan sponge scaffolds prepared by the TIPS and the modified TIPS process.

The addition of the nonsolvent in the chitosan solution was to make the homogeneous micropores and interconnectivity between macropores without any surface skin layer. The modified TIPS process is a solid-liquid-liquid separation process adding a nonsolvent for chitosan. The difference of the freezing temperature of each of the components induced the liquid-liquid and the liquid-solid phase separation via the solvent/nonsolvent/chitosan solution. The pore diameters of the CS scaffold prepared by the TIPS process were mainly about 80$\sim$100 μm. On the contrary, the pore diameters of the μ-CS scaffold could be controlled mainly within the range 4$\sim$50 μm, with a variation of the solvent/nonsolvent ratio in the modified TIPS process.[27]

Table 23.1. Mechanical properties of chitosan sponge scaffolds (n = 4).[27]

Scaffolds	Tensile strength kPa	Breaking elongation %	Young's modulus Pa
Dry state			
CS	35.46 ± 4.76	4.36 ± 0.42	2985.8 ± 282.3
μ−CS	20.48 ± 1.47	6.45 ± 0.65	1655.6 ± 271.2
Hydrated state			
CS	12.65 ± 1.53	26.24 ± 1.20	118.2 ± 2.9
μ−CS	12.32 ± 1.91	53.56 ± 1.85	64.8 ± 5.7

The mechanical properties of the chitosan sponge scaffolds measured by the INSTRON universal testing machine is shown in Table 23.1.[27] The dry scaffolds were cut into rectangular strips that were 20 mm in length and 10 mm in width. The hydrated samples were prepared by immersing the dry samples with the same dimensions in a phosphate buffered saline (PBS) buffer at room temperature for three hours prior to strain to failure on the testing system. The μ-CS scaffold prepared by the modified TIPS process exhibited a larger value of breaking elongation, more elasticity, but less tensile strength than the CS scaffold prepared by the TIPS process.

23.3.2 *Cell Culture on Chitosan Sponge Scaffolds*

Primary human dermal fibroblasts (HDFs) from newborn foreskin (HDF-N, MCTT Co., Korea) at passage 2 and 6 were used. The cultured HDFs were inoculated into chitosan sponge scaffolds at a concentration of 7×10^4 cell/mL/well. The inoculated HDFs were cultured in a medium (DMEM:F12 = 3:1) including 10% FBS in an incubator under 95% humidity at 37°C with 5% CO_2.

Figure 23.3 shows scanning electron microscope (SEM) micrographs of the HDFs cultured on chitosan sponge scaffolds after seven days of culture. The HDFs attached and proliferated well on the surface of the CS and μ-CS chitosan sponge scaffolds. The HDFs were shown in the cross section of the μ-CS scaffold prepared by the modified TIPS process, while the HDFs were not shown in the cross section of the CS scaffold prepared by the TIPS process. The results of

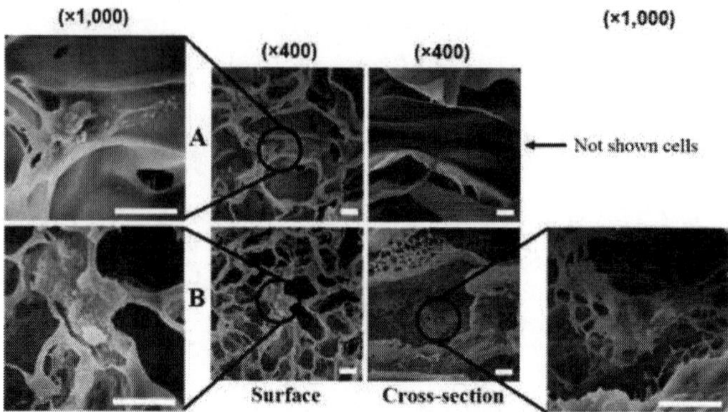

Figure 23.3. SEM micrographs of the HDFs cultured on chitosan sponge scaffolds after 7 d of culture. HDFs in the surface/cross section of (a) the CS scaffold prepared by the TIPS process and (b) the μ-CS scaffold prepared by the modified TIPS process. (Scale bar = 15 μm).

the 3-(4,5-dimethyl-2-thiazolyl)-2,5-diphenyl-2H-tetrazolium bromide (MTT) assay were that the initial cell adhesion on the μ-CS scaffold was 190% higher than on the CS scaffold. The proliferation rate of HDFs in the μ-CS scaffold was 1.82-fold that in the CS scaffold after three days of culture.[8]

The μ-CS chitosan sponge scaffold in which interconnectivity between pores was improved, compared with the CS scaffold, had the specific surface area enough for cell attachment and tissue ingrowth, facilitating a uniform distribution of cells and adequate transport of nutrients and cellular waste products.

23.4 Chitosan Bead Scaffolds

Bead-type scaffolds can be a large surface to attach and proliferate cells and easily injected into the body with a syringe if they are prepared in submicro size. Roh and Kwon fabricated pure chitosan bead-type scaffolds having various microstructures by an extended TIPS process.[26] Wu *et al.* reported that a three-dimensional alginate microbead platform was coated with cartilaginous ECM components to emulate the chondrogenic microenvironment, and the microbead system promoted bone marrow–derived MSC proliferation and protein deposition resulting in the formation of bigger aggregates compared with conventional pellet culture.[28] Park *et al.* reported a double-bead microsphere device to deliver bioactive molecules to cells. These double-beaded PLGA microsphere constructs containing dexamethasone and dehydroepiandrosterone show promise as coatings for implantable biomedical devices to improve biocompatibility and ensure *in vivo* performance.[29] Park and Kim also developed chitosan bead scaffolds having interconnective micropores between macropores by the modified TIPS process.[7] This method is detailed below.

23.4.1 *Preparation of Chitosan Bead Scaffolds*

Chitosan (75.6% deacetylated) was dissolved in 1% acetic acid to give 1% and 2% (w/v) solution. Ten milliliters of 20% (v/v) n-butanol/1% acetic acid solution was dropped into 10 mL of 2% chitosan solution. Final volumes of n-butanol were 10% in the 1% chitosan solution. The solution was homogenized to obtain the n-butanol/chitosan mixture and was then poured into 5 mL syringes with needles. The 18G and 23G needles were used for the 10% n-butanol solution and the 1% chitosan solution, respectively. Drops of the chitosan solution were extruded manually from the syringes into a beaker containing a cooling medium. Methylene chloride (MC), an MC/ether mixture, and liquid nitrogen (LN_2) were used as the cooling media. The temperature of the cooling medium was controlled at –15°C, –30°C, and –70°C by a refrigerator, or into LN_2 (–196°C) directly. After five minutes for LN_2 and three hours for other cooling media, the completely solidified chitosan beads were moved to a cooled, freeze-dried glass vessel for freeze-drying. The solidified chitosan beads were freeze-dried at –70°C for six hours. The density of the cooling medium at the set temperature should be only slightly less than that of the chitosan solution. Therefore, a 32% MC (in ether) solution was used as the cooling medium for 10% n-butanol (in 1% chitosan) solution at –15°C, –30°C, and –70°C.

To fabricate beads with a larger pore size, the cooling rate was controlled slowly in the process. Drops of the chitosan solution were extruded manually from the syringes into a Teflon 100-well plate (homemade) containing MC at room temperature (a drop per well).

Table 23.2. Preparation conditions and designations of chitosan bead scaffolds.[7]

Sample	Chitosan	Butanol	Acetic acid	Temperature	Cooling media
CS-MC15	1%	-	1%	−15°C	MC
CS-MC30	1%	-	1%	-30°C	
CS-MC70	1%	-	1%	-70°C	
CS-LN2	1%	-	1%	−196°C	LN_2
CS-RT70	1%	-	1%	RT→-70°C	MC
10B-MCE15	1%	10%	1%	-15°C	
10B-MCE30	1%	10%	1%	-30°C	MC/ether mixture
10B-MCE70	1%	10%	1%	-70°C	
10B-LN2	1%	10%	1%	−196°C	LN_2
10B-RT70	1%	10%	1%	RT→-70°C	MC

Abbreviation: RT, room temperature.

After five minutes, the 100-well plate containing drops of chitosan solution in MC was moved to a −70°C deep-freezer. This procedure slows the cooling rate of chitosan solution drops than that of chitosan solution drops in a −70°C cooling media directly. After three hours, the solidified chitosan beads were moved to a cooled, freeze-dried glass vessel for freeze-drying. The solidified chitosan beads were freeze-dried at −70°C for six hours. The porous chitosan beads were washed serially with 100%, 70%, and 50% alcohol for three hours each and then washed with deionized water until the filtrate reached neutral pH. The designation of each sample is listed in Table 23.2.

The photographs of chitosan bead scaffolds are shown in Fig. 23.4. The chitosan bead scaffolds were round shaped and about 2 mm in diameter uniformly. Figure 23.5 shows morphologies of the surface and cross section of the chitosan bead scaffold. The macropore size of the beads was controlled by changing of the temperature of the cooling medium and was increased by increasing the temperature.

The macropore size of the CS-RT70 and 10B-RT70 chitosan bead scaffolds prepared by the slowest cooling rate in the process was the largest of the prepared chitosan bead scaffolds. The bead scaffolds prepared by the modified TIPS process, showed good interconnected micropores between macropores in scaffold

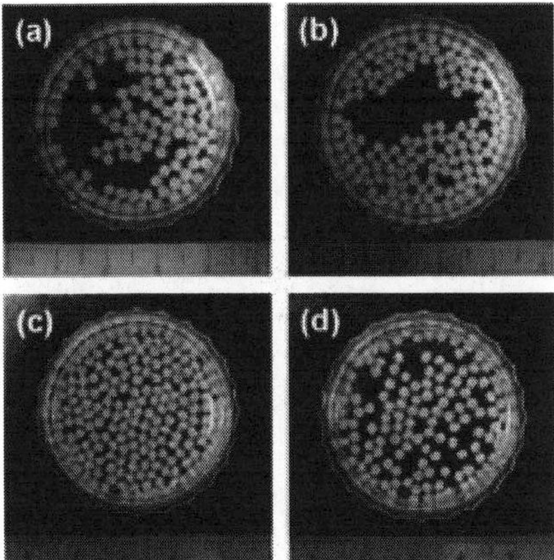

Figure 23.4. Photographs of chitosan bead scaffolds. (a) CS-LN2, (b) 10B-LN2, (c) CS-RT70, and (d) 10B-RT70.[7]

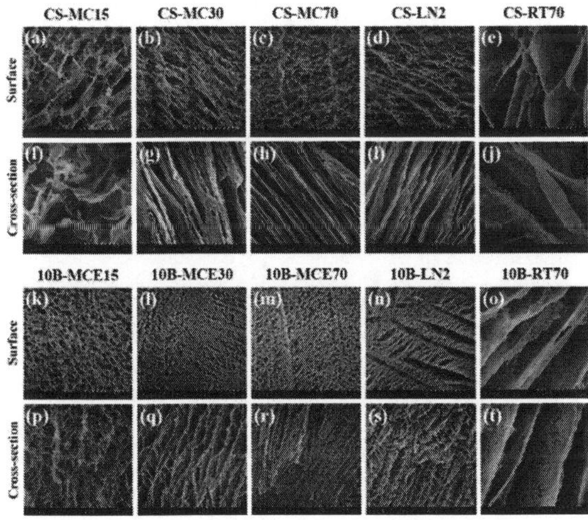

Figure 23.5. SEM micrographs of chitosan bead scaffolds prepared at various temperatures and in various cooling media. (a–e, k–o) Surface and (f–j, p–t) cross section of chitosan bead scaffolds (magnification ×400). The designation of each sample is listed in Table 23.2.[7]

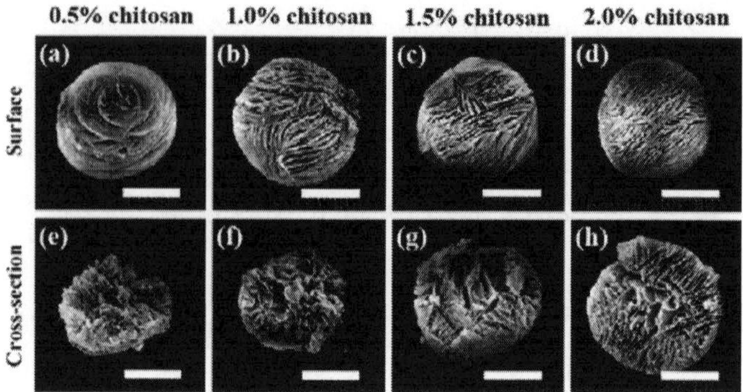

Figure 23.6. SEM microographs of chitosan bead scaffolds prepared in various chitosan concentrations. (a, e) 0.5%, (b, f) 1.0%, (c, g) 1.5%, and (d, h) 2.0% chitosan in 1% acetic acid solution. (a–d) Surface and (e–h) cross section of chitosan bead scaffolds (magnification ×25, scale bar = 1.2 mm).[7]

architecture. Figure 23.6 shows morphologies of the surface and cross section of the chitosan bead scaffolds prepared in various chitosan concentrations. The microstructure of the chitosan bead scaffolds became looser with a decrease in the chitosan concentration.

23.4.2 *Cell Culture on Chitosan Bead Scaffolds*

Chondrocytes from rabbit costal cartilage at passage 3 were used and inoculated into the chitosan bead scaffolds at a concentration of 2.4×10^4 cells/100 μL/well (7 beads/well). The inoculated chondrocytes were cultured in a medium (MSCGM BulletKit, Lonza, Walkersville, MD, USA), including basic fibroblast growth factor (b-FGF) (R&D Systems, Inc.) (1 ng b-FGF in 1 mL media) in an incubator under 95% humidity at 37°C with 5% CO_2.

Figure 23.7 shows the morphologies of chondrocytes cultured on the surface of the chitosan bead scaffolds after 14 days of culture.

The cultured chondrocytes and ECM were shown on the surface of the bead scaffolds.

In the results of cellular viability and proliferation by the MTT assay, initial cell proliferation in the bead scaffolds with micropores was higher than that in the bead scaffolds without micropores.[7] The safrain-O staining images of the bead scaffolds showed that the

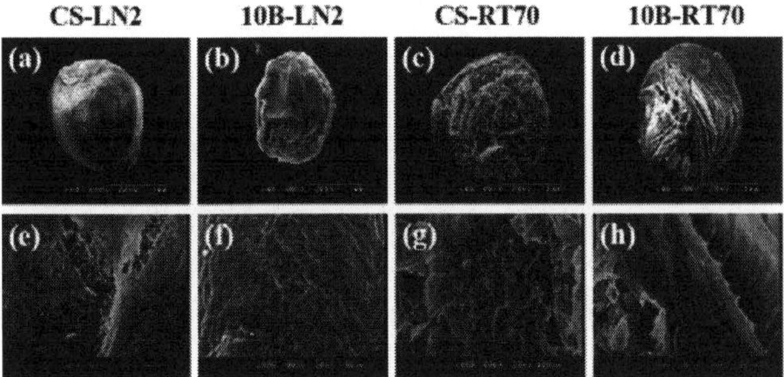

Figure 23.7. SEM micrographs of the surface of the chitosan bead scaffolds cultured on chondrocytes after 14 d of culture. (a–d) Entire morphology (magnification ×40) and (e–h) magnified images (magnification ×400).[7]

proliferation of chondrocytes was exhibited only on the surface of the CS-LN2 and 10B-LN2 bead scaffolds having small-size macropores but exhibited on both the surface and in the center of the CS-RT70 and 10B-RT70 bead scaffolds having large-size macropores.[7]

23.5 Chitosan Hydrogels

Biodegradable thermo-gelling hydrogels that undergo a sol-gel transition with increasing temperature were studied by many researchers in tissue engineering.[30] The advantages of the injectable hydrogels are that they can fill any shape of a defect, may incorporate drugs and growth factors by simple mixing, and do not require invasive surgery.[31] A block copolymer of poly-(ethylene oxide-propylene oxide-ethylene oxide) (Pluronic, known as Poloxamer)[32] and the chitosan/β-glycerol phosphate (β-GP) system are well known as thermogels.[33] Hoemann *et al.* reported that the thermo-gelling hydrogels consisting of chitosan and β-GP and can support *in vitro* and *in vivo* accumulation of cartilage matrix by primary chondrocytes, while persisting in osteochondral defects at least one week *in vivo*.[34] Kim *et al.* also reported that the bone formation from rat muscle-derived stem cells (rMDSCs) using an injectable *in situ*-forming chitosan gel *in vivo* was examined, and the rMDSCs survived well on the hydrogel created by the *in vitro* and *in vivo in situ*–forming chitosan gel, indicating that *in situ* gel-forming chitosan was a suitable substrate for the attachment and proliferation of rMDSCs.[35] BioSyntech developed hydrogels called BST-Gel®, which are liquid at low temperature and solid at human body temperature. BST-Gel® based on chitosan and β-GP is biocompatible, muco-adhesive, and safely biodegradable, with controllable residence times, allowing for a degradation time adapted to the specific application. BioSyntech is currently conducting a pivotal clinical trial on BST-CarGel® in Canada, Spain, and Korea.[36]

23.5.1 *Preparation of Chitosan Hydrogels*

Chitosan (86% deacetylated) was dissolved in 0.1M HCl to give 2.0, 2.6, and 3.0% (w/v) solutions, and β-GP was dissolved in water to give a 45% (w/v) solution. To produce the experimental mixtures, 1 mL of the β-GP solution was added dropwise to 1 mL of each of the chitosan solution and the mixtures were mixed. Final volumes of chitosan were 1.0, 1.3, and 1.5% in 22.5% β-GP solution. The obtained liquid solutions were translucent and homogeneous.

Figure 23.8 shows the storage modulus versus temperature curves for chitosan/β-GP thermo-gelling hydrogels containing different concentrations of chitosan by a rheometer equipped with a 40 mm 1° steel cone plate (AR2000ex, TA Instruments). The gelling temperature was measured to 26°C in case of 1.5% chitosan

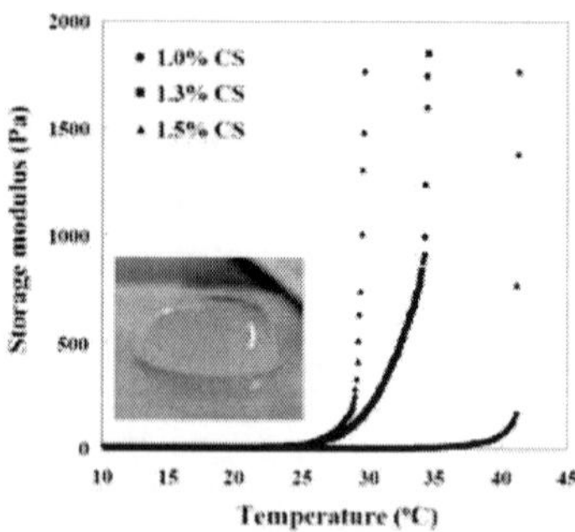

Figure 23.8. Storage modulus versus temperature curves for chitosan/β-GP thermo-gelling hydrogels containing different concentrations of chitosan. The inserted image is the chitosan/β-GP hydrogel at 37°C.

concentration and decreased with increasing of chitosan concentration. The inserted image in Fig. 23.8 shows the chitosan/β-GP hydrogel at 37°C.

23.5.2 *Cell Culture on Chitosan Hydrogels*

Primary HDFs from newborn foreskin (HDF-N, MCTT Co., Korea) at passage 6 were used. The chitosan was dissolved in 0.1M HCl to give a 2.0% (w/v) solution, and the β-GP was dissolved in water to give a 45% (w/v) solution. The chitosan, β-GP solution, and the media containing HDFs were mixed by a volume ratio of 1:0.8:0.2, respectively. The seeding cell density was 5×10^5 cells per 100 μL of thermo-gelling solution. The inoculated HDFs were cultured in a medium (DMEM:F12 = 3:1) including 10% FBS in an incubator under 95% humidity at 37°C with 5% CO_2. Figure 23.9 shows that the left and right images are a thermo-gelling hydrogel and hematoxylin and eosin (H&E) staining of a hydrogel cultured on HDFs for three days of culture, respectively. The result of H&E staining showed that the cultured HDFs well distributed in the hydrogel.

The optimum cell concentration in the chitosan/β-GP hydrogel was measured by $5 \times 10^5 \sim 1 \times 10^6$ cells per 100 μL of thermo-gelling solution.

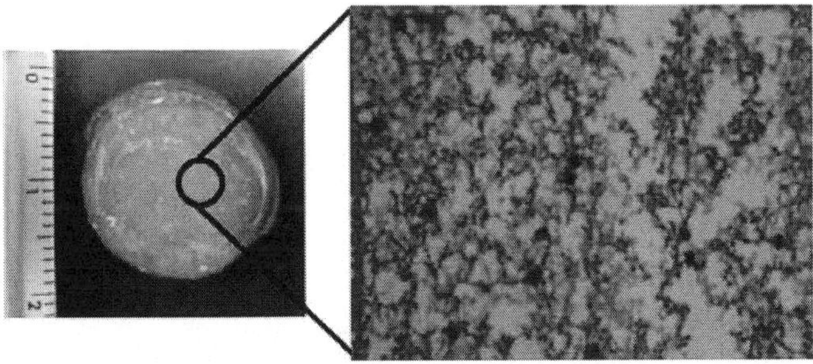

Figure 23.9. The thermo-gelling hydrogel cultured on primary HDFs (left image) and H&E staining of the hydrogel cultured on HDFs for 3 d of culture (right image).

23.6 Conclusions and Outlook

Tissue engineering, which is the restoration or replacement of lost or damaged tissue or organs with transplantation of engineered tissue, has come into the spotlight. During the past decade, many researchers have developed various artificial tissue, including skin, cartilage, bone, liver, and nerve, in tissue engineering. Chitosan has great potential as a scaffold material by the presence of reactive functional groups, gel-forming ability, high adsorption capacity, biodegradability, intrinsic antibacterial nature, minimal foreign body reaction, and nontoxicity. The scaffolds can be controlled by their size and shape, pore morphology, and mechanical property for cells and application sites in the body. Chitosan has the ability to be prepared in various geometries and forms, such as porous sponges, beads, and thermo-gelling hydrogels, suitable for cell ingrowths and phenotype. However, the scaffolds based on chitosan require improvement of the mechanical property for application in complicated tissues and organs, and many studies for the control mechanism of the interaction with various cells and chitosan must be carried out. Many *in vivo* studies also have to be investigated since the majority of the studies have been carried out *in vitro*.

In recent years, HemCon® Bandage, a hemostatic dressing based on chitosan for the external temporary control of severely bleeding wounds, was developed by HemCon Medical Technologies, Inc., in USA, and was approved by the FDA. HemCon Medical Technologies, Inc., commercialized HemCon® dental dressings, KytoStat™ bandages, ChitoFlex® hemostatic dressings, and ChitoFlex® surgical dressings for battlefield, emergency, laboratory, office, and home use.[25] BioSyntech also developed BST-DermOn™, BST-CarGel®, BST-Ossifil™, and BST-Disc™ for wound healing, cartilage repair, bone-filling product, and restoration disc volume using injectable thermo-gelling hydrogels (BST-Gel®) consisting of chitosan and β-GP. Chitosan is expected to be approved by the FDA for implants since BioSyntech is currently conducting a pivotal clinical trial on its products in Canada, Spain, and Korea.[36] Therefore, it is expected that chitosan will be used in scaffolds and cell delivery vehicles in clinical medicine before long.

Acknowledgments

The authors would like to acknowledge the financial support from the Ministry of Knowledge Economy (MKE) (Research grant 10035291, Republic of Korea) and the Ministry of Education, Science and Technology (MEST) (Research grant 2010-0005408, Republic of Korea).

References

1. C. Shi, Y. Zhu, X. Ran, M. Wang, Y. Su, and T. Cheng, *J. Surg. Res.*, **133**, 185 (2006).
2. S. V. Madihally and H. W. T. Matthew, *Biomaterials*, **20**, 1133 (1999).
3. D. M. Alberto, S. Michael, and V. R. Makarand, *Biomaterials*, **26**, 5983 (2005).
4. G. Khang, M. S. Kim, B. H. Min, I. Lee, J. M. Rhee, and H. B. Lee, *Tissue Eng. Regen. Med.*, **3**, 376 (2006).
5. S. J. Park and C. H. Kim, *Tissue Eng. Regen. Med.*, **4**, 471 (2007).
6. C. H. Kim, Y. J. Choi, S. J. Lee, Y. J. Gin, and Y. Son, *Key Eng. Mater.*, **342–343**, 181 (2007).
7. S. J. Park and C. H. Kim, *Tissue Eng. Regen. Med.*, **5**, 697 (2008).
8. H. J. Chun, G. W. Kim, and C. H. Kim, *J. Phys. Chem. Solids*, **69**, 1573 (2008).
9. T. Freier, H. S. Koh, K. Kazazian, and M. S. Shoichet, *Biomaterials*, **26**, 5872 (2005).
10. F. Shahidi and R. Abuzaytoun, *Adv. Food Nutr.*, **49**, 93 (2005).
11. G. A. F. Roberts, Eds., *Chitin Chemistry* (Macmillan, London, 1992).
12. M. N. Kumar, *React. Funct. Polym.*, **46**, 1 (2000).
13. E. Khor and L. Y. Lim, *Biomaterials*, **24**, 2339 (2003).
14. K. Kurita, T. Sannan, and Y. Iwakura, *Makromol. Chem.*, **178**, 3197 (1977).
15. A. Lahiji, A. Sohrabi, D. S. Hungerford, and C. G. Frondoza, *J. Biomed. Mater. Res.*, **51**, 586 (2000).
16. K. M. Varum, M. M. Myhr, R. J. Hjerde, and O. Smidsrød, *Carbohydr. Res.*, **299**, 99 (1997).
17. D. Ren, H. Yi, W. Wang, and X. Ma, *Carbohydr. Res.*, **340**, 2403 (2005).
18. Y. H. Youn, C. H. Kim, Y. J. Choi, Y. J. Gin, and Y. Son, *Key Eng. Mater.*, **342–343**, 185 (2007).

19. R. A. A. Muzzarelli, *Carbohyd. Polym.*, **76**, 167 (2009).
20. J. L. Cuy, B. L. Beckstead, C. D. Brown, A. S. Hoffman, and C. M. Giachelli, *J. Biomed. Mater. Res. A*, **67**, 538 (2003).
21. M. Mochizuki, Y. Kadoya, Y. Wakabayashi, K. Kato, I. Okazaki, M. Yamada, T. Sato, N. Sakairi, N. Nishi, and M. Nomizu, *FASEB J.*, **17**, 875 (2003).
22. C. Y. Hsieh, S. P. Tsai, D. M. Wang, Y. N. Chang, and H. J. Hsieh, *Biomaterials*, **26**, 5617 (2005).
23. J. M. Dang, D. D. Sun, Y. Shin-Ya, A. N. Sieber, J. P. Kostuik, and K. W. Leong, *Biomaterials*, **27**, 406 (2006).
24. J. H. Cho, S. H. Kim, K. D. Park, M. C. Jung, W. I. Yang, S. W. Han, J. Y. Noh, and J. W. Lee, *Biomaterials*, **25**, 5734 (2004).
25. http://www.hemcon.com
26. I. J. Roh and I. C. Kwon, *J. Biomater. Sci. Polym. Edn*, **13**, 769 (2002).
27. J. I. Lim, G. W. Kim, J. Na, I. S. Noh, Y. Son, and C. H. Kim, *Key Eng. Mater.*, **342–343**, 65 (2007).
28. Y. N. Wu, Z. Yang, J. H. P. Hui, H. W. Ouyang, and E. H. Lee, *Biomaterials*, **28**, 4056 (2007).
29. K. Park, J. S. Park, D. G. Woo, H. N. Yang, H. M. Chung, and K. H. Park, *Biomaterials*, **29**, 2490 (2008).
30. M. S. Kim, H. Hyun, G. Khang, and H. B. Lee, *Macromolecules*, **39**, 3099 (2006).
31. A. Gutowska, B. Jeong, and M. Jasionowski, *Anat. Rec.*, **263**, 342 (2001).
32. M. Malmsten and B. Lindman, *Macromolecules*, **25**, 5446 (1992).
33. G. Molinaro, J. C. Leroux, J. Damas, and A. Adam, *Biomaterials*, **23**, 2717 (2002).
34. C. D. Hoemann, J. Sun, A. Légaré, M. D. McKee, and M. D. Buschmann, *Osteoarthr. Cartilage*, **13**, 318 (2005).
35. K. S. Kim, J. H. Lee, H. H. Ahn, J. Y. Lee, G. Khang, B. Lee, H. B. Lee, and M. S. Kim, *Biomaterials*, **29**, 4420 (2008).
36. http://www.biosyntech.com

Part VI

NOVEL FABRICATION METHODS FOR SCAFFOLD

Chapter 24

CONTROLLING A CELLULAR NICHE IN SCAFFOLD DESIGNS FOR EPITHELIAL TISSUE ENGINEERING

Zhilian Yue,[a] Yan-Ru Lou,[a] Nur Aida Abdul Rahim,[b] and Hanry Yu[a–f*]

[a] *Institute of Bioengineering and Nanotechnology, A*STAR, The Nanos, #04-01, 31 Biopolis Way, 138669 Singapore*
[b] *Department of Physiology, Yong Loo Lin School of Medicine, National University of Singapore, #03-03, 2 Medical Drive, 117597 Singapore*
[c] *NUS Graduate School for Integrative Sciences and Engineering, Centre for Life Sciences (CeLS), #05-01, 28 Medical Drive, 117456 Singapore*
[d] *Singapore-MIT Alliance, National University of Singapore, E4-04-10, 4 Engineering Drive 3, 117576 Singapore*
[e] *NUS Tissue-Engineering Programme, DSO Labs, National University of Singapore, 117597 Singapore*
[f] *Department of Mechanical Engineering, Massachusetts Institute of Technology, Cambridge, Massachusetts 02139, USA*
*hanry_yu@nuhs.edu.sg; hyu@ibn.a-star.edu.sg

Epithelial tissues have sophisticated spatially and temporally defined biochemical and structural features, which are the prerequisites for functioning tissues. These include extracellular matrix (ECM) proteins and growth factors, both in terms of presence as well as presentation, topographical cues that direct polarity, mechanical properties of the underlying substratum, which

Handbook of Intelligent Scaffolds for Tissue Engineering and Regenerative Medicine
Edited by Gilson Khang

www.panstanford.com

regulate the interplay between cell-cell and cell-matrix interactions, and fluidic effects that govern mass transfer as well as applied physical stresses. Focus has been placed on quantitatively defining and controlling the microenvironmental cues to regulate the functional and structural features of epithelial tissue constructs. These requirements present special challenges to the engineering of innovative scaffolds for tissue engineering. Over the past decades, significant inroads have been made into tackling these challenges, and improvements in the field present a trend and a path to more innovative engineering efforts. The ultimate goal is to converge at an approach that drastically improves the performance of such constructs in liver, kidney, intestine, esophagus, and lung tissue engineering applications that facilitate tissue repair.

24.1 Introduction

Epithelial tissues such as liver, lung, esophagus, intestine, and kidney serve as a filterlike processing barrier prior to nutrients, water, air, and other essential components of life being allowed access to the cells in the rest of the body. For example, nutrients in food are given access to the blood when in the intestine, after which the digested components then enter the liver via basolateral domains of hepatocytes for further processing. These functions of epithelial tissues require them to exhibit several important morphological features (Fig. 24.1):

- The epithelial cells are typically polarized with one side (basolateral domain) taking up raw materials and the other side (apical domain) exporting processed precuts.
- The epithelial cells tightly interact forming tight junctions to separate the basolateral and apical domains so that the raw and processed contents do not mix.
- The surfaces of epithelial tissues are highly convoluted to exhibit a large surface area for interaction with external materials at nanoscales.
- ECMs are typically scarce in the form of thin layers like the basement membrane.

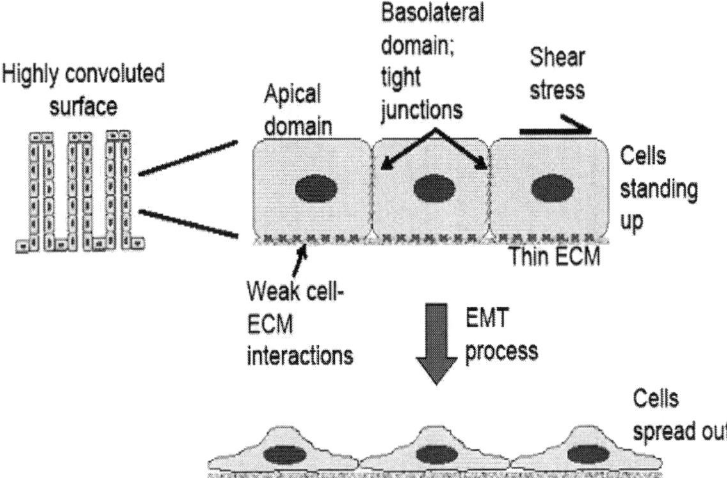

Figure 24.1. Various morphological features of generic epithelial tissue organization compared with mesenchymal cells.

- Interactions between cells and the ECM are generally weak such that there are no observable stress fibers or excessive focal adhesions in cells with the ECM.
- The cells are reacting and functioning normally within a narrow range of fluidic shear.

Thus, most epithelial cells in natural-tissue environments are standing up with strong cell-cell interaction and weak cell-ECM interaction.

A typical example of the breakdown in epithelial organization is the phenomenon of epithelial-mesenchymal transition (EMT) of certain cell types or cancer cells. In epithelia of mammary gland, cancer cells go through EMT upon induction into invasive phenotypes.[1] The epithelial morphology of a cluster of well-aligned cells transforms into the mesenchymal morphology typically seen in fibroblastic cells cultured on hard polystyrene cell culture plastic.[2] These characteristics of epithelial tissue require special attention to the microenvironment outside the cells when designing tissue engineering scaffolds for the repair and stimulation of epithelial tissue regeneration.

The purpose of scaffolds either for *in vitro* or *in vivo* applications of engineered epithelia is to allow restoration of the natural functions of these tissues. Scaffolds should be soft enough to limit mechanical tension to cells so that the cells are not stretched and forced to exhibit a mesenchymal morphology. Ideally the scaffold's mechanical properties are tunable to cater to subtle variations in mechanical requirements of different epithelial tissues. Scaffolds should also be macroporous even in hydrated states such that in *ex vivo* bioreactors and applications a certain range of fluidic shear can be introduced and for *in vivo* implant applications vasculature infiltration can occur. Scaffold-cell interfaces should have nanosized features, be easily modified with different chemical ligands to react with receptors on cellular surfaces, and be controllable spatially, such that the cells experience cues that restore their naturally polarized structures and functions.

Over the past decade, we and other groups have developed a range of cell culture technologies to individually control the extracellular microenvironments of epithelial tissues, such as the development of elastomeric biodegradable materials for esophagus,[3] hydrogels for three-dimensional (3D) presentation of the growth factors,[4,5] sandwich cultures for spatial presentation of polarity cues,[6] and microfluidic devices to recapitulate the mechanical shear of the liver.[7] Here we aim to summarize some of these individual efforts to control the microenvironments of the tissue-engineered epithelia. We also introduce efforts to develop scaffold materials and constructs that combine these considerations to potentially yield future generations of tissue-engineered epithelia whose functions can match those of native epithelia for specific applications *in vitro* and *in vivo*.

24.2 Native Extracellular Microenvironment for Epithelial Cells

Epithelial tissue covers the whole internal and external surface of the body. It is made up of one or more layers of epithelial cells. These cells adhere tightly with little or no intercellular substance. Based on the cell morphology and spatial organization, epithelial tissues can be classified into different types: simple squamous epithelium (e.g., lung, heart, and blood vessels), simple cuboidal epithelium (e.g., kidney tubules and ducts of glands), simple columnar epithelium (e.g., gastrointestinal tract and gallbladder), ciliated columnar epithelium (e.g., upper respiratory tract and uterus), glandular epithelium (e.g., endocrine glands), stratified squamous epithelium (e.g., epidermis and esophagus), and transitional epithelium (e.g., urinary bladder). Epithelial cells have diverse functions in absorption, secretion, sensory system, transport, and protection.

24.2.1 *Extracellular Matrix and Growth Factors*

Epithelial tissues are usually separated from the underlying connective tissues by a basement membrane. The basement membrane is a specialized ECM found in most tissues. It serves as a barrier and provides structural support for the epithelial cell layer.[8] Its constitutive components bind to cell membrane receptors to initiate outside-in signaling.[9] The molecular compositions of all basement membranes are not static, varying from tissue to tissue and during development and disease. All basement membranes contain laminins, type IV collagen, heparan sulfate proteoglycans, and entactin-1/nidogen-1.[10] Type IV collagen is the most abundant protein in basement membranes.[8] Matrigel, a solublized basement membrane derived from mouse sarcoma, is commonly used in cell culture. Laminin-1 trimer is a major component of Matrigel. Compared with other organs, such as kidney, pancreas, and lung, the liver has some unique features. Mature hepatocytes lack a basement membrane, and the ECM constitutes only a minor part of the normal liver.[11] In the space of Disse, between endothelial cells and hepatocytes, there is only little matrix, in which fibronectin is the most abundant component. In addition, collagen types I, III, IV, V, and VI are also present in the space of Disse.[11] This is different from other epithelial tissues in which collagen type I is restricted to the stroma and not directly in contact with epithelial cells. A better understanding of tissue-specific basement membrane compositions and structures will help design suitable scaffolds for various epithelial tissue engineering.

The basement membrane sequesters growth factors and cytokines and consequently influences their local concentration and biological activity.[12,13] It is now clear that many growth factors and cytokines can not only bind with high affinity to their signaling receptors but also bind with lower affinity to matrix molecules, such as proteoglycans, on cell surfaces or within the ECM. For example, heparan sulfate proteoglycans bind to hepatocyte growth factor (HGF)[14] and in turn stabilize HGF oligomers, which then facilitate the dimerization and activation of the HGF-signaling receptor c-Met in hepatocytes.[15] The involvement of the ECM was also found in the action of other growth factors, such as basic fibroblast growth factor[16] and vascular endothelial growth factor.[17] In some cases, only certain isoforms can bind to specific ECM molecules,[18–20] which therefore can lead to localization of certain growth factors in special region. Binding of growth factors to the ECM can also form long-range gradients or limit growth factor diffusion to one direction.[12] Knowledge of growth factor–ECM interactions will certainly provide guidance in scaffold design to tether growth factors in an appropriate spatial manner.

ECM molecules and growth factors are produced by different types of cells in the epithelial tissues. In the liver, hepatocytes produce heparan sulfate proteoglycans, endothelial cells produce laminin, and hepatic stellate cells produce type IV collagen and laminin, whereas activated hepatic stellate cells produce types I and III collagen.[21] HGF is a paracrine factor, being secreted by non-parenchymal cells and active in hepatocytes. Therefore, co-culturing is an effective means to temporally and spatially provide biological cues for epithelial cells.

The ECM is dynamic during development[22] and is not uniform across the adult epithelial tissues. Laminin, types III and IV collagen, and hyaluronans are predominant in fetal and neonatal livers, while fibronectin is dominant in the adult livers.[11] There is a gradient in the adult liver matrix composition from the portal triad to the central vein, with more type IV collagen, laminin, and heparan sulfate proteoglycans in the periportal area and more type I collagen, fibronectin, and heparin proteoglycans in the pericentral area.[23] This gradient provides a stem cell niche, influencing self-renewal and differentiation of hepatic progenitors: hepatic stem cells and hepatoblasts.[24,25] When progenitors are used in epithelial tissue engineering, different strategies must be used to present the stem cell microenvironment.

24.2.2 *Epithelial Polarity, Differentiation, and Function*

The cell surface attaching to the basement membrane is called the basal surface. The opposite side facing the external environment, or the lumen, is called the apical (free) surface. This cellular environment generates apical-basal polarity. In all epithelial tissues, the apical domain is separated from the basolateral region by tight junctions. Epithelial polarity is initiated by cell-cell and cell-matrix interactions[26] and reinforced and maintained by polarized distribution of proteins and lipids.[27] Molecules that establish epithelial polarity include Cdc42, PAK1, PI3K, PTEN, Rac, and the PAR proteins.[28] A normal apical-basal polarity is essential for epithelial cells to perform diverse functions. The simple columnar epithelium of the mammalian intestine and kidney is the most studied polarized tissue. It is made up of interconnected tubular networks. In contrast, the liver consists of interconnected single-layer cell sheets of hepatocytes. The hepatocytes' plasma membrane is divided into three domains: the basal (sinusoidal) domain, the site for transmembrane proteins that recognizes specific matrix components and is responsible for exchanging metabolites with blood; the lateral domain, the site for cell-cell adhesion via intermediate junctions and desmosomes and intercellular communication via gap junctions; and the apical (bile canalicular) domain, the surface of hepatocytes for secreting bile acid and detoxification products.[29,30] These microscale bile canaliculus assemble into 3D interconnected channels in liver tissue.

To obtain functional cells in epithelial tissue engineering, it is important to establish and maintain proper membrane polarity. Renal tubular cells have been used as a model in the study of epithelial polarity. To mimic the *in vivo* epithelial tissue structure, the formation of cysts in 3D culture systems has been investigated. Wang *et al.* found that cell-cell and cell-substratum contact has a distinct role in the development of renal epithelial polarity.[31] Recent studies further demonstrated that E-cadherin-mediated cell-cell adhesion triggered renal cell polarity.[32] ECM proteins also affect the formation of epithelial polarity. β1-integrin-mediated Rac1 activation leads to laminin assembly, which in turn orients epithelial apical polarity.[33,34] Madin-Darby canine kidney (MDCK) renal tubular cells grown on top of type I collagen gel lost their polarity and exhibited some of the characteristics of mesenchymal cells.[35] The loss of apical-basal polarity and the gain of front-back polarity is a morphogenetic process called EMT. EMT plays an important role in embryonic development[36] and a number of diseased states, including fibrogenesis[37,38] and tumor metastasis.[39] A recent study described that hepatocytes cultured on dried collagen undergo EMT and become resistant to transforming growth factor-β (TGF-β)-induced apoptosis.[40] An *in vivo* study showed TGF-β1-induced alveolar epithelial cell mesenchymal transition, which is mediated by matrix protein fibronectin but not laminin/collagen mixtures.[41] Therefore, EMT must be well controlled in engineering epithelial tissues by providing appropriate ECM molecules or growth factors.

It is now known that ECM-mediated cell morphological change selectively stimulates gene transcription. The expression of albumin, a liver-specific gene, is increased when hepatocytes grown on an appropriate ECM substratum exhibit cuboidal, differentiated cell morphology.[42] Tensile property of the substratum appears critical in regulating cell shape and differentiation. Collagen gel promotes hepatocyte differentiation, while dried nongel collagen promotes de-differentiation.[42] The formation of epithelial polarity is accompanied by proper functions. One of the major liver functions is transcellular bile secretion, which is driven by the polarized expression of distinct transport systems on the apical and basal domains of hepatocytes. In cholestatic liver injury, hepatocytes become de-differentiated and lose their characteristic surface transport polarity.[43,44]

24.2.3 *Other Physical Factors: Biomechanics and Microfluidics*

The ECM contains collagen fibers that resist tension and compression. While the epithelial sheet itself is a two-dimensional (2D) structure, the whole tissue microenvironment is 3D. The combined architecture of the basal membrane and ECM presents cells with 3D topographical cues that include surface roughness on the nano- and microscale,[45] curvature both at the micro- and macroscale,[46] and isotropy, as in the parallel or random alignment of underlying collagen fibers.[46]

The combination of structural proteins interwoven with proteoglycans also provides cells with a local and a global mechanical environment. Variations in composition and architecture determine their stiffness, measured variously as elasticity, viscoelasticity, or poroelasticity.[47] Central corneal tissue has a stiffness of 8.6 MPa and is able to distribute applied loads with great precision in order to remain nearly perfectly spherical and aberration free,[48] while liver tissue has a stiffness of 500 Pa.[49] Thus it is clear that cells require specific stiffness cues in order to maintain proper physiological functions and that their stiffness-sensing mechanisms allow them to promote or inhibit cellular functions accordingly. This is further demonstrated in the changes that occur in cells in response to changes in tissue stiffness, resulting in a remodeling of the tissue. Taking liver tissue as an example, hepatic stellate cells (HSCs), which are activated during tissue injury, produce collagen type I, which in turn causes the HSC microenvironment to increase in stiffness, exacerbating the injury response.[50] On a larger scale, the stiffness of the organs is directly related to matrix and cellular organization at the microscale. The skin is an example of thin, organized sheets that are strong in biaxial tension[51]; cylindrical blood vessels can resist circumferential tension[52]; lung structure and the ability to tolerate complex physical forces, including expansive pressures and pulmonary blood flow, is attributed to the connective tissue network[53]; wide intercellular spaces of intestinal epithelium in the villi actcount for the balance between outward interstitial pressure and inward muscle pull.[54]

By virtue of their location lining internal cavities and lumen, epithelial tissues are subjected to shear stresses resulting from fluid flow. Atheroprotective shear stresses experienced by endothelial cells in healthy arteries is >15 dyne/cm^2, whereas atherosclerotic arteries experience a lower shear stress of <4 dyne/cm^2 which results in modification of endothelial gene expression and phenotype, exacerbating the problem.[55]

Interstitial flow is also present in epithelial tissues. This is due to the existence of vasculature within tissues, which sets up flows on the order of 0.1–1 μm/s.[56] External stresses exerted on tissues also contribute to this effect, even though the stresses are temporary.[57] The fluidics of the tissue play an important part in establishing local gradients of growth factors and cytokines within tissues, as convection and diffusion interplay with protein secretion and uptake. From this it is obvious that tissue behavior is greatly affected by the mechanical parameters of the tissue environment.

The above-mentioned biomechanical and fluidic effects are important considerations when engineering scaffolds for epithelial tissue. The scaffold should have an optimum stiffness for the target tissue type and allow sheetlike assemblies of cells that interact in a 3D fashion and whose surfaces are accessible by fluid flow.

24.3 Engineering an Extracellular Microenvironment for Epithelial Cells

24.3.1 *The State of Art*

Polymeric matrices have been developed as cell support in various epithelial tissue engineering covering transplantation, extracorporeal devices, and *in vitro* models for pathophysiological research and drug screening. They are composed of synthetic, natural, or hybrid materials. The matrices developed so far are either in 2D or 3D configurations, in various forms, including membranes, fibers, microcarriers, microcapsules, hydrogels, and macroporous scaffolds. Generally speaking, they have been designed based on the following criteria:

(a) Controlled cell-matrix interactions
(b) Controlled homotypic and heterotypic cell-cell contacts
(c) Retention of differentiated phenotypes and functions by the cells and possibly proliferation
(d) Biocompatibility and biodegradability for implantation applications
(e) Biomechanical integrity to the targeted tissues/organs
(f) Efficient delivery of nutrients and oxygen and removal of metabolite waste

24.3.1.1 2D plastic substrata

The monolayer culture of epithelial cells onto flat substrata remains one common approach in epithelial tissue engineering, especially in retinal epithelium engineering[58] and bioartificial kidney devices.[59,60] The substrata can be tissue culture surfaces or polymeric membranes, generally coated/conjugated with ECM proteins or cell adhesion molecules. The structures and compositions of these bioadhesive molecules are critical for cells to reestablish *in vivo*-like cell-cell and cell-matrix interactions to maintain cell polarity and differentiated functions.

Primary hepatocytes are one major cell type in cell-based therapies in hepatic tissue engineering and drug metabolism and toxicity studies. How to maintain their phenotypic stability *in vitro* after isolation has been a subject of intensive studies. The spatial geometry of the ECM in hepatocyte culture proves to be important.[29] While single-layer cultured hepatocytes on type I collagen typically lose differentiated functions within the first week of culture, the overlay with a second layer of collagen can restore the polarity and function. Hepatocytes cultured in a sandwich configuration adopt *in vivo*-like morphology and intracellular organization with a functional bile canaliculi network. Their characteristic liver-specific functions can be retained up to several months.[61] A collagen overlay can also cause the reorganization of other types of epithelial cells such as MDCK and normal murine mammary gland with the formation of lumen,[62] highlighting the importance of spatial and temporal presentation of polarity cues on epithelial cell culture.

Hepatocytes cultured on soft and/or weakly adhesive substrata self-assemble into closely associated multicellular spheroids that possess structural polarity and functional bile canalicular channels delimited by tight junctions.[63] The channels are highly interconnected, mimicking the 3D nature of bile canaliculi in native liver tissue. Examples of these substrata include agarose, poly(hyroxyethyl methacrylate), Matrigel, positively charged surface and liver-derived proteoglycan-coated/galactosylated surfaces, etc.[64] Compared with the 2D monolayer, hepatocyte spheroids maintain up-regulated liver-specific functions for a long term, in terms of albumin production, urea synthesis, and cytochrome P450 and glucuronidation activity. Both sandwich-cultured and spheroid-cultured hepatocytes have been used intensively in bioartificial liver-assisted devices (BLAD) and *in vitro* studies of drug toxicity and metabolism.[61]

Cell sheet technology has been developed by Okano's group targeting for the regeneration of cell-dense tissues.[65] Culture surfaces are grafted with poly(*N*-isopropylacrylamide) and serve as temporal supports for a monolayer culture. The thermo-sensitive nature of culture surfaces allows for harvest of intact cell sheets with maintained differentiated functions, attributable to the preservation of cell surface proteins, cell-cell interactions, and deposited ECM. Three-dimensional tissues could be constructed by stacking individual cell sheets together, free of scaffold support.[65,66]

24.3.1.2 3D polymeric scaffolds

Three-dimensional gels of ECM proteins remain one popular culture system for epithelial cells, especially for glandular epithelial cells.[67,68] For example, breast epithelial cells cultured in Matrigel form cystlike spheroids with a central lumen and distinct apical and basolateral polarity, recapitulating certain structural features of glandular epithelium *in vivo,* whereas malignant breast epithelial cells only form random aggregates in 3D culture. These 3D models can be used in the mechanistic studies of tumor initiation and progression.[69]

A broad range of 3D polymeric scaffolds have been developed for hepatocyte culture, as exemplified in Table 24.1. Among them, microcapcules and hydrogels are of particular interest, where hepatocytes grow as aggregates or spheroids with intimate cell-cell contacts. Chia *et al.* reported a microencapsulation procedure for hepatocytes using complex coacervation in a physiological environment.[70] The microcapsules of functional hepatocytes are $\sim$150 μm, with a 2$\sim$3 μm outer layer of polyanions, and an inner layer of modified positively charged collagen to provide suitable cell-matrix support. The microcapsules are permeable to small molecules up to albumin, allowing efficient exchange of nutrient, oxygen, growth factors, and metabolites, while preventing attack by immunoglobulins of the immune systems. Their applications in hepatocyte transplantation and BLAD warrant future study.

Hydrogels are attractive matrices for cell culture, due to their tissuelike properties such as hydrophilicity and mechanical integrity to soft tissues. Polyethylene glycol (PEG)-based hydrogels have been

Table 24.1. Typical examples of 3D polymeric matrices in hepatic tissue engineering.

Type of scaffolds	Structure of core materials	Preparation method	References
Macroporous biodegradable, sponges	PLLA,PLGA	Particulate leaching	77–79
Microcarriers	Dextran with conjugated type I collagen	Commercial available	80,81
Microcapsules	Alginate/polylysine Poly(hydroxylethyl methacrylate-co–methyl methacrylate-co-methylacrylic acid)/methyl ated collagen	Polyelectrolyte complexation	70,82,83
Nanofibers	Poly(ε-caprolactone-co-ethyl ethylene phosphate), chitosan, PuraMatrix	Electrospinning	84,85
Hydrogels	Chitosan, alginate, alginate/chitosan, PEG, PuraMatrix	*In situ* gelation, freeze-drying,	73–75,86

Abbreviations: PLLA, poly(L-lactid acid); PLGA, poly(DL-lactic-*co*-glycolic acid).

used for the encapsulation of hepatocyte aggregates in a 3D context via *in situ* gelation.[71] *In situ* gelation often faces a dilemma in engineering large tissue constructs, between providing adequate matrix support and maximizing transport of nutrient, gas, and metabolite wastes. The capability of entrapped cells to migrate is also an issue to be addressed. Recent progress has evidenced the efforts in combining *in situ* gelation with microfabrication to form multilayered structures to tackle these issues.[72] Alternatively, hepatocytes can grow in preformed, macroporous hydrogels to form spheroids in a spatially distributed manner. One typical example is macroporous alginate hydrogels.[73–75] The macroporous structure of scaffolds in a physiological relevant condition, in terms of porosity, pore size, and interconnectivity, together with other physiochemical properties, is critical to the cellular functions of hepatocyte spheroids. One advantage of the hydrogel-borne spheroids is that the formation of spheroids within well-defined pore-size polymeric scaffolds could limit spheroid sizes to a better extent than in cell suspension and 2D substrata.[75] Compact spheroids with sizes over 100 μm suffer from limited mass transfer, resulting in necrotic cores and a rapid decline in cellular functions.[76]

24.3.2 *Spatial and Temporal Presentation of Extracellular Cues in Scaffolds for Liver Tissue Engineering*

In this section we use the liver as a model to discuss scaffold design, given its structural and functional complexity among epithelial tissues. While the early studies on polymer matrices of hepatocytes was focused more on the "carrier" nature and examination of their *in vivo* performances, growing efforts have been made to the engineering of biomaterials to optimize hepatocyte microenvironments *in vitro*. Incorporation of the biological principles, gleaned from the fundamental studies in hepatocyte culture, co-culture, and hepatogenesis, into scaffold design to create liver-specific materials has become and will continue to be one major research focus in hepatic tissue engineering. The development of stem cell technology stimulate research interest in the engineering of 3D niches for controlled differentiation of hepatocyte-like cells using human embryonic stem cells, human bone marrow mesenchymal stem cells, and liver progenitors,[71,87] bringing a new dimension to this field.

In a sandwich culture, the matrix composition affects hepatocyte morphology and polarity. A sandwich culture containing Matrigel restores markers lacking in a conventional collagen sandwich culture, such as the gap junction protein connexin 32 and epidermal growth factor (EGF) receptor (sinusoidal protein), though long-term albumin function is not affected.[88] The expression of connexin 32 can also be induced by adding soluble heparan sulfate proteoglycan into a collagen sandwich culture.[89]

Hepatocytes attach onto galactose-containing substrata via galactose-asialoglycoprotein receptor (ASGPR)-mediated interaction and tend to aggregate into spheroids with maintained liver-specific functions. Presentation of galactose functionality in polymeric matrices represents a promising strategy to improve hepatocyte attachment and functions. It has been applied successfully to almost all the forms of synthetic matrices, including membranes,[90,91] nanofibers,[84,85] microcapsules,[92,93] sandwich cultures,[6] and hydrogels.[94] Here are two examples. Hepatocytes cultured on porous galactosylated nanofibers (~760 nm) of poly (ε-caprolactone-co-ethyl ethylene phosphate) (PCLEEP) form flattened and smaller aggregates (20–100 μm) that engulf the nanofibers.[84] By contrast, spheroids formed on the smooth functional PCLEEP membrane are 50–300 μm and easily detached from the surface upon agitation. This observation highlights the importance of the nanosized topography of PCLEEP fibers on promoting cell-matrix interaction. A prespheroid monolayer configuration has been reported by Du *et al.* on a hydrid Arg–Gly–Asp (RGD)/galactose substratum based on polyethylene terephthalate (PET).[95] The monolayer of hepatocytes exhibits 3D cell behavior with high levels of liver-specific functions yet is attached effectively to the hydrid membrane. A synthetic sandwich culture was also established between the hydrid and galactosylated PET membranes, with improved mass transfer and differentiated functions compared with a collagen sandwich culture.[6] Both methods, nanofiber immobilization of spheroids and prespheroid monolayer configuration, may facilitate the application of hepatocyte spheroids in a microplate-based, high-throughput xenobiotics-screening system.

E-cadherin is a homotypic cell-cell adhesion receptor mediating cell-cell interaction. A 2D co-cultured cell-based E-cadherin presentation promotes differentiation of hepatocytes.[96–99] Substrate-based presentation of acellular E-cadherin induced 3D morphogenesis, resulting in 3D hepatocyte aggregation with high degrees of differentiated functions.[100] Incorporation of exogenous E-cadherin into scaffold design may be an interesting approach to produce favorable environments for the maintenance of differentiated phenotype and functions of hepatocytes *in vitro.*

Heterotypic cell-cell interactions are integral to the maintenance of hepatocyte functions.[101] The conventional co-culture concept has been introduced to various 3D culture configurations to produce more *in vivo*–like microenvironments for hepatocytes. For example, enhanced liver-specific functions were observed in the co-culture with NIH3T3 in alginate and alginate/galatocylated chitosan hydrogels,[102] alginate/galactosylated chotosan/heparin hydrogels,[94] and 3D microcapsules formed with a hybrid natural/

synthetic matrix.[103] An enhanced co-culture effect on hepatocyte function was observed, imposed by the 3D microcapsule environment. Furthermore, biochemical l assays revealed an important role of TGF-β1 regulation on the functional improvement. This finding motivated the development of a simplified co-culture mimic, 3D-microfluidic cell culture configuration of only hepatocytes via *in situ* controlled release of active TGF-β1 from gelatin microspheres.[5] Cell-cell interaction can also be regulated by the modulation of cell-matrix interaction.[95]

Controlled temporal/spatial presentation of soluble factors in 3D matrices is essential to the engineering of more *in vivo*–like microenvironments for hepatocyte differentiation and functions. The recent decade has seen tremendous progress in the development of controlled release technology.[4] Integration of this technology with scaffold material design and fabrication technology will certainly aid in the development of tissue-engineered liver tissue constructs. A few pioneer studies demonstrated an improved survival rate in hepatocyte transplantation by incorporating microspheres into porous scaffolds to deliver hepatotrophic factors[104] and vascular endothelial growth factor.[105]

24.3.3 *Biomechanical Issues*

Encouraging epithelial cells to achieve polarity depends a lot on the physical environment in which the cells are cultured. Culturing hepatocytes on stiff plastic or glass encourages the cells to assume a spreading morphology as they are exerting forces to match the stiffness of their substrate.[49] They also lose polarity. Polarity can be regained by culturing hepatocytes on a softer substrate, for example, Matrigel,[106] to promote a more rounded morphology and encourage cell-cell contact formation. A sandwich configuration, as previously presented, of a porous membrane placed on top of a hepatocyte monolayer on collagen I–coated plastic, where biochemical and biomechanical cues are given to both the basal and the apical layer, is also able to promote polarity.[6]

Scaffold design and scaffold materials, both natural and synthetic ECM, provide a range of controllable parameters for regulation of the biomechanical microenvironment presented to cells. Natural type I collagen can be gelled at various concentrations and pHs to produce constructs with varying stiffness for endothelial cell culture.[107] Matrigel is rich in laminin, collagen IV, and heparan sulfate proteoglycans and forms more compliant gels compared with type I collagen, as does fibrin.[57] However, it is difficult to isolate and quantify the added effects of other biochemical parameters on cellular behavior, as ECM proteins are inherent to these scaffolds from the outset.

Discounting biochemical factors, an ideal scaffold material should provide handles for controlling material stiffness, strut size and length, connectivity, and pore size, thus resulting in the ability to regulate bulk elasticity and porosity. Even if cells sit on a scaffold with nanofeatures, cells can assess the stiffness of the base polymer itself; struts that are on the order of the size of cells essentially present cells with a 2D instead of a 3D surface; strut length and connectivity determine pore size, which influences shear stresses, interstitial flow, and ability to migrate. Previously presented constructs by and large cater for this, including established polymers such as hydrogels and polyglycolide (PGA), polymers under clinical investigation such as polyorthoester and polyhydroxyalkanoate,[108] and novel biomaterials such as self-assembling peptide chains RAD16.[109]

A few examples where these factors are taken into account are given here. Hepatocytes cultured in a PEG hydrogel and layered into a 3D architecture improved hepatocyte survival and liver-specific function.[72] PGA-urothelial cells composites that were implanted into mice and retrieved at extended times demonstrated multilayered sheetlike structure formation.[110] RAD16-I has been used for endothelial sprout formation studies[111] as well as to foster tissue-like function of hepatocyte progenitor cells.[112]

Besides providing an optimal *in vitro* biomechanical microenvironment for cells, designs geared towards tissue implants should also match the mechanical properties of the tissue at sight of implantation. For long-term *in vitro* culture and implants, scaffold degradation properties are also important as part of the tissue-remodeling process, which results in changes in biomechanical properties and determines the tissue functionality and implant longevity. Scaffold integrity should be maintained for as long as it takes cellular constructs to produce their own ECM and assemble replacement architecture.[108]

24.3.4 *Fluid Dynamics and Mass Transfer*

In the majority of current tissue engineering applications, diffusion governs the primary means of mass transport, bringing gases, nutrients, proteins, cells, and waste products into and/or out of the constructs.[113] While this might be sufficient for survival, functional parameters can be improved in some applications by incorporating some forms of fluid flow. Consider the limitation of large, cellular spheroids with diameter more than 1 mm, which are generally organized into viable cells on the rim that surround a hypoxic, necrotic center.[114] The presence of fluid flow in the form of shear stress, as well as interstitial flow within tissue engineering constructs, is important in two ways. Firstly, the flow-induced stresses itself, at physiological values, promotes good cellular organization and function, as previously described, and secondly, fluid flow facilitates mass transfer of oxygen and soluble nutrients and removal of waste products. Scaffold porosity determines the magnitude of shear stresses experienced by cells. Porosity along with surface charges also factor into the mass transport properties of a scaffold. Besides steric hindrance, it is important to consider the interaction between the scaffold and charged nutrients, as repulsive forces could prevent nutrients to access cells, while attractive forces could sequester a large pool of nutrients to the scaffold itself.[113] One way to precisely control microscale flow profiles is to utilize microfluidic systems where fluid is applied through small 1–100 μm channels to 2D or 3D cellular constructs cultured within the system.[57]

Endothelial cells plated in a monolayer within a macrofluidic device and subject to laminar shear stress produce angiogenic sprouts within collagen I gels whose invasion distance was proportional to the magnitude of shear stress.[115] Interstitial flow causes hepatocytes to aggregate and form a functional mass[116] and induces interconnected tube formation of endothelial cells cultured in natural and synthetic gels.[111] In terms of mass transfer properties, a perfusion culture of hepatocytes has been shown to dramatically improve functional activity, such as albumin secretion and urea genesis rates, as well as morphological stability, including tight junctions, glycogen storage, and bile canaliculi.[117] Various methods can be used to overcome mass-transfer limitations of *in vitro* tissue-engineered constructs, including utilizing spinner-flask, hollow-fiber, or direct perfusion bioreactors, and rotating-wall vessels.[118]

Another aspect to consider is that the problem of fluid dynamics and mass transfer lends itself to mathematical modeling and computational tools. A multitude of software such as FEMLAB (Comsol, USA), AutoCAD (Autodesk, USA), and FLUENT (ANSYS, USA) is currently available for designing bioreactors with optimized and controlled flow profiles and flow rates, as well as for modeling engineered constructs for accurate prediction of its fluid dynamics.[57]

24.4 Applications and Outlook

Epithelial tissue engineering represents a special class of tissue engineering. The rationale behind all engineering efforts is to control spatial and temporal presentation of polarity cues via the use of polymeric scaffolds, which allow reestablishment of key structural features and consequent natural functions of tissue-engineered constructs. The polarity cues include the ECM, growth factors, biomechanical stimuli, fluidic effects, etc., which regulate the fine interplay of cell-matrix and cell-cell interactions and cell metabolite removal. In this chapter, starting from a detailed understanding of the structure, function, and native extracellular environment of epithelial cells, we discussed the key parameters in scaffold design for epithelial tissue engineering. We also summarized recent efforts to engineer the extracellular niches of liver tissue constructs. These findings will benefit the development of scaffold-aided epithelial tissue constructs such as kidney, intestine, esophagus, and lung.

Focusing specifically on the liver, our lab has developed a number of hepatocyte culture platforms for hepatotoxicity drug testing. These include 3D prespheroid monolayer culture, synthetic sandwich culture, and perfusion culture and microfludic devices, etc., where liver-specific phenotype and functions are maintained *in vitro*. We also develop 3D cellulosic macroporous hydrogels,

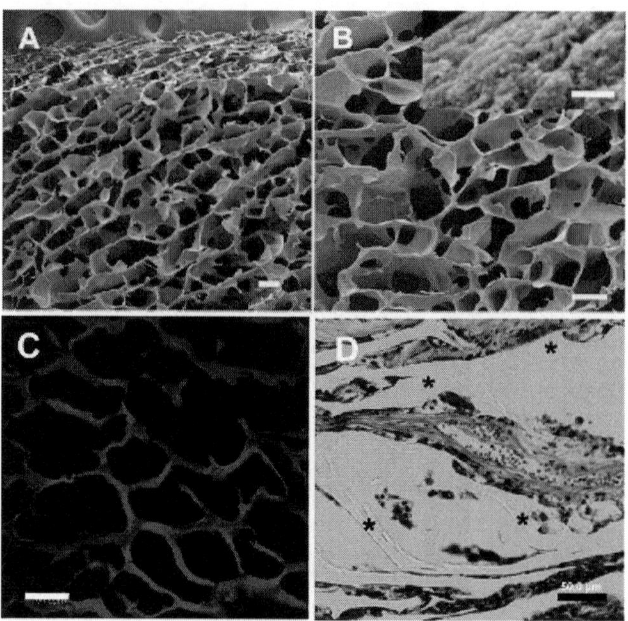

Figure 24.2. Macroprous cellulosic hydrogels. (a and b) SEM micrographs of scaffolds in the dry state; scale bar: 100 μm. (b) The nanofeature of a scaffold; scale bar: 1 μm. (c) Confocal micrograph of a hydrated scaffold; scale bar: 100 μm. (d) Masson's trichrome staining of scaffolds after subcutaneous implantation for 4 wk, showing collagen in blue; cytoplasm, muscle, and erythrocytes in red; and nuclei in black. * highlights the porous scaffold. Scale bar: 50 μm. The implant integrates with surrounding tissue, showing minimal foreign-body response and blood vessel formation with erythrocytes around. *Abbreviation*: SEM, scanning electron microscopy. See also Color Insert.

taking account of the important considerations discussed in previous sections for scaffold design (Fig. 24.2). The scaffolds exhibit combined macro- and nanofeatures, interconnected macroporous structures in hydrated states, with controlled mechanical integrity to soft tissues. Subcutaneous implantation shows integration with surrounding tissue with minimal foreign-body response. In addition, their surface chemistry allows facile introduction of ECM cues to regulate cell-cell and cell-cell matrix interactions and biodegradation profiles. These macroporous hydrogels can be promising candidates in the generation of tissue-engineered liver constructs and other epithelia *in vitro* and *in vivo*.

The integrated approach to construct a precision-engineered scaffold for epithelial tissue engineering would find other applications in lung, esophagus, intestine, pancreas, and kidney tissue regeneration as well. In the lung, sheets of polymer membrane-supported cell constructs with fine structural and functional integrity would be used to seal leakages in the airways to prevent lung collapse due to infection-induced lung damage. In intestine and esophagus, multilayered constructs with multiple cell types are being developed to repair the digestive functions; and kidney glomeruli or proximal tubules are being reconstructed for eventual treatment of kidney failure.

The cell source remains one major bottleneck in liver tissue engineering. Due to the instability of adult hepatocytes, hepatic progenitors are becoming an alternative. Hepatic progenitors display both proliferation and differentiation capacity. A few studies have demonstrated differentiation and maturation of hepatic progenitors in PEG hydrogels,[71] macroporous alginate hydrogels,[119] porous PLA scaffolds,[120] and alginate microspheres.[87] Cell-cell contact seems to be critical in initiation of polarity and subsequent differentiation and functionality,[71,119] while stronger cell-matrix contact prevents hepatic differentiation.[119] Because of the bipotential differentiation capacity of hepatoblasts, both hepatocytes and cholangiocytes can be derived, which will eventually form the complex liver tissue with bile duct structure.[87] Soluble factors perform better in 3D porous PLLA scaffolds.[120] However, scaffolds capable of regulating progenitor cell differentiation and maturation are still lacking. With the detailed understanding of stem cell biology, integration of novel engineering approaches with stem cell technology is becoming one major direction in epithelial tissue engineering.

Acknowledgments

Nur Aida Abdul Rahim is a recipient of the Global Enterprise for Micro-Mechanics and Moledular Medicine (GEM4) postdoctoral fellowship.

References

1. J. Yang, S. A. Mani, J. L. Donaher, S. Ramaswamy, R. A. Itzykson, C. Come, P. Savagner, I. Gitelman, A. Richardson, and R. A. Weinberg, *Cell*, **117**, 927 (2004).
2. M. Shtutman, E. Levina, P. Ohouo, M. Baig, and I. B. Roninson, *Cancer Res.*, **66**, 11370 (2006).
3. Y. Zhu, K. S. Chian, M. B. Chan-Park, P. S. Mhaisalkar, and B. D. Ratner, *Biomaterials*, **27**, 68 (2006).
4. M. Biondi, F. Ungaro, F. Quaglia, and P. A. Netti, *Adv. Drug Deliv. Rev.*, **60**, 229 (2008).
5. C. Zhang, S. M. Chia, S. M. Ong, S. Zhang, Y. C. Toh, D. van Noort, and H. Yu, *Biomaterials*, **30**, 3847 (2009).
6. Y. Du, R. Han, F. Wen, S. Ng San San, L. Xia, T. Wohland, H. L. Leo, and H. Yu, *Biomaterials*, **29**, 290 (2008).
7. Y. C. Toh, C. Zhang, J. Zhang, Y. M. Khong, S. Chang, V. D. Samper, D. van Noort, D. W. Hutmacher, and H. Yu, *Lab Chip*, **7**, 302 (2007).
8. V. S. LeBleu, B. Macdonald, and R. Kalluri, *Exp. Biol. Med. (Maywood)*, **232**, 1121 (2007).
9. C. ffrench-Constant and H. Colognato, *Trends Cell Biol.*, **14**, 678 (2004).
10. A. C. Erickson and J. R. Couchman, *J. Histochem. Cytochem.*, **48**, 1291 (2000).
11. A. Martinez-Hernandez and P. S. Amenta, In Eds. M. A. Zern and R. L. M., *Extracellular Matrix: Chemistry, Biology, and Pathobiology with Emphasis on the Liver* (Marcel Dekker, 1993), pp. 255.
12. J. Taipale and J. Keski-Oja, *FASEB J.*, **11**, 51 (1997).
13. J. Schlessinger, I. Lax, and M. Lemmon, *Cell*, **83**, 357 (1995).
14. S. Ashikari, H. Habuchi, and K. Kimata, *J. Biol. Chem.*, **270**, 29586 (1995).
15. T. F. Zioncheck, L. Richardson, J. Liu, L. Chang, K. L. King, G. L. Bennett, P. Fugedi, S. M. Chamow, R. H. Schwall, and R. J. Stack, *J. Biol. Chem.*, **270**, 16871 (1995).
16. A. Yayon, M. Klagsbrun, J. D. Esko, P. Leder, and D. M. Ornitz, *Cell*, **64**, 841 (1991).
17. T. Miralem, R. Steinberg, D. Price, and H. Avraham, *Oncogene*, **20**, 5511 (2001).
18. M. Pekny, A. Ostman, A. Hermansson, M. Nister, C. H. Heldin, and B. Westermark, *Growth Factors*, **10**, 77 (1994).

19. R. A. Pollock and W. D. Richardson, *Growth Factors*, **7**, 267 (1992).
20. Z. Poltorak, T. Cohen, R. Sivan, Y. Kandelis, G. Spira, I. Vlodavsky, E. Keshet, and G. Neufeld, *J. Biol. Chem.*, **272**, 7151 (1997).
21. J. P. Iredale and M. J. Arthur, *Gut*, **35**, 729 (1994).
22. W. P. Daley, S. B. Peters, and M. Larsen, *J. Cell Sci.*, **121**, 255 (2008).
23. E. Schmelzer, R. E. McClelland, A. Melhem, L. Zhang, H. L. Yao, E. Wauthier, W. S. Turner, M. E. Furth, D. Gerber, S. Gupta, and L. M. Reid, In Eds. C. S. Potten, R. B. Clarke, J. Wilson, and A. G. Renehan, *Tissue Stem Cells* (Taylor & Francis, 2006), pp. 161.
24. R. McClelland, E. Wauthier, J. Uronis, and L. Reid, *Tissue Eng. Part A*, **14**, 59 (2008).
25. L. Zhang, N. Theise, M. Chua, and L. M. Reid, *Hepatology*, **48**, 1598 (2008).
26. C. Yeaman, K. K. Grindstaff, and W. J. Nelson, *Physiol. Rev.*, **79**, 73 (1999).
27. K. Simons and A. Wandinger-Ness, *Cell*, **62**, 207 (1990).
28. F. Martin-Belmonte and K. Mostov, *Curr. Opin. Cell Biol.*, **20**, 227 (2008).
29. J. C. Dunn, M. L. Yarmush, H. G. Koebe, and R. G. Tompkins, *FASEB J.*, **3**, 174 (1989).
30. S. C. Stamatoglou and R. C. Hughes, *FASEB J.*, **8**, 420 (1994).
31. A. Z. Wang, G. K. Ojakian, and W. J. Nelson, *J. Cell Sci.*, **95**(Pt 1), 137 (1990).
32. L. N. Nejsum and W. J. Nelson, *J. Cell Biol.*, **178**, 323 (2007).
33. L. E. O'Brien, T. S. Jou, A. L. Pollack, Q. Zhang, S. H. Hansen, P. Yurchenco, and K. E. Mostov, *Nat. Cell Biol.*, **3**, 831 (2001).
34. W. Yu, A. Datta, P. Leroy, L. E. O'Brien, G. Mak, T. S. Jou, K. S. Matlin, K. E. Mostov, and M. M. Zegers, *Mol. Biol. Cell*, **16**, 433 (2005).
35. A. Zuk, K. S. Matlin, and E. D. Hay, *J. Cell Biol.*, **108**, 903 (1989).
36. E. D. Hay, *Dev. Dyn.*, **233**, 706 (2005).
37. R. Kalluri and E. G. Neilson, *J. Clin. Invest.*, **112**, 1776 (2003).
38. S. L. Friedman, *Gastroenterology*, **134**, 1655 (2008).
39. J. Yang and R. A. Weinberg, *Dev. Cell*, **14**, 818 (2008).
40. P. Godoy, J. G. Hengstler, I. Ilkavets, C. Meyer, A. Bachmann, A. Muller, G. Tuschl, S. O. Mueller, and S. Dooley, *Hepatology*, **49**, 2031 (2009).
41. K. K. Kim, M. C. Kugler, P. J. Wolters, L. Robillard, M. G. Galvez, A. N. Brumwell, D. Sheppard, and H. A. Chapman, *Proc. Natl. Acad. Sci. U S A*, **103**, 13180 (2006).

42. C. M. DiPersio, D. A. Jackson, and K. S. Zaret, *Mol Cell Biol.*, **11**, 4405 (1991).
43. M. Trauner and J. L. Boyer, *Physiol. Rev.*, **83**, 633 (2003).
44. L. Wang and J. L. Boyer, *Hepatology*, **39**, 892 (2004).
45. N. W. Karuri, S. Liliensiek, A. I. Teixeira, G. Abrams, S. Campbell, P. F. Nealey, and C. J. Murphy, *J. Cell Sci.*, **117**, 3153 (2004).
46. V. Vogel and M. Sheetz, *Nat. Rev. Mol. Cell Biol.*, **7**, 265 (2006).
47. B. P. Chan and K. W. Leong, *Eur. Spine J.*, **17 Suppl 4**, 467 (2008).
48. C. R. Ethier, M. Johnson, and J. Ruberti, *Annu. Rev. Biomed. Eng.*, **6**, 249 (2004).
49. P. C. Georges, J. J. Hui, Z. Gombos, M. E. McCormick, A. Y. Wang, M. Uemura, R. Mick, P. A. Janmey, E. E. Furth, and R. G. Wells, *Am. J. Physiol. Gastrointest. Liver Physiol.*, **293**, G1147 (2007).
50. R. G. Wells, *Hepatology*, **47**, 1394 (2008).
51. Y. C. Fung, *Biomechanics: Mechanical Properties of Living Tissues* (Springer, NY, 1981).
52. T. Azuma and S. Oka, *Am. J. Physiol.*, **221**, 1310 (1971).
53. B. Suki, S. Ito, D. Stamenovic, K. R. Lutchen, and E. P. Ingenito, *J. Appl. Physiol.*, **98**, 1892 (2005).
54. Y. Hosoyamada and T. Sakai, *Anat. Embryol. (Berl.)*, **210**, 1 (2005).
55. A. M. Malek, S. L. Alper, and S. Izumo, *JAMA*, **282**, 2035 (1999).
56. S. R. Chary and R. K. Jain, *Proc. Natl. Acad. Sci. U S A*, **86**, 5385 (1989).
57. L. G. Griffith and M. A. Swartz, *Nat. Rev. Mol. Cell Biol.*, **7**, 211 (2006).
58. L. Lu, M. J. Yaszemski, and A. G. Mikos, *Biomaterials*, **22**, 3345 (2001).
59. Y. Sato, M. Terashima, N. Kagiwada, T. Tun, M. Inagaki, T. Kakuta, and A. Saito, *Tissue Eng.*, **11**, 1506 (2005).
60. H. Zhang, F. Tasnim, J. Y. Ying, and D. Zink, *Biomaterials*, **30**, 2899 (2009).
61. Y. Nahmias, F. Berthiaume, and M. L. Yarmush, *Adv. Biochem. Eng. Biotechnol.*, **103**, 309 (2007).
62. H. G. Hall, D. A. Farson, and M. J. Bissell, *Proc. Natl. Acad. Sci. U S A*, **79**, 4672 (1982).
63. S. F. Abu-Absi, J. R. Friend, L. K. Hansen, and W. S. Hu, *Exp. Cell Res.*, **274**, 56 (2002).
64. N. Koide, K. Sakaguchi, Y. Koide, K. Asano, M. Kawaguchi, H. Matsushima, T. Takenami, T. Shinji, M. Mori, and T. Tsuji, *Exp. Cell Res.*, **186**, 227 (1990).

65. J. Yang, M. Yamato, T. Shimizu, H. Sekine, K. Ohashi, M. Kanzaki, T. Ohki, K. Nishida, and T. Okano, *Biomaterials*, **28**, 5033 (2007).
66. K. Ohashi, T. Yokoyama, M. Yamato, H. Kuge, H. Kanehiro, M. Tsutsumi, T. Amanuma, H. Iwata, J. Yang, T. Okano, and Y. Nakajima, *Nat. Med.*, **13**, 880 (2007).
67. G. Y. Lee, P. A. Kenny, E. H. Lee, and M. J. Bissell, *Nat. Methods*, **4**, 359 (2007).
68. T. R. Sodunke, K. K. Turner, S. A. Caldwell, K. W. McBride, M. J. Reginato, and H. M. Noh, *Biomaterials*, **28**, 4006 (2007).
69. J. Debnath and J. S. Brugge, *Nat. Rev. Cancer*, **5**, 675 (2005).
70. S. M. Chia, K. W. Leong, J. Li, X. Xu, K. Zeng, P. N. Er, S. Gao, and H. Yu, *Tissue Eng.*, **6**, 481 (2000).
71. G. H. Underhill, A. A. Chen, D. R. Albrecht, and S. N. Bhatia, *Biomaterials*, **28**, 256 (2007).
72. V. Liu Tsang, A. A. Chen, L. M. Cho, K. D. Jadin, R. L. Sah, S. DeLong, J. L. West, and S. N. Bhatia, *FASEB J.*, **21**, 790 (2007).
73. R. Glicklis, L. Shapiro, R. Agbaria, J. C. Merchuk, and S. Cohen, *Biotechnol. Bioeng.*, **67**, 344 (2000).
74. M. Dvir-Ginzberg, I. Gamlieli-Bonshtein, R. Agbaria, and S. Cohen, *Tissue Eng.*, **9**, 757 (2003).
75. M. Dvir-Ginzberg, T. Elkayam, E. D. Aflalo, R. Agbaria, and S. Cohen, *Tissue Eng.*, **10**, 1806 (2004).
76. R. Glicklis, J. C. Merchuk, and S. Cohen, *Biotechnol. Bioeng.*, **86**, 672 (2004).
77. L. G. Cima, D. E. Ingber, J. P. Vacanti, and R. Langer, *Biotechnol. Bioeng.*, **38**, 145 (1991).
78. D. J. Mooney, K. Sano, P. M. Kaufmann, K. Majahod, B. Schloo, J. P. Vacanti, and R. Langer, *J. Biomed. Mater. Res.*, **37**, 413 (1997).
79. P. M. Kaufmann, S. Heimrath, B. S. Kim, and D. J. Mooney, *Cell Transplant*, **6**, 463 (1997).
80. A. A. Demetriou, A. Reisner, J. Sanchez, S. M. Levenson, A. D. Moscioni, and J. R. Chowdhury, *Hepatology*, **8**, 1006 (1988).
81. J. Rozga, F. Williams, M. S. Ro, D. F. Neuzil, T. D. Giorgio, G. Backfisch, A. D. Moscioni, R. Hakim, and A. A. Demetriou, *Hepatology*, **17**, 258 (1993).
82. Z. H. Cai, Z. Q. Shi, M. Sherman, and A. M. Sun, *Hepatology*, **10**, 855 (1989).
83. C. Yin, S. Mien Chia, C. Hoon Quek, H. Yu, R. X. Zhuo, K. W. Leong, and H. Q. Mao, *Biomaterials*, **24**, 1771 (2003).

84. K. N. Chua, W. S. Lim, P. Zhang, H. Lu, J. Wen, S. Ramakrishna, K. W. Leong, and H. Q. Mao, *Biomaterials*, **26**, 2537 (2005).
85. Z. Q. Feng, X. Chu, N. P. Huang, T. Wang, Y. Wang, X. Shi, Y. Ding, and Z. Z. Gu, *Biomaterials*, **30**, 2753 (2009).
86. S. Wang, D. Nagrath, P. C. Chen, F. Berthiaume, and M. L. Yarmush, *Tissue Eng. Part A*, **14**, 227 (2008).
87. N. Cheng, E. Wauthier, and L. M. Reid, *Tissue Eng. Part A*, **14**, 1 (2008).
88. P. V. Moghe, F. Berthiaume, R. M. Ezzell, M. Toner, R. G. Tompkins, and M. L. Yarmush, *Biomaterials*, **17**, 373 (1996).
89. F. Berthiaume, P. V. Moghe, M. Toner, and M. L. Yarmush, *FASEB J.*, **10**, 1471 (1996).
90. L. Ying, C. Yin, R. X. Zhuo, K. W. Leong, H. Q. Mao, E. T. Kang, and K. G. Neoh, *Biomacromolecules*, **4**, 157 (2003).
91. S. Tobe, Y. Takei, K. Kobayashi, and T. Akaike, *Biochem. Biophys. Res. Commun.*, **184**, 225 (1992).
92. S. Ng, Y. N. Wu, Y. Zhou, Y. E. Toh, Z. Z. Ho, S. M. Chia, J. H. Zhu, H. Q. Mao, and H. Yu, *Biomaterials*, **26**, 3153 (2005).
93. S. J. Seo, T. Akaike, Y. J. Choi, M. Shirakawa, I. K. Kang, and C. S. Cho, *Biomaterials*, **26**, 3607 (2005).
94. S. J. Seo, Y. J. Choi, T. Akaike, A. Higuchi, and C. S. Cho, *Tissue Eng.*, **12**, 33 (2006).
95. Y. Du, S. M. Chia, R. Han, S. Chang, H. Tang, and H. Yu, *Biomaterials*, **27**, 5669 (2006).
96. T. Takehara, K. Matsumoto, and T. Nakamura, *J. Biochem.*, **112**, 330 (1992).
97. T. Nakamura, Y. Nakayama, and A. Ichihara, *J. Biol. Chem.*, **259**, 8056 (1984).
98. T. Nakamura, K. Yoshimoto, Y. Nakayama, Y. Tomita, and A. Ichihara, *Proc. Natl. Acad. Sci. U S A*, **80**, 7229 (1983).
99. T. A. Brieva and P. V. Moghe, *Biotechnol. Bioeng.*, **76**, 295 (2001).
100. E. J. Semler, A. Dasgupta, and P. V. Moghe, *Tissue Eng.*, **11**, 734 (2005).
101. S. N. Bhatia, U. J. Balis, M. L. Yarmush, and M. Toner, *FASEB J.*, **13**, 1883 (1999).
102. S. J. Seo, I. Y. Kim, Y. J. Choi, T. Akaike, and C. S. Cho, *Biomaterials*, **27**, 1487 (2006).
103. S. M. Chia, P. C. Lin, and H. Yu, *Biotechnol. Bioeng.*, **89**, 565 (2005).
104. M. K. Smith, K. W. Riddle, and D. J. Mooney, *Tissue Eng.*, **12**, 235 (2006).

105. A. Kedem, A. Perets, I. Gamlieli-Bonshtein, M. Dvir-Ginzberg, S. Mizrahi, and S. Cohen, *Tissue Eng.*, **11**, 715 (2005).
106. R. Coger, M. Toner, P. Moghe, R. M. Ezzel, and M. L. Yarmush, *Tissue Eng.*, **3**, 375 (1997).
107. N. Yamamura, R. Sudo, M. Ikeda, and K. Tanishita, *Tissue Eng.*, **13**, 1443 (2007).
108. D. W. Hutmacher, *J. Biomater. Sci. Polym. Ed.*, **12**, 107 (2001).
109. S. Zhang, T. C. Holmes, C. M. DiPersio, R. O. Hynes, X. Su, and A. Rich, *Biomaterials*, **16**, 1385 (1995).
110. A. Atala, *Curr. Urol. Rep.*, **2**, 83 (2001).
111. A. L. Sieminski, A. S. Was, G. Kim, H. Gong, and R. D. Kamm, *Cell Biochem. Biophys.*, **49**, 73 (2007).
112. C. E. Semino, J. R. Merok, G. G. Crane, G. Panagiotakos, and S. Zhang, *Differentiation*, **71**, 262 (2003).
113. J. L. Drury and D. J. Mooney, *Biomaterials*, **24**, 4337 (2003).
114. R. M. Sutherland, B. Sordat, J. Bamat, H. Gabbert, B. Bourrat, and W. Mueller-Klieser, *Cancer Res.*, **46**, 5320 (1986).
115. H. Kang, K. J. Bayless, and R. Kaunas, *Am. J. Physiol. Heart Circ. Physiol.*, **295**, H2087 (2008).
116. R. Sudo, S. Chung, I. K. Zervantonakis, V. Vickerman, Y. Toshimitsu, L. G. Griffith, and R. D. Kamm, *FASEB J.*, (2009).
117. M. J. Powers, D. M. Janigian, K. E. Wack, C. S. Baker, D. Beer Stolz, and L. G. Griffith, *Tissue Eng.*, **8**, 499 (2002).
118. I. Martin, D. Wendt, and M. Heberer, *Trends Biotechnol.*, **22**, 80 (2004).
119. M. Dvir-Ginzberg, T. Elkayam, and S. Cohen, *FASEB J.*, **22**, 1440 (2008).
120. S. Hanada, N. Kojima, and Y. Sakai, *Tissue Eng. Part A*, **14**, 149 (2008).

Chapter 25

BIOLOGICAL IMPLICATIONS OF POLYMERIC SCAFFOLDS FOR BONE TISSUE ENGINEERING DEVELOPED VIA SOLID FREEFORM FABRICATION

Andrew B. Yeatts and John P. Fisher*

Fischell Department of Bioengineering, University of Maryland, College Park, MD 20742, USA

*jpfisher@umd.edu

Clinically viable cell-based bone tissue engineering strategies require the development of advanced scaffold-manufacturing techniques. Stereolithography (SLA), three-dimensional printing (3DP), selective laser sintering (SLS), and fused deposition modeling (FDM) all are advanced solid freeform fabrication (SFF) techniques with bone tissue engineering applications; however, each presents a unique set of benefits and drawbacks. These benefits and drawbacks relate to the ability of the scaffolds to support cell proliferation, migration, and differentiation. SFF scaffolds are manufactured with a controlled architecture, but all methods reviewed have significant limitations with regard to the type of polymer used and resolution that can be obtained, both of which have great biological implications on cell and tissue interaction

Handbook of Intelligent Scaffolds for Tissue Engineering and Regenerative Medicine
Edited by Gilson Khang

www.panstanford.com

with the scaffold. This review assesses these limitations, along with the utilities of each method and the effect these have on the ability to create an effective bone tissue engineering polymeric scaffold using SFF.

25.1 Introduction

Tissue engineering approaches provide a promising alternative treatment of bone injuries, and with advances in the field of scaffold manufacture through SFF, bone tissue engineering is closer to becoming a viable clinical option. This chapter will overview common SFF techniques but will focus on evaluating the biological implications of each individual process. The biological implications, referring to growth, proliferation, differentiation, and signaling of cells seeded on or growing into the scaffold, are influenced by the surface roughness, porosity, pore size, and material properties of the scaffold. SFF techniques are able to be used to manufacture scaffolds to successfully incorporate these factors to varying degrees, but also have restrictions incurred through the mechanical process.[1,2] SFF techniques have been used to fabricate 3D bone tissue engineering scaffolds from materials commonly used in tissue engineering, such as poly(propylene fumarate) (PPF), poly(D,L-lactic-co-glycolic acid) (PLGA), and polycaprolactone (PCL).[3–8] Optimizing SFF techniques would greatly improve the clinical viability of bone tissue engineering techniques as these scaffolds could be readily and quickly made to be both reproducible and customizable to a defect.

25.1.1 *The Need for Bone Tissue Engineering*

Every year over six million bone injuries occur in the United States and approximately one million bone-grafting procedures are performed.[9] The source of bone for these grafts is either from the patient's own body in the case of an autograft or from a cadaver in the case of an allograft. Unfortunately both of these methods have significant disadvantages. The incidence of medical complications arising after surgery involving an autograft from the iliac crest are nearly 30%, as pain and morbidity at the donor site often occurs.[10]

Allografts are subject to rejection by the body, an immune response and disease transmission.[10,11] Since these traditional means of treating bone injuries are often ineffectual, a tissue engineering approach to replace damaged bone represents a promising alternative. A tissue engineering approach involves seeding and growing a cell source on a scaffold and implanting the scaffold and cells into the injury site.[12] The interaction of the scaffold with the seeded cells and the surrounding tissue at the implant site is vital to the success or failure of the tissue engineering treatment. The optimal tissue engineering scaffold should promote seeded cell proliferation and osteogenesis, support osteoconduction (bone ingrowth from surrounding tissue), be biodegradable, provide mechanical support, and be able to be shaped to fill the injury site.[13,14] In order to be used for clinical applications the scaffold should be manufactured quickly and easily with a highly reproducible design. The manufacturing technique of SFF or rapid prototyping provides a way to accomplish this as the scaffold is manufactured to defined design specifications layer by layer without the use of a mold.[15] This chapter will focus on the factors influencing the biological interactions of cells and tissue with the scaffold and how SFF techniques impact these interactions.

25.1.2 *Benefits of Scaffolds Developed via Solid Freeform Fabrication*

SFF techniques create a structure, layer by layer, to manufacture a 3D scaffold with controlled architecture. Producing scaffolds with a controlled architecture is vital for bone tissue engineering as scaffold properties such as porosity, pore size, and pore interconnectivity greatly influence osteogenic growth and scaffold mechanical properties.[16] Realization of this technology would lead to the production of scaffolds that are cost effective, easy to manufacture, and highly reproducible.[15] Design specifications for SFF scaffolds are typically inputted into a computer control system, allowing for a specific architecture. This process could be combined with images of a bone defect to create a scaffold that is customizable to the defect.[15,17,18] Much work has been done in developing bone tissue engineering scaffolds using SFF techniques, but relatively little work has been done in evaluating the biological implications of these scaffolds. This chapter seeks to review that work and speculate on the ability of each of these methods to develop effective bone tissue engineering scaffolds.

25.2 Stereolithography

25.2.1 *Scaffold-Manufacturing Process*

Becoming commercially available over 20 years ago, SLA is perhaps the most popular SFF method in use today. SLA utilizes a UV laser that selectively photopolymerizes a liquid polymer material (Fig. 25.1).[15,19–21] To develop a scaffold using SLA, a liquid polymer is placed in a vat and a platform is positioned just below the surface of the liquid polymer. The polymer is then cross-linked using a laser, which is guided by a mirror to cross-link the polymer to the shape specified. The photopolymerization only takes place at surface of the polymer. Once a layer is complete, the platform is lowered one layer and recoated with the liquid polymer and subsequent layers are cross-linked, following the same method. After completion of the scaffold, it is removed from the vat, excess polymer is removed, and the scaffold is placed in a UV oven to complete the curing process. SLA has the ability to create custom-shaped scaffolds with

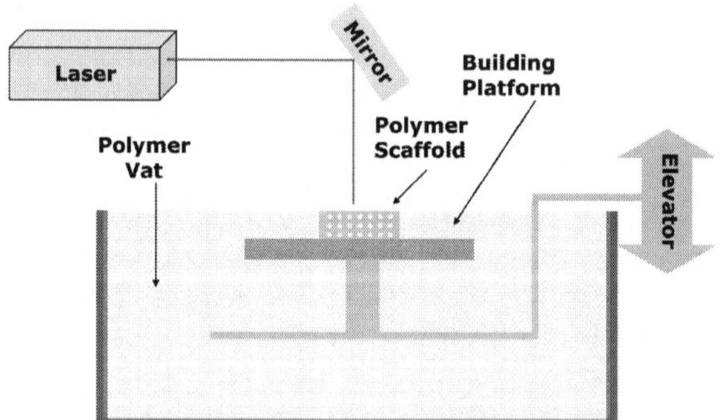

Figure 25.1. A schematic of the stereolithographic process.[1,20] The mirror-guided laser cross-links the polymer at the surface of the vat and can move in the x and y directions. The elevator lowers the completed scaffold one layer in the z direction.

resolutions as small as 20 μm. SLA was first used to aid in the treatment of bone injuries by oral and maxillofacial surgeons to create a model of complex cranial injuries.[17,22] Surgeons utilized computed tomography (CT) to develop 3D displays of the injury site and then fabricated a detailed model to aid in surgical planning. Tissue engineers have now adapted this idea to use SFF to create polymer scaffolds based on CT images of an injury site in order to create a platform for bone tissue engineering that will fit precisely in a defect. Several groups have evaluated the effectiveness of scaffolds developed through SLA for cell-based tissue engineering (Table 25.1). Research has shown that SLA can be used to make PPF scaffolds with controlled architectures.[3,4,23–27] Commonly used in bone tissue engineering applications, PPF has strong mechanical properties, as well as being biodegradable.[23,28–33] The ability of SLA to be used in the manufacture of PPF scaffolds provides great utility for the process in bone tissue engineering, though research has shown that further biological evaluations must be completed in order to optimize the manufacturing process of the scaffolds.[4]

Similar in principle to SLA, micro-SLA has been developed to create scaffolds with even smaller architectures by focusing the laser beam more precisely. This technique allows for resolutions in the 500 nm range, and it has been demonstrated that highly structured PPF scaffolds could be developed using this technique and that the scaffolds would support the growth of pre-osteoblasts.[24,25] Despite these promising developments, SLA has several drawbacks to its effective use in bone tissue engineering. Many common biomaterials used in bone tissue engineering cannot be used in SLA as high

Table 25.1. Scaffolds manufactured using SLA and used for bone tissue engineering purposes.

Scaffold material	Properties	References
PPF	Biodegradable, biocompatible, small pore size	3, 4, 24, 25
PEO and PEGDA	Low mechanical properties, not biodegradable, cell incorporation	37
HA	Indirect scaffold formation, osteoconductive	34–36

temperatures are associated with the removal of the unreacted polymer. Also many scaffolds will shrink as a result of curing, leading to changes in the controlled architecture and possible cracking of the scaffolds. Shrinking on the order of 25% following curing has been reported in the literature.[24] Despite these drawbacks SLA represents a promising technique to manufacture scaffolds with controlled architecture for bone tissue engineering.

25.2.2 *Biological Implications*

Several studies have been completed to test *in vitro* cell growth and *in vivo* bone growth within scaffolds manufactured by SLA.[24,25,34–38] In a study designed to test the ability of the SLA-developed scaffolds to support bone growth *in vivo*, porous hydroxyapatite (HA) scaffolds with a porosity of 40% were developed using an indirect SLA approach.[35,36] These scaffolds were created with two different architectures, orthogonal and radial, by developing a mold using SLA and forming a scaffold around it. The scaffolds were implanted into porcine mandibles, and bone ingrowth was evaluated by scanning electron microscopy (SEM) and toluene blue staining. The amount of bone ingrowth was quantified using histomorphometry, and the architecture of the scaffold was shown to influence the amount of the regenerated bone as the orthogonal scaffolds had a larger bone growth area than the radial scaffolds. Scaffold architecture was also shown to influence bone geography, as the orthogonal HA scaffolds formed an interpenetrating matrix, while the bone in the radial scaffolds formed in an intact piece in the center of the scaffold. The shape of the new bone greatly influences the mechanical properties of the regenerated tissue, and thus it is important to develop scaffolds with controlled architecture to induce proper regenerated bone formation.[39,40] Most solid freeform scaffold fabrication techniques involve seeding cells onto the surface of the scaffold. Though this can be effective, it is difficult to achieve a uniform distribution of cells throughout the scaffold using this method as the cell seeding is influenced by diffusion and architecture of the scaffold. In another study utilizing SLA a method was devised to encapsulate cells directly into the scaffold using a photopolymerizable poly(ethylene oxide) (PEO) and poly(ethylene glycol) diacrylate (PEGDA) hydrogel.[37] In the study a Chinese hamster ovary cell line was used to demonstrate the viability of the method. Cells were suspended in the photopolymer solution before exposure to the laser and were encapsulated upon polymerization. The cells remained viable throughout the encapsulation procedure. The scaffolds used in this particular study lacked the mechanical properties for effective use in hard-tissue applications, like bone, but the ability to encapsulate cells in a scaffold could be applied to bone tissue engineering to create a uniform cellular seeded scaffold with a controlled architecture and cell distribution.

Scaffold material must be liquid and photopolymerizable for use in SLA, thus greatly limiting the scaffold material choice. Despite this the ability to be used in PPF scaffold fabrication makes SLA a promising method to produce bone tissue engineering scaffolds. More cell studies are needed to fully assess the cell interactions with SLA-fabricated scaffolds, but SLA has been shown to create scaffolds with feature architecture and porosity small enough for bone tissue engineering purposes. The osteoconduction of scaffolds fabricated via indirect SLA has been demonstrated, but the indirect approach used to create these scaffolds adds an extra manufacturing step and could lead to uncontrolled microarchitecture in the scaffolds. This makes the indirect SLA approach less clinically viable than a direct approach. The possibility of the incorporation of cells and biomolecules directly into the scaffold is another important attribute of SLA. Growth factors such as bone morphogenetic protein-2 (BMP-2) and transforming growth factor-beta 2 (TGF-β2) have been shown to have great effects on the osteoinduction and osteoconduction of bone tissue engineering scaffolds.[41–43] Direct incorporation of these heat-sensitive materials into scaffolds cannot be achieved by SFF methods that require high heat such, as sintering. Incorporation of growth factors as well as cells through SLA could create dynamic bone tissue engineering scaffolds with a uniform cell population, growth factors to enhance osteogenesis, and controlled, reproducible scaffold architecture. The use of PPF, possible direct incorporation of biomolecules, and nanoscale resolution makes SLA one of the most promising SFF techniques for bone tissue engineering, yet additional cell studies using SLA are needed to realize this promise.

25.3 Three-Dimensional Printing

25.3.1 *Scaffold-Manufacturing Process*

Three-dimensional printing (3DP) utilizes a similar technology as an inkjet printer to create 3D scaffolds (Fig. 25.2). In order to develop a scaffold via 3DP, a computer-aided drawing (CAD) representation of the scaffold is inputted. An inkjet printer head then travels over a thin powder bed of the material forming the scaffold. As the inkjet printer head moves over the powder layer, it deposits a binding solution that fuses the powder in the shape dictated by the CAD. After a layer is completed, the powder bed is lowered a layer from the printer head, either through the lowering of the building platform or through the raising of the printer head. Another layer of powder is then deposited on the building platform, and the printer head deposits more binding solution to continue to form the scaffold. This process is repeated until the entire 3D structure is completed.[15,20,21,32,44,45] The unreacted powder then must be removed either by brushing or blowing. Removal of the unreacted polymer can be problematic in highly porous scaffolds as polymer powder becomes trapped in the pores. Despite this many researchers utilize 3DP to create tissue engineering scaffolds

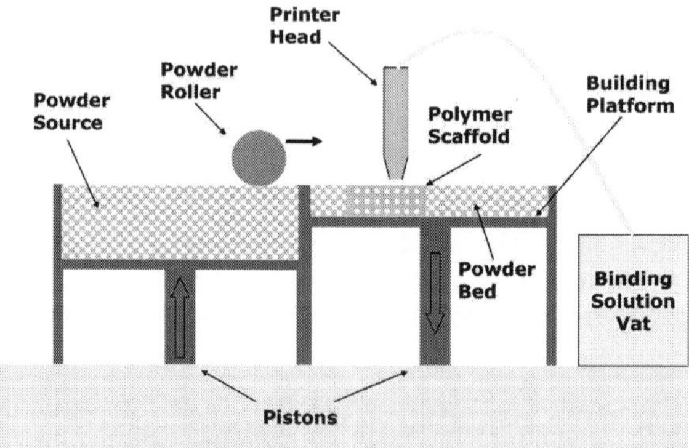

Figure 25.2. A schematic of the 3DP process.[1,20] The printer adds a binding solution in the x and y directions to form the scaffold. The building platform is lowered by a layer thickness, and the powder roller adds additional powder to achieve three dimensionality.

Table 25.2. Scaffolds manufactured using 3DP and used for bone tissue engineering purposes.

Scaffold material	Properties	References
HA	Osteoconductive, high mechanical strength, minimal inflammatory response	5, 48, 49, 56–58
PLGA	Biodegradable, osteoconductive, minimal inflammatory response	5
PLA	Biodegradable	50, 54, 59

(Table 25.2).[46–52] Porous HA scaffolds have been created using 3DP with a minimum pore size of 450 μm and a minimum wall thickness of 330 μm.[49] Though these results were promising a possible drawback of 3DP is highlighted, as the size of the pores achieved were not as small as the optimal pore size for bone tissue engineering, which is approximately 300 μm.[53] Just as indirect approaches were utilized with SLA, bone tissue engineering scaffolds have been manufactured using an indirect 3DP approach. In this approach molds were created using 3DP, and PLGA scaffolds was cast around them. Porogen leaching was then used to achieve small pore sizes.[47] This work showed that 3DP could be an effective tool for scaffold design but that other techniques may be necessary to achieve the porosity desired. Other problems associated with 3DP include limited scaffold material choice, required processing of many scaffold materials before use in 3DP, and possible shrinking of scaffold during curing.

25.3.2 *Biological Implications*

A significant drawback to the use of 3DP for fabrication of bone tissue engineering scaffolds is the inability of this process to be used to fabricate scaffolds with small pore sizes. The indirect 3DP approach overcomes this limitation as it allows for a controlled macroarchitecture, which can be custom made to fit a defect, while the necessary microarchitecture can be attained through the use of a secondary method such as porogen leaching or emulsion. These indirect 3DP scaffolds have a significant advantage over scaffolds made through traditional manufacturing techniques as they have increased mechanical properties, controlled global architecture that aids in bone growth and vascularization, and reproducibility of global architecture. Significant drawbacks to this technique include the extra steps associated with the indirect procedure, which include designing a mold, casting a polymer into the mold, and then adding a microstructure through a traditional fabrication technique.[54] In a study utilizing a slightly different 3DP approach poly(L-lactic acid) (PLA) scaffolds were created by mixing PLA with a salt of different pore sizes and molded into a 3D architecture using 3DP and the salt was leached out to create pores.[50] These porogen-leaching 3DP scaffolds were made in two different void fractions (75% and 90%) and pore sizes ranging from 90 μm to 120 μm. To test cellular response to this scaffold fabrication method, three different cell types (canine dermal fibroblasts, vascular smooth muscle cells, and microvascular epithelial cells) were seeded onto the scaffolds and the viability and extracellular matrix deposition of the cells was measured. The results demonstrated that the 75% void fraction groups were less optimal for supporting cellular infiltration than the 90% void fraction groups.[50] This study highlights the need for strictly controlled architectures but also highlights a weakness of SFF methods utilizing porogen leaching, as the pore sizes were not uniform in the scaffold and were larger than the size of the salt used.

A benefit of controlled interconnectivity of scaffold pores in addition to enhancing cellular infiltration is the added utility to use these scaffolds in a perfusion bioreactor system. Perfusion bioreactor systems enhance *in vitro* cell culture by increasing nutrient transfer and replicating *in vivo* mechanical stresses, but these systems require pore interconnectivity to allow fluid to flow through the scaffold.[55] Successfully integration of scaffolds formed by 3DP have been demonstrated in the literature as fluid was easily perfused through the 3DP scaffolds and greater interior cell growth was observed in scaffolds in the perfusion system.[48] *In vivo* performance of scaffolds developed by 3DP was also assessed in the literature.[5,56] Scaffolds fabricated using the commercially available TheriForm™ 3DP process were made from PLGA of two different molecular weights and combined with 20% tricalcium phosphate (TCP). Scaffolds incorporated two different architectures: one with macroscopic channels and another with a microscopic porosity gradient. The scaffolds were implanted into rabbit cranial defects and bone ingrowth was measured by histology using Stevenel's Blue and van Gieson's Picro Fuschin (SVG stain). Histomorphometric analysis was used to quantify new bone area, and results demonstrated that the high-molecular-weight PLGA scaffolds with macroscopic channels had a higher new bone area than both the scaffolds without channels and the defect without any scaffold. The lower-molecular-weight PLGA groups rapidly hydrolytically degraded and collapsed before the eight-week study was completed. The higher-molecular-weight PLGA scaffolds were shown to be osteoconductive and only mildly inflammatory as significant bone growth was observed after eight weeks, as shown by histology. The scaffolds guided bone growth down the channels, and the porosity gradient of the scaffold had an effect on the type of tissue that was produced, as only soft tissue was produced outside the radial channels of the scaffold. These results are important as they demonstrated that scaffold channels, porosity, and pore size influence bone growth. In another study HA scaffolds were formed using 3DP in two architectures, channeled and porous and porous, and the scaffolds once again implanted into rabbit cranial defects. Using histology extensive amount of bone growth was observed in the channeled scaffolds and the channels were shown to guide bone growth from different areas. The channeled scaffolds guided bone formation from the edges of the defect into the center of the scaffold. The HA scaffolds were shown to induce more bone formation than the PLGA-TCP scaffolds. These studies not only demonstrated the efficacy of using scaffolds designed through 3DP in a tissue engineering strategy to treat cranial defects but showed that the osteoconduction of a scaffold can be greatly influenced by its architecture. Scaffold material properties and architecture must both be taken into account when designing scaffolds for bone tissue engineering.

Three-dimensional printing has several disadvantages that hinder its use to manufacture porous scaffolds. In highly porous scaffolds unreacted powder becomes trapped in the pores and is difficult to remove. Second, the pore size is limited by the size of the printer head and particle size, which make it difficult to achieve scaffolds with optimal porosities to enhance bone formation. Void spaces within fused particles have been shown to create a microarchitecture within 3DP scaffolds, which may influence cell behavior and degradation of the scaffold, but this microarchitecture is uncontrolled.[15,56] The material choice for 3DP is limited as most biomaterials for tissue engineering purposes are not available in powder. Despite this studies have successfully demonstrated 3DP as a viable method to create a scaffold that supports new bone formation in a critical-sized defect. Even with these valuable studies, only a limited amount of research has been done to assess the *in vitro* and *in vivo* osteoinduction and osteoconduction of 3DP-fabricated scaffolds, and considerably more work needs to be done to demonstrate the utility of 3DP for cell-based bone tissue engineering applications.

25.4 Selective Laser Sintering

25.4.1 *Scaffold-Manufacturing Process*

Like 3DP, scaffolds designed via SLS begin with a powder material (Fig. 25.3). The powder material is placed on a flat bed and

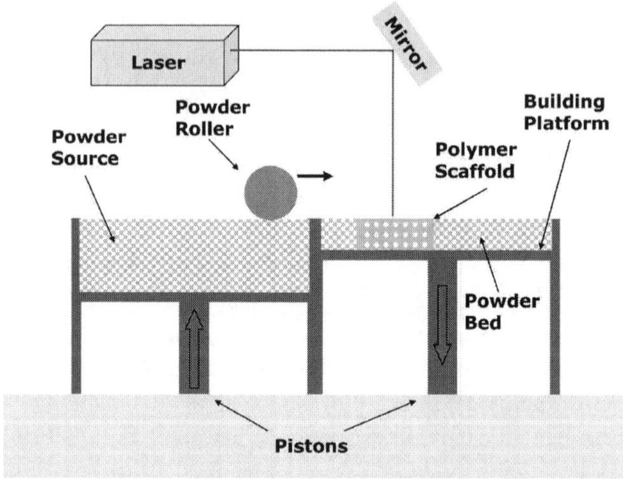

Figure 25.3. A schematic of the SLS process.[1,20] The polymer in the powder bed is sintered by the laser, which can be moved in the x–y direction. The building platform is lowered by a layer thickness, and the powder roller adds additional powder to achieve three dimensionality.

then heated to just below the glass transition temperature. A powerful infrared laser is then directed at the powder material, and it heats the powder to just above the glass transition temperature, sintering the particles together. The laser fuses the particles in a pattern dictated by an inputted computer drawing. Once a layer is completed, the powder bed is lowered one-layer thickness and the laser begins to fuse a new layer. The fusing of a new layer results in the bonding of the particles of the new layer to the previously completed layer. The powder that remains provides structural support and is removed after the scaffold is complete. Like 3DP, problems are encountered with removing unused powder from highly porous scaffolds. Techniques such as air pressure, solvents, and vibration are utilized in addition to traditional brushing to aid in removing the powder.[15,20,21,60] SLS can be used with a fairly large range of materials, and many common polymers and ceramics used for bone tissue engineering applications have been successfully utilized in scaffolds produced by SLS, including PCL, polyetheretherketone, HA, poly(vinyl alcohol) (PVA) , PLA, Nylon-6, and PLGA (Table 25.3).[6,18,61–70] Scaffolds made from PCL, a readily available, bioresorbable polymer used in bone tissue engineering, were successfully fabricated into 3D scaffolds and shown to have mechanical properties similar to trabecular bone. Despite the strong mechanical properties of this scaffold its minimum feature size was 1.75 mm, a lower resolution than what is optimal for bone tissue engineering scaffolds. The SLS-fabricated scaffold feature size is dictated by the particle size as it is in 3DP. Also like 3DP, overall porosity of the fabricated scaffolds is reduced by unremoved polymer particles from the pores.[6] Though improved techniques may lead to a smaller pore size and increased removal of loose powder, the pore size is limited

Table 25.3. Scaffolds manufactured using SLS and used for bone tissue engineering purposes.

Scaffold material	Properties	References
PCL	Bioresorbable, maintain mechanical strength	6, 63, 66
Nylon	Does not biodegrade, biocompatible	68
PLGA and HA	Osteoconductive, biodegradable	61, 66
PVA and HA	Biodegradable	64, 66, 72

by the beam of the laser used, which has a typical beam thickness of 400 μm. This creates a technological limitation to the resolution achieved via SLS.

25.4.2 *Biological Implications*

Scaffolds developed by indirect SLS have been shown to induce bone growth when seeded with bone marrow stromal cells.[67] The scaffolds in this study were formed from a mold constructed via SLS, and two experimental groups were created (smooth and textured), as well as a nonmolded control. The use of a mold provided for a highly controlled macrostructure but the potential for variability in the microstructure of the scaffolds. This microstructure was shown to have a significant effect on the seeded cells as increased cell attachment was observed on surfaces of the scaffold with textured surfaces (grooves $\sim$25 μm deep and $\sim$40 μm wide). Cell density was higher on the textured outer surface of the scaffold, revealing that the textured surface influenced the cell distribution. In addition areas of the scaffold with textured surfaces demonstrated an increased formation of budding bone, as shown by histology, though the total amount of new bone area was not different between the smooth and textured scaffolds. Two different scaffold properties likely caused the increased bone budding on the textured surface. First, increased surface texture leads to a greater surface area for cells to adhere and therefore influences cell distribution. Second, calcium and phosphate have been shown to enhance bone formation; therefore it is hypothesized that the increased degradation of the scaffold at areas of surface texture leads to increased local concentrations of calcium and phosphate and enhanced *in vitro* bone formation. This study demonstrated an effective use of cell-seeded bone tissue engineering scaffolds developed by SLS and provided evidence for the need for scaffolds with controlled architecture. In another study involving SLS and bone tissue engineering, a significant amount of *in vivo* bone formation was shown four weeks following seeding in an SLS-manufactured scaffold with primary human fibroblasts infected with a BMP-7 gene that causes the cells to produce bone.[6] The cell-seeded scaffolds were implanted subcutaneously into mice, and bone growth was detected using μCT and hematoxylin and eosin staining. Bone growth was observed inside the pores of the scaffolds, and the bone mineral density of the new bone was measured and found to be approximately 500 mg/cm^3, which is in the middle of the typical bone mineral densities of normal cortical and trabecular bone. These SLS-developed scaffolds had compressive moduli of approximately 60 MPa and yield strengths in the range of 2–3 MPa, which is within the range of normal trabecular bone. Mechanical properties of the scaffold are important aspects for two primary reasons. First, the scaffold must support the normal loading of the tissue. If not, excessive deformation of the scaffold and the cells seeded on it will inhibit proper bone formation. Second, bone remodeling is naturally influenced by mechanical forces; thus proper mechanical properties of the scaffold may enhance both the amount of bone ingrowth and osteoinduction of the scaffold.[1,14,71] In a final aspect of the study, a porous PCL condyle scaffold was created from CT data taken from a condyle of a minipig. The fabricated scaffolds replicated the actual condyle and thus demonstrated an important proof of the principle of creating a custom-made bone tissue engineering scaffold from patient CT data.

SLS is hampered by many of the same shortcomings as 3DP, including unreacted powder in pores and limitations on the pore size, depending on powder size. Changing SLS parameters such as laser power and altering particle processing can overcome some of these limitations and give the user more control, but some technical limitations still exist.[2] Gaps resulting from irregular particles can create porosity, but this porosity is not completely controllable.[61] Resolution is also limited by the laser dimensions. Unreacted powder has been shown to have negative effects on cell survival and reduce the overall porosity of the scaffold.[68] SLS often involves high heat, which makes inclusion of biomolecules directly into the scaffold impossible. The biomaterial needs to be readily available in a highly controlled powder form, which can restrict the materials that can be used. On the positive side SLS does not require the use of potentially harmful solvents, and since the polymer is sintered using heat, a binding agent is not required, which reduces the risk of any harmful substance remaining in the scaffold. Relatively few cell-based tissue engineering studies utilizing SLS have been completed, perhaps because many studies have focused on the effect

SLS parameters have on scaffold design. Despite this proof of principle, work has been done with SLS, demonstrating its potential to be used to create an effective cell-seeded bone tissue engineering scaffold from image data of a defect site that will support osteogenesis of the seeded cells.[6]

25.5 Fused Deposition Modeling

25.5.1 *Scaffold-Manufacturing Process*

FDM utilizes an extrusion-based technique to form 3D scaffolds. An FDM machine consists of a building platform and an extrusion nozzle, in which a filament of the scaffold material is fed (Fig. 25.4). The extrusion nozzle heats the scaffold material to just above its melting point and deposits the material on the building platform, where it immediately solidifies. The extrusion nozzle then moves horizontally to deposit a layer of the scaffold. After the layer is complete the building platform is lowered by a layer thickness and subsequent layers are built on top of the existing layer to create the 3D scaffold. Materials used in FDM must be thermoplastics as the material needs to be readily melted and solidified.[20] Materials commonly used for bone tissue engineering purposes in FDM include PCL and polypropylene (PP), as well as composite scaffolds of these polymers with HA and TCP (Table 25.4).[7,8,73–78]

Scaffolds developed via FDM from PCL were shown to support growth of osteoblast-like cells. These scaffolds had a minimum channel size of 160 μm and porosity as high as 77%.[8,77] FDM is able to achieve higher porosity and a smaller pore size than 3DP and SLS without trapping particles within the pores since the scaffold is built from an extruded polymer rather than a particle bed. PP-TCP composite scaffolds manufactured using FDM were also developed for bone tissue engineering applications. Samples were formed having a pore size of 160 μm and porosity of 36% to 52%. The failure strength was shown to be comparable to that of cancellous bone.[74] Limitations of FDM primarily result from the limited amount of scaffold materials that can be used and the elevated temperatures associated with the melting process.

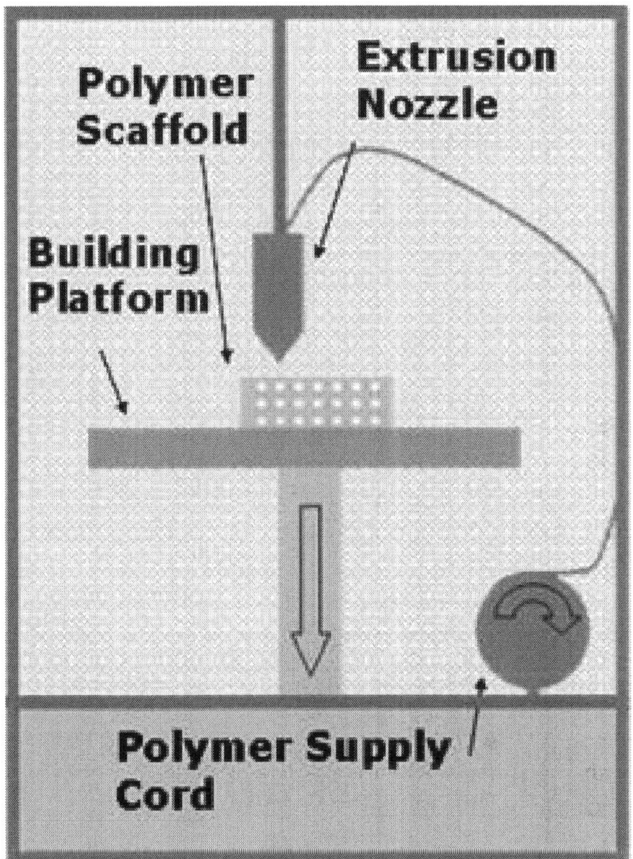

Figure 25.4. Schematic of the FDM process.[1,20] An extrusion nozzle forms a scaffold in the x–y direction. Three dimensionality is achieved by movement of the platform in the z direction.

25.5.2 *Biological Implications*

Several studies have been done to evaluate biological interactions of cells supported by scaffolds fabricated using FDM. In one such study, PCL scaffolds were fabricated using FDM for the co-culture of osteoblasts and chondrocytes. Cells in these scaffolds were shown to maintain their phenotype, and the approach represents a viable strategy for the repair of osteochondral defects.[7] Osteoblastic phenotype was determined by alkaline phosphatase (ALP) and

Table 25.4. Scaffolds manufactured using FDM and used for bone tissue engineering purposes.

Scaffold material	Properties	References
PCL, HA-PCL	Support osteoblast proliferation and phenotype, bioresorbable	7, 8, 73, 75–81
PLA, PLLA	Biodegradable	75, 82
PP, PP-TCP	Not biodegradable, mechanically strong, support cell growth	74

osteocalcin expression as well as visualization of the extracellular matrix using SEM. Scaffolds for this tissue engineering strategy require support of the growth, interaction, and extracellular matrix production of both osteoblasts and chondrocytes. The scaffolds were constructed with a pore size from 300 to 568 μm, a porosity of $\sim$60%, and a partition in the scaffold to separate the bone and cartilage components. Results demonstrated the scaffolds to be conducive for cell communication, growth, and attachment, as shown by confocal microscopy and SEM to visualize cells. The scaffolds became completely filled with cells and matrix, which could be attributed to the interconnected scaffold architecture and the large available surface area. Higher ALP expression was observed in the osteochondral scaffold seeding with chondrocytes and osteoblasts than the control, indicating the controlled architecture scaffolds developed using FDM could potentially be used for osteochondral tissue engineering. Another important aspect to mediating the biological interaction between the cells and the scaffold is the ability of the scaffold to be incorporated with bioactive factors. In another study utilizing PCL, scaffolds fabricated using FDM were shown to be effective as a delivery system for recombinant human BMP-2 (rhBMP-2), a growth factor shown to enhance both osteoinduction and osteoconduction.[79,80] The loaded rhBMP-2 was shown to have a loading efficiency as high as 73% and to remain bioactive. The bioactivity of the rhBMP-2 postrelease was verified by incubating human osteoblasts for four days with the released rhBMP-2 and comparing rhBMP-2 ALP levels to controls. The rhBMP-2 was shown to be released from the scaffolds in both a burst and a pattern profile, depending on the loading concentration. Scaffolds were seeded with canine osteoblast cells, and the scaffolds with rhBMP-2 were shown to induce higher osteocalcin expression than control groups, but even scaffolds without rhBMP-2 were shown to sustain osteogenic expression, as shown by osteocalcin expression and mineral content. This demonstrated the feasibility of using FDM-fabricated PCL scaffolds as a platform for cell-based tissue engineering and a delivery system for rh-BMP-2. In this study rhBMP-2 was not incorporated into the scaffold during manufacture as the FDM process involves high heat; rather rhBMP-2 was loaded directly on the scaffold after fabrication. In a separate study demonstrating the mechanical properties of FDM-fabricated PCL scaffolds, honeycomb-shaped scaffolds with a completely interconnected pore structure were subjected to compressive testing.[8] The scaffolds were shown to have a yield strength of 2.3 MPa and a compressive stiffness of 29.4 MPa at physiological conditions. The mechanical properties were shown to change based on the lay-down pattern of the scaffold. Scaffolds were seeded with human fibroblasts and osteoblast-like cells to evaluate the efficacy of using the scaffolds in cell-based tissue engineering strategies. Immunohistochemistry analysis and confocal microscopy revealed the cells had proliferated throughout the scaffolds and deposited an extracellular matrix. This study demonstrated that FDM could be used to develop a tissue engineering scaffold suitable for cell seeding and that the scaffold had sufficient mechanical properties to serve as a bone tissue engineering scaffold.

Like many other SFF techniques, FDM is limited by the materials that can be used, as the polymer must be melted and resolidified. Because of the high heat associated with the polymer melting, biomolecules cannot be directly incorporated into the scaffold. FDM is able to achieve higher porosities than SLS and 3DP, but as pore size is decreased it becomes technically difficult to maintain high porosity. Like SLS, FDM does not require the use of a binding agent or solvent, which is advantageous in avoiding toxic contamination in the scaffold. However, some scaffolds may require a secondary support material, which could potentially be harmful to cells. Several cell studies have demonstrated the ability of FDM-fabricated scaffolds to support osteoblastic cell phenotype, verifying that FDM represents a promising tool for the development of scaffolds with a controlled architecture.

25.6 Conclusions

Osteoinduction and osteoconduction are greatly influenced by scaffold architecture. Because of this clinical bone tissue engineering scaffolds require that scaffolds be made with properties to enhance bone formation and that these scaffolds be reproducible. SLA, 3DP, SLS, and FDM all represent promising methods to achieve this. SLA has been shown to generate scaffolds with the highest porosity and lowest pore size, but the requirement of a liquid photopolymerizable polymer greatly restricts material choice. Three-dimensional printing may have utility in some aspects of bone tissue engineering, but porosities over 50% are difficult to attain and the pore size is also limited. SLS is also unable to achieve porosities over 50% and is very limited in the use of polymeric scaffolds. FDM is able to be used to fabricate scaffolds with high porosities, but is only able to be used with a limited amount of materials. Manufacturing methods that are unable to achieve scaffold architectures required of bone tissue engineering may be better utilized in an indirect approach, with a secondary method to achieve the necessary pore size and porosity. These approaches have the benefit of creating a scaffold with a controlled macroarchitecture, while still achieving high-resolution microarchitectures, but they add an extra manufacturing step and still leave variation in scaffold design. Cell-based studies using various SFF techniques have been completed, but more work must be completed to fully understand the biological implications of these scaffolds. As the optimal bone tissue engineering scaffold has not yet been found, additional research needs to be done to utilize these promising SFF techniques in the manufacturing of a successful bone tissue engineering scaffold.

References

1. S. J. Hollister, *Nat. Mater.*, **4**(7), 518 (2005).
2. K. F. Leong, C. M. Cheah, and C. K. Chua, *Biomaterials*, **24**(13), 2363 (2003).
3. M. N. Cooke, J. P. Fisher, D. Dean, C. Rimnac, and A. G. Mikos, *J. Biomed. Mater. Res. Part B-Appl. Biomater.*, **64B**(2), 65 (2003).

4. K. W. Lee, S. F. Wang, B. C. Fox, E. L. Ritman, M. J. Yaszemski, and L. C. Lu, *Biomacromolecules*, **8(4)**, 1077 (2007).
5. T. D. Roy, J. L. Simon, J. L. Ricci, E. D. Rekow, V. P. Thompson, and J. R. Parsons, *J. Biomed. Mater. Res. Part A*, **66A**(2), 283 (2003).
6. J. M. Williams, A. Adewunmi, R. M. Schek, C. L. Flanagan, P. H. Krebsbach, S. E. Feinberg, S. J. Hollister, and S. Das, *Biomaterials*, **26**(23), 4817 (2005).
7. T. Cao, K. H. Ho, and S. H. Teoh, *Tissue Eng.*, **9**, S103 (2003).
8. D. W. Hutmacher, T. Schantz, I. Zein, K. W. Ng, S. H. Teoh, and K. C. Tan, *J. Biomed. Mater. Res.*, **55**(2), 203(2001).
9. C. T. Laurencin, A. M. Ambrosio, M. D. Borden, and J. A. Cooper, Jr., *Annu. Rev. Biomed. Eng.*, **1**, 19 (1999).
10. S. Gitelis and P. Saiz, *J. Am. Coll. Surg.*, **194**(6), 788 (2002).
11. A. S. Mistry and A. G. Mikos, *Adv. Biochem. Eng. Biotechnol.*, **94**, 1 (2005).
12. R. Langer, *Tissue Eng. Mol. Ther.*, **1**(1), 12 (2000).
13. E. Alsberg, E. E. Hill and D. J. Mooney, *Crit. Rev. Oral Biol. Med.*, **12**(1), 64 (2001).
14. S. F. Yang, K. F. Leong, Z. H. Du, and C. K. Chua, *Tissue Eng.*, **7**(6), 679 (2001).
15. D. W. Hutmacher, M. Sittinger and M. V. Risbud, *Trends Biotechnol.*, **22**(7), 354 (2004).
16. S. J. Hollister, R. D. Maddox and J. M. Taboas, *Biomaterials*, **23**(20), 4095 (2002).
17. H. Anderl, D. Z. Nedden, W. Muhlbauer, K. Twerdy, E. Zanon, K. Wicke, and R. Knapp, *Br. J. Plastic. Surg.*, **47**(1), 60 (1994).
18. E. N. Antonov, V. N. Bagratashvili, S. M. Howdle, A. N. Konovalov, V. K. Popov, and V. Y. Panchenko, *Laser Phys.*, **16**(5), 774 (2006).
19. D. T. Pham and R. S. Gault, *Int. J. Mach. Tools Manuf.*, **38**(10–11), 1257 (1998).
20. P. J. Bartolo, H. A. Almeida, R. A. Rezende, T. Laoui, and B. Bidanda, in *Virtual Prototyping and Bio Manufacturing in Medical Applications,* Ed. B. Bidanda and P. J. Bartolo (Springer, New York, 2008), p. 149.
21. D. W. Hutmacher, T. Woodfield, P. Dalton, and J. Lewis, C. A. Van Blitterswijk, in *Tissue Engineering*, Ed. C. A. Van Blitterswijk, (Elsevier, London, 2008), p. 403.
22. J. S. Bill, J. F. Reuther, W. Dittmann, N. Kubler, J. L. Meier, H. Pistner, and G. Wittenberg, *Int. J. Oral Maxillofac. Surg.*, **24**(1), 98 (1995).

23. J. P. Fisher, J. W. M. Vehof, D. Dean, J. P. C. M. van der Waerden, T. A. Holland, A. G. Mikos, and J. A. Jansen, *J. Biomed. Mater. Res.*, **59**(3), 547 (2002).
24. P. X. Lan, J. W. Lee, Y. J. Seol, and D. W. Cho, *J. Mater. Sci.-Mater. Med.*, **20**(1), 271 (2009).
25. J. W. Lee, P. X. Lan, B. Kim, G. Lim, and D. W. Cho, *Microelectr. Eng.*, **84**(5–8), 1702 (2007).
26. J. W. Lee, P. X. Lan, B. Kim, G. Lim, and D. W. Cho, *J. Biomed. Mater. Res. Part B-Appl.Biomater.*, **87B**(1), 1 (2008).
27. M. D. Timmer, C. G. Ambrose and A. G. Mikos, *J. Biomed. Mater. Res. Part A*, **66A**(4), 811 (2003).
28. J. P. Fisher, T. A. Holland, D. Dean, P. S. Engel, and A. G. Mikos, *J. Biomater. Sci. Polym. Ed.*, **12**(6), 673 (2001).
29. J. P. Fisher, T. A. Holland, D. Dean, and A. G. Mikos, *Biomacromolecules*, **4**(5), 1335 (2003).
30. J. P. Fisher, T. A. Holland, D. Dean, P. S. Engel, and A. G. Mikos, *J. Biomater. Sci. Polym. Ed.*, **12**(6), 673 (2001).
31. F. K. Kasper, K. Tanahashi, J. P. Fisher, and A. G. Mikos, *Nat. Protoc.*, **4**(4), 518 (2009).
32. M. W. Betz, D. M. Yoon, and J. P. Fisher, in *Topics in Bone Biology*, Eds. F. Bronner M. C. Farach-Carson, and A. G. Mikos (Springer, New York, 2006), p. 81.
33. D. M. Yoon and J. P. Fisher, Polymeric scaffolds for tissue engineering applications, In *CRC's Biomedical Engineering Handbook: Tissue Engineering and Artificial Organs,* Ed. J. D. Bronzino (CRC Press, Boca Raton, 2006), p. 37-1.
34. T. M. G. Chu, J. W. Halloran, S. J. Hollister, and S. E. Feinberg, *J. Mater. Sci.-Mater. Med.*, **12**(6), 471 (2001).
35. T. M. G. Chu, S. J. Hollister, J. W. Halloran, S. E. Feinberg, *Rep. Med.: Growing Tissues Organs*, **961**, 114 (2002).
36. T. M. G. Chu, D. G. Orton, S. J. Hollister, S. E. Feinberg, and J. W. Halloran, *Biomaterials*, **23**(5), 1283 (2002).
37. B. Dhariwala, E. Hunt and T. Boland, *Tissue Eng.*, **10**(9–10), 1316 (2004).
38. J. Jansen, F. P. W. Melchels, D. W. Grijpma, and J. Feijen, *Biomacromolecules*, **10**(2), 214 (2009).
39. L. D. Carbonare and S. Giannini, *J. Endocrinol. Invest.*, **27**(1), 99 (2004).
40. C. H. Turner, *Osteopor. Int.*, **13**(2), 97 (2002).

41. P. C. Bessa, M. Casal and R. L. Reis, *J. Tissue Eng. Regen. Med.*, **2**(2–3), 81 (2008).
42. M. W. Betz, J. F. Caccamese, D. P. Coletti, J. J. Sauk, and J. P. Fisher, *J. Biomed. Mater. Res. A*, 2008.
43. D. Dean, *et al*, *Tissue Eng.*, **11**(5–6), 923 (2005).
44. E. Sachs, M. Cima, P. Williams, D. Brancazio, and J. Cornie, *J. Eng. for Ind.-Trans. Asme*, **114**(4), 481 (1992).
45. D. K. Dimitrov, Schreve and N. de Beer, *Rapid Prototyp. J.*, **12**(3), 136 (2006).
46. A. Park, B. Wu and L. G. Griffith, *J. Biomater. Sci. Polym. Ed.*, **9**(2), 89 (1998).
47. M. Lee, J. C. Y. Dunn and B. M. Wu, *Biomaterials*, **26**(20), 4281 (2005).
48. B. Leukers, H. Gulkan, S. H. Irsen, S. Milz, C. Tille, M. Schieker, and H. Seitz, *J. Mater. Sci.-Mater. Med.*, **16**(12), 1121 (2005).
49. H. Seitz, W. Rieder, S. Irsen, B. Leukers, and C. Tille, *J. Biomed. Mater. Res. Part B-Appl. Biomater.*, **74B**(2), 782 (2005).
50. J. Zeltinger, J. K. Sherwood, D. A. Graham, R. Mueller, and L. G. Griffith, *Tissue Eng.*, **7**(5), 557 (2001).
51. K. W. Lee, S. F. Wang, L. C. Lu, E. Jabbari, B. L. Currier, and M. J. Yaszemski, *Tissue Eng.*, **12**(10), 2801 (2006).
52. C. X. F. Lam, X. M. Mo, S. H. Teoh, and D. W. Hutmacher *Mater. Sci. Eng. C-Biomimetic Supramol. Syst.*, **20**(1–2), 49 (2002).
53. V. Karageorgiou and D. Kaplan, *Biomaterials*, **26**(27), 5474 (2005).
54. J. M. Taboas, R. D. Maddox, P. H. Krebsbach, and S. J. Hollister, *Biomaterials*, **24**(1), 181 (2003).
55. I. Martin, D. Wendt and M. Heberer, *Trends Biotechnol.*, **22**(2), 80 (2004).
56. T. D. Roy, J. L. Simon, J. L. Ricci, E. D. Rekow, V. P. Thompson, and J. R. Parsons, *J. Biomed. Mater. Res. Part A*, **67A**(4), 1228 (2003).
57. L. Ciocca, F. De Crescenzio, M. Fantini, and R. Scotti, *Computer. Med. Imaging Graphics*, **33**(1), 58 (2009).
58. J. L. Simon, E. D. Rekow, V. P. Thompson, H. Beam, J. L. Ricci, and J. R. Parsons, *J. Biomed. Mater. Res. Part A*, **85A**(2), 371 (2008).
59. R. A. Giordano, B. M. Wu, S. W. Borland, L. G. Cima, E. M. Sachs, and M. J. Cima, *J. Biomater. Sci. Polym. Ed.*, **8**(1), 63 (1996).
60. S. F. Yang, K. F. Leong, Z. H. Du, and C. K. Chua, *Tissue Eng.*, **8**(1), 1 (2002).

61. R. L. Simpson, F. E. Wiria, A. A. Amis, C. K. Chua, K. F. Leong, U. N. Hansen, M. Chandraselkaran, and M. W. Lee, *J. Biomed. Mater. Res. Part B-Appl. Biomater.*, **84B**(1), 17 (2008).
62. W. Y. Zhou, S. H. Lee, M. Wang, W. L. Cheung, and W. Y. Ip, *J. Mater. Sci.-Mater. Med.*, **19**(7), 2535 (2008).
63. B. Partee, S. J. Hollister and S. Das, *J. Manuf. Sci. Eng.-Trans.Asme*, **128**(2), 531 (2006).
64. C. K. Chua, K. F. Leong, K. H. Tan, F. E. Wiria, and C. M. Cheah, *J. Mater. Sci.-Mater. Med.*, **15**(10), 1113 (2004).
65. F. E. Wiria, K. F. Leong, C. K. Chua, and Y. Liu, *Acta Biomaterialia*, **3**(1), 1 (2007).
66. K. H. Tan, C. K. Chua, K. F. Leong, C. M. Cheah, W. S. Gui, W. S. Tan, and F. E. Wiria,, *Biomed. Mater. Eng.*, **15**(1–2), 113 (2005).
67. C. E. Wilson, J. D. de Bruijn, C. A. van Blitterswijk, A. J. Verbout, and W. J. A. Dhert, *J. Biomed. Mater. Res. Part A*, **68A**(1), 123 (2004).
68. S. Das, S. J. Hollister, C. Flanagan, A. Adewunmi, K. Bark, C. Chen, K. Ramaswamy, D. Rose, and E. Widjaja, *Rapid Prototyp. J.*, **9**(1), 43 (2003).
69. K. H. Tan, C. K. Chua, K. F. Leong, C. M. Cheah, P. Cheang, M. S. Abu Bakar, and S. W. Cha, *Biomaterials*, **24**(18), 3115 (2003).
70. I. Gibson and D. P. Shi, *Rapid Prototyp. J.*, **3**(4), 129 (1997).
71. D. W. Hutmacher, *Biomaterials*, **21**(24), 2529 (2000).
72. F. E. Wiria, C. K. Chua, K. F. Leong, Z. Y. Quah, M. Chandrasekaran, and M. W. Lee, *J. Mater. Sci.-Mater. Med.*, **19**(3), 989 (2008).
73. D. Rohner, D. W. Hutmacher, T. K. Cheng, M. Oberholzer, and B. Hammer, *J. Biomed. Mater. Res. Part B-Appl. Biomater.*, **66B**(2), 574 (2003).
74. S. J. Kalita, S. Bose, H. L. Hosick, and A. Bandyopadhyay, *Mater. Sci. Eng. C-Biomimetic Supramol. Syst.*, **23**(5), 611 (2003).
75. S. H. Hsu, H. J. Yen, C. S. Tseng, C. S. Cheng, and C. L. Tsai, *J. Biomed. Mater. Res. Part B-Appl. Biomater.*, **80B**(2), 519 (2007).
76. H. Chim, D. W. Hutmacher, A. M. Chou, A. L. Oliveira, R. L. Reis, T. C. Lim, and J. T. Schantz, *International J. Oral Maxillofac. Surg.*, **35**(10), 928 (2006).
77. I. Zein, D. W. Hutmacher, K. C. Tan, and S. H. Teoh, *Biomaterials*, **23**(4), 1169 (2002).
78. J. T. Schantz, A. Brandwood, D. W. Hutmacher, H. L. Khor, and K. Bittner, *J. Mater. Sci.-Mater. Med.*, **16**(9), 807 (2005).
79. B. Rai, S. H. Teoh, K. H. Ho, D. W. Hutmacher, T. Cao, F. Chen, and K. Yacob, *Biomaterials*, **25**(24), 5499 (2004).

80. B. Rai, S. H. Teoh, D. W. Hutmacher, T. Cao, and K. H. Ho, *Biomaterials*, **26**(17), 3739 (2005).
81. W. Swieszkowski, B. H. S. Tuan, K. J. Kurzydlowski, and D. W. Hutmacher, *Biomol. Eng.*, **24**(5), 489 (2007).
82. Z. Xiong, Y. N. Yan, R. J. Zhang, and L. Sun, *Scripta Materialia*, **45**(7), 773 (2001).

Chapter 26

CONSIDERATIONS ON THE STRUCTURE OF BIOMATERIALS FOR SOFT- AND HARD-TISSUE ENGINEERING

Hideaki Kagami,[a] Hideki Agata,[a] Makoto Satake,[b] and Yuji Narita[c]

[a] *Tissue Engineering Research Group, Division of Molecular Therapy, Advanced Clinical Research Center, the Institute of Medical Science, the University of Tokyo 4-6-1, Shirokanedai, Minato-ku, Tokyo, 108-8639, Japan*

[b] *Integrative Technology Research Institute, Teijin Limited, Hino, Tokyo, Japan*

[c] *Department of Cardiac Surgery, Nagoya University, Graduate School of Medicine, Nagoya, Aichi, Japan*

kagami@ims.u-tokyo.ac.jp; agata@ims.u-tokyo.ac.jp; ma.satake@teijin.co.jp; ynarita@med.nagoya-u.ac.jp

In the field of tissue engineering, various novel biomaterials have been developed in order to achieve regeneration of target tissues. Scaffolds for a specific tissue require specific characteristics such as mechanical properties, biocompatibility, and degeneration period. In terms of mechanical properties, one of the ultimate goals of these materials is to mimic naturally existing scaffolds such as tissue matrices. However, the physiological nature of naturally existing tissues is still difficult to mimic by most artificially generated biomaterials. Controlling the degeneration period is

Handbook of Intelligent Scaffolds for Tissue Engineering and Regenerative Medicine
Edited by Gilson Khang

www.panstanford.com

also an important research interest for scaffold development. The degradation process *in vivo* is quite different from that *in vitro*. Furthermore, the degradation period is important, not only for the smooth replacement with regenerating tissue, but also for the long-term stability of the regenerated tissue. This idea is especially important for hard-tissue engineering, such as bone. In this chapter, the required properties of scaffolds for soft- and hard-tissue engineering are introduced and are intended to be a reference for designing novel scaffolds. There is a broad range of target tissues, so this chapter focuses on small-caliber blood vessels as an example of soft-tissue engineering and bone as an example of hard-tissue engineering. From a clinical point of view, several key features of scaffolds for these tissues are discussed and possible approaches to achieve these requirements are reviewed.

26.1 Introduction

To achieve efficient tissue engineering, various novel biomaterials and processing procedures have been developed. Recent developments in materials science have exponentially increased, and novel biomaterials have been introduced almost every year as a result. Scaffolds for tissue engineering should be designed from a clinical point of view since the ultimate goal of the materials is for clinical applications. Effective communication between material scientists and clinical doctors is one of the key factors for the development of successful biomaterials. Mechanical properties, biocompatibility, and degeneration period are other factors to be considered, which should also fulfill the clinical application requirements. In this chapter, we focus on two tissues, small-caliber blood vessels and bone. Through our preclinical *in vivo* and clinical studies, several key features of scaffold materials for these tissues are discussed, which aim to fill the gap between material scientists and clinical doctors.

Small-caliber blood vessels were selected as an example of soft-tissue engineering since generation of competent scaffolds for small-caliber blood vessels has been a challenge.[1] Despite enormous efforts to mimic natural blood vessels, the nature of the currently

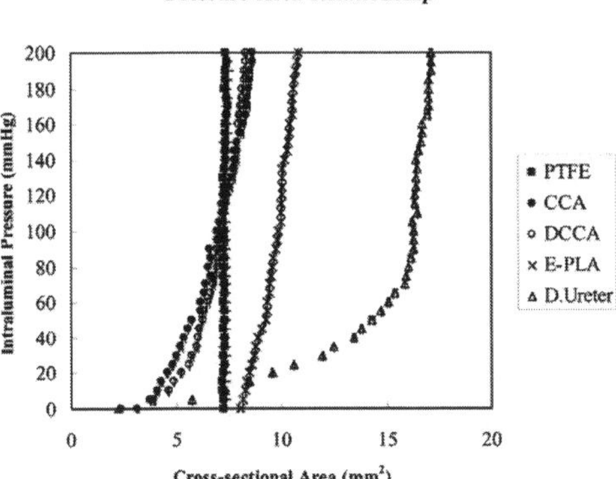

Figure 26.1. Pressure cross-sectional area relationship of each material. Polytetrafluoroethylene (PTFE) (■) showed a noncompliant response and almost no change in the cross-sectional area in response to intraluminal pressure. Canine common carotid artery (CCA) (•) and decellularized CCA (DCCA) (○) exhibited compliant responses, in a so-called J-shaped curve. Elastin gel combined with polylactic acid nanofiber tube (E-PLA) (×) evidenced cross-sectional area changes in response to an increase of intraluminal pressure, although the differences were relatively small compared with that of CCA, DCCA, or decellularized canine ureter (D.Ureter) D.Ureter (Δ) also showed a compliant response marked by more distensibility at a lower pressure range than CCA. and DCCA. Cross-sectional areas are expressed as mean values ± standard deviations (from Ref. 2, Murase *et al.*, 2006).

available materials is still not satisfactory (Fig. 26.1).[2] In order to overcome these difficulties, various approaches, including our own, are described.

Bone tissue engineering has been widely tested in the fields of orthopedic and oral surgery and has already been applied in clinics. In particular, regeneration of alveolar bone for dental implant surgery has attracted much attention since only a small volume of regenerated bone is required in most of the cases, which is advantageous in tissue engineering, and there is an increasing demand by elderly patients for dental implant surgery. Currently, animal-derived bioartificial bone substitutes have been widely used, but risks such as infection and immunoreactions against animal-derived proteins have not been completely eliminated. Although the use of artificial bone substitutes has increased, animal-derived artificial bone may not be accepted as an equal alternative of autologous bone by most clinical doctors at present. Accordingly, the development of alternatives such as tissue-engineered bone is of immediate importance. From recent clinical studies, knowledge about the clinical fate of biomaterials has accumulated.[3–6] In this chapter, generation of scaffolds for bone tissue engineering, as an example of hard tissue, will also be considered. The required and favorable nature of scaffold materials for bone tissue engineering is discussed with reference to those findings obtained from recent clinical studies including at our own facility.

26.2 Material Design for Soft-Tissue Engineering: Small-Caliber Vascular Grafts

26.2.1 *Tissue Engineering for Cardiovascular Surgery*

Artificial organs, including vascular graft prostheses, have been investigated for more than 100 years in the field of cardiovascular surgery. Although state-of-the-art cardiovascular prostheses show increased patient survival rate, have minimized morbidity, and improved patient quality of life, currently available cardiovascular prostheses are still far from perfect due to immunogenicity (foreign-body reaction), thrombogenicity, lack of long-term durability, low patency rate of small-caliber prostheses, and growth inability. Tissue-engineered vascular grafts are a new promising alternative since they are expected to generate viable autologous tissue so that these shortcomings might be overcome. Despite such expectations, development of tissue-engineered vascular grafts is still in its early phase largely due to the lack of suitable scaffold materials. In general, two different scaffold materials have been used: (i) decellularized (natural) tissue matrix scaffolds and (ii) biodegradable synthetic polymer scaffolds. The latter can also be used as a slow-release device for various growth factors/cytokines since they are biodegradable.[7] In this section, we focus on tissue-engineered, small-caliber vascular grafts, which are one of the most challenging targets for tissue engineering. In the last section, the prospects of scaffolds for cardiovascular tissue engineering are also discussed.

26.2.2 *Decellularized Tissue Scaffolds for Tissue-Engineered Small-Caliber Vascular Grafts: Methodology, Biocompatibility, and Mechanical Properties*

A decellularized technique has been developed to reduce immunogenicity and/or calcification of bioprostheses after implantation.[8,9] Advantages of using decellularized matrix for tissue engineering have been reported by many researchers.[10] The most favorable advantage of the decellularized scaffold is that it can be used to generate a completely identical structure of the target organ/tissue even if the structure is extremely complicated.

Various cell extraction methods, including detergent treatments, enzymatic digestion, and a combination of detergent and enzyme, have been reported for decellularization. Triton X-100,[11] sodium deoxycholate,[12] and sodium dodecyl sulfate (SDS)[13] are used as detergents, and trypsin[14] is used as an enzyme to disassociate the cells. As a unique method for decellularization, Fujisato *et al.* reported the use of ultrahigh pressure.[15] Each method has advantages and disadvantages, and optimization is necessary for each organ/tissue. For example, we found that esophagus treated with sodium deoxycholate showed superior mechanical properties, maintenance of extracellular matrix (ECM), and lower DNA content than that treated with Triton X-100.[16] Conversely, the degree of decellularization and maintenance of the matrix were best in the Triton X-100–treated ureters, while Triton X-100–treated and sodium deoxycholate–treated ureters had lower remnant DNA content and immunogenicity than the other treatments[17] (Fig. 26.2).

In vitro and *in vivo* biocompatibility of the decellularized scaffold was satisfactory (Fig. 26.3). Endothelial cells were easy to seed, and they adhered to the inner surface of the decellularized scaffold. The seeded cells functioned *in vivo* because cell-seeded grafts were found to have anti-thrombogenicity. Biocompatibility of the decellularized scaffolds depends on the native ECM, which may contribute

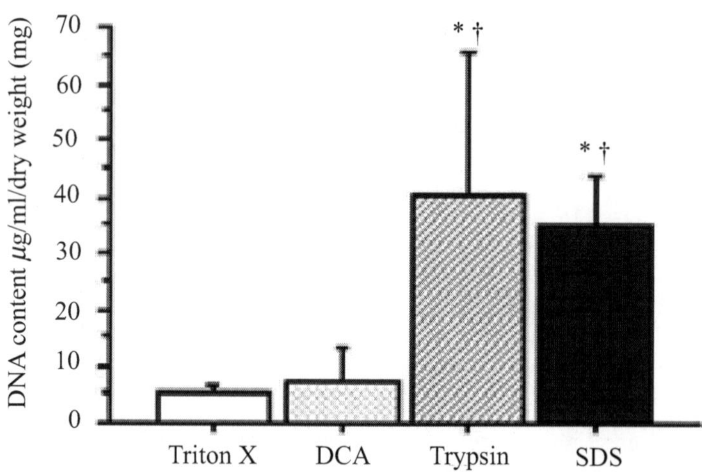

Figure 26.2. Residual DNA content of decellularized scaffolds prepared by four different methods. Triton X-100–treated and sodium deoxycholate–treated decellularized scaffolds had a significantly lower DNA content than SDS-treated and trypsin-treated scaffolds. DCA, deoxycholate. $^{*}P < 0.05$ versus Triton X, $^{\dagger}P < 0.05$ versus DCA (from Ref. 16, Narita *et al.*, 2008).

to cell adhesion and proliferation. The mechanical properties, for example, compliance, tensile strength, burst strength, and suture retention strength, of decellularized scaffolds were similar to those of the native tissue[17,18] (Fig. 26.1). Accordingly, it was suggested that cellular components may not play a major role in the mechanical properties of the decellularized scaffold but that the extracellular matrices could contribute to the mechanical strength and stiffness.

Potential shortcomings of decellularized scaffolds for small-caliber vascular grafts are long-term durability, aneurysmal formation, calcification, foreign-body (immunological) reactions due to residual allogenic or xenogenic proteins, and shortage of the donor if allogenic material is required. Sharp *et al.* demonstrated a case of aneurysmal formation using decellularized tissue matrix (SynerGraft, Cryolife) after peripheral arterial bypass surgery.[19] Other reports state that xenogenic decellularized matrix remnants contain xenogenic protein and may cause immunological foreign-body reactions.[20] Therefore, further studies will be necessary to confirm the safety and feasibility of decellularized scaffolds.

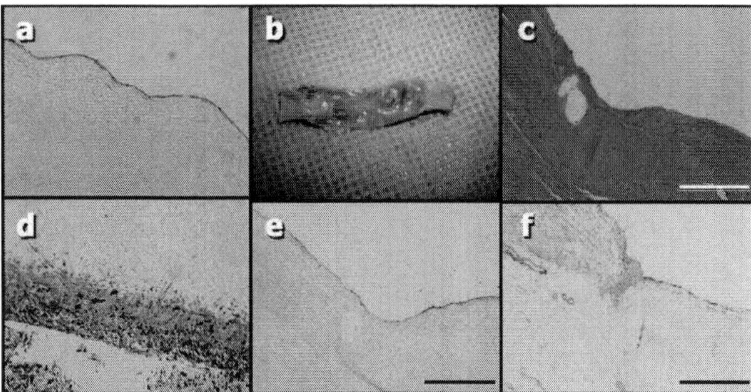

Figure 26.3. *In vitro* and *in vivo* biocompatibility of decellularized scaffolds. Endothelial cells (EC) and myofibroblasts were seeded inside and outside a decellularized ureter, respectively, and were cultured for 5 days. Immunohistochemical staining showed factor VIII–positive cells in a monolayer (a) and multilayered α-smooth muscle actin-positive cells (d) inside and outside the decellularized scaffold. Decellularized scaffolds removed 24 weeks after implantation (canine carotid arterial replacement model). The macro findings of EC-seeded grafts showed no thrombus formation and revealed a smooth and glistening surface (b). The section of the anastomotic site was stained with hematoxylin-eosin (c). Histological findings of decellularized scaffolds revealed complete re-endothelialization by immunohistochemical staining of factor VIII (e). However, the absence of a smooth muscle layer in the wall of decellularized ureters was confirmed by immunohistochemical staining with α-smooth muscle actin (f). Scale bar = 200 μm (from Ref. 17, Narita *et al.*, 2008). See also Color Insert.

26.2.3 *Biodegradable Synthetic Polymer Scaffolds for Tissue-Engineered Small-Caliber Vascular Grafts*

Scaffolds with biodegradable synthetic polymers can be manufactured artificially, which enables low-cost production compared with natural scaffolds such as decellularized scaffolds. We have developed a scaffold with electrospun nanoscaled fibers for cardiovascular tissue engineering, including small-caliber vascular grafts (Fig. 26.4). Scaffolds based on nanofibers offer great advantages for tissue engineering. They mimic the ECM (50–500 nm diameter fibers) and serve as a three-dimensional matrix for growing cells.[21–24] It is known that nanoscaled fibers also affect cellular

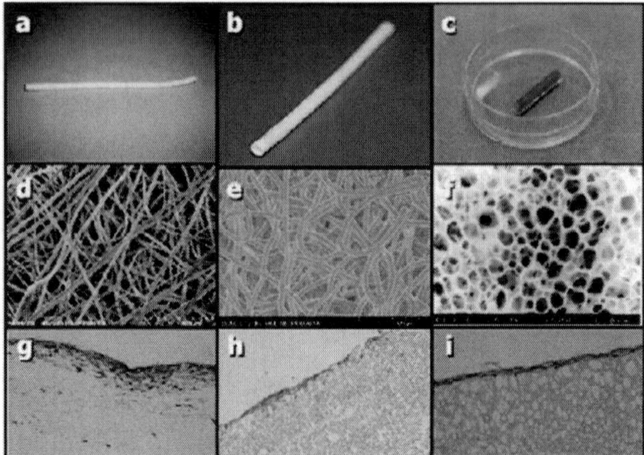

Figure 26.4. Appearance, SEM images, and biocompatibility of electrospun scaffolds with an inner diameter of 3 mm for small-caliber vascular grafts. PLA scaffolds (a, d, g), PCL scaffolds (b, e, h), and PLA with elastin scaffolds (c, f, i). SEM showed that PLA scaffolds had many crevices (d) and cells entered the wall of the scaffold (g) when stained histologically. On the other hand, PCL and PLA with elastin did not have many crevices (e, f), so the cells attached only to the surface of the scaffolds (h, i). *Abbreviations*: SEM, scanning electron microscopy; PLA, polylactic acid; PLC, poly(ε-caprolactone). See also Color Insert.

adhesion, morphology, proliferation, endocytotic activity, motility, and gene expression of various cell types.[25]

26.2.4 *How to Create Scaffolds for Tissue-Engineered Small-Diameter Vascular Grafts Using Electrospun Nanofibers*

We have generated four types of scaffolds for tissue-engineered, small-caliber vascular grafts using electrospun nanofibers.

Preparation for polylactic acid (PLA) or poly(ε-caprolactone) (PCL) tubelike scaffolds: A solution consisting of PLA or PCL and methylene chloride was loaded into a 1 mL syringe equipped with a blunt-end 18-gauge needle. The needle was clamped to the positive electrode of an electrospinning apparatus, and a voltage range of 10–15 kV was applied. The charged polymer was spun toward the circular, cylinder-like counterelectrode, which was rotating at 60 rpm. The fibrous material collected on the counterelectrode formed a tubelike structure (Fig. 26.3a,b,d,e).

Preparation for a complex of PLA and water-soluble elastin[26]: A tubelike complex of PLA and water-soluble elastin was generated. In brief, water-soluble elastin was added to deionized water. A water-soluble cross-linking agent was added to the solution and stirred for a few minutes at room temperature. Triethylamine was then added to the solution and stirred for another few minutes. The mixture was poured into a cylindrical template already installed into the PLA tube and left to stand for two days until a gel had formed. The gel was then copiously rinsed with deionized water. Finally, a milky-white cylindrical molded article, which was a complex between the PLA tube and water-soluble elastin, was obtained. The complex was sterilized by autoclaving for all subsequent experiments (Fig. 26.3c,f).

Preparation of a three-layered electrospun tube: An artery is composed of three layers: intima, media, and adventitia. The intima is composed of endothelial cells and the ECM. The media is composed of smooth muscle cells and elastic tissues. The adventitia is made up of fibroblasts and connective tissues. Since the anatomical structure of the three-layered artery may be important to support the mechanical properties of the vessel, we tried to mimic the three-layered architecture using electrospinning techniques.

An electrospun tube was fabricated as follows: First, a mixture consisting of poly(D,L,-lactide-*co*-glycolide) (PLGA; 50/50) and poly(D,L-lactide-*co*-caprolactone) (PLCA; 76/24) was spun toward the circular, cylinder-like counterelectrode, which was rotating at 60 rpm. The fibers were discharged in the air using a neutralization apparatus during development of the tube. Since the speed for fiber spinning was reduced by neutralization, the fibers lay down softly on the counterelectrode and were rolled up to create the intima layer. For the media layer, PLCA was spun in a similar manner to give an elastic texture. Finally, a mixture of PLGA and PLCA were spun to generate the outermost layer (Fig. 26.5).

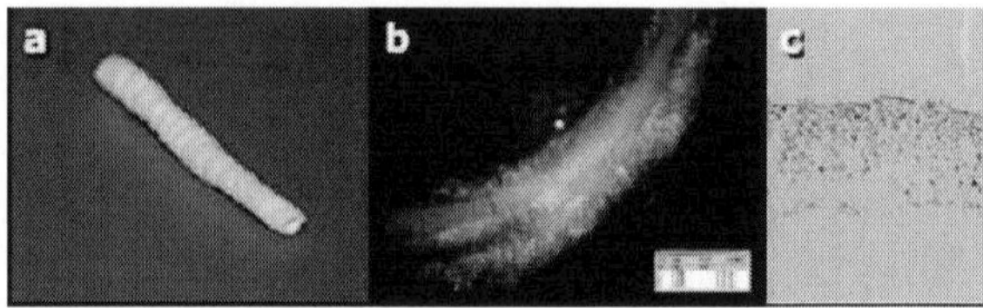

Figure 26.5. Appearance, loupe image, and biocompatibility of a three-layered electrospun scaffold. Histological findings revealed that three-layered scaffolds have good biocompatibility compared with PCL and PLA with elastin scaffolds. See also Color Insert.

26.2.5 *Biocompatibility and Mechanical Properties of Electrospun Synthetic Scaffolds*

To evaluate *in vitro* biocompatibility of electrospun nanofiber scaffolds, "cell-seeding tests" were performed. Cells were simply seeded onto the inside and the outside of the grafts and were evaluated histologically. When PLA was used as a material for electrospun nanofiber scaffolds, cells could populate the inside of the scaffold wall and form good adhesions to the scaffold (Fig. 26.3g). On the other hand, when a mixture of PCL, PLA, and elastin was used, cells attached only to the surface of the scaffold wall and did not populate the inside of the wall (Fig. 26.3h,i). The results from scanning electron microscopy (SEM) showed ultrastructural differences between the scaffolds made by these two materials (Fig. 26.3d–f). Although addition of PCL improved the mechanical strength of the scaffold (details of scaffold mechanical strength are described below), the fibers with PCL tended to attach to each other, which reduced the space between the fibers (Fig. 26.3e). Since the size of electrospun nanofibers is relatively small, this morphological change significantly affected the mobility of the cells.

Next, we hypothesized that a combination of PLA, which has good biocompatibility, and PCL, which has excellent mechanical strength, could form a satisfactory scaffold. Arteries have a three-layered structure, so we decided to create three-layered scaffolds to satisfy both mechanical strength and biocompatibility. In the case of three-layered scaffolds, the layers on both the inside and outside of the wall were made of PLA so that infiltration of the cells was satisfactory (Fig. 26.5). When generating electrospun nanofiber scaffolds,

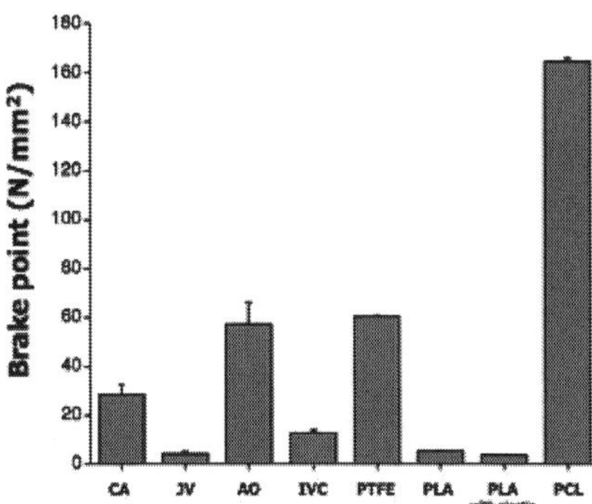

Figure 26.6. Maximum tensile strength of electrospun scaffolds and native tissues. PCL scaffolds were 3 times stronger than the aorta and PTFE. *Abbreviations*: CA, carotid artery; JV, jugular vein; AO, aorta; IVC, inferior vena cava; PTFE, polytetrafluoroethylene graft.

slight changes in materials could significantly affect nanoscaled structures. Since adhesion and infiltration of the cells depends on the structure of the scaffold, the difference in material constituents could affect biocompatibility of the electrospun nanofiber scaffold.

The tensile strength of the scaffold was measured, and the break points (maximum tensile strength) of the PLA, PLA with elastin, and PCL were compared with canine carotid artery, jugular vein, inferior vena cava, aorta, and polytetrafluoroethylene (PTFE) tubes. The tensile strength of PCL was three times that of the aorta and PTFE. However, the tensile strength of PLA and PLA with elastin was more fragile than that of the vein (Fig. 26.6).

Compliance tests were performed by our original compliance measurement system (Fig. 26.1). This system is able to observe pressure and cross-sectional diameter or area relationship of the prostheses.[2] The cross-sectional area of PLA with elastin increased in response to high-pressure load, but the response was different than the native artery. In fact, the pressure-area relation curve of PLA with elastin was approximately linear. On the other hand, PCL and PLA showed a noncompliant response and almost no cross-sectional area change, which was similar to the PTFE tube. It is important to emphasize the fact that compliance tests by our original system can measure the compliance of material in conditions similar to native blood vessels since pulsatile flow was generated, which created liquid movement inside of the scaffold. Unfortunately, most of the compliance tests were performed with a simpler system due to the lack of adequate equipment. However, our results indicate that the results cannot recapitulate the conditions *in vivo*, which may mislead evaluation of scaffold materials and development of novel materials. Compliance is an important factor in designing scaffolds for tissue-engineered, small-caliber vascular grafts since a compliance mismatch is thought to affect prognosis. Such a mismatch may cause neointimal hyperplasia at anastomotic sites, which, in turn, can lead to thrombosis and occlusion even at early stages after bypass surgery.[27]

We performed rat carotid arterial replacement using electrospun scaffolds to evaluate their function *in vivo*. Grafts were patent for 12 weeks, and no thrombus formation was observed. However, scaffold materials remained, and neither the smooth muscle layer, which is an important component of the vessel, nor regeneration of elastic fibers was observed (Fig. 26.7). When biodegradable materials

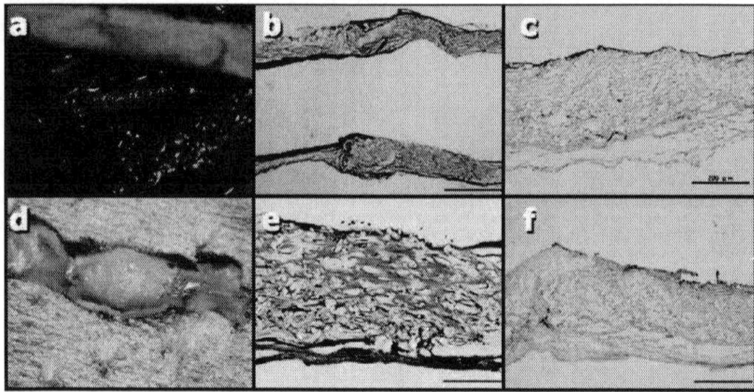

Figure 26.7. *In vivo* findings of electrospun scaffolds (PCL). Rat carotid arterial replacement was performed (a). The scaffold was removed 12 weeks after implantation. Grafts were in place for 12 weeks, and no thrombus formation was observed (d). Elastica van Gisson staining revealed that collagen was growing in the scaffold wall, but there was little elastin in the scaffold compared with native artery (b, e). Immunohistochemical staining for factor VIII showed single-layered endothelial cells covering the inside of the graft (c). However, α-smooth muscle actin staining revealed that the smooth muscle layer, which is important for vascular strength, was not regenerated (f). Scale bar = 200 μm (c and f) and 500 μm (b and e). See also Color Insert.

for vascular scaffolds are used *in vivo*, optimization of the balance of material absorption and tissue regeneration is the most important and most difficult issue. Development of biodegradable elastomers, which do not cause compliance mismatches at early stages of implantation, is required.

26.2.6 *Prospects of Designing a Scaffold for Cardiovascular Tissue Engineering*

The requirements for scaffold materials for cardiovascular tissue engineering are excellent biocompatibility, adequate mechanical properties, good texture that has easy surgical handling and optimized timing of material absorption, and tissue regeneration. To improve scaffold materials, various manipulations have been tested such as ECM coating (e.g., collagen, fibronectin), drug coating (e.g., heparin, Argatroban), growth factor linking (e.g., basic fibroblast growth factor (bFGF), vascular endothelial growth factor (VEGF), chemical treatment (e.g., argon-plasma), and peptides (e.g., the Arg-Gly-Asp [RGD] sequence). It is known that drugs can be capsulated directly into electrospun nanoscaled fibers, and these systems showed sustained release of the drug.[28,29] The use of electrospun fibers as drug carriers holds promise for future biomedical applications, especially scaffolds for tissue engineering. We developed a novel, controlled drug delivery device to prevent anastomotic stricture using electrospinning nanofibers, which were composed of biodegradable polymers and Tacrolimus.[25] In the future, intellectually multifunctional scaffolds may be developed for cardiovascular tissue grafts.

26.3 Scaffold Design for Hard-Tissue Engineering: Alveolar Bone

26.3.1 *Bone Reconstruction/Regeneration in Orthopedic and Oral Applications*

Bone defects that do not heal spontaneously require bone reconstruction. Although autologous bone grafting is the current gold standard for reconstruction of relatively large bone defects, autografts may cause donor site morbidity.[30] To avoid morbidity, several types of ceramic-based bone substitutes, which have similar mineral compositions to natural bone tissue, have been developed. Among them, synthetic hydroxyapatite (HA) and tricalcium phosphate (TCP) are the most commonly used ceramics as alternatives to autologous bone.[31] Block-type ceramics have proven useful for reconstruction of segmental bone defects, which are the major targets in orthopedic applications.[32] In contrast, relatively small and complicated-shape bone defects (alveolar bone) are the major targets in oral applications.[33] Since block-type bone substitutes are difficult to apply for small and complicated-shape bone defects, ceramic granules have been used. However, reconstruction of these defects with granules of HA and TCP is more difficult than block-type materials and inefficient partly due to the geometry of the substratum, which will directly affect the osteoconductive properties of the materials.[34] Accordingly, granules of natural materials such as allogenic/xenogenic freeze-dried bones are preferable for use in alveolar bone regeneration since these natural materials have better osteoconductive properties than synthetic HA and TCP.[33,35] However, use of allogenic/xenogenic bones cannot completely eliminate possible contamination risk (e.g., prions, viruses, zoonosis) or potential for immunological reactions against allogenic/xenogenic proteins.

Tissue engineering is an interdisciplinary approach to regenerate tissue through integration of cell biology and biomaterial/biomedical sciences. Since one of the concepts for tissue engineering is based on enhancement of the natural healing process of tissues, it can be considered an ideal therapeutic option for treating various tissue defects. For bone regeneration, tissue engineering is considered less invasive than autologous bone transplantation and more efficient than artificial bone substitutes.[36] Accordingly, bone tissue engineering has been studied extensively and has even reached the stage of clinical application in some facilities, including our own. Among various target diseases for bone tissue engineering, application to alveolar bone (oral application) has attracted much attention because currently available granule-type bone substitutes for use in alveolar bone regeneration do not possess osteoinductive properties. Furthermore, there is an increasing demand for bone regeneration in patients who undergo dental implant placement. Our group has investigated the safety and efficacy of alveolar bone tissue engineering using autologous bone marrow stromal cells (BMSCs) and β-TCP granules as a scaffold. The interim results of this clinical study showed the feasibility of alveolar bone tissue engineering (Asahina *et al.*, manuscript under review). This clinical study also provided us information about the fate of transplanted scaffolds in the human body, which should help us to consider which scaffolds are truly useful for future bone tissue engineering.

26.3.2 *Ideal Ceramic Scaffolds for Bone Tissue Engineering from a Clinical Point of View*

Scaffolds for bone tissue engineering should act as a template for new bone growth and are expected to be eventually replaced by autologous bone tissue.[31] Therefore, the biodegradable properties of ceramics are of substantial importance when ceramics are applied for bone tissue engineering (Table 26.1). In this respect,

Table 26.1. Scaffolds used in current clinical trials are described.

Currently used ceramic-scaffold for cell-based bone tissue engineering		
Target tissue/area	Scaffold	References
large bone diaphysis defect	HA block (porosity:60% or 80% pore size: 613 or 430 μm)	marcacci et al.[44]
Upper and lower jaw bone defects	HA particles (porosity: 65%: pore size:?)	meijer et al.[45]
Infrabony periodontal bone defects	HA granules (porosity: ?; pore size:?)	yamamiya et al.[45]
Sinus floor augmentation	60%HA/40%TCP cubes (porosity:?; pore size: 300–500 μm)	Shayeteh et al.[46]
Maxillary defects	β-TCP granules (porosity: 65%; pore size:?)	Mesimaki et al.[46]
Femoral head defects	β-TCP granules (porosity: 75%; pore size: 100–400μm)	Kawate et al.[47]

synthetic HA is not an ideal ceramic for bone tissue engineering since HA does not degrade properly in the human body.[32] In contrast, TCP is a degradable calcium phosphate ceramic; thus, TCP is considered to be a better ceramic scaffold for bone tissue engineering. However, the osteoconductive properties of TCP are known to be less than those of HA.[37] In order to improve the osteoconductive properties of TCP, several biphasic calcium phosphate ceramics (HA/TCP) have been developed.[37,38] Although the optimal composition of HA/TCP remains controversial, this approach seems promising to develop a more reliable ceramic scaffold for bone tissue engineering.

Porosity of the ceramic scaffold has a great influence on the efficacy of bone tissue engineering since it directly affects cell adhesion, migration, and proliferation of the osteogenic cells.[31,39,40] Anatomically, cortical bone has 3%–12% porosity, while trabecular bone has porosity in the range of 50%–90%.[41] Since the primary target of bone tissue engineering is the regeneration of trabecular bone and ceramic scaffolds should have enough strength to provide physical support for the cells, 65%–75% porosity might be ideal for ceramic scaffolds. Pore diameter is also known to influence cell migration, proliferation, and eventually the ability to support bone regeneration. The minimum pore size to regenerate bone is generally considered to be 50–100 μm.[38,40] Accordingly, β-TCP granules, which have 75% porosity and a pore size ranging from 100 to 400 μm, were utilized in our clinical trials. Interconnectivity of the pore and geometry of the scaffold are also known to influence the efficacy of bone tissue engineering, and these factors might cause a difference in osteoconductive properties between block- and granule-type ceramic scaffolds.[34] However, the influence of these factors on osteogenic cells has not been well investigated.

Although the development of novel biomaterials is a rapidly expanding area of science, basic understanding of cell-to-material interactions should be carefully considered to develop an ideal ceramic scaffold for bone tissue engineering. From a clinical point of view, degradability, composition, porosity, pore size, interconnection of the pore, and geometry are factors that need to be considered.

26.3.3 *Fate of Transplanted Scaffolds in the Human Body: A Clinical Study of Alveolar Bone Tissue Engineering Using Bone Marrow Stromal Cells and β-TCP*

Among the biomaterials used for bone tissue engineering, the fate of ceramic-based materials has been relatively well described since these materials can be used alone as bioartificial bone substitutes such as β-TCP.[3–6] However, the degradation process of β-TCP *in vivo* is much different from that *in vitro*. Transplants are immersed in an aqueous solution with ions and enzymes at various pHs. Furthermore, cells such as osteoclasts are known to interact with the materials and play some role during the degradation process. Biodegradation of TCP is considered to be mediated in two different ways: (1) resorption by osteoclasts and (2) dissolution by interstitial fluid.[4] Interestingly, when TCP was transplanted alone without cells, Zerbo *et al.* reported that osteoclastic activity did not precede osteogenic activity and most of the degradation may have happened as a result of a local drop in pH because of the production of acidic by-products and poorly developed vascularlization in the regenerating tissue.[4]

Ceramic-based materials have been used not only as bioartificial bone substitutes, but also as scaffolds for bone tissue engineering. Since only a limited number of publications are available on clinical bone tissue engineering, information about the fate of β-TCP as a scaffold is rather limited. However, cell-to-material interactions should play important roles during the degradation process of the scaffolds since the materials are transplanted with cells.

We have performed a clinical study of alveolar bone regeneration using autologous BMSCs and β-TCP granules as a scaffold. BMSCs were harvested from the iliac crest under local anesthesia and cultured with α-minimum essential medium (MEM) containing 10% autoserum and antibiotic/antimicotic reagents (Fig. 26.8). Nonadherent cells were discarded. After osteogenic induction for seven days, adherent cells were detached from the flasks and suspended in platelet-rich plasma (PRP), which was turned into a gel using autologous thrombin. The gel was then mixed with β-TCP granules as a scaffold and transplanted to the sinus floor and alveolar ridge to obtain enough bone volume to support the dental implant.

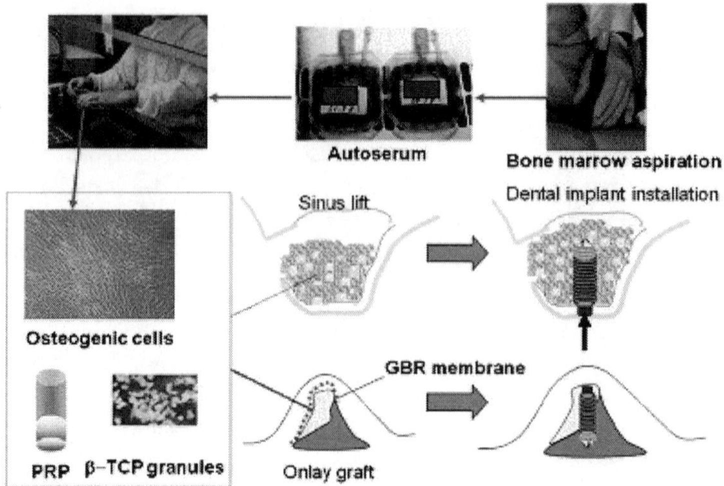

Figure 26.8. Schematic figure showing the procedure for alveolar bone tissue engineering. BMSCs were harvested from the iliac crest under local anesthesia and cultured in a cell culture facility. Nonadherent cells were discarded. At the time of surgery, cells were detached from flasks and suspended in PRP, which was turned into a gel using autologous thrombin. The gel was then mixed with β-TCP granules as a scaffold and transplanted into the sinus floor and/or alveolar ridge. See also Color Insert.

The results showed that bone regeneration using autologous BMSC-derived osteogenic cells was feasible (Asahina *et al.*, manuscript under review).

In this study, six months after cell transplantation, bone biopsies were performed using a trephine bur at the site of implant placement. Histology of the regenerated bone was analyzed. Newly formed bone was observed adjacent to the scaffold as well as between the scaffolds (Fig. 26.10). The available tissue sample from patients was from only one time point, so the time course of scaffold degradation could not be analyzed. However, there might be two differential types of scaffold degradation, as reported previously.[4] When the newly formed bone was adjacent to the scaffolds, it presented as brushlike borders, which may support the idea that β-TCP granules had started to degrade due to resorption by osteoclastic cells prior to bone regeneration (Fig. 26.9). Some of the scaffold seemed like it was degraded spontaneously but not

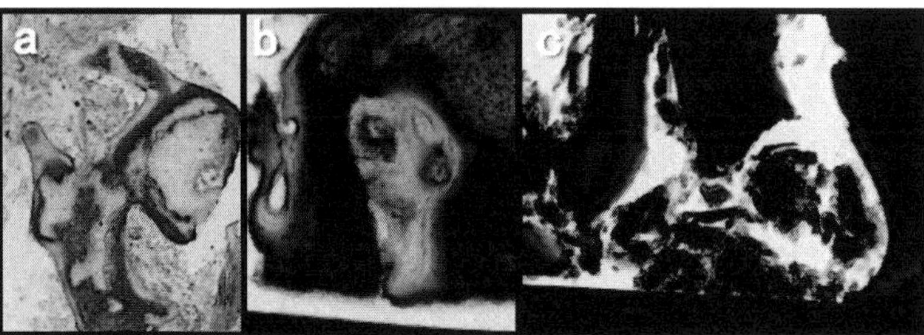

Figure 26.9. Histology of regenerated bone. Six months after cell transplantation, a bone biopsy was performed using a trephine bur at the site of implant placement. Nondecalcified tissue sections were cut, and the sections were stained with Villanueva-Goldner stain. Newly formed bone was observed adjacent to the scaffold, and the scaffolds seemed like they were absorbed by osteoclastic cells (a, b). Bone also formed between the scaffolds. In these cases, it seemed like the scaffolds were degraded spontaneously (c). See also Color Insert.

actively by osteoclastic cells since cells were not observed close to the degrading scaffold (Fig. 26.9). Actual tissues display a combination of two types of degradation. Compared with the transplantation of bone substitute alone, the presence of new bone adjacent to the scaffold was more frequent, which may reflect the role of transplanted cells. However, quantitative analyses were not performed, and these observations require further investigation.

26.3.4 *Considerations for Designing Scaffolds for Clinical Bone Tissue Engineering*

Factors to consider for successful scaffold materials include biocompatibility, degradation time, and mechanical properties. For bone tissue engineering scaffolds, morphology is important. In terms of ceramic-based scaffolds, porous structure is important and a certain size of pore is essential for osteoconductivity, as described above. In terms of fibrous scaffolds, fiber diameter also affects the success of bone regeneration. Alveolar bone defects do not bear large physiological loads until after the implant placement. Accordingly, mechanical strength of the scaffold may not be critical for dental implants. However, mechanical strength is important for most orthopedic applications. The shape of the scaffold is also important. Bone defects in alveolar bone are small and complex in shape and so are difficult to treat with block-type scaffolds. On the other hand, segmental bone defects may require solid, block-type scaffolds to support the mechanical strength.

The process of clinical bone tissue engineering can be divided into at least three phases: generation of tissue grafts for transplantation, scaffold degradation as bone regeneration occurs, and normal remodeling of the newly formed bone. The initial phase of bone tissue engineering is the generation of tissue-engineered grafts, which includes cell seeding, culture, induction, and transplantation (Fig. 26.10). Cultured cells can be seeded directly onto the scaffold, induced into osteogenic cells, and then transplanted. Alternatively, cells cultured in flasks can be induced into osteogenic cells and detached from the flask at the time of surgery. These are then mixed with the scaffold for transplantation. For the granular-type scaffolds,

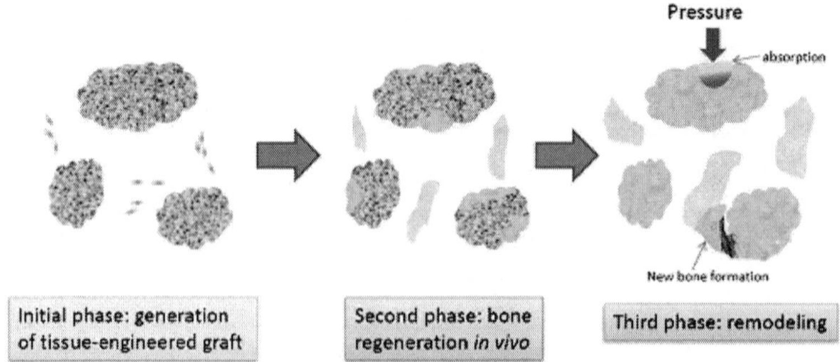

Figure 26.10. The process of clinical bone tissue engineering can be divided into at least three phases. The initial phase of bone tissue engineering is generation of tissue-engineered grafts. Autologous BMSCs are seeded on scaffolds before or after osteogenic induction. The scaffolds with cells are then transplanted. The second phase is the process of bone regeneration. During this phase, scaffolds are degraded as new bone formation occurs, which continues until the scaffolds are completely degraded. The third phase is the process of bone remodeling after bone regeneration and scaffold absorption. The volume of regenerated bone is regulated by coordinated bone resorption and formation, which are affected by various local and systemic factors such as pressure from surrounding soft tissues and physiological loads from the dental implant.

it is beneficial to culture cells on scaffolds rather than mixing them just prior to the operation.[42] Accordingly, the scaffolds should be highly biocompatible to allow cell attachment, growth, and differentiation of the cells. Since the induction period is two to three weeks, the materials should also be stable for this culture/induction period.

After cell transplantation, the scaffolds are degraded as new bone formation occurs, which continues until the scaffold is completely degraded. This is the phase of actual bone regeneration. When β-TCP granules are used, bone regeneration may continue for up to one year. During this stage, the scaffold seems to be directly degraded by resorption by osteoclastic cells. Compared with the dissolution by interstitial fluid, this process might be favorable since morphology of the tissue is maintained. To accelerate this active resorption process of the scaffold, the scaffold should mimic bone and be presented as a target tissue for osteoclasts. Ceramic-based materials might be advantageous since osteoclastic activities were confirmed from *in vivo* studies.

The third phase starts when the scaffold is completely degraded and the newly formed bone enters the normal remodeling process. For clinical cases, stability of the regenerated bone for a longer period is important. However, factors which affect stability are not well known. It is known that mechanical stimuli such as dental implant placement may help to maintain the volume of regenerated bone. Importantly, the shape of hard tissue should be compatible with the covering soft tissues. Long-term changes of atrophic alveolar bone shape result in a soft-tissue environment that conforms. Accordingly, immediate morphological changes after bone regeneration may lead to increased pressure on the regenerated bone and cause unexpected absorption. To prevent late absorption of the regenerated bone, the resorption of the scaffold should be long enough until the soft tissue surrounding the newly formed bone becomes stabilized. From this point of view, scaffolds with a relatively short degradation time, such as collagen, have disadvantages, and their application is limited to areas without pressure from adjacent soft tissues. Conversely, if scaffold degradation takes too long, the bone regeneration process may be delayed and the mechanical strength of the regenerating tissue may not be enough to support the dental implant. This is one of the reasons that TCP is preferred for dental implants. One ideal approach for these two differential and contradictory scaffold properties is to implement composite strategies rather than one simple scaffold.

26.3.5 *Prospective Novel Biomaterials for Hard-Tissue Engineering*

Currently used biomaterials for clinical tissue engineering have been used for many years, which is not surprising. Unfortunately, evaluation of newly developed materials takes time and the time lag between material development and clinical application cannot be avoided. However, it is logical to discuss the future of biomaterials as these may replace currently used materials. Since various novel materials are described elsewhere in this book, we have focused on several materials that may potentially overcome the current problems discussed in this chapter.

26.3.5.1 Composite and combined materials

As stated above, successful scaffolds for bone tissue engineering should have two different properties: early resorption to support immediate regeneration of bone and long-term presence to resist pressure from soft tissues and retain their original shape. Accordingly, one ideal approach is to make a composite of two or more different materials, which would play different roles during the regeneration period. A combination of natural polymers with HA and more complex forms of composite such as collagen-PLA-HA have been reported (reviewed by Hutmacher *et al.*[43]).[44–47]

Tissue engineering usually utilizes one type of cell. However, recent studies have shown the benefits of combining two different cell types. Although bone tissue engineering using BMSCs is feasible, the size of the regenerated bone is limited. Grafted cells require oxygen and nutrition to survive, and early neovascularization is considered essential for successful bone tissue engineering.[3] Development of an efficient neovascularization method to sustain engineered transplants is therefore clinically important. For this purpose, endothelial cells (EC) or endothelial progenitor cells (EPC) have been combined with BMSCs, which enhanced vascularization of the regenerated tissue.[48]

26.3.5.2 Growth factor incorporation into scaffolds

Results from clinical studies have shown that the process of bone regeneration is relatively long and may continue more than one year after cell transplantation. The time for cell transplantation is not a goal but the beginning of more complex phenomena taking place in the body, including cell-to-cell and/or cell-to-biomaterial interactions. Unfortunately, the fates of the transplanted cells and the detailed regenerating process have not been well documented, especially in humans. Transplanted cells play an important role initially, but there are likely many other factors playing roles in later stages. The transplants (transplanted cells and scaffolds) interact with host-derived cells, which affect degradation of the scaffold. The process could also be regulated by local growth factors/cytokines.

An exciting approach for novel biomaterials for bone tissue engineering is the generation of slow-release drug delivery systems for growth factors/cytokines.[49] These are released from the scaffold to establish a favorable environment for bone regeneration, rather than a permissive effect of the location. Growth factors can be directly immersed in ceramic-based scaffolds. When the scaffold is a composite of synthetic polymers, the polymer can function as a slow-release device by controlling the rate of degradation of the polymer. The factors are expected to coordinate with transplanted cells to regenerate bone. Investigations into the mechanism should facilitate this approach in the future.

26.4 Conclusions and Outlook

This article has focused on scaffold design almost exclusively from a clinical point of view, and we selected small-caliber blood vessels and alveolar bone as examples of tissue engineering. It is one of our aims to emphasize the importance of effective discussions between material scientists and clinical teams.

Recent developments in translational and clinical studies should be carefully considered in order to design scaffolds. Materials for small-caliber vascular grafts are still a challenge, and the development of novel materials, with mechanical properties similar to natural blood vessels without loss of their biocompatibility and controllable degradation time, is eagerly anticipated. Scaffold materials for bone tissue engineering have already been applied in clinical studies, although there is a need for more efficient materials. This may depend on investigation of more detailed mechanisms of bone regeneration.

Various scaffolds other than ceramic-based biomaterials are available for bone tissue engineering. Information about novel biomaterials for bone tissue engineering can be found in Parts II, III, VI, and VII and also in recent reviews.[43,50,51]

Acknowledgments

The authors would like to thank Professor Izumi Asahina for clinical studies on bone tissue engineering and Dr. Hiroaki Kaneko for developing the electrospun nanofiber scaffolds. This work was supported in part by a grant for Research on Human Genome and Tissue Engineering from the Ministry of Health, Labor and Welfare of Japan to HK, a Japanese grant-in-aid for scientific research from the Japan Society for the Promotion of Science to HK, and by a grant-in-aid from TES Holdings Co. Ltd (Tokyo, Japan). We are grateful to Olympus-Terumo Biomaterials (Tokyo, Japan) for generously providing the β-TCP granules.

References

1. L. Bordenave, P. Menu, and C. Baquey, *Expert Rev. Med. Devices*, **337** (2008).
2. Y. Murase, Y. Narita, H. Kagami, K. Miyamoto, Y. Ueda, M. Ueda, and T. Murohara, *ASAIO J.*, **450** (2006).
3. J. Wiltfang, K. A. Schlegel, S. Schultze-Mosgau, E. Nkenke, R. Zimmermann, and P. Kessler, *Clin. Oral. Impl. Res.*, **213** (2003).

4. I. R. Zerbo, A. L. J. J. Bronckers, G. de Lange, and E. H. Burger, *Biomaterials*, **1445** (2005).
5. Z. Suba, D. Takacs, D. Matusovits, J. Barabas, A. Fazekas, and G. Szabo, *Clin. Oral. Impl. Res.*, **102** (2006).
6. L. Cordaro, D. D. Bosshardt, P. Palattella, W. Rao, G. Serino, and M. Chiapasco, *Clin. Oral. Impl. Res.*, **796** (2008).
7. M. Mutsuga, Y. Narita, A. Yamawaki, M. Satake, H. Kaneko, Y. Suematsu, A. Usui, and Y Ueda, *J. Thorac. Cardiovasc. Surg.*, **703** (2009).
8. J. M. Malone. K. Brendel, R. C. Duhamel, and R. L. Reinert, *J. Vasc. Surg.*, **181** (1984).
9. S. G. Lalka, L. M. Oelker, J. M. Malone, R. C. Duhamel, M. A. Kevorkian, B. A. Raper, J. C. Nixon, K. J. Etchberger, M. C. Dalsing, and D. F. Cikrit, *Ann. Vasc. Surg.*, **108** (1989).
10. J. P. Stegemann, S. N. Kaszuba, and S. L. Rowe, *Tissue Eng.*, **2601** (2007).
11. A. Bader, T. Schilling, O. E. Teebken, G. Brandes, T. Herden, G. Steinhoff, and A. Haverich, *Eur. J. Cardio-Thorac Surg.*, **279** (1998).
12. S. Cayan, C. Chermansky, N. Schlote, N. Sekido, L. Nunes, R. Dahiya, and E. A. Tanagho, *J. Urol.*, **798** (2002).
13. C. Booth S. A. Korossis, H. E. Wilcox, K. G. Watterson, J. N. Kearney, J. Fisher, and E. Ingham, *J. Heart Valve Dis.*, **457** (2002).
14. G. Steinhoff, U. Stock, N. Karim, H. Mertsching, A. Timke, R. R. Meliss, K. Pethig, A. Haverich, and A. Bader, *Circulation*, **III-50** (2000).
15. T. Fujisato, K. Minatoya, S. Yamazaki, Y. Meng, K. Niwaya, A. Kishida, T. Nakatani, and S. Kitamura, In Eds. H. Mori and H. Matsuda, *Cardiovascular Regeneration Therapies Using Tissue Engineering Approaches* (Springer Tokyo, Tokyo, 2005), p. 83.
16. M. Ozeki, Y. Narita, H. Kagami, N. Ohmiya, A. Itoh, Y. Hirooka, Y. Niwa, M. Ueda, and H. Goto, *J. Biomed. Mater. Res. A*, **771** (2006).
17. Y. Narita, H. Kagami, H. Mastunuma, Y. Murase, M. Ueda, and Y. Ueda, *J. Artif. Organs*, **91** (2008).
18. B. S. Conklin, E. R. Richter, K. L. Kreutziger, D. S. Zhong, and C. Chen, *Med. Eng. Phys.*, **173** (2002).
19. M. A. Sharp, D. Phillips, I. Roberts, and L. Hands, *Eur. J. Vasc. Endovasc. Surg.*, **42** (2004).
20. E. Rieder, G. Seebacher, M. T. Kasimir, E. Eichmair, B. Winter, B. Dekan, E. Wolner, P. Simon, and G. Weigel, *Circulation*, **2792** (2005).
21. Y. Zhang, C. T. Lim, S. Ramakrishna, and Z. M. Huang, *J. Mater. Sci. Mater. Med.*, **933** (2005).

22. W. J. Li, C. T. Laurencin, E. J. Caterson, R. S. Tuan, and F. K. Ko, *J. Biomed. Mater. Res.*, **613** (2002).
23. S. Liao, B. Li, Z. Ma, H. Wei, C. Chan, and S. Ramakrishna, *J. Biomed. Mater.*, **R45** (2006).
24. N. Ashammakhi, A. Ndreu, L. Nikkola, I. Wimpenny, and Y. Yang, *Regen. Med.*, **547** (2008).
25. S. G. Kumbar, R. James, S. P. Nukavarapu, and C. T. Laurencin, *Biomed. Mater.*, **R45** (2008).
26. E. Kitazono, H. Kaneko, T. Miyoshi, and K. Miyamoto, *J. Synth. Org. Chem. Jpn.*, **514** (2004).
27. W. M. Abbott, J. Megerman, J. E. Hasson, G. L'Italien, and D. F. Warnock, *J. Vasc. Surg.*, **376** (1987).
28. E. R. Kenawy, G. L. Bowlin, K. Mansfield, J. Layman, D. G. Simpson, E. H. Sanders, and G. E. Wnek, *J. Control. Rel.*, **57** (2002).
29. J. Zeng, X. Xu, X. Chen, Q. Liang, X. Bian, L. Yang, and X. Jing, *J. Control. Rel.*, **227** (2003).
30. H. Agata, I. Asahina, Y. Yamazaki, M. Uchida, Y. Shinohara, M. J. Honda, H. Kagami, and M. Ueda, *J. Dent. Res.*, **79** (2007).
31. J. R. Jones, In Eds. L. Hench and J. Jones, *Biomaterials, Artificial Organs and Tissue Engineering* (Woodhead Publishing Limited, UK, 2005), p. 201.
32. M. Mastrogiacomo, A. Muraglia, V. Komlev, F. Peyrin, F. Rustichelli, A. Crovace, and R. Cancedda, *Orthod. Craniofac. Res.*, **277** (2005).
33. L. Cordaro, D. D. Bosshardt, P. Palattella, W. Rao, G. Serino, and M. Chiapasco, *Clin. Oral Implants. Res.*, **796** (2008).
34. S. P. van Eeden and U. Ripamonti, *Plast. Reconstr. Surg.*, **959** (1994).
35. Z. Schwartz, M. Goldstein, E. Raviv, A. Hirsch, D. M. Ranly, and B. D. Boyan, *Clin. Oral Implants Res.*, **204** (2007).
36. H. Agata, N. Watanabe, Y. Ishii, N. Kubo, S. Ohshima, M. Yamazaki, A. Tojo, and H. Kagami, *Biochem. Biophys. Res. Commun.*, **353** (2009).
37. T. L. Arinzeh, T. Tran, J. Mcalary, and G. Daculsi, *Biomaterials*, **3631** (2005).
38. Y. H. Hsu, I. G. Turner, and A. W. Miles, *J. Mater. Sci. Mater. Med.*, **2319** (2007).
39. P. Kasten, I. Beyen, P. Niemeyer, R. Luginbühl, M. Bohner, and W. Richter, *Acta Biomater.*, **1904** (2008).
40. J. Isaac, J. C. Hornez, D. Jian, M. Descamps, P. Hardouin, and D. Magne, *J. Biomed. Mater. Res. A*, **386** (2008).

41. G. A. Renders, L. Mulder, L. J. van Ruijven, and T. M. van Eijden, *J. Anat.*, **239** (2007).
42. M. Uchida, H. Agata, H. Sagara, Y. Shinohara, H. Kagami, and I. Asahina, *J. Biomed. Mater. Res. A*, **84** (2009).
43. D. W. Hutmacher, J. T. Schantz, C. X. F. Lam, K. C. Tan, and T. C. Lim, *J. Tissue. Eng. Reg. Med.*, **245** (2007).
44. S. Liao, W. Wang, Uo M, S. Ohkawa, T. Akasaka, K. Tamura, F. Cui, and F. Watari, *Biomaterials*, **7564** (2005).
45. Z, Li, L Yubao, Y. Aiping, P. Xuelin, W. Xuejiang, and Z. Xiang, *J. Mater. Sci. Mater. Med.*, **213** (2005).
46. S. S. Liao and F. Z. Cui, *Tissue Eng.*, **73** (2004).
47. S. S. Liao, F. Z. Cui, W. Zhang, and Q. L. Feng, *J. Biomed. Mater. Res. B Appl. Biomater.*, **158** (2004).
48. K. Usami, H. Mizuno, K. Okada, Y. Narita, M. Aoki, T. Kondo, D. Mizuno, J. Mase, H. Nishiguchi, H. Kagami, and M. Ueda, *J. Biomed. Mater. Res. A*, **730** (2009).
49. M. Biondi, F. Ungaro, F. Quaglia and P. A. Netti, *Adv. Drug Deliv. Rev.* **229** (2008).
50. M. Patel and J. P. Fisher, *Pediatric Res.*, **497** (2008).
51. D. Howard, L. D. Buttery, M. Shakesheff, and S. J. Roberts, *J. Anat.*, **66** (2008).

Chapter 27

MECHANO-ACTIVE SCAFFOLDS

Sang-Heon Kim,[a] Youngmee Jung,[a] Young Ha Kim,[b] and Soo Hyun Kim[a*]

[a]*Biomaterials Research Center, Korea Institute of Science and Technology, P.O. Box 131, Cheongryang, Seoul, Korea*
[b]*Department of Materials Science & Engineering, Gwangju Institute of Science and Technology, 261 Cheomdan-gwagiro, Buk-gu, Gwangju, Korea*
*soohkim@kist.re.kr

Biological processes are regulated through mechanical stimuli as well as biochemical interactions. The process by which mechanical stimuli are sensed and transmitted to the nucleus to induce changes in cell morphology and phenotype is also not clearly understood. Nevertheless, tissue engineering research takes mechanical stimuli into consideration, particularly in efforts to engineer components of the cardiovascular system and articular cartilage. Strategies concerning the mechanical environment of cells or tissues have been termed "mechano-active tissue engineering." Mechano-active scaffolds have employed elastic materials in mechano-active vascular and cartilage tissue engineering. Natural polymers and biodegradable polymers were studied for design of mechano-active scaffolds. The poly-(L-lactide-*co*-caprolactone) (PLCL) copolymer is composed of a soft matrix of ε-caprolactone moieties and hard domains containing additional

Handbook of Intelligent Scaffolds for Tissue Engineering and Regenerative Medicine
Edited by Gilson Khang

www.panstanford.com

L-lactide units, and exhibits a rubber-like elasticity in its physically cross-linked structure. This very elastic PLCL copolymer has also been fabricated as a macroporous scaffold for tissue engineering applications. PLCL has been fabricated for microporous scaffolds using a variety of techniques such as extrusion-particulate leaching, gel spinning, freeze drying, and electrospinning. This chapter discusses some of the recent insight into the fabrication and application of PLCL scaffolds for tissue engineering.

27.1 Introduction

It is well known that mechanical stimulation regulates the specialized structures and functions of mammalian cells, tissues, and organs. Mechanical stresses primarily originate either from tension that is caused by cells themselves or from the extracellular matrix (ECM) through cell adhesion. Significant recent progress has been achieved through studies of cardiovascular tissues and skeletal tissues, such as bone and articular cartilage, in mechanical, stress-related cell biology.[1–3]

The field of tissue engineering has progressed to develop means of regenerating damaged tissues and organs using cells and scaffolds. The fundamental role of a scaffold is not only to provide a temporary substrate on which transplanted cells can adhere but also to maintain mechanical integrity during the healing process and to deliver appropriate mechanical signals to adherent cells that ultimately comprise dynamic physiological systems.[4–7] Within tissue engineering, strategies concerning the mechanical environment of cells or tissues have emerged, and the resulting field has been termed "mechano-active tissue engineering." Specific examples include the development of elastic, mechano-active scaffolds that transmit mechanical stimulation to cells or protect cells from mechanical forces for use in vascular and cartilage tissue engineering. The mechanical properties of native blood vessels and of cartilage provide researchers with key design elements to apply in properly developing a scaffold that will function under conditions similar to those of the native tissues.

Natural polymers such as collagen have been studied as an elastic scaffold created by cross-linking these polymers with chemicals such as glutaldehyde, carbodiimide, and divinylsulfone.[8,9] Although chemical cross-linking increases the elasticity of the scaffolds, these chemicals also may be cytotoxic.[10,11] On the other hand, polyglycolic acid (PGA) fibers were cross-linked with poly(L-lactic acid) (PLA) to form a synthetic, biodegradable polymer scaffold for mechano-active tissue engineering.[12] However, this synthetic polymer scaffold exhibited significant, permanent deformation under cyclic mechanical strain conditions.[13–16] The PLCL copolymer is composed of a soft matrix of ε-caprolactone moieties and hard domains containing additional L-lactide units, differs a lot in their mechanical properties depending on the monomer content, and exhibits a rubberlike elasticity in its physically cross-linked structure.[17–19] In numerous studies, a PLCL copolymer prepared from 50 wt% of L-lactide and 50 wt% of ε-caprolactone was found to be highly elastic and has been fabricated as a macroporous scaffold for tissue engineering application.[20]

In this chapter, we review the recent insight into PLCL scaffolds for tissue engineering, focusing on blood vessel and cartilage regeneration.

27.2 Mechano-Active Scaffolds

27.2.1 *Elastic Biodegradable PLCL Copolymer*

A mechano-active polymer, a PLCL copolymer, was synthesized by the ring-opening polymerization of L-lactide and ε-caprolactone in the presence of $Sn(Oct)_2$ (Fig. 27.1). PLCL is composed of a soft matrix of ε-caprolactone moieties and hard domains of L-lactide units. The monomers used in this system differ greatly in mechanical properties and time to reach complete mass loss; however, once physically cross-linked in specific monomer ratios, the copolymer system exhibits a rubberlike elasticity.[18] The structure of PLCL was identified by ^1H NMR spectra in $CDCl_3$ (Fig. 27.1). The methine protons of the lactide unit appear as two singlets at δ 5.1–5.2 (a, a') as a result of the sequence distribution of the lactyl and caproyl units,

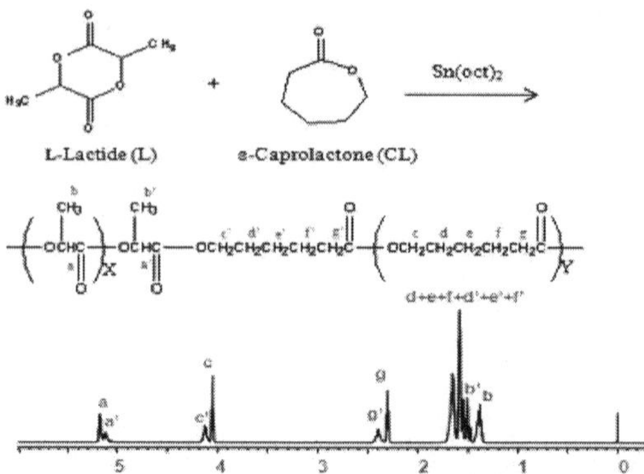

Figure 27.1. Chemical synthesis and ^{1}H-NMR spectrum of PLCL (50:50).

while the methylene protons of the caproyl unit adjacent to the ester group appear at δ 4.0–4.2 (c, c') and δ 2.3–2.5 (g, g'). The copolymer compositions synthesized by these relative intensities almost identically correspond to the initial feed compositions. PLCL prepared from 50 wt% of L-lactide and 50 wt% of ε-caprolactone is highly elastic and has been fabricated for microporous scaffolds using a variety of techniques such as extrusion-particulate leaching, gel spinning, gel pressing, freeze drying, and electrospinning. The mechanical properties of PLCL scaffolds were measured and compared with those of PLGA scaffolds; PLCL scaffolds fabricated with 60% porosity exhibit a strain of 500%, and scaffolds of 90% porosity can be extended to 200%. PLCL scaffolds with 90% porosity show 100% recovery at near-100% strain. In addition, PLCL scaffolds can be easily twisted and bent. In contrast, PLGA scaffolds largely deform and are broken even at strains as low as 20%. To further examine the elastic properties of PLCL scaffolds, scaffolds with varying porosity were subjected to cyclic strain at 10% amplitude and 1 Hz frequency for 27 days in a culture medium.[19] PLCL scaffolds of all tested porosities maintain excellent elasticity even in the hydrolytic medium over a 27-day experimental time course. Degradation tests show that the degradation rate of PLCL scaffolds is somewhat faster *in vivo* than *in vitro*, and this may be explained by enzymatic degradation possibly playing a role in degradation in the body. In addition, the CL moieties degrade faster than the LA units in PLCL scaffolds, although their hydrophilicities are in opposite order. This behavior appeared more prominently *in vivo*, probably indicating that amorphous regions composed of primarily CL units are first to be attacked by water, which can penetrate into the amorphous regions easier than into the hard domains that are composed primarily of LA units.

27.2.2 *Tubular PLCL Scaffold*

A tubular PLCL scaffold was fabricated by a particulate-leaching/extrusion method (Fig.27.2) for vascular tissue engineering.[18] In conduit PLCL scaffolds that are fabricated by the particulate leaching/extrusion, scanning electron microscopy (SEM) micrographs reveal extensive pores that are roughly spherical in shape. The pores appear to form interconnected networks, and the porosities and pore sizes of the scaffolds can be varied

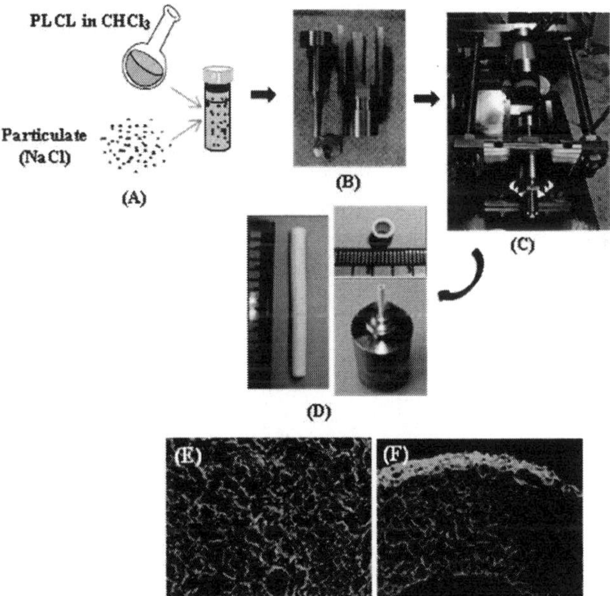

Figure 27.2. Fabrication of extruded tubular PLCL scaffolds. (a) Preparation of PLCL/NaCl mixture; (b) extrusion mold; (c) piston extrusion tool; (d) tubular scaffold extruded (I.D.: 4 mm, O.D.: 5 mm). SEM images of tubular PLCL scaffolds prepared by a particulate/extrusion method (e, f). See also Color Insert.

independently by adjusting the fractions and sizes of particles used (Fig. 27.2e,f). We have used tubular PLCL scaffolds fabricated by the extrusion-particulate leaching technique as mechano-active scaffolds for the application of cultured, small-diameter blood vessels. However, these extruded PLCL scaffolds are limited in mechanical strength, and vessel tensile properties are fundamental factors with which to evaluate scaffolds for vascular tissue engineering applications. In particular, the tensile strength of scaffolds is an important element for a successful vascular graft because vascular grafts must have adequate strength to resist rupture or excessive dilation when subjected to pulsatile pressures during implantation.

Therefore, a seamless tubular fibrous scaffold was fabricated from PLCL (50:50, molar ratio) to overcome the limited mechanical strength of extruded PLCL scaffolds by a custom-made gel-spinning molding device[21] (Fig. 27.4A). The gel-spinning molding device includes three separate drivers that make a cylindrical shaft turn on its axis and orbit and concurrently move up and down. A viscous PLCL solution in $CHCl_3$ was injected through a nozzle into a methanol bath, as shown in Fig. 27.4B. The injection of the PLCL solution into the methanol resulted in the solidification of PLCL as a fiber form, due to exchange of methanol in $CHCl_3$ and methanol in PLCL. The PLCL fibers were subsequently processed to produce a seamless tubular scaffold on the rotating cylindrical shaft, which was connected with the gel-spinning molding device, in the methanol bath. The distinctive moving of the cylindrical shaft in the methanol bath allowed the fibers to be spun and fabricated without excessive aggregation. The fibrous, tubular PLCL scaffolds showed good mechanical strength and cell adhesion and proliferation efficiencies compared with PLCL tubular scaffolds fabricated by the extrusion/particulate-leaching method.

27.2.3 *Seamless Double-Layered Scaffold*

In a different strategy, we used a gel-spinning molding technique to fabricate a double-layered tubular scaffold with high mechanical strength and elasticity. Our aim was to develop a seamless tubular PLCL scaffold that would resist rupture or leakage under high pressure to be applied for the implantation of

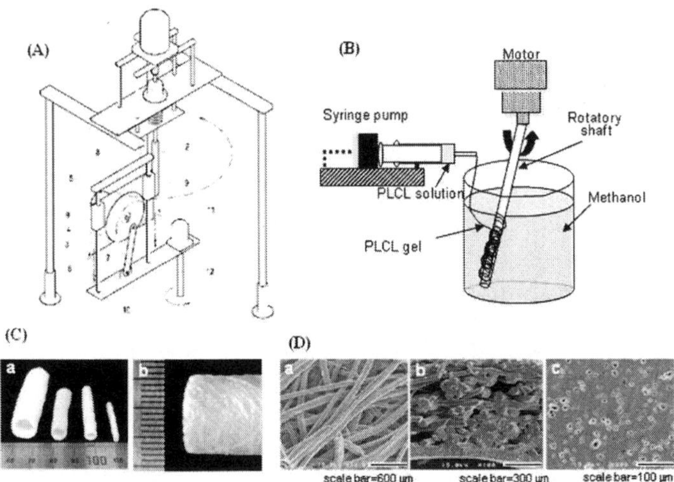

Figure 27.3. Device (A) for gel-spinning molding technique (B) and appearance (C) of tubular fibrous PLCL scaffolds prepared by the technique. SEM images of a double-layered PLCL scaffold (D, a: outer surface, b: cross-section, and c: luminal surface).

blood vessels, such as the aorta, rather than for use in *in vitro* culture for tissue reconstruction.[22] Fibrous PLCL scaffolds were enforced with a microporous inner layer for blocking of blood leakage during implantation due to the large pore space on the interior of nonreinforced PLCL scaffolds. Cylindrical shafts were dip-coated with a PLCL/salt mixture and coupled to a gel-spinning molding device for the formation of the outer fibrous structure. The morphology of double-layered scaffold is shown in (Fig. 27.3D). The individual fibers of the outer layer were found to be crossed and/or fused together with neighboring fibers (Fig. 27.3D-b). These cross-linked fibrous networks cause the spun PLCL scaffolds to be open-pore structures and well-interconnected between pores (Fig. 27.3D-a). An inner layer (Fig. 27.3D-c), which was a membrane with separated pores, was shown to be well fused with the outer fibrous networks (Fig. 27.3D-b). The scaffolds exhibited 550-670% elongation-at-break. The elastic features of the scaffolds are shown in (Fig. 27.4A). The scaffolds exhibited a complete rubber-like elasticity being recovered

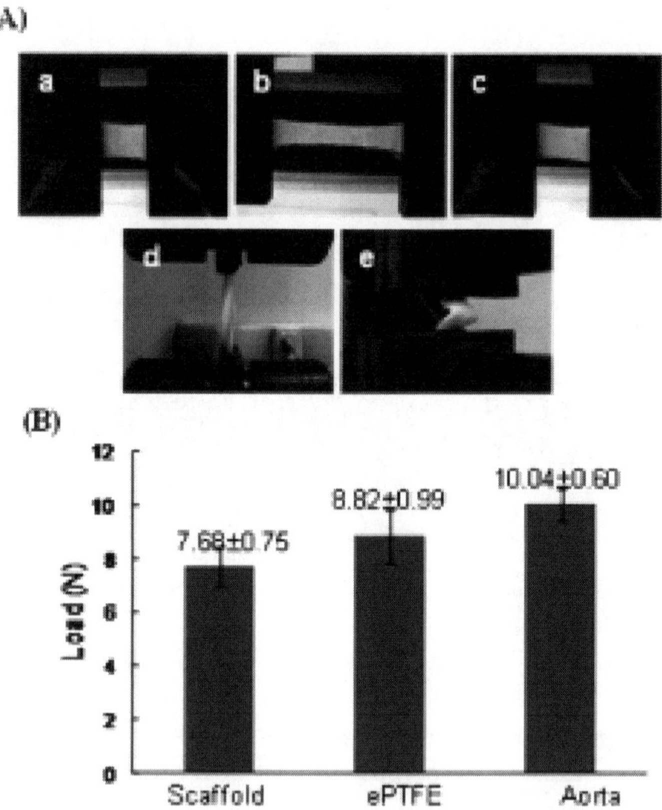

Figure 27.4. Physical properties of double-layered scaffolds. Elastic properties of double-layered scaffolds. (A) Scaffolds (a) were extended at 250% of initial length (b) with 3MPa for 5 sec and recovered (c) by releasing the load. Scaffolds were twisted (d) and folded (e) as through testing via a tensile system clamp. (B) Suture retention strength of double-layered scaffold, ePTFE, and canine abdominal aorta.

(Fig. 27.4A-c) after 250% extension (Fig. 27.4A-b) of the initial length (Fig. 27.4A-a) for 5 sec. The PLCL scaffolds were easily twisted (Fig. 27.4A-d) and folded (Fig. 27.4A-e). The suture retention strength of the scaffold was measured to be 7.68±0.75 N (n = 5), as shown in (Fig. 27.4B), compared with the strengths of ePTFE (8.82±0.99 N) and canine abdominal aorta (10.04±0.60 N). The burst pressures were more than 900 mmHg.

27.2.4 *Sheet-Form PLCL Scaffold*

A sheet-form PLCL scaffold with 80% porosity and pores ranging from 300–500 μm in size was fabricated for cartilage tissue engineering using a gel-pressing method.[23] The surface topology and cross-sectional images obtained by SEM reveal that the PLCL scaffolds have a homogeneously interconnected and open-pore structure without a skin layer. These properties will improve cell-seeding efficiency, cell ingrowth, and, eventually, cartilage regeneration. We showed that tensile properties of sheet-form PLCL scaffolds can be controlled by changing the initial monomer composition of PLCL. Tensile tests show that scaffolds that are fabricated from PLCL-1 (50:50 in LA:CA) have a lower tensile modulus (0.31 MPa) and a higher elongation at break (520%) than scaffolds that are fabricated from PLCL-2 (60:40 in LA:CA), which possess a modulus of 0.73 MPa and an elongation of 136%. Both PLCL films and scaffolds exhibit a completely rubberlike elasticity and show almost complete recovery (over 94%) up to the tensile strain before a break occurs. These elastic properties suggest that the PLCL scaffolds are suitable for use in cartilage tissue engineering. Because the mechanical properties of PLCL scaffolds can be controlled by changing the monomer content, these materials can be applied in a range of applications related to cartilage tissue engineering that depend on specific mechanical properties for ingrowth and eventual cartilage regeneration. PLCL-1 and PLCL-2 films have very high strains at break (1,100% and 800%, respectively), low stresses at break (6.7 MPa and 7.5 MPa, respectively), and very low tensile moduli (0.015 MPa and 0.014 MPa, respectively).

27.3 Mechano-Active Tissue Engineering

27.3.1 *Vascular Tissue Engineering*

Blood vessels are dynamic tissues with high elasticity and strength suited to withstand both the flow of blood and the associated pressure. Blood vessel walls are comprised of three layers: the tunica intima, the tunica media, and the tunica adventitia. The intima consists of a single layer of endothelial cells (EC) that are in direct contact with the blood flow and are supported by a subendothelial layer containing collagen fibers. Between the intima and the media is the internal elastic lamina, a layer of cross-linked elastin fibers.[24] The media contains smooth muscle cells (SMCs) that are subjected to the pulsatile load experienced by blood vessel walls. Numerous studies have reported that SMC phenotype and alignment are regulated significantly by mechanical stimulation, such as cyclic strain and pulsatile flow, in two-dimensional or three-dimensional culture systems.[25–27] Under mechanical stimulation, confluent SMCs orient perpendicular to applied strain and highly express SMC markers such as SM α-actin, myosin heavy chain, and caldesmon. In efforts to engineer blood vessels using bioreactor systems, interactions between ECs and SMCs (e.g., EC adhesion to and lining of the inner lumen in contact with cultured SMCs) are improved under the proper mechanical stimuli. In addition, cyclic strain has been shown to induce stem cell differentiation into SMCs.[28]

Due to the need for elastic properties in mechano-active tissue engineering, we have developed PLCL as a mechano-active scaffold material for vascular tissue engineering. Specifically designed PLCL scaffolds were seeded with SMCs labeled with CM-DiI and implanted into nude mice to investigate tissue compatibility, SMC growth, and *in vivo* scaffold degradation behavior.[29,30] Immunohistochemical analyses of extracted scaffolds demonstrated that SM α-actin gradually increased in the implanted scaffolds (Fig. 27.5a,c). The fluorescence was concentrated in each cell that was seeded in the early implantation period, but also extended and dispersed into the surrounding areas as the seeded cells extended and/or cells from surrounding areas grew into the implanted scaffolds (Fig. 27.5b,d). The implanted PLCL scaffolds degraded at a proper rate, while SM tissues were regenerated in the scaffolds. Thus, PLCL scaffolds exhibit good biocompatibility for SMCs and a proper *in vivo* degradation rate.

To investigate the effects of mechanical stimulation on the proliferation and phenotype of SMCs adhered to PLCL scaffolds, an extruded tubular scaffold was utilized as a three-dimensional cell culture substrate for SMCs under pulsatile strain and shear stress conditions.[18] We hypothesized that a radial distention would

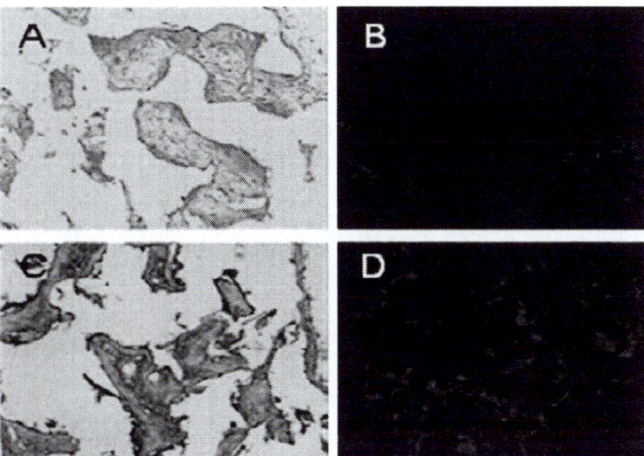

Figure 27.5. *In vivo* evaluation of SMC-seeded PLCL scaffolds. CM-DiI-tagged SMC-seeded PLCL scaffolds were subcutaneously implanted in nude mice for 2 (a, c) and 8 weeks (c, d). (a, c) Immunohistochemical staining for SM a-actin and (b, d) CM-DiI detection.

induce the phenotype of SMCs *in vitro* to be similar to that of SMCs *in vivo*. Aortic SMCs were seeded onto PLCL scaffolds and subjected to pulsatile strain in culture in pulsatile perfusion bioreactors. SMCs proliferate and eventually cover the surface of the scaffolds over eight weeks in pulsatile perfusion bioreactors. In control, static culture, cells grow much slower and the scaffold surface was not completely covered at eight weeks. Pulsatile strain enhances SMC proliferation and collagen production. In additional to collagen, elastin is a major ECM component secreted by functional SMCs in native blood vessels. The elastin content of transplanted PLCL scaffolds was examined as a measure of SMC differentiation in a mechanically active culture system. Elastin expression is greater in the scaffolds exposed to mechanical stimuli relative to that in static constructs.[31] Western blot analysis demonstrates that the expression of SM α-actin is upregulated by 2.5-fold in SM tissues reconstructed under the mechano-active conditions compared with that in tissues grown in static conditions.[18] The study demonstrates that tissue engineering of SM tissues *in vitro* using pulsatile perfusion bioreactors and elastic PLCL scaffolds leads to enhanced tissue development and differentiated SMC phenotype.

It was previously stated that tubular PLCL scaffolds were fabricated as a mechanically active artificial blood vessel by an extrusion-particulate leaching technique. However, these extruded PLCL scaffolds caused problems to be improved in respect of cell-seeding efficiency and cell ingrowth in the interior of the scaffold. We have developed a collagen/SMC-incorporated PLCL scaffold to facilitate small-diameter vascular tissue formation, using collagen gel as a substrate to provide the microporous PLCL scaffold, which is charge of mechanical support, with high surface-to-volume ratio.[32] Collagen gel can contain pores large enough to accommodate living cells and create pores into which living cells may penetrate and proliferate. A method for the preparation of a collagen/cells-incorporating scaffold is presented in Fig. 27.6a. A collagen solution that can be

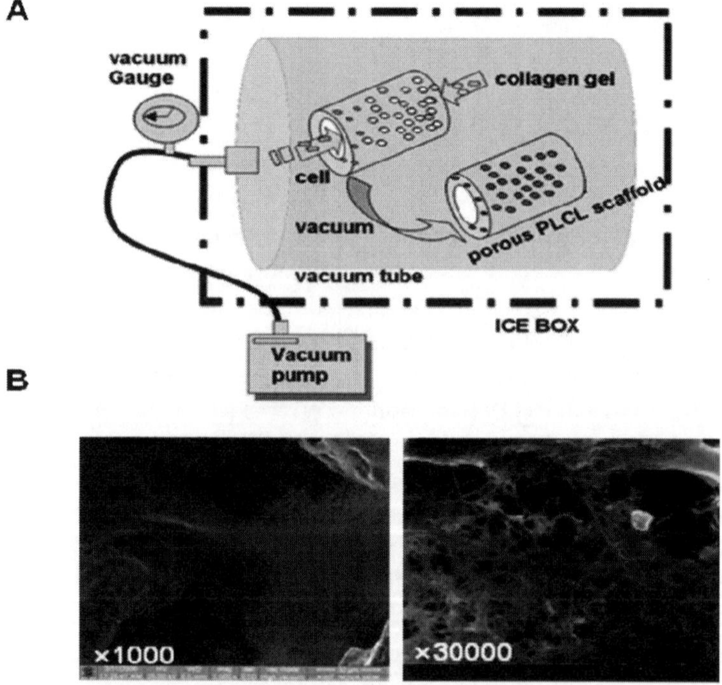

Figure 27.6. Vacuum seeding method. Experimental setup of a vacuum seeding system (a) and SEM images of collagen gel fibers infiltrated into the inside of scaffolds (b).

gelled is not easily infiltrated into the interior pores of scaffold due to high viscosity. In order to make a collagen/SMCs mixture infiltrate into the inside of PLCL scaffold, a decompressing device was used, as shown in Fig. 27.6a. SEM examinations showed that gelled collagen, which was observed as a fibrous network (Fig. 27.6b), was incorporated in the inside of the scaffolds under the decompressed condition. Cell adhesion and proliferation rate increased in collagen/SMC-incorporated tubular PLCL scaffolds compared with the scaffolds in which only SMCs were seeded. From SEM image and histological analysis, we further found that SMCs grew in the inside as well as on the surface of collagen/SMC-incorporated scaffolds and the cells continued to grow as a monolayer on collagen fibers. Four weeks after culture, the pores of the inner lumen, inside surface, and outer surface were covered with grown SMCs, and tissue-like structure was found in collagen/SMC-incorporated constructs (Fig. 27.7d), whereas the pores of those were kept open in SMC-seeded constructs (Fig. 27.7a). In 4′-6-diamidino-2-phenylindole (DAPI) staining, most cells were grown on the surface of scaffolds in SMC-seeded scaffolds (Fig. 27.7b). On the other hand, cells were found to be grown evenly in the inside as well as on the inner/outer surface in collagen/SMC-incorporated scaffolds (Fig. 27.7e). The content of elastin was examined to evaluate differentiation

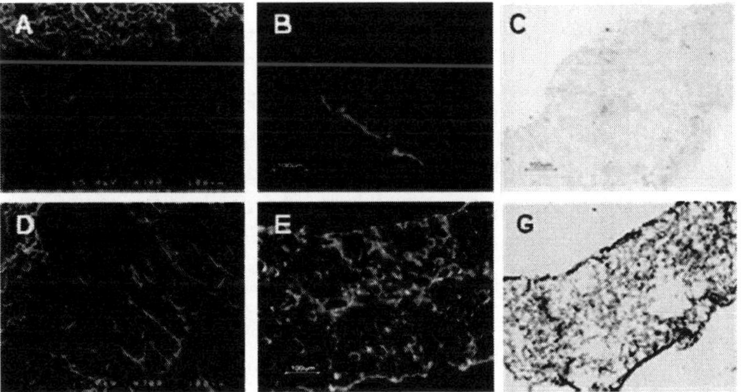

Figure 27.7. SEM micrographs (a, d), DAPI staining (b, e), and elastin staining (c, g) of tissue-engineered grafts cultured for 4 weeks. (a–c) Unmodified PLCL scaffolds and (d–f) collagen-incorporated PLCL scaffolds. See also Color Insert.

activity of SMCs cultured in the collagen/SMC-incorporated PLCL scaffold compared with those cultured in the SMC-seeded PLCL scaffold. Figure 27.7 illustrated the stained network (reddish brown) of elastin in the scaffolds with SMCs cultured during four weeks. Elastin was much more expressed in collagen/SMC-incorporated constructs (Fig. 27.7g) than in SMC-seeded constructs (Fig. 27.7c). In particular, much more elastin was observed in the collagen/SMC-incorporated scaffold having 50–100 μm pores than in any other scaffolds. A collagen/SMC-incorporated PLCL scaffold may support SMC growth and functions and can be used as a scaffold for tissue engineering to facilitate small-diameter vascular tissue formation, although we should further analyze the cell functions in the system.

The seamless double-layered tubular scaffold was fabricated by a novel gel-spinning technique, implanted into a canine abdominal aorta, and characterized through examination of mechanical and biological properties for blood vessel reconstruction. Angiography and ultrasonography showed that the implanted scaffolds kept their patency one year after operation (unpublished data).

27.3.2 *Cartilage Tissue Engineering*

The physiological characteristics of articular cartilage, whose main role is to bear high stress and friction loads in the body, are avascular, aneural, and alymphatic.[33,34] In particular, the lack of blood supply or a source of undifferentiated cells that promote wound healing is the major factor in little or no intrinsic capacity for cartilage repair in response to injury.[36–38] As a result, even minor lesions or large cartilage defects may lead to progressive damage and joint degeneration.[38,39]

Although there have been many studies on treatments for cartilage defects using autologous chondrocytes transplantation, bone marrow stimulation therapy, and even joint transplantation, there still remains unsolved problems such as the formation of fibrocartilage, a lower mechanical strength of newly formed tissue, and limited repair capacity.[40–43] Recently, tissue-engineered cartilage using scaffolds and cells has been considered for improved treatment of cartilage defects.[44–46]

Articular hyaline cartilage is subjected to particularly complex loads that affect its development and maintenance in the body.[44] Hence, mechanical stimulation associated with normal bodily functions is crucial in properly re-forming articular cartilage with tissue engineering. During hyaline cartilage engineering, scaffolds do not just provide a temporary substrate to which transplanted cells can adhere; they also play an important role in maintaining mechanical integrity during the healing process and delivering appropriate mechanical signals to adherent cells in dynamic physiological systems.[47–50] Much previous work has been devoted to evaluating scaffolds composed of various biodegradable polyesters, including PGA, PLA, poly(ε-caprolactone) (PCL), and their copolymers.[21,51–53] In particular, the mechanical properties of the PLCL copolymer can be altered by changing the monomer content, and the physically cross-linked structure of PLCL exhibits a rubberlike elasticity.[54] Successfully generating functional engineered cartilage requires using a mechano-active scaffold that can transmit mechanical signals to its adherent cells in physiologically dynamic bodily environments.

To deliver the required mechanical signals associated with the surrounding biological environment of cartilage, we fabricated a sponge-type microporous scaffold from the elastic PLCL copolymer by a gel-pressing method. Xie *et al.* evaluated microporous PLCL scaffolds in an *ex vivo* system for mechano-active, scaffold-based cartilage tissue engineering. A new cell-seeding technique was devised to improve chondrocyte distribution into and viability in microporous PLCL scaffolds under compression force-induced suction. They successfully delivered cells to microporous PLCL scaffolds with high cell-seeding efficiency, cell viability, and homogeneous cell distribution under multiple cycles of suction. Xie *et al.* focused on how chondrocytes respond biomechanically to various modes of compressive loading, specifically loading frequency, loading duration per cycle, loading period, and continuous or intermittent compression, in mechano-active PLCL scaffolds. The mRNA expression of ECM molecules associated with cartilage is observed in chondrocyte-containing PLCL scaffolds cultured at three different loading frequencies: 0.05, 0.1, and 0.5 Hz. Type II collagen expression is the highest in scaffolds cultured at 0.1Hz, whereas little difference is observed in aggrecan mRNA levels at these frequencies. Continuous, dynamic compression results in decreased mRNA levels of the cartilage-dependent ECM components, aggrecan and type II collagen. An intermittent loading (24-hour cycle of loading and unloading) program maintains high levels of ECM component mRNA expression. Continuous, dynamic compression also causes the release of sulfated glycosaminoglycan (S-GAG) from the chondrocyte-containing PLCL scaffolds into the surrounding medium.[55,56] Thus, excessive mechanical stress is not favorable for mRNA expression and protein accumulation, both of which influence cartilage formation. An appropriate mechanical stimulation program is required for the construction of functional cartilage using mechano-active scaffolds *in vitro*.

To evaluate the suitability of microporous PLCL scaffolds for mechano-active cartilage tissue engineering, chondrocyte-seeded PLCL scaffolds were cultured for 10 days or 25 days under continuous, compressive deformation of 5% strain at 0.1 Hz using a compressive-mode bioreactor (Fig. 27.8) or under control static conditions and subsequently implanted subcutaneously into nude mice.[25] The collagen and GAG content of mechanically stimulated scaffolds increases significantly over 10 days in culture compared with that in static-cultured scaffolds. Histological analysis shows that mechanically stimulated implants formed mature and well-developed cartilaginous tissue, as evidenced by the presence of chondrocytes within lacunae and the abundant accumulation of S-GAGs. However, more unhealthy lacunae shapes and hypertrophic forms are observed in the implants that have been mechanically stimulated for 24 days compared with those stimulated for 10 days. This result suggest that the proper periodical application of dynamic compression can encourage the maintenance of chondrocyte phenotype and enhanced GAG production, which will improve the function of cartilaginous tissue constructed both *in vitro* and *in vivo*. Previously, although there were many studies for formation of the articular cartilage with polymer scaffolds and engineering the constructs with mechanical stimulation, fibrous cartilage formation was indicated due to the mechanical properties of rigid polymer scaffolds.[45,57–59] In mechano-active cartilage tissue engineering, the materials for scaffolds need to recover completely after deformation in order for the effective transfer of the periodic mechanical

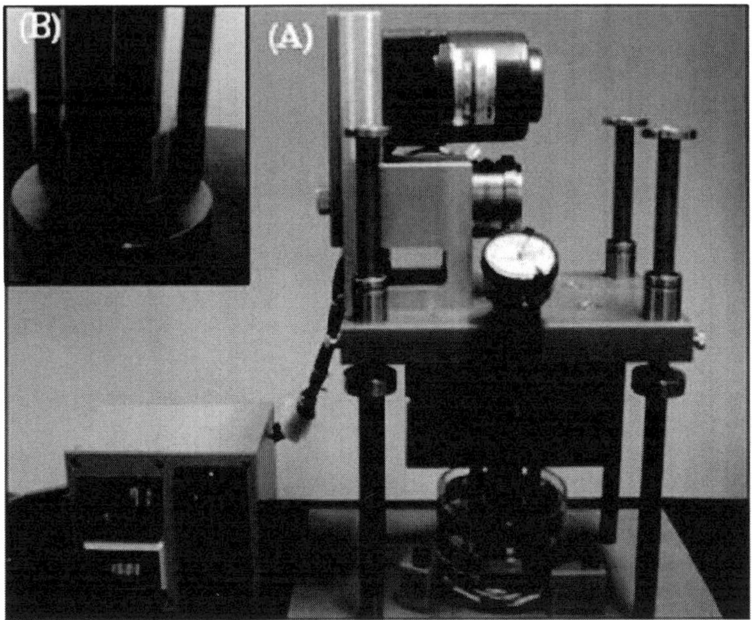

Figure 27.8. A bioreactor in compressive mode (a) and the units of the cradle of scaffolds and the plunger for compressive forces (b).

stimulation. For effectively engineering the constructs, we used the biodegradable PLCL scaffolds with complete rubberlike elasticity. It is believed that the elasticity of the PLCL scaffolds might have contributed to the transfer of the compressive forces to the seeded chondrocytes onto the scaffolds.

We compared the mechanical properties of elastic PLCL scaffolds with those of conventional rigid polymer scaffolds, such as PLA and PLGA scaffolds.[60] The PLCL scaffolds possess a completely rubberlike elasticity, are easily twisted and bent, and exhibit an almost complete (over 97%) recovery from applied strain (up to 500%), while the control PLA scaffolds show little recovery after strain. We evaluated their abilities to promote cartilaginous tissue formation and cartilage regeneration *in vitro*, in nude mice, and in a rabbit cartilage defect model. We seeded scaffolds with rabbit chondrocytes, cultured them *in vitro*, and subcutaneously implanted them into nude mice for up to eight weeks. *In vitro* and *in vivo*

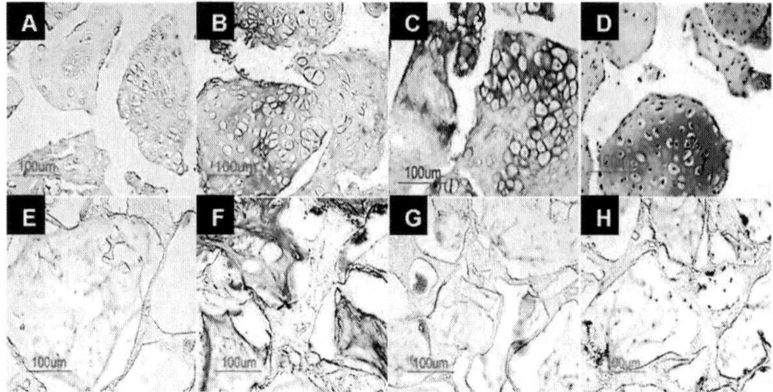

Figure 27.9. Histological studies of the cell-scaffold constructs explanted from nude mice at 8 weeks (tissue sections). The sections were stained with H&E (a, e), Masson's trichrome (b ,f), Alcian blue (b, g), or safranin O (d, h). The images represent the PLCL-chondrocyte constructs (a–d) and the PLA-chondrocyte constructs (e–h). (The full color images can be found in the online version of this article.) See also Color Insert.

accumulation of ECM on the cell-PLCL constructs demonstrates that these scaffolds can not only sustain but significantly enhance chondrogenic differentiation. Moreover, the mechanical stimulation of the dynamic *in vivo* environment promotes deposition of the chondral ECM onto implanted PLCL scaffolds (Fig. 27.9). In contrast, on the PLA scaffolds, most of the chondrocytes de-differentiate and form fibrous tissues. In the rabbit defect model, animals implanted with PLCL scaffolds exhibit significantly enhanced cartilage regeneration compared with those that receive an empty control or PLGA scaffold. These results indicated that the mechano-active PLCL scaffolds effectively deliver mechanical signals associated with biological environments to adherent chondrocytes, suggesting that these elastic PLCL scaffolds can successfully be used for cartilage regeneration.

Consequently, in mechano-active cartilage regeneration, it is important that the implants that are inserted to the defect sites not only maintain their mechanical integrity but also deliver mechanical signals to adherent cells in the body. In order to effectively transfer these mechanical signals, the scaffold material must completely recover from deformations induced by various body forces. With this point in mind, we developed an elastic, biodegradable PLCL scaffold and tested it in several cartilage regeneration models. From the results, we could confirm that the mechano-active PLCL scaffolds could effectively deliver the mechanical signals of the surrounding biological environment to the adherent chondrocytes.

27.4 Conclusions and Outlook

Mammalian cells and tissues reside in mechanically dynamic microenvironments in the body. Inappropriate physical loads upon tissues result in tissue deformation. Numerous studies have confirmed that proper mechanical stimuli applied to cells or tissues are involved in maintaining cell/tissue morphology and inducing specialized functions. Mechanical stimuli have recently been applied to regenerate functional tissues, in particular, tissues in the cardiovascular system and articular cartilage. To reconstruct functionally active SM tissues and chondrogenic tissues that are comparable to the native tissues, the *in vitro* re-creation of the *in vivo* mechano-active microenvironments may be necessary in the tissue engineering process. Tissue engineering scaffolds play a crucial role in delivering mechanical stresses from the extracellular environment to the cells responsible for tissue formation. PLCL is a rubberlike, elastic, biodegradable polymer that has developed as a key material for mechano active tissue engineering. In both *in vitro* and *in vivo* studies, PLCL scaffolds have been beneficial in sustaining SMC phenotype and chondrogenic differentiation in vascular tissue engineering and cartilage tissue engineering, respectively, presumably through effectively transmitting mechanical signals through dynamic microenvironments. PLCL will be an excellent candidate material for mechano-active scaffolds that deliver mechanical stresses to cells and tissues via a transducer.

Acknowledgments

This study was supported in part by a grant from the Korea Health 21 R&D Project, Ministry of Health & Welfare (MOHW) (A050082).

References

1. F. Wernig and Q. Xu, *Prog. Biophys. Mol. Biol.*, **78**, 105 (2002).
2. J. Klein-Nulend, R. G. Bacabac, and M. G. Mullender, *Pathologie. Biologie.*, **53**, 576 (2005).
3. C. T. Hung, D. R. Henshaw, C. C. Wang, R. L. Mauck, F. Raia, G. Palmer, P. H. Chao, V. C. Mow, A. Ratcliffe, and W. B. Valhmu, *J. Biomech.*, **33**, 73 (2000).
4. S. D. Waldman, C. G. Spiteri, M. D. Grynpas, R. M. Pilliar, J. Hong, and R. A. Kandel, *J. Bone Joint Surg.*, **85A**, 101 (2003).
5. E. M. Darling and K. A. Athanasiou, *Ann. Biomed. Eng.*, **31**, 1114 (2003).
6. L. Lu, X. Zhu, R. G. Valenzuela, B. L. Currier, and M. J. Yaszemski, *Clin. Orthop. Relat. Res.*, **391**, 251 (2001).
7. R. K. Aaron, D. M. Ciombor, S. Wang, and B. Simon, *Ann. N. Y. Acad. Sci.*, **1068**, 513 (2006).
8. M. E. Nimni, D. Cheung, B. Strates, M. Kodama, and K. Skeikh, *J. Biomed. Mater. Res.*, **21**, 741 (1987).
9. L. H. Olde Damink, P. J. Dijkstra, M. J. van Luyn, P. B. van Wachem, P. Nieuwenhuis, and J. Feijen, *Biomaterials*, **17**, 765 (1996).
10. L. L. Huang-Lee, D. T. Cheung, and M. E. Mimni, *J. Biomed. Mater. Res.*, **24**, 1185 (1990).
11. M. J. van Luyn,, P. B. van Wachem, L. H. Olde Damink, P. J. Dijkstra, J. Feijen, and P. Nieuwenhuis, *Biomaterials*, **13**, 1017 (1992).
12. B. S. Kim and D. J. Mooney, *J. Biomech. Eng.*, **122**, 210 (2000).
13. S. C. Woodward, P. S. Brewer, F. Montarned, A. Schindler, and D. G. Pitt, *J. Biomed. Mater. Res.*, **19**, 437 (1985).
14. C. G. Pitt, M. M. Gratzl, G. L. Kimme, J. Surles, and A. Schindler, *Biomaterials*, **17**, 215 (1996).
15. P. van der Valk, A. W. van Pelt, H. J. Busscher, H. P. de Jong, C. R. Wildevuur, and J. Arends, *J. Biomed. Mater. Res.*, **17**, 807 (1983).
16. T. Nakamura, Y. Shimizu, Y. Takimoto, T. Tsuda, Y. Li, T. Kiyotani, M. Teramachi, S. Hyon, Y. Ikada, and K. Nishiya, *J. Biomed. Mater. Res.*, **42**, 475 (1998).
17. S. I. Jeong, B. S. Kim, Y. M. Lee, K. J. Ihn, S. H. Kim, and Y. H. Kim, *Biomacromolecules*, **5**, 1303 (2004).
18. S. I. Jeong, J. H. Kwon, J. I. Limb, S. W. Cho, Y. Jung, W. J. Sung, S. H. Kim, Y. H. Kim, Y. M. Lee, B. S. Kim, C. Y. Choi, and S. J. Kim, *Biomaterials*, **26**, 1405 (2005).

19. S. I. Jeong, S. H. Kim, Y. H. Kim, Y. Jung, J. H. Kwon, B. S. Kim, and Y. M. Lee, *J. Biomater. Sci. Polym. Ed.*, **15**, 645 (2004).
20. A. G. Mikos, Y. Bao, L. G. Cima., D. E. Ingber, J. P. Vacanti, and R. Langer, *J. Biomed. Mater. Res.*, **27**, 183 (1993).
21. S. H. Kim, J. H. Kwon, M. S. Chung, E. Chung, Y. Jung, S. H. Kim, and Y. H. Kim, *J. Biomater. Sci. Polym. Ed.*, **7**, 1359 (2006).
22. S.-H. Kim, E. Chung, S.-H. Kim, Y. Jung, Y. H. Kim, and S. H. Kim, *J. Biomater. Sci. Polym. Ed.*, In press.
23. Y. Jung, S. H. Kim, H. J. You, S.-H. Kim, Y. H. Kim, and B. G. Min, *J. Biomater. Sci. Polym. Ed.*, **19**, 1073 (2008).
24. T. V. How, *Cardiovascular Biomaterials* (Springer-Verlag, London, 1992), p. 1.
25. M.-J, Qu, B. Liu, H. Q. Wang, Z.-Q. Yan, B.-R. Shen, and Z.-L. Jiang, *J. Vasc. Res.*, **44**, 345 (2007).
26. G. R. Houtchens, M. D. Foster, T. A. Desai, E. F. Morgan, and J. Y. Wong, *J. Biomech.*, **41**, 762 (2008).
27. F. Opitz, K. Schenke-Layland, W. Richter, D. P. Martin, I. Degenkolbe, T. Wahlers, and U. A. Stock, *Ann. Biomed. Eng.*, **32**, 212 (2004).
28. J. S. Park, N. F. Huang, K. T. Kurpinski, S. Patel, S. Hsu, and S. Li, *Front. Biosci.*, **12**, 5098 (2007).
29. B. S. Kim, S. I. Jeong, S. W. Cho, J. Nikolovski, D. J. Mooney, S. H. Lee, O. J. Jeon, T. W. Kim, S. H. Lim, Y. S. Hong, C. Y. Choi, Y. M. Lee, S. H. Kim, and Y. H. Kim, *J. Microbiol. Biotechnol.*, **13**, 841 (2003).
30. S. I. Jeong, B. S. Kim, S. W. Kang, J. H. Kwon, Y. Jung, Y. M. Lee, S. H. Kim, and Y. H. Kim, *Biomaterials*, **25**, 5939 (2004).
31. J. I. Lim, S.-H. Kim, S. H. Kim, and Y. H. Kim, *Biomat. Res.*, **10**, 154 (2006).
32. I. S. Park, S.-H Kim, Y. H. Kim, I. H. Kim, and S. H. Kim, *J. Biomater. Sci. Polym. Ed.*, In press.
33. O. Démarteau, M. Jakob, D. Schäfer, M. Heberer, and I. Martin, *Bioreology*, **40**, 331 (2003).
34. M. D. Buschmann, Y. A. Gluzband, A. J. Grodzinsky, and E. B. Hunziker, *J. Cell Sci.*, **108**, 1497 (1995).
35. L. A. Solchaga, V. M. Goldberg, and A. I. Caplan, *Clin. Orthop. Relat. Res.*, **391S**, S161 (2001).
36. E. B. Hunziker, *Osteoarthritis Cartilage*, **10**, 432 (2002).
37. A. J. Almarza and K. A. Athanasiou, *Ann. Biomed. Eng.*, **32**, 2 (2004).

38. F. Guilak, D. L. Butler, and S. A. Goldstein, *Clin. Orthop. Relat. Res.*, **391**, S295 (2001).
39. S. C. Ghivizzani, T. J. Oligino, P. D. Robbins, and C. H. Evans, *Phys. Med. Rehabil. Clin. N. Am.*, **11**, 289 (2000).
40. E. B. Hunziker and L. C. Rosenberg, *J. Bone. Joint. Surg. Am.*, **78A**, 721 (1996).
41. M. V. Risbud and M. Sittinger, *Trends Biotechnol.*, **20**, 351 (2002).
42. M. Brittberg, A, Lindahl, A. Nilsson, C. Ohlsson, O. Isaksson, and L. Peterson, *N. Engl. J. Med.*, **331**, 889 (1994).
43. A. K. Lynn, R. A. Brooks, W. Bonfield, and N. Rushton, *J. Bone Joint Surg. Br.*, **86B**, 1093 (2004).
44. R. L. Smith, D. R. Carter, and D. J. Schurman, *Clin. Orthop. Relat. Res.*, **427**, S89 (2004).
45. O. Demarteau, D. Wendt, A. Braccini, M. Jakob, D. Schafer, M. Heberer, and I. Martin, *Biochem. Biophy. Res. Comm.*, **310**, 580 (2003).
46. M. D. Buschmann, Y. J. Kim, M. Wong, E. Frank, E. B. Hunziker, and A. J. Grodzinsky, Arch. *Biochem. Biophys.*, **366**, 1 (1999).
47. S. D. Waldman, C. G. Spiteri, M. D. Grynpas, R. M. Pilliar, J. Hong, and R. A. Kandel, *J. Bone Joint Surg.*, **85A**, 101 (2003).
48. E. M. Darling and K. A. Athanasiou, *Ann. Biomed. Eng.*, **31**, 1114 (2003).
49. L. Lu, X. Zhu, R. G. Valenzuela, B. L. Currier, and M. J. Yaszemski, *Clin. Orthop. Relat. Res.*, **391**, S251 (2001).
50. R. K. Aaron, D. M. Ciombor, S. Wang, and B. Simon, *Ann. N. Y. Acad. Sci.*, **1068**, 513 (2006).
51. G. A. Ameer, T. A. Mahmood, and R. Langer, *J. Orthop. Res.*, **20**, 16 (2002).
52. R. D. Coutts, R. M. Healey, R. Ostrander, R. L. Sah, R. Goomer, and D. Amiel, *Clin. Orthop. Relat. Res.*, **391**, S271 (2001).
53. T. B. Woodfield, J. M. Bezemer, J. S. Pieper, C. A. van Blitterswijk, and J. Riesle, *Crit. Rev. Eukaryot. Gene Expr.*, **12**, 209 (2002).
54. J. Xie, M. Ihara, Y. Jung, I. K. Kwon, S. H. Kim, Y. H. Kim, and T. Matsuda, *Tissue Eng.*, **12**, 449 (2006).
55. J. Xie, Z. Han, S. H. Kim, Y. H. Kim, and T. Matsuda, *Tissue Eng.*, **13**, 29 (2007).
56. Y. Jung, S. H. Kim, S.-H. Kim, Y. H. Kim, J. Xie, T. Matsuda, and B. G. Min, *J. Biomater. Sci. Polym. Ed.*, **19**, 61 (2008).
57. K. Uematsu, K. Hattori, Y. Ishimoto, J. Yamauchi, T. Habata, Y. Takakura, H. Ohgushi, T. Fukuchi, and M. Sato, *Biomaterials*, **26**, 4273 (2005).

58. J. Chang, X. C. Ma, D. L. Ma, X. T. Li, and D. L. Xia, *J. Biomed. Mater. Res. B Appl. Biomater.*, **71**, 313 (2004).

59. M. Honda, N. Morikawa, K. Hata, T. Yada, S. Morita, M. Ueda, and K. Kimata, *Biomaterials*, **24**, 3511 (2003).

60. Y. Jung, M. S. Park, J. W. Lee, Y. H. Kim, S.-H. Kim, and S. H. Kim, *Biomaterials*, **29**, 4630 (2008).

Chapter 28

REINFORCED SCAFFOLD FOR TISSUE ENGINEERING

Young-Kwon Seo[a] and Jung-Keug Park[a,b*]
[a] *Dongguk University Research Institute of Biotechnology, Dongguk University, Seoul, 100-715, Korea*
[b] *Department of Medical Biotechnology, Dongguk University, Seoul, 100-715, Korea*
*jkpark@dongguk.edu

Wound healing is achieved via the synergistic effect of cells, growth factors, and scaffolds, all of which play important roles. Tissue engineering provides a promising solution for the repair of tissue defects. Scaffolds provide substrates to which cells can attach during the initial phase of repair that degrade after the completion of wound healing. Various natural materials, including collagen, hyaluronan, silk, and synthetic degradable polymers such as polyglycolic acid (PGA), polylactic acid (PLA), and polycaprolactone (PCL), have been investigated extensively for their ability to support the growth of cells. However, scaffolds constructed from these compounds have various problems associated with their mechanical properties and biocompatibility. To obtain good repair, scaffolds should not only possess properties that support cell adhesion, growth, and differentiated function, but must also create a stable, three-dimensional structure with sufficient porosity and mechanical properties to facilitate repair. Specifically, a

Handbook of Intelligent Scaffolds for Tissue Engineering and Regenerative Medicine
Edited by Gilson Khang

www.panstanford.com

tissue-engineered scaffold should provide adequate mechanical strength during the initial stage of repair and be composed of a highly porous structure that provides an ideal environment for the migration and proliferation of cells. Additionally, the scaffold should be designed to withstand handling and suturing during surgical implantation.

28.1 Introduction

Although various types of biomaterials have been developed and utilized for the repair of damaged tissue and organs, the biocompatibility of these materials is of major concern.[1–4] Synthetic biomaterials such as PGA and poly-L-lactic acid (PLLA) have commonly been used in clinical applications. Synthetic materials may afford better plasticity and controlled biodegradation than grafts prepared from natural extracellular matrix (ECM) materials, but they are also less biocompatible. However, natural ECM materials such as gelatin, alginate, chitosan, hyaluronate, collagen, amniotic membrane, and small intestine submucosa (SIS)[5,6] may be less suitable for use in tissue repair in many instances because of their weaker mechanical properties. For example, the mechanical properties of PGA are better than collagen, but the biocompatibility of PGA is not as good. Specifically, PGA has been shown to increase the proliferation of T cells and subsequently induce an inflammatory reaction.[7] As a result, many researchers have designed composite or reinforced scaffolds to overcome these shortcomings associated with PGA.

The scaffold design for tissue regeneration is one of the key technologies used in tissue engineering. Scaffolds should mimic the structure and biological function of the ECM. Collagen sponges are highly porous, with interconnected pore structures that enable the effective infiltration of cells and supply of oxygen and nutrients to the cells. However, a drawback associated with using a collagen sponge as a scaffold for cell proliferation and differentiation is its poor mechanical strength. To overcome this problem, attempts have been made to reinforce or combine sponges with other materials.[8]

Early reinforced composite scaffolds were developed for use in urological surgery in 1987. Because the scaffolds required biocompatibility and strong mechanical properties, natural and synthetic polymers were combined. These reinforced scaffolds were composed of Vicryl (Polyglactin) and collagen sponges or films. Many researchers utilized these reinforced scaffolds to repair urinary tract defects as well as bladder and kidney surfaces. These reinforced composite scaffolds were shown to be biodegradable, prevent the leakage of urine, and to be readily replaced by collagenous scar tissue lined with aurothelium.[9–12]

Recently, a collagen sponge tube reinforced with copoly(L-lactide/ε-caprolactone) was designed for urethral reconstruction. The results of animal tests demonstrated that this reinforced scaffold was slightly fibrotic but completely epithelialized and that it supported the regeneration of smooth muscle layers without fistulae or stenoses.[13]

Furthermore, reinforced scaffolds have been used for hard-tissue regeneration. A collagen scaffold reinforced with chitin fibers was designed, and cell proliferation *in vitro* and bone growth *in vivo* were evaluated. The results of these tests indicated that the cell population and bone growth were superior to that obtained when collagen scaffolds without chitin fibers were used. Further, the results of these studies suggested that the suppression of shrinkage of the sponges in the presence of chitin fibers maintained the interspace, resulting in increased cell migration and bone formation.[14,15] A fibrin scaffold reinforced with PGA fibers was also utilized for hard tissue regeneration (Fig. 28.1).[16]

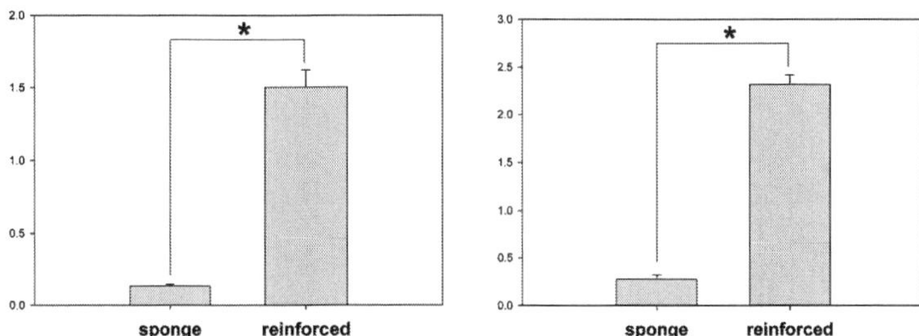

Figure 28.1. Comparison of UTS (a) and comparison of compression tests (b) between the collagen sponge (sponge) and reinforced collagen sponge (reinforced) after cell culture. (Seo *et al.*, Biotechnol. Bioprocess Eng., 1998). *Abbreviation*: UTS, ultimate tensile strength.

Moreover, a reinforced scaffold composed of Vicryl and a collagen sponge were used as dural substitutes because these compounds should prevent the leakage of cerebrospinal fluid, seal fistulae, prevent cortical adhesion, be soft and easy to suture, resist infection, produced a minimal inflammatory response, and (if restorable) allow regeneration of the host dura to occur.[17]

28.2 Biocompatibility of Reinforced Composite Scaffolds

Most scaffolds derived from nonautologous sources will cause some degree of foreign-body response after implantation *in vivo*. Host responses to scaffolds are evaluated by morphologic and histologic examination of the tissue at the implant site and are therefore related to the biocompatibility of the scaffold. Because reinforced composite scaffolds are composed of various biomaterials, it is necessary to closely evaluate the biocompatibility of reinforced composite scaffolds.

Many studies have been conducted to evaluate the effects of inflammatory cells on various scaffolds. Several groups have reported the potential for the use of collagen, silk, and biodegradable polymers, such as PGA in PLA for scaffolds. Over the past 20 years, silk and a variety of degradable synthetic materials, including PGA, have been used as biomaterials. In general, various cell lines have been utilized for the purpose of evaluating the cell compatibility of scaffolds *in vitro*.[18] However, Mathew *et al.* revealed that, although polyurethane showed a high level of cell compatibility when cultured with immortal cell lines such as L292 *in vitro*, it led to the migration of inflammatory cells such as neutrophils, monocytes, and macrophages toward the implantation site when it was implanted into rat muscles.[19] Some studies have tested the biocompatibility of various biomaterials by culturing inflammatory cells, such as monocytes, lymphocytes, and macrophages, with various biomaterials *in vitro* to assess the growth and synthesis of various cytokines. Peripheral blood leukocytes have been used extensively as models to assess the potential tissue response to inflammatory agents, such as biomaterials, microorganisms, and allografts.[20–25] Biocompatibility is typically evaluated by cell line culture *in vitro*; however, if the scaffold has a hydrophobic surface, the cell line cannot be attached. In a recent study, T lymphocytes were cultured with the scaffold and then assayed for proliferation and cytokine secretion.[26–28] Such *in vitro* studies are related immune reactions, which are more important than cell line growth because these reactions are directly related to the inflammatory cell-induced degradation of the implanted scaffold *in vivo*.

In general, silk or synthetic polymers have low cell compatibility. Therefore, many studies have been conducted to modify the surface of polymers to improve cell attachment and proliferation. For example, silk and PGA scaffolds were coated with collagen, which is a well known protein that has good biocompatibility and cell compatibility. The initial attachment of mesenchymal stem cells (MSCs) to the PGA scaffold was superior to the initial attachment of cells to the silk scaffold, but the average cell density did not differ significantly between groups following MSC culture. However, the PGA scaffold showed a greater increase in peripheral blood mononuclear cell (PBMC) proliferation than the silk scaffold *in vitro*. Additionally, the PGA scaffold resulted in increased levels of interleukin (IL)-1β and interferon-gamma (IFN-γ) when compared with the silk scaffold.[29] Taken together, these findings suggest that the results observed when *in vitro* inflammatory related cell cultures were used were more closely related to the *in vivo* results of implantation than the results of *in vitro* fibroblast-like or MSC cultures.

28.3 Reinforced Composite Scaffold for Bioartificial Tissue

Tissue regeneration using tissue engineering techniques is designed to replace the function and action of the damaged tissue or to repair original tissue before it reaches its defect point. Synthetic, nonbiodegradable scaffolds currently available play a role in mechanical function for a short period of time but lead to increased inflammatory reactions. Indeed, biodegradable scaffolds developed to date do not concurrently possess satisfactory biocompatibility and physical properties. The ideal scaffold use for tissue engineering must be biocompatible, nontoxic, and strong.

28.3.1 *Bioartificial Ligament and Tendon*

A new tissue engineering strategy has recently been attempted in ligament and tendon reconstruction.[30] The ideal ligament and tendon replacement scaffold should be biodegradable, porous and biocompatible, show adequate mechanical strength, and promote the formation of ligament or tendon tissue and blood vessels. Several groups have reported the potential construction of ligament and tendon scaffolds using a carbon copolymer, collagen, silk, and biodegradable polymers, such as PGA and PLA.[31–37]

The use of composite anterior cruciate ligament (ACL) prostheses, which are composed of a collagen inductor and a synthetic fiber (Dacron®), was investigated as a possible means of ACL reconstruction. Composite ACL prostheses were implanted in 10 sheep models for six months. The tibial portion showed an osseous ingrowth around the braid, and numerous vessels were observed. In that study, a synthetic fiber functioned as a support for mechanical strength, and the use of a collagen scaffold resulted in the migration of cells and vessels.[31] However, because the Dacron tube used in that study was a nonbiodegradable polymer, it induced long-lasting inflammation at the site of implantation. The mechanical properties and biocompatibility of the carbon-PLA-polycapralactone (carbon-copolymer) composite ligament prosthesis make it suitable for use as a scaffold material for the repair and reconstruction of human ligaments and tendons.[32] However, long-term studies indicate that the carbon fiber slowly breaks up at the site of implantation and subsequently appears in the regional lymph nodes.[33]

Therefore, a natural polymer, collagen, was used as a scaffold material in an attempt to reduce the inflammatory reaction. However, rapid degradation is often a problem with collagen.[34,36] As a result, reinforced composite scaffolds were fabricated by embedding parallel collagen fibers within a PLA polymer, but the ligament ruptured in 20% of the implants.[36]

The results of the studies conducted to date indicate that a new biocompatible biomaterial with strong mechanical properties and slow degradation is needed. A number of studies have recently been conducted to evaluate the possibility of ligament reconstruction using silk or PGA.[37] The results of recent studies conducted using silk materials were satisfactory in terms of both biocompatibility

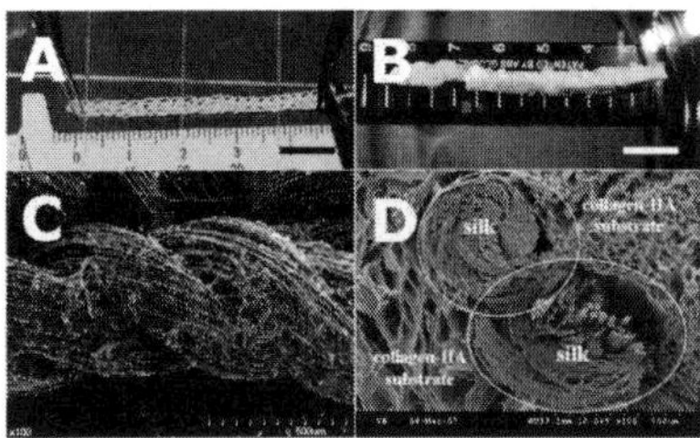

Figure 28.2. Gross appearance (a, b) and scanning electron microscopy (c, d) of the silk scaffold (a, c) and reinforced collagen-hyaluronan scaffold with silk thread (b, d). (Circles: Cross-section of silk thread). Scale bar = a, b: 1,000 μm; c, d = 500 μm. Magnification = c, d ×100. (Seo *et al.*[40])

and physical properties,[38] but use of the silk material alone did not enable sufficient attachment or growth of cells.

Recently, a reinforced composite silk scaffold was designed and found to have the mechanical properties of a silk material, while enabling increased adhesion and proliferation of cells by lyophilized collagen-hydroxyapatite (HA) substrates. The collagen-hyaluronan substrates in the reinforced silk scaffold also led to increased rates of cell migration when compared with silk. Moreover, these substrates induced angiogenesis, which is essential for the initial phase of repair of a damaged ligament (Figs. 28.2 and 28.3).[39,40]

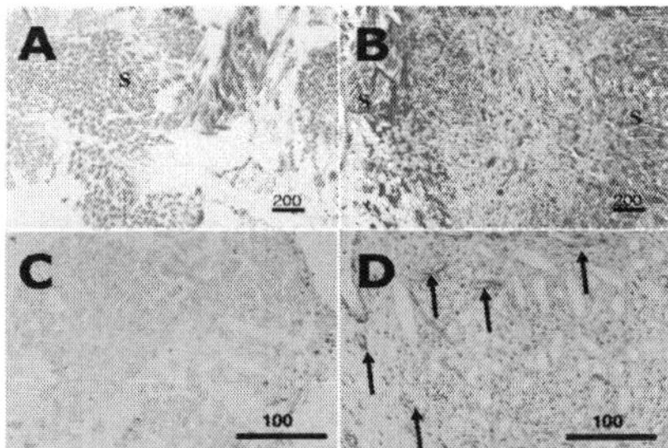

Figure 28.3. Histological comparison of the silk scaffold (a, c) and composite silk scaffold (b, d). The cross section was performed in the center of the silk and composite scaffolds. (a, b: s = silk fiber; B: Blue color = new synthesis collagen; H: arrow: blood vessels). Masson Trichrome stain: (a, b) = ×40, scale bar = 200 μm. CD31 stain: (c, d) = ×200, scale bar = 100 μm. (Seo *et al.*[40]) See also Color Insert.

28.3.2 *Bioartificial Bone*

Bone fracture or injuries occur for various reasons, including degenerative, surgical, and traumatic processes. Such injuries compromise the quality of life. There are some critical-size defects above which bone will not regenerate and will instead require clinical repair. When using a biodegradable material for tissue repair, the biocompatibility and toxicity of both the scaffold itself and by-products of its degradation and subsequent metabolites must all be considered.[41] Various synthetic and natural materials have been used to facilitate the repair of bone, including PGA, PLLA, polylactic-co-glycolic acid (PLGA), PCL, poly(ethylene-co-vinyl alcohol), HA, tricalcium phosphate (TCP), demineralized bone matrix (DMB), SIS, collagen, chitosan, and hyaluronate.[42–59] Such materials are used either alone or in mixtures, but the mechanical properties of the porous or blended scaffold are still much lower than that of natural sponge bone.

Regarding scaffolds used to repair bone, it is generally agreed that a highly porous structure and a large surface area are conductive to cell growth and that a satisfactory scaffold should possess biocompatibility and satisfactory mechanical properties. Indeed, only materials that possess proper mechanical properties can enable the scaffold to maintain its shape and structure after being in implanted in the defect.

Li *et al.* attempted to increase the mechanical strength of scaffolds using nano-HA and collagen sponge reinforced with chitosan fiber. It has been reported that chitin induces fibroblasts to release IL-8, which is involved in the migration and proliferation of fibroblasts and vascular endothelial cells.[60,61] Chitosan fibers can enhance the compressive strength of the scaffold by about four times, and a human marrow MSC culture revealed that the reinforced scaffold was more cytocompatible than a scaffold without reinforcement.[14] Additionally, bone regeneration was evaluated after implantation by radiography, histology, bone mineral density, and mechanical strength. The results revealed that only the chitin reinforced implants showed nearly perfect regeneration at 15 weeks after operation.[15] Reinforcement of PGA fiber resulted in a collagen sponge having increased compression *in vitro* and *in vivo*. In addition, when rat MSCs were seeded into scaffolds, more attached cells were observed in the collagen sponge reinforced with PGA fiber than in collagen sponge that did not contain PGA fiber.[8,62,63]

Recent studies have been conducted to evaluate the use of chitosan, calcium phosphate cement (CPC), and biodegradable polymers, such as VicrylTM mesh or poly-glactin mesh. The knitted mesh effectively improved the load-bearing behavior of scaffolds, and in the case of chitosan and CPC scaffolds reinforced with mesh, the strength increased by about tenfold when compared with unreinforced CPC scaffolds. These studies demonstrate that a PCL scaffold reinforced with PLLA fiber could be useful for bone tissue engineering.[64–66]

28.3.3 *Bioartificial Vessel*

The use of several synthetic vascular conduits has been evaluated, but these have proven to be thrombogenic to some extent and have no growth potential. Additionally, these synthetic materials eventually require replacement because of either severe stenosis at the site of conduct implantation due to calcification or complications from size discrepancies that occur as the patients grow.[67]

Using the classical tissue engineering precedent, cells can be seeded onto a biodegradable scaffold that provides sites for cell attachment and space for neotissue regeneration.[68]

Various artificial vessels constructed of biodegradable materials have also been evaluated. Some researchers have constructed tube-shaped reinforced scaffolds composed of L-lactide and ε-carprolactone (PCL) with PGA nonwoven fabric sheets.[67,69]

These tissue-engineered reinforced scaffolds were surgically implanted into dogs after being cultured with autolgous vascular myofibroblasts and smooth muscle cells for one week. The implanted tissue-engineered scaffolds showed no evidence of stenosis, dilatation, or thrombus, and their overall gross appearance was similar to that of native veins.[67]

Recent studies were conducted to evaluate the use of collagen, a PGA mesh, and a PLA mesh. These scaffolds were fabricated by compounding a collagen sponge with a PGA knitted mesh that was reinforced on the outside with woven PLA (Fig. 28.4).

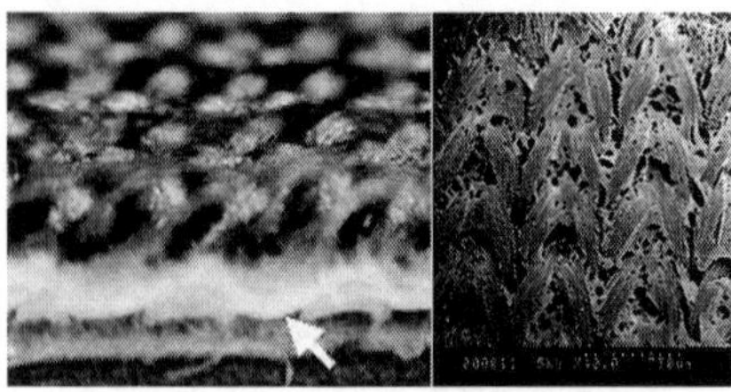

Figure 28.4. Formation of a biodegradable scaffold reinforced with woven a PLA mesh (arrow) cross-linked with collagen sponge (a). Scanning electron microscopy image of the tissue-engineered patch shows the uniformly distributed and interconnected pore structure (pore size 50–150 μm) of the collagen sponge (b) (magnification 40x). (Watanabe *et al.*[67])

The scaffolds were then grafted into the porcine descending aorta, the porcine pulmonary arterial trunk, or the canine ventricular outflow tract. The results of this study demonstrated that there was no thrombus formation and this scaffold provided good *in situ* regeneration.[70] When combined with the results of the aforementioned studies, these findings indicate that these reinforced scaffolds can be used as surgical scaffolds for the repair of vascular defects.

28.3.4 *Bioartificial Tracheae*

Tissue-engineered tracheal tissue requires a scaffold that can maintain its specific shape and size *in vivo*. Many types of scaffolds have been used in attempts to repair tracheal defects, but these have had limited success because of graft ischemia and inflammatory reactions leading to anastomotic dehiscence and stenosis.[71] Additionally, tissue-engineered tracheae have been found to be incapable of providing enough mechanical strength, which led to stenotic complications. Therefore, scaffolds used to generate tracheae should be designed to maintain their shape during the initial surgery and provide enough mechanical strength to prevent collapse.[72]

Fabricating a PLGA scaffold in a specific shape involves coating a fibrous PGA mesh with solutions of PLA and then allowing the solvent to evaporate so that PLA is deposited on the mesh. However, despite the increased mechanical properties, it is unlikely that the properties of PLA will induce the adhesion of cells and formation of tissue.[73] Wu *et al.* attempted to fabricate tissue-engineered tracheae using a PLGA nonwoven mesh coated with collagen, seeded with chondrocytes, and implanted in nude mice. After eight weeks of implantation, the constructs were similar to hyaline cartilage in gross appearance.[74] In addition, Kojima *et al.* evaluated a tissue-engineered trachea using cell and reinforced scaffolds. Specifically, they implanted PGA nonwoven meshes seeded with nasal chondrocytes subcutaneously in nude mice. After six weeks, epithelial cells were injected into the cartilaginous cylinders following removal of the mold. After an additional six weeks, gross inspection of this tissue-engineered trachea demonstrated that the regenerated tissue was similar to the native tracheal cartilage and had patency, rigidity, and a cylindrical shape.[75] Additionally, tracheal scaffolds composed of a gelatin-coated polypropylene mesh cultured with epithelial cells were found to induce stenosis in dogs.[71] Furthermore, Lin *et al.* evaluated reinforced scaffolds composed of PCL and a type II collagen sponge for tissue-engineered tracheae. To accomplish this, the chondrocytes were seeded onto the type II collagen sponge and then implanted subcutaneously into nude mice. The constructs were found to be strong enough to retain their tubular shape, and the gross appearance and histological analyses of the reinforced scaffold revealed that they were similar to native trachea (Fig. 28.5).[72]

The results of the studies described above demonstrate that reinforced PLGA, a PGA nonwoven mesh, polypropylene, and PCL can be used as scaffolds for tracheae because they possess adequate mechanical properties and facilitate increased cell adhesion and growth (Table 28.1).

28.3.5 *Bioartificial Skin*

Poly(2-hydroxyethyl methacrylate) (pHEMA) reinforced with nylon was developed in studies conducted to evaluate artificial skin.[76] This reinforced system increased the mechanical properties of the skin replacement, but reinforcement with a nondegradable polymer was problematic because it prevented the removal of the polymer after implantation.

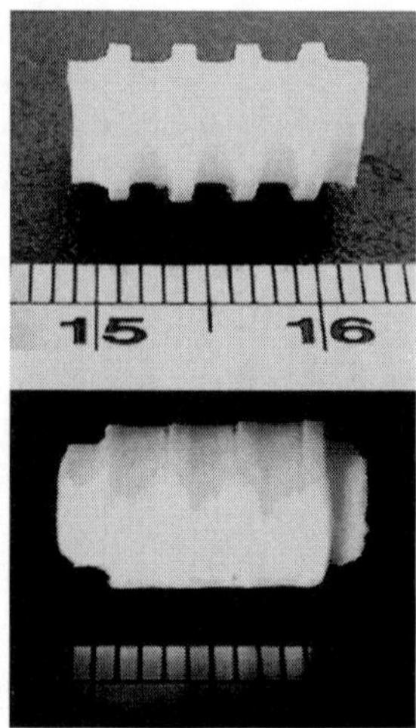

Figure 28.5. Photographs of the PCL tracheal scaffold: (a) the PCL frame and (b) the PCL frame with the type II collagen sponge in the grooves. (Lin *et al.*[73])

Table 28.1. Development of various reinforced composite scaffolds.

Application	Type of reinforced scaffold	References
Ligament/tendon	Dacron tube + inductor (collagen sponge)	31
	Carbon fiber + PLA/PCL coating	32
	Collagen thread + PLA coating	36
	Silk thread + RGD coating	37
	Silk mesh + silk sponge	38
	Silk thread + collagen/hyaluronan sponge	40
Bone	PLLA fiber + polycaprolactone	65
	Chitin fiber + collagen/nano-hydroxyapatite	Li *et al.*, 2005
	PGA fiber + collagen sponge	63
	Vicryl^{TM} mesh + chitosan/calcium phosphate cement	64
Vessel	PGA nonwoven fabric + L-lactide/PCL	67
	PGA/PLA mesh + collagen sponge	70
Trachea	Polypropylene mesh + gelatin coating	71
	PLGA fiber + collagen coating	72
	PCL + type ± collagen sponge	73
Skin	Nylon mesh + poly 2-hydroxyethyl methacrylate hydrogel	76
	PGA fiber + collagen sponge	77
	Collagen thread + collagen sponge	78

Abbreviation: RGD, arginine–glycine–aspartic acid.

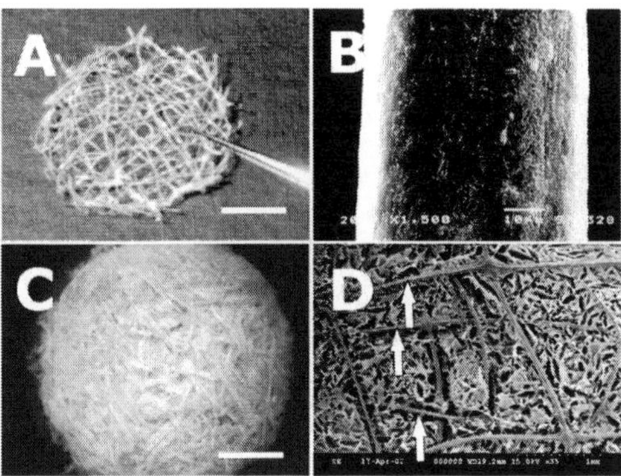

Figure 28.6. Optical microscopic photographs of a collagen mesh (a, diameter = 15 mm) and a reinforced collagen sponge with collagen mesh (c). Scanning electron microscopy of a collagen thread (b) and a reinforced collagen sponge with a collagen mesh (a). Original magnification: (b) ×1500; (d) ×35, scale bar = (b) 10 μm: (d) 1,000 μm. (Seo *et al.*, Biotechnol. Bioprocess Eng. 1998)

Recently, Okochi *et al.* attempted to increase wound healing through enhanced structural stability using collagen sponges. A collagen sponge reinforced with PGA was developed, which increased hair formation and growth when compared with a typical collagen sponge (Fig. 28.6). Epidermal cysts and ectopic hairs formed on the collagen sponge without PGA, and the volume of the sponge decreased as a result of humidity after seeding of the cell suspension. This resulted in the pores of the collagen sponge shrinking, which may have inhibited the movement of the transferred cells and led to the formation of epidermal cysts forming at the grafted site and the induction of cell leakage and loss.

However, the collagen sponge reinforced with PGA underwent less shrinkage, thereby maintaining the size of the pore structures so that the grafted cells could be retained.[77] Recently, a novel type of reinforced scaffold that had high biocompatibility and strong mechanical properties was designed. This collagen sponge was reinforced with a collagen mesh using a combination of lyophilization and cross-linking methods, which resulted in a structure composed of collagen threads and the optimal materials for prevention of an inflammatory reaction.[78]

28.4 Conclusion and Outlook

Previous studies demonstrated that growth factors and the ECM are important, but it is also apparent that the stability of the three-dimensional structure is essential. The collagen or hyaluronan substrates played an important role in increased cell attachment and proliferation. These substrates are very well-known biocompatible and cytotropic natural materials that contain RGD sequences; therefore, they have been utilized in a number of clinical treatments, including the construction of artificial skin, bone, and cartilage, and in injectable materials. However, the physical properties of these substrates were very poor, thus they were insufficient to support the mechanical load from these materials. Therefore, composite scaffolds reinforced with PGA, PLA, PCL monomers or polymers, silk, and chitin were developed. These composite scaffolds were found to have strong mechanical and physical properties. Taken together, these findings indicate that tissue reconstruction using a degradable scaffold is feasible. However, the synthetic scaffold system alone is not suitable for repair and reconstruction, but reinforced composite scaffolding can be very useful for repairing various tissues *in vivo*.

Future studies will concentrate on improving both the cell migration, proliferation, differentiation, and the divided area of the scaffold for co-culture.

Acknowledgments

This work was supported by a grant from the Korean Health 21 R&D Project, Ministry of Health and Welfare, Republic of Korea (0405-BO01-0204-0006).

References

1. G. H. Altman, F. Diaz, C. Jakuba, T. Calabro, R. L. Horan, J. C. Helen, J. Richmond and D. L. Kaplan *Biomaterials* **401** (2003).
2. G. H. Altman, R. L. Horan, H. H Lu, J. Moreau, I. Martin, J. C. Richmond and D. L. Kaplan, *Biomaterials*, **4131** (2002)

3. J. A Cooper, H. H Lu, F. K. Ko, J. W. Freeman, and C. T. Laurencin, *Biomaterials*, **1523** (2005).
4. L. Meinel, S. Hofmann, V. Karageorgiou, C. Kirker-Head, J. McCool, G. Gronowicz, L. Zichner, R. Langer, G. Vunjak-Novakovic, and D. L. Kaplan, *Biomaterials*, **147** (2005).
5. J. Hodde, *Tissue Eng.*, **295** (2002)
6. K. Kawai, S. Suzuki, Y. Tabata, Y. Ikada, and Y. Nishimura, *Biomaterials*, **489** (2000).
7. S. D. Gorham, M. J. Monsour, and R. Scott, *Urol. Res.*, **53** (1987)
8. H. Hosseinkhani, M. Hosseinkhani, F. Tian, H. Kobayashi, and Y. Tabata, *Tissue Eng.*, **11** (2007).
9. M. J. Monsour, R. Mohammed, S. D. Gorham, D. A. French, and R. Scott, *Urol. Res.*, **235** (1987)
10. R. Scott, R. Mohammed, S. D. Gorham, D. A. French, M. J. Monsour, A. Shivas, and T. Hyland, *Br. J. Urol.*, **26** (1988)
11. C. G. Gemmell, S. D. Gorham, M. J. Monsour, F. McMillan, and R. Scott, *Urol. Res.*, **381** (1988).
12. R. Scott, S. D. Gorham, M. Aitcheson, S. P. Bramwell, M. J. Speakman, and R. N. Meddings, *Br. J. Urol.*, **421** (1991)
13. I. Kanatani, A. Kanematsu, Y. Inatsugu, M. Imamura, H. Negoro, N. Ito, S. Yamamoto, Y. Tabata, Y. Ikada, and O. Ogawa, *Tissue Eng.*, **2933** (2007)
14. X. Li, Q. Feng, W. Wang, and F. Cui, *J. Biomed. Mater. Res. B Appl. Biomater.*, **219** (2006)
15. X. Li, Q. Feng, X. Liu, W. Dong, and F. Cui, *Biomaterials*, **1917** (2006)
16. A. Hokugo, T. Takamoto, and Y. Tabata, *Biomaterials*, **61** (2006)
17. N. Meddings, R. Scott, R. Bullock, D. A. French, T. A. Hide, and S. D. Gorham, *Acta. Neurochir. (Wien).*, **53** (1992).
18. W. S. Yang, H. W. Roh, W. K. Lee, and G. H. Ryu, *J. Biomater. Sci. Polym. Ed.*, **151** (2006).
19. A. S. Mathew, K. Sreenivasan, P. V. Mohanan, T. V. Kumary, and M. Mohanty, *Trends. Biomater. Artif. Organs*, **115** (2006).
20. J. Y. Wang, D. T. Tsukayama, B. H. Wicklund, and R. B. Gustilo, *J. Biomed. Mater. Res.*, **655** (1996).
21. O. Gonzalez, R. L Smith, and S. B. Goodman, *J. Biomed. Mater. Res.*, **463** (1996).
22. K. M. Miller, and J. M. Anderson, *J. Biomed. Mater. Res.*, **713** (1998).
23. A. Kesisoglou, J. C. Knowles, and I. Olsen, *J. Mater. Sci.*, **1189** (2002).

24. K. Kohilas, M. Lyons, R. Lofthouse, C. G. Frondoza, R. Jinnah, and D. S. Hungerford, *J. Biomed. Mater. Res.*, **95** (1999).
25. K. M. Miller, and J. M. Anderson, *J. Biomed. Mater. Res.*, **911** (1989).
26. J. Y. Wang, D. T. Tsukayama, and B. H. Wicklund, *J. Biomed. Mater. Res.*, **655** (1996)
27. A. Kesisoglou, J. C. Knowles, I. Olsen, *J. Mat. Sci. Mat. Med.*, **1189** (2002)
28. K. M. Miller, V. Rosecapara, and J. M, anderson, *J. Biomed. Mater. Res.*, **1007** (1989)
29. Y. K. Seo, H. H. Yoon, Y. S. Park, K. Y. Song, W. S. Lee, and J. K. Park, *Cell Biol. Toxicol.*, (2008).
30. G. Vunjak-Novakovic, G. A. Altman, and D. L. Kaplan, *Ann. Rev. Biomed. Eng.*, **131** (2004)
31. D. Huguet, J. Delecrin, and G. Daculsi, *J. Mater. Sci. Mater. Med.*, **67** (1997)
32. A. B. Weiss, M. E. Blazina, and H. Alexander, *Clin. Orthop. Relat. Res.*, **77** (1985)
33. D. H. R. Jenkins, *J.Bone. Joint. Surg. Br.*, **520** (1978)
34. M. Chvapil, D. P. Speer, and D. H. King, *J. Biomed. Mater. Res.*, **313** (1993)
35. J. F. Cavallaro, P. D. Kemp, and K. H. Kraus, *Biotec. Bioeng.*, **781** (1994)
36. M. G. Dunn, L. D. Bellincampi, and J. P. Zawadsky, *J. Appl. Polym. Sci.*, **1423** (1997)
37. J. Chen, G. H. Altman, and D. L. Kaplan, *J. Biomed. Mater. Res.*, **559** (2003)
38. H. Liu, H. Fan, Y. Wang, S. L. Toh, and J. C. Goh, *Biomaterials*, **662** (2008).
39. H. Fan, H. Liu, S. L. Toh, and J. C. Goh, *Biomaterials*, **1017** (2008).
40. Y. K. Seo, H. H. Yoon, K. Y. Song, S. Y. Kwon, H. S. Lee, Y. S. Park, and J. K. Park, *J. Orthop. Res.*, **495** (2009).
41. D. A. Wahl, and J. T. Czernuszka, *Eur. Cell Mater.*, **43** (2006).
42. Y. M. Lee, S. H. Nam, Y. J. Seol, T. I. Kim, S. J. Lee, Y. Ku, I. C. Rhyu, C. P. Chung, S. B. Han, and S. M. Choi, *J. Periodontol.*, **865** (2003).
43. S. L. Ishaug-Riley, G. M. Crane-Kruger, M. J. Yaszemski, and A. G. Mikos, *Biomaterials*, **1405** (1998).
44. B. Inanc, A. E. Elcin, and Y. M. Elcin, *Tissue Eng.*, **257** (2006).
45. K. Matsumura, S. H. Hyon, N. Nakajima, H. Iwata, A. Watazu, and S. Tsutsumi, *Biomaterials*, **4817** (2004).
46. K. G. Marra, J. W. Szem, P. N. Kumta, P. A. DiMilla, and L. E. Weiss, *J. Biomed. Mater. Res.*, **324** (1999).
47. Y. C. Fu, H. Nie, M. L. Ho, C. K. Wang, and C. H. Wang, *Biotechnol. Bioeng.*, **996** (2008).

48. P. Q. Ruhé, H. C. Kroese-Deutman, J. G. Wolke, P. H. Spauwen, and J. A. Jansen, *Biomaterials*, **2123** (2004).
49. M. J. Hoffer, D. J. Griffon, D. J. Schaeffer, A. L. Johnson, and M. W. Thomas, *Vet. Surg.*, **639** (2008).
50. S. Honsawek, D. Dhitiseith, and V. Phupong *J. Med. Assoc. Thai.*, **S189** (2006).
51. D. C. Moore, H. A. Pedrozo, J. J. Crisco 3rd, and M. G. Ehrlich, *J. Biomed. Mater. Res*, **259** (2004).
52. C. V. Rodrigues, P. Serricella, A. B. Linhares, R. M. Guerdes, R. Borojevic, M. A. Rossi, M. E. Duarte, and M. Farina, *Biomaterials*, **4987** (2003).
53. I. Tcacencu and M. Wendel, *J. Mater. Sci. Mater. Med.*, **2015** (2008).
54. Y. Xiao, H. Qian, W. G. Young, and P. M. Bartold, *Tissue Eng.*, **1167** (2003).
55. T. Nakahara, T. Nakamura, E. Kobayashi, M. Inoue, K. Shigeno, Y. Tabata, K. Eto, and Y. Shimizu, *Tissue Eng.*, **153** (2003).
56. Y. F. Zhang, X. R. Cheng, Y. Chen, B. Shi, X. H. Chen, D. X. Xu, and J. Ke, *J. Biomater. Appl.*, **333** (2007).
57. F. Zhao, Y. Yin, W. W. Lu, J. C. Leong, W. Zhang, J. Zhang, M. Zhang, and K. Yao, *Bioamterials*, **3227** (2002).
58. L. S. Liu, A. Y. Thompson, M. A. Heidaran, J. W. Poser, and R. C. Spiro, *Biomaterials*, **1097** (1999).
59. J. Kim, I. S. Kim, T. H. Cho, K. B. Lee, S. J. Hwang, G. Tae, I. Noh, S. H. Lee, Y. Park, and K. Sun, *Biomaterials*, **1830** (2007).
60. Y. Okamoto, M. Watanabe, K. Miyatake, M. Morimoto, Y. Shigemasa, and S. Minami, *Biomaterials*, **1975** (2002).
61. T. Mori, M. Okumura, M. Matsuura, K. Ueno, S. Tokura, Y. Okamoto, S. Minami, and T. Fujinaga, *Biomaterials*, **947** (1997).
62. H. Hosseinkhani, M. Hosseinkhani, F. Tian, H. Kobayashi, and Y. Tabata, *Biomaterials*, **5089** (2006).
63. H. Hosseinkhani, Y. Inatsugu, Y. Hiraoka, S. Inoue, and Y. Tabata, *Tissue Eng.*, **1476** (2005).
64. H. H. Xu, J. B. Quinn, S. Takagi, and L. C. Chow, *Biomaterials*, **1029** (2004).
65. V. Guarino, F. Causa, P. Taddei, M. di Foggia, G. Ciapetti, D. Martini, C. Fagnano, N. Baldini, and L. Ambrosio, *Biomaterials*, **3662** (2008).
66. A. S. Von Gonten, J. R. Kelly, and J. M. Antonucci, *J. Mater. Sci. Mater. Med.*, **95** (2000).
67. M. Watanabe, T. Shin'oka, S. Tohyama, N. Hibino, T. Konuma, G. Matsumura, Y. Kosaka, T. Ishida, Y. Imai, M. Yamakawa, Y. Ikada, and S. Morita, *Tissue Eng.*, **429** (2001).

68. T. Shinoka and C. Breuer, *Yale J. Biol. Med.*, **161** (2008).
69. S. H. Lim, S. W. Cho, J. C. Park, O. Jeon, J. M. Lim, S. S. Kim, and B. S. Kim, *J Biomed Mater Res Part B*, **537** (2008).
70. S. Iwai, Y. Sawa, S. Taketani, K. Torikai, K. Hirakawa, and H. Matsuda, *Ann. Thorac. Surg.*, **1821** (2005).
71. J. Kim, S. W. Suh, J. Y. Shin, J. H. Kim, Y. S. Choi, and H. Kim, *J. Thorac. Cardiovasc. Surg.*, **124** (2004).
72. W. Wu, X. Feng, T. Mao, X. Feng, H. W. Ouyang, G. Zhao, and F. Chen, *Br. J. Oral Maxillofac. Surg.*, **272** (2007).
73. C. H. Lin, J. M. Su, and S. H. Hsu, *Tissue Eng. Part C Methods.*, **69** (2008).
74. J. M. Moran, D. Pazzano, and L. J. Bonassar, *Tissue Eng.*, **63** (2003).
75. K. Kojima, L. J. Bonassar, A. K. Roy, H. Mizuno, *J. Cortiella C. A. Vacanti*, **823** (2003).
76. C. D. Young, J. R. Wu, and T. L. Tsou, *Biomaterials*, **1745** (1998).
77. M. Itoh, Y. Hiraoka, K. Kataoka, N. H. Huh, Y. Tabata, and H. Okochi, *Tissue Eng.*, **818** (2004).
78. Y. K. Seo, H. H. Youn, C. S. Park, K. Y. Song, and J. K. Park, *Biotechnol. Bioproc. Eng.*, **745** (2008).

Chapter 29

THREE-DIMENSIONAL SHAPE CONTROL OF IMPLANT DEVICES

Ung-il Chung,* Hideto Saijoh, Kazuyo Igawa, Yuki Kanno, Yoshiyuki Mori, and Tsuyoshi Takato

Department of Bioengineering, the University of Tokyo Graduate Schools of Engineering and Medicine & Division of Tissue Engineering, the University of Tokyo Hospital 7-3-1 Hongo, Bunkyo-ku, Tokyo 113-0033, Japan

*tei@bioeng.t.u-tokyo.ac.jp

The performance of the scaffolds holds a key to the realization of tissue engineering/regenerative medicine in clinical settings. We have focused on the vital role of the dimensional compatibility of the scaffolds. By controlling the three-dimensional (3D) shape of the scaffolds, we have significantly improved the performance of the artificial bones, which have good dimensional compatibility, resultant reduction in the operation time and corresponding invasiveness, and resultant speedy union with the host bone tissues. We conclude that 3D shape control is vital to the performance of the scaffolds. We propose that it is worth considering at least once to attempt to control the 3D shape of the scaffold by optimizing the design and fabrication method, before using expensive and high-risk growth factors and cells.

Handbook of Intelligent Scaffolds for Tissue Engineering and Regenerative Medicine
Edited by Gilson Khang

www.panstanford.com

29.1 Introduction

To bring tissue engineering/regenerative medicine into reality, it is crucial to sufficiently advance and combine the three pillars, that is, cells, signaling molecules, and scaffolds.[1] However great the advancement in the basic research on the first two may appear to be in the recent years, cells and signaling molecules have limitations to being clinically applied without the help of scaffolds. In addition, in some tissues, including the skeletal system, the scaffolds alone appear to be able to provide treatment to some extent, provided that the performance of the scaffolds is sufficiently improved.

Demand characteristics for the tissue engineering scaffolds have distinctive features in comparison with those for other industrial ones.[2] The demand characteristics for the former are defined not only by the properties of the scaffold materials but also by the interactions of the scaffold materials with the living body, while those for the latter are mainly defined by the properties of the materials. The demand characteristics can be categorized into at least six items: mechanical strength, workability, biocompatibility, biodegradability, biosafety, and regeneration inductivity (Table 29.1). The first two items can be well predicted by the information on the scaffold materials, thanks to the great achievement of materials science. The third, fourth, and fifth items can never be fully predicted by the study of the scaffold materials alone at this moment. We can obtain information on these three only by studying the interactions of the scaffold materials with the living body; the study usually consists of

Table 29.1. Demand characteristics for tissue engineering scaffolds.

Items	Major source of prediction
1. Mechanical strength	Material property
2. Workability	Material property
3. Biocompatibility	Interaction with living body
4. Biodegradability	Interaction with living body
5. Biosafety	Interaction with living body
6. Regeneration inductivity	Biology

implanting the scaffold materials in animals and subsequently observing the reactions of the animals to the materials. Information on the last item cannot be obtained without conducting biological research. The last item is a desirable property, that is, the activity of the scaffolds to influence the proliferation and differentiation of transplanted and host cells to actively induce regeneration. If realized, it will dramatically enhance the performance of the scaffolds.

In the current paper, we would like to take up bone tissue engineering as an example, discuss ways to optimize the performance of scaffolds, and point out a pivotal role of the 3D shape control of the scaffolds.

29.2 Current Status of Artificial Bones

Reconstructive surgery using autograft, allograft, and artificial bones has been performed to treat craniofacial bone deformities.[3] An autograft is superior to the other techniques in function and engraftment because it is a bone tissue derived from the same individual, containing live cells and growth factors. An autograft has both the ability to facilitate bone regeneration (osteoconductivity) and the ability to actively induce bone regeneration (osteoinductivity) and is speedily fused and integrated to the bone of the implantation site. However, because this process requires highly invasive bone collection surgery from healthy sites, donor site morbidity often occurs with the quantity of the autograft being limited.[4] Although an allograft is usually collected from cadavers and thus is free from the invasiveness to the recipient and less restricted in quantity, it runs a biological risk of contamination by pathogens as well as an ethical risk associated with illegal body trading.[5] Because an allograft is usually heat-treated and kept frozen in order to reduce immunological reactions, no live cells are present and growth factors are inactivated to some extent. Therefore, its osteoconductivity and osteoinductivity are inferior to those of an autograft. These factors cause an allograft to fail more often than an autograft. In addition, common to the both procedures, the graft has to be manually carved to fit to deformities during surgery. This process is often time consuming and laborious, being associated with low precision and a poor aesthetic outcome.[6] These situations call for the development of the artificial bones that can replace autografts and allografts.

Currently, among the three major biomaterials (polymers, metals, and ceramics), ceramics is most widely used; calcium phosphates are the most popular materials for artificial bones.[7,8] The choice of the calcium phosphates is reasonable because approximately 70% of bone in our body is made of the calcium phosphates, amounting to about 1.7 kg for a male weighing 60 kg.[9] Therefore, their biocompatibility and biosafety were, in a sense, already tested in the living body. The calcium phosphates are metabolized and degraded by the endogenous system for bone remodeling, although the speed of degradation varies depending on their particle size and form. The calcium phosphates are naturally osteoconductive.[9] This unique combination of the excellent material properties suitable for the artificial bones gives the calcium phosphates a big edge over the other materials.

The artificial bones made of the calcium phosphates are superior to autografts and allografts in biosafety, unlimited quantity, and low invasiveness because they are made from limestone and mineral phosphates, are thus free from contamination by pathogens, and do not require bone collection from donor sites.[10] However, the artificial bones that have been developed and used so far in clinical settings come in relatively simple, uniform shapes, requiring cumbersome and low-precision shape adjustment by surgical operators during surgery. In addition, these artificial bones are usually sintered after fabrication in order to increase mechanical strength and thereby withstand surgical procedures.[11,12] The postfabrication sintering process, however, causes contraction in size and often decreases biodegradability by increasing the calcium phosphates in glass state.[13] Therefore, novel artificial bones are in need that have better shape compatibility to deformities, appropriate mechanical strength without the postfabrication sintering, and biodegradability. We have conceived that the key to meeting all demands may lie in the innovation of the 3D fabrication technology.[14]

29.3 3D Fabrication Technologies and Their Comparison

There are three major 3D fabrication methods in industries: machining, casting, and layer manufacturing (Table 29.2).[15,16] Maxillofacial bone deformities tend to have complex external shapes, including overhangs and internal structures such as interconnecting pores suitable for blood vessel and cell invasion.[17] Machining is a method to remove excess parts from bulk materials using cutting tools. For example, a sculpture is made by cutting from wood and stone. Currently, most sophisticated machining in industries is achieved by the five-axis machining center. Although machining is good at the reproduction of surface shape and at the custom-made production, it has difficulties in fabricating overhangs and complex internal structures because of the limited accessibility of the cutting tools to these structures.

Casting is a method to produce objects by pouring molten materials into a shaped mold and allowing it to cool and solidify. Casting is used to shape materials, including glass, plastics, metals, and alloys. Although casting is good at the reproduction of substantially complex shapes, including some overhangs, it is unable to reproduce complex internal structures. It is suitable for the production of the same shape in a large quantity, not for custom-made production.

Layer manufacturing is also called rapid prototyping or solid free-form fabrication.[18,19] In this method, virtual designs created by computer-aided design (CAD) are transformed into a collection of thin-slice data. The manufacturing machines then read in the data and lay down thin layers of materials such as liquid, powder, and sheet, which are sequentially fused to create the final shape. Layer

Table 29.2. Three major 3D fabrication methods and their characteristics.

	Strong points	Weak points
1. Machining	Custom-made production	Overhangs Internal structures
2. Casting	Fairly complex shapes	Custom-made production
3. Layer manufacturing	Any shape Custom-made production	Available to limited biomaterials

manufacturing enables to create practically any shapes or geometric features, including overhangs and complex internal structures. The method is most suitable for custom-made production. Therefore, layer manufacturing is superior to the other two 3D fabrication methods in manufacturing medical implants.[19]

Within layer manufacturing, there are several major submethods that have distinct features, using different materials.[15] They include stereolithography, fused deposition modeling, selective laser sintering, and inkjet printing. The current stereolithography method uses photosensitive resins and photoinitiators that are usually not biocompatible or biodegradable.[20] Furthermore, the method produces radicals that are toxic to human bodies. The fused deposition modeling method uses thermoplastics that are usually not biocompatible or biodegradable and require a support structure in some cases.[21] Selective laser sintering uses thermoplastic resins and metal powders at high temperature, which causes size contraction.[22] By the inkjet printing technology, in contrast, biocompatible and biodegradable materials can be processed at room temperature, and porous scaffolds with controlled internal structures with high resolution can be manufactured.[23] As we already argued, calcium phosphates have a great advantage as biomaterials for artificial bones over metals and polymers because of their excellent combination of good material properties.[9] Thus, among the different submethods of layer manufacturing, inkjet printing seems to have the greatest potential to directly fabricate calcium phosphate materials.

29.4 Inkjet Printing Technology

In inkjet printing, powder material is first spread by a roller onto a thin sheet. Then, an inkhead prints a curing solution and draw pictures on the sheet. The thin sheet of the powder material then becomes solidified where the curing solution is printed. In the next step, another thin sheet of the powder material is newly spread on top of the solidified layer, and the process is repeated until the final shape is created.[23] Despite efforts by a number of researchers, there has been no success in directly manufacturing clinically useful custom-made implants from calcium phosphates by the inkjet printing technology, due to the technical difficulties in keeping dimensional precision and mechanical strength at once. As we mentioned above, the postfabrication sintering process is usually used to increase the mechanical strength of the implants made from calcium phosphates; however, postfabrication sintering inevitably causes contraction in size by up to as much as 27%, compromising the greatest advantage of the inkjet printing technology, that is, the excellent shape control.[24] Postfabrication sintering also compromises biodegradability by increasing the crystallinity of the calcium phosphates, leading to poor resorption by osteoclasts.[25–27] Therefore, it is necessary to find a way to fabricate the calcium phosphates without postfabrication sintering. We have focused on the finding that the addition of water turns α-tricalcium phosphate (TCP) powder into hydroxyapatite and solidifies it.[7] Although this finding was paid attention to by other groups and adopted as a fabrication principle, their final products failed to exhibit sufficient mechanical strength for clinical use, calling for postfabrication sintering.[28] By systematically going over fabrication parameters, we succeeded in optimizing the diameter of the α-TCP particles and the viscosity and pH of the curing solution for inkjet printing so that, without postfabrication sintering, we could obtain the mechanical strength sufficient for surgical handling and non-weight-bearing applications in dogs.[23]

Based on this technological breakthrough, we started fabricating custom-made artificial bones from α-TCP powder using the inkjet printing technology and implanted them in 10 patients with maxillofacial deformities.[29] The artificial bones had good dimensional compatibility to deformities due to the excellent shape control by the fabrication method, thereby requiring minimum dimensional adjustment for several minutes before implantation and minimum fixation by suturing at several points after implantation. In contrast, the conventional artificial bones typically require at least an hour for dimensional adjustment and another hour for fixation. Thus, the inkjet-printed custom-made artificial bones helped dramatically reduce the operation time and the corresponding invasiveness to the patients. The initial clinical evidence of bone union between the artificial bones and host bone tissues was seen as early as at 6 months, and all the patients had partial union at 12 months. This relatively speedy union may have occurred for the following reasons. First, the inkjet-printed custom-made artificial bones had superior dimensional compatibility, which might have helped facilitate bone regeneration from host tissues by bringing them in close proximity to the inkjet-printed custom-made artificial bones. Second, the inkjet-printed custom-made artificial bones were not sintered and therefore likely had better biodegradability, enhancing bone formation. Third, the inkjet-printed custom-made artificial bones had macroscopic pores, which might have helped facilitate cell and blood vessel invasions. All the patients were satisfied with aesthetic facial appearance. Based on these findings, a large-scale clinical trial of the inkjet-printed custom-made artificial bones is now underway.

29.5 Conclusions and Outlook

By controlling the 3D shape of the scaffolds, we succeeded in dramatically improving the performance of the artificial bones. Using the growth factors and cells is interesting and eye catching; however, it is extremely expensive, is associated with various risks that cannot be ignored, and is therefore far from clinical applications. Before thinking of combining the scaffolds with the growth factors and cells, we should seriously ask ourselves whether there may be anything left to be done on the scaffolds themselves. In my opinion, there appears to be much room for improvement of the 3D shape control of the scaffolds. If such improvement is done, much of the problem will be solved for a large number of the patients, bringing us closer to the realization of tissue engineering/regenerative medicine in clinical settings. For the minority of the patients who need more regeneration inductivity, they will be treated in combination with the growth factors, and if it is unsuccessful, the treatment using exogenous cells will be finally considered. In order to achieve success in real clinical settings, we find it crucial to structure the treatment strategy stepwise from the easiest and inexpensive toward the most difficult and expensive.

Using the same principle, we have started improving the performance of granule-type artificial bones. The conventional granule-type artificial bones are manufactured by pulverizing porous block-type artificial bones, and therefore the size and shape of the artificial bones are heterogeneous. This heterogeneity in size and shape leads to heterogeneous spatial packing in the aggregate and consequent generation of mechanically and biologically deleterious dead space. By fabricating the granules of the same size and shape by injection molding, we have discovered that these problems can be solved. The details will be published elsewhere in the near future.

Acknowledgments

This work was supported by Grant-in-Aid for Scientific Research (A) #20240043 from the Japan Society for the Promotion of Science and by Terumo Life Science Foundation.

References

1. R. P. Lanza, R. Langer, and J. Vacanti, Eds., *Principles of Tissue Engineering* (Academic Press, San Diego, 2000).
2. L. Moroni, J. R. de Wijn, and C. A. van Blitterswijk, *J. Biomater. Sci. Polym. Ed.*, **19**(5), 543 (2008).
3. S. W. Herrin and, P. Ocharoen, *Orthod. Craniofac. Res.*, **8**, 174 (2005).
4. P. Tessier, H. Kawamoto, D. Matthews, J. Posnick, Y. Raulo, J. F. Tulasne, and S. A. Wolfe, *Plast. Reconstr. Surg.*, **116**, 6S (2005).
5. B. L. Eppley, W. S. Pietrzak, and M. W. Blanton, *J. Craniofac. Surg.*, **16**, 981 (2005).
6. M. Hallman and A. Thor, *Periodontology*, **47**, 172 (2008).
7. M. Hatoko, H. Tada, A. Tanaka, S. Yurugi, K. Niitsuma, and H. Iioka, *J. Craniofac. Surg.*, **16**, 327 (2005).
8. S. Tomita, S. Molloy, L. E. Jasper, M. Abe, and S. M. Belkoff, *Spine*, **29**, 1203 (2004).
9. S. V. Dorozhkin and M. Epple, *Angew. Chem. Int. Ed. Engl.*, **41**, 3130 (2002).
10. E. Fischer-Brandies and E. Dielert, *J. Oral Implantol.*, **12**, 40 (1985).
11. B. L. Eppley, *J. Craniofac. Surg.*, **13**, 650 (2002).

12. H. Tada, M. Hatoko, A. Tanaka, M. Kuwahara, K. Mashiba, S. Yurugi, H. Iioka, and K. Niitsuma, *J. Craniofac. Surg.*, **13**, 287 (2002).
13. S. Karashima, A. Takeuchi, S. Matsuya, K. I. Udoh, K. Koyano, and K. Ishikawa, *J. Biomed. Mater. Res. A*, **88**, 628 (2009).
14. W. Y. Yeong, C. K. Chua, K. F. Leong, and M. Chandrasekaran, *Trends Biotechnol.*, **22**, 643 (2004).
15. S. J. Hollister, *Nat. Mater.*, **4**, 518 (2005).
16. P. F. Jacobs, Ed., *Rapid Prototyping & Manufacturing: Fundamentals of Stereolithography* (McGraw-Hill, NY, 1992).
17. I. Ono, K. Abe, S. Shiotani, and Y. Hirayama, *J. Craniofac. Surg.*, **11**, 527 (2000).
18. E. Sachlos and J. T. Czernuszka, *Eur. Cell Mater.*, **5**, 29 (2003).
19. S. M. Peltola, F. P. Melchels, D. W. Grijpma, and M. Kellomaki, *Ann. Med.*, **40**, 268 (2008).
20. K. W. Lee, S. Wang, B. C. Fox, E. L. RitmanL, M. J. Yaszemski, and L. Lu, *Biomacromolecules*, **8**, 1077 (2007).
21. D. W. Hutmacher and S. Cool, *J. Cell Mol. Med.*, **11**, 654 (2007).
22. M. H. Smith, C. L. Flanagan, J. M. Kemppainen, J. A. Sack, H. Chung, S. Das, S. J. Hollister, and S. E. Feinberg, *Int. J. Med. Robot.*, **3**, 207 (2007).
23. K. Igawa, M. Mochizuki, O. Sugimori, K. Shimizu, K. Yamazawa, H. Kawaguchi, K. Nakamura, T. Takato, R. Nishimura, S. Suzuki, M. Anzai, U. Chung, and N. Sasaki, *J. Artif. Organs*, **9**, 234 (2006).
24. F. C. Fierz, F. Beckmann, M. Huser, S. H. Irsen, B. Leukers, F. Witte, O. Degistirici, A. Andronache, M. Thie, and B. Müller, *Biomaterials*, **29** 3799 (2008).
25. C. L. Camire, P. Nevsten, L. Lidgren, and I. McCarthy, *J. Biomed. Mater. Res. B Appl. Biomater.*, **79**, 159 (2006).
26. M. Yamada, M. Shiota, Y. Yamashita, and S. Kasugai, *J. Biomed. Mater. Res. B Appl. Biomater.*, **82**, 139 (2007).
27. T. Okuda, K. Ioku, I. Yonezawa, H. Minagi, Y. Gonda, G. Kawachi, M. Kamitakahara, Y. Shibata, H. Murayama, H. Kurosawa, and T. Ikeda, *Biomaterials*, **29**, 2719 (2008).
28. A. J. Ambard and L. Mueninghoff, *J. Prosthodont.*, **15**, 321 (2006).
29. H. Saijo, K. Igawa, K. Kanno, Y. Mori, K. Kondo, K. Shimizu, S. Suzuki, D. Chikazu, M. Iino, M. Anzai, N. Sasaki, U. Chung, and T. Takato, *J. Artif. Organs*, (2009).

Chapter 30

NOVEL FABRICATION AND CHARACTERIZATION OF PORE-SIZE-GRADIENT SCAFFOLDS BY A CENTRIFUGATION TECHNIQUE

Se Heang Oh and Jin Ho Lee*

Department of Advanced Materials, Hannam University
461-6 Jeonmin Dong, Yuseong Gu, Daejeon, Korea
*jhlee@hnu.kr

It is well recognized that the pore size of polymer scaffolds plays an important role for cell growth and tissue regeneration: different kinds of cells were shown to have different optimal pore size ranges in the scaffolds for effective cell growth. So, if the tissue scaffold with a pore size gradient (i.e., the scaffold with gradually increasing pore size along one direction) can be prepared, it will become a powerful tool for basic studies of the interactions between cells or tissue and scaffolds with different pore sizes since the effect of pore size can be effectively examined using one scaffold. In recent years, several techniques have been used to fabricate porous polymer scaffolds having a three-dimensional (3D) pore structure. However, it is not possible from those techniques to fabricate scaffolds with a pore size gradient.

Handbook of Intelligent Scaffolds for Tissue Engineering and Regenerative Medicine
Edited by Gilson Khang

www.panstanford.com

Recently, we developed a new technique to fabricate pore-size-gradient, cylindrical scaffolds by a simple centrifugation. In this technique, the pore size ranges of the scaffold can be easily controlled by adjusting the centrifugal speed. The pore-size-gradient, cylindrical scaffolds were fabricated with alginate, chitosan, and poly-ε-caprolactone (PCL) by a centrifugation technique, and their characterizations in terms of scaffold pore sizes were discussed in this chapter.

30.1 Introduction

The control of a 3D pore structure is of great importance for the development of scaffolds for tissue engineering. Highly porous scaffolds with an interconnected pore structure are desirable in many cases to facilitate cell seeding and adhesion, secretion of extracellular matrices, and eventual tissue regeneration. Numerous investigations, including scaffold fabrication,[1] surface modification,[2–4] and a bioreactor system[5] for the development of the scaffolds, which can provide a desirable environment for cell growth, have been actively conducted. It is well recognized that the pore size of scaffolds plays an important role for cell binding, migration, and ingrowth[6,7] and tissue ingrowth and regeneration.[8,9] A number of cell types exhibit a preference for binding to scaffolds with pore sizes significantly larger than the characteristic cell size, often utilizing a characteristic bridging mechanism where adjacent cells act as support structures to assist bridging large pores. Generally, it was reported that a large pore size or porosity of the scaffold can allow effective nutrient supply, gas diffusion, and metabolic waste removal but lead to low cell attachment and intracellular signaling, while a small pore size or porosity can provide opposite properties of above.[6,9,10] Many researchers have reported optimum pore size ranges for the different kinds of cells or tissues, for example, the pore sizes of $\sim$5 μm for neovascularization,[11] 5 $\sim$ 15 μm for fibroblast ingrowth,[12] $\sim$ 20 μm for hepatocyte ingrowth,[9] 20 $\sim$ 125 μm for skin regeneration,[13] 70 $\sim$ 120 μm for chondrocyte ingrowth,[14] 40 $\sim$ 150 μm for fibroblast binding,[15] 45 $\sim$ 150 μm for liver tissue regeneration,[16] 60 $\sim$ 150 μm for vascular smooth muscle cell binding,[10] 100 $\sim$ 300 μm for

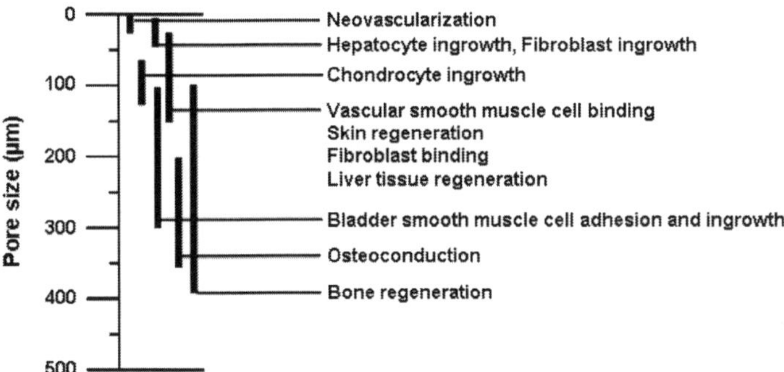

Figure 30.1. Suggested optimum pore size ranges for different cells or tissues (Ref. 22).

bladder smooth muscle cell adhesion and ingrowth,[17] $100 \sim 400$ μm for bone regeneration,[18–21] and $200 \sim 350$ μm for osteoconduction[8] (summarized in Fig. 30.1), depending of the porosity as well as scaffold materials used. Nevertheless of these investigations, more systematic evaluation on optimum pore size ranges of scaffold has been still required because of the absence of well-defined scaffolds with a wide range of pore sizes. Thus, if the scaffold with a pore size gradient can be prepared, it will become a powerful tool for basic studies of the interactions between cells or tissues and scaffolds with different pore sizes since the effect of pore size can be effectively examined using one scaffold. Several techniques, including particulate leaching, phase separation, freeze-drying, gas foaming, fiber bonding, and 3D printing have been used to fabricate porous polymer scaffolds having a 3D pore structure.[1] Among them, the particulate leaching and freeze-drying methods, which can control the pore size of scaffolds by porogen size and freezing temperature, respectively, have been commonly chosen to evaluate of the pore size effect.[18,23] However, it is not possible from those techniques to fabricate scaffolds with a pore size gradient. Also it is troublesome that the scaffolds with different pore sizes are separately fabricated, and thus low reproducibility of pore size distribution is still a limitation. Recently, 3D printing[24] and spinning[25] techniques have been tried to fabricate scaffolds with a pore size gradient.

Recently, we developed a novel scaffold fabrication technique (centrifugation technique) that can prepare cylindrical scaffolds with a pore size gradient (i.e., a gradually increasing pore size along the longitudinal direction). The pore-size-gradient, cylindrical scaffolds were made by both natural (alginate and chitosan) and synthetic (PCL) polymers, which are widely used as cell carrier or scaffold materials for tissue engineering applications. The scaffold was simply obtained by the centrifugation of a cylindrical mold containing fibril-like polymers and the following fibril bonding process, if needed. During the centrifugation, a porous scaffold having a pore size gradient was formed in the cylindrical mold by the gradual increment of the centrifugal force along the cylindrical axis. And also, the pore size ranges of the scaffold could be easily controlled by adjusting the centrifugal speed.

In this chapter, the fabrications of pore-size-gradient, cylindrical scaffolds made by alginate, chitosan, and PCL and their characterizations in terms of scaffold pore sizes are discussed.

30.2 Fabrication of Pore-Size-Gradient Scaffolds

30.2.1 *Materials*

Sodium alginate (also called algin or alginic acid sodium salt; medium viscosity, Mw 80,000 $\sim$ 120,000, Sigma, USA), chitosan (low molecular weight, degree of acetylation 15% $\sim$ 25%; Fluka, Switzerland), and PCL (Mw 42,500; Aldrich, USA) were used to fabricate pore-size-gradient scaffolds. Sodium alginate was washed several times with 65% methanol to remove impurities and vacuum-dried. Pluronic F127 ($EG_{99}PG_{65}EG_{99}$, Mw 12,500; BASF, USA) as a hydrophilic additive of PCL was used as received. Calcium chloride ($CaCl_2$; Oriental Chemical Industry, Korea) as a cross-linking agent of sodium alginate, NaOH (Oriental Chemical Industry, Korea) as a precipitation agent of chitosan, and all other chemicals were analytical grade and were also used as received. Water was ultrapure grade ($<$18 mΩ) supplied from a Milli-Q purification system (Millipore Co., USA).

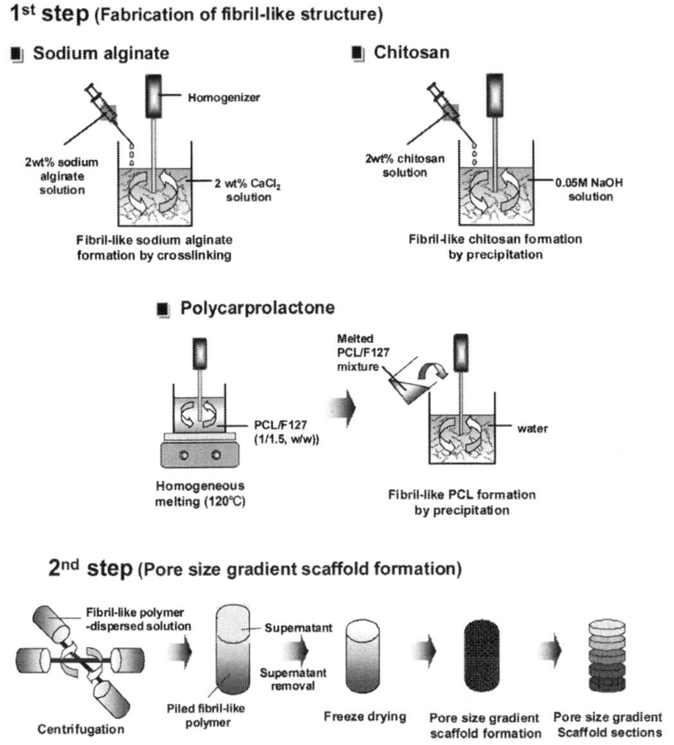

Figure 30.2. Schematic diagrams showing the fabrication process of pore-size-gradient, cylindrical scaffolds by the centrifugation method (Ref. 22).

30.2.2 *Pore-Size-Gradient Alginate Scaffolds*

To prepare alginate cylindrical scaffolds having a pore size gradient, 2 wt% alginate solution was slowly dropped into a 2 wt% $CaCl_2$ solution with vigorous agitation using a homogenizer (20,000 rpm; HG-3000, SMT, Japan) to obtain fibril-like cross-linked alginate structures (Fig. 30.2). The prepared fibril-like alginates were washed in excess water to remove residual $CaCl_2$. Then the fibril-like alginate-suspended aqueous solution was poured into a polypropylene (PP) cylindrical mold (inner diameter 14 mm; height 7.5 cm), and the mold was centrifuged (3,000 rpm) for five minutes. During the centrifugation, the fibril-like alginates were accumulated in the cylindrical mold with a range of pore size gradients by the gradual increment of the centrifugal force along the cylindrical axis. After supernatant removing and freeze-drying, a pore-size-gradient alginate scaffold (height ~5.0 cm) was formed. The prepared alginate scaffold was further treated by soaking in 1 wt% chitosan solution to bond alginate fibrils to each other via anion complex and thus to strengthen the mechanical properties of the scaffold as well as to prevent the disentanglement of fibrous structures in a cell culture medium or in the body. The prepared cylindrical alginate scaffold was sectioned along the longitudinal direction by cutting the scaffold using a blade after being frozen in liquid nitrogen to characterize it in terms of scaffold pore sizes (section thickness 5 mm).

30.2.3 *Pore-Size-Gradient Chitosan Scaffolds*

Chitosan solution (2 wt% in 0.1 M acetic acid) was slowly dropped into 0.05 M NaOH (nonsolvent for chitosan) with vigorous agitation using a homogenizer (20,000 rpm) to obtain fibril-like chitosan precipitates. The prepared fibril-like chitosans were washed in excess water to remove residual NaOH (Fig. 30.2). Then the fibril-like chitosan-suspended solution was poured into a cylindrical PP mold, and centrifugation was carried out, same as the fabrication method for the alginate scaffold above. During the freeze-drying, the somewhat sticky chitosan fibrils were bonded to each other.

30.2.4 *Pore-Size-Gradient PCL Scaffolds*

PCL and Pluronic F127 (1/1.5, w/w) were heated together to 120°C with vigorous agitation using a homogenizer (20,000 rpm) until a homogeneous molten phase was formed. The Pluronic F127 blended with PCL helps to reduce the viscosity of molten PCL and also provides the hydrophilicity in the prepared PCL scaffold. The PCL/Pluronic F127 mixture was slowly dropped into water (nonsolvent for PCL) with vigorous agitation using a homogenizer (20,000 rpm) to obtain fibril-like PCL precipitates (Fig. 30.2). The prepared fibril-like PCLs were washed in excess water to remove residual Pluronic F127. Then the fibril-like PCL-suspended solution was poured into a cylindrical PP mold, and centrifugation was carried out, same as the fabrication method for the alginate or chitosan scaffold above. The prepared PCL scaffold was further treated by a thermal fibril-bonding process for strengthening of mechanical properties; the scaffold was heat-treated nearby the PCL melting temperature (59°C, 1 hr) to bond accumulated PCL fibrils to each other.

30.3 Characterization of Pore-Size-Gradient Scaffolds

30.3.1 *Measurements of Pore Sizes and Porosity*

Surface morphologies of the prepared cylindrical alginate, chitosan, and PCL scaffold sections (thickness 5 mm) along the longitudinal direction were observed by scanning electron microscopy (SEM; Model S-3000N, Hitachi, Japan) operated at an accelerating voltage of 15 kV. Before morphology observations, the specimens were coated with platinum using a sputter coater (SC 7680, Quorum Technologies, UK) under an argon atmosphere. The average pore sizes of the scaffold section surfaces were measured using an image analysis program (i-solution, IMT, Korea) from the SEM photographs. The porosity of the scaffold sections was measured using a specific gravity bottle (Hubbard, Hanil, Korea) based on Archimedes' principle.[26] Briefly, the porosity of the scaffold was determined as follows:

$$\text{Porosity (\%)} = [(W_2 - W_3 - W_s)/\rho_e]/[(W_1 - W_3)/\rho_e] \times 100$$

where, W_1 = specific gravity bottle weight filled with ethanol; W_2 = specific gravity bottle weight including ethanol and scaffold; W_3 = specific gravity bottle weight taken out ethanol-saturated scaffold from W_2; W_s = scaffold weight; ρ_e = density of ethanol; and thus $(W_1 - W_3)/\rho_e$ = total volume of the scaffold including pores, $(W_2 - W_3 - W_s)/\rho_e$ = pore volume in the scaffold.

Figure 30.3 shows the sectional morphologies of the cylindrical alginate, chitosan, and PCL scaffolds fabricated by the centrifugation method. It was observed that the pore size and porosity of the scaffolds gradually increased along the cylindrical axis, regardless of polymer types used, owing to the gradual increment of the centrifugal force along the longitudinal axis in the cylindrical mold containing fibril-like polymers during the

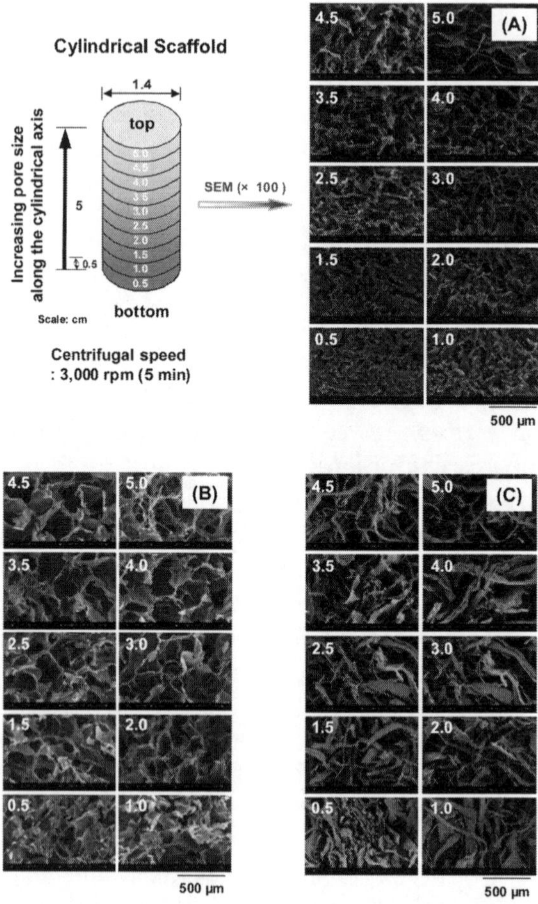

Figure 30.3. SEM photographs of cylindrical scaffold section surfaces along the longitudinal direction. (a) Alginate, (b) chitosan, and (c) PCL scaffolds (× 100) (Ref. 22).

centrifugation. Since the centrifugal force increases with the increasing radius of gyration (position dependent) at the same centrifugal speed, bottom positions of the cylinder (longer radii of gyration) cause more dense accumulation of the polymer fibrils, resulting in smaller pore sizes as well as lower porosities, while the top positions (shorter radii) have larger pore sizes and higher porosities. As the centrifugal speed, 3,000 rpm, was applied to the cylindrical mold

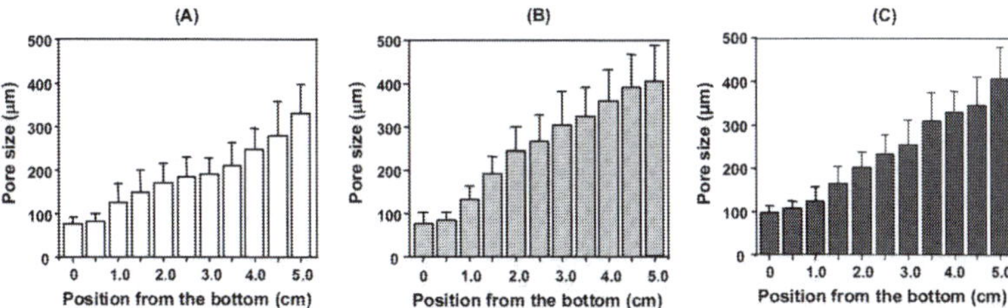

Figure 30.4. Pore size distributions of cylindrical scaffold sections along the longitudinal direction. (a) Alginate, (b) chitosan, and (c) PCL scaffolds (n =3) (Ref. 22).

for five minutes, the pore size and porosity ranges in the prepared scaffold along the cylindrical axis were observed from 80 ~ 88 μm to 310 ~ 405 μm (Fig. 30.4) and from 80% ~ 83% to 93% ~ 94% (Fig. 30.5), respectively, depending on the polymers used. The pore size ranges could be also controlled by adjusting centrifugal speed; for example, 180 ~ 325 μm for 1,000 rpm and 135 ~290 μm for 2,000 rpm were observed for the alginate scaffold (not shown).

30.3.2 *Measurements of Mechanical Properties*

The mechanical properties of the scaffold sections (thickness 5 mm) were measured by a biaxial tensile test equipment designed by our laboratory for small scaffold samples, which was attached in an Instron machine (AG-5000G, Shimadzu, Japan) with a 50 kg_f load cell.[27] A rod with a ball-shaped (diameter 6 mm) tip was hammered vertically at a crosshead speed of 1 mm/min on the

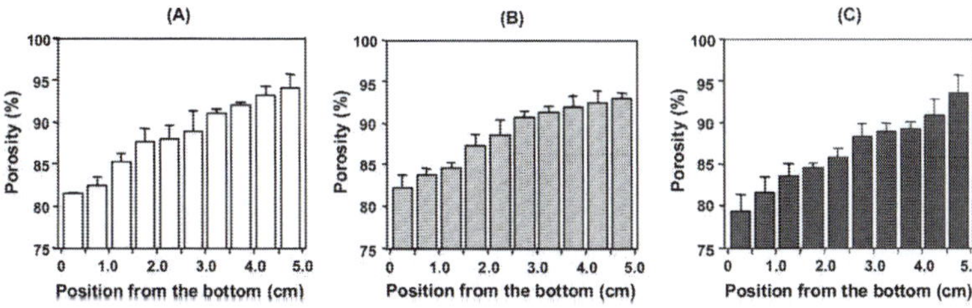

Figure 30.5. Porosity distributions of cylindrical scaffold sections along the longitudinal direction. (a) Alginate, (b) chitosan, and (c) PCL scaffolds (n = 3) (Ref. 22).

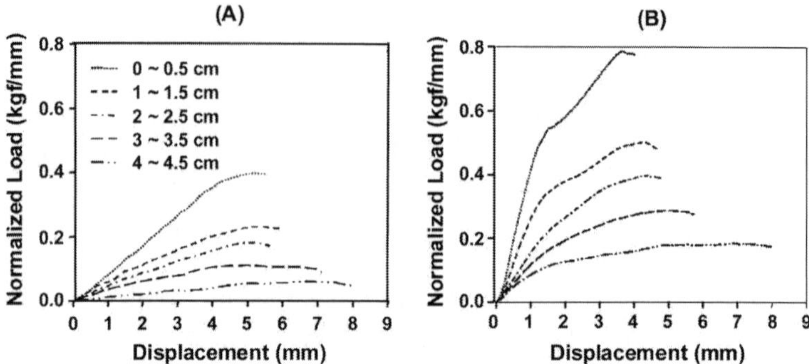

Figure 30.6. Normalized load displacement curves of cylindrical alginate scaffold sections along the longitudinal direction. (a) Nontreated and (b) chitosan-treated scaffold sections (Ref. 22).

specimen placed on a mold-type grip, and the load displacement curves were obtained. The load values were normalized by dividing with the thickness of scaffold (load/scaffold thickness, kg_f/mm). Figure 30.6 compares the mechanical strengths of alginate scaffold sections with/without chitosan treatment. The mechanical strength of the scaffold sections along the longitudinal direction gradually decreased with increasing pore size and porosity, as expected (bottom positions of the cylindrical scaffold have smaller pore sizes as well as lower porosities, as discussed earlier). The chitosan-treated scaffold sections (Fig. 30.6b) showed dramatically increased mechanical strengths compared with the scaffold sections without the treatment (Fig. 30.6a).

It seemed that the positively charged chitosan provides the ion complex with the negatively charged alginate fibrils (i.e., ion complex between $-NH^{3+}$ of chitosan and $-COO^{-}$ of alginate), resulting in the fibril bonding and the increased mechanical strengths. The chitosan treatment to alginate scaffolds may provide an additional benefit of improving cell compatibility. The mechanical strengths of chitosan scaffolds showed a similar trend to the alginate scaffold: the mechanical strength of the scaffold sections along the longitudinal direction gradually decreased with the increasing pore size and porosity (Fig. 30.7), but the strengths were lower than the alginate scaffold sections even though their pore sizes and porosities were almost the same.

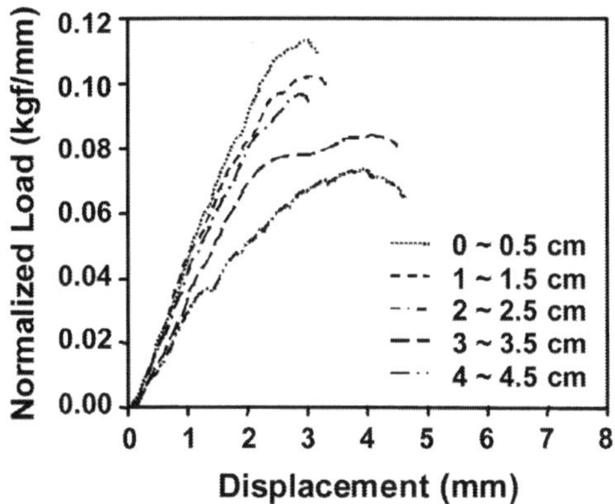

Figure 30.7. Normalized load displacement curves of cylindrical chitosan scaffold sections along the longitudinal direction (Ref. 22).

For PCL scaffolds, heat treatment near the PCL melting temperature (59°C) was conducted during the scaffold fabrication for fibril bonding between adjacent PCL by slight melting, leading to the prevention of the disentanglement of fibrous structures and to the improvement of mechanical strength of the scaffold.

As shown in Fig.30.8a, the mechanical property of the scaffold increased by the heat treatment (i.e., fibril-bonding process). The dimension or pore structure of the scaffold was little affected by the

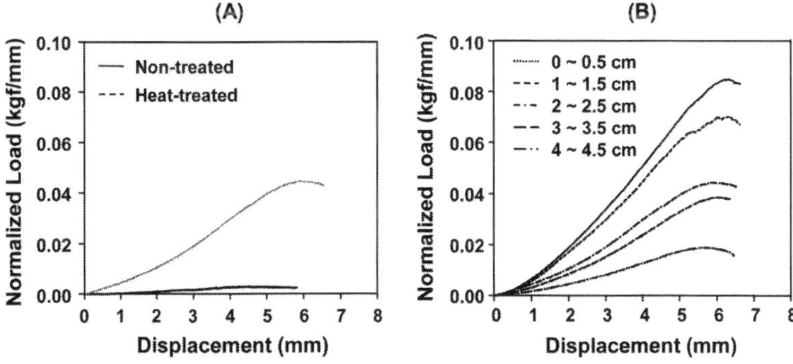

Figure 30.8. Normalized load displacement curves of cylindrical PCL scaffold sections. (a) Comparison of scaffold sections with/without heat treatment at the position of 2.0 $\sim$ 2.5 cm and (b) heat-treated scaffold sections along the longitudinal direction (Ref. 28).

heat treatment condition used. Figure 30.8b shows that the mechanical strength of the scaffold sections (after heat treatment) along the longitudinal direction gradually decreased with the increasing pore size and porosity, similarly to the alginate or chitosan scaffold case. It was recognized from our preliminary studies that all scaffold sections (even the largest-pore-sized scaffold section; position from the bottom, 4.0–5.0 cm) have sufficient mechanical property for *in vitro* cell culture and *in vivo* animal studies.

30.3.3 *Evaluation of Wettability*

The alginate, chitosan and PCL gradient scaffold sections were observed for their wettability or water absorbability, which is an essential factor for cell culture and homogeneous tissue regeneration. For this, a cell culture medium was dropped on top of the scaffold, and the time required for complete absorption of the medium into the scaffold was measured. When the culture medium was dropped to the scaffold sections, the medium was completely wetted into the alginate and chitosan scaffold sections for all the pore size ranges since they are natural polymers having hydrophilicity (Fig. 30.9a).

However, it was not wetted at all into the control PCL scaffold, owing to its hydrophobic character (Fig. 30.9b). Hydrophilization of the PCL scaffold sections (by the addition of Pluronic F127, as treated in this study) allowed the scaffolds to be completely wetted within a few seconds, similar to the alginate or chitosan scaffolds (Fig. 30.9c). The fast wetting of scaffolds into the cell culture medium are highly desirable for tissue engineering applications because the cells can be directly seeded and cultured in these hydrophilized scaffolds without any prewetting treatments.

30.3.4 In vitro *Cell Interactions*

Selected sections of the pore-size-gradient, cylindrical chitosan and PCL scaffolds (diameter 14 mm; thickness 2 mm; average pore sizes ~80, ~190, ~270, and ~400 μm [chitosan] and ~88, ~200, ~310, and ~405 μm [PCL]) sterilized by ethylene oxide (EO) were placed

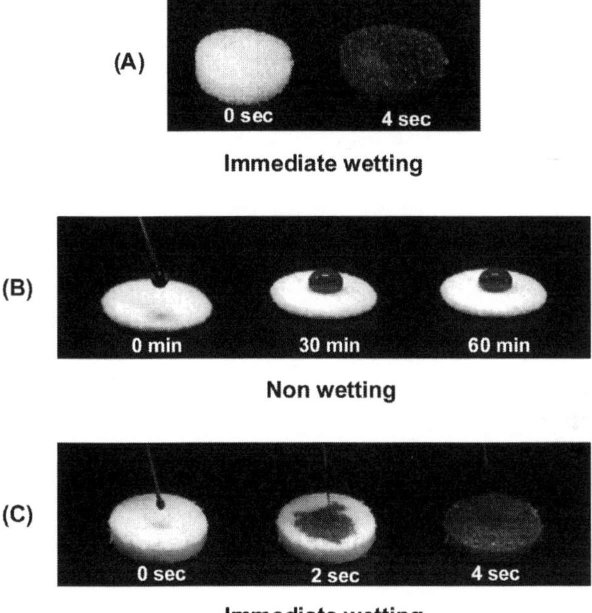

Figure 30.9. Photographs demonstrating the wettability of a cell culture medium into (a) alginate or chitosan, (b) PCL (no hydrophilized), and (c) PCL (hydrophilized by Pluronic F127 addition) scaffold sections at the position of 2.0 ~ 2.5 cm (Ref. 22).

into the 12-well polystyrene (PS) dishes (Corning, USA). Three different types of cells, fibroblasts (NIH/3T3 mouse embryo fibroblasts, Korean Cell Line Bank, Korea), chondrocytes (immortalized human costal chondrocyte cell line C-28/12, Beth Israel Deaconess Medical Center, USA), and osteoblasts (MG-63 human osteosarcoma, Korean Cell Line Bank), were used as model cells to estimate the pore size effect on cell interactions. A cell suspension in the Roswell Park Memorial Institute medium (RPMI, for fibroblasts) or Dulbecco's Modified Eagle's Medium (DMEM, for chondrocytes and osteoblasts) containing 10% fetal bovine serum (FBS, Gibco, USA), 0.1% gentamicin sulfate (Sigma, USA), and 1% penicillin G (Sigma) was seeded onto the each scaffold section (cell density 1×10^6 cells/specimen). The cell suspensions could be easily wetted into the scaffold sections, owing to their hydrophilic character, and thus directly seeded without any prewetting treatment. The cell-seeded scaffold sections were maintained for two hours at 37°C in an incubator with humidified 5% CO_2 atmosphere for cell adhesion to the scaffold sections. Then the scaffold sections were transferred to new 12-well PS dishes, the culture medium was added to the culture plate (3.0 mL/well), and the cells in the scaffold sections were cultured up to eight weeks with mild shaking (about 50 rpm). The culture medium was changed into a fresh one every day during the cell culture periods. The cells were cultured in the scaffold sections for given periods (1, 7, 14, 28, and 56 days), and the viable cell numbers in each specimen were estimated by the 3-(4,5-dimethylthiazol-2-yl)-2,5-diphenyltetrazolium bromide (MTT) assay method, as shown in Figs. 30.10 and 30.11. The pore-size-gradient scaffold sections showed over 90% cell-seeding efficiencies (after the cell-seeded specimens were maintained for one day at 37°C in a CO_2 incubator for cell adhesion), indicating that the scaffolds fabricated by the centrifugation method satisfied the requirements for cell scaffolds, that is, highly porous and interconnected pore structures. The cells were grown in the scaffold sections without significant differences up to 14 days, regardless of cell types, pore sizes, and scaffolds. Thereafter, the different types of cells showed different growth behaviors in the scaffolds; the chondrocytes and osteoblasts showed better cell growth in the scaffold sections having larger pore sizes, while the fibroblasts showed best cell growth in the scaffold sections having pore size about 200 μm. Generally the cells need the pores to be large enough to allow them to migrate into the pores of the scaffold and to allow effective nutrient supply and metabolic waste removal, which are essential for effective cell growth, but to be small enough to establish a sufficiently high surface area for efficient binding to the scaffold and cell-cell interactions for better cell growth, as discussed earlier. The exact mechanism of different growth behavior of different types of cells in the scaffolds with different pore sizes is not clear yet. However, the result of Figs. 30.10 and 30.11 suggest that the chondrocytes and osteoblasts may prefer larger pores (better conditions for efficient transport of nutrients or metabolites) for their growth, while the fibroblasts seem to prefer smaller pores (more surface areas for cell attachment and signaling).

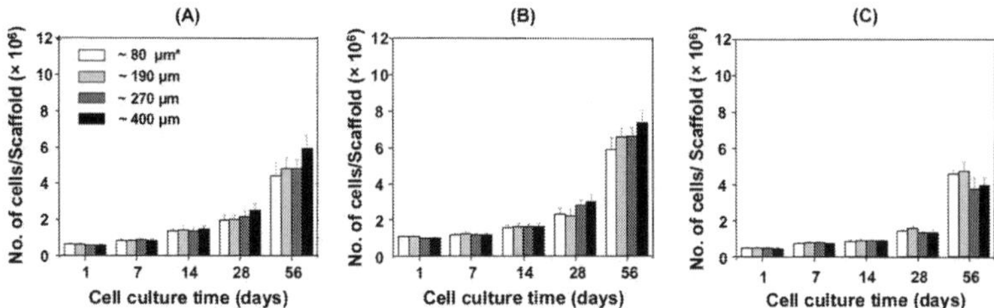

Figure 30.10. Number of cells adhered and grown in pore-size-gradient chitosan scaffold sections; (a) chondrocytes, (b) osteoblasts, and (c) fibroblasts. The cell numbers were determined from the MTT assay calibration curve (*, average pore size; $n = 3$) (Ref. 29).

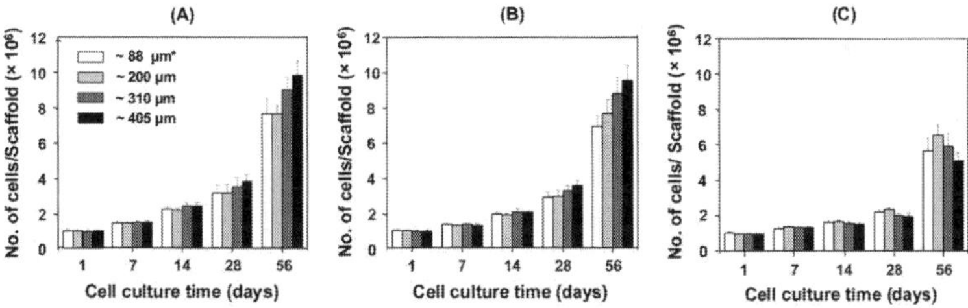

Figure 30.11. Number of cells adhered and grown in pore-size-gradient PCL scaffold sections; (a) chondrocytes, (b) osteoblasts, and (c) fibroblasts. The cell numbers were determined from the MTT assay calibration curve (*, average pore size; $n = 3$) (Ref. 28).

30.3.5 In vivo *Tissue Interactions*

To evaluate the pore size effect on *in vivo* tissue interactions, the selected sections of the pore-size-gradient, cylindrical PCL scaffold were implanted into the skull bone defects of a rabbit. For this, healthy and skeletally mature male New Zealand white rabbits with weight between 2.0 and 2.5 kg were selected and cared for according to the Korean Food and Drug Administration guidelines for the care and use of laboratory animals. The animals were housed in individual cages. Surgery was performed under general anesthesia. The anesthesia was induced by an intramuscular injection of ketamine hydrochloride (10 mg/kg, Ketala®, Yuhan Co., Korea) and 2% xylazine hydrochloride (2 mg/kg, Rumpun®, Byely Co., Korea). To reduce the perioperative infection risk, the rabbits received antibiotic prophylaxis (gentamycine, 4 mg/kg). The animals were placed in a ventral position for the insertion of the implants. The dorsum of the skull was shaved, washed, and disinfected with povidone-iodine. A midsagittal incision was made through the skin. The skin and subcutaneous tissue were separated from the periosteum by using blunt dissection. A second, longitudinal incision was made through the periosteum, which was the elevated and carefully dissected from the underlying skull bone. After exposure of the parietal calvaria, four full-thickness skull defects were made by using an 8 mm trephine hand instrument (Leibinger, Germany). Subsequently, the scaffold sections (diameter 8 mm; thickness 2 mm; average pore sizes $\sim$88, $\sim$200, $\sim$310, and $\sim$405 μm) without cells were inserted into the defects randomly. After inserting the scaffold sections, the periosteum was closed over the implant by using a 5-0 chromic catgut suture, and the skin was closed with use of a 5-0 nylon suture. At 4, 8, and 16 weeks postimplantation, euthanasia was performed with an overdose of pentobarbital sodium. A total of nine rabbits (three rabbits per period) were used for the implantation of scaffold sections. For histological study, the implants with surrounding cranial tissue were removed, fixed with 4% formaldehyde in phosphate buffered saline (PBS), and then decalcified in 10% formic acid. After dehydration in a graded series of ethanol, the specimens were embedded in paraffin wax and cut into 5 μm transverse sections in the center of the bony defects. These sections were stained with hematoxyline and eosin for the observation by light microscopy (Model BX50F4, Olympus, Japan). Throughout the observation periods, no significant differences were observed in tissue reaction within the defects. In all specimens, some inflammatory cells were observed inside the pores; however, the inflammatory response was limited and decreased for the implants at 8 and 16 weeks compared with 4 weeks of implantation. Analysis of the scaffold sections via light microscopy revealed various levels of bone formation at 4, 8, and 16 weeks postimplantation. At 4 weeks' implantation, it was observed that the scaffold section having $\sim$310 μm pore size showed faster new bone formation than other group sections (Fig. 30.12). These features were more obvious after 8 and 16 weeks' implantation. In the scaffold section having $\sim$310 μm pore size, new bone formation was more extensive inside the pores and progressed

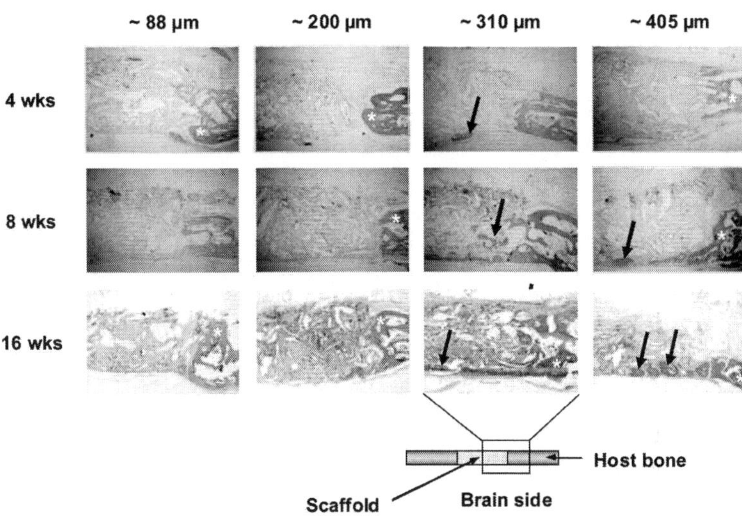

Figure 30.12. Histological sections of PCL scaffold sections with different pore sizes (× 100; *, host bone; arrow, new bone) (Ref. 28).

into the center of the scaffold. The scaffold section having ~405 μm pore size also showed new bone formation (but less than the scaffold section with ~310 μm pore size); however, in the other group sections having less than ~310 μm pore size, very limited bone growth behavior was observed. It seems that the pore size of about 310 μm is sufficient for bone tissue ingrowth in our experiment system, even though the result of this preliminary animal study was a little different from the *in vitro* osteoblast growth behavior: the osteoblasts showed better cell growth in the scaffold section having ~405 μm pore size than the scaffold section with ~310 μm pore size (see Fig. 30.11b). Actually, with respect to bone regeneration, there are some inconsensus regarding the optimal scaffold pore sizes. Hulbert *et al.*[21] reported that the scaffolds with pore sizes larger than 150 μm were suitable for bone reorganization and ingrowth, while Robinson *et al.*[18] reported that the bone formation and ingrowth occurred in scaffolds within a 350 μm pore size, but the scaffolds with a pore size less than 200 μm had no bone ingrowth.

It was also reported that excellent bone ingrowth occurred in scaffolds with pore sizes larger than 400 μm.[19,20] It seems that the optimal pore sizes for bone ingrowth and regeneration are highly dependent on pore structures as well as scaffold materials used.

30.4 Conclusions

We fabricated cylindrical scaffolds with gradually increasing pore size along the longitudinal direction by the centrifugation method. The pore size (or porosity) of the scaffolds increased along the cylindrical axis by the gradual increment of the centrifugal force. The prepared alginate, chitosan, and PCL scaffolds had the pore size (porosity) ranges along the cylindrical axis from 80 $\sim$ 88 μm to 310 $\sim$ 405 μm (from 80% $\sim$ 83% to 93% $\sim$ 94%), as the centrifugal speed of 3,000 rpm was applied. From the *in vitro* cell culture (chondrocytes, osteoblasts, and fibroblasts) and *in vivo* animal study (rabbit cranial defect model), it was observed that the cells and tissues were shown to have different pore size ranges in the scaffolds for effective cell growth and tissue regeneration: $\sim$400 μm pore size for chondrocytes and osteoblasts, $\sim$200 μm pore size for fibroblast growth, and $\sim$310 μm pore size for new bone formation. The pore-size-gradient scaffolds fabricated by the centrifugation method can be a good tool for the systematic studies of the interactions between cells or tissues and scaffolds with different pore sizes.

Acknowledgments

This work was supported by a grant from the Korea Research Foundation (KRF-2003-041-D00212).

References

1. T. S. Karande, J. L. Ong, and C. M. Agrawal, *Ann. Biomed. Eng.*, **32**, 1728 (2004).
2. H. Shin, S. Jo, and A. G. Mikos, *Biomaterials*, **24**, 4353 (2003).
3. B. D. Ratner and S. J. Bryant, *Ann. Rev. Biomed. Eng.*, **6**, 41, (2004).
4. S. Wang, W. Cui, and J. Bei, *Anal. Bioanal. Chem.*, **381**, 547 (2005).

5. K. Bilodeau and D. Mantovani, *Tissue Eng.*, **12**, 1 (2006).
6. T. G. Van Tienen, R. G. J. C. Heijkants, P. Buma, J. H. de Groot, A. J. Pennings, and R. P. H Veth, *Biomaterials*, **23**, 1731 (2002).
7. M. C. Wake, C. W. Patrick, and A. G. Mikos, *Cell Transplant.*, **3**, 339 (1994).
8. K. Whang, K. E. Healy, D. R. Elenz, E. K. Nam, D. C. Tsai, C. H. Thomas, G. W. Nuber, F. H. Glorieux, R. Travers, and S. M. Sprague, *Tissue Eng.*, **5**, 35 (1999).
9. S. Yang, K. F. Leong, Z. Du, and C. K. Chua, *Tissue Eng.*, **7**, 679 (2001).
10. J. Zeltinger, J. K. Sherwood, D. A. Graham, R. Mueller, and L. G. Griffith, *Tissue Eng.*, **7**, 557 (2001).
11. J. H. Brauker, V. E. Carr-Brendel, L. A. Martinson, J. Crudele, W. D. Johnston, and R. C. Johnson, *J. Biomed. Mater. Res.*, **29**, 1517 (1995).
12. J. J. Klawitter and S. F. Hulbert, *J. Biomed. Mater. Res. Symp.*, **2**, 161 (1971).
13. I. V. Yannas, E. Lee, D. P. Orgill, E. M. Skrabut, and G. G. Murphy, *Proc. Natl. Acad. Sci. U S A*, **86**, 933 (1989).
14. D. J. Griffon, M. R. Sedighi, D. V. Schaeffer, J. A. Eurell, and A. L. Johnson, *Acta Biomater.*, **2**, 313 (2006).
15. A. K. Salem, R. Stevens, R. G. Pearson, M. C. Davis, S. J. Tendler, C. J. Roberts, P. M. Williams, and K. M. Shakesheff, *J. Biomed. Mater. Res.*, **61**, 212 (2002).
16. S. S. Kim, H. Utsunomiya, J. A. Koski, B. M. Wu, M. J. Cima, J. Sohn, K. Mukai, L. G. Griffith, and J. P. Vacanti JP, *Ann. Surg.*, **228**, 8 (1988).
17. C. Danielsson, S. Ruault, M. Simonet, P. Neuenschwander, and P. Frey, *Biomaterials*, **27**, 1410 (2006).
18. B. P. Robinson, J. O. Hollinger, E. H. Szachowicz, and J. Brekke, *Otolaryngol. Head Neck Surg.*, **112**, 707 (1995).
19. L. D. Zardiackas, D. E. Parsell, L. D. Dillon, D. W. Mitchell, L. A. Nunnery, and R. Poggie, *J. Biomed. Mater. Res.*, **58**, 180 (2001).
20. J. D. Bobyn, G. J. Stackpool, S. A. Hacking, M. Tanzer, and J. J. Krygier, *J. Bone Joint Surg. Br.*, **81**, 907 (1999).
21. S. F. Hulbert, F. A. Young, R. S. Mathews, J. J. Klawitter, C. D. Talbert, and F. H. Stelling, *J. Biomed. Mater. Res.*, **4**, 433 (1970).
22. S. M. Lim, I. K. Park, S. H. Oh, and J. H. Lee, *Biomater. Res.*, **10**, 145 (2006).
23. F. J. O'Brien, B. A. Harley, I. V. Yannas, and L. J. Gibson, *Biomaterials*, **26**, 433 (2005).
24. T. B. F. Woodfield, C. A. Van Blitterswijk, J. De Wijn, T. J. Sims, A. P. Hollander, J. Riesle, *Tissue Eng.*, **11**, 1297 (2005).

25. B. A. Harley, A. Z. Hastings, I. V. Yannas, and A. Sannino, *Biomaterials*, **27**, 866 (2006).
26. J. Yang, G. Shi, J. Bei, S. Wang, Y. Cao, Q. Shang, Q. Yang, and W. Wang, *J. Biomed. Mater. Res.*, **62**, 438 (2002).
27. S. H. Oh, S. G. Kang, E. S. Kim, S. H. Cho, and J. H. Lee, *Biomaterials*, **24**, 4011 (2003).
28. S. H. Oh, I. K. Park, J. M. Kim, and J. H. Lee, *Biomaterials*, **28**, 1664 (2007).
29. S. M. Lim, S. H. Oh, I. K. Park, and J. H. Lee, *Key Eng. Mater.*, **342**, 285 (2007).

Chapter 31

SOLID FREEFORM FABRICATION METHOD APPLIED TO TISSUE SCAFFOLDS

Dong-Woo Cho,[a,d*] Jin Woo Lee,[b] Jong Young Kim,[c] and Tae-Yun Kang[a]

[a] *Department of Mechanical Engineering, POSTECH, P.O. Box 790-784,403, PIRO, Hyoja-dong, Nam-gu, Pohang, Korea*
[b] *Department of Nanoengineering, The University of California, San Diego*
[c] *Department of Mechanical Engineering, Andong National University*
[d] *Division of Integrative Biosciences and Biotechnology, POSTECH, P.O. Box 790-784,403, PIRO, Hyoja-dong, Nam-gu, Pohang, Korea*
*dwcho@postech.ac.kr

Until recently, tissue engineering scaffolds were produced by conventional fabrication methods such as particulate leaching, solvent casting, fiber bonding, and phase separation/inversion. However, these methods are not satisfactory in terms of three-dimensional (3D) freeform fabrication of the inner/outer architecture and the control of pores, porosity, and interconnectivity. To fully realize and account for the 3D environment, attempts have been made to design and fabricate scaffolds using computer-aided design (CAD) and computer-aided manufacturing (CAM). As a result, several solid freeform fabrication (SFF) technologies, including stereolithography (SL), fused deposition modeling (FDM), 3D printing (3DP), and selective laser sintering (SLS), recently have been developed. Because these methods use

Handbook of Intelligent Scaffolds for Tissue Engineering and Regenerative Medicine
Edited by Gilson Khang

www.panstanford.com

computers for design and fabrication, SFF technologies have the merit of being able to fabricate 3D scaffolds as designed, which enables their standardization. Using these technologies, scaffolds that are customized to each individual patient can be produced. If designs for new materials, optimal scaffold fabrication systems, and enhanced studies of cell physiology (i.e., optimal cell adhesion, proliferation, and vascularization) can be developed, SFF technologies will become an important aspect of tissue engineering research in the near future.

31.1 Introduction

Enhanced realization of the 3D freeform environment is becoming necessary to reinforce the viability of cells and the ability to regenerate tissues. Therefore, there have been many attempts to design and fabricate tissue engineering scaffolds using CAD and CAM. As a result, several SFF technologies, including SL, FDM, 3DP, and SLS have been developed. Because these methods use computer programs to design and fabricate the scaffolds, their inner and outer architectures, including the pore shape, porosity, and the interconnectivity of the scaffolds, can be more tightly controlled. Additionally, SFF technologies are able to fabricate 3D scaffolds as designed, which enables their standardization. This removes experimental errors caused by variability in the inner architecture from one scaffold to another, thus improving the reliability of the experiment. Further, we can use these technologies to fabricate scaffolds that are customized to each individual patient. The remainder of this chapter describes the basic characteristics of these technologies and gives examples of their application to tissue engineering.

31.2 SFF Methods Applied to Scaffolds

31.2.1 *Stereolithography*

SL was independently developed by both Kodama and Marutani of Japan around the same time period.[1,2] 3-D System Corporation was the first to realize an economical SL system. In SL, an ultraviolet

(UV) laser beam irradiates the surface of a UV-curable liquid photopolymer, causing it to solidify. Many scanned UV laser lines are overlapped to fabricate a given cross-sectional area. These cross-sectional areas are stacked together to form the desired 3D shape. Microstereolithgraphy (MSTL) was developed from SL and uses a specific laser beam that is a few μm in diameter. The laser beam is used to solidify a very small area of the photopolymer using a focusing lens (Fig. 31.1). MSTL is the highest-resolution SFF technology. MSTL also has an advantage over other fabrication technologies in that it enables the fabrication of 3D freeform structures at the nanometer and micrometer scale. Further, the inner architecture can be precisely controlled by utilizing CAD/CAM technology. As a result of these features, this technology offers great potential for the fabrication of adequate scaffolds for tissue engineering. However, this method requires photocurable materials to fabricate a structure, and this material limitation is a large barrier for scaffold fabrication. Several approaches may overcome this limitation, including developing photocurable biomaterials and combining methods that use various other biomaterials.

31.2.1.1 Photopolymer scaffold

Photocurable biomaterials such as polypropylene fumarate (PPF) and materials based on trimethylene carbonate (TMC) and gelatin have been developed for use in SL scaffold fabrication.

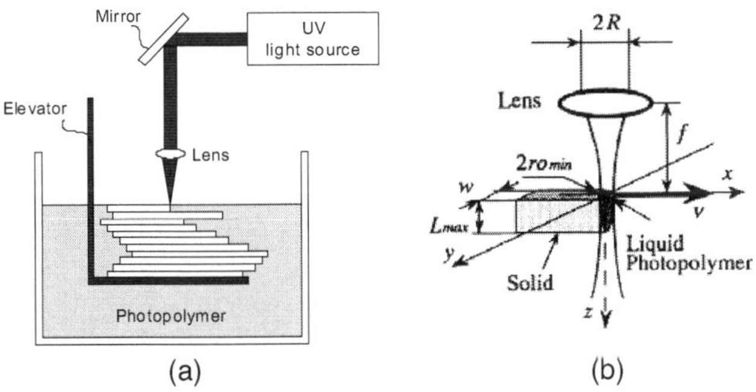

Figure 31.1. (a) Fundamental principle of SL technology[3] and (b) theoretical shape of the solidified polymer.[4]

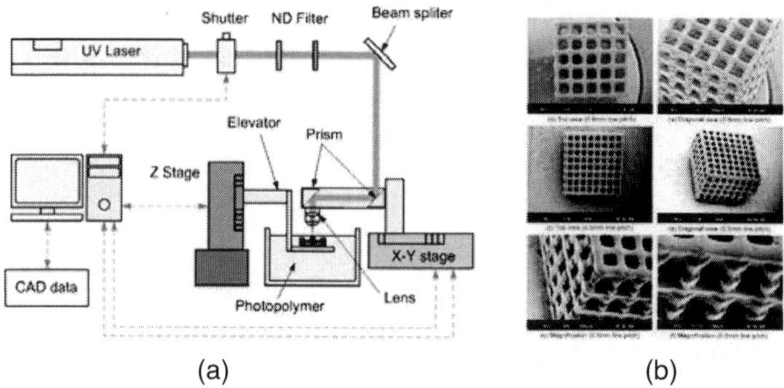

Figure 31.2. (a) MSTL system and (b) SEM image of 3D high-accuracy scaffolds.[7] *Abbreviation*: SEM, scanning electron microscopy.

PPF was first synthesized in 1988 by Sanderson *et al.* through the transesterification of diethyl fumarate and propylene glycol with a para-toluene sulfonic acid catalyst.[5] The mechanical properties of synthetic PPF are similar to those of trabecular bone. Since the first trial of UV polymerization using a conventional SL system by A.G. Mikos,[6] several similar attempts have been made to use this technique for tissue engineering.

Lee *et al.* reported a series of studies on the fabrication of high-precision scaffolds for tissue engineering applications using PPF and MSTL technologies. These fabricated scaffolds are extremely precise (to a few tenths of a micron), and these levels of precision cannot be realized with any other conventional fabrication method (e.g., salt-leaching and gas-forming methods). In addition, the pores are completely interconnected with each other.[7] The mechanical properties and cell affinities of the fabricated scaffolds were analyzed (Fig. 31.2)[7,8] revealing that the cell adhesion and proliferation capability were substantially enhanced through surface modification processes such as biomimetic apatite coating and the attachment of various peptides.[9] Another recent study reported the fabrication of scaffolds that gradually secrete bone morphogenetic protein-2(BMP-2) during degradation.[10]

Lee *et al.* analyzed the solidification characteristics and manufacturability of PPF according to the amount of diethyl fumarate (DEF) that was added to PPF as a cross-linking agent (Fig. 31.3).[11]

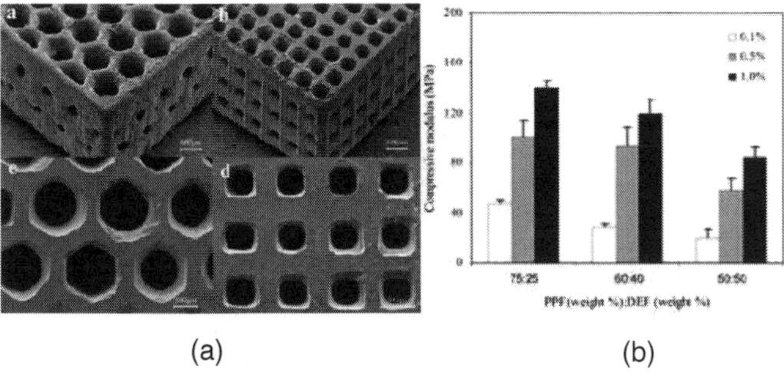

Figure 31.3. (a) Fabricated scaffolds using PPF/DEF resin and (b) compressive modulus of a cylindrical sample of PPF/DEF resin.[11]

TMC, which can be cross-linked by ring-opening polymerization, becomes a TMC-based photopolymer via combination with initiators such as trimethylolpropane (TMP), ε-caprolactone (CL), or various molecular weights of polyethylene glycol (PEG). Kwon *et al.* fabricated two-dimensional (2D) patterns and 3D structures with TMC/TMP and TMC/PEG and observed the effects of the materials on cell adhesion and proliferation (Fig. 31.4).[12] Further Lee *et al.* fabricated a 3D scaffold with TMC/TMP using the MSTL system and applied the scaffold in cartilage regeneration (Fig. 31.5).[13]

Schuster *et al.* developed an acrylate-based monomer formulation consisting of a biodegradable-basis monomer derived from gelation hydrolysate. These studies demonstrated that a 3D shape could be fabricated using this material. In addition, they mixed various reactive diluents such as urethanedimethacrylate (UDMA), diisobutylacrylamide (DBA), dipentaerythritol pentaacrylate (PPA), pentaerythritol triacrylate (PTA), trimethylolpropane triacrylate

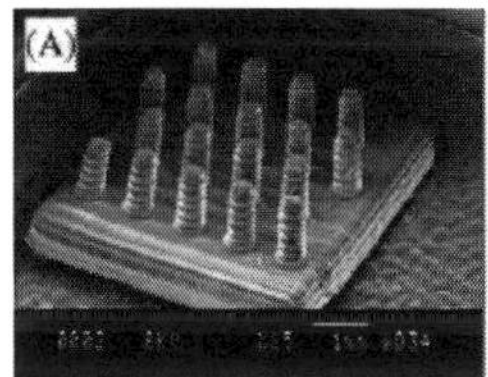

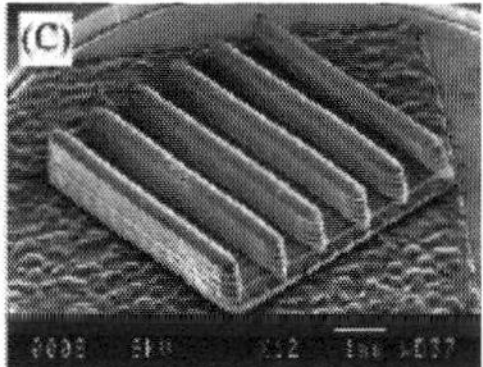

Figure 31.4. SEM image of (a) micropillar array, (b) microcorn array, and (c) microbank array from TMC/PEG200.[12]

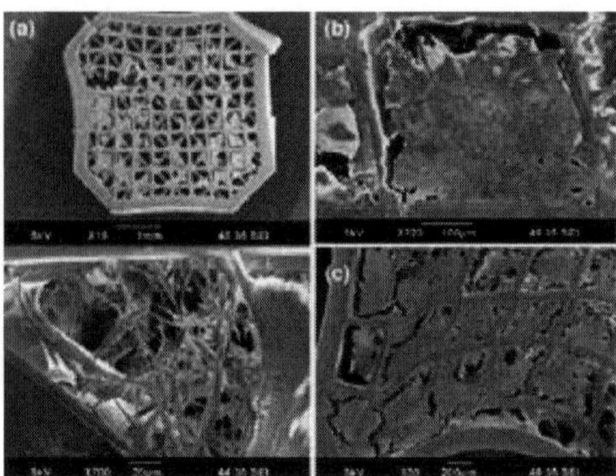

Figure 31.5. Cultured chondrocytes of scaffolds (a) 3 days after seeding, (b) 3 weeks after seeding, and (c) 2 weeks after seeding.[13]

(TTA), and glycol diacrylate (P3-A) with the gelatin hydrolysate to endow photo-polymerization capabilities. They evaluated the cell adhesion and proliferation capabilities of osteoblast-like cells (MG63) using the material to characterize cytocompatibility.[14]

Most photocurable polymers do not have mechanical properties that are adequate for these applications. Because the addition of an appropriate amount of minerals often improves mechanical properties, Lee *et al.* mixed PPF/DEF with hydroxyapatite (HA) to enhance the mechanical strength of the biomaterial as compared with existing polymers.[15]

A scaffold can be fabricated with ceramic materials using SL and a sintering process. In this method, photocurable polymers are blended with bioceramic materials such as HA or tri-calcium phosphate (TCP) to make a slurry. Next, the prepared slurry is transferred into the 3D negative mold cavity made using SL. Lastly, the mold undergoes a sintering process in which very high temperatures are applied. During this time the polymer material is selectively removed, and bonding between the ceramic powders is induced. Chu *et al.*[16] used commercialized epoxy resin to make a negative mold and applied the sintering process to fabricate a high-strength scaffold made of HA (Fig. 31.6). They also successfully made various shapes of scaffolds and demonstrated that the technology would be useful for tissue engineering applications. Woesz *et al.*[17] introduced the fabrication of CP scaffolds and described the proliferation and

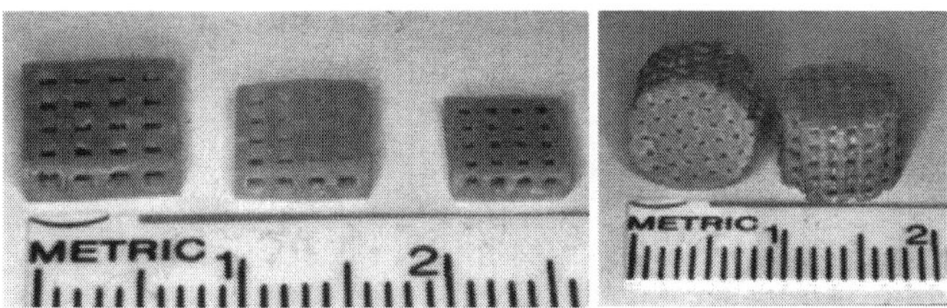

Figure 31.6. HA scaffolds.[16]

differentiation of the MC3T3-E1 cells (pre-osteoblasts) seeded in the fabricated scaffold.

In addition to using a negative mold, direct fabrication from the slurry (without the filling process) using MSTL is being studied. This approach allows for the easier fabrication of a structure because it does not require the molding process. Cho *et al.* introduced this fabrication method and showed that the fabricated structures that are produced have very complicated shapes (Fig. 31.7).

31.2.1.2 Biopolymer scaffold

Biopolymers, including synthetic polymers and natural polymers, cannot be used in the sintering process. Therefore, several researchers have introduced new methods for the fabrication of 3D porous structures composed of biopolymers. These methods make it possible to process various biomaterials, including synthetic and natural polymers.

Jiankang *et al.* created 3D chitosan/gelatin scaffolds using MSTL.[18] Specifically, high-resolution MSTL technology was used to

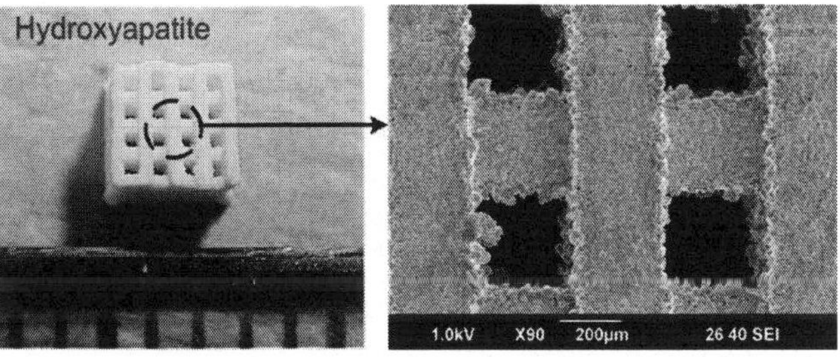

Figure 31.7. Fabricated HA scaffolds.[15]

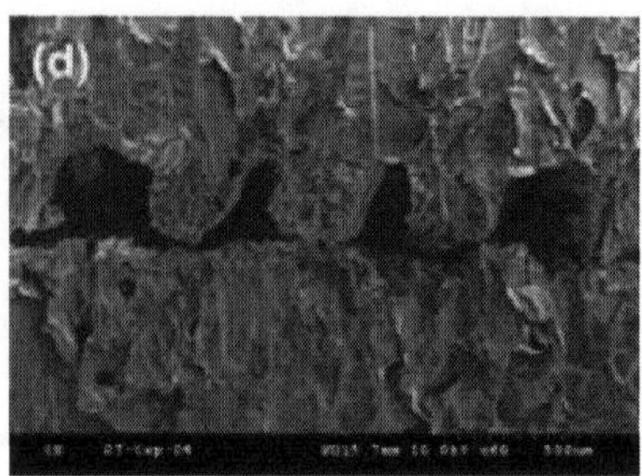

Figure 31.8. SEM image of a fabricated chitosan/gelatin structure.[23]

fabricate a master pattern for the 2D Polydimethylsiloxane (PDMS) mold. Chitosan/gelatin material was cast and freeze-dried on the mold, resulting in a 2D shape composed of natural polymer. A 3D structure was then created by assembling the 2D layers (Fig. 31.8). Cell culture results using hepatocytes were also presented.

In contrast, Schuster *et al.* developed a soluble photopolymer for use in SL that had high photoreactivity and mechanical strength[19] and could be removed by an alkali solution. A biocompatible copolymer was molded using the developed resin. M. Vallet-Regi used the etching properties of a commercialized photopolymer[20,21]; a 3D mold was fabricated using Accrura SI10 from 3-D Systems, Inc., and a biomaterial was injected. The mold was then etched using a 2 M NaOH solution. Kang *et al.* studied the molding process for various biomaterials, such as poly-ε-caprolactone (PCL), bone cement, and poly(lactic-co-glycolic) acid (PLGA), for tissue engineering. The soluble photopolymers proposed by Schuster[19] and high-resolution MSTL technologies[22] were used to fabricate 3D scaffolds, and the lost-mold shape-forming process was exploited using various biomaterials (Fig. 31.9).[24,25]

31.2.2 *Fused Deposition Modeling*

FDM is an SFF process commonly used in mechanical system design and manufacture. The technology was developed by S. S. Crump in the late 1980s and was successfully commercialized in 1990. A traditional FDM machine consists of a head-heated liquefier

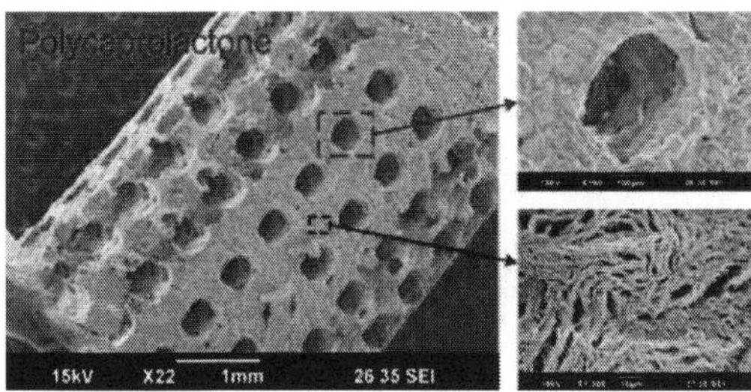

Figure 31.9. Fabricated 3D porous scaffolds by indirect-SFF method based on MSTL.[24]

attached to a carriage moving in the horizontal *xy* plane. The main function of the liquefier assembly is to heat and extrude the filament materials through a nozzle directly onto the build platform following a pre-programmed path, which is based on CAM. If one layer is fabricated, the machine moves down one step in the *z* direction to deposit the next layer. The final products can be produced in a layer-by-layer process. By changing the direction of material deposition for consecutively deposited layers and the spacing between the material roads, scaffolds with highly controllable pore geometry and complete pore interconnectivity can be manufactured.

In comparison with other SFF techniques, the FDM method does not require any solvent and offers great ease and flexibility in the handling and processing of materials. The use of a filament-modeling material also reduces the time required in the heating compartment and allows for continuous production without the need for replacing feedstock. However, one drawback of the FDM technique is the need for preformed fibers with uniform size and material properties to feed through the rollers and nozzle. In addition, its application to biodegradable polymers, excluding PCL, are rarely reported Therefore, various modified FDM processes were developed and evaluated to overcome the limitations of the traditional FDM process in tissue scaffold fabrication.[26–34]

Hutmacher *et al.* investigated the mechanical properties and responses of cells cultured on PCL scaffolds designed and fabricated using commercialized FDM (Stratasys Inc., Eden Prairie, MN).[29] Two types of PCL scaffolds with an area of 32.0 (length) × 25.5

(width) × 13.5 mm (height) and 61% porosity were manufactured using different patterns (0°/60° /120° and 0°/72°/144°/36°/108°), to give a honeycomb-like pattern of triangular and polygonal pores, respectively. The scaffolds with a 0°/60°/120° lay-down pattern had a compressive stiffness and yield strength of 41.9 ± 3.5 and 3.1 ± 0.1 MPa, respectively. In comparison, the scaffolds with a 0°/72°/144°/36°/108° lay-down pattern had a compressive stiffness and yield strength of 20.2 ± 1.7 and 2.4 ± 0.1 MPa, respectively. *In vitro* studies over a period of three to four weeks showed that PCL scaffolds are biocompatible with human fibroblasts and periosteal cell culture systems. This result shows that FDM allows for the design and fabrication of highly reproducible biodegradable 3D scaffolds with a fully interconnected pore network.

Zein *et al.* demonstrated the efficacy of the commercialized FDM technique for scaffold design and fabrication. Scaffolds with regular geometrical honeycomb pores were manufactured with pore/channel sizes ranging from 160 to 700 μm and with porosities of 48%–77%.[30] The mechanical properties of these scaffolds were found to be generally dependent on porosity, regardless of the lay-down pattern and channel size. These results are in agreement with theoretical concepts on the structure–property relationships of porous solids.

Cao *et al.* demonstrated successful *in vitro* co-culturing of osteoblasts and chondrocytes on PCL scaffolds for more than 50 days using a commercialized FDM apparatus.[31] Rectangular-shaped, honeycomb-like scaffolds were fabricated with a three-angle lay-down pattern (0°/60°/120°). The porosity ranged from 60% to 65%, and the pore size ranged from 300 to 580 μm. The 10 × 10 × 3.2 mm PCL scaffold was partitioned vertically into two halves with a gap between them. One-half (the bone compartment) of the partitioned scaffold was designated for bone marrow stromal cell (BMSC) seeding, and the other half (the cartilage compartment) was designated for chodrocyte seeding. It was reported that both osteoblasts and chodrocytes produce a rich extracellular matrix (ECM) in their respective scaffold compartments. At the interface region, a mixture of cell types was observed. Therefore, it was demonstrated that 3D porous PCL scaffolds produced by FDM are biocompatible, as

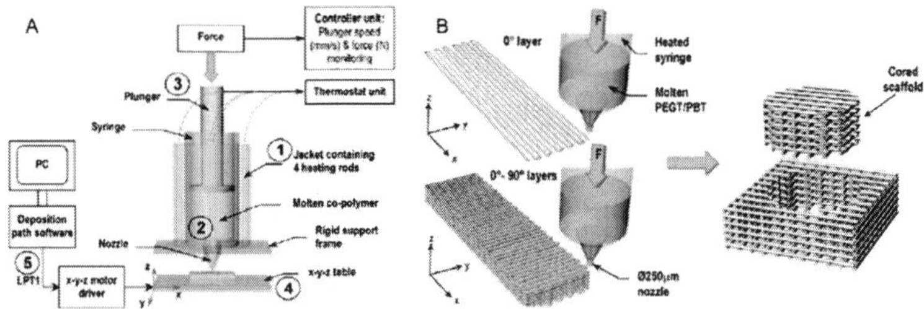

Figure 31.10. (a) The fiber deposition device and (b) the 3D deposition process where 250 μm PEGT/PBT fibers are successively laid down in a computer-controlled pattern (0°–90° orientation shown). Scaffolds are subsequently cored from the deposited bulk material.

evidenced by extensive cell adhesion and proliferation. Further, they are suitable for *in vitro* osteochondral constructs in tissue engineering.

Woodfield *et al.* developed and characterized a fiber deposition technique for producing 3D PEG-terephthalate (PEGT)-poly(butylene terephthalate) (PBT) block copolymer scaffolds with a 100% interconnecting pore network for the engineering of articular cartilage (Fig. 31.10).[32] By varying the PEGT/PBT composition, porosity, and pore geometry, 3D-deposited scaffolds were produced with an equilibrium modulus and a dynamic stiffness ranging from 0.05 to 2.5 MPa and 0.16 to 4.33 MPa, respectively. The 3D-deposited scaffolds supported the rapid attachment of bovine chondrocytes and tissue formation following dynamic culture *in vitro* and subcutaneous implantation in nude mice, as demonstrated by the presence of articular cartilage extracellular matrix (ECM) constituents, glycosaminoglycan (GAG), and type II collagen throughout the interconnected interior pore volume. Similar results were achieved with respect to the attachment of expanded human articular chondrocytes, resulting in a homogenous distribution of viable cells after five days of dynamic seeding.

Wang *et al.* developed the precision extruding deposition (PED) system.[33] In contrast to the conventional FDM process, which requires the use of precursor filaments, the PED process directly extrudes scaffolding materials in a granulated or pellet form without filament preparation and freeform deposits according to the

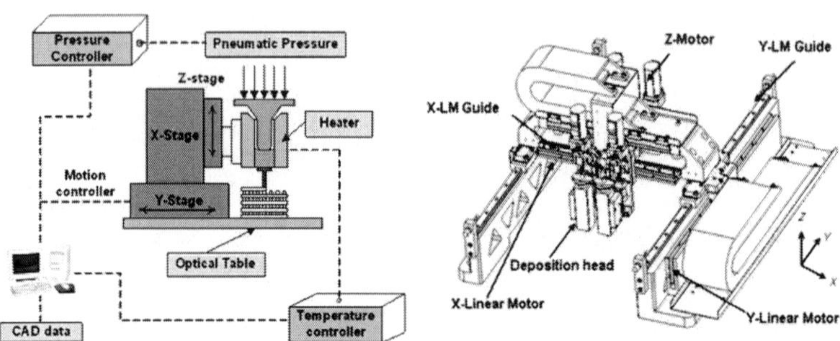

Figure 31.11. (a) Schematic diagram of the developed MHDS and (b) image of the designed and manufactured MHDS.

designed microscale features. The typical pore size of fabricated PCL scaffold ranges from 200 to 300 μm, which is near the optimal size suggested for bone tissue scaffold applications. The compressive modulus of scaffolds ranges from 150 to 200 MPa. The preliminary results of biological experiments demonstrate the biocompatibility of the PED process and PCL materials.

Kim *et al.*[26,34] described the adhesion of human BMSCs and the proliferation characteristics of various scaffolds that consist of biodegradable materials and that were fabricated using a multihead deposition system (MHDS) developed by the authors (Fig. 31.11). The MHDS may be superior to other commercialized systems using the FDM process in that it can conveniently and quickly fabricate various hybrid scaffolds. The MHDS enables the fabrication of 3D tissue scaffolds with a resolution of several tens of microns (Fig. 31.12). In addition, multideposition heads installed in the MHDS permit rapid fabrication and manufacture of blended scaffolds with various biomaterial compositions. PCL, PLGA, blended PCL/PLGA, and blended PCL/PLGA/TCP scaffolds, which had the same inner/outer architecture, were manufactured adequately and had perfectly interconnected pores. The mechanical testing and cell proliferation results show that the blended PCL/PLGA/TCP scaffold was superior to other scaffolds. The feasibility of this application to tissue engineering of SFF-based 3D scaffolds fabricated using MHDS was confirmed through scaffold fabrication, mechanical testing, and cell interaction evaluation.

Figure 31.12. SEM images of scaffolds (7.4 × 7.4 × 3.2 mm) fabricated by MHDS. (a) A 3D PCL scaffold and (b) a 3D PLGA scaffold.

31.2.3 *3D Printing*

The process of 3DP is an SFF technique used for the rapid manufacture of accurate and complex 3D microstructures from predesigned patterns of the final structures. The 3DP technology was developed and commercialized at the Massachusetts Institute of Technology (MIT) in 1995. Fundamentally, 3DP is also a layer-by-layer fabrication process, in which the sliced 2D pattern of a computer model is printed on a fresh layer of powder via an inkjet print head. Successive 2D profiles are then printed on a new layer of powder until the whole 3D model is completed. After the binder has dried in the powder bed, the finished component is retrieved and the unbound powder is removed. The 3DP process is one of the most investigated SFF techniques for scaffold fabrication.[35–42]

An advantage of 3DP for scaffold fabrication is that commercial inkjet printers can be easily reconstructed to print cell-encapsulated natural polymers at precise positions inexpensively and with high throughputs. In addition, the scaffold can be manufactured easily in an ambient environment. Therefore, bio-printers, which are based on the traditional 3DP apparatus, have been developed to perform the computer-assisted deposition of natural materials, including bioactive molecules, biomolecules, and viable cells. However, if the scaffold is designed to be porous, one problem of the powder-supported and powder-filled structure is the difficulty of removing internal unbounded powder. In addition, the resolution of the scaffold is limited by the fixed nozzle size and the low position accuracy of the inkjet printer.

Lee *et al.* fabricated scaffolds with large pore sizes and fine features using commercially available 3DP systems.[40] They reported that direct 3DP, in which the final scaffold materials are utilized during the 3DP process, imposes limitations on pore size and shape complexity, among other limitations. Therefore, this study presented an indirect 3DP protocol, in which molds were printed and the final materials were cast into the mold cavity to overcome the limitations of the direct technique. PLGA scaffolds with a diameter of 500 μm and a height of 1 mm were produced by solvent casting into plaster molds, followed by particulate leaching and mold removal using deionized water. The uniform distribution of intestinal epithelial cells (IEC6) on the scaffold after seven days suggests the presence of highly interconnected pores within the scaffolds that allow cells seeded at the top surface to settle evenly within the scaffolds. The cell density within the villi-shaped regions of the scaffolds increased with culture time, whereas the cell density within the region below the villi remained low. The observed heterogeneous cell distributions might be due to 1) limitations on the diffusion of oxygen and nutrients into the interior of the scaffolds, 2) cell migration, or 3) the presence of isolated voids within the scaffold. To investigate the SFF capabilities of 3DP with common medical image data, anatomically shaped zygoma scaffolds with 300–500 μm interconnected pore sizes were produced and characterized.

Mondrinos *et al.* established a drop-on-demand printing (DDP) process, which uses thermoplastic, porogen-based injection molding manufacturing methods. This process was demonstrated to efficiently and reproducibly fabricate porous PCL and PCL-CP composite scaffolds with pore sizes as small as 200 μm (Fig. 31.13).[41] The compressive strengths of the 90:10 and 80:20 PCL–CP scaffolds were 19.5 $\pm$ 1.4 and 24.8 $\pm$ 1.3 MPa, respectively, according to ASTM standards. The compressive strengths of pure polycaprolactone (PCL) scaffolds were 2.77 $\pm$ 0.26 MPa. Cytocompatibility tests using human embryonic palatial mesenchymal (HEPM) cells indicate that all porogen-based scaffolds facilitate attachment and support the proliferation of HEPM cells *in vitro*. In addition, it was suggested that the presence of CP in the PCL–CP composite enhanced the proliferation of HEPM cells and reduced spreading in favor of multilayer assembly. The advantage of the porogen-based process is the ability to use multiple biomaterials for injection molding with a single porogen using dual inkjet heads.

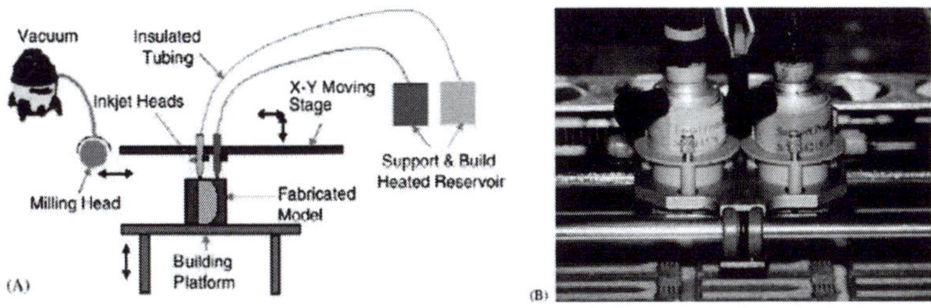

Figure 31.13. Drop-on-demand 3DP system: (a) diagram of DDP system components and (b) digital photograph of inkjet print heads in this DDP system, Solidscape Modelmaker II. See also Color Insert.

Xu *et al.* used a commercial thermal printer, which was composed of an HP desktop printer (HP 550C) and an HP 51626a ink cartridge, to deposit Chinese hamster ovary (CHO) cells and hydrogel-based materials for tissue engineering.[42] Analyses of the CHO cells showed that more than 90% of cells were not damaged during nozzle dispensing. A complete ring of CHO cells was observed on soy agar biopaper after day 7. The cells continued to proliferate until a complete circle was formed by day 25. Therefore, cell-printing technologies based on the inkjet printer will offer researchers a cost-effective tool to rapidly fabricate cell patterns and tissuelike structures.

31.2.4 *Selective Laser Sintering*

SLS is a technology that uses a CO_2 laser beam to sinter thin layers of polymer, ceramic, or composite (polymer/ceramic/multiphase metal) powders, thus forming solid 3D objects. During the SLS process, the laser beam is scanned selectively over the surface of a powder bed, following the cross-sectional profiles of the slice data generated by the computer program. The irradiation of the laser beam raises the powder temperature at the irradiation point, and sintering occurs just beyond the glass transition temperature, causing the particles to fuse together to form a solid structure. The solid structure sinks to bottom of the powder, and new layers of powder are deposited by a roller.[43] Next, layers are fabricated directly on top of the previously sintered layers. During fabrication, the structure is supported and embedded by the surrounding unprocessed

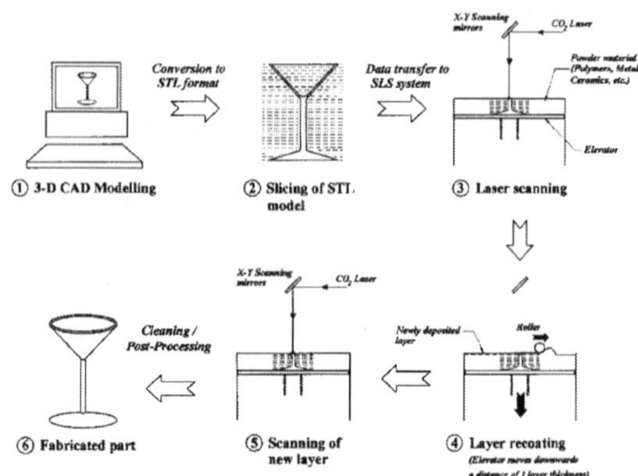

Figure 31.14. Schematic layout of the SLS process.[46]

powders and must be extracted from the powder bed after fabrication (Fig. 31.14). Because the powders are subjected to low compaction forces during their deposition to form new layers, fabricated structures are usually porous.

SLS is an effective method for creating complex-shaped prototypes and allows for the design of anatomically shaped scaffolds with defined pore sizes, porosity, and topographies. In fact, SLS-fabricated scaffolds possess the best mechanical properties among all SFF technologies, and it is the preferred fabrication process for producing complex porous ceramic matrices suitable for implantation in bone defects. Various materials, including polymers, ceramics, metallic powder, and their composites, can be processed using the SLS system. However, because the SLS technology involves high processing temperatures, the technology is limited to the processing of thermally stable polymers.[44–46] Further, the pores created in SLS-fabricated scaffolds are dependent on the particle size of the powder stock used; typical pore sizes encountered in SLS-fabricated structures are limited to smaller pore size ranges (<50 μm).

Lee *et al.* first fabricated ceramic bone implants using SLS technology and CP powders of various Ca/P ratios in 1993.[47] The powders were coated with an intermediate polymer binder, poly(methyl methacrylate-co-n-butyl methacrylate) emulsion copolymer, containing 30% (v/v) polymer emulsion and ceramic powders.[48] The pore sizes of the SLS-fabricated ceramic parts were reported to measure approximately 50 μm and were well interconnected. They also evaluated the SLS-fabricated CP oral implants in both the rabbit and dog models.[49]

Rimell *et al.* reported the fabrication of clinical implants using a simplified SLS apparatus and ultra-high-molecular-weight polyethylene (UHMWPE).[45] In their study, solid linear continuous bodies could be formed; however, material shrinkage caused problems during the production of sheetlike structures. The material exposed to the laser beam was shown to undergo degradation, that is, chain scission, cross-linking, and oxidation. It was concluded that the development of improved starting powders with increased densities are required for the application of this technology to the fabrication of UHMWPE devices.

Huang *et al.* designed a novel scaffold with a 3D branching and joining flow-channel network that was composed of multiple tetrahedral units (4 mm edge length) to engineer implantable liver tissues.[50] To fabricate this network, biodegradable PCL and 80% (w/w) NaCl salt particles serving as a porogen were thoroughly mixed and applied in an SLS process. The fabricated scaffold had a high porosity (89%) with pore sizes of 100–200 μm and 3D flow channels (Fig. 31.15). To evaluate its biocompatibility, human hepatoma Hep G2 cells were seeded into the scaffold using avidin–biotin (AB) binding and cultured in a perfusion system for nine days. The results demonstrated that such 3D flow channels are essential to the cells' growth and function. Williams *et al.*[51] also used PCL for scaffold material.

Wiria *et al.* fabricated scaffolds using biocomposite materials consisting of PCL and HA.[52] Biocomposite blends with different percentage weights of HA were physically blended and sintered to assess their suitability for fabrication via SLS. Cell culture experiments showed that Saos-2 cells were able to live and replicate on the fabricated scaffolds. Tan *et al.* also used nondegradable polyetheretherketone (PEEK)/HA powder blends to fabricate 3D scaffolds on a commercial SLS machine.[46] To assess their suitability for SLS processing, different weight percentage (w/w) compositions of

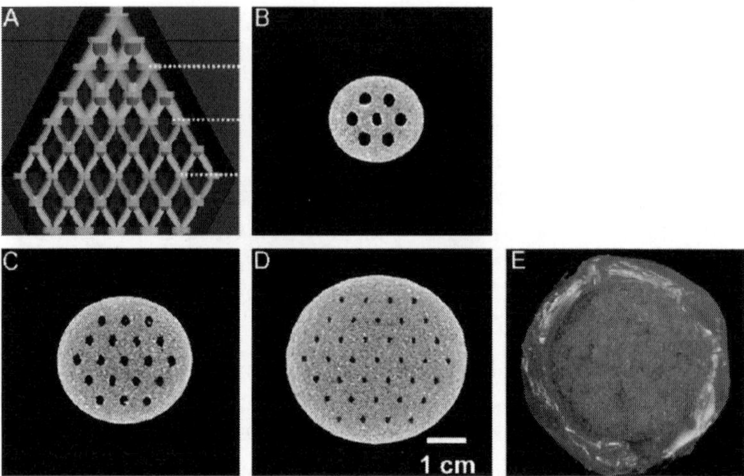

Figure 31.15. (a) CAD design from front view of a scaffold, (b–d) micro-computed tomography at the bottoms of second, fourth, and sixth layers of CAD data, and (e) observation of culture medium flow of a cell-free half-scaffold.

physically mixed PEEK/HA powder blends were sintered by varying the laser power and temperature settings.

Goodridge *et al.* evaluated the biological performance of a porous apatite–mullite glass-ceramic to determine its potential as a bone replacement material.[53] Direct contact and extract assays were used to assess the cytotoxicity of the material. A pilot animal study, in which the material was implanted into rabbit tibiae for four weeks, was also performed to assess *in vivo* bioactivity. They reported that the material produced by SLS did not show any acute cytotoxic effects using either contact or extract methods.

Popov *et al.* have developed surface SLS (SSLS) for bioactive and bioresorbable scaffold fabrication based on a modified and enhanced SLS procedure.[54–56] SSLS is initiated by melting only the polymer particle surface but not the poly(D,L-lactic) acid (PLA) powder, which does not absorb ($\sigma_{ab} < 0.1$ cm^{-2}) near-infrared (λ = 0.97 lm) laser radiation. The laser light is absorbed by a small quantity of homogeneously distributed biocompatible carbon black (CB) microparticles that are added to the PLA on the surface of

Figure 31.16. Photomicrographs of SSLS-PLA generated scaffolds. Optical image of a mandibular scaffold generated by SSLS-PLA (a), prototype SSLS-PLA scaffolds (b), SEM microphotograph of individual SSLS-PLA scaffold structures (c), *in vitro* cell growth characterization on SSLS-PLA; representative images of adherence and viability of fetal femur-derived cells on SSLS-PLA scaffolds as evidenced by labeling of the cell viability marker CMFDA and negligible EH-1 (necrosis) marker in extended culture, 7 days postseeding onto SSLS-PLA scaffolds (d). Bare SSLS-PLA scaffold in comparison to (e) fetal femur-derived cell-seeded scaffolds, which expressed high activity of alkaline phosphatase (f) 7 days postseeding. Original magnifications 100 (d–f). Scale bars = 5 mm (c), 50 lm (c) and 500 lm (d).[56]

each particle. This process leads to localized surface heating. Altering the laser intensity and laser beam scanning speed enables the reproducible fabrication of 3D polymer scaffolds with specific shapes and internal structures (Fig. 31.16).

Hollander *et al.* reported a direct laser forming (DLF) technology, which enables prompt modeling of metal parts with high bulk density.[57] They tested DLF-produced material on the basis of the titanium alloy Ti–6Al–4V because of its applicability as a hard tissue biomaterial. They also investigated mechanical and structural properties. Finally, they cultured human osteoblasts on nonporous and porous blasted DLF-Ti–6Al–4V specimens to study the morphology, vitality, proliferation, and differentiation of the cells.

31.3 Summary

This chapter has described various 3D scaffold fabrication methods based on SFF technologies. These technologies provide great advantages over conventional scaffold fabrication methods in the control of pore size, porosity, and interconnectivity. Because these methods are designed and fabricated by a computer, SFF technologies have the merit of being able to fabricate a 3D scaffold as designed, thus enabling the standardization of 3D scaffolds. Additionally, patient-customized scaffolds can be fabricated using these technologies. The improvement of scaffold fabrication systems and medical image systems such as computed tomography and magnetic resonance imaging can lead to rapid progress; however, the development of adequate biocompatible and biodegradable materials for each SFF system is required for the fabrication of the scaffolds for tissue engineering applications. It is also essential to study the interaction between cells and scaffold materials. Therefore, future developments in SFF require the design of new materials, optimal scaffold fabrication systems, and enhanced studies of cell physiology, including optimal cell adhesion, proliferation, and vascularization. Given the concentration of these efforts, SFF technologies will become an important aspect of tissue engineering research in the near future.

Acknowledgments

This work was supported by the National Research Foundation of Korea (NRF) grant funded by the Korea government (MEST) (No. 2010-0018294).

References

1. H. Kodama, *Rev. Sci. Instrum.*, **52**(11), 1770 (1981).
2. T. Nakai and Y. Marutani, *Conference on Laser and Electro-Optics* 1986, San Francisco, ME-2, (1986).
3. H. W. Kang, I. H. Lee, and D.W. Cho, *Tran. ASME: J. Manuf. Sci. Eng.*, **126**(4), 766 (2004).

4. T. Nakamoto, K. Yamaguchi, P. A. Abraha, and K. Mishima, *J. Micromech. Microeng.*, **6**(2), 240 (1996).
5. J. E. Sanderson, *US Patent*, **4722948**, 1 (1998).
6. M. N. Cooke, J. P. Fisher, D. Dean, C. Rimnac, and A. G. Mikos, *J. Biomed. Mater. Res.*, **64B**(2), 65 (2002).
7. J. W. Lee, P. X. Lan, B. Kim, G. B. Lim, and D. W. Cho, *J. Biomed. Mater. Res. Part B: Appl. Biomater.*, **87B**(1), 1 (2008).
8. J. W. Lee, P. X. Lan, B. Kim, G. B. Lim, and D. W. Cho, *Proc. of the 32nd Int'l Conf. on Micro-and Nano-engineering*, **84**(5–8), 1702 (2007).
9. P. X. Lan, J. W. Lee, Y. J. Seol, and D. W. Cho, *J. Mater. Sci.: Mater. Med.*, **20**(1), 271 (2009).
10. J. W. Lee, N. T. Anh, K. S. Kang, Y. J. Seol, and D. W. Cho, *2008 TERMIS-EU Chapter Meeting*, Porto, Portugal, **June 22** (2008).
11. K.-W. Lee, S. Wang, B. C. Fox, E. L. Ritman, M. J. Yaszemski, and L. Lu, *Biomacromolecules*, **8**(4), 1077 (2007).
12. I. K. Kwon and T. Matsuda, *Biomaterials*, **26**, 1675 (2005).
13. S. J. Lee, H. W. Kang, J. K. Park, J. W. Rhie, S. K. Hahn, and D. W. Cho, *Biomed. Microdevices*, **10**(2), 233 (2008).
14. M. Schuster, C. Turecek, F. Varga, H. Lichtenegger, J. Stampfl, and R. Liska, *Appl. Surf. Sci.*, **254**, 1131 (2007).
15. J. W. Lee, P. X. Lan, Y. J. Seol, and D. W. Cho, *2007 TERMIS-EU Chapter Meeting*, London, UK, **September 4–7**, 1745 (2007).
16. T. M. G. Chu, J. W. Halloran, S. J. Hollister, and S. E. Feinberg, *J. Mater. Sci.: Mater. Med.*, **12**, 471 (2001).
17. A. Woesz, M. Rumpler, J. Stampfl, F. Varga, N. Fratzl-Zelman, P. Roschger, K. Klauschofer, and P. Fratzl, *Mater. Sci. Eng. C*, **25**, 181 (2005).
18. H. Jiankang, L. Dichen, L. Yaxiong, Y. Bo, Z. Hanxiang, L. Qin, L. Bingheng, and L. Yi, *Acta Biomater.*, **5**(1), 453 (2009).
19. M. Schuster, R. Infuhr, C. Turecek, J. Stampfl, F. Varga, and R. Liska, *Monatshefte fur Chemie*, **137**, 843 (2006).
20. M. V. Cabanas, J. Pena, J. Roman, and M. Vallet-Regi, *J. Biomed. Mater. Res. Part A*, **78A**(No.3), 508 (2006).
21. S. Sanchez-Salcedo, A. Nieto, and M. Vallet-Rgi, *Chem. Eng. J.*, **137**, 62 (2008).
22. P. F. Jacobs, *Soc. Manuf. Eng.* (1992).
23. H. Jiankang, L. Dichen, L. Yaxiong, Y. Bo, L. Bingheng, and L. Qin, *Polymer*, **48**, 4578 (2007).

24. H. W. Kang, T. Kang, and D. W. Cho, *Micro/Nanoeng. 2008*, Athens, Greece, **241**, September 15–18 (2008).
25. H. W. Kang and D. W. Cho, *TERMIS-AP 2007*, Tokyo, Japan, **December 3–5** (2007).
26. J. Y. Kim, E. K. Park, S. Y. Kim, and J. W. Shin, and D. W. Cho, *J. Micromech. Microeng.*, **18**, 055027 (2008).
27. A. Yamada, F. Niikura, and K. Ikuta, *J. Micromech. Microeng.*, **18**, 025035 (2008).
28. Z. Xiong, Y. Yan, S. Wang, R. Zhang, and C. Zhang, *Scripta Mater.*, **46**, 771 (2002).
29. D. W. Hutmacher, T. Schantz, I. Zein, K. W. Ng, S. H. Teoh, and K. C, *J. Biomed. Mater. Res.*, **55**, 203 (2001).
30. I. Zein, D. W. Hutmacher, K. C. Tan, and S.-H. Teoh, *Biomaterials*, **23**, 1169 (2002).
31. T. Cao, K. H. Ho, and S.-H. Teoh, *Tissue Eng.*, **9**, 103 (2003).
32. T. B. F. Woodfield, J. Malda, J. D. Wijn, F. Peters, J. Riesle, and C. A. V. Blitterswijk, *Biomaterials*, **25**, 4149 (2004).
33. F. Wang, L. Shor, A. Darling, S. Khalil, W. Sun, S. Guceri and A. Lau, *Rapid Prototyp. J.*, **10**, 42 (2004).
34. J. Y. Kim, J. J. Yoon, E. Y. Park, D. S. Kim, S. Y. Kim, and D. W. Cho, *Biofabrication*, **1**, 015002 (2009).
35. S. Limpanuphap and B. Derby, *J. Mater. Sci., Mater. Med.*, **13**, 1163 (2002).
36. H. Seitz, W. Rider, S. Irsen, B. Leukers, and C. Tille, *J. Biomed. Mater. Res.*, **74B**, 782 (2005).
37. A. Khalyfa, S. Vogt, J. Weisser, G. Grimm, A. Rechtenbach, W. Meyer, and M. Schnabelrauch, *J. Mater. Sci., Mater. Med.*, **18**, 909 (2007).
38. C. X. F. Lam, X. M. Mo, S. H. Teoh, and D. W. Hutmacher, *Mater. Sci. Eng. C*, **20**, 49 (2002).
39. V. Mironov, T. Boland, T. Trusk, G. Forgacs and R. R. Markwald, *Trends Biotechnol.*, **21**, 157 (2003).
40. M. Lee, J. C. Y. Dunn and B. M. Wu, *Biomaterials*, **26**, 4281 (2005).
41. M. J. Mondrinos, R. Dembzynski, L. Lu, V. K. C. Byrapogu, D. M. Wootton, P. I. Lelkes, and J. Zhou, *Biomaterials*, **27**, 4399 (2006).
42. T. Xu, J. Jin, C. Gregory, J. J. Hickman, and T. Boland, *Biomaterials*, **26**, 93 (2005).
43. B. K. Paul, and S. Baskaran, *J. Mater. Process. Technol.*, **61**, 168 (1996).
44. R, Landers, U. Hubner, R. Schmelzeisen, and R. Mulhaupt, *Biomaterials*, **23**, 4437 (2002).

45. J. T. Rimell and P. M. Marquis, *J. Biomed. Mater. Res.*, **53**, 414 (2000).
46. K. H. Tan, C. K. Chua, K. F. Leong, C. M. Cheah, P. Cheang, M. S. Abu Bakar and S. W. Cha, *Biomaterials*, **24**, 3115 (2003).
47. G. Lee and J. W. Barlow, *Proc. Solid Freeform Fab. Symp.*, Austin, TX, **August 9–11**, 376 (1993).
48. N. K. Vail, J. J. Beaman, D. L. Bourell, H. L. Marcus, and J. W. Barlow, *J. Appl. Polym. Sci.*, **52**(6), 789 (1994).
49. G. Lee, J. W. Barlow, W. C. Fox, and T. B. Aufdermorte, *Proc. Solid Freeform Fab. Symp.*, Austin, TX, **15** (1996).
50. H. Huang, S. Oizumi, N. Kojima, T. Niino, and Y. Sakai, *Biomaterials*, **28**, 3815 (2007).
51. J. M. Williams, A. Adewunmi, R. M. Schek, C. L. Flanagan, P. H. Krebsbach, S. E. Feinberg, S. J. Hollister, and S. Das, *Biomaterials*, **26**, 4817 (2005).
52. F.E. Wiria, K.F. Leong, C.K. Chua, and Y. Liu, *Acta Biomater.*, **3**, 1 (2007).
53. R. D. Goodridge, D. J. Wood, C. Ohtsuki, and K. W. Dalgarno, *Acta Biomater.*, **3**, 221 (2007).
54. V. K. Popov, E. N. Antonov, V. N. Bagratashvili, A. N. Konovalov, and S. Howdle, *Proc. Boston. V.EXS-1*, **F5**(4), 1 (2003).
55. V. K. Popov, E.N. Antonov, V.N. Bagratashvili, A. L. Ivanov, A.N. Konovalov, J. J. A. Barry, and S. M. Howdle, *4th Int. Conf. Mater. Process. Prop. Perform.*, Ibaraki, Japan, Singapore: Institute of Materials (East Asia), 152 (2005).
56. J. M. Kanczler , S.-H. Mirmalek-Sani, N. A. Hanley, A. L. Ivanov , J. J.A. Barry, C. Upton, K. M. Shakesheff, S. M. Howdle, E. N. Antonov, V. N. Bagratashvili, V. K. Popov, and R. O.C. Oreffo, *Acta Biomater.*, **5**(6), 2063 (2009).
57. D. A. Hollander, M. von Walter, T. Wirtz, R. Sellei, B. Schmidt-Rohlfing, O. Paar, and H.-J. Erli, *Biomaterials*, **27**, 955 (2006).

Chapter 32

NOVEL MICROSPHERES FOR PROLONGED CELL SURVIVAL

Sing Muk Ng,[a,b] Jeung Soo Huh,[b] Syed Izhar Haider Abdi,[c] and Jeong Ok Lim[a,c,d*]

[a] *Bio-Medical Research Institute, Kyungpook National University School of Medicine, Daegu, Republic of Korea*
[b] *Department of Materials Science and Metallurgy, Kyungpook National University, Daegu, Republic of Korea*
[c] *Department of Biomedical Science, Kyungpook National University School of Medicine, Daegu, Republic of Korea*
[d] *Joint Institute for Regenerative Medicine, Kyungpook National University Hospital, Daegu, Republic of Korea*
*jolim@knu.ac.kr

The development of novel biomaterials and its practical applications have been the subject of much research in the field of scaffolds for regenerative medicine. Classical paradigms in this discipline have demonstrated the success of producing scaffold biomaterials that facilitate tissue growth and provide structural support for cells and the delivery of bioactive molecules. The next challenge is in engineering biomaterials with innovative designs, offering smart and special functionalities which fulfill the optimum requirements for cell growth and tissue regeneration. This chapter aims to review one of such frontier efforts, particularly in the development of novel microspheres having smart capability in the controlled release of oxygen for efficient tissue engineering. Highlights will be on the current status of such development, the

Handbook of Intelligent Scaffolds for Tissue Engineering and Regenerative Medicine
Edited by Gilson Khang

www.panstanford.com

state of the art in fabricating oxygen-producing microspheres, in lab evaluation protocols, and its possible integration into tissue engineering applications. The discussion is based on real practice involving related technology and methodology for developing oxygen-releasing microspheres (ORMs) via encapsulating hydrogen peroxide in poly(lactide-*co*-glycolide) (PLGA) microspheres.

32.1 Introduction

The ultimate goal of tissue engineering is to assemble cells into a structural form together with specific biological functionalities in replacing diseased or damaged tissue caused by wear, trauma, neoplasm, or congenital deformity.[1] In general, this is done by collecting cells from a patient, expanding their number, and growing the cells into tissue using a suitable scaffold. In some cases, the process includes the need for cellular and tissue patterning, microcirculation development, and uses of appropriate external stimuli, such as chemical, biological, mechanical, or electrical, to ensure the appropriate biofunctional and mechanical properties are restored within the newly formed tissue. The more advanced stage in this area includes the reconstruction of complex organs and their transplantation into animals or humans. In this engineering process, key factors that ensure the success fall on the ability to control optimum conditions required for cell survival and the advancement of technology in producing suitable biomaterial scaffolds, which enable cell migration, adhesion, and differentiation to occur. These two factors seem to be treated independently at the early days but have now slowly merged into one subject, in which scaffolds in parallel operate as an artificial extracellular matrix (ECM) and provide the optimum needs and specific functions for cell survival (e.g., nutrients, salts, oxygen, growth factors, etc.). Later, this area of interest became known as intelligent scaffolds and generated huge research interest, especially from the aspect of their novel production, potential applications, and the prospect for commercialization.

One of the possible routes to fabricate intelligent scaffold is via the incorporation of functionalized materials into the main matrix of a scaffold. Others include the use of a hybrid material formed between organic-synthetic matrices.[2] The functionalized materials act as a smart terminal within the scaffold, having the ability to interact dynamically with the cell-growing environment and subsequently control the condition at its optimum. In this aspect, microspheres encapsulated with either chemical compounds or bioactive reagents or both can be a good candidate to serve the purpose. Along with the small-size factor that makes integration work simple, there are several strong driving reasons for the selection of using microspheres. First, microspheres allow the encapsulation of various ingredients, and their release from the microcapsule can be carefully controlled. Such works have been well demonstrated, especially in the area of drug delivery for encapsulating different kinds of drugs, and the microspheres manage to target specific treatment sites with controlled release of dosage.[3–5] Second, microsphere systems increase the life span of an active constituent due to the direct protection from the shell of the microspheres. This ensures prolonged effect or continuous activity of the ingredient without being deactivated by potential inhibitors from the surrounding or during the delivery route to the targeted area.[6,7] Third is the flexibility in the choice of materials and techniques that can be used to prepare the microspheres. Varieties of matrices either synthetic or natural are employed for this purpose, eliminating the effect of incompatibility between the microspheres and the active ingredients or with the main matrix of the scaffold during its integration for tissue engineering.[8] The choices of methods to prepare microspheres are also diverse, and one of the commonly reported is the double emulsion solvent evaporation technique, which is technically simple, has no instrumentation involved, and is economical in cost.[9–11] Finally, microspheres have been technically and practically proven to be compatible either *in vivo* or *in vitro* in biological systems.[12–14] This avoids the interfering effect that may cause serious malfunction to cell activities when blended into the scaffold or used individually for tissue engineering purposes. The advantages and flexibilities simplify the work of producing novel microspheres as the effort can focus directly on the smart property, manufacturing methodology, and its integration for applications.

In dealing with more complex systems such as a large-volume implant, the issue of stable cell survival and function often became a major concern. Basically this is due to the practical limitation in supplying all the required nutrients to the cells on time. Of all, oxygen is one of the most important elements for cell survival, yet it remains the toughest challenge so far in designing a suitable mechanism to supply a sufficient amount to cells. This is due to the gas nature of oxygen and its low solubility in aqueous solution ($\approx$7.8 mg L^{-1} at 25°C)[15], even lower in culture media having various solutes.[16] As a result, direct supply often damages cells due to sparging,[17] bubble formation,[18] and forming,[19] unlike other nutrients that can dissolve directly into the media of the cells. In the absence of oxygen, cells of living organisms undergo anaerobic metabolism that generates a large amount of lactic acid and subsequently harms or kills the cells with the highly acidic condition.[20] Under normal physiological conditions, oxygen is delivered to the body tissues via blood circulation that comes under the control of the respiratory system. The concentration of oxygen at the tissues is tightly maintained at an optimum level in ensuring healthy, growing cells. Such a complex system is elegant in nature but has turned into one of the biggest limitations in the field of tissue engineering as the formation of blood vessels naturally is a slow process. As tissue starts to grow thicker, cells imbedded more than 200 μm from the oxygen source will face necrosis before even the blood vessels are formed, due to oxygen diffusion limitations.[21,22] The incorporation of a system which mimics the natural respiratory function is still far too complex to be made realistic using the current engineering level, although measures are undergoing, especially in the area of vascularization. Therefore, is still an advantage to use scaffolds in providing oxygen to cells for tissue engineering and possibly via incorporating smart microspheres into scaffolds that release oxygen in a controlled manner.

32.2 Current Status and Development in Supplying Oxygen for Tissue Engineering

The issue of the insufficient supply of oxygen for the cell survival during the tissue engineering process has been known for a long time but remains problematic. Numerous efforts have been made to overcome this limitation over the years, which include the use of artificial oxygen carriers, enhancement and speeding up of the vascularization in tissues, and the attempts at fabricating oxygen-generating biomaterials. All these efforts have improved the supply efficiency of oxygen within the engineering system but to some extent have not fully succeeded in achieving survival of a clinically applicable large tissue mass. This section gives some overview of and insight into these efforts.

32.2.1 *The Use of Artificial Oxygen Carriers*

In nature, the supply of oxygen to cells is accomplished using the circulation of blood as a carrier. Hemoglobin (Hb) is the component in blood that is responsible for the role of binding oxygen at the lung, delivering it, and releasing it to cells over the entire body for the respiratory process. The mechanisms involved have been intensively investigated by researchers and well documented.[23,24] Nevertheless, the development of such a system is still currently impossible due to innovation and technology limitations. However, it has inspired tenacious and successful efforts in mimicking such a system, especially in developing novel liquid artificial oxygen carriers based on synthetic Hb and perfluorocarbons (PFCs).

Cell-free Hb itself tends to dissociate into its individual $\alpha\beta$ dimmer-dimmer that are rapidly eliminated from the circulatory system and reduces the effective lifetime besides causing kidney damage. In addition, cell-free Hb lacks its essential 2,3-diphosphoglycerate (2,3-DPG) affector, preventing the release of oxygen to the tissues.[25] Therefore, modifications are made to increase the stability besides optimizing its performance towards the delivery of oxygen. Modifications can be made via direct cross-linking between Hb, known as the acelullar type, or the encapsulation of Hb into a secondary matrix, known as the cellular type.[26] The Hb-modified carriers have been studied to oxygenate tissues in overcoming the problem of stroke, sickle cell crisis, and cardiac arrest.[27] Besides, the potentials are also being explored to be used in overcoming major surgical bleeding, hemorrhagic shock, and clinical efficacy in blood sparing.[28]

PFCs are another kind of commonly used oxygen carriers. PFCs are fluorine-substituted linear, cyclic, or polycyclic anthropogenic hydrocarbons that have total chemical, biological, and enzymatic inertness, low toxicity toward biological systems, thermostability, and very strong hydrophobicity and lipophobicity.[23,29] Uniquely, PFC liquids can dissolve large volumes of respiratory and other nonpolar gases, having the solubility decreasing in the order $CO_2 > O_2 > CO > N_2$. PFCs have the solubility of oxygen typically in the range of 1,000–1,400 mg L^{-1} compared to just around 8.0 mg L^{-1} for water.[30–32] All these factors make it suitable to be employed as an oxygen carrier for biomedical applications. In aerobic cell culture studies, PFCs have been used to substitute the limitations caused by conventional aeration systems, a method where air or a mixture of gases is forced through the volume of a bioreactor.[33] By abolishing the aeration system, the intensive physical interference caused by air movement and its mixing with the culture medium is eliminated, thus avoiding the generation of shear forces that can trigger the phenomenon known as hydrodynamic (turbulent) shear cell stress.[34]

32.2.2 *Induction and Enhancement of Vascularization*

In nature, tissues rely on blood vessels for the supply of oxygen and nutrients. The formation of new blood vessels is required when tissue grows beyond the diffusion limit of oxygen, and this is also applicable for tissue-engineered constructs. Experiences have shown that *in vitro* culture of a larger-volume construct is still possible with the aid of specially designed bioreactors, but after implantation, the supply of oxygen and nutrients to the implant is often limited by diffusion processes. The spontaneous invasion of blood vessels from the host will occur in response to signals that are secreted by the implanted cells due to hypoxia.[35] However, the rate is often too slow, and normally, spontaneous vascular growth is around several tenths of micrometers per day, indicating that complete vascularization of an implant of several millimeters needs weeks.[36] This could not support the need required by the implant, leading to failure in the tissue regeneration effort.

Having this understanding, an approach has been taken to induce, enhance, and speed up the formation of blood vessels in the implant and the merging process of blood vessels with the host system. Three processes have been distinguished during blood vessel formation, which are vasculogenesis, angiogenesis, and arteriogenesis.[37] Vasculogenesis is the *de novo* vessel-forming process that takes place during early embryonic development. Endothelial cells differentiate from their precursors and proliferate within avascular tissue to form a capillary network. Angiogenesis is the morphogenic process of new blood capillaries emerging from preexisting vessels. The process comprises matured vessel changes, pericyte detachment, ECM degradation and remodeling, proliferation, and migration and assembly of endothelial cells (EC) into tubule structures. Finally comes arteriogenesis, a process of structural enlargement and remodeling of preexisting small arterioles into larger vessels. Under well-controlled pathological and physiological conditions, angiogenesis is regulated by a number of mediators, either pro-angiogenic or anti-angiogenic factors.[38,39] This means the mediators can be used as tuning parameters in inducing, controlling, and speeding up the whole angiogenesis process. Thus in the interest of supplying oxygen to tissue, the use of angiogenic factors permits enhanced diffusion of oxygen across the newly formed tissue. This is made possible by the stimulation of blood vessel development and maturation by different growth factors such as vascular endothelial growth factor and basic fibroblast growth factor.[40,41] Besides, new microvascular networks in tissue substitutes have been engineered using vascular endothelial cells and stem cells or by creating arteriovenous shunt loops. With the rapid formation of vessels and the merging with the sounding readily available respiratory system, the possibility of necrosis has been reduced.

32.2.3 *The Utilization of Oxygen-Generating Biomaterials*

There are attempts to supply a sufficient amount of oxygen for tissue generation via a direct method, using chemically generated oxygen that is incorporated within scaffolds. It is known that peroxide salts such as sodium percarbonate and calcium peroxide will get dissociated into hydrogen peroxide upon dissolving in water,[42] and subsequently the hydrogen peroxide decomposes into water and oxygen. Such kind of work has been investigated by researchers from the Wake Forest University by employing peroxide salts to generate oxygen that is directly supplied for cell regeneration purpose.[43,44] In their work, an oxygen-generating salt was directly incorporated in the scaffold constructed using biodegradable polymer. Their most recent study showed such scaffolds manage to extend cell viability under hypoxic conditions for at least 10 days, allowing increased cell survivability while the neovascularization mechanism is being established after implantation.[44]

32.3 Oxygen-Releasing Microspheres (ORMs)

32.3.1 *The State of the Art*

Nature often offers examples of systems having smart materials and working mechanisms, which can be used as learning models in attempts to produce innovations that fulfill the needs of daily life. The mimicking effort no doubt is challenging but gives the most promising potential of success in application for various areas, especially those biologically related. The same approach is required in the interest of developing ORMs for efficient cell survival. Planning and implementation should be based upon concrete scientific knowledge that can make integration between connecting working mechanisms work, while having smooth flow of operational steps in a chronological manner. In addition, there is a need to consider those basic aspects that are related in such development, as summarized in Table 32.1, based on practicality, relying on knowledge and experience supports, hands-on facilities, and financial status. A simple method involving low-cost materials but achieving the objective of producing microspheres that release oxygen and are safe to biological systems is preferable.

As listed in Table 32.1, several main factors are required to be considered in designing novel ORMs. The first effort is in identifying suitable, clean oxygen sources that can be encapsulated and release oxygen upon being activated. These can be chemical compounds that decompose into products having oxygen or biological substances that undergo bioactivity such as photosynthesis, which generates oxygen. The direct usage of oxygen gas is seen as unpractical in this

Table 32.1. Basic aspects involved in and considered during the development of ORMs.

Process involved in developing ORMs
1. Identification of a suitable oxygen-generating source
2. Selection of materials that can be used as building blocks for ORMs
3. Selection of a suitable technique for encapsulating the oxygen source
4. The study of oxygen-releasing profiles.
5. Incorporation of a suitable agent to decompose the source (if required)
6. Storage and integration of ORMs into applications

case as the encapsulation of the gas is often difficult and gives low loading efficiency. If the gas is to be compressed in achieving higher concentration for encapsulation, a shell of ultra strength is required to sustain the pressure and is often unachievable by the commonly used biodegradable matrices. Having the oxygen source identified, the next factor to be considered is the materials that can be used to form the shell of the microspheres. Whether synthetic or natural, the material chosen should be biocompatible, pose no interference reaction with the other ingredients, and be preferably biodegradable. The utilization of hybrid materials having high suitability for tissue engineering tasks, formed from the homogenous blending of different matrices, is also a good option. In addition, microspheres can be formed via combination of different materials as long as the layers are attaching to each other. Upon having all the ingredients, a suitable method can be chosen based on the capability of forming microspheres and at the same time encapsulating the oxygen source. Efficient encapsulation should be achieved without decomposing the source during the process. There are many techniques reported for the encapsulation of compounds into microspheres, especially in drug delivery systems that can be readily modified or improved.[3,45] An ideal technique should be simple and highly reproducible, use nontoxic materials, involve no sophisticated instrumentations, and have a short processing time. The microspheres produced require evaluation for their oxygen-releasing profile, and if necessary, an activating mechanism to transform the source into oxygen will be integrated. The activation may be induced by mechanical force, thermal heat, electromagnetic radiation, catalyst, etc. In having a more elegant system, the activation mechanism is designed to be used as a switch in controlling the oxygen release time and rate. Such steps are very important to ensure the correct dosage and rate are supplied to the targeted applications. Finally, a proper storage method should be considered for the microspheres that are not being used immediately to avoid loss of oxygen. One good option will be the direct freezing of microspheres under a super-low temperature using liquid nitrogen to reduce the release rate of the oxygen source. Besides, the direct incorporation into other formats such as scaffolds to be used immediately is another option to replace the storage process.

Having the understanding of the state of the art for the development of ORMs, an attempt has been made to produce ORMs using hydrogen peroxide (H_2O_2) as an oxygen source that is encapsulated within a biodegradable matrix. The main concept behind the system is the production of oxygen upon the chemical decomposition of H_2O_2 that is released from the core of the microspheres. Encapsulation stabilizes H_2O_2 and induces a controlled release for prolonged generation of oxygen in sustaining optimum cell survival. In contrast, the direct usage of free H_2O_2 generates oxygen in high level of concentration at once and can cause a serious toxic effect to cells instead of facilitating their survival.[46] The use of H_2O_2 was based on several considerations besides its well-known chemistry.[47] The decomposition process of H_2O_2 is rather simple, consisting of a single-step reaction, and produces only oxygen and water as by-products. This eliminates the possibility of having harmful by-products generated from the chemical reaction. Even the by-products produced may not be toxic, such as metal cations in some cases, but accumulation can increase the concentration, leading to a high possibility of an interfering effect towards the biological system. Besides, H_2O_2 is commercially available, easy to obtain, and considerably economical. However, there are also some drawbacks in using H_2O_2, which you need to be aware of and improve to avoid failure in producing ORMs for biological applications. H_2O_2 decomposes easily under normal conditions and requires extra precaution to avoid excess loss during the handling. Besides, H_2O_2 is harmful to cells when in direct contact and is often used as an antiseptic to kill bacteria.[48] Therefore, assurance is required that only oxygen molecules reach the cells during the tissue reconstruction process but not free H_2O_2. In this aspect, a high standard of engineering work is needed to create a microsphere shell that acts as a barrier between the cells and H_2O_2 and can immediately decompose the H_2O_2 once released to avoid potential leakage.

The decomposition rate of H_2O_2 is well known theoretically to be affected by several parameters such as a change of temperature, a change of pH, or the existence of impurities that act as a catalyst.[49,50] These affecting variables can be utilized as smart controlling keys for the releasing rate of oxygen once incorporated within the microsphere. For instance, the use of different catalysts and the amount added can vary the decomposition rate of H_2O_2, and when these parameters are fixed, a known, controlled releasing rate can be achieved. A more advanced example will be the use of pH changes from the surrounding to induce the decomposition rate. This is made possible by first incorporating a pH-sensitive catalyst to the shell of microspheres. Some catalysts lose their activity once transformed into different formats such as in a complexed form but will resume their activity when free, back to their initial condition by some physical factors. As mentioned earlier, the acidity of the cell environment increases under anaerobic conditions due to the formation of lactic acid. The drop of pH can dynamically induce dissociation of the complexed catalysts into their free form having normal activity. When this occurs, H_2O_2 will be decomposed to generate more oxygen for the cells' need. As this is a responsive mechanism, the level of oxygen can be maintained dynamically at the required level, not over or under dosage.

32.3.2 *Materials as Building Blocks of Microspheres*

In developing smart microspheres, there is a diverse selection of choices in materials readily to be employed, depending on the objective, suitability, and intended applications. Different kinds of materials such as organic (synthetic and biologically derived or naturally obtained) and inorganic (ceramic based) have been investigated and well reported in comprehensive reviews.[1,2,51] In general, biologically derived materials have excellent physiological activities such as selective cell adhesion, biodegradability, and low risk of viral infection, antigenicity, or deterioration. On the other hand, synthetic materials offer versatility in designing an exogenous ECM with specific properties such as porosity and mechanical strength that can be manufactured reproducibly on a large scale.

Among the available candidate biomaterials, extensive research concentrates on PLGA, a synthetic-type material classified under hydrolytically degradable polymers. PLGA undergoes bulk erosion through hydrolysis of the ester bonds with the rate of degradation that depends on a variety of parameters, including the lactic and glycolic acids ratio and the molecular weight, shape, and structure of the matrix.[52] The major popularity of PLGA is attributed in part to its approval by the Food and Drug Administration (FDA) for use in humans, its good processibility, which enables fabrication of a variety of structures and forms, and controllable degradation rates. Besides, PLGA demonstrates good cell adhesion and proliferation, making it a potential candidate for tissue engineering applications.[53] Under such considerations, the use of PLGA is also one of the ideal choices as the main matrix for the shell of ORMs. In this case, PLGA is inert towards H_2O_2 and can be easily fabricated into micron-sized particles having H_2O_2 encapsulated. PLGA plays a role in shielding the H_2O_2 from being decomposed by possible catalysts from the surroundings, while at the same time becoming the barrier to avoid harmful direct contact of the free H_2O_2 with cells. Slow release of the H_2O_2 gives sufficient time for its complete decomposition into oxygen.

Sometimes the issue of incompatibility of physical properties between ingredients used in producing ORMs arises under unavoidable circumstances. As a consequence, the microsphere loses the smart capability in creating an optimum level of oxygen for cell survival. This issue also exists in the system of using H_2O_2 encapsulated in PLGA for oxygen generation, where two main ingredients chosen have different physical properties. PLGA is highly hydrophobic, while H_2O_2 is highly hydrophilic. Thus the possibility of incomplete H_2O_2 encapsulation may occur due to the strong expel force generated during the *meta stable* stage of the synthesis, a stage when all ingredients are still in the liquid phase with high mobility. Phase separation can easily occur, failing the incorporation of H_2O_2 within the solidified PLGA microsphere. To overcome such a problem, two possible steps can be employed—creating better compatibility between ingredients and modifying the preparation methodology that can minimize the expel force during the meta stable stage. The latter effort on methodology alteration will be discussed in the next section.

The addition of an extra ingredient having an intermediate property from these extremes is a possible approach to increase the physical compatibility between ingredients. For instance, in the study carried out, the addition of polyvinyl alcohol (PVA) has found to increase the encapsulation efficiency of H_2O_2 in PLGA microspheres. PVA acts as a surfactant that stabilizes H_2O_2 emulsion within the organic PLGA phase, creating a hydrophilic microenvironment within the hydrophobic polymer phase during the meta stable stage. Thus this reduces the expel force and promotes effective encapsulation. However, in this case, further increment in the hydrophilicity of PLGA by blending in methoxy polyethylene glycol (m-PEG) shows no significant improvement on the encapsulation efficiency, although such an approach is being reported suitable for the encapsulation of several other compounds.[54,55] PEG is the hydrophilic segment widely used to change the physicochemical properties of PLGA in achieving better encapsulation for these highly hydrophilic compounds.[56,57] The failure may be due to the small size of H_2O_2 that can diffuse out quickly from a more hydrophilic shell consisting of the blended matrices during the meta stable stage. The observation has demonstrated the importance of first understanding the nature of the ingredients chosen and the whole working mechanism involved before trying to tune its physical compatibility to the optimum using a secondary addictive.

The design of an elegant ORM sometimes requires the combination of various working mechanisms that operate in a chronological manner. Each working step is interrelated, and the sequence of the processes must be followed, such as the rate of release for the oxygen source matches the decomposition rate and the flow of the source is from the reservoir to the activation region, not the opposite direction. It is also important for the source to stay stable at the reservoir and only get decomposed when required. Under considerations based on the differences in role and activation time, integration of these functionalities within one same segment of matrix without them interacting before required is often difficult. Thus, the option of using a combination of different matrices each having the required functionality can be an easy solution. In this case, matrices are engineered into different segments or layers by clear boundaries according to their role. For instance, the source will be encapsulated at the core and then coated with another layer having the activating catalyst. Such design allows the generation of oxygen in a controlled manner during application, while the loss of the source will be minimized during production as the activating agent is not in contact directly with the source. Materials chosen for each segment may be of different types but must be able to attach together boundary to boundary, while being ready to be grafted with the required functionality.

A successful attempt has been demonstrated in mixing PLGA microspheres encapsulated with H_2O_2 into alginate beads, in which the alginate was first grafted with catalase. Alginate is chosen as the second coating layer as its backbones consist of a carboxylic group that is suitable to be grafted with catalase, besides showing no toxic effect, being biocompatible, and being well employed in encapsulated bioreagents.[58,59] In addition, an alginate solution can be easily formed into a hydrogel using a divalent cation such as barium (II) or calcium (II) ions. PLGA microspheres, although from different kinds of materials, were found to be evenly distributed within the alginate bead, as observed clearly from the image taken under an optical microscope (Fig. 32.1).

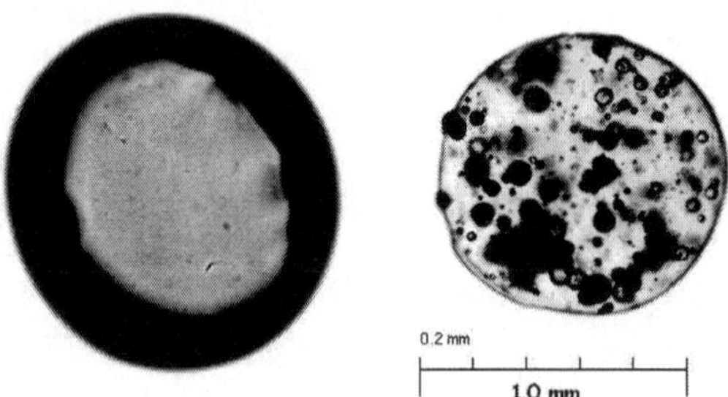

Figure 32.1. H_2O_2-encapsulated PLGA microspheres that are evenly distributed within an alginate matrix that has been grafted with catalase to decompose the H_2O_2. Black spots in the right bead were the PLGA microspheres, while the alginate hydrogel was transparent under the optical microscope. The left microbead is the control, produced using alginate without the addition of PLGA microspheres.

In such a system, the decomposition of H_2O_2 only occurs once it is released from the PLGA segment and enters the alginate layer. The decomposition rate depends on the controlled release rate of the H_2O_2 from the PLGA and the concentration of catalase grafted on the alginate. The use of two layers avoids serious losses of H_2O_2 during the ORM production as H_2O_2 does not have direct contact with catalase.

32.3.3 *Techniques for Producing ORMs*

32.3.3.1 Double-emulsion and solvent evaporation technique

The solvent evaporation method has been used extensively to produce microspheres encapsulated with various compounds such as drugs,[60–62] peptides,[63] and bioreagents.[64,65] PLGA is one of the most frequently used materials in this technique. Microencapsulation by the solvent evaporation technique can be done via various protocols, and the selection for the best option readily depends on the property of the compounds used.[11] Therefore, this technique is also a suitable alternative in producing ORMs to encapsulate the oxygen-releasing source into a polymeric matrix with a micrometer-ranged capsule. As the oxygen sources often consist of water-soluble compounds, the double-emulsion-type option made up of a few major steps can be employed.[9–11] First is the dispersion of the oxygen-generating source into an organic solvent that is a predissolved polymeric matrix. The state of the art falls on the unique interphase formation between immiscible aqueous and organic layers, turning the aqueous phase into fine emulsion droplets. This emulsion mixture is then dispersed into an aqueous solution forming a secondary emulsion, which is commonly known as a water-oil-water ($W_1/O/W_2$) double emulsion. Under such a process, the first aqueous droplets will be entrapped in the core of the polymer, while physically the polymer is shaped into small, fine spheres due to the secondary dispersion. The microspheres are usually stabilized using a suitable surfactant. The organic solvent will be extracted from the dispersed phase by a continuous phase, transforming the droplets into solid microspheres that are free from the toxic organic solvent. Finally is the harvesting and drying of the microspheres from the continuous phase. In real practice, all these process-engineering steps may vary from case to case, depending on the ingredients involved. Therefore, optimization is needed for different systems, and steps can be modified in particular to achieve the encapsulation efficiency required.

In encapsulating H_2O_2, a series of modifications and improvements on the standard technique are required. Although microencapsulation of water-soluble compounds is not something new and has been commonly reported, to our best knowledge, no studies have pursued a similar thought on molecules having a weight less than 100 g mol^{-1}. Theoretically, the idea seems to give no significant variations in achieving the final encapsulation efficiency as those reported for larger compounds; however, a practical study using H_2O_2 reviews that the outcome was not even near to the expectation. No significant existence of H_2O_2 was observed from the PLGA microspheres prepared using the conventional standard method. The major reason is related to the high diffusion coefficient of H_2O_2 due to its small size, leading to failure of encapsulation as leaching occurs seriously into the continuous phase. In order to avoid the leaching, a thermodynamic factor was considered to create a condition where the diffusion will occur back to the microspheres instead of out or be at least stabilized during the meta stable stage. One of the possible approaches is via inducing a *backward concentration gradient*, where H_2O_2 is added into the continuous phase and the concentration is controlled above or is the same as the one added in the microspheres. Under such circumstances, H_2O_2 diffuses from the high-concentration region of the continuous phase toward the low-concentration region of the microsphere core. Effective encapsulation can be achieved after the hardening of the microspheres (Fig. 32.2). As far as cost, safety, and handling precautions are concerned, the volume of the continuous phase needs to be reduced as a counterbalance of the high concentration of H_2O_2 used. Having reduced volume, the stirring time for solvent evaporation is extended for complete organic solvent removal, while increasing the concentration of the surfactant to stabilize emulsion droplets to avoid coalescence.[66] No obvious differences in size and shape were observed from the batch of microspheres prepared from the modified method compared with the standard protocol.

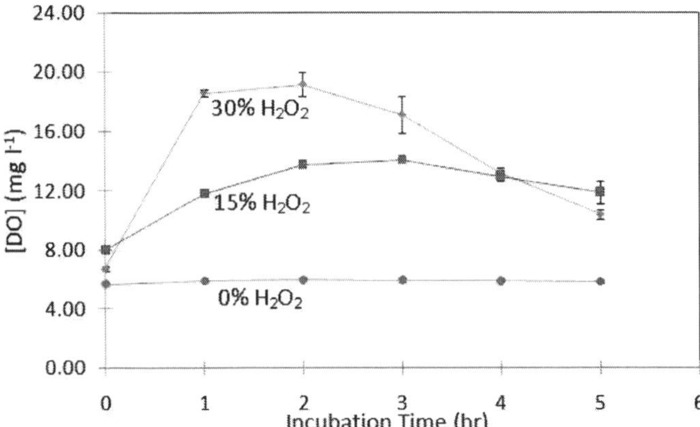

Figure 32.2. The effect of the addition of H_2O_2 on the continuous phase during the synthesis of microspheres. Encapsulation efficiency is dependent on the concentration of H_2O_2 added to the continuous phase due to the backward concentration gradient. Percentage values stated on the graph represent the concentration of H_2O_2 added to the continuous solution during synthesis.

The pH condition of the continuous phase is also an influencing parameter during the synthesis, which needs to be controlled. During the encapsulation of H_2O_2, a slight basic condition produces a smoother surface of microspheres compared with the lower pH that seems to produce tiny holes and uneven-shaped spheres (Fig. 32.3).

Although H_2O_2 is stable at a neutral condition, while it usually decomposes in a highly alkaline medium, this observation was not

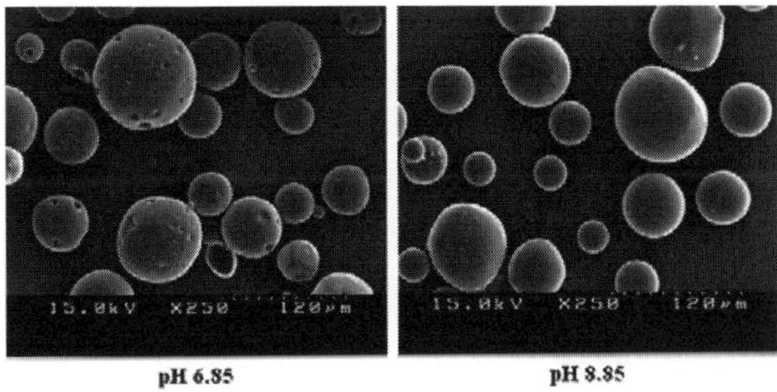

Figure 32.3. The effect of pH at continuous phase during synthesis towards the morphology of PLGA microspheres encapsulated with H_2O_2.

the case. Rather, a lower pH shows greater decomposition, as evidenced from a scanning electron microscopy (SEM) micrograph. A possible reason may be related to the presence of PVA in the continuous phase that undergoes de-protonation at high pH, leaving the negatively charged terminal to form cheating complexes with impurities. This avoids the decomposition of H_2O_2 by these impurities, which act as a catalyst. Oppositely, protonation of PVA occurs at a lower pH and subsequently causes the dissociation of possible present complexes of impurities. As this occurs, the impurities may diffuse through the polymer layer and decompose the H_2O_2, forming a gas that will increase the porosity of the PLGA matrix. This indicates less H_2O_2 is present in the final microspheres harvested due to decomposition, which is not preferable.

The freeze-drying process is normally performed in many microencapsulation works of drugs to remove excess water content via low temperature and pressure. It is also been reported by Kim and Park[56] as having a critical effect towards the sustained release of drugs. The freezing process can cause microcracks on the polymer matrix, while the removal of water produces microchannels, both of which favor drug release. However, in this study, H_2O_2 was chosen as the encapsulation molecule, which has almost similar physical properties as water. Under such circumstances, H_2O_2 will be removed along with water molecules during the drying process. In the same study by Kim and Park,[56] the freeze-drying process has reported to remove almost 98% of the water content. Thus, it is not surprising that PLGA microspheres show no evidence of the existence of H_2O_2 after being freeze-dried for six hours. Realizing this factor, the preparation process of consecutive batches of microspheres has skipped the freeze-drying process, but they have instead just been dried overnight under flowing air on a clean bench. The freezing process was found to cause no significant effect towards the encapsulation efficiency of H_2O_2.

32.3.3.2 Functionalization of matrices selected as building blocks

In tissue engineering, regenerative medicine, and other biomedical fields, the modification of potential crude biomaterials is often required to improve their characteristics towards better cell adhesion, proliferation, viability, and enhanced ECM secretion functions. This is usually done by functionalizing the matrices with bioactive species or chemical compounds. In this aspect, emphasis is usually on the surface engineering as this is the barrier that will be in contact directly with the cells. As this area covers an important key factor in the success of utilizing a biomaterial, intensive research and reviews on this subject matter have been undertaken. In general, it is reported that such an approach consists of three steps, which are the modification step involving different techniques, assessment and characterization of the materials, and a biocompatibility test through biologically related studies.[66] Modification for functionalizing purposes can be carried out with different synthetic routes such as grafting, adsorption, and blending, depending on the material suitability. General principles and techniques used for the modification will not be further discussed in this section as that information can be readily obtained from the literature.[56,66,67] Such modification is also required for the ORMs, depending on the final intended application, and the standard protocols from the literature can be used.

Particularly in developing ORMs, there is an extra need for functionalizing the matrix with the capability of decomposing the free H_2O_2 into oxygen and water besides the common properties required as tissue scaffolds. Various catalysts for decomposing H_2O_2, chemical or biological, can be used as a coating layer on the microsphere surface to serve this purpose. However, the mobility of the catalyst needs to be confined, and it should not be released into the surrounding, which may cause serious contamination. In this case, a biological catalyst, specifically catalase, can be employed and immobilized on the biomaterials matrix. Catalase is chosen due to its specific activity towards H_2O_2 decomposition with a fast turnover, ease to be obtained with considerably low cost, and most importantly the possibility to carry out immobilization covalently or physically into a matrix.

Covalent immobilization is more preferable as the chances of contaminating the biological surrounding will be lowered and the immobilization protocols for catalase are well reported.[68–70] One possible covalent method is via the use of glutaraldehyde as a cross-linking agent between amine-amine groups.[71,72] However, in this study, the covalent immobilization was performed using the well-established carbodiimide chemistry. In this approach, carbodiimide is used to activate carboxylic from a matrix and the activated terminal will form a covalent bond in the presence of an amine group from the catalase. N-hydroxysuccinimide is added to stabilize the activated species for better covalent bond formation. Alginate will be a good candidate for the immobilization as the backbone contains a carboxylic group for the grafting to take place. After the immobilization, the alginate solution was formed into a hydrogel using calcium (II) ions, and PLGA microspheres were incorporated within the alginate. Observation shows that the catalase retains its activity in decomposing H_2O_2 after immobilization and also after being transformed into the hydrogel format, although the decomposition rate was slightly slower compared with the free catalase.

32.3.3.3 Instrumentations for the preparation of microspheres

The involvement of some simple instrumentation for the production of ORMs can ensure the batches obtained in a more standard fashion and repeatable manner. Besides reducing the deviation in the microspheres' physical properties, the involvement of instrumentation also opens a possible step in mass production and commercialization. In the first stage of H_2O_2 encapsulation, involvement of instrumentations is not required as the double-emulsion and solvent evaporation method is simple and reproducible and can be scaled up easily for mass production. However, there are other options of microsphere production that involve instrumentations, such as the spray-drying technique.[73]

In the study of producing ORMs, the later stage of alginate bead production having PLGA microspheres imbedded can employ instrumentation to obtain better and standard beads. For instance, the use of a syringe injected by hand produces uneven-sized alginate beads ranging from 2 to 6 mm. When a microinjector was attached to the syringe for a fixed injection rate, beads having an evenly distributed size of around 2 mm were obtained. In an effort of reducing the size into the micrometer range, further improvement was made by applying high voltage on the needle head having the setup as

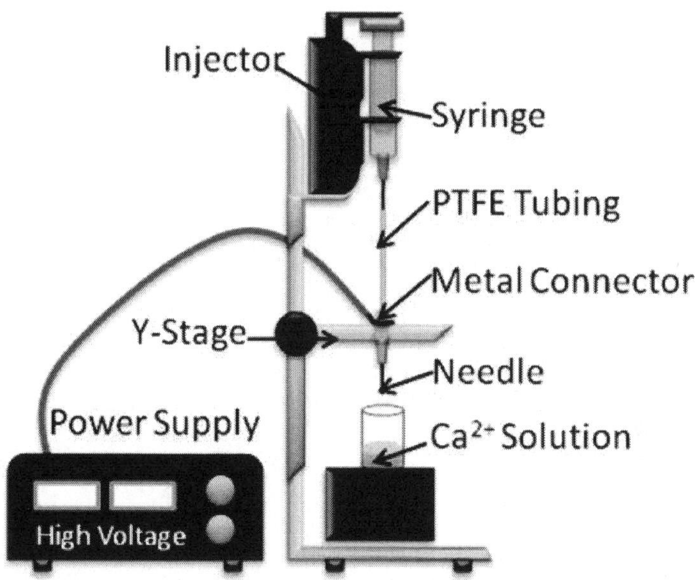

Figure 32.4. Instrumentation used for the production of alginate microbeads that employed a high voltage in obtaining even and standard microbeads. The syringe has the maximum capacity of 5.0 mL, and the power can supply a voltage up to 20 KV. The position of the calcium (II) ion solution is at a fixed distance of 10 cm from the tip of the needle, and the injection rate is set at 10 mL hour^{-1}.

illustrated in Fig. 32.4. The applied voltage ranging from 15 to 20 KV generates an electrostatic repulsion force on the inner-side alginate solution that subsequently causes it to expel out of the needle head. The outlet of the needle head was faced towards a target bottle containing a solution of calcium (II) ions. This has successfully produced even-sized alginate beads having a size below 1 mm in a very fast rate and conveniently. Besides, the PLGA microspheres were also homogenously distributed within the alginate matrix (Fig. 32.4).

32.3.4 *Evaluation of the Oxygen-Releasing Profile*

32.3.4.1 Direct observation

The simplest evaluation way on the release of oxygen from ORMs is via qualitative observation by the naked eye under an optical

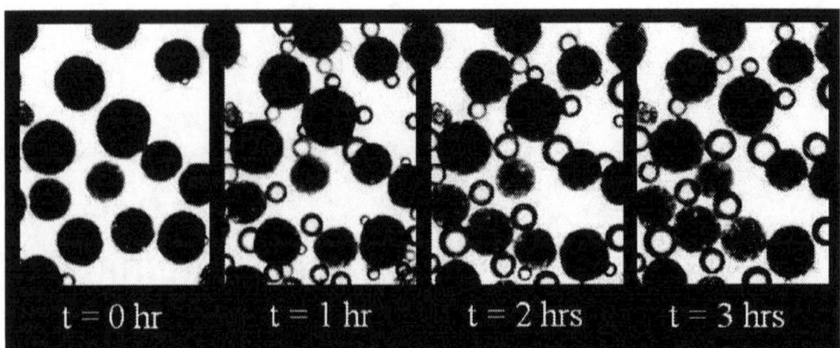

Figure 32.5. Microsphere images taken under an optical microscope, showing the formation of bubbles. The bubble size increases with time, indicating the continuous release of oxygen. Bubbles are seen as white cycles, and the microbeads are the opaque spheres.

microscope. In this study, the ORMs were sandwiched between two glass slides, and after that, water was carefully introduced through the edge. During this process, entrapment of air bubbles was avoided. As the oxygen was produced, bubble formation was observed on the microspheres' surface, and when images were taken under a periodic period, the growing size of the bubbles entrapped between the slides indicated the continuous release of oxygen. An example of such images observed is shown in Fig. 32.5. Although this method provides no concrete data on the concentration and rate of oxygen production, direct observation can give very convincing evidence in proving ORMs are releasing oxygen, as intended.

32.3.4.2 Quantitative analytical approach

Besides qualitative observation, systematic and precise analytical measurements are required in evaluating the release profiles of oxygen from the microspheres. This approach generates quantitative concrete values that are used directly to analyze and compare the performance of microsphere batches prepared either using the same or different materials, methods, or production settings. In addition, the microspheres can be categorized according to the oxygen-releasing amount and rate, which is a useful guideline for assigning or predicting their suitability for some applications, as some require more concentration than others.

Practically, analytical measurement of the oxygen level can be performed via different standard analytical methods, either directly or indirectly, using conventional laboratory techniques, optical spectroscopy, or the use of a commercially available portable oxygen meter.

In the conventional method, microspheres will be placed in dry or incubated in a solution inside an airtight reaction vessel having an outlet that is connected with tubing. The other end of the tubing will be inserted into a burette inlet that is first filled with water and placed backward in a water reservoir. The oxygen produced from the microspheres will replace the water inside the burette, and the amount can be recorded directly from the granulated marks on the burette. This method is simple and cheap but is, however, subjected with higher errors and lower accuracy. It will be also problematic if the amount of oxygen produced is too low and insignificant in causing any water replacement. More fine and small-scaled apparatuses will be required when dealing with a low oxygen production rate.

Another option for a simple, sensitive, and accurate measurement will be via the use of standard optical spectroscopy. ORMs can be incubated in an airtight standard cuvette containing a solution with a fluorescent dye that is sensitive to oxygen quenching, such as ruthenium complexes[74,75] or erythrosin B.[76] Under continuous monitoring using a spectrofluorometer, the decrease of fluorescent intensity upon being quenched by the oxygen produced will be recorded. The quantitative value on the oxygen concentration can be derived from a calibration curve obtained using various suitable quenching models.[77] Measurements recorded using this method are very reliable but require handling by trained personal and the use of expensive optical instrumentation.

A simplified version of the spectroscopy method will be the use of commercially available oxygen sensors. These sensors are very common nowadays and can be purchased easily at a comparatively low price. Oxygen concentration is recorded directly without the need of conversion, no calibration curve, and no extra chemical. The handling of such sensors is also often simple, and effort is only needed on setting up a controlled study with a good plan that can evaluate the oxygen-releasing profile for the ORMs.

32.3.4.3 Biologically related study

Currently, the main intention of developing novel ORMs is to be applied in biologically related areas such as in tissue engineering for efficient cell survival. Therefore, the microspheres not only should release oxygen as planned but also should be biocompatible in a biological environment. Thus, the microspheres need to be tested with a series of studies in ensuring the safety aspect is fulfilled and the existence will not affect the normal activity of cells. Commonly, a cell culture study will be used for the initial evaluations before upgrading into more complex systems, if necessary, including *in vivo* transplant. For the ORMs described in this work, an initial cell culture study was performed using neural N2a cells. Initial results show that cells can proliferate and maintain the viability well in the presence of microspheres. Although quantitative counting of cells has not been carried out, the images obtained under the microscope can significantly show the difference in the cell numbers. The number of cells when incubated with microspheres was observed to be more compared with the control well in the absence of microspheres (Fig. 32.6).

In the more advanced stage, an indirect biological related study can be planned accordingly to be used as another powerful tool to evaluate and verify the release of oxygen from the ORMs. For

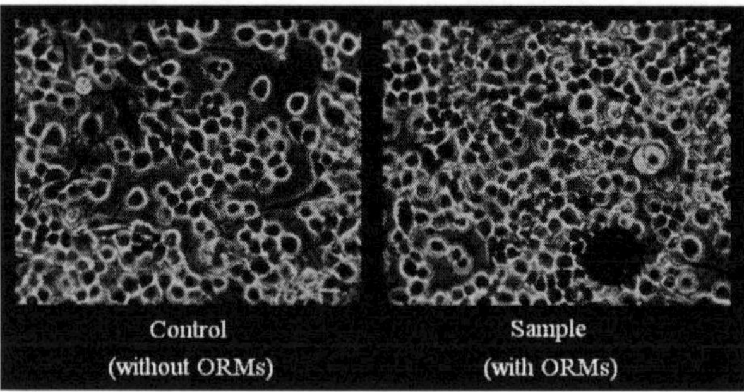

Figure 32.6. Cell images recorded after 24 h incubation under normal condition. The total volume of the medium was fixed at 3.0 mL. The control well was not added with ORMs, while the sample well was added with 0.2 mg of ORMs. The dark spot observed on the sample well was the microsphere added.

instance, a comparison of cell density for culture seeded under normoxic (21% O_2, 5% CO_2) and hypoxic (<1% O_2) conditions in the presence and absence of microspheres over a period of time can be made. Any significant density differences of cells between the cultures will contribute towards the effort in drawing conclusions over the oxygen-releasing phenomena. Theoretically, under hypoxic condition, the oxygen level is very low and cells usually will not survive long under such conditions. With the addition of microspheres under such conditions, if the cells can survive, this gives an indication that oxygen is supplied by the microspheres. In contrast, if cells fail to survive and show the same result as the cultures without microspheres, this indicates a failure. In this type of a comparison study, all conditions set between the control and the sample need to be identical and well controlled. Such kind of an evaluation method has been practically used in the work by Oh *et al.*[44]

32.4 ORMs in Applications for Efficient Cell Survival

Despite the success in developing ORMs, the next important aspect is to utilize the microspheres for efficient cell survival effectively in applications. The microspheres should be able to release the oxygen to cells at the targeted place and time, without being wasted or targeted on the wrong area. Thus, this reflects the need for designing elegant delivery mechanisms, integrated on a suitable platform that acts as a bridge to merge the microspheres into a real biological system. Based on the literature, there are several potential options that can be employed and further modified in achieving this task. This section will discuss some of the methods with suggestions based on the authors' opinions.

32.4.1 *Direct Integration in Scaffolds*

The integration of microspheres encapsulated with active agents into scaffolds for tissue engineering purposes has been widely reported.[12,78–80] The motivation is to combine the strengths and advantages, both of conventional scaffolds and encapsulated microspheres, into a new class of a composite biomaterial in fulfilling the needs in the area of tissue engineering. An encapsulated microsphere by itself has portrayed tremendous characteristics such as controlled release of active reagents and the capability to encapsulate various compounds of biological importance, but nevertheless these advantages fail to be utilized directly as a scaffold due to its physical constraint. This is because a typical scaffold needs to meet several basic criteria[53]: an extensive network with interconnecting pores for cells migration, multiplication, and attachment; channels to provide oxygen and nutrients and also to remove waste products; biocompatibility with a high affinity for cells to attach and proliferate; the right shape and size as desired by the surgeon; and appropriate mechanical strength and biodegradation profile.

The concept of a composite scaffold utilizing ORMs in tissue engineering can be adopted. The ORMs can be integrated homogenously as part of the scaffold within the main matrix. The microspheres act as an active microreactor that produces oxygen for the cells, while cells can attach, proliferate and differentiate, and form an ECM on the specially designed scaffold. In this approach, there is also a possibility to separate the oxygen-producing mechanisms into both the matrices—encapsulation of the oxygen source such as H_2O_2 in microspheres and the grafting of decomposition reagents such as catalase on the scaffold. As the source is released at a controlled rate, it will get decomposed when in contact with the scaffold matrix.

Integration of the microspheres can be done via various methods. One of the ways is by direct mixing of the microspheres into the scaffold material during the construction of the three-dimensional (3D) network while the material is still in solution form. Upon being molded into the desired shape and size of the scaffold, the microspheres will be evenly distributed within the matrix of the scaffold to release oxygen for cell survival. Other approaches include the entrapment technique, where microspheres are physically spread in between the networks of the scaffold. An example of that will be the electrospinning of nanofibers into a requested design (e.g., sheet, capillary, layered, etc.), and at the same time, ORMs are spread evenly during the electrospinning process. By doing this, microspheres will be knitted in between the porous fiber networks, and cells can grow directly on the microspheres' surface, ensuring a sufficient and high level of oxygen supply.

32.4.2 *Oxygen-Generating Reservoir*

The small size of the microsphere requires extra handling caution when used for tissue engineering applications so as to avoid the interference effect that may be caused by its high mobility. Unless integrated within a fixed position, as discussed earlier, the microspheres will have the possibility of being washed away by the present biofluid to other sensitive areas and subsequently cause a serious contamination problem. For example, if the ORMs are used to provide oxygen at a wound for faster tissue regeneration, without integration, the microspheres may go into the bloodstream via injured or cut vessels and get circulated to other parts of the body. This can cause serious problems and the malfunction of those organs due to the presence of the microspheres.

In overcoming such a problem, direct usage of microspheres to the targeted area can be avoided, and instead a delivery system can be employed that can supply oxygen to the cells. This means the ORMs will act as a reservoir to generate oxygen without direct contact with the cells. In this case, the medium on the targeted area is constantly circulated through the microspheres that are confined within membrane barriers to saturate its oxygen level. Upon reaching the cells, diffusion of oxygen occurs from the highly concentrated fluidic phase to the cells. Such an approach is more suitable to be applied for an *in vitro* system such as those using microfluidic channels for the supply of nutrients to cells.

32.4.3 *Medicine to Oxygenate Tissues*

The failure in supplying enough oxygen to organs causes serious damage to their functionality and subsequently can paralyze a normally operating biological system. One of the well-known examples is stroke, caused by the blockage of the cerebral microvasculature and resulting in decreased blood and nutrient supply postocclusion to the brain tissue.[81] This ischemic and hypoxic environment leads to several necrotic tissue events, including the loss of adenosine triphosphate (ATP) production, elevated levels of intracellular calcium, etc., that ultimately lead to neuronal and cerebrovascular cell death.[82] Under such conditions, it will be useful to employ the ORMs as a medicine used to oxygenate tissue that is under hypoxic conditions. Due to its small size, the ORMs can be directly injected to the affected area and act as a temporary generator for oxygen supply before the problem of vessel blockage is overcome via operation or other methods. This is seen to be a better option compared with the use of oxygen-rich fluids as the oxygen supply can be generated continuously at a constant rate with a considerably higher concentration for a longer duration. As for this type of application, the ORMs should be specially designed to be extremely fine in size and having fast degradation and to be eliminated (microspheres or the degraded products) from the area to avoid other harmful effects that may arise due to their long-term existence at the area. With this taken into consideration, harmful effects caused by the existence of microspheres, as pointed out in the previous section, will be eliminated and serve the purpose of supplying oxygen to needed areas, playing the role of blood.

32.5 Conclusion

This chapter has focused almost exclusively on the frontier research trend in current biomaterial development, which is the effort in producing ORMs for efficient cell survival. All rounded aspects concerning this trend such as the current status, experimental approaches, and the possible integrations into applications have been discussed. In order to make the scope more specific, a large section of the discussion has been made based on real practice, which was the development of H_2O_2-encapsulated PLGA microspheres that generate oxygen. Several suggestions, potential approaches, and possible modifications of standard protocols for producing ORMs have been given based on the authors' opinions and the laboratory observations obtained during the study.

Currently, there are still a lot of research potentials in the area of developing ORMs. These include the research of new concepts for generating oxygen from microspheres and the innovation of a novel methodology used in the production of microspheres. Besides, the possible future applications for such smart materials are diverse, such as in biologically related fields, regenerative medicine, and biosensor platform technology.

Acknowledgments

This research was supported by the Conversing Research Center Program through the National Research Foundation of Korea (NRF) funded by the Ministry of Education, Science and Technology (grant number 2010K001133).

References

1. O. Arosarena, *Curr. Opin. Otolaryngol Head Neck Surg.*, **233** (2005).
2. Francesco Rosso, G. Marino, A. Giordano, M. Barbarisi, D. Parmeggiani, and A. Barbarisi, *J. Cell. Physiol.*, **465** (2005).
3. S. Freiberg and X. X. Zhu, *Int. J. Pharm.*, **1** (2004).
4. J. K. Vasir, K. Tambwekar, and S. Garg, *Int. J. Pharm.*, **13** (2003).
5. R. Herrero-Vanrell and M. F. Refojo, *Adv. Drug Del. Rev.*, **5** (2001).
6. S. Martins, B. Sarmento, E. B. Souto, and D. C. Ferreira, *Carbohydr. Polym.*, **725** (2007).
7. H. Tamber, P. Johansen, H. P. Merkle, and B. Gander, *Adv. Drug Del. Rev.*, **357** (2005).
8. B. Ríhová, *Adv. Drug Del. Rev.*, **157** (1996).
9. P. B. O'Donnell and J. W. McGinity, *Adv. Drug Del. Rev.*, **25** (1997).
10. S. Freitas, H. P. Merkle, and B. Gander, *J. Control. Rel.*, **313** (2005).
11. M. Li, O. Rouaud, and D. Poncelet, *Int. J. Pharm.*, **26** (2008).
12. H. Tan, J. Wu, L. Lao, and C. Gao, *Acta Biomaterialia*, **328** (2009).
13. F. Buket Basmanav, G. T. Kose, and V. Hasirci, *Biomaterials*, **4195** (2008).
14. Y. Kimura, M. Ozeki, T. Inamoto, and Y. Tabata, *Biomaterials*, **2513** (2003).
15. P. A. Ruffieux, U. von Stockar, and I. W. Marison, *J. Biotechnol.*, **85** (1998).
16. A.S.B.K.W.D.D. G. Quicker, *Biotechnol. Bioeng.*, **635** (1981).
17. S. K. W. Oh, A. W. Nienow, M. Al-Rubeai, and A. N. Emery, *J. Biotechnol.*, **245** (1992).
18. J. Wu, *J. Biotechnol.*, **81** (1995).

19. S. Zhang, A. Handa-Corrigan, and R. E. Spier, *J. Biotechnol.*, **289** (1992).
20. Z. Wang, S. Wang, Y. Marois, R. Guidoin, and Z. Zhang, *Biomaterials*, **7387** (2005).
21. I. P. Torres Filho, M. Leunig, F. Yuan, M. Intaglietta, and R. K. Jain, *Proc. Natl. Acad. Sci. U. S. A.*, 2081 (1994).
22. J. Folkman, *Cancer Cell*, **113** (2002).
23. R. Kocian and D. R. Spahn, *Best Pract. Res. Clin. Anaesthesiol.*, **63** (2008).
24. W. A. Eaton, E. R. Henry, J. Hofrichter, and A. Mozzarelli, *Nat. Struct. Biol.*, **351** (1999).
25. S. Sivan and N. Lotan, *Biomol. Eng.*, **83** (2003).
26. A. M. Piras, A. Dessy, F. Chiellini, E. Chiellini, C. Farina, M. Ramelli, and E. Della Valle, *Biochim. Biophys. Acta, Proteins Proteomics*, **1454** (2008).
27. D. J. Cole, L. McKay, W. K. Jacobsen, J. C. Drummond, and P. M. Patel, *Artif. Cells, Blood Substit. Biotechnol.*, **95** (1997).
28. A. Schubert, *Semin. Anesth., Perioper. Med. Pain*, **78** (2001).
29. J. K. Aronson, Ed., Oxygen-carrying blood substitutes, in *Meyler's Side Effects of Drugs: The International Encyclopedia of Adverse Drug Reactions and Interactions* (Elsevier: Amsterdam, 2006), p. 2653.
30. K. C. Lowe, *J. Fluor. Chem.*, **19** (2002).
31. C. N. Murray and J. P. Riley, *Deep Sea Res. Oceanogr. Abs.*, **311** (1969).
32. M. W. Grenfell, *Fluoropolymers: Properties* (Springer, NY, 1999).
33. M. Pilarek and K. W. Szewczyk, *Biochem. Eng. J.*, **38** (2008).
34. R. Pohorecki, J. Baldyga, A. Ryszczuk, and T. Motyl, *Biochem. Eng. J.*, **147** (2001).
35. F. J. Giordano and R. S. Johnson, *Curr. Opin. Genet. Dev.*, **35** (2001).
36. E. L. C. Eliot and R. Clark, *Am. J. Anat.*, **251** (1939).
37. J. Rouwkema, N. C. Rivron, and C. A. van Blitterswijk, *Trends Biotechnol.*, **434** (2008).
38. D. Bouïs, Y. Kusumanto, C. Meijer, N. H. Mulder, and G. A. P. Hospers, *Pharmacol. Res.*, **89** (2006).
39. M. Nomi, A. Atala, P. D. Coppi, and S. Soker, *Mol. Aspects Med.*, **463** (2002).
40. P. Divya, P. R. Sreerekha, and L. K. Krishnan, *Biomol. Eng.*, **593** (2007).
41. M. H. Sheridan, L. D. Shea, M. C. Peters, and D. J. Mooney, *J. Control. Rel.*, **91** (2000).
42. A. McKillop and W. R. Sanderson, *Tetrahedron*, **6145** (1995).
43. B. S. Harrison, D. Eberli, S. J. Lee, A. Atala, and J. J. Yoo, *Biomaterials*, **4628** (2007).

44. S. H. Oh, C. L. Ward, A. Atala, J. J. Yoo, and B. S. Harrison, *Biomaterials*, **757** (2009).
45. V. R. Sinha, A. K. Singla, S. Wadhawan, R. Kaushik, R. Kumria, K. Bansal, and S. Dhawan, *Int. J. Pharm.*, **1** (2004).
46. J. Han and J.-J. Zhong, *Enzyme Microb. Technol.*, **498** (2003).
47. C. Samanta, *Appl. Catal. A: General*, **133** (2008).
48. M. P. Gaikowski, J. J. Rach, and R. T. Ramsay, *Aquaculture*, **191** (1999).
49. R. J. Watts, D. D. Finn, L. M. Cutler, J. T. Schmidt, and A. L. Teel, *J. Contam. Hydrol.*, **312** (2007).
50. N. Frikha, E. Schaer, and J.-L. Houzelot, *Thermochimica Acta*, **47** (2006).
51. L. S. Nair and C. T. Laurencin, *Prog. Polym. Sci.*, **762** (2007).
52. J. M. Anderson and M. S. Shive, *Adv. Drug Del. Rev.*, **5** (1997).
53. C. Liu, Z. Xia, and J. T. Czernuszka, *Chem. Eng. Res. Des.*, **1051** (2007).
54. D. Mallardé, F. Boutignon, F. Moine, E. Barré, S. David, H. Touchet, P. Ferruti, and R. Deghenghi, *Int. J. Pharm.*, **69** (2003).
55. G. Ruan and S.-S. Feng, *Biomaterials*, **5037** (2003).
56. K. Kato, E. Uchida, E.-T. Kang, Y. Uyama, and Y. Ikada, *Prog. Polym. Sci.*, **209** (2003).
57. J. K. Jackson, T. Hung, K. Letchford, and H. M. Burt, *Int. J. Pharm.*, **6** (2007).
58. P. de Vos, M. M. Faas, B. Strand, and R. Calafiore, *Biomaterials*, **5603** (2006).
59. W. R. Gombotz and S. Wee, *Adv. Drug Del. Rev.*, **267** (1998).
60. W.-J. Shih, Y.-H. Chen, C. -J. Shih, M. -H. Hon, and M. -C. Wang, *J. Alloys and Compd.*, **826** (2007).
61. T.-W. Chung, Y.-Y. Huang, and Y.-Z. Liu, *Int. J. Pharm.*, **161** (2001).
62. J.-H. Lee, T. G. Park, and H.-K. Choi, *Int. J. Pharm.*, **75** (2000).
63. C. Fude, C. Dongmei, T. Anjin, Y. Mingshi, S. Kai, Z. Min, and G. Ying, *J. Control. Rel.*, **310** (2005).
64. C. Sturesson, P. Artursson, R. Ghaderi, K. Johansen, A. Mirazimi, I. Uhnoo, L. Svensson, A. Ann-Christine, and J. Carlfors, *J. Control. Rel.*, **377** (1999).
65. Q. Wei, W. Wei, R. Tian, L.-Y. Wang, Z.-G. Su, and G.-H. Ma, *J. Colloid Interface Sci.*, **267** (2008).
66. Z. Ma, Z. Mao, and C. Gao, *Colloids Surf. B: Biointerfaces*, **137** (2007).
67. H. Chen, L. Yuan, W. Song, Z. Wu, and D. Li, *Prog. Polym. Sci.*, **1059** (2008).
68. M. Yoshimoto, H. Sakamoto, and H. Shirakami, *Colloids Surf. B: Biointerfaces*, **281** (2009).

69. M. Alves Da Silva, M. H. Gil, A. P. Piedade, J. S. Redinha, A. M. O. Brett, and J. M. C. Costa, *J. Polym. Sci. Part A: Polym. Chem.*, **269** (1991).
70. N. Patel, M. C. Davies, M. Hartshorne, R. J. Heaton, C. J. Roberts, S. J. B. Tendler, and P. M. Williams, *Langmuir*, **6485** (1997).
71. S. Akkus Çetinus and H. Nursevin Öztop, *Enzyme Microb. Technol.*, **889** (2003).
72. A. M. Eberhardt, V. Pedroni, M. Volpe, and M.L. Ferreira, *Appl. Catal. B: Environm.*, **153** (2004).
73. A. Gharsallaoui, G. Roudaut, O. Chambin, A. Voilley, and R. Saurel, *Food Res. Int.*, **1107** (2007).
74. M. Florescu and A. Katerkamp, *Sens. Actuators B: Chem.*, **39** (2004).
75. B. C. Towe, S. Flechsig, and G. Spaulding, *Biosens. Bioelectron.*, **799** (1996).
76. R. T. Bailey, F. R. Cruickshank, G. Deans, R. N. Gillanders, and M. C. Tedford, *Anal. Chim. Acta*, **101** (2003).
77. A. Mills, *Sens. Actuators B: Chem.*, **69** (1998).
78. X. Niu, Q. Feng, M. Wang, X. Guo, and Q. Zheng, *Polym. Degrad. Stab.*, **176** (2009).
79. A. Jaklenec, E. Wan, M. E. Murray, and E. Mathiowitz, *Biomaterials*, **185** (2008).
80. F.-M. Chen, Y.-M. Zhao, H.-H. Sun, T. Jin, Q.-T. Wang, W. Zhou, Z.-F. Wu, and Y. Jin, *J. Control. Rel.*, **65** (2007).
81. C. Warlow, C. Sudlow, M. Dennis, J. Wardlaw, and P. Sandercock, *Lancet*, **1211** (2003).
82. K. S. Mark and T. P. Davis, *Peptides*, **1965** (2000).

Chapter 33

EMULSION TEMPLATING

Elizabeth Cosgriff-Hernández
Biomedical Engineering Department, Texas A&M University, 3120 TAMU, College Station, Texas 77843-3120, USA
cosgriff.hernandez@tamu.edu

Advancements in tissue engineering are revolutionizing our capability to repair and replace damaged tissue by augmenting the body's native regenerative processes. Biomaterial scaffolds play a pivotal role in tissue regeneration by restoring function and serving as a template for necessary cellular interactions. The fabrication process dictates scaffold properties such as architecture, surface chemistry, biodegradation, and strength. Emulsion templating is a relatively new method for the production of high-porosity scaffolds via template polymerization of high-internal-phase emulsions (HIPEs). PolyHIPEs show great promise as bone scaffolds due to the tremendous control of scaffold morphology, potential as an injectable system, and superior mechanical properties. This chapter reviews several studies that utilize emulsion templating to generate bone scaffolds with highly interconnected pores.

33.1 Introduction

Musculoskeletal conditions have an enormous impact on the quality of life and remain one of the leading reasons that patients seek

Handbook of Intelligent Scaffolds for Tissue Engineering and Regenerative Medicine
Edited by Gilson Khang

www.panstanford.com

medical care. Orthopedic conditions such as sprains, fractures, arthritis, back pain, and bone tumors result in 136.8 million ambulatory health care visits, more than 3 million hospitalizations, 245 billion dollars in medical costs, and 488 million days of restricted work activity each year. Overall, musculoskeletal ailments comprise more than 14% of health care expenditures in the United States.[2] A commonly encountered challenge for reconstructive surgeons is treatment of large bony defects resulting from traumatic injury, tumor resection, degenerative diseases, and congenital deformities. Although small defects in bone are effectively repaired naturally, the body is often incapable of repairing these large defects.[5,6] Nationwide Inpatient Statistics show that over 1,100,000 surgical procedures involving the partial excision of bone, bone grafting, and inpatient fracture repair were performed in 2004 at an estimated total cost of over $5 billion, illustrating the importance of addressing the functional, aesthetic, and social impact of such defects.[7] An important area of musculoskeletal research has focused on biocompatible orthopedic implants for the treatment of nonunion defects, replacement of diseased tissue, and maxillofacial surgery.

Surgeons have long been interested in producing a synthetic bone-grafting material that can accelerate the healing process, integrate well with the surrounding tissue, and later remodel to resemble native bone. During the past century, a wide variety of alloplastic materials were utilized in reconstruction therapies, including celluloid, aluminum, gold, vitallium, tantalium, stainless steel, titanium, methyl methacrylate (MMA) resins, polyethylene, silicone elastomers, and hydroxyapatite ceramics. Drawbacks and complications inherent in the use of current alloplastic materials include inadequate tissue integration, limited biodegradability, and stress shielding.[8–16] Biologically active implants such as demineralized bone matrix obtained from allogeneic or xenogeneic sources have shown promise as reconstructive materials because of their high osteoinductivity, but drawbacks include the risks of disease transmission as well as cost.[8,14] The current gold standard for new bone-grafting materials continues to be autogenous bone as a result of its potential for growth and remodeling as well as its ability to osseointegrate. Although often used, these treatments have several limitations and have reported failure rates as high as 30%. Autologous bone harvesting has anatomical limitations, donor site pain and morbidity due to infection, and excessive resorption.[5,6] The limitations of current surgical techniques reveal the need for significant improvement in methods to promote bone regeneration. Development of tissue engineering strategies that can be used to repair critical-size bone defects would thus be invaluable.

33.2 Bone Tissue Engineering

Engineered bone grafts provide a means to regenerate bone without the complications associated with traditional transplants.[8,17,18] Tissue engineers attempt to repair or regenerate damaged tissue by using living cells or attracting endogenous cells to aid in tissue formation. In the traditional tissue engineering paradigm, cells are seeded on a scaffold composed of a synthetic polymer or natural material and appropriate bioactive agents are added to aid cell differentiation and tissue growth (Fig. 33.1).[19] Tissue engineers utilize biomaterial scaffolds to sustain functionality during regeneration and to serve as a template for necessary cellular interactions. In bone constructs, scaffolds are seeded with osteoblasts or mesenchymal cells prior to implantation to establish new centers for

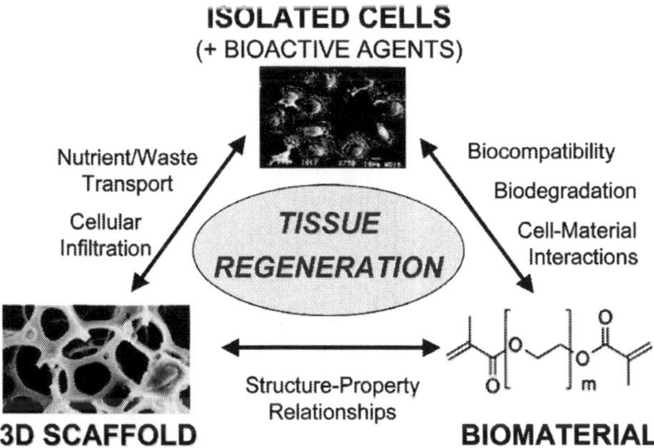

Figure 33.1. Tissue engineering paradigm.

bone formation.[19,20] The construct may be matured *in vitro* prior to implantation or implanted directly.

Tissue engineering scaffolds typically consist of biodegradable materials that will integrate fully with the host tissue.[19] In addition to the choice of a suitable biomaterial, the success of tissue engineering constructs depends on the three-dimensional (3D) architecture of the scaffold. An interconnected porous structure enables cellular migration and proliferation, vascularization, and the transport of nutrients and metabolic waste.[10,13,20,21] The mechanical properties of the scaffold are especially important in orthopedic applications. Adequate support similar to native bone is necessary to allow for continued loading and prevent stress-shielding effects.[10,11,17,22] Thus, the advancement of bone tissue engineering strategies is strongly dependent on the development of high-porosity scaffolds that meet these key requirements.

33.3 Scaffold Fabrication

Tissue engineers have developed numerous strategies to create 3D bone scaffolds using a wide range of fabrication techniques.[1,4,10,17,20,23–27] The fabrication process dictates scaffold properties such as architecture, surface chemistry, biodegradation, and strength. One of the primary challenges is to generate a scaffold with high porosity that retains sufficient mechanical strength for orthopedic applications. High porosity is necessary for the proper influx of nutrients and removal of waste products during tissue regeneration. The most common techniques to produce porous scaffolds include particulate leaching, gas foaming, and fiber bonding (Fig. 33.2).[1,21,26,27]

Many of these strategies provide exceptional architecture control; however, irregular defect sites require fabrication of complex polymer molds to replicate the contours of the defect. *In situ*-forming scaffolds can fill irregular-shaped defects, improve contact between the scaffold and the surrounding tissue, and eliminate the need for costly molding techniques.[28,29] Poly(methyl methacrylate) (PMMA) bone cements are perhaps the most widely used injectable material in orthopedics; however, PMMA is nondegradable, which

Figure 33.2. Scanning electron micrographs of porous scaffolds fabricated with (a) electrospinning,[1] (b) salt leaching,[3] and (c) gas foaming.[4] Reprinted with permission from Ref. 1; copyright 2006 Mary Ann Liebert, Inc.; Ref. 3; copyright 2002 Wiley; Ref 4; copyright 2006 Elsevier.

may impede bone healing.[28] Recently, investigators have developed highly cross-linked, degradable networks such as poly(propylene fumarates) (PPFs) and polyanhydrides that can be formed *in situ* using either thermal or photoinitiated cross-linking.[30–33] Although biodegradable and injectable, these materials lack the porosity necessary to repair critical-size defects. In contrast, *in situ*–curing hydrogels have sufficient mass transport properties but lack the mechanical strength necessary for orthopedic applications. A scaffold fabrication method that is injectable and porous, yet retains high mechanical strength, would provide a significant improvement over current methods.

Emulsion templating is a relatively new method for the production of high-porosity scaffolds, which involves the template polymerization of HIPEs.[34] These heterogeneous liquid-liquid emulsions are characterized by an internal, droplet phase volume fraction of at least 0.74. This limiting value represents the hexagonal packing of the spherical droplets. Increasing the internal phase volume beyond this value results in deformation of the droplets into a polyhedral shape.[35] When the external, continuous phase is composed of a polymerizable monomer, it may be polymerized to form a rigid, high-porosity foam (Fig. 33.3).

Researchers at Unilever Research Port Sunlight Laboratory (Cheshire, UK) first coined the term "polyHIPEs" to describe these polymeric foams.[34,36] Initially polyHIPE foams were primarily closed-pore architectures with the internal aqueous phase trapped

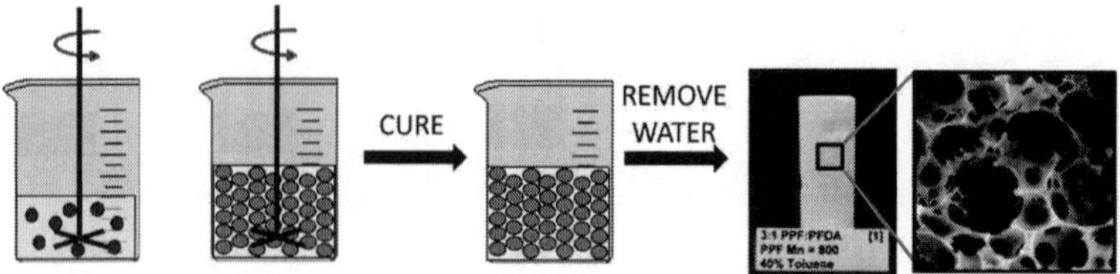

Figure 33.3. Schematic of the polyHIPE process.

within the cellular structure. In 1982, Unilever scientists generated open-pore polyHIPEs using low-hydrophile-lipophile-balance (HLB) surfactants with the styrene-divinylbenzene system.[34] The open-pore structure facilitated the rapid removal of the aqueous dispersed phase and resulted in low-density foams of less than 0.1 g/cm^3.[37] A wide range of porosities (75%–99%), pore sizes (5–100 μm), and closed- or open-pore morphologies can be achieved by varying the HIPE composition.[34,35,37–43] The windows between pores form when the thin film separating droplets opens due to polymer shrinkage during polymerization. Thus, the open- or closed-pore morphology is related to the thickness of this film and the amount of polymer shrinkage.[37] Such materials have been investigated for a variety of uses, including solid-phase synthesis and catalyst supports, aerosol filters, and substrates for porous Ni electrodes.[34,35,39,40,44]

An attractive characteristic of these new polymeric foams for tissue engineering is the high level of control over the structural architecture. Experimental parameters may be controlled by varying such factors as composition, mixing speed, and temperature to generate either open- or closed-cell polyHIPE foams (Fig. 33.4).

Porosity (75%–99%) and pore size (5–100 μm) can also be tailored by tuning the HIPE composition.[34,37,42–44] Furthermore, prior to polymerization, the HIPE has a whipped-cream or mayonnaise consistency that is well within the viscosity range of injection. The homogeneous and spherical pores of the polyHIPE also eliminate stress concentrators that can limit mechanical properties as in salt-leached scaffolds.[45] Overall, the combination of injectability, high porosity, and superior mechanical properties make polyHIPEs excellent candidates for tissue engineering scaffolds.[36]

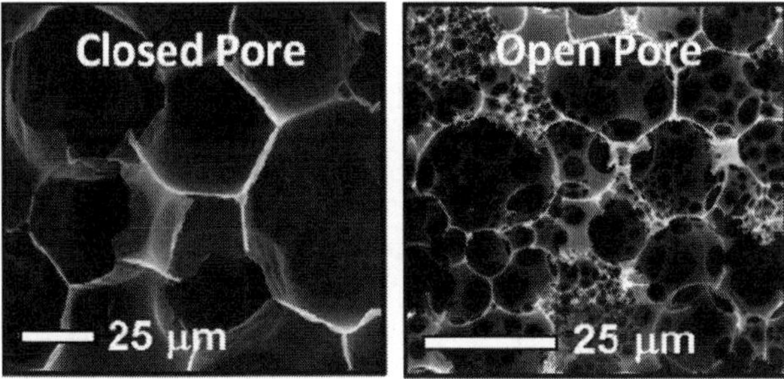

Figure 33.4. SEM of polyHIPEs with open and closed pore structures. Reprinted with permission from Ref. 37; copyright 2007 American Chemical Society.

33.4 PolyHIPEs as Bone Scaffolds

The first polyHIPE investigated as a bone scaffold was based on styrene and divinylbenzene.[36] However, the styrene-based system is nonbiodegradable, which led to the investigation of polyHIPEs based on poly(ε-caprolactone-*co*-styrene), poly(lactic acid-*co*-styrene), dextran, pullulan, and gelatin.[23,46–51] Although high-porosity polyHIPEs were prepared with these formulations, their application to bone tissue engineering strategies are restricted due to the limited biodegradability of the copolymer systems and the low mechanical strength of the natural polymers. Fumarate-based polyHIPEs were developed in the Mikos group using toluene as a diluent to reduce viscosity sufficiently to permit emulsion formation.[37] However, osteoblast growth and mineralized matrix production on these semi- and fully degradable polyHIPEs has yet to be established.

33.4.1 *Nondegradable: Styrene-Based PolyHIPEs*

Akay *et al.* first investigated polyHIPE scaffolds to support the growth of osteoblasts *in vitro*.[36] PolyHIPE foams based on styrene with a divinylbenzene cross-linker were prepared (Fig. 33.5). Elevated temperatures (up to 80°C) during emulsification were utilized to generate a broader range of pore sizes. Rat osteoblasts were seeded on polyHIPE scaffolds with three average pore sizes

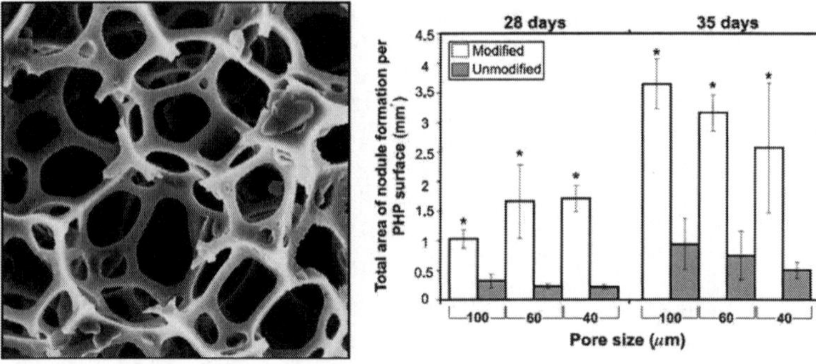

Figure 33.5. Mineralized nodule formation on modified and unmodified styrene-based polyHIPEs after 28 and 35 days of rat osteoblasts cells culture. Reprinted with permission from Ref. 36; copyright 2004 Elsevier.

(40, 60, and 100 mm) that were either unmodified or modified with hydroxyapatite. Osteoblasts formed multicellular layers on the polyHIPE surface and migrated up to a depth of 1.4 mm after 35 days in culture. Retention of the osteoblastic phenotype was confirmed with reverse transcriptase-polymerase chain reaction (RT-PCR), and osteogenesis markers were examined with histological staining. There were few phenotypic differences between cells cultured on scaffolds of different pore sizes. However, the overall penetration of osteoblasts into the polymer was significantly greater in the 100 μm scaffold as compared to the other sizes investigated. The hydroxyapatite coating enhanced the penetration of cells into the polymer, cellular proliferation, and the area of mineralized nodule formation as compared to the unmodified controls (Fig. 33.5). This first study demonstrates that polyHIPE scaffolds can be utilized as osteoconductive bone scaffolds with excellent architectural control.

33.4.2 *Semidegradable: Polyester-Based PolyHIPEs*

Busby *et al.* reported the first preparation of polyHIPE foams containing polyester macromers by free radical homo- or copolymerization.[23,46] First, vinyl groups were introduced onto poly(ε caprolactone) (PCL) and poly(lactic acid) PLA diols. Diluents (styrene, MMA, or toluene) were utilized to reduce the viscosity of the polyester macromer to enable HIPE formation. Polymerization of the continuous phase of HIPEs resulted in low-density foams with pore diameters ranging from 5 to 100 μm. The effect of the diluent type on foam morphology was investigated with scanning electron microscopy (SEM). In particular, the effect of the diluent type (reactive: styrene or MMA; unreactive: toluene) and the ratio of polyester to diluents was examined. Although styrene and MMA yielded stable HIPEs up to 60 wt% polyester content, the addition of toluene resulted in soft foams that collapsed upon drying. Regardless of the diluent type, only low levels of vinyl-terminated polester yielded microcellular foams. PolyHIPEs with polyester concentrations higher than 40 wt% displayed characteristics of phase separation. In general, 20 wt% polyester polyHIPEs had pore sizes suitable for tissue engineering scaffolds (Fig. 33.6). Cytocompatibility studies on the polester-based foams using whole chicken embryo explants, rat skin explants, or individual human skin cells indicated promising cell adhesion, spreading, and proliferation up to six days.

Lumelsky *et al.* utilized two approaches for incorporating PCL into polyHIPEs.[50] Similar to Busby *et al.*, the first approach copolymerized vinyl-terminated PCL with styrene.[23] This approach yielded a typical polyHIPE structure with voids on the order of tens of microns. The second approach generated semi-interpenetrating polymer networks (semi-IPN) of a PCL diol and styrene. The relatively hydrophilic PCL diol destabilized the HIPE, which resulted in voids on the order of hundreds of microns. In contrast to the vinyl-terminated PCL, which was covalently immobilized in the

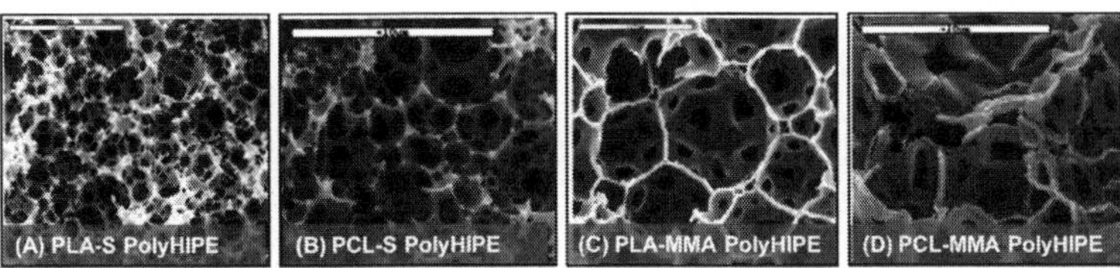

Figure 33.6. SEM of polyester-based polyHIPEs using styrene (S) and methyl methacrylate (MMA) as diluents. (a) 20:80 PLA:styrene, (b) 20:80 PCL:styrene, (c) 20:80 PLA:MMA, and (d) 20:80 PCL:MMA. Scale bars = 10 microns. Reprinted with permission from Ref. 23; copyright 2001 American Chemical Society; and Ref. 46; copyright 2002 Society of Chemical Industry.

polystyrene (PS) network, a significant amount of PCL diol was removed during purification of the polyHIPE. Initial cell culture studies indicated that the semi-IPN monoliths support the grown and differentiation of mouse skeletal cells.

33.4.3 *Fully Degradable: Fumarate-Based PolyHIPEs*

Although several studies have examined semidegradable polyHIPEs based on styrene or MMA copolymer systems, Christenson *et al.* reported the first fully biodegradable polyHIPE based on fumarate macromers.[37] PPF is an unsaturated polyester that biodegrades via ester hydrolysis into the biocompatible compounds fumaric acid and propylene glycol.[52] When reacted with the cross-linker propylene fumarate diacrylate (PFDA), this system forms a highly cross-linked network that imparts mechanical strength sufficient for its use in bone tissue engineering scaffolds. Previous studies with fumarate-based scaffolds have demonstrated its biocompatibility, biodegradability, and osteoconductivity.[52–54]

Similar to previous polyester-based polyHIPEs, toluene was utilized as a diluent to reduce the viscosity of the organic phase to enable HIPE formation. A range of polyHIPE scaffolds of different pore sizes and morphologies were generated by varying the diluent concentration (40%–60 wt%), cross-linker concentration (25–75 wt%), and macromer molecular weight ($M_n = 800–1000$ g/mol). Although some formulations resulted in macroporous monoliths (pore diameter > 500 μm), the majority of the polyHIPEs studied were rigid, microporous monoliths with average pore diameters in the range of 10–300 μm. Gravimetric analysis confirmed the porosity of the microporous monoliths as 80%–89% with most scaffolds above 84%. PolyHIPEs with pores greater than 50 μm had a closed-pore morphology, whereas scaffolds with pores smaller than 50 μm had an open-pore morphology (Fig. 33.7).

These two morphologies were attributed to differences in film thickness between the droplets at the gel point. It was concluded that the smaller droplet size resulted in a reduction of film thickness below the threshold value that results in interconnect formation. Unlike the vinyl-terminated PCL and PLA systems, the fumarate-based HIPEs generated rigid, microcellular foams without the

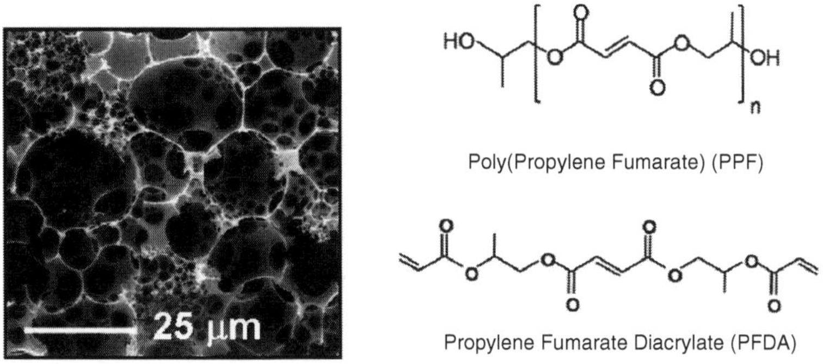

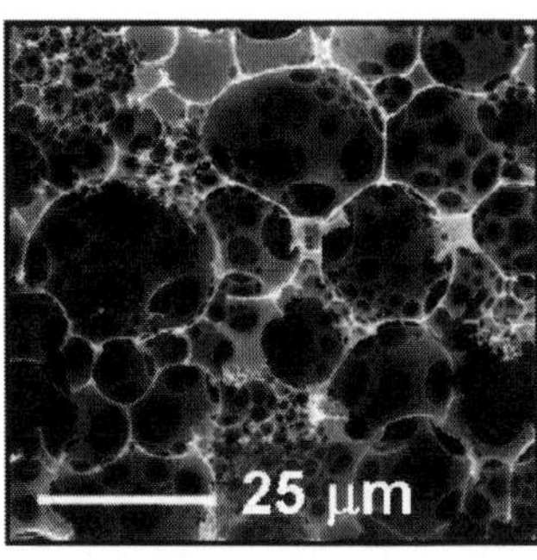

Poly(Propylene Fumarate) (PPF)

Propylene Fumarate Diacrylate (PFDA)

Figure 33.7. SEM image of fumarate-based polyHIPEs fabricated from PPF and PFDA. Reprinted with permission from Ref. 37; copyright 2007 American Chemical Society.

addition of a vinyl-based comonomer such as styrene. This resulted in the first fully biodegradable polyHIPE scaffold.

33.5 New Synthesis Routes

Vinyl monomers copolymerized with a divinyl cross-linker are the most commonly used systems to generate polyHIPEs.[23,36,46,55] The resulting monoliths are rigid and relatively brittle. In contrast, Lepine *et al.* reported the first preparation of polyHIPEs based on a polyurethane network.[56] This strategy was used to enhance the elasticity and toughness of the monoliths. Scaffolds of a pure polyurethane network were found to be too soft and underwent a high level of shrinkage. The addition of styrene and divinylbenzene resulted in the generation of rigid, microcellular IPNs. A difunctional comonomer, hydroxybutyl methacrylate, was then utilized to create covalent links between the two networks of the IPNs. As expected, the incorporation of a high concentration (40%) of a polyurethane (PU) elastomer into the PS rigid network decreased the modulus and increased the flexibility of the polyHIPE. However, a lower concentrations (20%) of PU resulted in monoliths with an increased modulus as compared to the PS network. The authors hypothesized that this effect was due to an increase in the proportion of cross-links, which was supported by a threefold increase in the modulus upon introduction of internetwork links between the PU and PS networks. More recently, Cosgriff-Hernandez *et al.* reported the generation of a poly(ester urethane urea) polyHIPE using similar addition reactions.[57] In this study, a PCL prepolymer was synthesized by end-capping PCL triols with hexane diisocyanate (HDI). During cure of the HIPE, the PCL prepolymer reacted with water and an excess of HDI to form a poly(ester urethane urea) network. Overall, these studies demonstrate the feasibility and utility of generating polyHIPEs using polyaddition reactions.

33.6 Conclusions and Outlook

The application of emulsion templating in the development of tissue engineering scaffolds is a relatively new area of orthopedic research. This chapter has focused on a review of polyHIPE scaffolds; however, the scientific developments reported above do not exhaust the emulsion-based research efforts in tissue engineering. For example, double emulsions have found great utility in the fabrication of drug-eluting microspheres. The preliminary investigations reviewed here indicate that these strategies have great potential to improve current orthopedic scaffold development. Significant advancements are still necessary to realize the full potential of these novel scaffolds in clinical use. In particular, cellular differentiation, proliferation, and mineralized matrix production on these new scaffolds must be established in both *in vitro* and *in vivo* environments. The mechanical properties and rate of biodegradation are also key material properties that will need to be explored in more depth. Overall, these initial studies foreshadow a bright future through the use of emulsion-templated scaffolds in bone reconstructive strategies.

References

1. Q. P. Pham, *Tissue Eng.*, **12** (2006).
2. *Orthop. Res. Soc. Newslett.*, **18** (2005).

3. M. S. Wolfe, *J. Biomed. Mater. Res.*, **61** (2002).
4. T. K. Kim, *Biomaterials*, **27** (2006).
5. M. Artico, *Surg. Neurol.*, **60** (2003).
6. Y. T. Konttinen, *Clin. Orthop. Related Res*, (2005).
7. *HCUP Nationwide Inpatient Sample (NIS)* Healthcare Cost and Utilization Project (HCUP): Agency for Healthcare Research and Quality: Rockville.
8. P. D. Costantino, *Facial Plast. Surg.*, **18** (2002).
9. T. M. Freyman, *Prog. Mater. Sci.*, **46** (2001).
10. V. Karageorgiou, *Biomaterials*, **26** (2005).
11. X. Liu, *Ann. Biomed. Eng.*, **32** (2004).
12. L. G. Mercuri, *J. Oral Maxillofac. Surg.*, **62** (2004).
13. A. G. Mikos, *Electr. J. Biotechnol.*, **3** (2000).
14. A. M. Pou, *Curr. Opin. Otolaryngol. Head Neck Surg.*, **11** (2003).
15. F. Rose, *Biochem. Biophys. Res. Commun.*, **292** (2002).
16. B. Seal, *Mater. Sci. Eng.*, **34** (2001).
17. F. R. A. J. Rose, *Biochem. Biophys. Res. Commun.*, **292** (2002).
18. R. Kenley, *Pharm. Res.*, **10** (1993).
19. R. Langer, *Science*, **260** (1993).
20. S. N. Bhatia, *Biomed. Microdevices*, **2** (1999).
21. M. Hacker, *Biomaterials*, **28** (2007).
22. V. I. Sikavitsas, *Biomaterials*, **22** (2001).
23. W. Busby, *Polym. Int.*, **51**, 154–164 (2002).
24. T. M. Freyman, *Prog. Mater. Sci.*, **46** (2001).
25. M. Hacker, *Biomaterials*, **28** (2007).
26. L. D. Harris, *J. Biomed. Mater. Res.*, **42** (1998).
27. A. S. Mistry, *J. Biomed. Mater. Res.* (2008).
28. J. A. Burdick, *Biomaterials*, **23** (2002).
29. E. Jabbari, *Biomacromolecules*, **6** (2005).
30. S. J. Peter, *J. Biomed. Mater. Res.*, 44 (1999).
31. S. J. Peter, *Biomaterials*, **21** (2000).
32. S. J. Peter, *Tissue Eng.*, **1** (1995).
33. D. Svaldi-Muggli, *J. Biomed. Mater. Res.*, **16** (1998).
34. P. Hainey, *Macromolecules*, **24** (1991).
35. N. R. Cameron, *Colloid Polym. Sci.*, **274** (1996).

36. G. Akay, *Biomaterials*, **25**, 3991–4000 (2004).
37. E. M. Christenson, *Biomacromolecules*, **8**, 3806–3814 (2007).
38. R. Butler, *Adv. Mater.*, **13** (2001).
39. N. R. Cameron, *J. Mater. Chem.*, **10** (2000).
40. N. R. Cameron, *J. Mater. Chem.*, **7** (1997).
41. J. M. Williams, *Langmuir*, **6** (1990).
42. J. M. Williams, *Langmuir*, **7** (1991).
43. J. M. Williams, *Langmuir*, **4** (1988).
44. R. J. Carnachan, *Soft Matter*, **2** (2006).
45. J. Zhang, *Polymer*, **46** (2005).
46. W. Busby, *Biomacromolecules*, **2**, 871–881 (2001).
47. P. Krajnc, *Macromol. Rapid Commun.*, **26** (2005).
48. M. Silverstein, *Polym. Eng. Sci.*, **41** (2001).
49. M. Silverstein, *Polymer*, **46** (2005).
50. Y. Lumelsky, *Macromolecules*, **41** (2008).
51. M. Silverstein, *Macromolecules*, **42** (2009).
52. S. He, *Polymer*, **42** (2001).
53. J. P. Fisher, *J. Biomed. Mater. Res.*, **59** (2002).
54. J. S. Temenoff, *Electr. J. Biotechnol.*, **3** (2000).
55. M. A. Bokhari, *J. Mater. Chem.*, **17** (2007).
56. O. Lepine, *Polymer*, **46** (2005).
57. E. M. Cosgriff-Hernandez, *PMSE Preprints (Amer. Chem. Soc., Div. Poly. Mater. Sci. Eng.* (2009).

Part VII

SCAFFOLD FOR TARGET ORGAN

Chapter 34

PGA FIBER FOR SOFT TISSUE ENGINEERING

Wei Liu and Yilin Cao

Department of Plastic and Reconstructive Surgery, Shanghai 9th People's Hospital, Shanghai Tissue Engineering Key Laboratory, Shanghai Jiao Tong University School of Medicine, 639 Zhi Zao Ju Road, Shanghai 200011, P.R. China

liuwei_2000@yahoo.com; yilincao@yahoo.com

34.1 Introduction

Treatment of various soft-tissue injuries and defects remains a major challenge in reconstructive surgery, partly because there are limited sources available for autologous tissue grafts. Soft-tissue injuries and defects, such as of skin, tendon, cartilage, or blood vessels, are common diseases in clinics. Currently, most of the reconstructive surgical procedures are involved in tissue transfer, which usually leads to donor site morbidity and functional disability. In contrast, the tissue engineering approach is able to generate a tissue graft by using expanded cells derived from the biopsied tissue and scaffold materials and thus to avoid donor site morbidity.

In 1993, Dr. Langer and Dr. Vacanti proposed the basic principle of tissue engineering, which employs degradable scaffold and seed cells to develop living tissue either *in vitro* or *in vivo*.[1] One of the

Handbook of Intelligent Scaffolds for Tissue Engineering and Regenerative Medicine
Edited by Gilson Khang

www.panstanford.com

key issues is the selection of a proper scaffold material. Generally, an ideal scaffold material should have the following characteristics in order to perform its function properly[2]:

1. Good biocompatibility: Besides the general requirements for biomaterials, such as they should be nontoxic, noncarcinogenic, noninflammatory, scaffolds should be able to support cell attachment, proliferation, matrix production, and even differentiation. Particularly, the degraded products should not be harmful to seeded cells.
2. Suitable biodegradability: After tissue formation, the scaffold should be able to completely degrade. In addition, the degradation rate should match the rate of cell growth and tissue formation. Furthermore, the degradation rate should be able to control according to the requirements of different types of tissues.
3. Three-dimensional porous structure: Generally speaking, the porosity should be above 90% with a high ratio of surface/volume so that the scaffold can provide a large surface area for cell attachment, growth, and matrix production and deposition and for nutrition and waste transportation and the access of neovascularization.
4. Good plasticity and mechanical property: An ideal scaffold should be able to be prefabricated into a certain shape and possess a certain level of mechanical property so that it can support tissue function before the engineered tissue can be remodeled and be mature enough to regain the normal mechanical strength.
5. Appropriate cell surface properties for cell–scaffold interaction: This is particularly important for maintaining normal cell phenotype or promoting cell differentiation in addition to cell attachment and growth.
6. Easy to manufacture: Low cost and ease for manufacture are generally required in order to fabricate scaffolds in a large scale for practical applications.
7. Sterilization: An ideal material should be easily handled for various types of sterilization without affecting its own characteristics.

Based on the source of materials, scaffolds can be divided into two groups, natural scaffolds and synthetic polymer scaffolds. The polyglycolic acid (PGA) scaffold falls into the category of synthetic polymers and was the first scaffold used for pioneering research of tissue engineering.[3–6] Different from natural scaffolds, synthetic polymers can be designed and manufactured for their exact degradability and degradation time, as well as their pore size and porosity and other physical and chemical properties. Thus the quality of synthetic scaffolds can be better controlled compared with natural scaffolds. This allows for reproduction in large quantities with similar characteristics. Additionally, due to the nature of artificially made substances, they can avoid the risk of transferring pathogens to the host, which possibly reside in human and animal tissues, such as prion and viruses. Among different synthetic polymers, PGA belongs to the group of poly (α-hydroxy acids).

The poly (α-hydroxy acids) are hydrolyzed by cleavage of their ester bonds in a water-containing environment, which results in a reduction of molecular weight, but not the total mass, of the scaffold. The degradation via hydrolysis usually reduces the molecular weight to around 5,000. Afterward, cellular degradation takes over the degradation process to further degrade the polymer into monomers, and they are finally metabolized into water and CO_2.[7] Due to this nature, poly (α-hydroxy acids) are considered one of the optimal synthetic scaffolds for tissue engineering.

One of the common physical forms for a PGA scaffold is the unwoven PGA fibers, which were originally reported as being a scaffold by Langer's group and Vacanti's group.[3–6] According to literature, unwoven PGA fibers have been widely applied to the engineering of various types of soft tissue, including skin, tendon, cartilage, corneal stroma, blood vessels, etc.[3–6,8] This chapter will provide a review of their application in soft-tissue engineering.

34.2 PGA Fibers for Tendon Engineering

Tendons are the connective tissue that links muscles to bones so that the tensile force created by muscles can be transmitted to bones for body movement. The main tendon extracellular matrix (ECM) is type I collagen, which is highly organized in a hierarchy of bundles that are aligned in a parallel fashion. This unique structure provides the unique biomechanical properties of tendon tissues. Therefore, the parallel alignment structure and strong mechanical property should be considered for tendon scaffold design. It was proposed that an ideal tendon scaffold should fulfill the following requirements[9]:

1. Biodegradability with adjustable degradation rate
2. Biocompatibility before, during, and after degradation
3. Superior mechanical properties and maintenance of mechanical strength during the tissue regeneration process
4. Biofunctionality: The ability to support cell proliferation and differentiation, ECM secretion, and tissue formation
5. Processability: The ability to be processed to form desired constructs of complicated structures and shapes, such as woven or knitted scaffolds, etc.

As early as 1994, Cao *et al.* performed a pioneer research study of tendon engineering by using unwoven PGA fibers as the scaffold for *in vivo* tendon engineering in a nude mouse model.[10] First, unwoven PGA fibers were arranged in a parallel fashion, and then tenocytes isolated from calf tendons were seeded onto the scaffold followed by *in vivo* implantation in the subcutaneous tissue of nude mice. After 12 weeks of implantation, tendonlike tissue formed and revealed longitudinally aligned collagen fibers. Afterward, the same scaffold was used for engineering tendons in immunocompetent animals.

In the first experiment, a hen claw was used as a model for tendon regeneration and repair inside a tendon sheath. First, autologous tenocytes were isolated and seeded onto unwoven PGA fibers and the cell-scaffold construct was *in vitro*–cultured for one week followed by *in vivo* transplantation to repair a 3 cm long defect of the flexor digitorum profundus tendon. At 14 weeks' postrepair, mature tendon tissue was formed when observed grossly (Fig. 34.1, *top*). Histologically, longitudinally aligned collagen fibers with a curving pattern could be observed as well (Fig. 34.1, *middle*) similar to that of the native tendon (Fig. 34.1, *bottom*).

More importantly, the engineered tendon reached 83% of the native tendon's tensile strength.[11] In the second experiment, a porcine model was used to perform the study. However, dermal

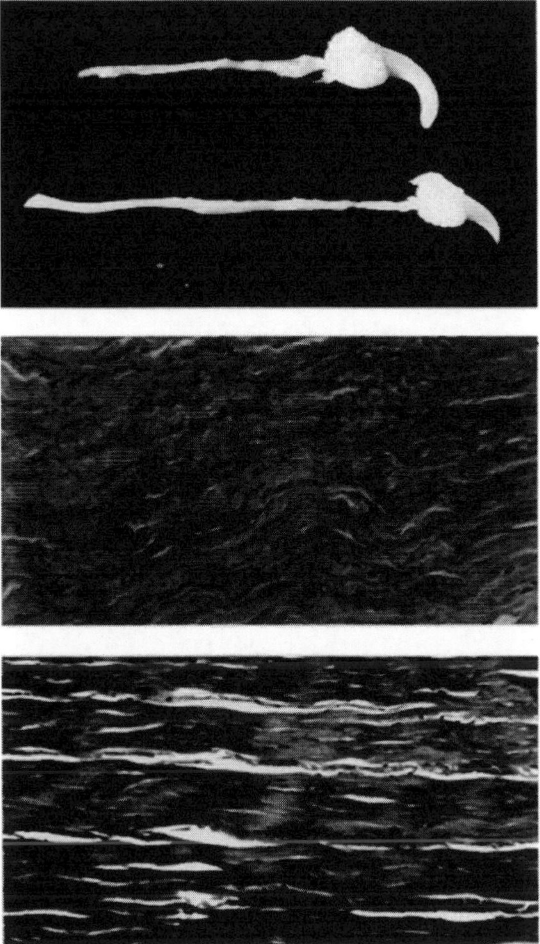

Figure 34.1. Gross view (*top*, between arrows) and histology of engineered tendon (*middle*) and natural tendon (*bottom*) at 14 weeks' postrepair. (Reprinted by permission from Ref. 11).

fibroblasts were used to replace tenocytes as seed cells to engineer the tendon and repair tendon defects. As similarly described, dermal fibroblasts were first isolated and *in vitro*-cultured and expanded, and then the cells were seeded onto unwoven PGA fibers for one week of *in vitro* culture. Scanning electron microscope examination revealed abundant matrix production by both

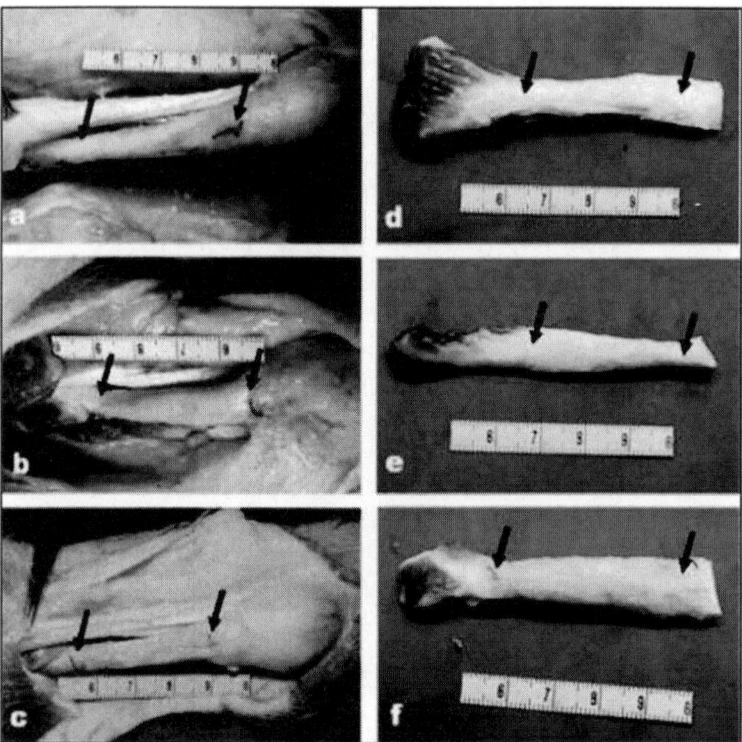

Figure 34.2. Repair of tendon defects by gross view of a fibroblast engineered tendon (a and d, between arrows), a tenocyte-engineered tendon (b and e, between arrows), and a formed tissue in control group 2 (c and f, between arrows) at 26 weeks. Note that both engineered tendons have a relatively smooth surface, whereas the control tissue has a rough surface. (a–c) Before harvesting; (d–f) after harvesting. (Reprinted by permission from Ref. 12).

dermal fibroblasts and tenocytes that were seeded onto the scaffold, indicating good cell compatibility between the cells and the scaffold. The prefabricated cell-scaffold construct was transplanted *in vivo* to repair a 3 cm long defect created on the flexor digitorum superficialis tendon. When examined at 26 weeks' postrepair, mature tendon tissue was formed, which was similar to the tendon tissue engineered by autologous tenocytes in a gross view (Fig. 34.2). Histologically, a similar tissue structure

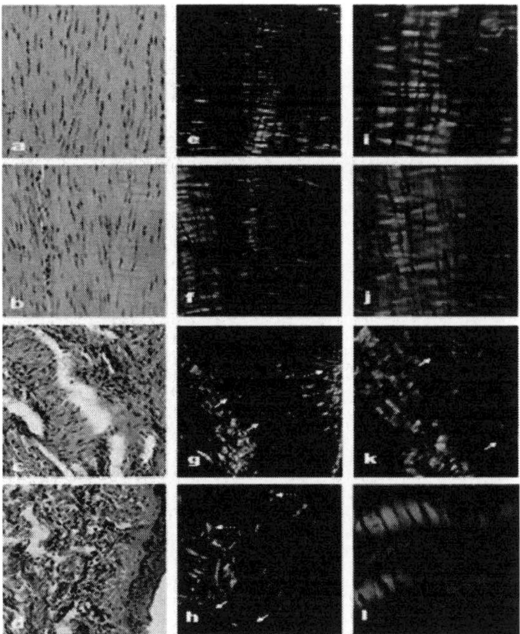

Figure 34.3. Histological finding of formed tissue at 26 weeks. H&E staining shows histological structures of fibroblast-engineered tendons (a), tenocyte-engineered tendons (b), a control tissue in control group 2 (c), (d) and normal pig skin. Collagen III (delicate collagen fibers with a light-green color) is detected only in the polarized images of control tissues (g and k) and in normal pig skin (h), as indicated by white arrows. In addition, collagen I (golden color, indicated by white, dotted arrows) is also detected in these tissues. In the polarized images of (e and i) fibroblast- and (f and j) tenocyte-engineered tendons and natural tendons (l), collagen I (golden color) is the predominant collagen type. Original magnification ×400 (i–k); ×200, all others. (Reprinted by permission from Ref. 12). *Abbreviation*: H&E, hematoxylin and eosin. See also Color Insert.

was observed among dermal fibroblast– and tenocyte-engineered tendons and native tendon tissues (Fig. 34.3). Importantly, a strong mechanical property was also achieved in engineered tendons with dermal fibroblasts.[12]

These two experiments showed the feasibility of using unwoven PGA fibers as a scaffold for *in vivo* tendon engineering, and the scaffold did fit the requirement of good biocompatibility,

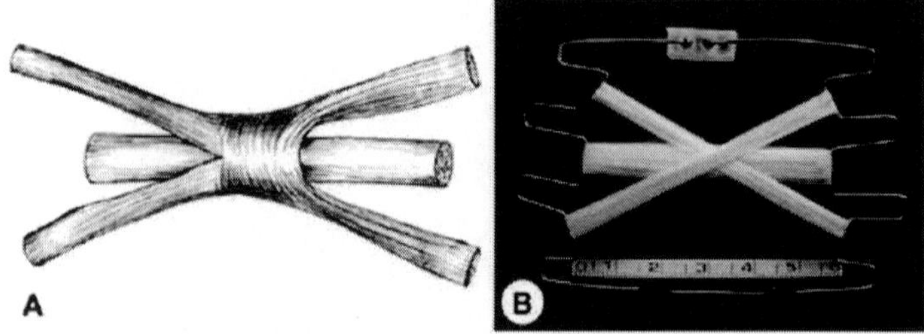

Figure 34.4. Design of extensor tendon complex scaffold. (a) Schema chart showing the central slip and two lateral bands. (b) PGA scaffold secured on a custom-made spring to mimic the complex structure. (Reprinted by permission from Ref. 13).

biodegradability with proper degradation rate, biofunctionality for supporting cell growth and matrix production, and processability. However, an obvious disadvantage of unwoven fibers is the lack of proper mechanical support during tendogenesis; therefore, an acellular small intestinal submucosa membrane was needed to wrap around the cell-loaded PGA fibers in order to enhance the mechanical property of the cell-scaffold constructs.[11,12]

In a recent study performed in our center, we tried to understand what would be the optimal condition for tissue-engineering-mediated tendogenesis.[13] In this study, long PGA fibers and human fetal extensor tenoctyes (isolated from a three-month-old aborted fetus donated by the patients for research only) were used to engineer an extensor tendon equivalent. The long PGA fibers were arranged to mimic the extensor tendon complex–like structure (Fig. 34.4) followed by cell seeding onto the scaffold.

After *in vitro* culture for six weeks, the cell-scaffold constructs were further divided into three groups: (1) *in vitro* culture with mechanical loading, (2) *in vivo* implantation without mechanical loading; and (3) *in vivo* implantation with mechanical loading by suturing the construct to fascia. And thus mouse movement can provide a natural dynamic loading. The results showed that human fetal cells could form an extensor tendon complex structure *in vitro* and become further matured *in vivo* by mechanical stimulation. In contrast to *in vitro*–loaded and *in vivo*–nonloaded tendons, *in vivo*–loaded tendons exhibited bigger tissue volume, better aligned

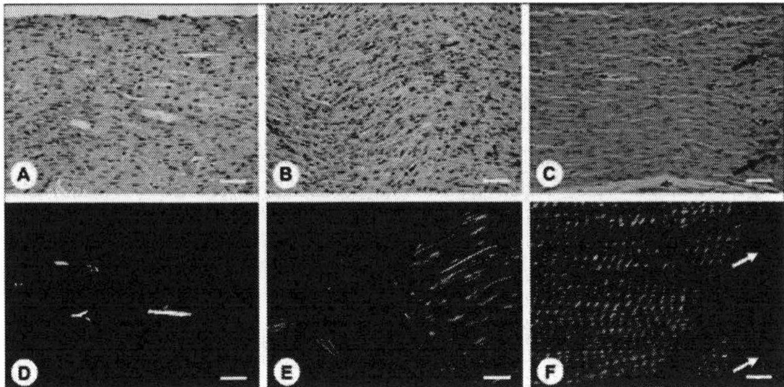

Figure 34.5. H&E staining (a–c) and polarized microscopic views (d–f) of *in vitro*-engineered tendon after total 12 weeks (a and d) and of *in vivo*-implanted tendons for 14 weeks without (b and e) and with (c and f) loading. Original magnifications: ×200; bar represents 50 μm for all. (Reprinted by permission from Ref. 13). See also Color Insert.

collagen fibers, a more mature collagen fibril structure with D-band periodicity, and stronger mechanical properties (Fig. 34.5).

These results indicate that *in vivo* mechanical loading via an *ex vivo* approach might be an optimal approach for engineering functional tendon tissue. Therefore, a reasonable strategy for engineered functional repair of tendon defects might be to generate a neo-tendon tissue first *in vitro* and then to implant *in vivo* for its further maturation and for carrying out its functions.

The physical form of polymer fibers may also affect their mechanical property and likely their degradation rate. A previous study performed in our center found that once in a woven fashion, PGA fibers could significantly enhance their mechanical strength. Additionally, the woven fibers also seemed to degrade more slowly compared with nonwoven fibers at the same time points.[14]

34.3 PGA Scaffold for Cartilage Engineering

Cartilage engineering is another major area of tissue construction. Engineering of hyaline cartilage to repair large, full-thickness defects of articular cartilage became the first target because the defect remains to be a major concern in clinical practice due to the lack of proper therapy.[15] In a porcine model, autologous articular

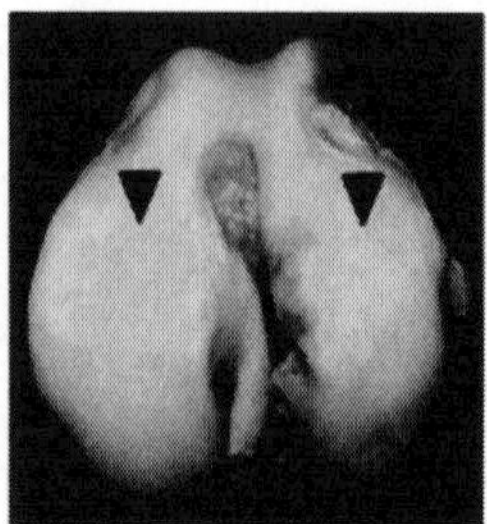 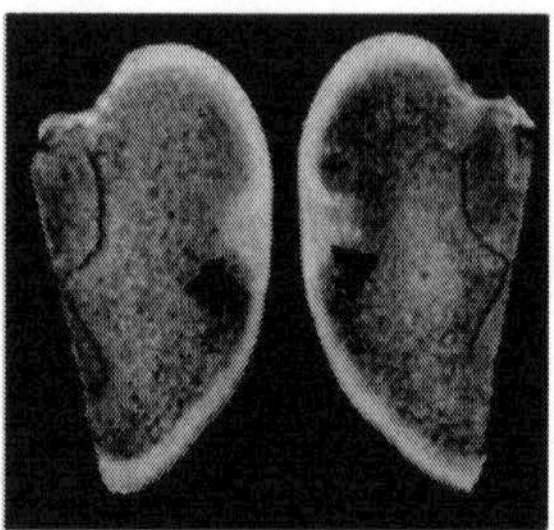

Figure 34.6. Gross view (*left*) and cross-sectional view (*right*) of repaired articular cartilage defects (arrowheads indicate the repaired defects). (Reprinted by permission from Ref. 8).

cartilage at a non-weight-bearing area was harvested from the knee joint of one side to isolate chondrocytes. On the other side, an 8 mm full-thickness articular cartilage defect deep in the underlying cancellous bone was created on both weight-bearing areas of medial and lateral femoral condyles. The isolated chondrocytes were mixed with 30% Pluronic F127 (BASF, Mount Oliver, NJ) at a final concentration of 5×10^7 cells/mL at 4°C. A 0.6 mL aliquot of cell–Pluronic solution was then mixed with 60 mg of PGA (Albany International Research, Albany, NY) and stored at 4°C until use. After the creation of the defect, the cell-scaffold construct containing PGA, Pluronic solution, and chondrocytes was then used to repair the defects in the experimental group. In the control group, the defects were either repaired with scaffold material alone or left unrepaired. Grossly, cartilage tissue was formed in the defect of experimental group as early as 4 weeks postrepair. Histological examination demonstrated the presence of hyaline cartilage tissue. At 24 weeks postrepair, gross examination revealed a complete repair of the defect by engineered cartilage, shown by a smooth articular surface indistinguishable from nearby normal cartilage. A cross section showed an ideal interface healing between the engineered cartilage and the adjacent normal cartilage (Fig. 34.6).

Histology of the tissue harvested from repaired defects further revealed a typical structure of cartilage lacuna and an ideal interface healing to adjacent normal cartilage, as well as to underlying cancellous bone (Fig. 34.7).

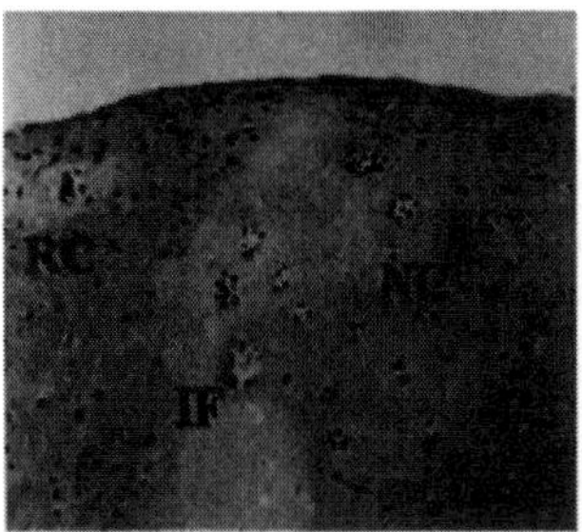

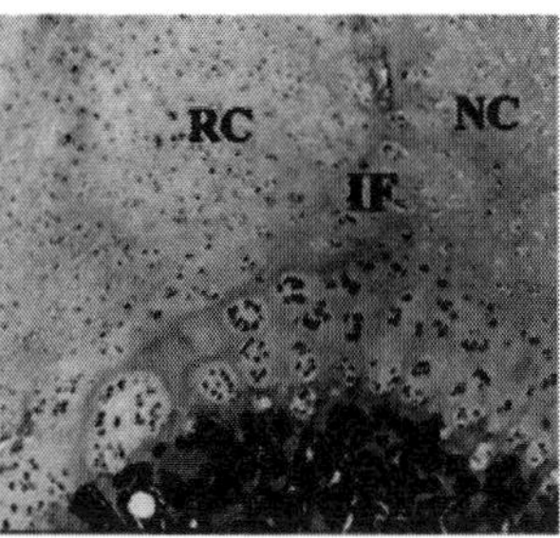

Figure 34.7. Histology demonstrates that engineered cartilage heals ideally with adjacent cartilage (*left*) and underlying cancellous bone (*right*). *Abbreviations*: NC, normal cartilage; RC, repaired cartilage; IF, interface healing; CB, cancellous bone. (Reprinted by permission from Ref. 8).

Moreover, the engineered cartilage exhibited enhanced ECM production and improved biomechanical properties, indicating that engineered cartilage resembles the native articular cartilage not only in morphology and histology but also in biochemical components and biomechanical properties.[15]

Recently, induced bone marrow stem cells (BMSCs) with a chondrogenic phenotype were used to repair articular cartilage defects in a similar model, except that the defects were created at non-weight-bearing areas.[16] To prepare the scaffold, a silicone rubber mold was created that contained a cylinder cavity with an 8 mm diameter and a 6 mm depth. Thirty milligrams of unwoven PGA fibers (Albany International Research, Albany, NY) were inserted into the cavity and 0.3 mL of 1.5% polylactic acid (PLA) (Sigma, St. Louis, MO) diluted in a dichloromethane solvent was added to maintain the scaffold shape. The scaffolds were then removed from the mold and sterilized by soaking in 75% alcohol and washed three times with phosphate buffered saline (PBS), followed by two washes with Dulbecco's Modified Eagle's Medium (DMEM). Chondrogenically induced or dexamethasone-treated BMSCs at passage 2 (1.5×10^7 in 0.3 mL) were evenly dropped onto PGA/PLA scaffolds respectively to form cell-scaffold constructs, and the constructs were cultured at 37°C in a humidified atmosphere of 5% CO_2 for four to five hours, which allowed complete adhesion of BMSCs to the scaffold. Inductive and dexamethasone-containing media were then added to the corresponding wells, and the media were changed every other day. PGA/PLA scaffolds alone (without cells) were cultured in inductive media as another control. After *in vitro* culture for one week, the constructs were implanted *in vivo* to repair the osteochondral defect.

Gross view and histology also demonstrated that the articular cartilage defect was fully repaired by engineered hyaline cartilage. Interestingly, an across section at the repair site showed that the underlying cancellous bone defect was repaired by bone tissue (Fig. 34.8) rather than the hyaline cartilage as shown in the previous study in which chondrocytes were used, indicating that BMSCs can differentiate into different cell types *in vivo*, possibly induced by different local environmental factors. Thus BMSCs may serve as a better cell source for repairing articular cartilage defect if it is associated with a defect of underlying bone tissue.[16]

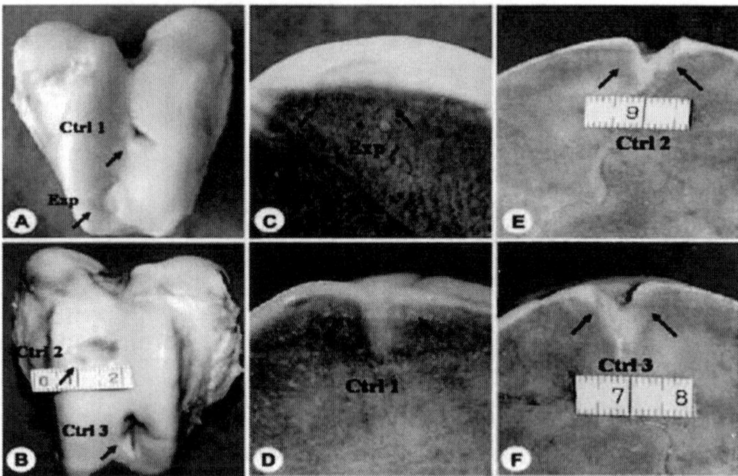

Figure 34.8. Gross and cross-sectional view of repaired defects at 6 months postrepair. Arrows indicate the repaired regions. The experimental defect exhibits a relatively regular surface (Exp, a), and the osteochondral defect is completely repaired with both engineered cartilage and bone when observed at the cross section (c). The repaired surface of control group 1 remains irregular (Ctrl 1, a), but the osteochondral defect is mostly repaired at the cross section (d). The defects in control group 2 (Ctrl 2, b, e) and control group 3 (Ctrl 3, b, f) remain largely unrepaired at both cartilage and bony layers. (Reprinted by permission from Ref. 16).

Recently, human BMSCs (hBMSCs) have also been used to explore the possibility of engineering cartilage tissue *in vitro* and their fate after *in vivo* implantation.[17] To pursue the goal, a PLA-coated unwoven PGA scaffold was prepared as previously described.[16] Seven milligrams of unwoven PGA fibers (Albany International Research, Albany, NY) were compressed into a cylinder shape with a 5 mm diameter and a 2 mm thickness. In order to solidify the scaffold shape, 50 μl of 1.5% PLA (Sigma) diluted in dichloromethane was used to coat the PGA fibers. The scaffolds were sterilized by soaking in 75% alcohol and washed three times with PBS, followed by two washes with DMEM. The hBMSCs at passage 2 (2.5×10^6 in 40 μl) were then evenly dropwised onto each scaffold. After incubation for four to five hours to allow for complete adhesion of the cells to the scaffold, the regular culture medium was added to cover the cell-scaffold construct. The constructs were then kept in an incubator at 37°C with 95% humidity and 5% CO_2.

The cell-scaffold constructs were chondrogenically induced from 4 to 12 weeks for *in vitro* chondrogenesis and then implanted subcutaneously into nude mice for 12 or 24 weeks. The engineered cartilages were then evaluated before and after implantation. Histological examination showed typical cartilage structure formation after 8 weeks of induction *in vitro*. However, part of the constructs became ossified after implantation when *in vitro* induction lasted for 8 weeks or less time. In contrast, when the cell-scaffold construct was *in vitro*-cultured to 12 weeks and longer, the *in vitro*-engineered cartilage became more mature with an obvious lacuna structure and more matrix production and deposition. More importantly, long-term *in vitro*-engineered cartilage maintained its cartilageous structure after *in vivo* implantation without being ossified (Figs. 34.9 and 34.10). These results indicate that a fully differentiated stage achieved by extended chondrogenic induction *in vitro* is necessary for hBMSCs to form stable ectopic chondrogenesis *in vivo*.[17]

The unwoven PGA fibers were also used for engineered repair of meniscus defects.[18] The meniscus was first harvested from one side to isolated meniscal fibrochondrocytes. A full-thickness meniscal defect of length 1 cm was then created on the other side, and the two ends of the defect were bridged by an acellular intestinal submucosa membrane that wrapped around the ends. Inside the membrane

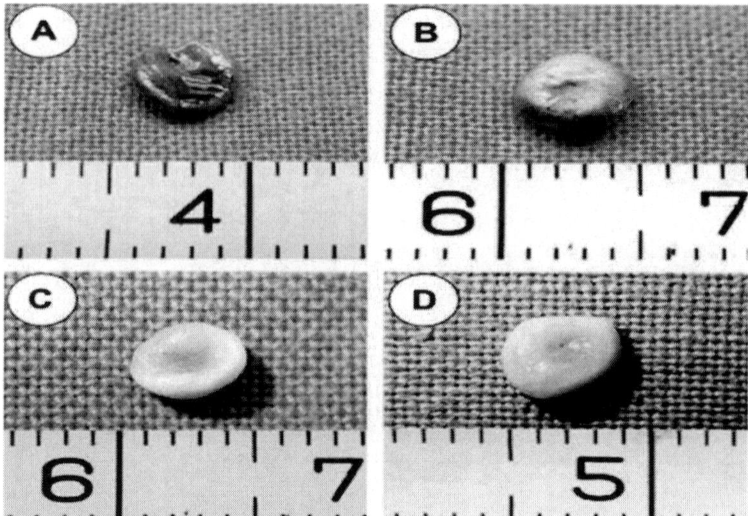

Figure 34.9. Gross view of engineered tissue after subcutaneous implantation. (a) The tissue engineered by 4 weeks of *in vitro* induction became obviously vascularized after 12 weeks of implantation. (b) Vascular invasion was also observable after 12 weeks of implantation in part of the surface of some implanted tissues engineered by 8 weeks of *in vitro* induction. The cartilages engineered by 12 weeks of *in vitro* induction exhibited mature cartilage appearance after 12 (c) or 24 (d) weeks of implantation. (Reprinted by permission from Ref. 17).

tube, unwoven PGA fibers mixed with polyethylene-polypropylene hydrogel and fibrochondrocytes were transplanted. In the control group, biomaterial alone was used. Grossly, the engineered tissue resembled the native meniscus in morphology, color, and texture at 25 weeks posttransplantation of the cell-scaffold construct. Histologically, the engineered meniscus displayed a typical structure of fibrocartilage tissue, similar to that of the native meniscus (Fig. 34.11).[8]

34.4 PGA Fibers for Skin Engineering

Engineered skin might be the most mature tissue-engineering product. Most of the skin products that are commercially available currently are made from ECM derived from either animals

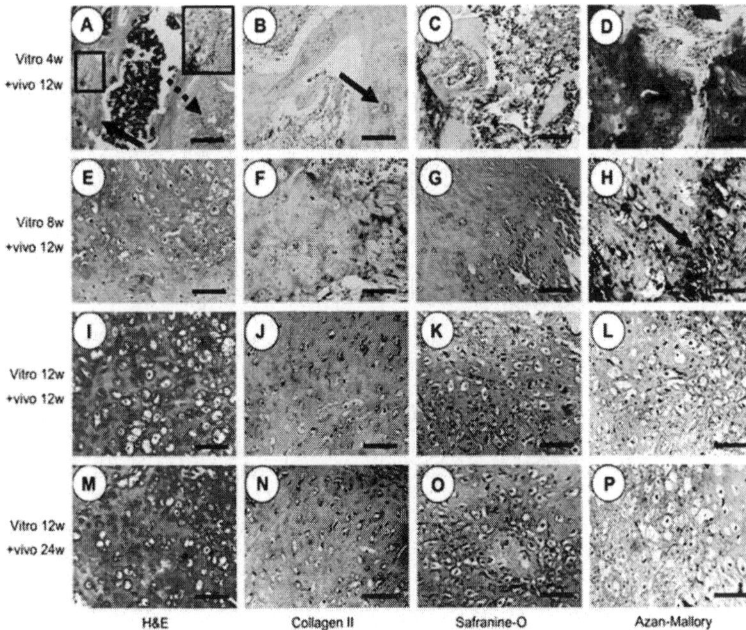

Figure 34.10. Histological structure and staining of engineered tissues after subcutaneous implantation. After 12 weeks of implantation, tissue engineered by 4 weeks of induction *in vitro* mainly formed ossified tissue (a, solid arrow) with strong positive staining of Azan-Mallory (d). Blood vessels (A, rectangle), fibrous tissue (a, dotted arrow), and positive staining of type II collagen (b, arrow) were also observed in a part of some specimens. No obvious positive staining of Safranine O was observed (c). The cartilage engineered by 8-week induction *in vitro* mainly formed cartilage-like tissue (e) with positive staining of type II collagen (f) and Safranine O (g) after implantation. Nevertheless, positive staining of Azan-Mallory could still be observed in the edge area of some specimens (h, arrow). For the cartilage engineered by 12-week induction *in vitro*, however, all the specimens showed cartilage-like tissue (i, m) with strong, positive staining of type II collagen (j, n) and Safranine O (k, o), as well as negative staining of Azan-Mallory (l, p) after 12 (i–l) or 24 (m–p) weeks of implantation, and neither ossified tissue nor fibrous tissue was observed. Scale bar = 100 μm. (Reprinted by permission from Ref. 17). See also Color Insert.

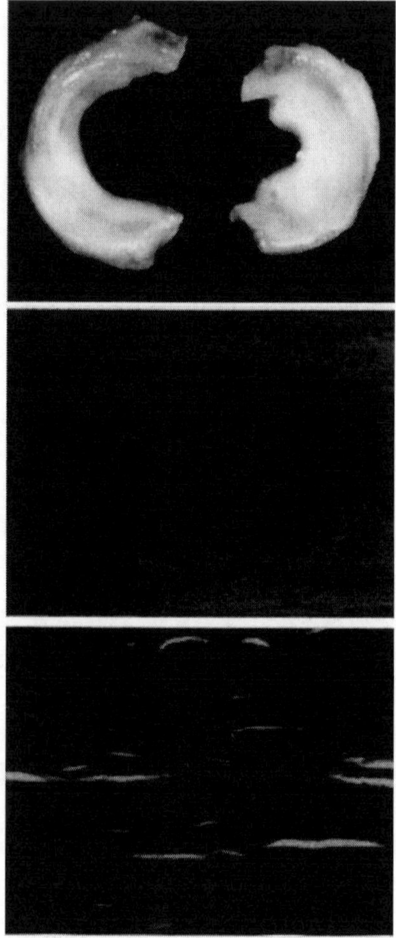

Figure 34.11. Gross view (*top*) and histology of engineered meniscus (*middle*) and natural meniscus (*bottom*) at 25 weeks postrepair (arrows indicate the engineered meniscus). (Reprinted by permission from Ref. 8).

or human beings. The concern for these products is the potential contamination of some unknown pathogens, which might become a threat to human health. Because of this concern, we focused our skin engineering on the use of synthetic material that is biodegradable and bioabsorbable. Because unwoven PGA fibers have been used to successfully engineer various types of tissues in our center, the same material has also been used in skin engineering. Briefly, autologous full-thickness skin (2 × 2 cm) was harvested, and keratinocytes and dermal fibroblasts were isolated and then separately seeded onto a sheet of unwoven PGA fibers with a polyethylene-polypropylene hydrogel. Two layers of cell-scaffold sheets were then overlapped to form a composite skin construct with a thickness of 2 mm. In a porcine model, two full-thickness wounds (4 cm in diameter) deep to the fascia were created at the dorsal aspect of the pigs, and a titanium isolation chamber was then inserted to prevent tissue in growth from adjacent skin. The wounds were repaired either with a composite skin construct as an experimental group or with biomaterial alone as a control group. The engineered skin was harvested at one, two, four, and eight weeks postrepair for histological examination. When examined grossly, neoskin formation was observed at two weeks postrepair in the experimental group. Although the wound was totally covered by the engineered skin in the experimental group at four weeks, mature, full-thickness skin was formed only in part of the wound area. However, the maturation completed uniformly at eight weeks, which resembled the morphology of native porcine skin (Fig. 34.12, *left*).

In contrast, only granulation tissue was observed in control wounds at different time points. Histologically, a double-layer structure was observed as early as one week postrepair. Interestingly, the rete ridges of the epidermis became enlarged and started to migrate deeply into dermal part of the engineered skin at two weeks and reached further deeper at four weeks. At eight weeks, these

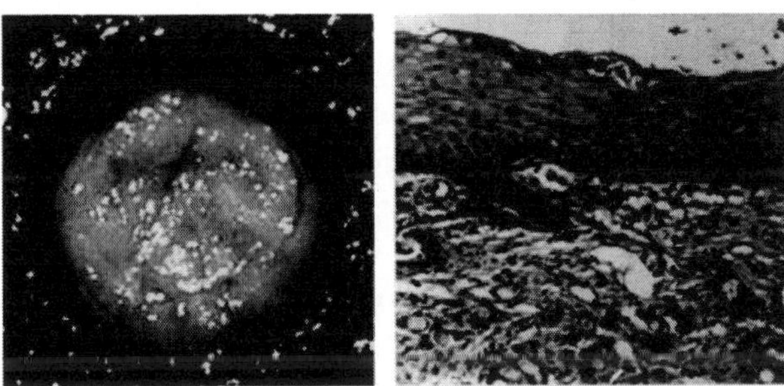

Figure 34.12. Gross view (*left*) and histology (*right*) of engineered full-thickness skin at 8 weeks. (Reprinted by permission from Ref. 8).

outreaching rete ridges retreated to where they originally resided and the histology appeared to be similar to that of normal skin (Fig. 34.12, *right*). Furthermore, histological and transmission electronic microscope examination revealed the presence of a basement membrane structure in the tissue-engineered skin. Likewise, only granulation tissue was observed in the control group histologically. Supported by the *in vivo* result, we are currently investigating the possibility of *in vitro* skin engineering using the same strategy. The preliminary results showed that a double-layer skin was formed after *in vitro* culture for two weeks, which contained both epidermis and dermis. At four weeks, the engineered skin became more mature by formation of stratified epidermis, deposition of abundant matrix in the dermis, and partial formation of a basement membrane. Thus, we propose that a synthetic biopolymer might be qualified as a scaffold for *in vitro* skin engineering.[8]

34.5 PGA Fibers for Corneal Stroma Engineering

Since the corneal stroma represents the major component responsible for corneal transparency, the primary effort was made to test the possibility of stroma engineering. To prepare the cell-scaffold construct, PGA fibers (Albany International Research, Mansfield, MA), 15 μm in diameter and weighing 5 mg, were made into plate constructs, with a diameter of 5 mm and a thickness of 1 mm. The constructs were soaked in 75% ethanol for one hour and washed with PBS three times, followed by washing with an F-12 medium containing 10% fetal bovine serum (FBS). The medium was removed afterward, and the constructs were air-dried for 30 minutes under ultraviolet light. Corneal stromal cells were collected and resuspended in a culture medium at a density of 5×10^7 cells/mL. In culture dishes, 1×10^7 cells were seeded evenly onto the PGA fibers, and the cell-scaffold constructs were kept in an incubator for four hours to allow complete adhesion of the cells to the PGA fibers. Ham's F-12 was then added, and the cell-scaffold constructs were incubated for seven days before *in vivo* transplantation. The culture medium was replaced twice per week.

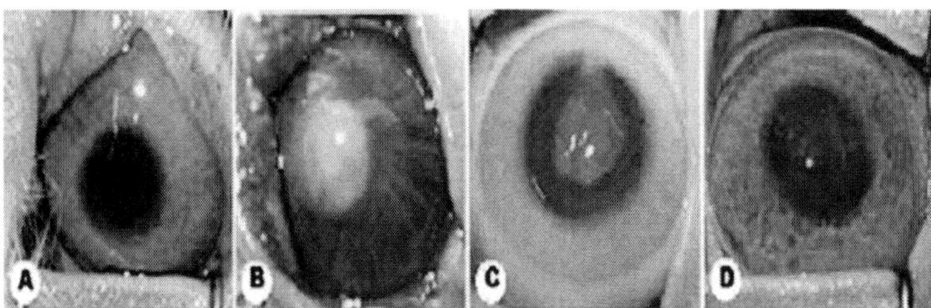

Figure 34.13. Gross observation shows that engineered tissue gradually becomes transparent over an 8-week period. (a) Preoperation, (b) postoperation, (c) 4 weeks postoperation, and (d) 8 weeks postoperation. (Reprinted by permission from Ref. 18).

After *in vivo* implantation for eight weeks, the construct that was opaque at the time of transplantation became a nearly transparent stroma at eight weeks posttransplantation (Fig. 34.13). The newly formed stroma has also been proven as an engineered stroma by the presence of green fluorescence protein (GFP)-labeled stromal cells. Histology of the engineered corneal stroma that was nearly transparent also showed a structure similar to that of the native counterpart (Fig. 34.14).[18]

In another similar study, dermal fibroblasts were used to explore the possibility of replacing stromal fibroblasts for corneal stroma engineering.[19] Again PGA fibers were used as the scaffold as similarly prepared.[18] Dermal fibroblasts were harvested from newborn rabbits, seeded onto biodegradable, unwoven PGA fibers, cultured *in vitro* for one week, and then implanted into adult rabbit corneas. After eight weeks of implantation, nearly transparent corneal stroma was formed, with a histological structure similar to that of its native counterpart (Fig. 34.15). The existence of cells that had been retrovirally labeled with GFP demonstrated the survival of implanted cells. In addition, all GFP-positive cells that survived expressed keratocan, a specific marker for corneal stromal cells, and formed fine collagen fibrils with a highly organized pattern similar to that of native stroma (Fig. 34.16). Interestingly, the condrocyte-seeded PGA scaffold formed opaque cartilage instead of transparent corneal stroma (Fig. 34.15). The results demonstrated that neonatal dermal fibroblasts could switch their phenotype in the

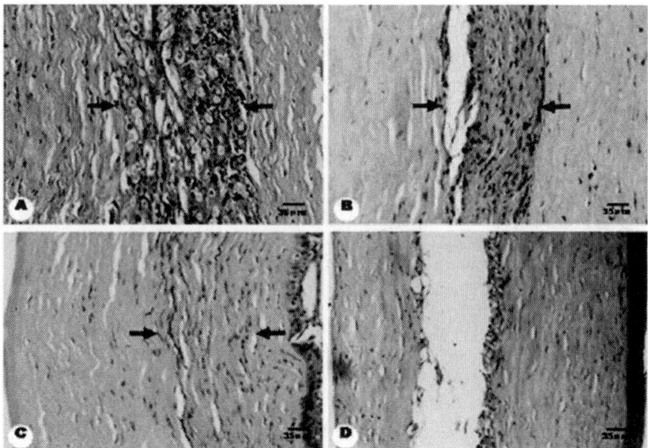

Figure 34.14. Sectional views of H&E-stained stroma. Implanted corneal stromal cell–PGA construct regions are indicated by arrows in each experimental group (a, 4 weeks; b, 6 weeks; c, 8 weeks postoperation), and implanted PGA-alone region in the control group is also shown (d, 8 weeks postoperation). Original magnification: (a–d) ×200. Scale bars: 35 μm.

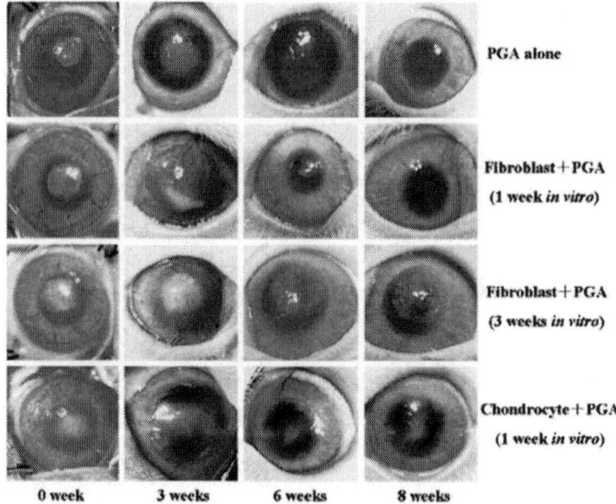

Figure 34.15. Gross view of the grafts in each group at 0, 3, 6, and 8 weeks postimplantation. At 8 weeks, the corneas become transparent in the PGA-alone group and in the group with implantation of fibroblast-PGA construct precultured *in vitro* for 1 week but not in the chondrocyte-PGA group and the group implantated with precultured fibroblast-PGA *in vitro* for 3 weeks. (Reprinted by permission from Ref. 19). See also Color Insert.

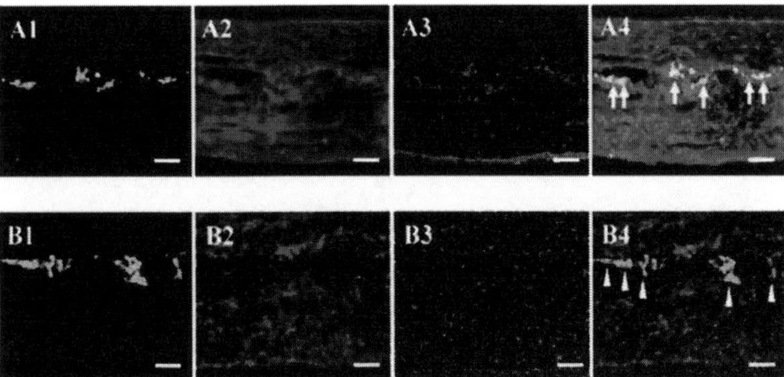

Figure 34.16. Phenotypic switch of dermal fibroblasts to corneal stromal cells. After 8 weeks of implantation, keratocan expression is observed in the group with the implantation of fibroblast-PGA constructs precultured *in vitro* for 1 week (a, arrow) but not in the group with 3-week precultured cell-PGA constructs (b, arrow head). a1, b1: GFP expression; a2, b2: keratocan staining; a3, b3: nuclei staining (Hoechst 33258); a4, b4: merged images of 3 colors. Bars: 50 μm. (Reprinted by permission from Ref. 19). See also Color Insert.

new tissue environment under restricted conditions. The functional restoration of corneal transparency using dermal fibroblasts suggests that they could be an alternative cell source for corneal stroma engineering.[19]

34.6 PGA Fibers for Blood Vessel Engineering

To prove the possibility that a neovascular structure containing both endothelium and a smooth muscle (SM) layer can be engineered, studies were performed first in a nude mouse model. Endothelial cells and smooth muscle cells (SMCs) were isolated from human neonate umbilical veins. After *in vitro* expansion, these cells were seeded onto unwoven PGA fibers wrapped around a silicone tube and then *in vitro*-co-cultured for one week followed by *in vivo* implantation into the subcutaneous tissue of a nude mouse. A scaffold tube alone was transplanted as a control. Engineered vessels were harvested at 2, 6, and 11 weeks postrepair for histology and immunohistochemical staining. Grossly, a tubular structure

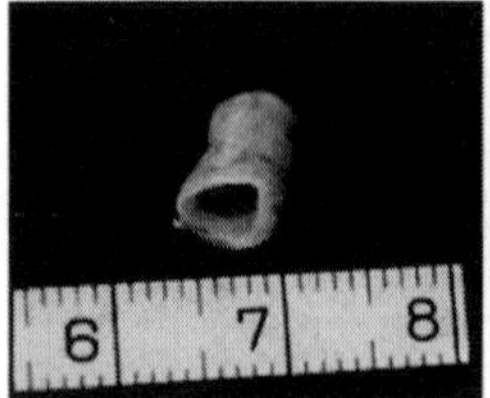

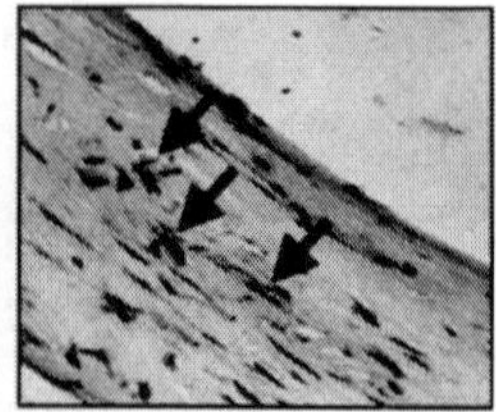

Figure 34.17. Gross view of engineered vessel at 6 weeks (*left*). Immunohistochemistry shows factor VIII–positive cells at the luminal surface (*middle*) and SM α-actin-positive cells in the vessel wall (*right*). Arrows indicate SM α-actin-positive cells. (Reprinted by permission from Ref. 8).

was formed in both groups at 2 weeks but disappeared in the control group at 6 weeks. Histologically, the implanted PGA fibers were mostly degraded at 6 weeks (Fig. 34.17). The neovascular structure formed in the experimental group contained the endothelium on its luminal surface and a tissue layer similar to the middle layer of a vessel. Immunohistochemical staining demonstrated that the endothelium lined at the luminal surface was positive for factor-VIII and the middle layer contained cells that stained positive for SM α-actin (Fig. 34.17). However, the neovascular tissue harvested at 11 weeks became atrophic compared with the tissue of 6 weeks, although trichrome staining showed that more SM fibers were formed in the 11-week tissue than in the 6-week tissue. This phenomenon suggests that mechanical stimulation might be an essential element for vessel engineering.

To enhance the mechanical property of engineered vessel wall tissue, an *in vitro* approach was employed using a bioreactor system.[20] To prepare a cell-scaffold construct, unwoven PGA fibers (Albany International Research Company, Albany, NY), 15 μm in diameter and 30 mg in weight, were made into a 40 × 30 × 2 mm mesh. The scaffold was soaked in 75% ethanol for one hour and washed three times with PBS, followed by incubating with DMEM containing 10% FBS for 10 minutes. The medium was removed afterward, and the scaffold was air-dried for 30 minutes under ultraviolet light before use. Canine SMCs were collected, resuspended in the culture medium at a density of 6×10^7 cells/mL, and 3×10^7 cells were then evenly seeded onto each PGA mesh in tissue culture

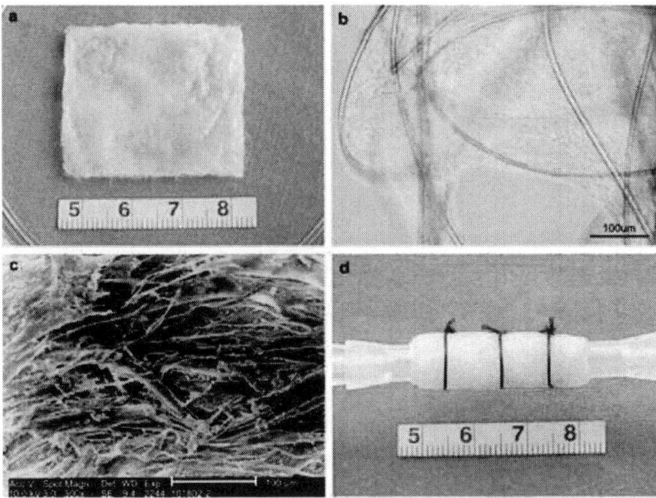

Figure 34.18. Cell-scaffold constructs cultured in a dish. (a) SMCs were seeded onto an unwoven PGA fiber mesh and cultured for 5 days in a dish. (b) Microscopic observation shows SMC growth on the PGA mesh at day 5. (c) Scanning electron microscopic view of ECM production by SMCs on PGA fibers at day 5. (d) The cell-PGA sheet was wrapped around a silicone tube in the culture chamber of a vessel reactor, secured by biodegradable sutures. (Reprinted by permission from Ref. 20).

dishes. Thereafter, the cell-scaffold constructs were kept in an incubator for four hours to allow for the complete adhesion of the cells to the PGA fibers. DMEM with 10% FBS was then added to cover the constructs. The cell-PGA sheets were incubated for another five days in the culture dishes.

Afterward, the SMC-PGA sheets were wrapped around the silicone tubes in the culture chamber of the vessel reactor and further secured by biodegradable sutures (Ethilon, Ethicon, Inc., USA) (Fig. 34.18). The loaded chamber was then filled with DMEM containing 10% FBS to cover the cell-scaffold construct, followed by connecting the culture chamber to the whole reactor. A pulsatile flow of sterile PBS was applied through the silicone tubes at a frequency of 75 beats/min. The flow rate was gradually increased and adjusted (between 70–80 mL/min) to reach a radial distension about 5% of the original diameter of the constructs. The culture was maintained at 37°C in the incubator for eight weeks with medium change twice a week. The cell-PGA constructs cultured without pulsatile radial stress were set up as a control.

After eight weeks, the SMC-PGA constructs were transformed into a tubular structure with excellent elasticity and a round lumen 6 mm in diameter when cultured under dynamic mechanical loading. In contrast, vessel walls derived from the static group revealed a collapsed lumen structure and a rough surface when pulsatile stimulation was not applied (Fig. 34.19).

Histology revealed the structure with multiple layers of SMCs and orientated orderly in the vessel walls of the dynamic group with collagenous fibers distributed evenly in the vessel wall and complete degradation of PGA fibers. On the contrary, randomly orientated SMCs together with disorganized collagenous fibers were observed in the static group (Fig. 34.20).

The above findings were further confirmed by the results of immunohistochemical staining for SM α-actin and calponin in the dynamic group. In contrast, fewer cells positive for SM α-actin and calponin with a disorganized pattern were observed in the

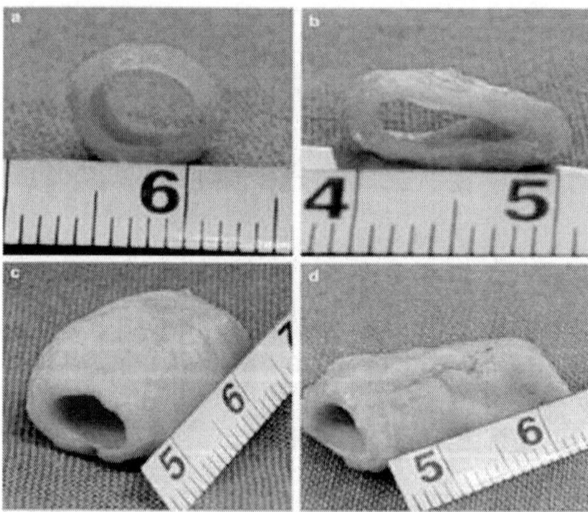

Figure 34.19. Gross view of engineered vessels. Vessel walls with a round lumen (6 mm in diameter) are formed after 4 (a) and 8 (c) weeks of culture with pulsatile stimulation. Vessel walls with a collapsed lumen and a rough surface are observed after 4 (b) and 8 (d) weeks of culture without pulsatile stimulation. (Reprinted by permission from Ref. 20).

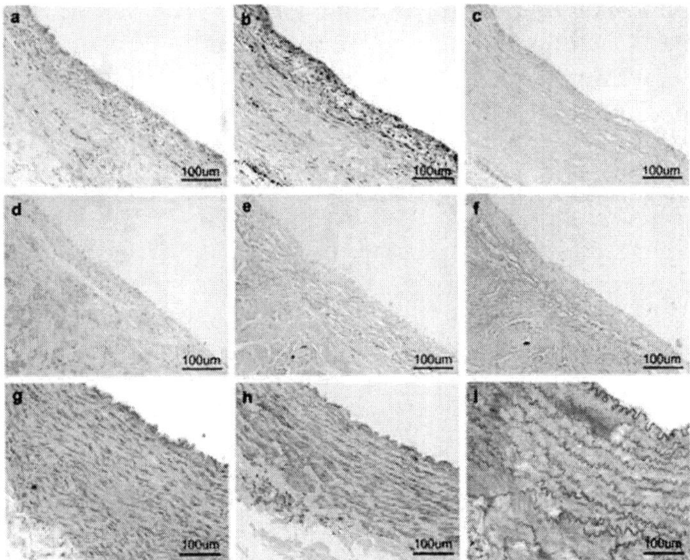

Figure 34.20. Histology of the engineered vessel walls after 8 weeks of culture with (a–c) or without (d–f) pulsatile stimulation. Native canine abdominal arteries are presented as references (g–i). H&E staining shows that PGA fibers are completely degraded and SMCs are in well-orientated layers in the dynamic group (a), while SMCs are orientated randomly in the static group (d). Masson staining shows well-organized collagenous fibers distributed evenly in the vessel wall of the dynamic group (b), but only disorganized collagenous fibers are present in the static group (e). Elastic fibers are neither found in the dynamic group (c) nor in the static group (f) by Gomori staining. Bars: 100 μm. (Reprinted by permission from Ref. 20). See also Color Insert.

walls of the static group. In addition, mechanical properties such as tensile strength, suture-holding retention strength, and burst pressure were significantly enhanced in the dynamic group compared with the control group. These results indicate that such an approach is suitable for engineering large-vessel wall tissue and for engineering other tissues with a muscular tubular structure.[20]

34.7 PGA Fibers for Engineering Peripheral Nerve Tissue

PGA fibers were also investigated for their application in peripheral nerve engineering. Schwann cells were isolated from the sciatic nerves of newborn Wistar rats and cultured in an F12/DMEM medium for *in vitro* expansion. The PGA fibers with a diameter of 15 μm were arranged longitudinally into a cord and were then seeded with Schwann cells suspended in a culture medium and co-cultured for one week. In adult Wistar rats, the sciatic nerve was exposed and a 15 mm long defect was created and then bridged with the cell-scaffold construct that was wrapped with a degradable biomembrane in an experimental group or bridged with a syngenic nerve graft as a positive control and bridged with PGA fibers alone or left unrepaired as a negative control. Engineered nerves were harvested at three months posttransplantation for histology and electrophysiological evaluation. The results showed that peripheral nerve function was much better recovered in the experimental and positive control groups than in the negative control group. The gained functions of the engineered nerve were demonstrated by the response of rat extremities to pain and temperature stimulation and by the maintaining of gastrocnemius weight. Grossly, the engineered nerve has a morphology similar to that of the normal peripheral nerve, whereas the PGA fibers were totally degraded in the negative control group. Electromyogram evaluation revealed that no statistically significant difference was found between the

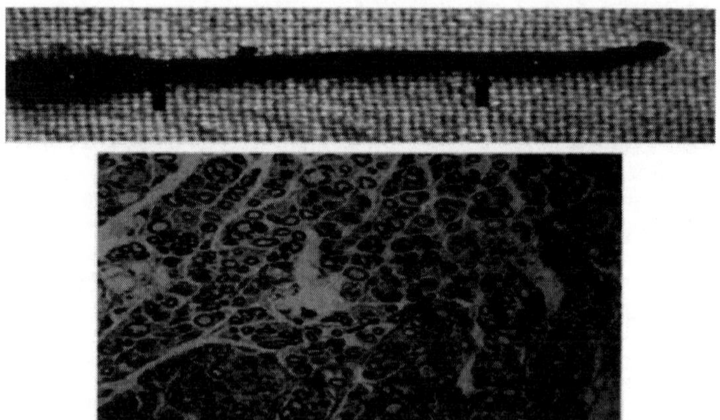

Figure 34.21. Gross view (*top*, between arrows) and histology of a tissue-engineered nerve (*bottom*). (Reprinted by permission from Ref. 8).

experimental and positive control groups in electroconductivity, the velocity, or the amplitude of action potentials. Histology also demonstrated that the level of axon regeneration in the tissue-engineered nerve was comparable to that of the positive control group (Fig. 34.21).[8]

34.8 Conclusion

This chapter reviewed the recent accomplishments in tissue reconstruction in our center. The results showed that it is possible to generate different types of soft tissues, such as tendon, cartilage, skin, cornea, vessel, and peripheral nerve, either *in vitro* or *in vivo* by using PGA fibers as the main scaffold, indicating PGA fibers remain one of the excellent candidates of scaffold materials for soft-tissue engineering, although they are considered conventional materials and have been used since the early stage of tissue engineering research. Future efforts to modify PGA fiber characters for reduced acidity of its degradation product, better regulation of its degradability, and enhanced mechanical property will be the future direction to better use PGA fibers as the scaffold for soft-tissue engineering.

Acknowledgments

These studies were supported by the National "973" and "863" Tissue Engineering Project Foundation, the Shanghai Science and Technology Development Foundation, the Key Laboratory Foundation of Shanghai Education Committee, and the National Natural Science Foundation. The authors would like to thank Drs. Yongtao Liu, Bin Chen, Dejun Cao, Feng Xu, Dan Deng, Bin Wang, Yanjie Zhang, and Hong Li for the contribution in tendon engineering; Drs. Guangdong Zhou, Yanchung Liu, Fuguo Chen, Yimin Cui, Tianyi Liu, and Kai Liu for the contribution in cartilage engineering; Drs. Guanghui Yang, Xie Cai, Jun Yang, and Chengliang Deng for the contribution in skin engineering; Drs. Zhicheng Xu and Yang

Liu for the contribution in blood vessel engineering; Drs. Xiaojie Hu and Yanqing Zhang for the contribution in cornea engineering; and Dr. Chuanchang Dai for the contribution in peripheral nerve engineering.

References

1. R. Langer and J. P. Vacanti, *Tissue Eng.*, **260**, 920 (1993).
2. J. M. Pachence and J. Kohn, *Biodegr. Polym.*, 263 (1997).
3. L. E. Freed, J. C. Marquis, A. Nohria, J. Emmanual, A. G. Mikos, and R. Langer, *J. Biomed. Mater. Res.*, **27**, 11 (1993).
4. J. P. Vacanti, M. A. Morse, W. M. Saltzman, A. J. Domb, A. Perez-Atayde, and R. Langer, *J. Pediatr. Surg.*, **23(1 Pt 2)**, 3 (1988).
5. L. G. Cima, J. P. Vacanti, C. Vacanti, D. Ingber, D. Mooney, and R. Langer, *J. Biomech. Eng.*, **113**, 143 (1991).
6. L. G. Cima, D. E. Ingber, J. P. Vacanti, and R. Langer, *Biotechnol. Bioeng.*, **38**, 145 (1991).
7. S. Commandeur, H. M. M. Van Beusekom, and W. J. Van Der Giessen, *J. Interven. Cardio.*, **19**, 500 (2006).
8. W. Liu , L. Cui, and Y. A. Cao, *Tissue Eng.*, **9 Suppl 1**, S17 (2003).
9. Y. Liu, H. S. Ramanath, and D. A. Wang, *Trends Biotechnol.*, **26**, 201 (2008).
10. Y. Cao, J. P. Vacanti, X. Ma, K. T. Paige, J. Upton, Z. Chowanski, B. Schloo, R. Langer, and C. A. Vacanti, *Transplant. Proc.*, **26**, 3390 (1994).
11. Y. Cao, Y. Liu, W. Liu, Q. Shan, S. D. Buonocore, and L. Cui, *Plast. Reconstr. Surg.*, **110**, 1280 (2002).
12. W. Liu, B. Chen, D. Deng, F. Xu, L. Cui, and Y. Cao, *Tissue Eng.*, **12**, 775 (2006).
13. B. Wang, W. Liu, Y. Zhang, Y. Jiang, W. J. Zhang, G. Zhou, L. Cui, and Y. Cao, *Biomaterials*, **29**, 2954 (2008).
14. X. Wei, P. H. Zhang, W. Z. Wang, Z. Q. Tan, D. J. Cao, F. Xu, L. Cui, W. Liu, and Y. L. Cao, *Key Eng. Mater.*, **288**, 7 (2005).
15. Y. Liu, F. Chen, W. Liu, L. Cui, Q. Shang, W. Xia, J. Wang, Y. Cui, G. Yang, D. Liu, J. Wu, R. Xu, S. D. Buonocore, and Y. Cao, *Tissue Eng.*, **8**, 709 (2002).
16. G. Zhou, W. Liu, L. Cui, X. Wang, T. Liu, and Y. Cao, *Tissue Eng.*, **12**, 3209 (2006).

17. K. Liu, G. D. Zhou, W. Liu, W. J. Zhang, L. Cui, X. Liu, T. Y. Liu, and Y. Cao, *Biomaterials*, **29**, 2183 (2008).
18. X. Hu, W. Lui, L. Cui, M. Wang, and Y. Cao, *Tissue Eng.*, **11**, 1710 (2005).
19. Y. Q. Zhang, W. J. Zhang, W. Liu, X. J. Hu, G. D. Zhou, L. Cui, and Y. Cao, *Tissue Eng. Part A*, **14**, 295 (2008).
20. Z. C. Xu, W. J. Zhang, H. Li, L. Cui, L. Cen, G. D. Zhou, W. Liu, and Y. Cao, *Biomaterials*, **29**, 1464 (2008).

Chapter 35

TISSUE ENGINEERING AND ANTI-AGING

Minoru Ueda
Department of Oral and Maxillofacial Surgery, School of Medicine, Nagoya University 65 Tsurumai-cho, Showa-ku, Nagoya Aichi, Japan
mueda@med.nagoya-u.ac.jp

35.1 Introduction

For the elimination of facial wrinkles and skin contour defects, injectable filler substances composed of commercially prepared materials are now widely available. Injectable soft-tissue substitutes provide an affordable nonsurgical alternative for the correction of facial signs of aging. The search for an ideal substitute for facial soft-tissue augmentation has been an ongoing effort for many years. The ideal soft-tissue filler should be nonantigenic, noninflammatory, stable after injection, nonmigratory, nontoxic, noncarcinogenic, biologically inert, long lasting but resorbable, and easy to apply.[1] In the past there have been three types of soft-tissue fillers: autologous, allogenic, and synthetic. Good results have been reported with autologous fat transplants, although resorption rates are high, postoperative down time is long, and better results can be achieved by repeated sessions. Injectable bovine collagen was the first material successfully used as a dermal filler, and the armamentarium now includes synthetic and other protein-based materials. However,

Handbook of Intelligent Scaffolds for Tissue Engineering and Regenerative Medicine
Edited by Gilson Khang

www.panstanford.com

these products have some limitations. For example, up to 6% of patients suffer hypersensitivity reactions to bovine collagen, which can manifest as granulomatous inflammation, necrosis, or abscess formation.[1] Rare systemic complications have been reported. The clinical effect of biological materials such as collagen is short lived because of their rapid degradation *in vivo*. For long-lasting treatment results, alternative procedures have been investigated. Synthetic materials such as silicone have the potential for adverse reactions, since artificial materials remain permanently in the body.

To overcome some of these limitations, an autologous living fibroblast culture technique was developed by Isolagen Technologies (Exton).[2] This technique may safely produce sustained improvements in contour defects without surgery and has a virtually zero risk of hypersensitivity reactions. This method involves a small postauricular punch biopsy, which is used to create an autologous fibroblast cell line through a specific culturing process. These numerically multiplied living autologous fibroblasts suspended in saline solution are then injected directly into the patient's dermis, where it is believed these cells create a continuous protein repair system. Recent studies have demonstrated objectively measured improvements in facial contour defects lasting at least 12 to 48 months. Indications include correction of facial lines, wrinkles, and scars. Disadvantages of this method include the necessity for a skin biopsy, high costs, long processing time (6 weeks), and the need to inject the material within 48 hours of receipt. Also, patients who expect immediate improvement may be dissatisfied because it takes time for the implanted fibroblasts to grow and ameliorate the treated area.[2]

Following the Isolagen method, we have developed the new method for the wrinkle treatment by using gingival fibroblasts.[1] The gingival fibroblast has been known to possess unique characteristics such as remodeling capabilities. In particular, gingival fibroblasts are considered responsible for the relatively scarless healing of oral mucosa compared with skin healing. Recently, it has been shown that gingival fibroblasts are a potential resource for not only autologous collagen but also the various growth factors that may affect repair and maintenance of dermal and superficial subcutaneous deficiencies. Gingival fibroblasts can be cultured from a small gingival segment. The biopsy sample is small (5 mm diameter), and the expected outcome includes low morbidity and minimal scar formation. In addition, gingival biopsy can be easily performed in a general dental practice, which may be advantageous for patients with dental implants.

To return to the subject, hyaluronic acid (HA) is a constituent of the ground substance of the normal dermis and has considerable water-binding capabilities, which influence dermal volume and compressibility.[3,4] It is a glycosaminoglycan composed of repeating dimeric units of D-glucuronic acid and N-acetyl-glucosamine, which provide a fluid matrix on which collagen and elastic fibers can develop. It is obtained from either avian or bacterial culture sources. Its derivatives that are used for soft-tissue augmentation are cross-linked to decrease proteolytic degradation rates. HA is highly suited for soft-tissue augmentation because it is insoluble, resists degradation, does not migrate, and retains a high water content. A major advantage of HA is due to the fact that its chemical structure is uniform throughout all living species, rendering a minimal chance of immunogenicity. Therefore, pretreatment allergic skin testing is not necessary. Also, unlike collagen products are colorless; therefore, they can be injected superficially without concern of discolorations showing through the skin. The effect of soft-tissue augmentation by HA appears immediately. The handling of these products is generally considered to be superior to bovine collagen because of easier flow rates. Additionally, they need not be refrigerated; however, if they are exposed to heat, monomers will form, which can contribute to inflammation.

In an effort to obtain results that last for a longer amount of time, a combination of cultured fibroblasts and stabilized HA has been developed. The aim of this article was to introduce the longevity of the filling effect, in terms of resorption time, of cultured fibroblasts with a HA matrix.

35.2 Materials and Method[1]

This Institutional Review Board (IRB)-approved study enrolled seven patients who were treated in our clinic. The study population consisted of patients with facial wrinkles mainly in the perioral region, such as nasolabial folds and lip wrinkles following implant treatment.

35.2.1 *Tissue Preparation*

Fibroblasts were obtained from the patient's buccal gingival tissue. Mucosal epithelium was removed with the aid of a dissecting microscope. The submucosal tissue sample (5 mm diameter) was treated for one hour at 37°C with 10 mg/mL collagenase (Wako) dissolved in Dulbecco's Modified Eagle's Medium (DMEM) (Invitrogen), containing 2.5 μg/mL amphotericin B (Sigma–Aldrich) and 50 μg/mL gentamicin sulfate (Sigma–Aldrich). After the resulting cell suspension was washed several times in phosphate buffered saline (PBS) (Invitrogen), it was resuspended in DMEM containing a 10% autologous serum prepared from the patient's peripheral blood, and supplemented with 2.5 μg/mL amphotericin B and 50 μg/mL gentamicin sulfate. The cells in the medium were then seeded into a 6-well tissue culture plate (Corning). For the histological evaluation, a piece of tissue was fixed with 4% paraformaldehyde and then embedded in paraffin. Paraffin sections were cut, de-paraffinized, and stained with hematoxylin and eosin.

35.2.2 *Cell Culture*

Cells were maintained in DMEM plus 10% autologous human serum and incubated at 37°C with 5% CO_2. The cells demonstrated a typical spindle shape, which was maintained throughout the cell culture (Fig. 35.1). When cells attained 80%–100% confluency, they were subcultured into flasks and expanded from T-75 (Greiner Bio-One) to T-225 flasks (Sumitomo Bakelite). For subculture, the cells were first washed in PBS and then treated for five minutes at 37°C with TrypLE Express (Invitrogen) before resuspension in a fresh medium. Cells were seeded at a density of 2–5 × 105 cells/mL and subcultured according to a schedule that accounted for the timing of the patient's treatments. The cell culture timetable was: P (passage) 0 cells cultured in 6-well plates (7–10 days), P1 cells in a T-75 flask (7–10 days), P2 cells in a T-225 flask (7–10 days), and P3 to P6 cells in

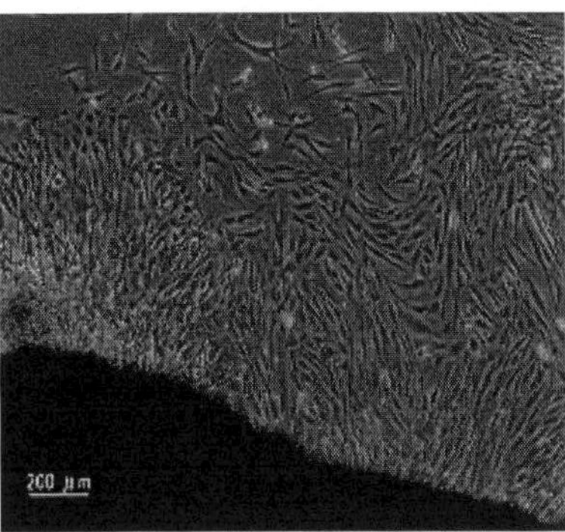

Figure 35.1. Phase contrast photomicrograph showing the morphology of the cultured gingival fibroblasts.

8–12 T-225 flasks (7–10 days). Cells cultured beyond the sixth passage were not used for treatments.

The characteristics of the cultured cells were checked by immunofluorescence microscopy for known fibroblast markers (Fig. 35.2). The cells were seeded in a 6-well plate (Greiner Bio-One) at a density of 2,500 cells/cm^2 and cultured until the cells became subconfluent. Cells were washed three times with PBS (Nissui Pharmaceutical Co.) and then fixed with 4% paraformaldehyde (Sigma-Aldrich) for 30 minutes. Cells were washed three times with PBS for 5 minutes and treated with a blocking reagent containing 2%

Figure 35.2. Dark-field photomicrographs showing results of indirect immunofluorescent staining of cultured gingival fibroblasts. Nuclei (blue) were visualized by counterstaining with DAPI. (a) Staining with anti-collagen type I and (b) staining with antivimentin. See also Color Insert.

BSA (Sigma–Aldrich) and 0.25% Triton X-100 (Sigma–Aldrich) in PBS. After removal of the blocking reagent, the cells were incubated with polyclonal antibodies against vimentin (1:100 dilution; Santa Cruz Biotechonology), cytokeratin 14 (AE1/AE3) (1:200 dilution; Chemicon International), type I collagen (1:1500 dilution; Rockland Immunochemicals), α-smooth muscle actin (Lab Vision), or CD90 (Thy-1) (1:100 dilutions; BD Pharmingen) for two hours at room temperature. The cells were washed three times with PBS and incubated with Alexa Fluor® secondary antibodies 466 mouse antihuman IgG (Invitrogen) or Alexa Fluor® 546 rabbit antihuman IgG (Invitrogen) for one hour at room temperature. After three washes with PBS, the cells were counterstained using a Vectashield mounting medium with 4′,6-diamidino-2-phenylindole (DAPI) (Vector Laboratories). Digital images were acquired using DP Controller 1.2.1 and DP Manager 1.2.1 (Olympus).

35.2.3 *Medium and Autologous Serum Preparation*

Autologous patient serum was prepared from 100–150 mL of peripheral blood. Peripheral blood was kept at room temperature for one hour and then centrifuged at 4°C. The supernatant was transferred to a fresh centrifuge tube and centrifuged again to remove any remaining blood cells. The resulting supernatant was sterile-filtered using a 0.22 μm pore tube top filter (Corning). The filtered serum was added to the culture medium containing antibiotics to a final serum concentration of 10%.

35.2.4 *Preparation of Cell Suspension and HA Admixture*

Gingival fibroblasts were harvested from 8–12 tissue T-225 culture flasks after being washed with PBS and treated with TrypLE Express (37°C, 5 min). The cells were washed twice with sterile saline (Otsuka Seiyaku, Tokyo, Japan), and resuspended in sterile saline to a final concentration of 1.0×10^7 cells/mL. The cell suspension and HA (1%, sodium hyaluronate, ARTZ®, Kaken Pharmaceutical Co. Ltd, Tokyo, Japan) was stored in 1.0 mL syringes until use.

35.2.5 *Safety Tests*

Before the beginning of cell culture, patients' sera (4 mL) was tested for hepatitis B surface (HBs) antigen (chemiluminescent immunoassay [CLIA]), HBs antibody (CLIA), hepatitis B core (HBc) antigen (CLIA), hepatitis C virus (HCV) antibody (RIA solid phase), human immunodeficiency virus (HIV) antigen and HIV antibody (enzyme-linked immunosorbent assays [ELISAs]), syphilis serology (rapid plasma regain [RPR]), syphilis serology (*Treponema pallidum* hemagglutination assay [TPHA]), human T-cell lymphotrophic virus (HTLV)-1 antibody (CLIA), parvovirus B19 DNA (polymerase chain reaction [PCR]), and mycoplasma antibody (ELISA). The patients were negative for all tests. Peripheral blood (3 mL) was also tested for white blood cell, red blood cell, and platelet counts; hemoglobin, hematocrit, mean corpuscular hemoglobin (MCV), and mean corpuscular hemoglobin concentration (MCHC).

Before cell culture, DMEM with 10% autologous serum (1 mL) underwent two safety checks. To test for contamination by bacteria and fungi, the medium was spread onto horse blood agar (Nissui Pharmaceutical) and the medium was also tested for mycoplasma. Mycoplasma DNA was extracted from the prepared medium using phenol/chloroform/isoamyl alcohol (PCI) (Sigma). Equal volumes of PCI were added to the medium (600 μL) and centrifuged at 15,000 revolutions/min for 5 minutes at room temperature. The supernatant was then mixed with 400 μL ice-cold 100% propan-2-ol (Wako) and centrifuged at 15,000 revolutions/min for 10 minutes. The pellet was then rinsed with 400 μL ice-cold 70% ethanol (Wako) and centrifuged at 15,000 revolutions/min for 5 minutes. The final pellet was dried for 3 minutes and then dissolved in 20 μL distilled water for PCR. The PCR mix (25 μL) contained 2.5 μL 10$\times$ buffer, 2 μL of each 10 mM dNTP, 0.5 μL of 10 mM of each primer (F1: s'-ACACCATGGGAGYTGGTAAT-3'; R1: s'-CTTCWTCGACTTYCAGACCCAAGGCAT-3'), and 0.1 μL of 5 units/μL Taq-DNA polymerase (Takara). Thirty-five cycles were run with the following conditions: 94°C for 1 minute (denaturation), 55°C for 30 seconds (annealing), and 72°C for 30 seconds (polymerization). Amplified DNA was separated on a 1.5% agarose gel and soaked in Tris/acetate/ethylenediaminetetraacetic acid (EDTA) (TAE) buffer containing 0.1 μg/mL ethidium bromide. The products were analyzed using the ChemiDoc XRS gel documentation system (Bio-Rad Laboratories).

Before cell injection, contamination was checked for again. The supernatant of each cell culture flask used was collected five days before cell harvest and tested as described for prepared the tissue culture medium.

On the day of injection, 5 μL of the cell suspension that had been prepared for injection was diluted 100-fold in normal saline solution and tested using the Endosafe PTS system (Charles River Laboratories) for endotoxin before injection. Samples containing more than 1.0 EU/mL were considered positive.

35.2.6 *Clinical Assessment of Aesthetic Improvement*

All satisfactory assessments were performed by the patient, and the following grading scale was used at 3, 6, 12, and 36 months after the first injection:

4. Completely satisfied
3. Satisfied
2. No remarkable change observed
1. Not satisfied

35.2.7 *Skin Replica and Analysis*

At the time of wrinkle induction and one week after the final injection of the gingival fibroblast and HA admixture, negative replicas of the skin surface were taken by using a silicon-based impression material, Flextime® (Heraeus Kulzes, New York). To obtain replicas of the wrinkles from the same skin area, the skin was marked using an oil-based marker pen. For ease of measurement, all replicas were cut into 1 cm square pieces, and the back of each replica was processed into a flat plane using the same impression material.

Light was directed at a 20-degree angle, and images were incorporated from replica using a charge-coupled device (CCD). The image of the negative replica was observed using a wrinkle analysis system skin visiometer SV 600 (Courage & Khazaka, Cologne,

Table 35.1. Improved parameters for wrinkles. R1 represents skin roughness; R2, maximum roughness; R3, average roughness; R4, smoothness roughness; and R5, arithmetic average roughness.

Parameters	Description
R1	Skin roughness
R2	Maximum roughness
R3	Average roughness
R4	Smothness roughness
R5	Arithmetic average roughness

Germany). The parameters used in the assessment of the skin wrinkles are listed in Table 35.1.

35.3 Result

Sixty-eight patients and two hundred and seventy-two sites were treated with live gingival fibroblast injections with HA (Table 35.2). The population was 1 male and 67 females. Age of the patients varied from 32 to 85 years. The mean age at the time of the first injection was 53.3 years. At 3 years' follow-up, the satisfaction score among treated patients was evaluated (Fig. 35.3, Table 35.2) at 3 years after injection, and the results were 3.1 in the nasolabial area (Fig. 35.4), 3.1 in the lip, 3.1 in the forehead (Fig. 35.5), and 2.8 in between the eyebrows (Fig. 35.6), respectively. The satisfaction score increased from 3.1, whereas the lateral eyelid to 4.00 within 6 months (Fig. 35.7).

When we measured the parameters for the wrinkles of replicas with the skin visiometers SV 600, injection of fibroblast and HA significantly reduced all parameters for wrinkles (Figs. 35.8 & 35.9).

No serious adverse events considered to be related to the study treatment were reported. Routine laboratory results, including hematology and chemistry, were unremarkable and showed no abnormalities or definitive trends. The longevity of the treatment effect continued for at least 3 years.

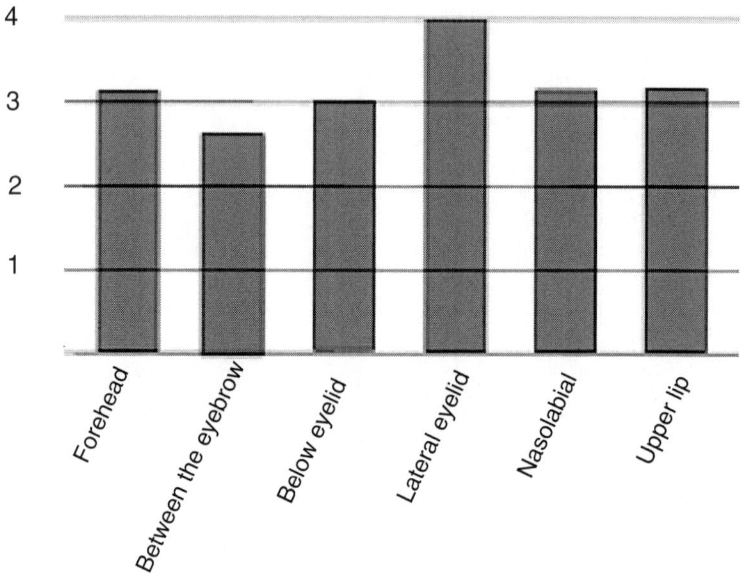

Figure 35.3. Results of satisfaction evaluation in sites.

Table 35.2. Patient data and results.

Patient Data		
Patient: 68 (1 male, 67 female)		
Age: 32–85 years old (Average: 53.3 years old)		
Observation Period: 36 month ~		
Results		
	Patient	Sites
4. Completely satisfied	34(50.0%)	106(38.9%)
3. Satisfied	17(25.0%)	84(30.6%)
2. No remarkable change observe	13(20.0%)	60(22.2%)
1. Not satisfied	4(5.0%)	22(8.3%)

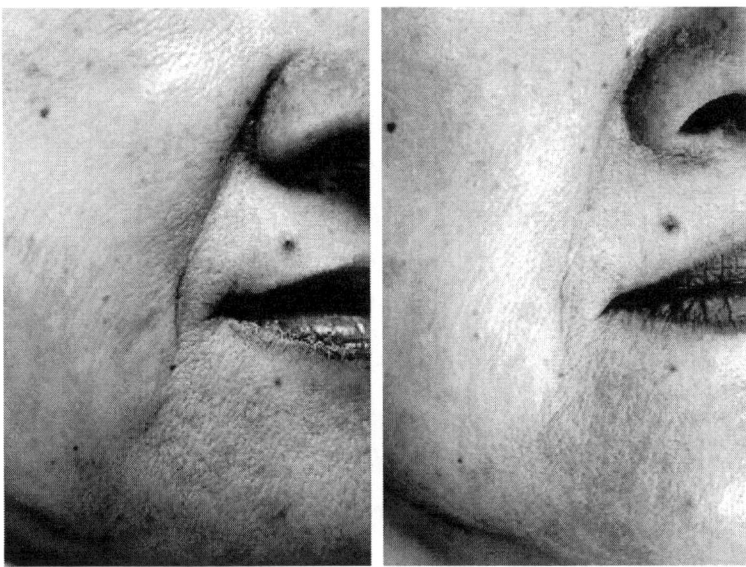

Figure 35.4. A 51-year-old woman, nasolabial area, 3 years after cell injection, satisfaction score 4.0, replica analysis, 70%. See also Color Insert.

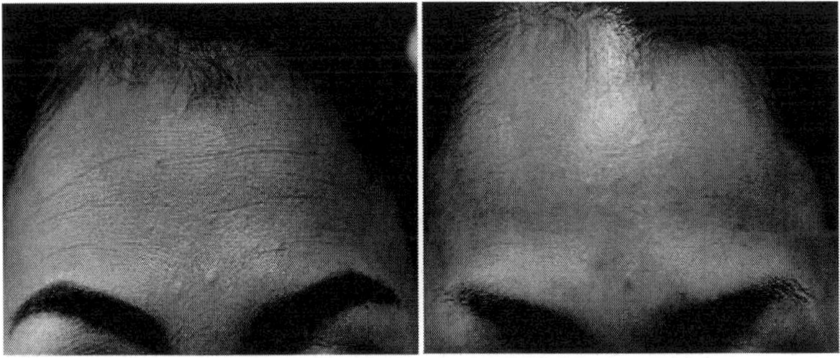

Figure 35.5. A 34-year-old woman, forehead area, 40 months after cell injection, satisfaction score 4.0, replica analysis, 65%. See also Color Insert.

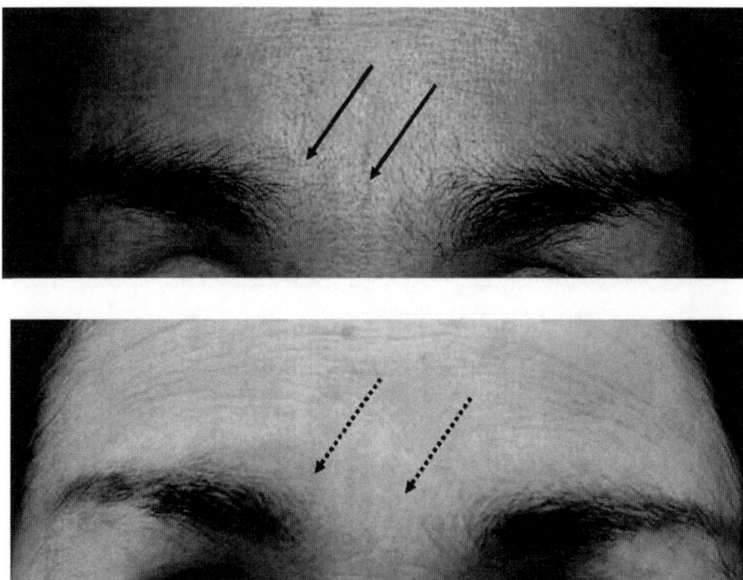

Figure 35.6. A 51-year-old woman, between the eyebrows, 1 year after injection, satisfaction score 2.8, replica analysis, 85%. See also Color Insert.

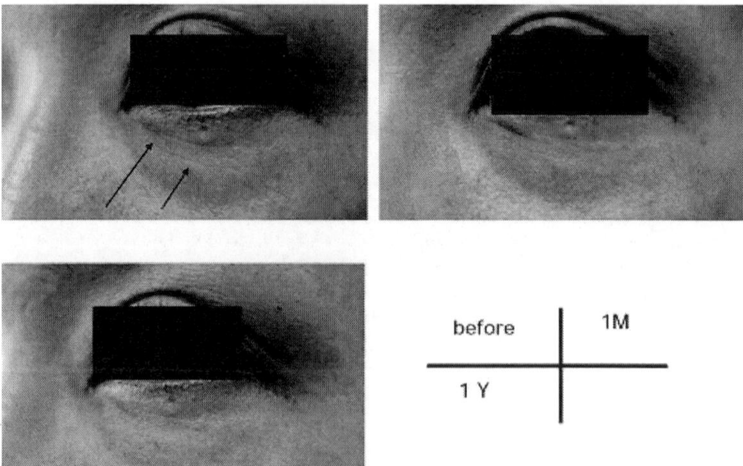

Figure 35.7. A 38-year-old woman, lower eyelid area, 38 months after cell injection, satisfaction score 3.5, replica analysis, 72%. See also Color Insert.

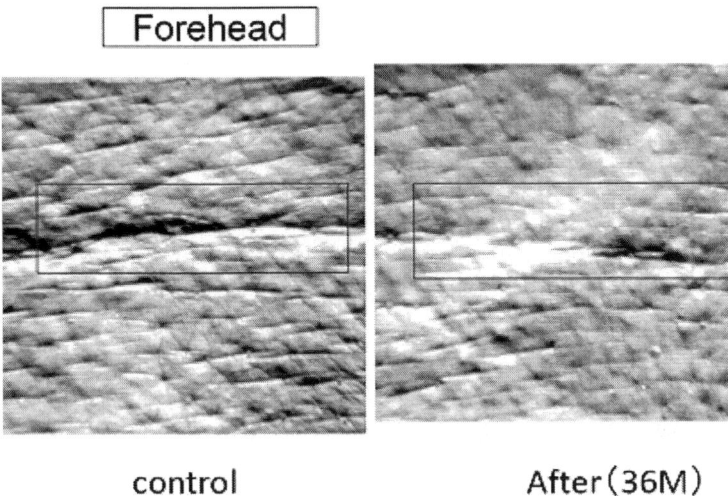

Figure 35.8. Evaluation of wrinkles by replica analysis after gingival fibroblast injection. Control (*left*), 1×10^7 cells' injection (*right*).

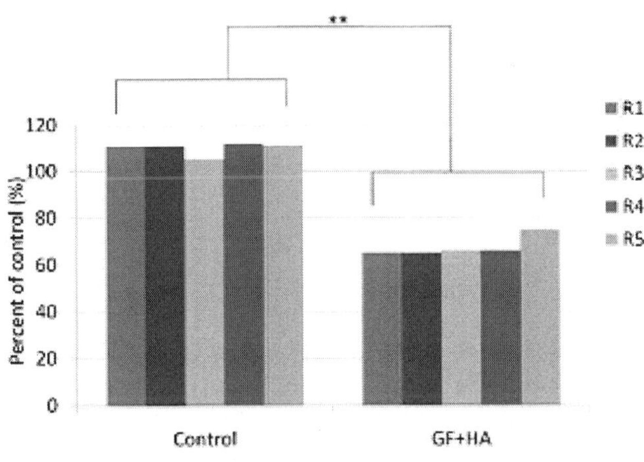

Figure 35.9. Results of replica analysis. See also Color Insert.

35.3.1 *Single-Case Report*

The protocol used in this study was approved by the Ethics Committee of the Nagoya University School of Medicine. Informed consent was obtained from the patient before tissue harvest. A 51-year-old female patient was referred to our hospital for the treatment of periodontal disease and dental implant. The patient suffered from severe periodontitis, and all her remaining upper teeth needed to be extracted. She had a strong gag reflex and could not wear a conventional denture. Accordingly, the patient presented with a typical concave facial profile after teeth extraction. She also exhibited a prominent rhytid (nasolabial fold) and irregularly shaped wrinkles on her upper lip, which became relatively deep over time. To correct masticatory dysfunction, a treatment with dental implant–based prosthesis was planned and eight dental implants were inserted to maxilla. Three months after implant installation, a second surgery was performed to connect an abutment to the fixtures. Since the patient complained not only of masticatory dysfunction but also of aesthetic problems, including deep and irregular wrinkles on the upper lip and a nasolabial groove, an additional treatment for soft-tissue augmentation using autologous fibroblast injection was planned.

At the time of second surgery, a 5 mm gingival biopsy specimen was excised from the buccal side of the left first molar of the patient, who received a local anesthetic. The gingiva was sutured with 4-0 vicryl. The autologous gingival fibroblasts were prepared and examined, as described in the Materials and Methods section, and the cells were prepared for injection at a final concentration of 1×10^7 cells/mL in saline.

One week after placement of the implant prosthesis, three fibroblast injection treatments were performed at two-week intervals. After an infraorbital nerve block was administered, the cell suspension was injected into both nasolabial grooves, as well as into the irregular skin wrinkles on the upper lip (Fig. 35.10). The skin showed faint redness with slight swelling in the area of the injected sites (Fig. 35.11). The sites were cooled with ice, and the redness disappeared within one hour after injection.

Appearance of the patient profile after teeth extraction was a typical concave shape with deep and irregular wrinkles on the upper

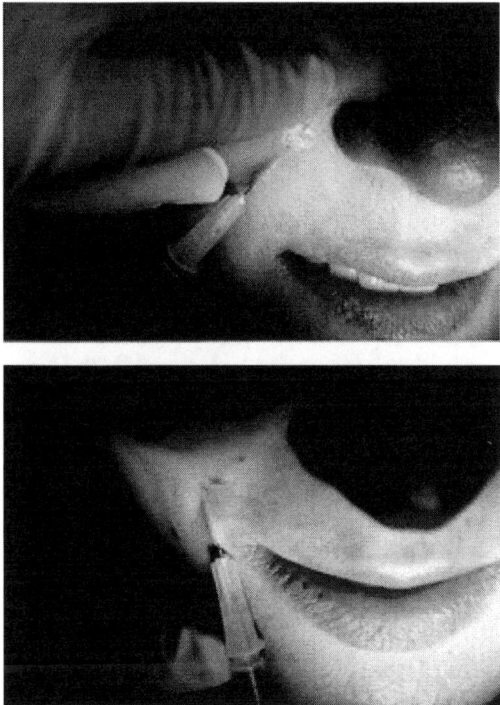

Figure 35.10. Injection of fibroblasts into perioral region. (a) Injection into the nasolabial groove and (b) injection into lip wrinkles. See also Color Insert.

lip, and clear nasolabial grooves were observed. The facial profile improved after the placement of dental implant–based prosthesis, though the aesthetic problem of the upper lip still remained. After serial injections of mucosal fibroblasts and prosthetic treatment, the facial profile, as well as the lip wrinkles, showed significant improvement. Subjective and objective improvement scores were obtained at every follow-up visit. Fair improvement (grade 3) was observed after 3 months and remained the same up to 12 months after treatment. The patient's subjective assessment was grade 2 (satisfied) 3 months after injection, and throughout the 1-year follow-up period. An intradermal injection of autologous gingival fibroblasts, together with the placement of a dental implant prosthesis, significantly improved the perioral skin texture and alleviated wrinkles.

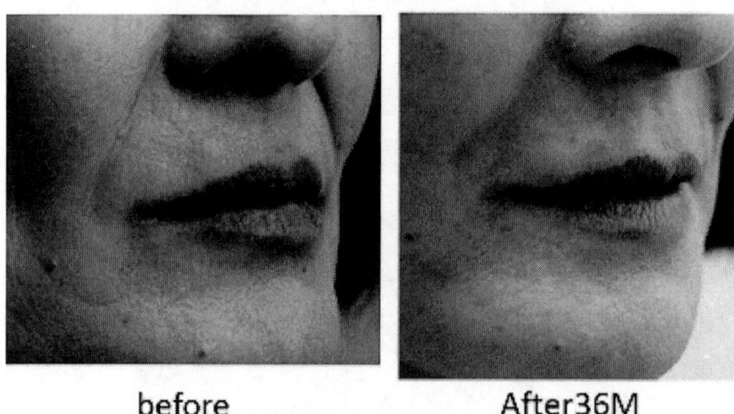

Figure 35.11. Appearance immediately after cell injection (*left*) and 36 months after injection (*right*). See also Color Insert.

35.4 Discussion

35.4.1 *Fibroblast and HA for Skin Rejuvenation*

The art of injection fillers began in the beginning of the twentieth century with a paraffin injection for augmentation after orchiectomy. The quest for more suitable filler materials continued over the years. A refined form of liquid silicone was introduced in the 1960s and was widely used until countless complications arose, and it was essentially banned. At the beginning of the 1980s, bovine collagen was introduced, and it soon became the gold standard against other newly introduced materials. Allergic reactions seen after the application of bovine collagen and its limited longevity have sustained the search for a more ideal filler material.[6]

Autologous fat, an abundant filler material, was popularized by Founrnier in the 1970s.[2] Resorption of 60% has been reported for autologous fat grafts. Newer injectable materials, such as autologous injectable human collagen (Autologen; Collagenesis Inc., Beverly, MA), developed from surgically excised skin, and dermalogen (Collagenesis, Inc.), manufactured from cadaveric skin, have been introduced.[6] Isolagen (Isolagen Technologies, Inc., Paramus, NJ) uses cultured and expanded autogous fibroblasts with saline solution; it is used as a filler material for dermis and subcutaneous tissue that have lost collagen.[6] Skin contains a relatively thick layer of dense, irregular connective tissue in the dermis, called the reticular or deep layer of the dermis. Collagen and other noncellular components (i.e., reticular and elastic fibers and ground substance) of connective tissue are synthesized by fibroblasts. The underlying mechanisms of skin repair by fibroblasts are still not fully understood.

We used gingival fibroblast for winkle treatment instead of dermal fibroblasts. In our method, a gingival rather than a skin biopsy can be easily achieved in a regular dental practice, using local anesthesia. The donor site can be sutured without any tension, heals within one week usually without scar formation, and is undetectable externally. These features of the gingiva demonstrate the advantages of using gingival (mucosal) fibroblasts for treatment. Also, the gingival autologous cell system does not require extensive surgical extraction—a small biopsy is performed on a nonvisible oral cavity, and the fibroblast cells are cultured, multiplying to millions. Unused cells can be stored in liquid nitrogen for new and repeat procedures. The characteristics of gingival fibroblasts are almost identical to those of skin fibroblasts, although there are several known differences. These differences include increased extracellular matrix (ECM) reorganizational ability, cellular migration, and experimental wound repopulation in comparison with skin fibroblasts.[7–9] The profile of growth factors is also known to be different, such that keratinocyte growth factor (KGF) is the growth factor that promotes epithelial cell turnover, and vascular endothelial growth factor (VEGF) is the growth factor that facilitates neovascularization at the injected site. These factors may contribute to the positive effects on the skin in the area of the injected sites.

To enhance the effect of gingival fibroblasts, we routinely use HA as a scaffold with the cells. In recent years, HA derivatives have been used as injectable filler materials. Unlike collagen, HA is identical in all species; therefore, it is biocompatible and does not cause a foreign-body reaction. To maintain the longevity, cross-linked HA preparations are used. HA has the ability to bind water and form hydrated polymers of high viscosity. It has an isovolemic pattern of degradation; as the concentration in the dermis decreases, its ability to bind water increases. Our method was based on the hypothesis that a long-lasting correction can be achieved with HA combined with cultured human fibroblasts compared with HA alone.

Aging is a complex process having pathologic similarities to skin wounds.[10] Fibroblasts play key roles in this process as they interact with keratinocytes, fat cells, and mast cells. They are the source of ECM proteins, glycoproteins, adhesive molecules, and various cytokines.[11] By supplying these molecules and supporting cell-to-cell interactions, fibroblasts contribute to the fibroblast-keratinocyte-endothelium complex that accelerates wound repair and maintains skin integrity and youthfulness. Conventional treatments of skin aging, such as laser and topical regimens, are mostly based on increasing ECM synthesis via fibroblast activation. Wound healing and skin rejuvenation from aging are a complex but orderly process and are orchestrated via cytokines and growth factors. These cytokines do not operate in isolation but rather interact with a variety of cytokines, as well as other regulatory proteins. Five major growth factor families have been studied with regards to the wound-healing process: transforming growth factor (TGF), insulin-like growth factor (IGF), platelet-derived growth factor (PDGF), epithelial growth factor (EGF), and fibroblasts growth factor. Growth factors released from platelets initiate this cascade, and continued secretion of growth factors by inflammatory cells, fibroblasts, and epithelial cells maintains this process.[12] The most significant skin rejuvenation process for aging is collagen remodeling, and the cell that is most important in the production of collagen remodeling is the dermal fibroblast. Collagen synthesis by fibroblasts is stimulated by a variety of growth factors, including IGF, EGF, interleukin-1, and tumor necrosis factor–α, but *in vivo* TGF-β appears to be the most potent stimulator. In summary,[13,14] it has been demonstrated that fibroblasts have an antiwrinkle effect, and it is enhanced by HA.

35.4.2 *Wrinkle Treatment in Dentistry*

Dental implants have revolutionized the field of dentistry. Due to their ability to preserve bone and provide proper emergence profiles, dental implants are usually the ideal treatment option in the aesthetic zone, which focuses on microaesthetics (anything inside the oral cavity). However, in the case of a patient who has a severely absorbed alveolar ridge, a prosthesis lacks a gum section, which may result in insufficient lip support.[15,16,17] This means that a patient with an implant-based prosthesis for the front teeth may develop facial concavity and skin wrinkles as the patient ages. Dentists provide patients with what they are more commonly seeking—a highly aesthetic appearance. There are other aspects of facial aesthetics—the so-called "macroaesthetic" profile that focuses on the overall facial appearance of the patient. Many patients are hyperaware of rhytides, or wrinkles, in the perioral region, which may detract from an overall pleasing appearance. A rhytide may be divided into dynamic and static conditions. Dynamic wrinkles occur during the contraction of muscles and are most prominent in the forehead and eye region of the face. These wrinkles are best treated by a paralyzing drug, such as botulinum toxin. Botulinum toxin (Botox, Allergan, Inc., Irvine, Calif) is a deadly poison produced by the *Clostridium botulinum* bacterium. Clinical applications include blepharospasm, strabismus, and hemipatial spasm. New clinical uses in the fields of cosmetic dermatology include its use for the management of hyperfunctional facial lines, most commonly in the regions of the glabella, periorbital crow's feet, and forehead lines.

Static wrinkles are present even when the face is relaxed. The most prominent static wrinkles are in the inferior third of the face, between the nose and the chin. This region is of primary importance to the dentist and will be addressed in this article.

Deep nasolabial folds, prominent "marionette" lines at the inferior corners of the mouth, or a flat, attractive appearance is a key factor in skin rejuvenation.

Our results indicate that a combination of dental implants and gingival fibroblasts with HA injections may provide an acceptable treatment for patients, especially those desiring better aesthetic results. Our initial experience with the autologous gingival fibroblast injection process indicates that it is probably capable of producing ongoing improvements in perioral lip wrinkles without the hypersensitivity complications and harvesting challenges associated with other treatments.

There were no signs of apoptosis, inflammation, or necrosis in any site, which was expected because the cultured cells were autologous.

The results of our morphologic and morphometric analyses suggest that cross-linked HA and fibroblasts lead to additional synthesis of extracellular components (i.e., fibers and ground substance) in connective tissue. This supports Yoon's results,[18–24] which showed that cross-linked HA combined with cultured human fibroblasts provides a shorter appearance of effect and a longer-lasting effect compared with fibroblasts alone.

The study demonstrates that cultured human gingival fibroblasts combined with HA can be a suitable, biocompatible, and long-lasting material and should be regarded as a new method in dermal renovation in dentistry.

35.5 Conclusion

Tissue engineering has opened a new era for more permanent filler materials by enabling combinations of cell elements with biodegradable polymer scaffolds. In our study, cross-linked HA was used as a biodegradable polymer scaffold for cultured human gingival fibroblasts, which are considered "biomaterials that heal." Living autologous gingival fibroblasts with HA have the potential to provide higher aesthetic results and satisfaction to patients. It opens a new window in the field of oral surgery.

References

1. M. Ueda, *Biochem. Soc.*, **11** (2007).
2. F. D. Sarosh, D. M. Carl, and Hom-Lay wang, *J. Oral Implantol.*, **191** (2007).
3. M. A. Biondi and S. C. Brown, *Phys. Rev.*, **1700** (1949).
4. E. S. Yoon, S. K. Han, and W. K. Kim, *Ann. Plast. Surg.*, **51**, 587 (2003).
5. J. B. A. Mitchell, *Phys. Rep.*, **215** (1990).
6. F. Duranti, G. Salti, B. Bovani, M. Calandra, and M. L. Rosati, *Dermatol. Surg.*, **24**, 1317 (1998).

7. J. Wm. McGowan, R. Caudano, and J. Keyser, *Phys. Rev. Lett.*, **1447** (1976).
8. D. Heinegard, S. Bjornsson, M. Morgelin, and Y. Sommarin, *London: Portland Press*, **113** (1998).
9. F. Duranti, G. Salti, B. Bovani, M. Calandra, and M. L. Rosati, *Dermatol. Surg.*, **24**, 1317 (1998).
10. M. Lee, *Phys. Rev.*, **A16**, 109 (1977).
11. S. Pollack, *J. Cutan. Med. Surg.*, **3(Suppl 4)**, S27 (1999).
12. P. Stephens, K. J. Davies, T. al-Khateeb, J. P. Shepherd, and D. W. Thomas, *J. Dent. Res.*, **75**, 1358 (1996).
13. J. B. A. Mitchell and C. Rebrion-Rowe, *Int. Rev. Phys. Chem.*, **201** (1997).
14. P. Stephens, K. J. Davies, and N. Occleston, *J. Dermatol.*, **144**, 229 (2001).
15. P. Stephens, S. Hiscox, H. Cook, W. G. Jiang, W. Zhiquiang, and D. W. Thomas, *Wound Repair Regen.*, **9**, 34 (2001).
16. R. E. Watson and C. E. Griffiths, *J. Cosmet. Dermatol.*, **4**, 230 (2005).
17. A. Le Pillouer-Prost, *J. Cosmet. Laser Ther.*, **5**, 232 (2003).
18. R. E. Fitzpatrick and E. F. Rostan, *J. Cosmet. Laser Ther.*, **5**, 25 (2003).
19. G. F. Pierce, T. A. Brown, and D. Mustoe, *J. Lab Clin. Med.*, **117**, 373 (1991).
20. A. Tallgren, *Acta Odontol. Scand.*, **28**, 251 (1970).
21. D. A. Atwood, *J. Periodontol.*, **50**, 11 (1979).
22. J. W. Unger, C. W. Ellinger, and J. C. Gunsolley, *J. Prosthet. Dent.*, **67**, 827 (1992).
23. C. Chaussain Miller, D. Septier, and M. Bonnefoix, *Clin. Oral. Investig.*, **6**, 39 (2002).
24. J. H. Chung, S. H. Youn, O. S. Kwon, K. H. Cho, J. I. Youn, and H. C. Eun, *J. Dermatol. Sci.*, **15**, 188 (1997).

Chapter 36

MATRICES FOR ZONAL CARTILAGE TISSUE ENGINEERING

Daisy Irawan, Dietmar Hutmacher,* and Travis Klein
Institute of Health and Biomedical Innovation, Queensland University of Technology 60 Musk Ave, Kelvin Grove, QLD 4059 Australia
*dietmar.hutmacher@qut.edu.au

Articular cartilage is a hydrated tissue with a zonal structure that allows for efficient joint articulation. The variations in cell, matrix, and functional properties from the articular surface are commonly lost in cartilage trauma or disease. Current treatments are not sufficient to recreate these variations and have been limited in their success. Hydrogel matrices offer an opportunity to provide cells with a three-dimensional environment that incorporates features of different zones and create zonal tissue-engineered cartilage. Natural hydrogels based on carbohydrates (including alginate and agarose) and proteins (including collagen and fibrin) allow for the viable encapsulation of chondrocytes and have been used to form zonal constructs with different cellular and extracellular matrix properties. Synthetic and semisynthetic matrices (including several based on poly(ethylene glycol) [PEG] and Extracel) also have been designed and implemented to offer modular control over numerous properties such as cell attachment, modulus, and degradability. The broad spectrum of hydrogel properties

Handbook of Intelligent Scaffolds for Tissue Engineering and Regenerative Medicine
Edited by Gilson Khang

www.panstanford.com

available will allow for optimization of conditions needed for engineering articular cartilage with functional zonal variations, with a final goal of treating a wide range of cartilage conditions.

36.1 Introduction

Normal articular cartilage is an avascular, aneural, alymphatic tissue with a zonal structure that allows for efficient low-friction load bearing in synovial joints. The noncalcified cartilage is commonly described as having three zones with distinct features and functions (Fig. 36.1). The superficial zone is soft in compression, allowing for even stress distribution, and the resident chondrocytes secrete a lubricant molecule, proteoglycan 4 (PRG4), which adheres to the surface and is abundant in the synovial fluid to aid in low-friction articulations.[1,2] The middle zone is stiffer in compression, due largely to the abundance of sulfated glycosaminoglycans. The deep zone integrates with the calcified cartilage and stiff subchondral bone through aligned collagen fibrils to maintain the integrity of the osteochondral junction.

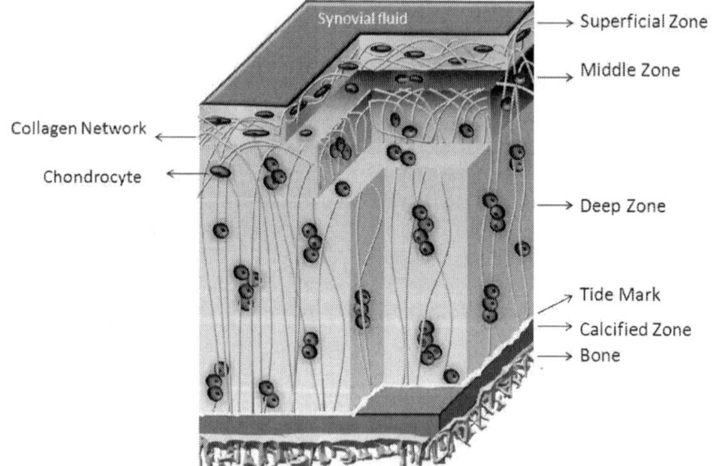

Figure 36.1. Zonal organization of articular cartilage. Differences are evident in cell morphology and function, as well as extracellular matrix content and organization with depth from the articular surface.

Unfortunately, cartilage has a poor natural regeneration capacity,[3] and this structure is commonly lost. Thus, articular cartilage damage, primarily osteoarthritis, causes severe pain, limits the ability to perform basic activities, and causes significant economic loss[4] in developed and developing countries. In the United States, this disease affects 27 million people,[5] and by 2015, it is predicted to lead to the replacement of 2 million knees and hips.[6] In Indonesia, this disease disables approximately 4.4 million people.[7]

Current efforts to restore cartilage function have many limitations in terms of cost, risk, availability of donor, success rate, and performance of the final tissue. Therefore, significant research has focused on developing tissue engineering–based cartilage treatments. Since 1987, autologuous chondrocyte transplantation (ACT),[8] the first cartilage tissue engineering approach applied clinically, has been used to treat full-thickness chondral defects in more than 12,000 patients. This method involves harvesting a small number of chondrocytes through microsurgery, propagating them in the laboratory, and then injecting the cell suspension underneath a sealed periosteal flap, or bilayered collagen membrane.[9] Since it is an autologous technique, this method has low risk of immune response. However, there are risks of chondrocyte leakage and unwanted periosteal response (e.g., hyperthrophy), and the surgical procedures are complex.[10] Further, the cells typically undergo dedifferentiation during expansion, as characterized by changes in morphology and gene expression, and the resultant tissues do not commonly attain the zonal structure that is functionally important in normal articular cartilage.[11]

Hydrogel-/matrix-based cartilage tissue engineering has gained interest, especially because of its capability to provide three-dimensional growth space, while maintaining chondrocyte viability and phenotype. Hydrogels are also interesting because of their high permeability (necessary for nutrient and metabolite transport), low toxicity (because of the small amount of materials being used), range of mechanical properties (Fig. 36.2), and capacity for functionalization.[12,13] There are several different types of hydrogel matrices that have been used for cartilage tissue engineering, that is, carbohydrate based, protein or amino acid based, and

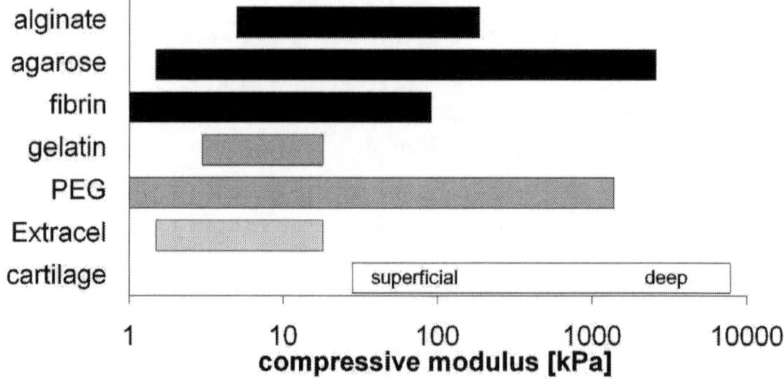

Figure 36.2. Range of compressive moduli in selected papers for common matrices used for cartilage tissue engineering approaches (alginate,[14] agarose,[15] fibrin,[16,17] gelatin,[18] PEG,[19*,20] and Extracel[21*]), as well as depth-dependent moduli in fetal[22] and adult[23] bovine articular cartilage. * indicates an estimate based on three times the shear modulus.

synthetic. Each of these matrices exhibits certain features of the articular cartilage matrix, which consists primarily of water (70%), glycosaminoglycans (10%), and proteins (20%, mainly collagen type II). With this broad range of matrices, possibilities exist for providing specific environments to generate/maintain zonal differences and to construct engineered tissues with zonal functions. In this chapter, we will describe and discuss several popular hydrogel systems used in cartilage tissue engineering and techniques to utilize these materials to generate advanced tissue-engineered cartilage that more closely mimics the zonal structure and function of native cartilage.

36.2 Carbohydrate-Based Matrices

One prominent feature of articular cartilage is the abundance of (sulfated) glycosaminoglycans in the extracellular matrix. These macromolecules, principally chondroitin-6-sulfate, but also chondroitin-4-sulfate, keratan sulfate, dermatan sulfate, and others, provide a fixed charge to the matrix and are responsible for the swelling and compressive properties of the tissue. In cartilage, they are mainly associated with aggrecan and hyaluronic acid to form massive proteoglycan aggregates. Also important is the variation in total sulfated glycosaminoglycan content (low at the surface) and the gradient of keratan sulfate (low at the surface[24]) and dermatan sulfate (high at the surface[25]). The extracellular environment established by these glycosaminoglycans is likely critical for the proper function of the resident chondrocytes. Thus, it is not entirely surprising that hydrogels based on carbohydrates with similar properties, primarily alginate and agarose, are successful for the maintenance of the chondrocyte phenotype and in the re-differentiation of chondrocytes that have de-differentiated in monolayer culture.[12,26,27]

While some concerns have been expressed regarding the safety of these natural polymers,[28] current review indicates that pure alginate is safe for implantation.[37] Also, human clinical trials using ultrapurified alginate-agarose constructs with autologous chondrocytes (Cartipatch) to repair focal cartilage defects have not shown any adverse events and have shown good functional improvement up to two years postoperation.[29]

36.2.1 *Alginate*

Alginate is an ionic linear polysaccharide constructed of (1,4)-linked β-D-mannuronic acid (M) and α-L-guluronic acid (G) monomers (Fig. 36.3a) arranged in block repeats such as MMMMM, GGGGGG, or GMGMGM, based on its source from brown seaweed (*Phaeophyceae*)[30] or bacteria.[31,32] In general alginate that has more G-blocks will be stiffer than one that is mixed (for example, MGMGMG).[33] While seaweed-based alginate is more popular than bacteria-based alginate (especially in the food industry[33]), due to its low cost, bacterial alginates, which can be produced in a more controlled manner, are worth consideration for tissue engineering purposes. Alginate is cross-linked to form a gel in the presence of divalent cations. The gelling structure of calcium-alginate is similar to an eggbox structure (Fig. 36.3b). Mainly, it involves entrapment of cations within two cavities formed by diguluronate so that the chains are "locked" to each other and a hydrogel is formed almost instantaneously.[34] The stiffness of the gel depends on the chain length, monomer sequence, molecular weight distribution, concentration of both the alginate and the cross-linking agent,[33] and the type of cross-linking cations ($Mg^{2+} < Ca^{2+} < Zn^{2+} < Sr^{2+} < Ba^{2+}$).[35]

For tissue engineering applications, cells are typically mixed with the alginate solution and cross-linked in a solution of ~100 mM $CaCl_2$. One interesting aspect of alginate is that it can be easily dissolved using a calcium chelator, such as sodium citrate, and the cells with associated matrix can be recovered and used to form constructs.[36] Alginate has shown excellent ability to retain the typical spherical chondrocyte phenotype and also to serve as a better model system of chondrogenesis than micromass pellets.[38] Alginate also can be functionalized with peptides, such as arginine–glycine–aspartic acid (RGD), to affect cell behavior, through carbodiimide chemistry.[39]

Alginate can be formed into hydrogel disks by mixing with $CaSO_4$ and cross-linking in a mold[40], partially dissolving in sodium citrate, and then combining gels by further cross-linking with $CaCl_2$ to form layered alginate constructs (Fig. 36.3c).[41] The shear stiffness and toughness of the layered constructs generally increased with time and were similar to or greater than single-layer constructs, indicating good integration using this method. Alginate can also be printed into arbitrary geometries using direct free-form fabrication techniques and could thus be used to create zonal constructs.[42]

Alternatively, alginate can be used to preculture zonal chondrocytes, and these cells with an associated matrix can then be used to form zonal constructs that do not contain exogenous matrices (e.g., alginate). These constructs maintained stratification and

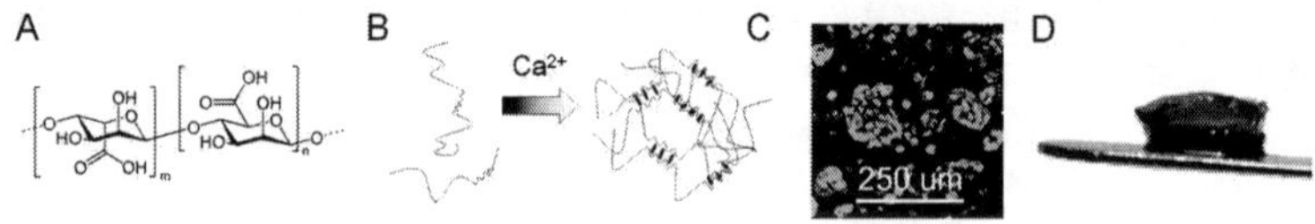

Figure 36.3. (a) Alginate is a linear polysaccharide composed of block repeats of (1,4)-linked β-D-mannuronic acid (M) and α-L-guluronic acid (G) monomers; (b) alginate cross-links in the presence of divalent cations, (c) maintains viable chondrocytes with a round morphology (confocal image at 8 weeks), and (d) can be formed into layered structures (dyed here with food color).

zonal function (PRG4 expression) *in vitro*[43] and can be mechanically induced to form various shapes.[44] However, these constructs rapidly lost their organization when implanted in femoral defects of minipigs,[45] likely due to insufficient mechanical properties upon implantation.

36.2.2 *Agarose*

Agarose is a neutral linear polysaccharide with repeating units of $(1\rightarrow3)$-β-D-galactopyranose-$1(1\rightarrow4)$-3,6-anhydro-α-L-galactoypyranose (Fig. 36.4a)[46] derived from agar, which is extracted from several species of *Rhodophyceae* (red seaweed). Compared with agar, which is widely consumed throughout the world, agarose is much more highly purified and more expensive. Agarose forms a thermopolymerizing physical hydrogel and is commonly used for electrophoresis, microbiology, plant tissue culture, and tissue engineering applications. Agarose powder is not soluble in cold water but must be heated above the gelatinization point, which ranges between 50°C–95°C for commercially available varieties (Sigma Catalog, 2009), to form a solution. After being fully hydrated, agarose will remain fluidic until it is near its setting temperature (Fig. 36.4b). The viscosity of hydrated agarose varies depending on its molecular weight, which can vary between batches. Low-viscosity agarose is easier to mix with a cell suspension and to be molded into specific shapes. Also, it is easier to make gel with a higher concentration of agarose. High-viscosity agarose is prone to entrap air bubbles during heating and may require degassing for several hours at 98°C.[46] Upon cooling to the setting temperature (8°C–42°C for various commercial varieties), the hydrogel forms.

Agarose is widely studied in the cartilage tissue engineering field, as cells remain viable and express chondrotypic genes in a three-dimensional system that allows for mechanical stimulation.[47,48] For cartilage tissue engineering applications, a low setting temperature is beneficial to ensure homogenous mixing with the cell suspension and to reduce cell death due to contact with high temperature. However, agarose with a low setting temperature tends to have low mechanical strength as well. A potential agarose for tissue engineering purpose, from *Gracilaria dura*, has a setting temperature of

≤ 35°C and mechanical strength of 2,200 g/cm^2 for 1% agarose.[49] This is relatively higher than agarose types VII and VIIA, which have been more frequently studied for cartilage tissue engineering.[47,50] While the unmodified agarose has shown potential, it is also possible to incorporate peptides such as RGD,[51] or growth factors[52] to modulate cell behavior.

Agarose has been used to construct tissue-engineered cartilage with zonal differences. When immature bovine chondrocytes were encapsulated in sequential layers of 2% and 3% agarose (copolymerized), initial differences in mechanical properties between layers were evident, although these differences diminished in cultures with 20% fetal bovine serum (FBS) (Fig. 36.4c).[53] Dynamic compression during bioreactor cultures (Fig. 36.4d) resulted in mechanical properties that switched from stiffer in the 3% agarose layer to stiffer in the 2% agarose layer after four weeks, due to increased mechanical stimulation and matrix accumulation in the 2% layer.[54] In a further study,[55] chondrocytes from different zones were encapsulated in sequential agarose layers of the same (2% in both layers) or different (3% and 2%) agarose concentrations to generate constructs with differences in cellular and/or mechanical properties. After 42 days of static culture, the glycosaminoglycan (GAG) and collagen content, as well as the compressive modulus, of the superficial layer were lower than the middle layer of the construct. Importantly, chondrocytes in the superficial layer expressed

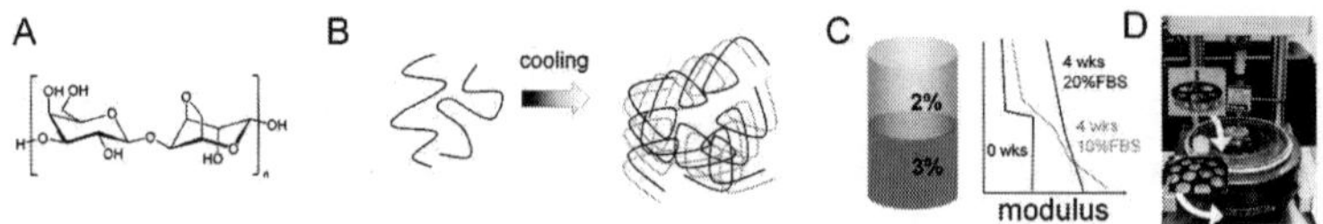

Figure 36.4. Agarose is a linear polysaccharide with (a) repeating units; (b) above the gelatinization point, hydrated agarose is a solution with random coils that forms a network of double helices upon cooling to its setting temperature; (c) Ng *et al.* formed bilayered agarose/chondrocyte constructs that varied in compressive modulus with depth; initial variations were affected by 4 weeks in culture, with a homogenization seen between layers in cultures with 20% FBS;[53] (d) compression bioreactors were used to stimulate layered constructs, resulting in increased stiffness in the 2% layer after 4 weeks.[54] See also Color Insert.

PRG4, while those in the middle layer did not, indicating biomimetic stratification.

36.3 Protein-Based Matrices

The other key component of the articular cartilage extracellular matrix (in addition to the proteoglycans) is the proteins, which make up the collagen network. This network is made primarily of collagen type II but also contains minor components of collagen type XI, IX, XI, and others. This cross-linked protein network resists the swelling pressure of the proteoglycans and provides the tensile properties.[56] The variations in collagen organization are important to the pronounced cartilage anisotropy and depth dependent properties. Protein-based hydrogel matrices can mimic parts of the natural cartilage matrix, including natural sites for cell attachment, enzymatic degradation, and incorporation into existing networks.

36.3.1 *Collagen or Gelatin*

As collagen is the most abundant protein in cartilage, it is not surprising that collagen gels have been investigated for cartilage tissue engineering purposes. Collagen gels can be formed by pH neutralization. Cells can be suspended in the collagen solution and incorporated into the hydrogel. However, collagen gels typically have the issue of cell-induced shrinkage due to the contractile forces of the cells and the weak mechanical properties of the gel. One method to overcome the problem of shrinkage is through incorporation of short collagen fragments in the gel.[57] Another interesting approach to improving the mechanical properties is through nonenzymatic glycation, which commonly occurs in the native cartilage matrix, in the solution before gel formation.[58] This treatment resulted in better glycosaminoglycan retention and modest increase in compressive modulus.

The type of collagen used can have a major influence on chondrogenesis, with collagen II hydrogels leading to better mesenchymal stem cell (MSC) chondrogenesis than collagen I hydrogels.[59] Regardless, collagen gels have shown good *in vivo* and clinical results when combined with progenitor cells and chondrocytes.[60,61] Importantly, for clinical applications, the telopeptides have been removed to reduce the chance of an adverse immune response.[62–64] Collagen gels have not been used for zonal cartilage studies, perhaps due to difficulty in controlling the dimensions of the gels, but would be interesting to study, especially if collagen fiber orientation can be controlled.

36.3.2 *Fibrin*

Among many different types of matrices, fibrin is the only one which is inherently produced by a human body for rapid tissue regeneration. During wound healing, fibrin clots are formed by thrombin-induced cleavage of fibrinogen and Factor XIII, resulting in Factor XIIIa (transglutaminase)-mediated cross-linking of the fibrinopeptides (Fig. 36.5). The fibrin clot is a porous structure, with 99.75% of void space, facilitating efficient nutrient exchange.[65] Cells bind to fibrin (which contains RGD sites), proliferate, and elaborate extracellular matrix, including collagen I.[66,67] Naturally, as cells invade and proliferate the fibrin matrix, they release proteases to

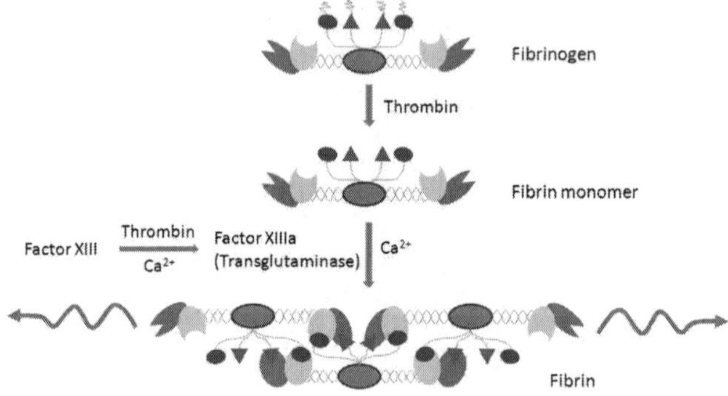

Figure 36.5. The polymerization mechanism of fibrin (adapted from Ref. 68). Thrombin cleaves peptide fragments from fibrinogen to form fibrin monomers. These monomers can form covalent bonds between monomers through a Factor XIIIa-mediated reaction in the presence of calcium to form a fibrous network that is useful for tissue engineering.

digest the fibrin while producing their own extracellular matrix to form the regenerated tissue.[67]

Fibrin hydrogels based on the natural fibrin clot are soft poroelastic materials[69] with mechanical properties that can be tailored. Human fibrin glue is used clinically as a two-component sealant (72–110 mg/ml fibrinogen, 500 IU/ml thrombin, purified from human plasma [Baxter]), making it attractive for regulatory reasons. Autologous fibrin can be alternatively prepared from blood, which contains ~5 mg/ml fibrinogen.[70] Finally, individual components of this system can be prepared from purified sources to better control the properties of the gel.

The mechanical strength of fibrin gels increases as the fibrinogen concentration increases; however, an overly dense network prevents cell proliferation.[71] The concentrations of fibrinogen and thrombin also influence the thickness and homogeneity of the fibrin strands. According to Zhao *et al.*, the optimum fibrinogen concentration is 20 mg/ml, with up to 40 mM $CaCl_2$ added to facilitate fibrin cross-linking. Meanwhile, thrombin concentration can be varied between 0.5–5 U/ml, depending on the required mechanical properties and reaction kinetics.[68] Generally, higher thrombin concentration results in thinner fibers, higher water content, and faster gel formation. At thrombin concentration >100 U/ml, fibrin will be formed within seconds.[68] This condition is not favorable for homogenous encapsulation of cells in a matrix of controlled shape, as is desired for *in vitro* cartilage tissue engineering.

One potential drawback of fibrin for use in cartilage tissue engineering is its rapid degradation. Fibrin made of 20 mg/mL of fibrinogen almost completely degrades in phosphate buffered saline (PBS) and only retains ~70% of its weight after six days in Dulbecco's Modified Eagle's Medium (DMEM).[68] Further, the presence of FBS, which inherently contains plasmin, in the media significantly increases the degradation rate.[68] To slow down the degradation process, a protease inhibitor such as aprotinin, tranexamic acid, or ε-aminocaproic acid, may be added to the culture medium.[16,66] However, according to Zhao *et al.*,[68] aprotinin only slightly reduces the degradation rate and is not completely effective during the first few days. Enhanced stability can be attained by producing fibrin gel with finer and denser fibrin strands. Promisingly, a systematic experiment with bovine fibrinogen showed that a stable fibrin gel can be produced by using 50 mg/mL fibrinogen, 2.5 U/mL thrombin, 20 mM $CaCl_2$, and pH 7.[16] This type of gel remains stable for 12 months, even without addition of fibrinolysis inhibitors.

Fibrin is potentially useful as an injectable matrix for cartilage tissue engineering, and it promotes collagen II synthesis by chondrocytes.[72] Fibrin can be used to encapsulate MSCs[73] or chondrocytes[16] by simply suspending the cells in fibrinogen and mixing this suspension with thrombin of equal volume in a dilution buffer containing $CaCl_2$. While not yet studied as a zonal construct, the adhesive nature of fibrin simplifies the construction and adhesion of multiple layers with different mechanical properties or cell phenotypes. An alternative approach would be to use fibrin microbeads (50–200 μm) as a three-dimensional environment for expansion[70] of chondrocytes or MSCs, prior to layering into a zonal construct.

36.4 Synthetic and Semisynthetic Matrices

The natural matrices listed above contain biological ligands that can either interact (e.g., fibrin) or minimally interact (e.g., alginate) with the cells and natural extracellular matrix produced by these cells. While the functionality of these hydrogels has been improved by incorporation of cell adhesion molecules, etc., they are still limited by batch-to-batch variations, lack of control over degradation, etc. To overcome the limitations of the natural hydrogels, several synthetic or semisynthetic modular hydrogel systems have been devised. These systems allow for fine control over the hydrogel properties (mechanical properties, cell and protein adhesion, degradability, etc.[74]), and their modularity opens the door for the development of constructs with systematic variations in properties known to vary in articular cartilage.

36.4.1 *Poly(Ethylene Glycol)*

The most frequently used base material for synthetic hydrogel systems in tissue engineering is PEG. This molecule is highly hydrophilic and resistant to protein and cell adhesion. These characteristics are an important starting point for fundamental studies of cell matrix interactions and cell behavior in modified materials because the data are not confounded by high nonspecific protein adhesion which influences the cell behavior. Further, several different chemistries can be used to incorporate polymerize and add functionalities into the hydrogel.

One system that has been studied extensively uses PEG end-functionalized with diacrylate (PEG-DA) and/or methacrylate (PEG-DM), which, when combined with a photoinitiator (1-[4-(2-Hydroxyethoxy)-phenyl]-2-hydroxy-2-methyl-1-propane-1 -one, also known as Irgacure 2959, CIBA Specialty Chemicals), can be photopolymerized. This cross-linking mechanism is advantageous in that it could be used as an injectable, *in situ* cross-linked treatment for cartilage defects.[75]

Using this system, the mechanical properties can be tuned over a large range; the compressive modulus was varied from 60 kPa to 500 kPa in photopolymerized gels with PEG-DM and a hydrolytically degradable poly(lactic acid) (PLA)-PEG-PLA dimethacrylate precursors by altering the precursor percentage from 10% to 20%.[76] The degradation of these constructs by hydrolysis was found to best match the extracellular matrix accumulation by encapsulated chondrocytes at an intermediate macromer content of 15%. These systems can incorporate biological molecules by first functionalizing the biomolecule with (meth)acrylate and reacting with PEG precursors. This technique has been used to generate gels functionalized with RGD peptides for cell adhesion, methacrylated chondroitin sulfate[77] covalently bonded with the PEG network, and physically entrapped collagen type I and hyaluronic acid.[78,79]

Interestingly, these functionalized gels showed differential effects on zonal chondrocytes, with superficial chondrocytes expressing PRG4 at the highest rate and deep chondrocytes expressing aggrecan and collagen II at the highest rate in PEG-CS gels.[79] Similar hydrogels have been used to form zonal constructs by sequential photopolymerization of PEG-DA gels containing chondrocytes from different zones (Fig. 36.6).[80] The immature bovine chondrocytes remained viable, remained within their respective layers, and retained biosynthetic differences over three weeks in culture.

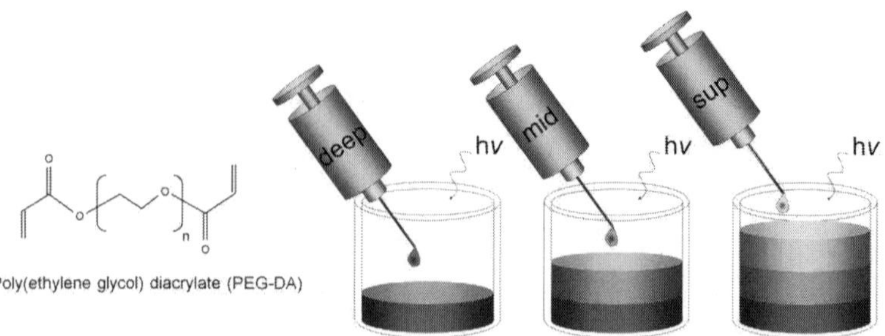

Figure 36.6. PEG-DA, when combined in solution with a photoinitiator, can be photopolymerized using ultraviolet light to form a hydrogel. Kim *et al.* encapsulated chondrocytes from different zones in the PEG-DA solution and sequentially photopolymerized the layers to form zonal constructs with biomimetic cellular organization.[80]

Interestingly, in another study, bilayered constructs attained higher compressive and shear moduli than homogeneous constructs, indicating potential synergistic effects of the co-culture conditions.[81]

PEG hydrogels using different polymerization approaches are also available and have potential in these areas. One such gel system forms by disulfinde bonding between multiarm PEG-vinyl sulfone and cysteines of synthetic peptides or proteins.[82] Gels form within 10 minutes after combining the precursor solutions using this system, and it is amenable to injectable delivery. The system is quite versatile, given the variety of peptides that can be designed and incorporated into the gel. For example, a matrix metalloproteinase (MMP)-sensitive sequence was used to make the hydrogel specifically degradable (instead of hydrolytically) and showed effects on chondrocyte proliferation, collagen network formation, and gene expression.[82]

Multiple functionalities (such as cell adhesivity) can be incorporated by adding different cysteine-containing peptides.[83] The mechanical properties can be controlled by the macromer content and number of arms on the starting PEG, and clearly tailoring the degradability has an influence on the mechanical properties of the final tissue. While these gels have not been used to form zonal cartilage constructs, differences observed using chondrocytes in this system indicate the capability of forming layers with different cell clustering and collagen organization, akin to native cartilage.

One further PEG-based system uses a natural cross-linking method found in wound healing so as to form synthetic, biomimetic hydrogels. This hydrogel is based on the natural fibrin clotting reaction (Fig. 36.4), invoking transglutaminase-catalyzed cross-linking of multiarm PEG molecules end-functionalized with glutamine- and lysine-containing synthetic peptides.[84,85] Functionalities such as MMP degradability can be incorporated into these peptides or as additional peptides containing lysine or glutamine. The gel is formed in minutes from liquid precursors and like the preceding gels is amenable to injectable delivery. Mechanical properties can be altered by changing the percentage of PEG, and varying these properties have been shown to influence chondrocyte secretion of glycosaminoglycans, as well as resultant mechanical properties.[86] These gels can be formed as separate disks and combined into a layered construct using a small amount of hydrogel precursor; although this layering technique works for at least three months *in vitro*, the interfacial strength may not be sufficient to maintain integrity at or following implantation into a cartilage defect.

36.4.2 *Extracel*

Semisynthetic matrices, which modularly combine natural carbohydrates and/or proteins with synthetic polymers, also have great potential for zonal cartilage tissue engineering. One such matrix, commercially available as Extracel, uses modified PEG (PEG-SS-DA) to cross-link thiol-modified carboxymethyl hyaluronic acid (CMHA-S) and thiol-modified gelatin (gelatin-DTPH).[87–89] Gels form in approximately three minutes with this system.

Degradation *in vitro* can be rapidly performed by adding up to 25 mM cysteine (such as *N*-acetyl cysteine) to break the disulfide bonds in the degradable PEG linker. This allows for easy and efficient recovery of cells and could thus be useful in cell expansion in a biomimetic environment.[90] *In vitro* and *in vivo*, these materials are also subject to enzymatic degradation, due to the incorporation of biological substrates, by MMPs and hyaluronidase.

Extracel gels are soft, relative to the articular cartilage matrix, with the shear modulus reaching up to only 5.5 kPa, in one instance.[21] Thus, while they should not be expected to directly replace articular cartilage with full load-bearing capacity, they may be expected to develop properties of the cartilage matrix with time in culture. Importantly, hyaluronic acid in the gel should be able to interact with link protein and aggrecan to form large proteoglycan aggregates that are integral to the natural cartilage matrix. While they have not been utilized thus far for zonal constructs, the ability to independently change the contents of hyaluronic acid and gelatin could provide specific properties of the different zones. These hydrogels mixed with cells can be sequentially polymerized to form layer structures, as has been shown in multilayered tubular constructs.[91]

36.5 Conclusions and Outlook

Hydrogel matrices come in a variety of basic types (e.g., carbohydrate based, protein based, synthetic), offering three-dimensional environments for cell encapsulation, growth, and differentiation. In addition to the variety of base materials, each type can be functionalized with biological molecules to give a range of responses from chondrogenic cells. These hydrogel matrices have been used in many *in vitro* and *in vivo* studies and have been used to a limited extent in clinical applications. The broad palette of hydrogel properties available for cartilage tissue engineering is particular interesting, given the natural variations in biochemical and biomechanical properties that are functionally important in articular cartilage. We now have building blocks to generate hydrogel/cell constructs *in vitro* with cellular, mechanical, and biochemical differences and are just beginning to utilize these systems for detailed studies on chondrocyte and progenitor cell chondrogenesis in zonal tissue-engineered cartilage. The literature to date indicates that biosynthetic and biomechanical differences can be maintained *in vitro* using zonal cells and differing matrix properties,[81,92] although not under all conditions.[54]

The materials highlighted here are representative of the most promising and most commonly used in cartilage tissue engineering. Many other hydrogel materials also have potential for cartilage tissue engineering, including chitosan,[93] hyaluronic acid,[94] peptide-amphiphiles,[95] poly(*N*-isopropylacrylamide), and poly(acrylic acid).[96] At this point, it is unclear whether an "ideal" matrix will be found; there may, however, be specific material characteristics that are "better" for specific regions of the cartilage and for each cell type. Combination of multiple hydrogels, such as fibrin-alginate[97] or gelatin-chitosan,[98] may be useful to tailor properties such as degradation and produce a protein-carbohydrate network that resembles the collagen-glycosaminoglycan network of native articular cartilage to a greater extent than single component hydrogels. Control over the organization of the matrix to replicate the organization of the extracellular matrix in articular cartilage, such as the orientation of the collagen fibers, remains to be addressed. Biomechanical stimulation through specifically designed bioreactors, or *in vivo* stimulation, will likely be needed to achieve such organization.

Work remains to be done to characterize cell responses to the different hydrogels and in the formation of constructs with zonal differences. Studies to date have focused primarily on multiple layering techniques to either form the second layer atop a prepolymerized layer or combine two prepolymerized gels using an intermediate layer as an adhesive. For the latter approach, natural "glues" such as fibrin or novel chondroitin sulfate-based adhesives[99] have great potential to form a strong interlayer bond.

More sophisticated approaches currently being investigated for producing zonal constructs include hydrogel printing[100] using techniques adapted from technologies used to produce polymer scaffolds for biomedical applications.[101] Incorporation of releasable growth factor gradients, by embedding microcarriers in a hydrogel, is another approach recently taken to generating constructs with zonal differences.[102] If typical zonal differences are attainable and maintainable *in vitro*, zonal tissue-engineered cartilage could be useful as an advanced model of cartilage behavior under various stimulatory and disease conditions. Finally, the effectiveness of prescribing zonal differences *in vitro* after implantation *in vivo* will be key to determining whether or not such zonal matrices will be beneficial in a clinical setting.

Acknowledgments

This work was supported by the Australian Research Council through the Discovery Projects scheme and the Australian Government through the AusAID scheme.

References

1. G. D. Jay, J. R. Torres, D. K. Rhee, H. J. Helminen, M. M. Hytinnen, C. J. Cha, K. Elsaid, K. S. Kim, Y. Cui, and M. L. Warman, *Arthritis Rheum.*, **56**, 3662 (2007).
2. T. A. Schmidt, N. S. Gastelum, Q. T. Nguyen, B. L. Schumacher, and R. L. Sah, *Arthritis Rheum.*, **56**, 882 (2007).
3. C. Chung and J. A. Burdick, *Adv. Drug Del. Rev.*, **60**, 243 (2008).
4. E. Loza, J. M. Lopez-Gomez, L. Abasolo, J. Maese, L. Carmona, and E. Batlle-Gualda, *Arthritis Rheum.*, **61**, 158 (2009).
5. R. C. Lawrence, D. T. Felson, C. G. Helmick, L. M. Arnold, H. Choi, R. A. Deyo, S. Gabriel, R. Hirsch, M. C. Hochberg, G. G. Hunder, J. M. Jordan, J. N. Katz, H. M. Kremers, and F. Wolfe, *Arthritis Rheum.*, **58**, 26 (2008).
6. S. Kim, *Arthritis Rheum.*, **59**, 481 (2008).
7. J. Soeroso, L. F. Dans, M. L. Amarillo, G. H. Santoso, and H. Kalim, *APLAR J. Rheumatol.*, **8**, 106 (2005).
8. M. Brittberg, A. Lindahl, A. Nilsson, C. Ohlsson, O. Isaksson, and L. Peterson, *N. Engl. J. Med.*, **331**, 889 (1994).
9. W. Bartlett, J. A. Skinner, C. R. Gooding, R. W. Carrington, A. M. Flanagan, T. W. Briggs, and G. Bentley, *J. Bone Joint Surg. Br.*, **87**, 640 (2005).
10. S. Marlovits, P. Zeller, P. Singer, C. Resinger, and V. Vécsei, *Eur. J. Radiol.*, **57**, 24 (2006).
11. T. J. Klein, J. Malda, R. L. Sah, and D. W. Hutmacher, *Tissue Eng. Part B Rev.*, (2009).
12. P. D. Benya and J. D. Shaffer, *Cell*, **30**, 215 (1982).
13. C. Domm, M. Schunke, K. Christesen, and B. Kurz, *Osteoarthr. Cartil.*, **10**, 13 (2002).
14. C. K. Kuo and P. X. Ma, *Biomaterials*, **22**, 511 (2001).
15. V. Normand, D. L. Lootens, E. Amici, K. P. Plucknett, and P. Aymard, *Biomacromolecules*, **1**, 730 (2000).

16. D. Eyrich, F. Brandl, B. Appel, H. Wiese, G. Maier, M. Wenzel, R. Staudenmaier, A. Goepferich, and T. Blunk, *Biomaterials*, **28**, 55 (2007).
17. G. M. Peretti, M. A. Randolph, V. Zaporojan, L. J. Bonassar, J. W. Xu, J. C. Fellers, and M. J. Yaremchuk, *Ann. Plast. Surg.*, **46**, 533 (2001).
18. H. A. Awad, M. Q. Wickham, H. A. Leddy, J. M. Gimble, and F. Guilak, *Biomaterials*, **25**, 3211 (2004).
19. S. C. Rizzi and J. A. Hubbell, *Biomacromolecules*, **6**, 1226 (2005).
20. S. J. Bryant and K. S. Anseth, *J. Biomed. Mater. Res.*, **59**, 63 (2002).
21. K. Ghosh, X. Z. Shu, R. Mou, J. Lombardi, G. D. Prestwich, M. H. Rafailovich, and R. A. Clark, *Biomacromolecules*, **6**, 2857 (2005).
22. T. J. Klein, M. Chaudhry, W. C. Bae, and R. L. Sah, *J. Biomech.*, **40**, 182 (2007).
23. C. C. Wang, N. O. Chahine, C. T. Hung, and G. A. Ateshian, *J. Biomech.*, **36**, 339 (2003).
24. M. Zanetti, A. Ratcliffe, and F. M. Watt, *J. Cell Biol.*, **101**, 53 (1985).
25. A. R. Poole, C. Webber, I. Pidoux, H. Choi, and L. C. Rosenberg, *J. Histochem. Cytochem.*, **34**, 619 (1986).
26. J. Bonaventure, N. Kadhom, L. Cohen-Solal, K. H. Ng, J. Bourguignon, C. Lasselin, and P. Freisinger, *Exp. Cell Res.*, **212**, 97 (1994).
27. F. Lemare, N. Steimberg, C. Le Griel, S. Demignot, and M. Adolphe, *J. Cell Physiol.*, **176**, 303 (1998).
28. H. Zimmermann, D. Zimmermann, R. Reuss, P. J. Feilen, B. Manz, A. Katsen, M. Weber, F. R. Ihmig, F. Ehrhart, P. Gessner, M. Behringer, A. Steinbach, L. H. Wegner, V. L. Sukhorukov, J. A. Vasquez, S. Schneider, M. M. Weber, F. Volke, R. Wolf, and U. Zimmermann, *J. Mater. Sci. Mater. Med.*, **16**, 491 (2005).
29. T. A. Selmi, P. Verdonk, P. Chambat, F. Dubrana, J. F. Potel, L. Barnouin, and P. Neyret, *J. Bone Joint Surg. Br.*, **90**, 597 (2008).
30. K. Clare, in *Industrial Gum: Polysaccharides and Their Derivatives*, Eds. R. L. Whistler and J. N. BeMiller (Academic Press, California, 1993).
31. U. Remminghorst and B. Rehm, *Biotechnol. Lett.*, **28**, 1701 (2006).
32. W. Sabra, A. P. Zeng, and W. D. Deckwer, *Appl. Microbiol. Biotechnol.*, **56**, 315 (2001).
33. A. D. Augst, J. K. Hyun, and D. J. Mooney, *Macromol. Biosci.*, **6**, 623 (2006).

34. Y. Fang, S. Al-Assaf, G. O. Phillips, K. Nishinari, T. Funami, P. A. Williams, and L. Li, *J. Phys. Chem. B*, **111**, 2456 (2007).
35. C. P. Reis, R. J. Neufeld, S. Vilela, A. n. J. Ribeiro, and F. Veiga, *J. Microencapsul.*, **23**, 245 (2006).
36. K. Masuda, R. L. Sah, M. J. Hejna, and E. J. Thonar, *J. Orthop. Res.*, **21**, 139 (2003).
37. H. Park and K. Lee, in *Natural-Based Polymers for Biomedical* Applications, Eds. R. Reis *et al.* (Woodhead Publishing, Cambridge, 2008), pp. 515–529.
38. I. H. Yang, S. H. Kim, Y. H. Kim, H. J. Sun, S. J. Kim, and J. W. Lee, *Yonsei Med. J.*, **45**, 891 (2004).
39. J. A. Rowley, G. Madlambayan, and D. J. Mooney, *Biomaterials*, **20**, 45 (1999).
40. S. C. Chang, J. A. Rowley, G. Tobias, N. G. Genes, A. K. Roy, D. J. Mooney, C. A. Vacanti, and L. J. Bonassar, *J. Biomed. Mater. Res.*, **55**, 503 (2001).
41. C. S. Lee, J. P. Gleghorn, N. Won Choi, M. Cabodi, A. D. Stroock, and L. J. Bonassar, *Biomaterials*, **28**, 2987 (2007).
42. D. L. Cohen, E. Malone, H. Lipson, and L. J. Bonassar, *Tissue Eng.*, **12**, 1325 (2006).
43. T. J. Klein, B. L. Schumacher, T. A. Schmidt, K. W. Li, M. S. Voegtline, K. Masuda, E. J. Thonar, and R. L. Sah, *Osteoarthr. Cartil.*, **11**, 595 (2003).
44. E. Han, W. C. Bae, N. D. Hsieh-Bonassera, V. W. Wong, B. L. Schumacher, S. Gortz, K. Masuda, W. D. Bugbee, and R. L. Sah, *Clin. Orthop. Relat. Res.*, **466**, 1912 (2008).
45. K. Chawla, T. J. Klein, B. L. Schumacher, K. D. Jadin, B. H. Shah, K. Nakagawa, V. W. Wong, A. C. Chen, K. Masuda, and R. L. Sah, *Tissue Eng.*, **13**, 1525 (2007).
46. V. Normand, D. L. Lootens, E. Amici, K. P. Plucknett, and P. Aymard, *Biomacromolecules*, **1**, 730 (2000).
47. M. M. Knight, T. Toyoda, D. A. Lee, and D. L. Bader, *J. Biomech.*, **39**, 1547 (2006).
48. E. M. Darling and K. A. Athanasiou, *Tissue Eng.*, **11**, 395 (2005).
49. R. Meena, A. K. Siddhanta, K. Prasad, B. K. Ramavat, K. Eswaran, S. Thiruppathi, M. Ganesan, V. A. Mantri, and P. V. S. Rao, *Carbohydr. Polym.*, **69**, 179 (2007).
50. A. Guaccio, C. Borselli, O. Oliviero, and P. A. Netti, *Biomaterials*, **29**, 1484 (2008).

51. Y. Luo, M. S. Shoichet, *Nat. Mater.*, **3**, 249 (2004).

52. Y. Aizawa, N. Leipzig, T. Zahir, and M. Shoichet, *Biomaterials*, **29**, 4676 (2008).

53. K. W. Ng, C. C. Wang, R. L. Mauck, T. N. Kelly, N. O. Chahine, K. D. Costa, G. A. Ateshian, and C. T. Hung, *J. Orthop. Res.*, **23**, 134 (2005).

54. K. W. Ng, R. L. Mauck, L. Y. Statman, E. Y. Lin, G. A. Ateshian, and C. T. Hung, *Biorheology*, **43**, 497 (2006).

55. K. W. Ng, G. A. Ateshian, and C. T. Hung, *Tissue Eng. Part A*, **15**, 1 (2009).

56. W. Wilson, C. C. van Donkelaar, B. van Rietbergen, and R. Huiskes, *J. Biomech.*, **38**, 1195 (2005).

57. E. Gentleman, E. A. Nauman, K. C. Dee, and G. A. Livesay, *Tissue Eng.*, **10**, 421 (2004).

58. R. Roy, A. L. Boskey, and L. J. Bonassar, *J. Orthop. Res.*, **26**, 1434 (2008).

59. D. Bosnakovski, M. Mizuno, G. Kim, S. Takagi, and M. Okumura, T. Fujinaga, *Biotechnol. Bioeng.*, **93**, 1152 (2006).

60. S. Wakitani, T. Goto, S. J. Pineda, R. G. Young, J. M. Mansour, A. I. Caplan, and V. M. Goldberg, *J. Bone Joint Surg.*, **76-A**, 579 (1994).

61. S. Wakitani, K. Imoto, T. Yamamoto, M. Saito, N. Murata, and M. Yoneda, *Osteoarthr. Cartil.*, **10**, 199 (2002).

62. M. Ochi, Y. Uchio, K. Kawasaki, S. Wakitani, and J. Iwasa, *J. Bone Joint Surg. Br.*, **84**, 571 (2002).

63. H. Yamaoka, H. Asato, T. Ogasawara, S. Nishizawa, T. Takahashi, T. Nakatsuka, I. Koshima, K. Nakamura, H. Kawaguchi, U. I. Chung, T. Takato, and K. Hoshi, *J. Biomed. Mater. Res. A*, **78**, 1 (2006).

64. U. Noth, L. Rackwitz, A. Heymer, M. Weber, B. Baumann, A. Steinert, N. Schutze, F. Jakob, and J. Eulert, *J. Biomed. Mater. Res. A*, **83**, 626 (2007).

65. J. W. Weisel, *Biophys. Chem.*, **112**, 267 (2004).

66. E. D. Grassl, T. R. Oegema, and R. T. Tranquillo, *J. Biomed. Mater. Res.*, **60**, 607 (2002).

67. B. Tawil, H. Duong, and B. Wu, in *Natural-Based Polymers for Biomedical Applications*, Eds. R. L. Reis *et al.* (Woodhead Publishing, Cambridge, 2008).

68. H. Zhao, L. Ma, J. Zhou, Z. Mao, C. Gao, and J. Shen, *Biomed. Mater.*, **31** (2008).

69. J. Noailly, H. Van Oosterwyck, W. Wilson, T. M. Quinn, and K. Ito, *J. Biomech.*, **41**, 3265 (2008).

70. R. Gorodetsky, *Expert Opin. Biol. Ther.*, **8**, 1831 (2008).
71. W. Ho, B. Tawil, J. C. Y. Dunn, and B. M. Wu, *Tissue Eng.*, **12**, 1587 (2006).
72. G. M. Peretti, J.-W. Xu, L. J. Bonassar, C. H. Kirchhoff, M. J. Yaremchuk, and M. A. Randolph, *Tissue Eng.*, **12**, 1151 (2006).
73. S. H. Park, S. R. Park, S. I. Chung, K. S. Pai, B.-H. Min, *Artif. Organs*, **29**, 838 (2005).
74. A. S. Gobin and J. L. West, *Faseb J*, **16**, 751 (2002).
75. J. Elisseeff, K. Anseth, D. Sims, W. McIntosh, M. Randolph, and R. Langer, *Proc. Natl. Acad. Sci. U S A*, **96**, 3104 (1999).
76. S. J. Bryant, R. J. Bender, K. L. Durand, and K. S. Anseth, *Biotechnol. Bioeng.*, **86**, 747 (2004).
77. S. J. Bryant, J. A. Arthur, and K. S. Anseth, *Acta Biomater.*, **1**, 243 (2005).
78. N. S. Hwang, S. Varghese, Z. Zhang, and J. Elisseeff, *Tissue Eng.*, **12**, 2695 (2006).
79. N. S. Hwang, S. Varghese, H. J. Lee, P. Theprungsirikul, A. Canver, B. Sharma, and J. Elisseeff, *FEBS Lett.*, **581**, 4172 (2007).
80. T. K. Kim, B. Sharma, C. G. Williams, M. A. Ruffner, A. Malik, E. G. McFarland, and J. H. Elisseeff, *Osteoarthr. Cartil.*, **11**, 653 (2003).
81. B. Sharma, C. G. Williams, T. K. Kim, D. Sun, A. Malik, M. Khan, K. Leong, and J. H. Elisseeff, *Tissue Eng.*, **13**, 405 (2007).
82. Y. Park, M. P. Lutolf, J. A. Hubbell, E. B. Hunziker, and M. Wong, *Tissue Eng.*, **10**, 515 (2004).
83. M. P. Lutolf, J. L. Lauer-Fields, H. G. Schmoekel, A. T. Metters, F. E. Weber, G. B. Fields, and J. A. Hubbell, *Proc. Natl. Acad. Sci. U S A*, **100**, 5413 (2003).
84. M. Ehrbar, S. C. Rizzi, R. Hlushchuk, V. Djonov, A. H. Zisch, J. A. Hubbell, F. E. Weber, and M. P. Lutolf, *Biomaterials*, **28**, 3856 (2007).
85. M. Ehrbar, S. C. Rizzi, R. G. Schoenmakers, B. S. Miguel, J. A. Hubbell, F. E. Weber, and M. P. Lutolf, *Biomacromolecules*, **8**, 3000 (2007).
86. T. J. Klein, S. C. Rizzi, J. C. Reichert, N. Georgi, J. Malda, W. Schuurman, R. W. Crawford, and D. W. Hutmacher, *Macromol. Biosci.*, **9**, 1049 (2009).
87. G. D. Prestwich, X. Z. Shu, Y. Liu, S. Cai, J. F. Walsh, C. W. Hughes, S. Ahmad, K. R. Kirker, B. Yu, R. R. Orlandi, A. H. Park, S. L. Thibeault, S. Duflo, and M. E. Smith, *Adv. Exp. Med. Biol.*, **585**, 125 (2006).
88. X. Z. Shu, S. Ahmad, Y. Liu, and G. D. Prestwich, *J. Biomed. Mater. Res. A*, **79**, 902 (2006).

89. X. Z. Shu, Y. Liu, Y. Luo, M. C. Roberts, and G. D. Prestwich, *Biomacromolecules*, **3**, 1304 (2002).
90. J. Zhang, A. Skardal, and G. D. Prestwich, *Biomaterials*, **29**, 4521 (2008).
91. V. Mironov, V. Kasyanov, X. Zheng Shu, C. Eisenberg, L. Eisenberg, S. Gonda, T. Trusk, R. R. Markwald, and G. D. Prestwich, *Biomaterials*, **26**, 7628 (2005).
92. K. W. Ng, G. A. Ateshian, C. T. Hung, *Tissue Eng. Part A*, **15**, 2315 (2009).
93. J. Berger, M. Reist, J. M. Mayer, O. Felt, and R. Gurny, *Eur. J. Pharm. Biopharm.*, **57**, 35 (2004).
94. D. L. Nettles, T. P. Vail, M. T. Morgan, M. W. Grinstaff, and L. A. Setton, *Ann. Biomed. Eng.*, **32**, 391 (2004).
95. K. L. Niece, C. Czeisler, V. Sahni, V. Tysseling-Mattiace, E. T. Pashuck, J. A. Kessler, and S. I. Stupp, *Biomaterials*, **29**, 4501 (2008).
96. A. Au, J. Ha, A. Polotsky, K. Krzyminski, A. Gutowska, D. S. Hungerford, and C. G. Frondoza, *J. Biomed. Mater. Res. A*, **67**, 1310 (2003).
97. C. Perka, R. S. Spitzer, K. Lindenhayn, M. Sittinger, and O. Schultz, *J. Biomed. Mater. Res.*, **49**, 305 (2000).
98. M. Risbud, J. Ringe, R. Bhonde, and M. Sittinger, *Cell Transplant.*, **10**, 755 (2001).
99. D. A. Wang, S. Varghese, B. Sharma, I. Strehin, S. Fermanian, J. Gorham, D. H. Fairbrother, B. Cascio, and J. H. Elisseeff, *Nat. Mater.*, **6**, 385 (2007).
100. N. E. Fedorovich, J. R. De Wijn, A. J. Verbout, J. Alblas, and W. J. Dhert, *Tissue Eng. Part A*, **14**, 127 (2008).
101. D. W. Hutmacher, M. Sittinger, and M. V. Risbud, *Trends Biotechnol.*, **22**, 354 (2004).
102. X. Wang, E. Wenk, X. Zhang, L. Meinel, G. Vunjak-Novakovic, and D. L. Kaplan, *J. Control. Release.*, **134**, 81 (2009).

Chapter 37

COLLAGEN-BASED SCAFFOLD FOR BONE TISSUE REGENERATION

Fu-Zhai Cui,* Zong-Gang Chen, and Xue Xia

Department of Materials Science and Engineering, Tsinghua University, 100084, Beijng, P.R. China

*cuifz@mail.tsinghua.edu.cn

This paper presents a review of collagen-based scaffolds for bone tissue regeneration by self-assembling. On the basis of discussing the component and hierarchical structure in natural bone tissue, the biomemitic fabrication of collagen-based bone tissue engineering scaffolds using self-assembled collagen mineralization and their current applications and development in bone tissue regeneration are described. Some basic questions of biomemitic fabrication concerning the composition, microstructure, crystallography, and organic–inorganic phase relationships have been reviewed. This kind of bone graft composite has already shown great promise and success in clinical applications, on account of its compositional and structural similarity to natural bone. It is suggested that future work in this should focus on both basic theoretical aspects, as well as the development of applications.

Handbook of Intelligent Scaffolds for Tissue Engineering and Regenerative Medicine

Edited by Gilson Khang

www.panstanford.com

37.1 Introduction

Bone is composed of osteocytes surrounded by a rigid, highly calcified organic matrix. Although bone has good mechanical properties, it often undergoes damage or suffers defects resulting from tumor reconstruction, chronic infection, or traumatic bone loss. There is a great need therefore for bone graft materials for various bone applications. Autologous bone grafting is widely accepted as the gold standard for the treatment of bone defects.[1] However, the supply of autografts is limited and donor site morbidity is also a concern along with the required prolonged operation times.[2] The alternative of allogeneic bone has potential risks of disease transmission and infection.[3] In order to avoid the problems associated with either autologous or allogeneic bone grafts, many synthetic bone graft materials, such as titanium alloy, ceramics, and polymers, have been used as bone-substitute materials during the past decades.[4] However, these substitute matrices have specific disadvantages in biocompatibility, degradability, osteogenic capability, and histochemical responses by the host tissue. It became necessary to find a promising alternative of autogenous bone for grafting indications. Bone tissue engineering is a recently developed, exciting approach for the repair of bone defects.[5,6] Recently, studies of three-dimensional scaffold materials have become a crucial element of bone tissue engineering. The three-dimensional scaffold materials were designed to mimic one or more of the bone-forming components of autografts, in order to facilitate the growth of vasculature into the material, and provide an ideal environment for bone regeneration.[7–9] Therefore, as a surrogate of the natural bone, an ideal bone tissue engineering scaffold should simulate both the components and the structure of natural bone. Previous studies have revealed that natural bone is a representative example with a typical hierarchically ordered organization. Bone tissues are mainly constructed from nanosized hydroxyapatite (HA) crystals and a collagen framework in which the crystals form, resulting in a highly complex but ordered mineral–organic composite material by self-assembling.[10] In this paper we have used the concept of self-assembling defined by Whitesides and Grzybowski,[11] that is, self-assembling is the autonomous organization of components into patterns or structures without human intervention. It is considered that self-assembling processes are common throughout nature and technology. Therefore we begin this paper with a discussion about the components and hierarchical structure in natural bone tissue by self-assembling and then give a description of work on biomemitic fabrication of bone tissue engineering scaffold materials using self-assembled collagen mineralization. Finally, we finish by reviewing the current applications and development of such scaffold materials.

37.2 Compositional and Structural Characteristics of Natural Bone

To produce an excellent bone substitute material, it is necessary to emphasize several compositional and structural characteristics of natural bone.

37.2.1 *Composition of the Natural Bone Matrix*

Bone is a complex tissue that is composed of some organic matrices upon which HA crystals are deposited. Ninety percent of the organic materials in the bone matrix are collagen I, with the remaining ten percent consisting of various glycoproteins and proteoglycans. Interspersed throughout the organic matrix are crystals of HA, which give the bone matrix its stiffness.[12]

37.2.2 *Hierarchical Structure of the Natural Bone Matrix by Self-Assembly*

Landis has put forward the microstructure model of the bone matrix, which is generally accepted.[13] According to the model, the interactions between the collagen molecules result in the characteristic quarter repeat of 67 nm and in a complex cross-striation banding, which results in assembly of the collagen triple helices into microfibrils, forming the supramolecular structures.[14–16] The packing of collagen molecules is that five tropocollagen molecules align longitudinally with an overlap of approximately a quarter of the molecular length to form a microfibril. This is the so-called quarter stagger, which is combined with the gap between successive macromolecules. It is responsible for the typical 67 nm periodical cross-striation patterns as observed by transmission electron microscopy (TEM), atomic force microscope (AFM), and X-ray diffraction (XRD) investigations.[15,17,18] Within collagen, deposition of HA mineral is drawn as occurring initially in the gap zones of the aggregated molecular arrays, but HA mineral can also be deposited independently in the overlap zones. The crystals are aligned with their respective crystallographic c-axes generally parallel to one another and to the long axis of the molecules and fibrils in which they are located.[13,19] In addition to crystal growth within the channels and gaps, nucleation and progressive mineralization on the surface of the collagen fibrils may also occur.[20] The microfibrils are then assembled into collage fibrils that may vary in thickness from 10 to 300 nm. These are further combined, oriented, and formed ordered structures with particular morphologies for tissues.

To better understand the complex bone architecture, several hierarchical models have been proposed. Weiner and Wagner have identified seven discrete levels of hierarchical organization in bone. We have given the seven hierarchical levels of organization of the zebra fish skeleton bone (Fig. 37.1).[21,22] Mann has presented a similar structural hierarchy containing six levels.[23]

As a consequence, natural bone is of a complex nanofibril system with an intricate hierarchical structure of mineralized collagen I by self-assembling. The assembly includes an orderly deposition of HA minerals within a type I collagen matrix. The crystallographic c-axis of the HA is oriented parallel to the longitudinal axis of the collagen fibril.[13,24,25]

37.3 Biomimetic Fabrication with Self-Assembled Collagen Mineralization

Natural bone is a three-dimensional composite with an intricate hierarchical structure of mineralized collagen fiber. Investigation and simulation of the hierarchical nanofibril structure in nature can offer some new ideas in the design and fabrication of new functional

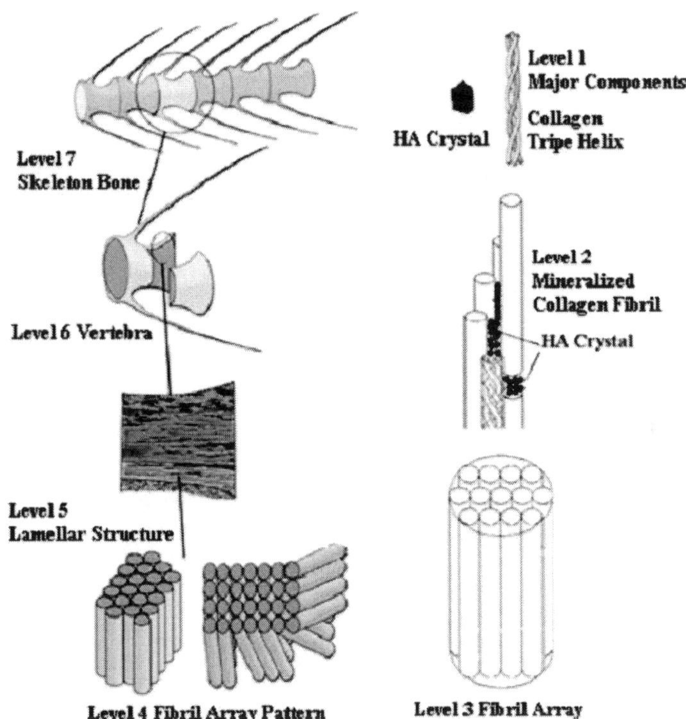

Figure 37.1. The seven hierarchical levels of organization of the zebra fish skeleton bone.[21,22] Level 1: HA crystals and part of a collagen fibril with the triple helix structure. Level 2: Mineralized collagen fibrils. Level 3: The array of mineralized collagen fibrils. Level 4: Two fibril array patterns of organization as found in the zebra fish skeleton bone. Level 5: The lamellar structure in one vertebra. Level 6: A vertebra. Level 7: Skeleton bone.

materials, such as tissue engineering scaffold materials and biomimetic engineering materials. The key steps in the composite synthesis of mineralized collagen are control of nucleation and growth of calcium phosphate minerals by collagen matrix and assembly of the nanofibril of mineralized collagen in aqueous media.[26–28]

37.3.1 *Mineralization Mechanism of Hydroxyapatite Crystals on Collagen Fibers*

The study of biomineralization is not only important to gain an understanding of how mineral-rich tissues are created *in vivo* but also because it is a great source of inspiration for the design of advanced materials.[29–34]

Previous investigations on the mineralization of HA crystals on collagen fibers have suggested that the binding of calcium ions on the negatively charged carboxylate groups of collagen is one of the key factors for the first-step nucleation of HA crystals.[35] Other partially negatively charged functional groups on collagen have previously been considered to be possible nucleation sites for HA crystals. It has been demonstrated, however, that carboxyl groups (–COOH) are the major nucleation sites for collagen fibrils. In a neutral solution, the carboxyl groups are ionized to COO-, which favors chelation of calcium ions. The carboxyl groups on the outside of the collagen threefold spiral are also one kind of site for collagen mineralization.

Zhang *et al.* have reported some investigations on the mineralization of calcium phosphate crystals on collagen fibers by using Fourier transform infrared spectroscopy (FTIR).[36] For the first time, they successfully showed that in the initial stage of collagen mineralization, collagen prefers to chelate calcium ions in solution and that these chelated ions subsequently form nucleation sites for calcium phosphate crystals. Moreover, in addition to their importance for the interfacial interaction between calcium ions and carboxyl groups, the carbonyl groups on the surface of collagen molecules also act as nucleation sites of calcium phosphate crystals.

In the recent research, it is reported that the conformation of collagen changes rapidly during the amorphous/crystalline conversion and crystal ripening. The results show the characteristics of cooperative interaction between calcium phosphate and collagen molecules, that is, collagen molecules control the nucleation of calcium phosphate through providing nucleation sites, while collagen molecules adopt different conformations to adapt the formation of biominerals.[37]

A recent review by Palmer *et al.*[38] also comprehensively presents protein-based mineralization, nonprotein biopolymers mineralization, synthetic polymers mineralization, and organoapatites mineralization. In addition, they still described several supramolecular systems that have been developed in their own laboratories, both as biomimetic mineralization models and as matrices for bone regeneration. Liao *et al.*[39] have investigated the fabrication of nano-HA/collagen/osteonectin composites for bone graft applications. The results demonstrated that ostonectin played a role in allowing crystal growth to proceed in an orderly way in bone. All these showed that noncollagenous proteins in natural bone also played a role in regulating mineralization of HA crystals in bone, which can help us to better understand the biomineralization mechanism of HA crystals.

In summary, preliminary studies have shown that collagen is an important structural agent to direct HA mineralization by proteins or polymers with bonelike organization. For these reasons, the form of bone can be an "organic matrix-mediated" mineralization process.

37.3.2 *Assembly of the Nano-Fibril of Mineralized Collagen*

Scientists have attempted to mimic the collagen-mineralization process *in vitro* in order to achieve a better understanding of the organizational structure in naturally occurring tissues in which the major organic matrix is collagen. Numerous studies about mineralized collagen have been reported, as described in the following paragraphs.[22]

Rhee[40] investigated the nucleation of Ca–P crystals through chemical interaction with collagen by soaking a collagen membrane in a supersaturated simulated body fluid solution. By combining the collagen fibril assembly and the calcium phosphate formation in a one-process step, Bradt *et al.*[41] obtained a homogeneously mineralized collagen gel consisting of a three-dimensional network of collagen fibrils covered with calcium phosphate. The initial precipitate, along with the fibril assembly, was amorphous calcium phosphate. This was then transformed into a crystalline apatite-like phase. The addition of polyaspartate to the reaction mixture was found to improve the attachment between the collagen fibrils and the calcium phosphate crystals. Goissis *et al.*[42] reported both *in vitro* and *in vivo* studies of biomimetic mineralization of charged collagen matrices. Their results showed that the calcium phosphate deposited in close resemblance to assembly of collagen fibrils *in vivo*. Additionally, the *in vitro* results suggested that amide hydrolysis may have introduced into the matrix signs for the controlled mineralization of collagen fiber. Amide hydrolysis was found from TEM investigations to occur near the overlap and gap zones.

Pederson and Ruberti[43] reported a strategy for exploiting temperature driven self-assembly of collagen and thermally triggered liposome mineralization to form a mineralized collagen composite.

Zhang *et al.* have also synthetically prepared nano-fibrils of mineralized collagen as a self-assembly model system, with the objective of evaluating the possibility of synthesizing materials with hierarchical structures similar to those found in nature (Fig. 37.2).[44] Firstly, they prepared solutions containing collagen, calcium and phosphate ions following a processing route described by Bradt *et al.*[45] and then changed either pH or temperature to induce the formation of collagen fibrils. TEM investigations of unstained samples at low magnification revealed that the composites formed consist of an intertwined assembly of collagen fibrils bundles more than 1 μm long (Fig. 37.2a). Each collagen fibril is surrounded by a

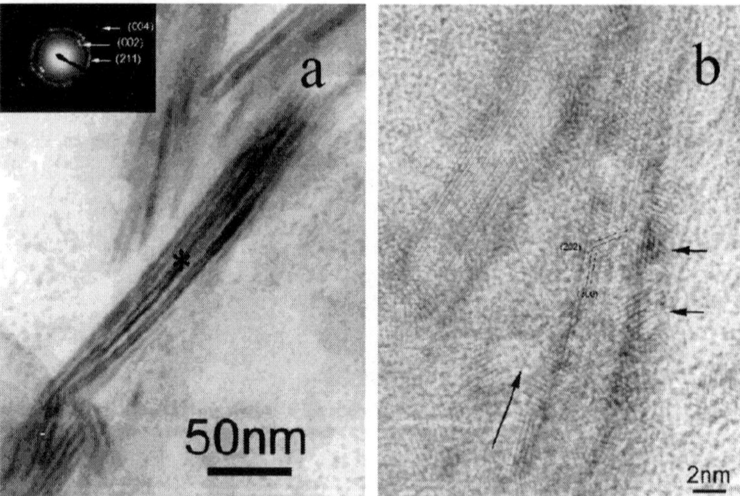

Figure 37.2. TEM image and HRTEM image of mineralized collagen fibrils.[44] (a) TEM image: The insert is a selected area electron diffraction pattern of the mineralized collagen fibrils. (b) HRTEM image: The long arrow indicates the longitudinal direction of the collagen fibrils. The two short arrows indicate two HA nanocrystals. *Abbreviation*: HRTEM, high-resolution transmission electron microscopy.

layer of HA nanocrystals grown on the surface of the collagen fibrils. Each mineralized bundle of collagen fibrils is 5.5–6.9 nm in diameter, which is thicker than the self-assembled collagen fibrils, implying that the self-assembled collagen nanofibrils act as the template for HA precipitation. The thickness of the HA crystal layer on the surface of the collagen fibrils was calculated to be 0.75–1.45 nm. Additionally, in order to discern the relative orientation of the HA crystals with respect to collagen fibrils, the mineralized collagen fibrils were analyzed by electron diffraction. The results finally demonstrate the preferential alignment of the HA crystallographic c-axis with the collagen fibril longitudinal axis. The HRTEM analysis of the parallel-aligned mineralized collagen fibrils revealed that a crystal lattice is seen not only on the side area of the collagen fibrils but also in the middle area and that the electron density on the surface of the collagen fibrils is higher than in the interior area (Fig. 37.2b). These findings indicate that HA crystals grown on the surface of the collagen surround the fibrils. Scanning electron microscopy (SEM) analysis showed that the mineralized collagen is a three-dimensional network of collagen fibrils on which crystals of calcium phosphate have settled. The mineralized collagen fibrils are aligned parallel to their longitudinal axes. The calcium phosphate nanocrystals can be seen to aggregate on the surface of collagen fibrils. This observation is in agreement with the TEM results (Fig. 37.3).

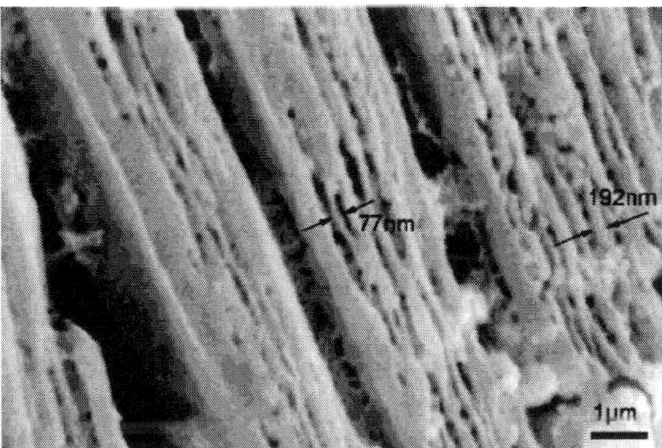

Figure 37.3. Scanning electron micrograph of mineralized collagen fibrils.[44]

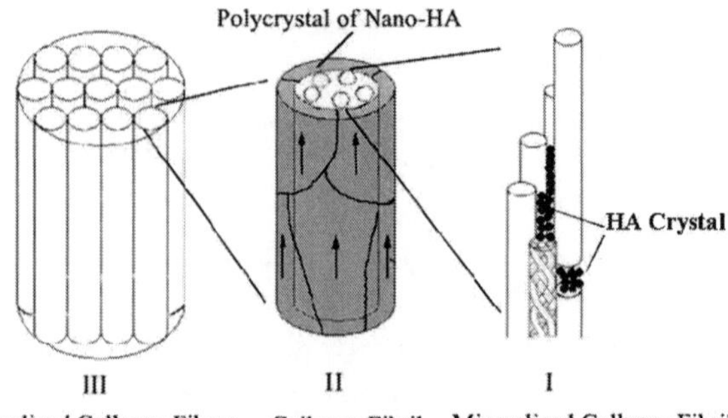

Figure 37.4. Schematic diagram of the hierarchical structure of self-assembled HA-collagen composite.[44] (I) The first level of the hierarchy is the organization of collagen molecules with the nano-HA crystals formed initially in the gap zones between the collagen fibrils. (II) The second level of the hierarchy is the organization of collagen fibrils with respect to HA crystals. The HA crystals are sheetlike and grow on the surface of these fibrils in such a way that their c-axes are oriented along the longitudinal axes of the fibrils, as indicated by the arrows in the figure. (III) The third level of the hierarchy is the organization of the mineralized collagen fibrils. The mineralized collagen fibrils align parallel to each other to form mineralized collagen fibers.

A schematic diagram of self-assembled HA/collagen composites comprising multiple levels of hierarchical organization has been depicted in this study. As shown in Fig. 37.4, the lowest level of this hierarchy is the organization of collagen molecules with some particulates of nano-HA crystals. The collagen fibrils are formed by self-assembly of collagen triple helices, and the HA crystals are formed initially in the gap zones between the collagens. Considering that the diameter of the collagen molecule is 1.5 nm, the diameter of the collagen fibrils in the five-stranded packing model should be approximately 4.0 nm, which is in agreement with TEM observations of self-assembled collagen fibrils. The second level of the hierarchy is the organization of these collagen fibrils with respect to the growth of the HA crystals. HA crystals grow on the surface of these fibrils in such a way that their c-axes are oriented along the longitudinal axes of the fibrils and surround the fibrils. This arrangement implies that the nucleation and growth of HA crystals are not random but rather are controlled by the fibrils themselves. The third level of the hierarchy is the organization of these mineralized collagen fibrils, which are aligned parallel to one another to form mineralized collagen fibers.

The nanoscale organization of the composites resembles that of HA crystals in mineralized tissue in which the HA crystals also align their c-axes with the longitudinal axes of the collagen fibrils. Such alignment is the most impressive characteristic of bone minerals. Development of novel self-assembled structures should therefore improve our understanding of collagen-mediated mineralization in other calcified tissues and point the way to the development of new functional materials for biomimetic engineering.[22] Moreover, these fundamental studies provide the basic theoretical support for the fabrication of HA/collagen composites and their application in bone regeneration,[46–48] both of which will be introduced in the next sections.

37.4 Synthesis and Application of Collagen-Based Scaffolds in Bone Tissue Regeneration

The principal concept of tissue engineering is to isolate a small biopsy of specific cells from a patient, to culture these cells on a scaffold material, and then transplant the cell-engineered scaffold into the defect site of the patient's body in order to direct new tissue formation into the scaffold, which will biodegrade over time. So an ideal bone scaffold should simulate the structure and composition of natural bone to promote early mineralization and support new bone formation while at the same time allowing for replacement by new bone. On the basis of this principle, the nano-HA/collagen-based composites synthesized by biomimetic strategy show great promise for bone reconstructive or regenerative surgery because of their compositional and structural analogy to natural bone.[49–51]

37.4.1 *Synthesis of Nano-HA/Collagen-Based Scaffolds*

By biomimetic strategy, many researchers have prepared HA-collagen composites showing great promise in clinical application because of their compositional and partly structural analogy to natural bone.[34,41,48,52–54]

The nano-HA, collagen-based composites can be prepared by directly mixing the nano-HA and collagen. But the weak interactions between HA and collagen make them a no-cooperation effect *in vivo* for bone defects repair. A recent review by Supova[55] comprehensively describes the problem of HA dispersion in polymer matrices. Another method involves coprecipitation of collagen fibrils and nano-HA spontaneously and is a promising route for achieving the same hierarchical structure in synthetic materials as in bone. It is also believed that the manufacture using biomimetic self-assembly of nanocomposite grafts with certain features of natural bone either compositionally or structurally may replicate the natural bone growth process.[34,40,48,51] This method was firstly reported in 1995.[51] Collagen is first dispersed in an acidic solution, and then the coprecipitation is induced by increasing the pH or by the addition of mixing agents. The direct nucleation of HA nanocrystals onto collagen fibers has been performed starting from an aqueous suspension of calcium solution and phosphate solution together with collagen solution of pH 9–10 at room temperature. A solution condition of high ion concentration and high pH value was employed. The obtained composite product was freeze-dried at 40 °C. The nanocrystals of HA elegantly aligned with their c-axis preferentially oriented along the collagen fibers. As this process mimics mineralization of natural bone to a certain extent, it is suggested that the HA/collagen nanocomposite can be used for bone repair and regeneration. Based on the previous research, a manufacturing route for assembly of a nano-HA/collagen composite has been designed in our lab.[56,57] Moreover, three-dimensional scaffolds have also been prepared with the addition a small amount of a poly(lactic acid) (PLA) polymer to improve the mechanical properties of the scaffold.[56] These materials are good examples that mimic natural bone to some extent at the supermolecular structure level. Their biological evaluation suggests that such bone-resembling composites are readily incorporated into the bone metabolism in a way similar to bone remodeling, instead of acting as permanent implants.[47,56]

37.4.2 *Applications and Development of Nano-HA/Collagen Scaffolds for Bone Tissue Engineering*

We have already developed osteogenic cells/nano-HA/collagen-based scaffold composites, and *in vitro* cellular functions of these materials have demonstrated that the scaffolds support cellular growth and related functions well and lead to new bone formation.[47] More recently, a three-dimensional bone-resembling scaffold prepared using nano-HA/collagen/PLA/osteoblasts composites has been developed.[58] This system supports cellular adhesion, proliferation, and migration. Interestingly, cells grown on this material were observed to penetrate deep into the matrix, to a depth of about 200–400 mm. Within a short period, probably due to both the compositional and structural similarity to natural bone, this material can provide a promising cell/scaffold for bone tissue engineering.[22]

Scaffolds loaded with growth factor have been shown to regulate cellular growth and related functions in a better way.[59] Growth factors can be effectively delivered to a bone defect through nanocomposites, and *in vivo* the efficacy of such methods has been evaluated.[56] The *in vivo* performance of the nanocomposite with recombinant human bone morphogenetic protein-2 (rhBMP-2) is better than that of the nanocomposite without rhBMP-2, which implies its efficacy as a good bone graft. The efficacy in bone regeneration of such a scaffold combined with bone marrow mesenchymal stem cells was also evaluated.[60] It was found that the implanted scaffold could enhance and accelerate bone formation in segmental defects in rabbits. These experimental results indicate that an effective bone graft should consist of an osteoconductive matrix in conjunction with osteogenic cells and osteoinductive growth factors with a structure, composition and physicochemical, mechanical, and biological features analogous to natural bone.[22]

Bone tissue engineering using nano-HA/collagen composites is still, however, in its infancy, although our knowledge in this area is expanding. Although *in vitro* and *in vivo* evidence strongly supports the effective use of the biomimetic nanocomposites as bone graft materials, further *in vivo* performance and clinical studies are needed to confirm their promise as effective graft materials for bone regeneration.

In our recent research, the nano-HA/collagen/PLA composites have been used to repair a 20 mm segmental defect in the radius of the beagle dog. The X-ray and histological observations demonstrated that the 20 mm segmental defect was completely healed within 24 weeks after surgery, as shown (Fig. 37.5). The results of the animal studies confirm the excellent biocompatibility, osteocompatibility, and bioactivity of the composites with surrounding tissues, and the implants were observed to stimulate the formation of new bone.

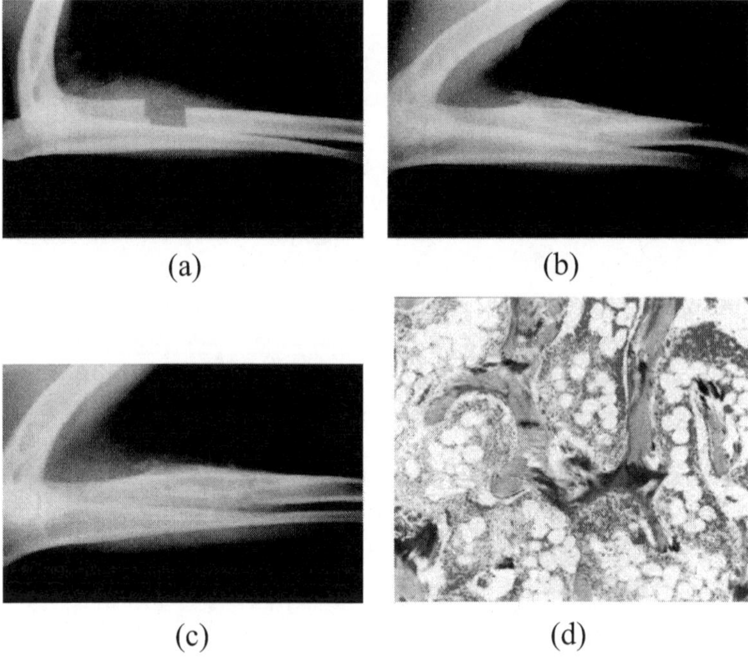

Figure 37.5. The X-ray (a, b, and c) and histological (d) observations of the nano-HA/collagen/PLA composites to repair a 20 mm segmental defect in a dog. (a) After surgery; (b) implant at 12 weeks; new bone and callus grew up, and cortical bones were connected completely; (c) implant at 24 weeks; bone callus decreased with the complete repair by new bone; (d) implant at 24 weeks; histology H&E staining; bone marrow appeared in the new bone tissue. *Abbreviation*: H&E, hematoxylin and eosin.

Nano-HA/collagen based scaffolds have now been successfully used for thousands of cases in clinical applications, including all the various types of hard-tissue repair. Yu *et al.* have reported 64 cases of posterolateral fusion in patients from 46 to 79 years old. In all cases the wound healed and no abnormity was found in local and systematic examinations during long-term follow-up.[61] Neither pseudoarthrosis nor complications of the pedicle screw system were found. The safety of nano-HA/collagen/PLA composites used in lumbar posterolateral fusion has already been proven. This study further shows that use of this composite, combined with autologous bone in posterolateral lumbar fusion, leads to results similar to those found for autologous bone. The composites can thus be used as a supplement to autologous bone.[22]

Collagens used in these HA-collagen composites are derived from animal tissues. They carry the risk of transmission of infectious agents.[62] The recent development of recombinant protein expression technology provides a reliable, predictable, and chemically defined source of purified, humanlike collagen multipeptide that is free of animal components. Recombinant collagens have been produced by transfected cells.[63–71] Purified recombinant collagens are capable of undergoing spontaneous alignment to form collagen fibrils and defined features that are characteristic of collagen. Recent studies[72,73] have revealed that recombinant, humanlike collagen has the same characteristics in the initial mineralization stage as natural collagen. Additionally it also can induce the deposition and direct the growth of HA nanocrystals *in vitro* in the form of self-assembly of nanofibrils of mineralized collagen resembling an extracellular matrix.[74] The research on the nano-HA/recombinant humanlike collagen/PLA (nHA/RHLC/PLA) scaffold also revealed that the scaffold possessed good biocompatibility, biodegradation, and osteoconductive property and the scaffold maintained the advantages of mineralized animal-sourced collagen, while avoiding potential transmission of infection.[75] It is a promising material for bone tissue engineering in the clinical repair of large bony defects.

With the development of nano-HA/collagen scaffolds for bone repair and regeneration, some advanced technologies and strategie, such as electrospinning and stem cell technologies, are also being introduced for the fabrication of bone-resembling material.

Liao *et al.*[76] have reviewed biomimetic electrospun nanofibers for tissue regeneration. Some researches on HA/collagen nanofiber composites for bone graft are carrying out by combining self-assembling with electrospinning methods.[77–79] The self-renewal and transdifferentiation of stem cells give us a promising way to enable tissue engineering for clinical application. Liao *et al.*[80] have reviewed stem cells and biomimetic materials strategies for tissue engineering. In-depth mechanisms on the signaling pathway of stem cells will provide guidance in engineering stem cell/scaffold construct for tissue regeneration. Progress in the understanding of self-renewal and directed differentiation of stem cells on biomimetic materials will also lead scientists to propose the possibility of cell-based therapies to treat diseases, including the use of stem cells in tissue engineering. In the future, more and more advanced technologies and strategies will be introduced to develop the ideal biomimetic-mineralized collagen composites for bone repair and regeneration. We can look forward to soon entering an exciting new era in the availability of biomimetic bone graft materials for the enhancement of orthopedic surgery.

37.5 Conclusions and Outlook

On the basis of discussing the component and hierarchical structure in natural bone tissue, we described the biomemitic fabrication of collagen-based bone tissue engineering scaffolds using self-assembled collagen mineralization and reviewed their current applications and development in bone tissue regeneration. Some basic questions concerning the composition, microstructure, crystallography, and organic–inorganic phase relationships have been reviewed in this article. Although great progress has been made in the research on the nanostructure of self-assembling, mineralized collagen composites in natural bone and the biomemitic fabrication of scaffold materials, there are still many open questions, including the formation of a hierarchical nanostructure, the control of morphology, the incorporation of foreign ions, the interaction with biomolecules, and the assembly of the organic and inorganic phases, which are still not well understood. Future research should lead to an improved understanding of collagen-mediated mineralization in other calcified tissues and point the way to the development of new functional materials for biomimetic engineering. This knowledge will also help in the treatment of widespread pathological calcifications, such as atherosclerosis, stone formation, or dental calculus. Further progress in the topic of collagen-based scaffolds can be expected to come from modern genetics, where gene structures are now known to control the dynamic self-assembly of both cellular and protein processes.

Acknowledgments

This work was supported by National Basic Research Program of China, Grant No. 2005CB623905, and by the National Natural Science Foundation of China, Grant No. 50830102.

References

1. J. M. Lane, E. Tomin, and M. P. G. Bostrom, *Clin. Orthop.*, **367S**, 107 (1999).
2. E. M. Younger and M. W. Chapman, *J. Orthop. Trauma*, **3**, 192 (1989).
3. T. W. Bauer and G. F. Muschler, *Clin. Orthop.*, **371**, 10 (2000)
4. L. L. Hench and J. Wilson, *Science*, **226**, 630 (1984).
5. U. A. Stock and J. P. Vacanti, *Annu. Rev. Med.*, **52**, 443 (2001).
6. R. Langer and J. P. Vacanti, *Science*, **260**, 920(1993).
7. T. Grenga, J. E. Zins, and T. W. Bauer, *Plast. Reconstr. Surg.*, **84**, 245 (1989).
8. H. Schliephake, F. W. Neukam, and D. Klosa, *Int. J. Oral. Max. Surg.*, **20**, 53(1991).
9. D. C. Tancred, B. A. O McCormack, and A. J. Carr, *Biomaterials*, **19**, 2303(1998).
10. S. Weiner and H. D. Wagner, *Annu. Rev. Mater. Sci.*, **28**, 271 (1998).
11. G. M. Whitesides and B. Grzybowski, *Science*, **295**, 2418 (2002).
12. D. L. Stocum, *Regenerative Biology and Medicine* (Science Press, Beijing, 2007), p. 230.

13. W. J. Landis, M. J. Song, A. Leith, L. McEwen, and B. F. McEwen, *J. Struct. Biol.*, **110**, 39 (1993).
14. E. Baer, A. Hiltner, and H. D. Keith, *Science*, **235**, 1015 (1987).
15. J. W. Smith, *Nature*, **219**, 157 (1968).
16. K. Kuhn, *Essays Biochem.*, **5**, 59 (1969).
17. W. J. Landis and F. H. Silver, *Comp. Biochem. Physiol.*, **A133**, 1135 (2002).
18. J. Ge, F. Z. Cui, X. M. Wang, and Y. Wang, *Mater. Sci. Eng.*, **C27**, 46 (2007).
19. W. Traub, T. Arad, and S. Weiner, *Proc. Natl. Acad. Sci. U S A*, **86**, 9822 (1989).
20. W. J. Landis, K. J. Hodgens, M. J. Song, J. Arena, S. Kiyonaga, M. Marko, C. Owen, and B. F. McEwen, *J. Struct. Biol.*, **117**, 24 (1996).
21. X. M. Wang, F. Z. Cui, J. Ge, and Y. Wang, *J. Struct. Biol.*, **145**, 236 (2004).
22. F. Z. Cui, Y. Li, and J. Ge, *Mat. Sci. Eng. R.*, **57**, 1 (2007).
23. S. Mann, *Biomineralization Principles and Concepts in Bioinorganic Materials Chemistry* (Oxford University Press, Oxford, 2001).
24. S. Weiner and W. Traub, *FEBS. Lett.*, **206**, 262 (1986).
25. F. Z. Cui, H. B. Wen, X. W. Su, and X. D .Zhu, *J. Struct. Biol.*, **117**, 204 (1996).
26. P. D. Yang, T. Deng, D. Y Zhao,. P. Y. Feng,; D. Pine, B. F. Chmelka, G. M. Whitesides, and G. D. Stucky, *Science*, **282**, 2244 (1998).
27. N. J. Mathers and J. T. Czernuszka, *J. Mater. Sci. Lett.*, **10**, 992 (1991).
28. K. I. Clarke, S. E. Graves, A. T. C. Wong, J. T. Triffitt, M. J. O. Francis, and J. T.Czernuszka, *J. Mater. Sci. Mater. Med.*, **107** (1993).
29. H. A. Lowenstam, *Science*, **211**, 1126 (1981).
30. S. Mann, *Nature*, **365**, 499 (1993).
31. S. Mann, D. D. Archibald, J. M Didymus, T. Douglas, B. R. Heywood, F. C. Meldrum, and N. J. Reeves, *Science*, **261**, 1286 (1993).
32. M. Fritz, A. M. Belcher, M. Radmacher, D. A. Walters, P. K. Hansma, G. D. Stucky, D. E. Morse, and S. Mann, *Nature*, **371**, 49 (1994).
33. S. I. Stupp and P. V. Braun, *Science*, **277**, 1242 (1997).
34. J. Aizenberg, J. C. Weaver, M. S. Thanawala, V. C. Sundar, D. E. Morse, and P. Fratzl, *Science*, **309**, 275 (2005).
35. M. Kikuchi, S. Itoh, S. Ichinose, K. Shinomiya, and J. Tanaka, *Biomaterials*, **22**, 1705 (2001).
36. W. Zhang, Z. L. Huang, S. S. Liao, and F. Z. Cui, *J. Am. Ceram. Soc.*, **86**, 1052 (2003).

37. F. Z. Cui, Y. Wang, Q. Cai, and W. Zhang, *J. Mater. Chem.*, **18**, 3835 (2008).
38. L. C. Palmer, C. J. Newcomb, S. R. Kaltz, E. D. Spoerke, and S. I. Stupp, *Chem. Rev.*, **108**, 4754 (2008).
39. S. S. Liao, M. Ngiam, C. K. Chan, and S. Ramakrishna, *Biomed. Mater.*, **4**, 1 (2009).
40. S. H. Rhee, J. D. Lee, and J. Tanaka, *J. Am. Ceram. Soc.*, **83**, 2890 (2000).
41. J. H. Bradt, M. Mertig, A. Teresiak, and W. Pompe, *Chem. Mater.*, **11**, 2694 (1999).
42. G. Goissis, S. V. Silva-Maginador, and V. Conceicao-Amaro-Martins, *Artif. Org.*, **27**, 437 (2003).
43. W. Pederson, J. W. Ruberti, and P. B. Messersmith, *Biomaterials*, **24**, 4881 (2003).
44. W. Zhang, S. S. Liao, and F. Z. Cui, *Chem. Mater.*, **15**, 3221 (2003).
45. J. Bradt, M. Mertig, A. Teresiak, and W. Pompe, *Chem. Mater.*, **11**, 2694 (1999).
46. Du, F. Z. Cui, Q. L. Feng, X. D. Zhu, and K. de Groot, *J. Biomed. Mater. Res.*, **42**, 540 (1998).
47. Du, F. Z. Cui, X. D. Zhu, and K. de Groot, *J. Biomed. Mater. Res.*, **44**, 407 (1999).
48. Du, F. Z. Cui, W. Zhang, Q. L. Feng, X. D. Zhu, and K. de Groot, *J. Biomed. Mater. Res.*, **50**, 518 (2000).
49. S. I. Stupp, G. C. Mejicano, and J. A. Hanson, *J. Biomed. Mater. Res.*, **27**, 289 (1993).
50. Y. Doi, T. Horiguchi, Y. Moriwaki, H. Kitago, T. Kajimoto, and Y. Iwayama, *J. Biomed. Mater. Res.*, **31**, 43 (1996).
51. R. Z. Wang, F. Z. Cui, H. B. Lu, H. B. Wen, C. L. Ma, and H. D Li, *J. Mater. Sci. Lett.*, **14**, 490 (1995).
52. K. S. Tenhuisen, R. I. Martin, M. Klimkiewicz, and P. W. Brown, *J. Biomed. Mater. Res.*, **A29**, 803 (1995).
53. K. I. Clarke, S. E. Graves, A. T. C. Wong, J. T. Triffitt, M. J. O. Francis, and J. T. Czemuszka, *J. Mater. Sci. Mater. Med.*, **4**, 107 (1993).
54. M. C. Chang, T. Ikoma, M. Kikuchi, and J. Tanaka, *J. Mater. Sci. Lett.*, **20**, 1199 (2001).
55. M. Supova, *J. Mater. Sci. Mater. Med.*, **20**, 1201 (2009).
56. S. S. Liao, F. Z. Cui, W. Zhang, and Q. L. Feng, *J. Biomed. Mater. Res.*, **B69**, 158 (2004).

57. S. Liao, F. Watari, M. Uo, S. Ohkawa, K. Tamura, W. Wang, and F. Cui, *J. Biomed. Mater. Res.*, **74B**, 817 (2005).
58. S. S. Liao, F. Z. Cui, and X. D. Zhu, *J. Bioact. Compat. Pol.*, **19**, 117 (2004).
59. X. B. Yang, R. S. Bhatnagar, S. Li, and R. O. Oreffo, *Tissue Eng.*, **10**, 1148 (2004).
60. S. Zhou, K. B. Zhao, Y. Li, and F. Z. Cui, *J. Bioact. Compat. Polym.*, **21**, 373 (2006).
61. X. Yu, L. Xu, and L. Y. Bi, *Orthop. J. Chin.*, **13**, 586 (2005).
62. C. L. Yang, P. J. Hillas, J. A. Baez, M. Nokelainen, J. Balan, J. Tang, R. Spiro, and J. W. Polarek. *Biodrugs*, **18**, 103 (2004).
63. M. Chen, F. K. Costa, C. R. Lindvay, Y. P. Han, and D. T. Woodley, *J. Biol. Chem.*, **277**, 2118 (2002).
64. S. Frischholz, F. Beier, I. Girkontaite, K. Wagner, E. Poschl, J. Turnay, U. Mayer, and K. von der Mark, *J. Biol. Chem.*, **273**, 4547 (1998).
65. J. Myllyharju, A. Lamberg, H. Notbohm, P. P. Fietzek, T. Pihlajaniemi, and K. I. Kivirikko, *J. Biol. Chem.*, **272**, 21824 (1997).
66. Lamberg, T. Helaakoski, J. Myllyharju, S. Peltonen, H. Notbohm, T. Pihlajaniemi, and K. I. Kivirikko, *J. Biol. Chem.*, **271**, 11988 (1996).
67. M. Nokelainen, H. Tu, A. Vuorela, H. Notbohm, K. I. Kivirikko, and J. Myllyharju, *Yeast*, **18**, 797 (2001).
68. D. D. Buechter, D. N. Paolella, B. S. Lelie, M. S. Brown, K. A. Mehos, and E. A. Gruskin, *J. Biol. Chem.*, **278**, 645 (2003).
69. Merle, S. Perret, T. Lacour, V. Jonval, S. Hudaverdian, R. Garrone, F. Ruggiero, and M. Theisen, *FEBS Lett.*, **515**, 114 (2002).
70. C. John, R. Watson, A. J. Kind, A. R. Scott, K. E. Kadler, and N. J. Bulleid, *Nat. Biotechnol.*, **17**, 385 (1999).
71. M. Tomita, H. Munetsuna, T. Sato, T. Adachi, R. Hino, M. Hayashi, K. Shimizu, N. Nakamura, T. Tamura, and K. Yoshizato, *Nat. Biotechnol.*, **21**, 52 (2003).
72. Y. Wang and F. Z. Cui, *Mater. Sci. Eng.*, **C26**, 635 (2006).
73. Y. Zhai and F. Z. Cui, *J. Cryst. Growth*, **291**, 202 (2006).
74. Y. Zhai, F. Z. Cui, and Y. Wang, *Curr. Appl. Phys.*, **5**, 429 (2005).
75. Y. Wang, F. Z. Cui, K. Hu, X. D. Zhu, and D. D. Fan, *J. Biomed. Mater. Res.*, **B86**, 29 (2008).
76. S. S. Liao, B. J. Li, Z. W. Ma, H, Wei, C. Chan, and S. Ramakrishna, *Biomed. Mater.*, **1**, 45 (2006).

77. M. Ngiam, S. Liao, A. J. Patil, Z. Y. Cheng, F. Y. Yang, M. J. Gubler, S. Ramakrishna, and C. K. Chan, *Tissue Eng.*, **A15**, 535 (2009).
78. J. H. Song, H. E. Kim, and H. W. Kim, *J. Mater. Sci. Mater. Med.*, **19**, 2925 (2008).
79. S. S. Liao, R. Murugan, C. K. Chan, and S. Ramakrishna, *J. Mech. Behav. Biomed. Mater.*, **1**, 252 (2008).
80. S. S. Liao, C. K. Chan, and S. Ramakrishna, *Mat. Sci. Eng.*, **C28**, 1189 (2008).

Chapter 38

SCAFFOLD CONSIDERATIONS FOR OSTEOCHONDRAL TISSUE ENGINEERING

Eric Farrell,[a,b*] Fergal J. O'Brien,[c] and Gerjo J. V. M. van Osch[a,b]

[a] *Department of Orthopaedics, Erasmus MC University Medical Centre Rotterdam, The Netherlands*
[b] *Department of Otorhinolaryngology, Erasmus MC University Medical Centre Rotterdam, The Netherlands*
[c] *Department of Anatomy, Royal College of Surgeons in Ireland, Dublin, Ireland*
*e.farrel@erasmusmc.nl

Osteochondral defects occur as a result of disease or acute injury and have very poor repair potential. Current approaches are usually a temporary solution, and the final fate of the patient is generally to undergo surgery and replacement of the joint with a prosthetic. This chapter will discuss the tissue engineering approaches to osteochondral repair from the perspective of the scaffold considerations needed to develop a viable substitute to joint replacement. Bone and cartilage are two very different tissues with unique characteristics and properties. For example, the avascular nature of cartilage compared with the highly vascularized bone creates a unique set of challenges for the tissue engineer. Mechanical properties and differences between the two tissues also must be carefully accounted for. Finally this chapter discusses current approaches to joint repair and postulates some potential future options.

Handbook of Intelligent Scaffolds for Tissue Engineering and Regenerative Medicine
Edited by Gilson Khang

www.panstanford.com

38.1 Introduction

38.1.1 *Tissue Engineering of the Bone Cartilage Interface*

Repair of the bone cartilage interface is a challenging task, one that is necessary not only in the combat of the degenerative joint disease osteoarthritis (OA) but also in the treatment of traumatic injuries. Osteochondral defects often occur in young people, particularly athletes, and these injuries have a poor spontaneous regeneration capacity. When a lesion of the cartilage occurs there is little self-repair potential in the tissue, making healing difficult. Recently it has become clear that the underlying bone must also be taken into consideration in such injuries as changes in the cartilage cause changes here, too, which, if untreated, can eventually lead to degenerative OA. OA is among the leading causes of disability in the elderly and forms a major burden to health care. About 30% of persons aged 65 and over are affected by knee or hip OA.[1] OA affects 27 million people in the United States alone and is becoming an increasingly prevalent disease. As with many tissues of the body, repair of osteochondral defects using autologous tissue is severely limited by tissue availability. To add to this problem, donor site morbidity and a difficulty matching the shape of the graft topology with the injured site compound the issue.[2] The ability to tissue-engineer replacement grafts would greatly reduce the socioeconomic burden associated with this type of disease/injury. Tissue engineering of the bone cartilage interface presents many problems. Some of these are encountered in other fields of tissue engineering, and some are unique. This chapter will discuss the considerations for the design of a successful tissue engineered osteochondral graft, the current state of the art, and some potential future prospects.

38.2 Joint Homeostasis

After trauma or in degenerative diseases such as OA, joint homeostasis is disturbed. The homeostasis in the joint is determined by biomechanical and biochemical factors. Although cartilage damage was the hallmark of OA for a long time, it is becoming more and more clear that changes in the subchondral bone and inflammation of the synovium also play a role in the disease process. Inflammation of the synovium, a well-recognized feature in OA but also following traumatic injury of the joint, can disturb homeostasis. High levels of synovial makers indicative of poor production of extracellular matrix and cartilage degradation are locally present in patients even one year after autologous chondrocyte implantation.[3] These inflammatory factors secreted by synovium can have harmful effects on the cartilage repair process.

The bone located directly beneath the cartilage (the subchondral bone) presumably plays an important role in the force distribution in the cartilage, resulting from joint loading, and plays a role in the biomechanical component of joint homeostasis. Changes in subchondral bone architecture could lead to a change in loading pattern, which is hypothesized to play a role in the development of OA. Radin *et al.* proposed the theory that a thicker subchondral bone plate has a reduced shock-absorbing capacity.[4] Several studies showed an increase[5] in the amount of subchondral bone in OA.[6–8] This would lead to increased stresses in the overlying cartilage layer, resulting in tears and progressive loss of the cartilage. However, until now it is still unclear whether the increase in thickness of the subchondral bone plate is a consequence or a cause of the disease. Contrary to the general belief that OA is accompanied by subchondral sclerosis, experimental studies with animals reported thinning of the subchondral bone plate in early phases of the disease that preceded sclerosis.[9–13] This thinning, caused by increased turnover, was due to osteoclastic resorption of the subchondral bone plate, sometimes even up to the calcified cartilage.

Changes in subchondral bone not only accompany OA but also are recognized after traumatic joint injuries such as a cruciate ligament rupture or a cartilage lesion. Their relation with progressive joint disease is under discussion, but severe bone bruises have been suggested as a precursor of early degenerative changes.[14] It is clear that in diseased joints subchondral bone changes occur. When the stable, well-functioning biomechanical system of cartilage and bone present in normal, healthy joints is disturbed, a myriad of cellular responses comes into play, linking changes in bone and cartilage. It is more and more accepted that the altered conditions in the bone should be restored at the time a cartilage repair procedure is performed.

38.3 Current and Recent Approaches to the Field of Osteochondral Tissue Engineering

Presently there are several clinical approaches to treat cartilage damage depending on severity, starting with pain medication, anti-inflammatory treatment, and physical therapy, eventually leading to surgery over time due to eventual worsening of the condition. The first surgical approach of choice involves microfracture or subchondral drilling. During these procedures holes are made through the subchondral bone to evoke bleeding and invasion of stem cells from the underlying bone marrow. This stimulates repair with fibrous cartilage. These techniques obviously affect processes in the subchondral bone[15] that may affect cartilage repair. Alternatively, for cartilage lesions, autologous chondrocyte transplantation is a clinically viable option. In this procedure, first described in patients in 1994 by Brittberg *et al.*,[16] chondrocytes are isolated from cartilage harvested from a less-weight-bearing site in the joint. After expansion of the number of chondrocytes in culture, the cells are transplanted in the cartilage defect as a suspension, under either a flap of periosteum or a collagen membrane attached to the cartilage defect edges. Newer-generation products use seeding of the cells on or in a scaffold material. This system has been coined matrix-induced autologous chondrocyte implantation (MACI). In these chondrocyte transplantation procedures the underlying bone is left untreated. However, the properties of this bone will have been changed due to the cartilage defect, as well as will change after the procedure because cartilage and bone are one entity and changes in one will evoke adaptational reactions in the other. Although the underlying bone initially did not receive attention, this is changing mainly due to the fast improvements in magnetic resonance imaging (MRI) techniques.[17]

In the case of larger osteochondral defects, osteochondral grafts can be used. These grafts can be either allografts (fresh or frozen) or autografts. Different techniques exist where only one or sometimes several small osteochondral tissue plugs are used to fill a larger defect (mosaicplasty). This approach has limited success because of difficulties with integration of the graft in the surrounding tissue. In the case of allografts, inflammation can be a serious problem. However, autografts have the disadvantage of creating large defects at the harvest site. To overcome these disadvantages, tissue engineering approaches are used to generate osteochondral transplants.

38.4 Functional Properties of Bone and Cartilage and the Important Differences Between Them

To understand the requirements for successful osteochondral graft manufacture, it is important to understand the roles of bone and cartilage and the differences between them with respect to load bearing, physical makeup, etc., in the joint region. Articular cartilage, also referred to as hyaline cartilage, is a complex tissue composed mainly of type II collagen, proteoglycans, water, and cells (chondrocytes). The interconnected collagen fibrils, composed mainly of collagen type II but also types IX and XI, lie parallel to the surface. Within the fiber matrix, aggrecan and hyaluronic acid combine to form highly negatively charged proteoglycan aggregates. This negative charge results in a large water retention capacity of the tissue, resulting in swelling, which is constrained by the collagen fibers. It is these properties that give cartilage its remarkable mechanical properties both in compression and in tension. Cartilage is subdivided into three distinct zones: the superficial, middle, and deep zones.[18] The outer, superficial zone faces into the joint space and articulates with the opposing joint surface. At this cartilage-cartilage interface components such as hyaluronan, proteoglycan 4, and surface-active phospholipids, produced by synovium, as well as cartilage, are important to provide boundary lubrication and keep joint homeostasis.[19] The superficial zone is where cell density is the highest. In the middle zone, cells are more sparse and rounded with the collagen fibers less aligned than in the superficial zone. Another change in morphology is seen in the deep zone. Here the fibers are larger and oriented perpendicular to the surface. The aggrecan concentration and compressive strength increase by an order of magnitude across the depth of the cartilage from the surface to the deep zone. The superficial zone is where cell density is the highest (Fig. 38.1).

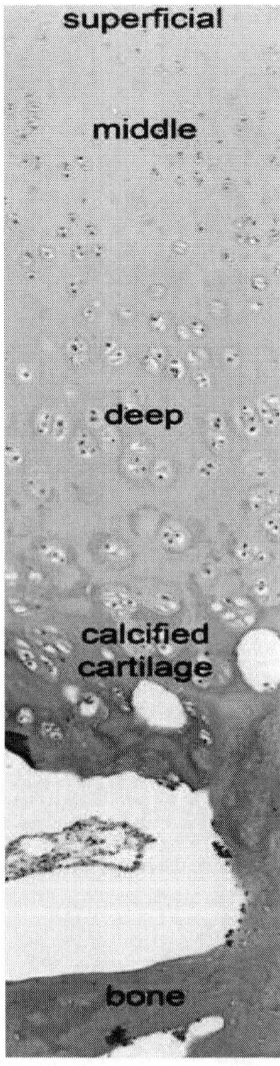

Figure 38.1. Histology of immature bovine cartilage showing the transitions from superficial- to deep-zone cartilage, the calcified cartilage layer, and the underlying bone.

Beneath the deep zone lies the calcified cartilage layer that forms the interface between bone and cartilage both morphologically and mechanically. Important to note is that cartilage is an avascular and aneural tissue comprising only a single cell type that resides in an oxygen-poor environment.

Bone itself is a complex tissue comprising several cell types (osteoblasts, osteocytes, osteoclasts, and bone marrow cells) and constantly undergoing remodeling. Osteoblasts are the bone-forming cells that are responsible for laying down the organic component of bone, which subsequently becomes mineralized. During development and bone formation some of these cells become trapped in their own matrix and differentiate further to become osteocytes. These cells are responsible for bone maintenance and also sensing of mechanical strain. Osteoclasts are the bone-resorbing cells of the body and play an integral role in tissue maintenance, working with the osteoblasts to remodel the bone in response to mechanical signals or lack thereof. According to Wolff's law, bones will adapt to the loads to which they are exerted, either increasing or decreasing their density and ability to withstand loads in a particular direction.[20] This is made possible by the interplay between the bone-forming osteoblasts and the bone-resorbing osteoclasts. Several modifications to this law have been proposed since, but the essence of it remains.[21–23] These same cells play a role in the alterations in the subchondral bone of an osteoarthritic or injured joint. Lying beneath the bone-cartilage interface is the subchondral bone, which is comprised of a thin layer of dense cortical bone. This bone is not heavily vascularized or porous compared with the trabecular (spongy) bone directly beneath it. The trabecular bone is filled with marrow and penetrated with blood vessels. Mechanically, bone is much stiffer than cartilage, with varying reports on the Young's moduli of the different bone regions but all in the giga-Pascal range. Cartilage by contrast measures in the tens or hundreds of mega-Pascals (for review see Ref. 24). The differences in mechanical properties enable cartilage to withstand and dampen impacts and to transfer the load to the underlying bone. Both cartilage and bone have a critical role to play in this process.

38.5 Vascularization and its Absence in Cartilage

Like most tissues of the body bone is a well-vascularized tissue. Cartilage, however, is avascular, with no blood vessels crossing the subchondral plate into the calcified cartilage layer. This offers a unique set of considerations for tissue engineers. Maintaining cartilage in an avascular state begins at development and is a continuing process throughout life. The formation of bone from a cartilage template via endochondral ossification (for review see Ref. 25) and its vascularization without the formation of blood vessels in the articular cartilage involves a delicate balance between pro- and anti-angiogenic factor release from chondrocytes at different stages of development. The two most important factors in this process are probably the anti-angiogenic factor chondromodulin 1 and the pro-angiogenic factor vascular endothelial growth factor (VEGF)-A. During development chondromodulin is heavily expressed in proliferating chondrocytes, but when these begin to hypertrophy its expression decreases with a concomitant increase in VEGF expression.[26] This combination leads to blood vessel invasion from the surrounding tissue, going hand in hand with matrix mineralization and bone formation. Other factors have also been identified as being critical for this process to take place, including high-mobility group box 1 protein (HMGB1) and collagen type X, both of which act as chemoattractants for blood vessels and bone-forming cells.[27,28]

Alongside chondromodulin, thrombospondin 1 and 2 and tissue inhibitor of metalloproteinases (TIMP)-2 and 3 are also expressed in developing cartilage but reduced in hypertrophic and calcified zones.[26,29,30] The presence of these factors in mature cartilage plays an important role in preventing vessel invasion from the underlying bone. VEGF-A plays a large role in the switch from vessel inhibition to invasion and is expressed in hypertrophic cartilage but not in proliferating cartilage.[31] Several knockout studies have been performed with regard to angiogenesis inhibitors (for review see Ref. 32). By expressing these various factors at different levels, a perfect homeostasis is maintained between avascular hyaline cartilage and the underlying vascularized bone tissue. Upsetting this balance may be one cause of the onset of OA.[33]

38.6 Scaffold Considerations for Osteochondral Tissue Engineering

Scaffolds for tissue engineering applications are designed to perform the following functions: (1) to encourage cell-material interactions, that is, cell attachment, differentiation, and proliferation, eventually leading to the deposition of the extracellular matrix; (2) to permit the transport of nutrients, wastes, and biological signaling factors to allow for cell survival; (3) to biodegrade at a controllable rate, which approximates the rate of natural tissue regeneration; and (4) to provoke a minimal immune and/or inflammatory response *in vivo*.[34] Scaffolds for osteochondral tissue engineering must fulfill all of these criteria; however, they must also address some additional challenges. In the field of tissue engineering, it is commonly postulated that scaffolds should mimic the natural architecture and mechanical properties of the tissues to be engineered. The great challenge with osteochondral tissue engineering is that, although a single-phase scaffold may promote healing in at least part of the defect, it is unlikely to result in completely adequate healing of the cartilaginous, transitional, and osseous parts of the tissue. Consequently, there is a clear requirement for a scaffold that more closely mimics the layered structure and gradient composition of osteochondral tissue.

Ideally, a scaffold for osteochondral repair would consist of distinct layers, specifically designed to closely mimic both the morphology and the composition of healthy anatomical tissue. However, developing such a biomaterial presents enormous challenges due to the different mechanical properties and levels of vascularization, in addition to the biological (cellular and matrix) composition of the overlying, superficial articular cartilaginous layer, the intermediate (transitional and radial zone) cartilage, the calcified cartilage, and the deeper subchondral bone. A range of scaffolds fabricated from natural polymers (e.g., collagen, glycosaminoglycan, hyaluronic acid, chitosan, fibrin, and alginate), synthetic polymers (e.g., polyglycolide acid [PGA], polylactides [poly-L-lactide (PLLA), poly-D-lactide (PDLA)], poly-lactic-co-glycolic acid [PLGA], polycaprolactone (PCL), and polyethylene-glycol [PEG]), ceramics (e.g., hydroxyapatite [HA] and calcium phosphates), hydrogels, devitalized bone, or combinations of these materials have been used to create osteochondral tissue-engineered grafts. Single-phase scaffolds have been employed to engineer both cartilage and bone layers; however, increasingly, research has begun to focus on multiphase scaffolds employing different material types, pore architecture, and porosities to better suit the biomechanical requirements of the cartilage and bone layers.[35,36] Most existing commercial scaffolds consist of a single phase and are not optimized for cartilage repair, resulting in poor healing of hyaline-like cartilage.[37]

A recent review by Martin *et al.* concisely described (some of) the different approaches for the fabrication of osteochondral composite constructs with specific considerations for scaffold type and cell source.[2] In this review they describe four construct options and four cell options as follows:

(a) A scaffold for the bone component but a scaffold-free approach for the cartilage component. In this approach, researchers generally made use of stiff bonelike scaffolds comprised of ceramic[38] or HA[39] or composites thereof. Upon these they seeded large volumes of cells, which would self-organize into a cartilage-like matrix.[40] The approach of using cells in high densities to generate cartilage-like tissue has been demonstrated independent of the bone component.[41]

(b) Different scaffolds for the bone and the cartilage components combined at the time of implantation. While combining two scaffolds might be easier than generation of a bilayer or multiphase construct, researchers are then faced with a second site where integration might pose a problem, namely, at the bone-cartilage interface. One of the reasons to attempt this approach was a lack of technology/capability to generate biphasic scaffolds. Nevertheless, some success was observed seeding and culturing the two phases separately and then suturing[35] or gluing them together afterward.[42] This was also demonstrated *in vivo*,[43] but integration with the surrounding cartilage tissue remained a problem, despite the relatively favorable environment of the rabbit knee.

(c) A single but heterogeneous composite scaffold. This approach might offer the best option of the four of these described approaches, enabling the researcher to carefully tailor the scaffold properties to the mechanical and physiological needs of both tissue types. Of course integration between the two materials is again a critical concern. This approach has shown promise *in vitro*,[44,45] but there is a need to properly characterize these treatment regimes in joint defects of large animals to fully assess repair potential.

(d) The fourth scaffold approach referred to by Martin *et al.* was to use a single homogenous scaffold for both components, as described by Cao *et al.*[46–48] Once again more research and better characterization is required to properly assess the best approach of those described here. To confound that issue, as also discussed by Martin *et al.*, these scaffolds have been (i) loaded with a single cell source having chondrogenic capacity, (ii) loaded with two cell sources having either chondrogenic or osteogenic capacities, (iii) loaded with a single cell source having both chondrogenic and osteogenic differentiation capacity, or (iv) used in a cell-free approach.

Within the field of tissue engineering several or all of the above-mentioned approaches have been attempted, most of them *in vitro* thus far. This is one classification system that could be used to describe the current approaches used, but there are also other considerations/approaches to consider, as will be discussed shortly.

Commercially, two products on the market which use a multilayered approach that are currently undergoing clinical trials include Smith & Nephew's TRUFIT product, which is proposed as a one-step arthroscopic procedure for repairing bone and articular cartilage defects, and Orthomimetic's Chondromimetic product, which proposes to support the body's natural repair mechanisms to encourage the simultaneous repair of both articular cartilage and the bone to which it is attached. Both of these products are bilayered with different material and properties in the parts aimed for bone and cartilage regeneration, respectively. TRUFIT implants are composed of a poly(D,L-lactide-*co*-glycolide) polymer, formed into a bilayered cylindrical scaffold that is absorbed by the body within six to nine months, while Chondromimetic is a highly porous, collagen I–based product within a mineral phase present in the region designed for implantation in the subchondral bone. Although neither mimics the complex anatomy of native osteochondral tissue, scaffolds such as these are undoubtedly a step in the right direction.

Collagen is a favored material in scaffold design for cartilage due to its biological properties.[49] Furthermore, as a normal constituent of the tissue, collagen has been shown to be a much more appropriate substrate for bone formation than biomaterials such as polystyrene, titanium, or PLA and PLGA, which are commonly used as implant materials.[50,51] The major disadvantage of collagen as a scaffold for bone regeneration is that it has relatively poor mechanical properties. However, the mechanical properties of collagen-based scaffolds can be improved through cross-linking mechanisms[52–54] or by allowing bone cells to produce osteoid on the scaffolds, followed by subsequent mineralization *in vitro*,[55] which also leads to improved mechanical properties. Although collagen-based scaffolds have also inferior properties compared with native cartilage, they have been implemented in the matrix-induced autologous chondrocytes implantation (MACI) technique, which has displayed promising results for the repair of cartilage.[56–58] However, type 1 collagen is the usual component of these scaffolds and, although type 2 collagen is the main constituent of articular cartilage, it is a component in few scaffolds, either on the market or currently in development. This is somewhat surprising. Studies by Nehrer *et al.*[59] have shown the preferential behavior of chondrocytes in type 2 collagen over type 1 collagen. A product currently undergoing commercial investigation, from the laboratory of one of the authors, is ChondroColl.[60] This is a highly porous, lyophilized collagen-based scaffold. In this product, three layers are combined into one three-dimensional construct, which aims to facilitate the healing of osteochondral tissue *in vivo*. This scaffold is produced using an "iterative layering" technique, resulting in interfacial adhesion between the scaffold layers and allowing the properties of the individual scaffold layers to be optimized for their respective roles in osteochondral tissue healing (Fig. 38.2). The scaffold thus offers a gradient composition and structure consisting of four bioactive and biocompatible constituents all found in the osteochondral tissue, namely, collagen type 1 and 2, glycosaminoglycans (GAG), and HA. The pore size and mechanical properties of the material are varied

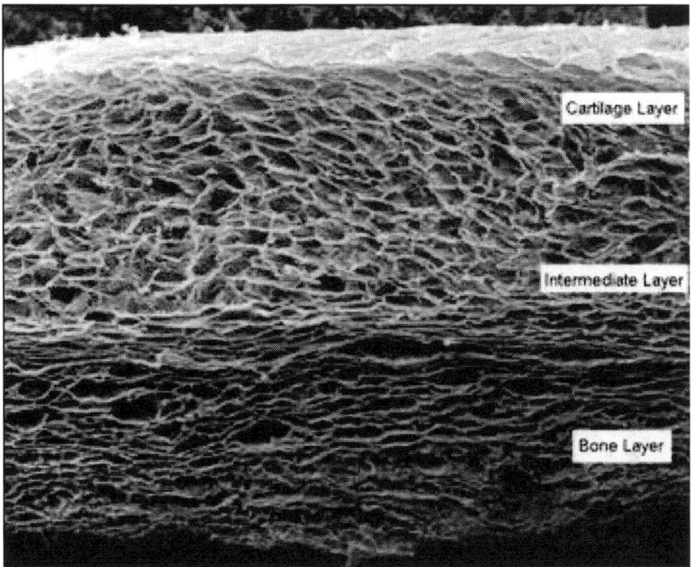

Figure 38.2. SEM image of ChondroColl, which is a 3-layered osteochondral scaffold composed of a bone layer (containing type 1 collagen and HA), an intermediate layer (containing type 1 collagen, type 2 collagen, and HA), and a cartilage layer (containing primarily type 2 collagen). *Abbreviation*: SEM, scanning electron microscopy.

throughout the structure to promote tissue formation by different cell types throughout the structure.

Previously we have shown[61] that scaffold pore size significantly affects cellular activity, and it is hypothesized that the optimal scaffold pore size is not only tissue specific but also cell specific. For example, it is hypothesized that scaffolds for bone tissue engineering should contain pores > 300 μm to enhance bone formation and vascularization,[62] while scaffolds for cartilage tissue engineering should employ a pore size < 200 μm.[63–65] It is thought that smaller pores sizes may encourage hypoxic conditions, which are more favorable for chondrogenesis. Therefore, by producing a scaffold with a different mean pore size in the osteo and chondral regions or by producing a scaffold with a gradient in mean pore size from the osteo to the chondral region, it might be possible to develop an optimal scaffold for osteochondral repair.

There are other distinct considerations with regard to the generation of a tissue-engineered osteochondral replacement construct. Firstly, fixation of the tissue-engineered construct into the defect site must be taken into account for several reasons. The most obvious is that the construct must remain in place if it is to function properly. Also, in order for proper integration to occur with the surrounding tissue, the construct must not be permitted to move excessively. Several methods are employed to fix constructs in place, the most common being gluing, usually using fibrin and suturing. Mechanical properties of the construct have already been mentioned in relation to the scaffold material and load bearing, but this must also be taken into account for the final tissue properties required and how this will affect integration with the surrounding tissue. This is one of the largest problems currently associated with repair of chondral and osteochondral defects. In osteochondral transplantation, poor integration at the cartilage and bone levels has been reported.[66] Poor integration prevents the structure from functioning as one biomechanical entity, and the high stress at the interface leads to further degeneration in time. In contrast to mature osteochondral grafts it is believed that the immature tissue of tissue-engineered constructs have better integration capacities.[67] Other options may be considered to improve integration upon implantation, such as enzymatic treatment of the wound area[68] or mechanical debridement or use of glues for initial fixation. Development of a construct with similar mechanical properties to the surrounding cartilage and underlying bone can only improve the chances of good initial integration.

Also to be considered is the fact that bone is a highly vascularized tissue, whereas cartilage remains avascular throughout life. Depending on the cell type chosen for the osteochondral construct, it is possible that blood vessels might invade from the lower osseous layer to the chondral layer. This could be prevented in a number of possible ways, from the placement of a degradable impermeable membrane between the two layers to allow for stabilization *in vivo* to the addition of gene vectors expressing anti-angiogenic factors to prevent invasion. This expression could also be programmed to be transient, switching off once the tissue has fully integrated and matured. A final consideration in designing a layered scaffold for osteochondral tissue repair is tailoring the degradation characteristics of the different layers. Ideally, an implanted scaffold will degrade *in vivo* and be replaced by a functional repair tissue. This needs to be accurately tailored in order to ensure that the rate of degradation does not exceed the rate of native tissue formation to the detriment of mechanical stability. A study by Solchaga *et al.*[69] found that faster-degrading scaffolds resulted in an increase in bone formation, while slow-degrading scaffolds prolonged the presence of cartilage tissue, delayed endochondral bone formation, and sustained a thicker cartilage layer. However, if bone formation was delayed, the lack of an appropriate mechanical support for the developing overlying cartilage resulted in cracks and discontinuities at the surface. The engineering of osteochondral tissues requires more attention to achieve a suitable graft solution to reproducibly provide regenerated, functional *in vivo* cartilage and bone tissues. It is also clear that the scaffold characteristics will greatly influence the mechanical environment to which cells are exposed once implanted into defects and therefore needs careful consideration during the design process.

38.7 Endochondral Ossification, a More Logical Approach for Osteochondral Tissue Engineering

The fields of tissue engineering and regenerative medicine (TERM) have improved noticeably in recent years with new developments and progression in bone and cartilage repair too numerous to discuss here (for review see Refs. 70–78). Despite this and the advances discussed earlier, very few clinical studies have been performed with regard to tissue-engineered bone with varying success[79] with even fewer studies at the bone-cartilage interface. There is currently a huge deficit in the available options for the millions of bone and joint replacement procedures performed every year worldwide. However, there are many factors that determine the success of a TERM-based approach to this issue, and some of these have yet to be addressed satisfactorily before a TERM treatment option becomes viable. Probably the most important issue of any TERM approach is that of vascularization. Researchers in every field are now aware that it is vital that constructs/treatments take vascularization into account. This is crucial at the bone-cartilage interface as discussed earlier. Ideally, rapid vascularization or even prevascularization of the bone component is desired without vascular penetration into the cartilage layers.

Tissue engineering of bone has, since its inception, focused on the generation of bone via the direct differentiation of mesenchymal stem cells (MSCs) to bone-forming cells, a process called intramembranous ossification. Normally, *in vivo*, this process is characterized by an increase in vascularity of the existing connective tissue present, prompting invasion of mesenchymal cells into the region. These cells then differentiate along the osteogenic route to become osteoblasts, which will lay down the organic component of bone, called osteoid. Over time this tissue becomes mineralized and cells either become trapped in this tissue further differentiating into osteocytes or retreat to the surface of the bone and remain as osteoblasts involved in bone formation and remodeling. This process, however, normally occurs only in a small proportion of special bones within the body, such as the skull. Most bones of the body, however, are formed via a process known as endochondral ossification, whereby a cartilage template is first generated, which is subsequently invaded by blood vessels (stimulated by this cartilage tissue) and mineralized to form bone. In this process, cartilage cells as they mature stop proliferating and enlarge. These "hypertrophic" chondrocytes begin to release pro-angiogenic factors and reduce their expression of anti-angiongenic factors, leading to vessel invasion along with invasion of mesenchymal and osteogenic cells. This occurs in parallel with mineralization of the tissue and formation of bone.

If one considers that one of the main reasons for failure of a tissue-engineered construct is that of core necrosis caused by hypoxia and lack of nutrient delivery and waste removal,[80–82] perhaps a complete rethink of the approach is required. It has become very clear that to successfully generate viable tissue-engineered bone, vascularization must be accounted for.[75,76,83–86] This may be done in a number of manners, for example, by generating tissue-engineered vessels around which bone could be generated.[85] The use of an *in vivo* bioreactor to grow bones in an ectopic location within the body, which would lead to generation of blood vessels concomitant with bone growth, would be another option.[78] Use of intelligent instructive scaffolds is also being increasingly explored, as discussed elsewhere. This can incorporate growth factors to induce vessel invasion or perhaps gene vectors to upregulate expression of VEGF, for example, thereby expediting vessel invasion following implantation. This type of research is at the cutting edge of current scientific approaches to bone and cartilage repair. Preconditioning MSCs to hypoxia prior to implantation is also being explored.[87–89]

There is, however, perhaps a much more logical approach to form bone. Why not induce bone formation by endochondral ossification? In this scenario a construct of chondrogenically primed MSCs or perhaps even a MSC-derived cartilage template would be implanted and allowed to progress further along the endochondral ossification route to form bone naturally. This approach has several advantages. First, MSCs are known to progress along the endochondral ossification route, naturally becoming hypertrophic, as occurs very often in cartilage tissue engineering approaches.[90–94] Second, chondrocytes usually reside in an environment of very low oxygen tension. Placing an unvascularized cartilage-like template into such an environment should not result in cell death, as would normally happen. Third, as explained earlier, endochondral ossification is characterized by attraction of vessels by the hypertrophic chondrocytes, which occurs simultaneously with bone formation. This is the opposite of the intramembranous approach usually used in tissue engineering, whereby bone formation is stimulated by an increase in vessel density, which has already occurred. A combination of this approach with new scaffolds could offer huge potential for the fields of both bone and osteochondral tissue engineering. Of course the challenge will be to maintain the chondral layer in an unhypertrophic, unvascularized state. However, with new instructive scaffolds and the surrounding cartilaginous and synovial environment, perhaps this will occur naturally. This approach is looking increasingly promising with recent research demonstrating the stability of implanted MSCs in chondral defects.[95] *In vitro* these same cells produced collagen type X and became hypertrophic; however, when implanted directly without *in vitro* culture, the same cells showed much lower expression of collagen type X and degrading enzymes indicative of the endochondral ossification process. Generation of stable cartilage in such a manner, using appropriate scaffolds, could enable this endochondral approach for osteochondral tissue engineering to succeed.

The field of osteochondral tissue engineering is a new and exciting one. One of the main driving forces behind where the next breakthrough comes from will certainly be the choice of next generation biomaterial, but other considerations, such as choice of cell type, should also play a role, as discussed here. Mimicking developmental approaches might be the simplest and most logical way to achieve complete repair in concert with an appropriate scaffold.

References

1. E. Odding, H. A. Valkenburg, H. J. Stam, and A. Hofman, *Eur. J. Epidemiol.*, **17(11)**, 1033 (2001).
2. L. Martin, S. Miot, A. Barbero, M. Jakob, and D. Wendt, *J. Biomech.*, **40(4)**, 750 (2007).
3. A. L. Vasara, Y. T. Konttinen, L. Peterson, A. Lindahl, and I. Kiviranta, *Clin. Orthop. Relat. Res.*, **467(1)**, 267 (2009).
4. E. L. Radin and R. M. Rose, *Clin. Orthop. Relat. Res.*, **213**, 34 (1986).
5. M. D. Grynpas, B. Alpert, I. Katz, I. Lieberman, and K. P. Pritzker, *Calcif. Tissue Int.*, **49(1)**, 20 (1991).
6. D. Bobinac, J. Spanjol, S. Zoricic, and I. Maric, *Bone*, **32(3)**, 284 (2003).
7. M. Ding, A. Odgaard, and I. Hvid, *J. Bone. Joint. Surg. Br.*, **85(6)**, 906 (2003).
8. B. Li, D. Marshall, M. Roe, and R. M. Aspden, *J. Anat.*, **195 (Pt 1)**, 101 (1999).
9. Y. H. Sniekers, F. Intema, F. P. Lafeber, G. J. van Osch, J.P . van Leeuwen, H. Weinans, and S. C. Mastbergen, *BMC Musculoskelet. Disord.*, **9**, 20 (2008).
10. S. M. Botter, G. J. van Osch, J. H. Waarsing, J. S. Day, J. A. Verhaar, H. A. Pols, J. P. van Leeuwen, and H. Weinans, *Biorheology*, **43(3–4)**, 379 (2006).
11. S. M. Botter, G. J. van Osch, J. H. Waarsing, J. C. van der Linden, J. A. Verhaar, H. A. Pols, J. P. van Leeuwen, and H. Weinans, *Osteoarthr. Cartilage.*, **16(4)**, 506 (2008).
12. D. L. Batiste, A. Kirkley, S. Laverty, L. M. Thain, A. R. Spouge, and D. W. Holdsworth, *Osteoarthr. Cartilage*, **12(12)** , 986 (2004).

13. T. Hayami, M. Pickarski, Y. Zhuo, G. A. Wesolowski, G. A. Rodan, and T. Duong le, *Bone*, **38(2)**, 234 (2006).

14. A. Nakamae, L. Engebretsen, R. Bahr, T. Krosshaug, and M. Ochi, *Knee Surg. Sports Traumatol. Arthrosc.*, **14(12)**, 1252 (2006).

15. H. Chen, J. Sun, C. D. Hoemann, V. Lascau-Coman, W. Ouyang, M. D. McKee, M. S. Shive, and M. D. Buschmann, *J. Orthop. Res.*, (2009).

16. M. Brittberg, A. Lindahl, A. Nilsson, C. Ohlsson, O. Isaksson, and L. Peterson, *N. Engl. J. Med.*, **331(14)**, 889 (1994).

17. S. Trattnig, S. A. Millington, P. Szomolanyi, and S. Marlovits, *Eur. Radiol.*, **17(1)**, 103 (2007).

18. P. J. Yang and J. S. Temenoff, *Tissue Eng. Part B Rev.*, (2009).

19. T. A. Schmidt, N. S. Gastelum, Q. T. Nguyen, B. L. Schumacher, and R. L. Sah, *Arthr. Rheum.*, **56(3)**, 882 (2007).

20. J. Wolff, *Das Gesetz der Transformation der Knochen*, A. Hirschwald Ed. (Berlin 1892).

21. D. R. Carter and N. J. Giori, Effect of mechanical stress on tissue differentiation in the bony implant bed, In J. E. Davies, Ed., *The Bone-Biomaterial Interface* (University of Toronto Press: Toronto 1991), pp. 367–376.

22. F. Pauwels, *Biomech. Locom. Appar.*, **375**, (1980).

23. P. J. Prendergast, R. Huiskes, and K. Soballe, *J. Biomech.*, **30(6)**, 539 (1997).

24. L. A. McMahon, F. J. O'Brien, and P. J. Prendergast, *Regen. Med.*, **3(5)**, 743 (2008).

25. E. J. Mackie, Y. A. Ahmed, L. Tatarczuch, K. S. Chen, and M. Mirams, *Int. J. Biochem. Cell. Biol.*, **40(1)**, 46 (2008).

26. N. Taniguchi, K. Yoshida, T. Ito, M. Tsuda, Y. Mishima, T. Furumatsu, L. Ronfani, K. Abeyama, K. Kawahara, S. Komiya, I. Maruyama, M. Lotz, M. E. Bianchi, and H. Asahara, *Mol. Cell. Biol.*, **27(16)**, 5650 (2007).

27. G. Shen, *Orthod. Craniofac. Res.*, **8(1)**, 11 (2005).

28. Y. Hiraki, H. Inoue, K. Iyama, A. Kamizono, M. Ochiai, C. Shukunami, S. Iijima, F. Suzuki, and J. Kondo, *J. Biol. Chem.*, **272(51)**, 32419 (1997).

29. C. Shukunami, K. Iyama, H. Inoue, and Y. Hiraki, *Int. J. Dev. Biol.*, **43**, 39–49 (1999).

30. C. I. Colnot and J. A. Helms, *Mech. Dev.*, **100(2)**, 245 (2001).

31. H. P. Gerber, T. H. Vu, A. M. Ryan, J. Kowalski, Z. Werb, and N. Ferrara, *Nat. Med.*, **5(6)**, 623 (1999).

32. C. Shukunami, Y. Oshima, and Y. Hiraki, *Biochem. Biophys. Res. Commun.*, **333(2)**, 299 (2005).

33. T. Hayami, H. Funaki, K. Yaoeda, K. Mitui, H. Yamagiwa, K. Tokunaga, H. Hatano, J. Kondo, Y. Hiraki, T. Yamamoto, T. Duong le, and N. Endo, *J. Rheumatol.*, **30(10)**, 2207 (2003).
34. F. Lyons, S. Partap, and F. J. O'Brien, *Technol. Health Care*, **16(4)**, 305 (2008).
35. D. Schaefer, I. Martin, P. Shastri, R. F. Padera, R. Langer, L. E. Freed, and G. Vunjak-Novakovic, *Biomaterials*, **21(24)**, 2599 (2000).
36. J. F. Mano and R. L. Reis, *J. Tissue Eng. Regen. Med.*, **1(4)**, 261 (2007).
37. R. Dorotka, U. Windberger, K. Macfelda, U. Bindreiter, C. Toma, and S. Nehrer, *Biomaterials*, **26(17)**, 3617 (2005).
38. R. A. Kandel, M. Grynpas, R. Pilliar, J. Lee, J. Wang, S. Waldman, P. Zalzal, and M. Hurtig, *Biomaterials*, **27(22)**, 4120 (2006).
39. X. Wang, S. P. Grogan, F. Rieser, V. Winkelmann, V. Maquet, M. L. Berge, and P. Mainil-Varlet, *Biomaterials*, **25(17)** , 3681 (2004).
40. R. Tuli, S. Nandi, W. J. Li, S. Tuli, X. Huang, P. A. Manner, P. Laquerriere, U. Noth, D. J. Hall, and R. S. Tuan, *Tissue Eng.*, **10**, 1169–1179 (2004).
41. A. D. Murdoch, L. M. Grady, M. P. Ablett, T. Katopodi, R. S. Meadows, and T. E. Hardingham, *Stem Cells*, **25(11)**, 2786 (2007).
42. J. Gao, J. E. Dennis, L. A. Solchaga, A. S. Awadallah, V. M. Goldberg, and A. I. Caplan, *Tissue Eng.*, **7(4)**, 363 (2001).
43. D. Schaefer, I. Martin, G. Jundt, J. Seidel, M. Heberer, A. Grodzinsky, I. Bergin, G. Vunjak-Novakovic, and L. E. Freed, *Arthr. Rheum.*, **46(9)**, 2524 (2002).
44. R. M. Schek, J. M. Taboas, S. J. Segvich, S. J. Hollister, and P. H. Krebsbach, *Tissue Eng.*, **10(9–10)**, 1376 (2004).
45. J. K. Sherwood, S. L. Riley, R. Palazzolo, S. C. Brown, D. C. Monkhouse, M. Coates, L. G. Griffith, L. K. Landeen, and A. Ratcliffe, *Biomaterials*, **23(24)**, 4739 (2002).
46. T. Cao, K. H. Ho, and S. H. Teoh, *Tissue Eng.*, **9 Suppl 1**, S103 (2003).
47. A. Alhadlaq, J. H. Elisseeff, L. Hong, C. G. Williams, A. I. Caplan, B. Sharma, R. A. Kopher, S. Tomkoria, D. P. Lennon, A. Lopez, and J. J. Mao, *Ann. Biomed. Eng.*, **32(7)**, 911 (2004).
48. A. Fukuda, K. Kato, M. Hasegawa, H. Hirata, A. Sudo, K. Okazaki, K. Tsuta, Y. Shikinami, and A. Uchida, *Biomaterials*, **26(20)**, 4301 (2005).
49. T. Aigner and J. Stove, *Adv. Drug. Deliv. Rev.*, **55(12)**, 1569 (2003).
50. J. Green, S. Schotland, D. J. Stauber, C. R. Kleeman, and T. L. Clemens, *Am. J. Physiol.*, **268(5 Pt 1)**, C1090 (1995).
51. S. Shi, M. Kirk, and A. J. Kahn, *J. Bone. Miner. Res.*, **11(8)**, 1139 (1996).

52. M. G. Haugh, M. J. Jaasma, and F. J. O'Brien, *J. Biomed. Mater. Res: Part A* 2008.

53. A. J. Bailey, N. D. Light, and E. D. Atkins, *Nature*, **288(5789)**, 408–410 (1980).

54. I. V. Yannas and A. V. Tobolsky, *Nature*, **215(5100)**, 509–510 (1967).

55. E. Farrell, J. O'Brien F, P. Doyle, J. Fischer, I. Yannas, B. A. Harley, B. O'Connell, P. J. Prendergast, and V. A. Campbell, *Tissue Eng.*, **12**, 459–468 (2006).

56. W. Bartlett, J. A. Skinner, C. R. Gooding, R. W. Carrington, A. M. Flanagan, T. W. Briggs, and G. Bentley, *J. Bone Joint Surg. Br.*, **87**, 640–645 (2005).

57. S. Marlovits, G. Striessnig, F. Kutscha-Lissberg, C. Resinger, S. M. Aldrian, V. Vecsei, and S. Trattnig, *Knee Surg. Sports Traumatol. Arthrosc.*, **13(6)**, 451–457 (2005).

58. M. H. Zheng, C. Willers, L. Kirilak, P. Yates, J. Xu, D. Wood, and A. Shimmin, *Tissue Eng.*, **13**, 737–746 (2007).

59. S. Nehrer, H. A. Breinan, A. Ramappa, G. Young, S. Shortkroff, L. K. Louie, C. B. Sledge, I. V. Yannas, and M. Spector, *Biomaterials*, **18(11)**, 769–776 (1997).

60. J. P. Gleeson, T. J. Levingstone, and J. O'Brien F, Layered scaffold suitable for osteochondral defect repair. 2009.

61. F. J. O'Brien, B. A. Harley, I. V. Yannas, and L. J. Gibson, *Biomaterials*, **26**, 433–441 (2005).

62. V. Karageorgiou and D. Kaplan, *Biomaterials*, **26**, 5474–5491 (2005).

63. C. Chang, F. H. Lin, C. C. Lin, C. H. Chou, and H. C. Liu, *J. Biomed. Mater. Res. B. Appl. Biomater.*, **71**, 313–321 (2004).

64. D. J. Griffon, M. R. Sedighi, D. V. Schaeffer, J. A. Eurell, and A. L. Johnson, *Acta. Biomater.*, **2**, 313–320 (2006).

65. S. M. Vickers, L. S. Squitieri, and M. Spector, *Tissue Eng.*, **12**, 1345–1355 (2006).

66. R. P. Silverman, L. Bonasser, D. Passaretti, M. A. Randolph, and M. J. Yaremchuk, *Plast. Reconstr. Surg.*, **105**, 1393–1398 (2000).

67. B. Obradovic, I. Martin, R. F. Padera, S. Treppo, L. E. Freed, and G. Vunjak-Novakovic, *J. Orthop. Res.*, **19**, 1089–1097 (2001).

68. P. K. Bos, J. DeGroot, M. Budde, J. A. Verhaar, and G. J. van Osch, *Arthr Rheum.*, **46**, 976–985 (2002).

69. L. A. Solchaga, J. S. Temenoff, J. Gao, A. G. Mikos, A. I. Caplan, and V. M. Goldberg, *Osteoarthr. Cartilage*, **13**, 297–309 (2005).

70. K. Zaslav, B. Cole, R. Brewster, T. Deberardino, J. Farr, P. Fowler, and C. Nissen, *Am. J. Sports Med.*, (2008).
71. S. J. Roberts, D. Howard, L. D. Buttery, and K. M. Shakesheff, *Br. Med. Bull.*, (2008).
72. K. Pelttari, E. Steck, and W. Richter, *Injury* **39 Suppl 1**, S58–S65 (2008).
73. A. Schaffler and C. Buchler, *Stem. Cells.*, **25**, 818–827 (2007).
74. W. C. Pederson and D. W. Person, *Orthop. Clin. North Am.*, **38**, 23–25, (2007).
75. U. Kneser, D. J. Schaefer, E. Polykandriotis, and R. E. Horch, *J. Cell. Mol. Med.*, **10**, 7–19 (2006).
76. S. X. Hsiong and D. J. Mooney, *Periodontol 2000*, **41**, 109–122 (2006).
77. A. Yokoyama, I. Sekiya, K. Miyazaki, S. Ichinose, Y. Hata, and T. Muneta, *Cell. Tissue Res.*, 1–10 (2005).
78. M. M. Stevens, R. P. Marini, D. Schaefer, J. Aronson, R. Langer, and V. P. Shastri, *Proc. Natl. Acad. Sci. U S A*, **102)**, 11450–11455 (2005).
79. G. J. Meijer, J. D. de Bruijn, R. Koole, and C. A. van Blitterswijk, *Biomaterials*, **29**, 3053–3061 (2008).
80. D. J. Kelly and P. J. Prendergast, *Med. Biol. Eng. Comput.*, **42**, 9–13 (2004).
81. A. Salim, R. P. Nacamuli, E. F. Morgan, A. J. Giaccia, and M. T. Longaker, *J. Biol. Chem.*, **279**, 40007–40016 (2004).
82. E. Volkmer, I. Drosse, S. Otto, A. Stangelmayer, M. Stengele, B. C. Kallukalam, W. Mutschler, and M. Schieker, *Tissue Eng. Part A*, **14**, 1331–1340 (2008).
83. J. Rouwkema, J. de Boer, and C. A. Van Blitterswijk, *Tissue Eng.*, **12**, 2685–2693 (2006).
84. J. K. Leach, D. Kaigler, Z. Wang, P. H. Krebsbach, and D. J. Mooney, *Biomaterials*, **27**, 3249–3255 (2006).
85. U. Kneser, E. Polykandriotis, J. Ohnolz, K. Heidner, L. Grabinger, S. Euler, K. U. Amann, A. Hess, K. Brune, P. Greil, M. Sturzl, and R. E. Horch, *Tissue Eng.*, **12**, 1721–1731 (2006).
86. J. Oswald, S. Boxberger, B. Jorgensen, S. Feldmann, G. Ehninger, M. Bornhauser, and C. Werner, *Stem Cells*, **22**, 377–384 (2004).
87. X. Hu, S. P. Yu, J. L. Fraser, Z. Lu, M. E. Ogle, J. A. Wang, and L. Wei, *J. Thorac. Cardiovasc. Surg.*, **135**, 799–808 (2008).
88. M. H. Theus, L. Wei, L. Cui, K. Francis, X. Hu, C. Keogh, and S. P. Yu, *Exp. Neurol.*, **210**, 656–670 (2008).
89. X. Mao, Q. Zeng, X. Wang, L. Cao, and Z. Bai, *J. Huazhong. Univ. Sci. Technol. Med. Sci.*, **24**, 566–568 (2004).

90. K. Pelttari, A. Winter, E. Steck, K. Goetzke, T. Hennig, B. G. Ochs, T. Aigner, and W. Richter, *Arthr. Rheum.*, **54**, 3254–3266 (2006).

91. P. Zimmermann, S. Boeuf, A. Dickhut, S. Boehmer, S. Olek, and W. Richter, *Arthr. Rheum.*, **58**, 2743–275 (2008).

92. A. Dickhut, K. Pelttari, P. Janicki, W. Wagner, V. Eckstein, M. Egermann, and W. Richter, *J. Cell. Physiol.*, **219**, 219–226 (2009).

93. J. M. Jukes, S. K. Both, A. Leusink, L. M. Sterk, C. A. van Blitterswijk, and J. de Boer, *Proc. Natl. Acad. Sci. U S A*, **105**, 6840–6845 (2008).

94. E. Farrell, O. P. van der Jagt, W. Koevoet, N. Kops, C. J. van Manen, C. A. Hellingman, H. Jahr, F. J. O'Brien, J. A. Verhaar, H. Weinans, and G. J. van Osch, *Tissue Eng. Part A* (2008).

95. E. Steck, J. Fischer, H. Lorenz, T. Gotterbarm, M. Jung, and W. Richter, *Stem. Cells. Dev.* (2008).

Chapter 39

APPLICATION OF SCAFFOLDS FOR ARTIFICIAL SKIN IN REGENERATIVE MEDICINE

Hyun Ju Lim[a] **and Ho Yun Chung**[b]

[a] *Department of Advanced Organic Materials Science and Engineering, Kyungpook National University, 1370 Sankyuk-dong, Buk-gu, Daegu, Republic of Korea*
[b] *Department of Plastic & Reconstructive Surgery, Kyungpook National University, School of Medicine, 50, Samduck-dong, Joong-gu, Daegu, Republic of Korea*
witchzz@hanmail.net; hy-chung@knu.ac.kr

Tissue engineering and regenerative medicine is a rapidly growing and emerging as an interdisciplinary field in the area of biological and medical cooperation, which aims to regenerate new biological material for replacing diseased or damaged tissues or organs. Tissue engineering for skin, as the largest organ of the human body, has been developed from early periods. Nowadays, skin is readily accessible for direct tissue modulation, for example, gene therapy or stem cell therapy. And many researchers are approaching to replace scar or nonfunctional tissue using biological scaffolds for skin regeneration, such as collagen, gelatin, fibrin glue, chitosan, poly(lactic acid) (PLA), poly(lactic-co-glycolic acid) (PLGA), and so on. Furthermore, many tissue-engineered products are showing, such as Epicel®, Integra®, Alloderm®, Transcyte®, Apligraf®, and so on. And their clinical applications are reported.

Handbook of Intelligent Scaffolds for Tissue Engineering and Regenerative Medicine
Edited by Gilson Khang

www.panstanford.com

39.1 Introduction

The skin, the largest organ in the body, primarily serves as a protective barrier against the environment. Skin injuries, including burns and infections, are among the most complex and harmful physical injuries to evaluate and manage.[1] Loss of integrity of large portions of the skin may result in significant disability or even death. Skin replacement has been a challenging task for surgeons ever since the introduction of skin grafts by Reverdin in 1871.

The skin consists of the epidermis and the dermis, with a complex nerve and blood supply. Figure 39.1 shows a schematic of requirements to create a fully functional skin and the structure of the skin. The adult skin is composed of two distinct layers, the epidermis and the dermis. The thin epidermis comprises rapidly dividing keratinocytes, which produce keratin, a strong structural protein. The topmost layer of the epidermis is the stratum corneum, a thin watertight layer of dead, flattened cells. The thicker dermis, situated below the epidermis, is a complex association of fibroblasts and the extracellular matrix (ECM), accounting for the skin's mechanical integrity and elasticity. The skin's blood vessels, nerve fibers, and lymphatics all reside within this layer. The epidermis is nourished by diffusion of small molecules from the dermis. The skin's appendages,

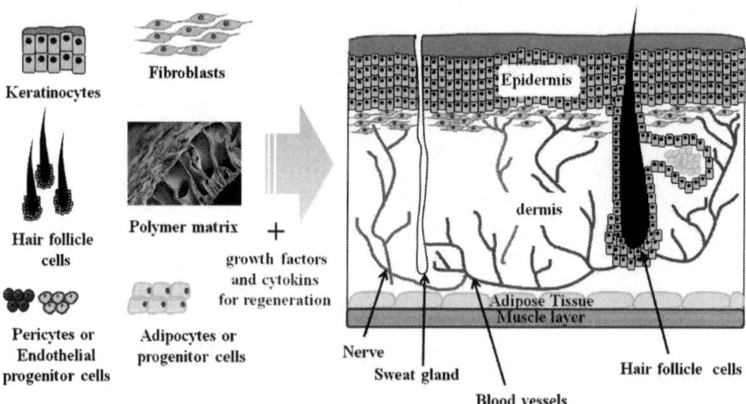

Figure 39.1. Schematic of requirements to create a fully functional skin. See also Color Insert.

namely, hair follicles and sweat glands, traverse both levels. Beneath the dermis is the hypodermis, a loose connective layer comprising primarily adipose tissue. These layers play an important role in protecting the body from any mechanical damage such as wounds and infection.[2]

Normal wound healing involves a complex and dynamic process involving soluble mediators, blood cells, ECM components, and resident cells that eventually result in the restoration of tissue integrity, but a superbly orchestrated series of overlapping processes result in a varying degree of functional and structural restoration. When successful, wound healing restores normal function with a well-organized minimal scar. But when the control mechanisms are abnormal, chronic wound or hypertrophic scar formation can occur. Wound healing occurs in interrelated and interdependent phases: inflammation, cell proliferation, and remodeling. During the initial stages of wound healing, a temporary repair is usually achieved in the form of a blood clot. Inflammatory cells, fibroblasts, and capillaries later invade the clot to form a contractile granulation tissue that draws the wound margins together; meanwhile, the cut epidermal edges migrate to cover the degenerated wound surface.[3] For improvement of wound-healing cases, a number of biomaterials are used as dressings or scaffolds that may use natural or artificial materials. Polymers, such as semipermeable polyurethane dressings, alginate dressing, and so on, are used as wound dressings for superficial wounds.[4–6] More complicated dressings vary from dermal replacements made from collagen and/or other proteins backed by a polymer layer to the biological dressing, which contains ECMs such as collagen, hyaluronic acid (HA), and fibrin with seeded cells.[7,8] Biomaterials for promoting wound healing have already been applied successfully in various clinical fields.[8]

Artificial skin research using tissue-engineered skin replacements such as cultured autologous and allogenic keratinocytes grafts, autologous or allogenic composites, acellular biological matrices, and cellular matrices, including biological substances such as fibrin sealant and various types of collagen and HA is an active field in regenerative medicine and has opened new options for clinicians to restore various skin defects.[9] In tissue-engineered skin substitutes, an autograft (biosynthetic skin substitutes and autologous cultures/noncultured skin engineering products) is the best replacement for lost skin; however, in clinical practice this is not always possible, particularly in large total-body surface area burns, as there is often an insufficient amount of skin for autografting available at the time of burn excision or the physiological condition of the patient precludes the harvesting of skin. Furthermore, most skin substitutes are expensive, and considerable experience is required to decide which material is appropriate for any given situation. These are allografts and xenografts. These can be used to provide temporary wound coverage, but there are issues with graft rejection, availability, cultural and ethical implications, and the possibility of disease transfer.[10] Tissue-engineered skin is one of the most advanced tissue constructs, yet it lacks several important functions such as skin appendages and immune cells, and it is difficult to recreate the entire human skin. Gene therapy can be used for better functional tissue and the healing process.[11] Recently, multipotent skin stem cells have been applied to promote tissue regeneration and to prevent scar formation.[12] For example, the application of hair follicle bulge cells (HFBCs) and dermal papilla cells (DPCs) was illustrated in an experiment.[13] The application of autologous or allogenic stem cells in combination with specific growth factors in artificial skin substitute may also help.[2]

39.2 Scaffolds: Biomaterials as One of Important Factors for Regeneration of Skin Tissue

In tissue engineering emerging at the end of the 1980s, two important factors are generally known, biological characteristics of cells and biocompatible scaffolds. When cells are used, donor tissue is dissociated into individual cells, which are expanded in culture, attached to a support matrix, and re-implanted after expansion. Among those factors, biocompatible scaffolds are important for maintaining tissue architecture. Only isolated cell have a limited capacity to reform their respective tissue structure, because they lack a template that guides in restructuring. Moreover, transplantation of large volumes of constructs cannot survive because of diffusion limitations that restrict interaction with the host environment for nutrients, gas exchange, and elimination of waste products. Therefore, implanted cells survive poorly under such conditions.[14]

Scaffolds used usually in therapeutic applications are made from natural or synthetic polymers that are often resorbed or degraded in the body. In degradable scaffolds, several key challenges exist. First, the primary generation of degradable polymers, widely used in regenerative medicine, was adapted from other surgical uses and has deficiencies in terms of mechanical and degradation properties.[15] The second major challenge is how to fabricate these polymers into scaffolds that have defined shapes and a complex porous internal architecture that can direct tissue growth.[16] In nondegradable scaffolds, which are rarely used for skin substitutes in these days, a major challenge is how foreign-body reactions are reduced. And also, these polymers and their degradation products as well as nondegradable polymers must also be nontoxic and nonimmunogenic upon implantation. The ideal synthetic biomaterial, such as wound dressing, and biologic skin substitutes should have the following characteristics, listed by Pruitt *et al.*[17,18]

- absence of antigenicity
- tissue compatibility
- absence of local or systemic toxicity
- impermeability for exogenous microorganisms
- water vapor transmission similar to normal skin
- rapid and sustained adherence to wound surface
- conformance to surface irregularities
- elasticity to permit motion of underlying tissue
- resistance to linear and shear stress
- tensile strength to resist fragmentation
- inhibition of wound surface flora and bacteria
- long shelf life, minimal storage requirements
- biodegradability (for permanent membranes)
- low cost
- minimization of nursing care of wound

- minimization of patient discomfort
- translucent properties to allow direct observation of healing
- reduction of healing time
- decreased rate of infection
- patient acceptance

Skin regeneration has been become the area to see the benefit of tissue engineering. In this field, a number of biomaterials are used as scaffolds that may use natural or synthetic materials. For examples, chitosan, gelatin, and HA are used as porous scaffolds[19]; collagen, polycaprolactone (PCL)[20] or fibrin-coated PCL[21] and polyurethane[6] are used as biofilms or matrices; and PLGA,[22] PLA, polyglycolic acid (PGA),[23] and their copolymers are used as biomeshes by microfabrication technology. Improvements in these biomaterials need to be optimized. In general, the most common approach is to create three-dimensional (3D) biodegradable scaffolds in the shape of the respective tissue. A successful 3D tissue engineering scaffold must have a highly porous structure and good mechanical stability. High porosity and an optimally designed pore size provide structural space for cells to accommodate and migrate to the inner stage and enable exchanging of nutrients and oxygen between the scaffold and the environment. The studies of 3D printing and electrospinning technologies are reported for accurate manufacture, creating materials of a defined pore size.[24–29] Biomimicking technology, such as adhesion peptides using Arg-Gly-Asp (RGD) sequences, which is important in integrin binding, or growth factors' incorporation into biomatrices with nanoscale surface manipulation, has been shown to enhance cell adhesion and migration.[30–32]

39.3 Clinical Applications of Tissue-Engineered Skin Products

Tissue-engineered biological dressings and skin substitutes offer great promise in the treatment of burns, chronic ulcers, donor site and other surgical wounds, and a variety of other dermatological conditions. Despite large potential benefits, tissue-engineered skin products have suffered setbacks in recent years and have garnered a considerably lower market share than commercial promoters anticipated. The reasons may be that the mechanism of action of these products is not universally agreed upon, that is, graft "take" is not usually considered to occur, and that these products do not specifically match a clinical treatment to an underlying pathology.[18] Currently available products approved by the Food and Drug Administration as matrices for wound repair include Epicel® (Genzyme Tissue Repair), Integra® (Integra Life Sciences), Alloderm® (Life Cell), Transcyte® (Advanced Biohealing), Apligraf® (Organogenesis), Orcel® (Ortec international), and so on. Table 39.1 shows examples of commercially created skin substitutes.

Table 39.1. Classification of commercialized skin substitutes.

Main component & commercial product name (Manufacturer)	Description	Characterization
Epidermal substitutes		
Epicel® (Genzyme Tissue Repair Corporation)	Cultured epidermal autograft grown from patient skin biopsy on petrolatum gauze backing	Deep dermal or full-thickness burns over a total-body surface area >= 30% coverage with little risk of rejection. 3 wk required to produce fragile confluent sheets.
Laserskin® (Fidia Advanced Biopolmers)	Subconfluent, autologous keratinocytes from skin biopsy in a perforated HA matrix	A less fragile delivery system for keratinocytes, 100% esterified HA membrane/cell interaction properties improve mechanical stability. Minimum of 3 wk required to expand keratinocyte population.
EpiDex® (Modex Therapeutiques)	Cultured, autologous outer root sheath hair follicle cells on silicon membrane	Using substitute to take up to 6 wk after harvesting, cells have increased proliferative capacity and can be cryopreserved for repeat applications. Success in chronic ulcer treatment. Product fragile.

(Contd.)

Table 39.1. (*Continued*)

Main component & commercial product name (Manufacturer)	Description	Characterization
BioSeed-S® (Biotissue Technologies)	Autologous human keratinocytes from patient skin biopsy on a fibrin matrix	After keratinocyte is growing for 2~3 wk, it is incorporated into a 3D fibrin matrix. This product can be applied to treat partial-thickness burns and chronic ulcers.
Myskin® (CellTran)	Subconfluent, autologous keratinocytes on a PVC polymer coated with a plasma-polymerized surface	PVC raises keratinocyte attachment and proliferation, providing a more stable delivery platform. Up to 14 d required for cell expansion. Repeated application needed for good clinical outcome.
CellSpray® (CellSpray Inc.)	Preconfluent, autologous keratinocytes delivered into a suspension for spray	Significant reduction in the use of pressure garments for scar management (down to <20% of previous usage). Repeated application needed.
Cryoskin® (Cell Tran)	Cryopreserved, confluent allogeneic keratinocytes delivered on chemical surface	Long shelf life, sterility data available at the moment of use, rapid availability. Pending MHRA (UK) approval. Repeated application needed.
Holoderm® (TegoScience)	Cultured, autologous keratinocytes from patient normal skin biopsy	Excellent wound repair ability, short growing term, low cost, no immune rejection, large area of wound coverage. Little normal skin from patient needed to obtain some cells.
Kaloderm® (TegoScience)	Cultured, allogeneic keratinocytes from same-kind-origin skin cells	No immune rejection, mass production can be obtained from same-kind-origin skin cells, minimized to make scar. Potential risk of cell viability, although none reported.
Dermal substitutes		
Alloderm® (Life Cell Corporation)	Processed cadaver allograft skin	Processing helps reduce antigenic components, successful in resurfacing full-thickness burns. Problems of graft rejection and disease transfer.

Table 39.1. (*Continued*)

Main component & commercial product name (Manufacturer)	Description	Characterization
Dermagraft® (Advanced Biohealing, Inc.)	Cryopreserved, allogeneic fibroblast-derived dermal matrix	Neonatal fibroblast rapidly and easily proliferated to produce some growth factors to aid in wound healing. Tends to have been used as a biological dressing with multiple applications, even if it does not clearly demonstrate rejection, and is very resistant to tearing.
Integra® (Johnson & Johnson)	Synthetic polysiloxane polymer; bovine type I collagen and GAGs	Encourages ingrowth of fibroblasts and epithelial cells. Epidermal equivalent replaced after 14–21 d with an autograft. Bovine collagen presents and antigenicity and disease risk.
Transcyte® (Advanced Biohealing, Inc)	Thin silicone layer over collagen-coated nylon mesh seeded with neonatal allogeneic fibroblasts	It's not marketed currently, but successfully used to treat second- and third-degree burns. Dermal fibroblasts secrete collagen; GAGs and growth factors to aid wound healing. Problems are nylon mesh (not biodegradable), rejection, and disease risk from fibroblasts.
EZ-Derm® (Brennen Medical, LLC)	Porcine-derived xenograft in which the collagen has been chemically cross-linked with an aldehyde	Temporary, protective barrier that allows the natural healing process to continue undisturbed, keeps body fluids in and bacteria out; aldehyde cross-linking adds strength and durability for extended wound coverage.
Repliform® (LifeCell Corporation)	Acellular human dermal allograft	Application for urological plastic surgery. Repeated application needed.
FortaFlex® (Organogenesis Inc.)	Acellular porcine small intestine submucosa	Application to treat full- and partial-thickness burns, venous and diabetic ulcers. Rejection and disease risk. Repeated application needed.

(*Contd.*)

Table 39.1. (*Continued*)

Main component & commercial product name (Manufacturer)	Description	Characterization
Biobrane® (UDL Laboratories Inc.)	Porcine collagen chemically bound to silicone/nylon membrane	Temporary covering of partial-thickness burns and wounds. Rejection and disease risk.
Hyalograft 3D® (Fidia Advanced Biopolymers)	Autologous fibroblasts on esterified HA matrix	Rapidly to proliferate keratinocyte, application for full- and partial-thickness wounds and management of diabetic foot ulcers. Repeated application needed.
Matriderm® (Skin&Health Care)	Acellular scaffold made with bovine collagen types I, III, V and elastin	Simultaneous use of this and split-thickness skin grafting in a single-step procedure. Reconstructed skin more elastic; visible reduction in the development of scar formation. Repeated application needed.
Permacol® (Tissue Sciences Laboratories)	Porcine-derived acellular dermal matrix	Nonimmunogenic due to processing to remove noncollagenous and cellular materials, supports host fibroblast infiltration and revascularization. Revascularization sometimes inefficient to support overlying epidermal graft.
Apligraf® (Organogenesis)	Human allogeneic neonatal keratinocytes (epidermis) and foreskin fibroblasts (dermis) in bovine type I collagen, ECM proteins, and cytokines	Many growth factors and cytokines to be included, and no major reported side effects. Improves granulation tissue deposition and no signs of rejection observed. Risk of chronic graft rejection and disease from allogeneic keratinocytes and fibroblasts. Requires repeated applications.
Bilayer substitutes		
OrCel® (Ortec International)	Human allogeneic neonatal keratinocytes (epidermis) and foreskin fibroblasts (dermis) in a bovine collagen sponge	Provides a favorable environment for host cell migration and provides a source of cytokines and growth factors, application for skin graft donor sites and mitten-hand surgery for epidermolysis bullosa. Not intended for use as a permanent skin replacement. Designed as a biological dressing.

Abbreviations: PVC, polyvinyl chloride; MHRA, Medicines and Healthcare Products Regulatory Agency; GAG, glycosaminoglycan.

However, at present, no manufactured skin substitute has provided an outcome consistently comparable with an autologous skin graft. There are problems that should be solved with currently available products, which are reduced vascularization, scarring, absence of differentiated structures, delay involved with cell culture, persistence of cells in heterologous grafts, biocompatibility, mechanical and handling properties, cell source, and development, safety, and product costs.[2] the clinical effect is often modest and sometimes not justifiable from a cost-benefit perspective. Nevertheless, these devices continue to find application, and improvements may facilitate broader use in the future. Skin replacement products are the most advanced, with a number of tissue-engineered wound care materials in several international communities. The potential effect of this field is that it offers novel solutions to the medical field for drug screening, genetic engineering, and tissue and organ replacements.

39.4 Conclusions and Outlook

It is hoped that less expensive and more clinically effective tissue-engineered skin products can be developed in the future. But, we should keep in mind the fact that skin pathology is a multivariable process that cannot necessarily be solved with a construct that histologically resembles the skin. A successful tissue-engineering technique requires intensive cooperation between clinicians, biologists, and engineers. Tissue engineering and regenerative medicine have given new therapeutic approaches that may overcome the drawbacks in current therapies.[33] The importance of available first-generation tissue-engineered products should not be discounted on account of modest clinical and commercial success. Although these products were not the success hoped for, an important proof-of-principle was demonstrated in regulatory, technical, manufacturing, logistical, and commercialization elements.

In the skin substitute field, a pathologic-based approach, improvement in quality, and more clinical-oriented construction will bring more similarity to native skin grafts. In this way, tissue-engineered skin substitutes may be a therapeutic method for regeneration rather than repair and the patients would be saved with regenerative medicine.

Acknowledgments

This work was supported by the Grant of the Korean Ministry of Education, Science and Technology (the Regional Core Research Program/Anti-aging and Well-being Research Center).

References

1. J. Lademann, *Skin Pha. Phy.*, **22** (2009).
2. A. D. Metcalfe and M. W. J. Ferguson, *J. R. Soc. Interface*, **4** (2007).
3. A. T. Singer and R. A. Clark, *N. Engl. J. Med.*, **341** (1999).
4. S. Kleczynski, T. Niedzwiecki, and M. Brzezinski, *Polym. Med.*, **16** (1986).
5. T. T. Phan, I. J. Lim, E. K. Tan, B. H. Bay, and S. T. Lee, *Cell Tissue Bank*, **6** (2005).
6. K. T. Holland, W. Davis, E. Ingham, and G. A. Gowland, *J. Hosp. Infect.*, **5** (1984).
7. M. V. Sefton and K. A. Woodhouse, *J. Cutan. Med. Surg.*, **3** (1998).
8. Y. F. Chin, *J. Nurs.*, **53** (2006).
9. F. Liu and J. Wu, *J. Biomed. Eng.*, **17** (2000).
10. C. Pham, J. Greenwood, H. Cleland, P. Woodruff, and G. Maddern, *Burns*, **33** (2007).
11. S. T. Andreadis, *Adv. Biochem. Eng. Biotechnol.*, **103** (2007).
12. G. Cotsarelis, *J. Invest. Dermatol.*, **126** (2006).
13. H. T. Wang, B. Chen, C. W. Tang, and D. H. Hu, *Zhonghua Shao Shang Za Zhi*, **23** (2007).
14. J. P. Vacanti, R. Langer, J. Upton, and J. J. Marler, *Adv. Drug Del. Rev.*, **33** (1998).
15. L. G. Griffith, *Ann. NY Acad. Sci.*, **961** (2002).
16. L. G. Griffith and M. A. Schwartz, *Nat. Rev. Mol. Cell. Biol.*, **7** (2006).
17. B. A. Pruitt and N. S. Levine, *Arch. Surg.*, **119** (1984).
18. M. Ehrenreich and Z. Ruszczak, *Tissue Eng.*, **12** (2006).
19. H. Liu, Y. Yin, and K. Yao, *J. Biomater. Appl.*, **21** (2007).

20. N. T. Dai, M. R. Willianson, N. Khammo, E. F. Adams, and A. G. Coombes, *Biomaterials*, **25** (2004).
21. H. L. Khor, K. W. Ng, A. S. Htay, J. T. Schantz, S. H. Teoh, and D. W. Hutmacher, *J. Mat. Sci. Mater. Med.*, **14** (2003).
22. K. W. Ng and D. W. Hutmacher, *Biomaterials*, **24** (2007).
23. W. Ryu, S. W. Min, K. E. Hammerick, M. Vyakarnam, R. S. Greco, F. B. Prinz, and R. J. Fasching, *Biomaterials*, **28** (2007).
24. V. Mironov, T. Boland, T. Trusk, G. Forgacs, and R. R. Markwald, *Trends Biotechnol.*, **21** (2003).
25. H. Seitz, W. Rieder, S. Irsen, B. Leukers, and C. Tille, *J. Biomed. Mater. Res. B Appl. Biomater.*, **74B** (2005).
26. K.W. Lee, S. Wang, L. Lu, E. Jabbari, B. L. Currier, and M. J. Yaszemski, *Tissue Eng.*, **12** (2006).
27. Y. K. Luu, K. Kim, B. S. Hsiao, B. Chu, and M. Hadjiargyrou, *J. Control. Rel.*, **89** (2003).
28. M. Li, M. J. Mondrinos, M. R. Gandhi, F. K. Ko, A. S. Weiss, and P. I. Lelkes, *Biomaterials*, **26** (2005).
29. M. C. McManus, E. D. Boland, D. G. Simpson, C. P. Barnes, and G. L. Bowlin, *J. Biomed. Mater. Res. A*, **82A** (2006).
30. J. A. Hubbell, *Swiss Med. Wkly.*, **136** (2006).
31. B. Li, J. Chen, and J. H. Wang, *J. Biomed. Mater. Res. A*, **79** (2006).
32. M. H. Fittkau, P. Zilla, D. Bezuidenhout, M. P. Lutolf, P. Human, and J. A. Hubbell, *Biomaterials*, **26** (2005).
33. Y. Ikada, *J. R. Soc. Interface*, **3** (2006).

Chapter 40

BIODEGRADABLE SCAFFOLDS FOR BONE REGENERATION

Yoichi Yamada
Center for Genetic and Regenerative Medicine, Nagoya University School of Medicine, Nagoya, Japan
yyamada@med.nagoya-u.ac.jp

Bone defects often occur due to various diseases, and the golden standard approach to the treatment is autogenous bone grafting. However, the bone collecting involved not only severe invasiveness with attendant donor site morbidity and complications but also the limited supply of autogenous material. For overcoming this problem, various artificial materials containing ceramic material, synthetic biological material (nanomaterial), and biomaterial have been developed. This chapter has focused on the use of biodegradable scaffolds, especially biodegradable ceramic composites (β-tricalcium phosphate [β-TCP]), nanofibers hydrogel peptide (PM), and injectable biomaterial—tissue-engineered bone (TEB). TEB consists of platelet-rich plasma (PRP) as the scaffold and signal molecules and mesenchymal stem cells (MSCs) as the isolated cells for effective bone regeneration applied by tissue engineering and the regenerative medicine concept. According to the animal study and clinical applications, these materials would

Handbook of Intelligent Scaffolds for Tissue Engineering and Regenerative Medicine
Edited by Gilson Khang

www.panstanford.com

be useful as a biodegradable scaffold for bone regeneration and lead to favorable results in therapeutic applications.

40.1 Introduction

A golden standard of clinical treatment for bone defects from tumor resection, congenital malformation, trauma, osteoporotic fractures, or periodontitis is autogenous bone grafting. However, bone extraction has severe invasiveness with attendant donor site morbidity and complications such as swelling, pain, and hemorrhage, is limited in supply, and is occasionally not suitable for the proposed reconstruction because of poor tissue quality or difficulty in shaping the graft. An approach to this problem focused on the development of various artificial materials, and it was thought that these materials containing ceramic materials, anorganic porous bovine-derived bone mineral scaffolds, synthetic biological material (nanomaterial), and biomaterial might be used instead of the autogenous bone.[1–3] However, most of these artificial materials alone were difficult to induce adequate bone formation and suffered from increased susceptibility to infection, an uncertain long-term interaction with the host's physiology.[4] And in the use of animal-derived materials, there are some problems that they may not have uniform quality, foreign substances may contaminate the material, and animal infection may occur. There is also currently an insufficient amount of information available on the essential properties of scaffolds for bone regeneration.

On the other hand, recently even in clinical applications, treatments based on tissue engineering and regenerative medicine (TERM) technology are expected to be of benefit from the viewpoint of reduced hospital days, donor site morbidity, and immune reactions compared with conventional grafts.[5–7] The concept is to regenerate tissues by transplanting stem cells with scaffolds made from natural or synthetic materials and appropriate signal molecules, and bone regeneration is regarded as a application of this concept (Fig. 40.1).

The selected approach involved the application of MSCs as the cells. The MSCs are easy to harvest by bone marrow aspiration

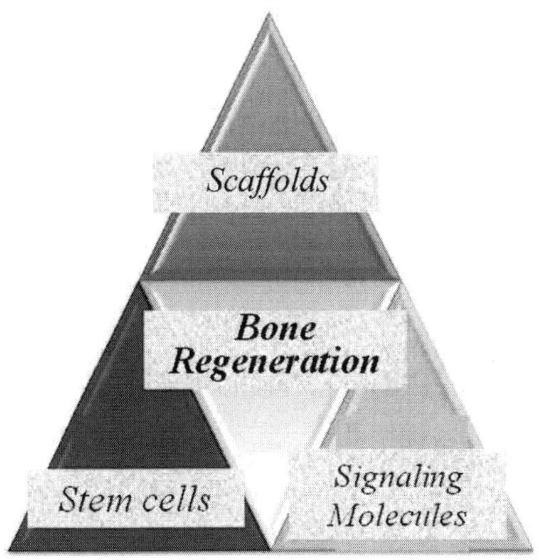

Figure 40.1. A concept of tissue engineering and regenerative medicine.

and can be cultured in the laboratory. They are multipotent and show active proliferation.[8] One of the TERM approaches has attempted to create new bone based on MSCs seeded onto porous ceramic scaffolds. The scaffolds in TERM also play an important role and the ideal biomaterial scaffolds, particularly in bone regeneration fields with complicated bone defect shapes, should have excellent plasticity for fitting into complex defect shapes, and the rate of absorption needs to be fast in order to avoid infection. Therefore in this chapter, we would show the possibility of applying biodegradable scaffolds for bone regeneration.

40.2 Biodegradable Scaffolds

40.2.1 *Biodegradable Ceramic Composite (β-TCP)*

Nonbiodegradable scaffolds such as a ceramic composite (hydroxyapatite [HA]) possess inherent problems, including infection, unexpected change of soft tissue overlying the graft, and a potential for late exposure. So attention was paid to biodegradable scaffolds. One of them was synthesized by materials with a three-dimensional structure, which provided a good environment for bone regeneration and its surface facilitated cell attachment, proliferation, and osteogenic differentiation.[9]

The pore average was 200 to 400 μm in diameter, the interconnection average was 60 μm, and the average void volume was 90%. These ceramic composites were disk shaped, 5 mm in diameter, and 4 mm in thickness. Following the concept of TERM, we investigated the bone regeneration ability of the scaffolds: β-TCP alone and β-TCP loaded with MSCs. The results showed by scanning electron microscopy (SEM) that a high percentage of open pores with a connection to each pore and the surface were covered by culture cells with a collagenous-like extracellular matrix (ECM) (Fig. 40.2). The result of bone formation ability showed that some mature bone together with cuboidal active osteoblasts was observed when alkaline phosphatase activity achieved a peak at two weeks.[10] And it confirmed that the bone areas increased in a time-dependent fashion at one, two, four, and eight weeks after implantation. However, empty β-TCP blocks alone did not show any bone formation in the pore area at their weeks, and only fibrous tissues were observed (Fig. 40.3).[10] On the other hand, β-TCP loaded with MSCs has excellent osteogenic characteristics and supports its potential in tissue engineering to repair bone defects. However, it is difficult for new bone to invade the defect for the lack of completely interconnected pores, slow resorption of the ceramic, and lack of osteoinductive properties, and in addition these delivery substances do not have good plasticity and the cellular implantation procedure was complicated by problems associated with the delivery vehicles. Optimally, these should combine with an appropriate rate of biodegradability with the capacity for the respective cells to multiply. But it is difficult to apply the block matrix for the field with the complicated form in oral maxillofacial surgery and orthopedic surgery; therefore next we paid attention to the injectable scaffolds with plasticity.

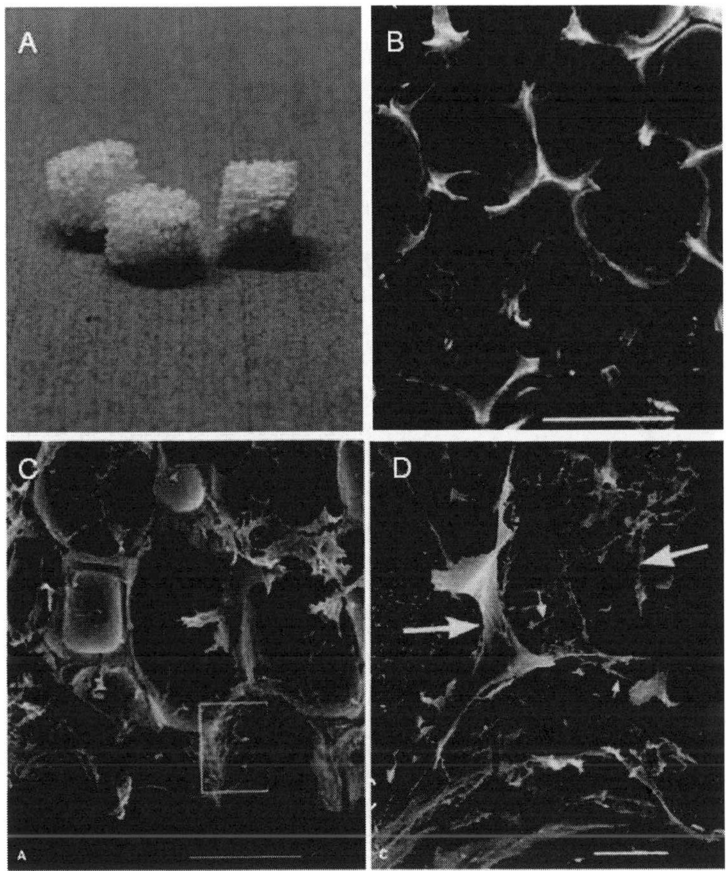

Figure 40.2. β-TCP scaffold. (A) Macro view of β-TCP matrices. (B–D) SEM photomicrograph of a cross section of β-TCP alone (B) or of a composite subcultured for 20 days before implantation. (C) Bar = 500 μm; (D) bar = 20 μm). Small arrow indicates a collagenous-like fiber. Large arrow indicates a cultured cell.[10]

40.2.2 *Nanofibers Hydrogel Peptide*

Nanofibers hydrogel peptide, PuraMatrix^{TM} (PM, 3-DMatrix, Ltd., Tokyo, Japan), is synthesized by chemical peptide methods and has similarities to the fibers and pore sizes found in the ECM. It is a dynamic, organized nanocomposite that not only provides mechanical support for embedded cells but also interacts with cells

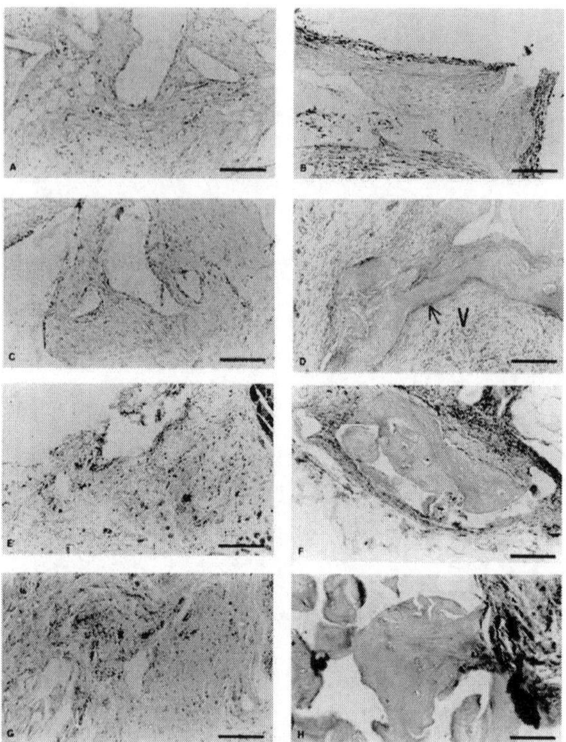

Figure 40.3. Typical histological finding of MSCs loaded in β-TCP scaffolds and empty β-TCP at 1, 2, 4, and 8 weeks after implantation (bar = 100 μm, hematoxylin & eosin staining).[10] Empty β-TCP (A) 1 week after implantation, (C) 2 weeks after implantation, (E) 4 weeks after implantation, and (G) 8 weeks after implantation. MSCs loaded in β-TCP scaffolds (B) 1 week after implantation, (D) 2 weeks after implantation (V indicates vasculature, and arrow indicates osteoblast lining), (F) 4 weeks after implantation, and (H) 8 weeks after implantation.

and promotes and regulates cellular functions such as adhesion, migration, proliferation, and differentiation and is consequently involved in three-dimensional morphogenesis.[11]

PM is also a synthetic biological material formed through the assembly of ionic self-complementary peptides consisting of a 16–amino acid sequence, Ac-RADARADARADARADA-$CONH_2$, which is called RADA16-I and is composed of alanine, arginine, and

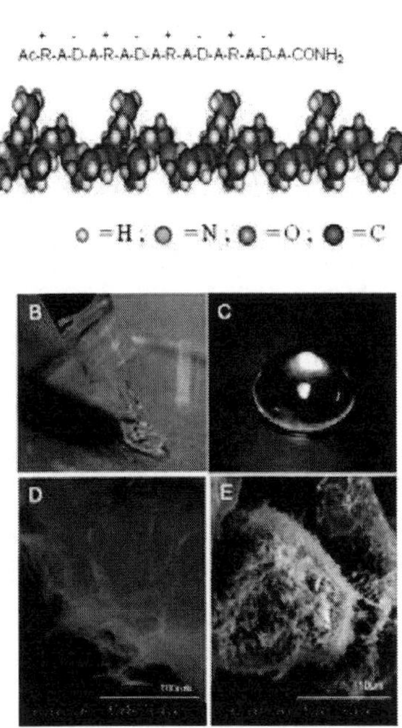

Figure 40.4. Images of PM and PM/dog MSCs (dMSCs).[13] (A) Molecular model of the peptide RADA16-I included in PM. A; alanine. R; arginine. D; aspartic acid. – and + refer to the negatively and positively charged residues. (B, C) Photograph of PM hydrogel at normal temperature. (D, E) SEM images of PM. The PM surface appears like a sheet [D]. dMSCs are sandwiched in PM sheets at a higher magnification [E]. A scale bar on each image is included: 100 nm (D); 10 μm (E). See also Color Insert.

aspartic acid and has greater than 99.5% water content (standard amino acids content 1% w/v)[12] and good fluid characteristics (Fig. 40.4).[13] The peptides have regular repeating units of positively charged residues (arginine) and negatively charged residues (aspartic acid) separated by hydrophobic residues (alanine) and contain 50% charged residues and are characterized by their periodic repeats of alternating ionic hydrophilic and uncharged hydrophobic amino acids.[14] The peptides interact individually by self-complementary and amphiphilic properties and are self-assembled. Thus, PM possesses good plasticity, absorption, and biocompatible properties and is devoid of animal-derived pathogens and antigens.[15] The self-assembled three-dimensional microenvironment structure is also similar to natural ECM, and PM may be useable as a scaffold because of these nanoscale morphological features that may help to control cell behavior.[13–16]

Indeed, in our animal study,[13] PM and PM/dMSCs were implanted into dog mandible bone defects. Macroscopic findings showed that these scaffolds had almost completely disappeared without infection after implantation. Implanted and nonimplanted control regions were collected after two, four, and eight weeks for histological examination. The cavities filled with PM/dMSCs resulted in new bone formation, which was manifested in a tubular pattern and by abundant vascularization after four weeks. The control, PM, and PM/PRP groups did not exhibit appreciable bone formation (Fig. 40.5). The bone-regenerating ability of all implants was assessed by measuring the cortical and medullary bone surface areas without native bone by image analysis. Adding PM alone to the cavity did not significantly increase the cortical or medullary bone surface area compared with the control at two, four, and eight weeks. In contrast, the PM/dMSCs groups showed a significant increase in the surface area at all three time points compared with the control and PM groups at two, four, and eight weeks.[13]

However, the quality of PM needs to be improved because it is too soft to keep the form of bone formation. Taken together, it is suggested that PM might be useful as a scaffold of bone regeneration in cell therapy, and these results would lead to an effective treatment method for bone defects.

40.2.3 *Injectable Tissue-Engineered Bone*

We have attempted to regenerate bone with TERM in a significant osseous defect, more effective with minimal invasiveness and good plasticity and nonimmunoreactive, and to provide a clinical alternative to autogenous bone grafts.[17–20] In the study, we used MSCs as the isolated cells and PRP as the growth factors and scaffold. The new tissue-engineered technology we developed is called

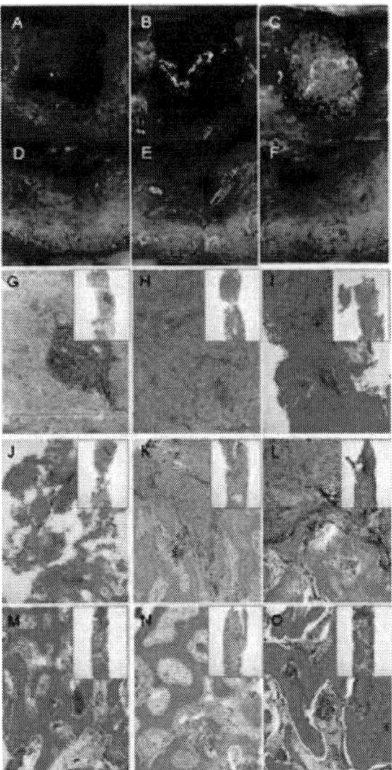

Figure 40.5. Macroscopic and histological evaluation of controls, PM, and PM/dMSCs implantations at each time point.[13] (A) Bone defects were prepared with a trephine bar. Controls were defects only. (B, C) The defects were filled by PM (B), PM/dMSCs (C). (D) At 8 weeks, bone regeneration in the controls (defect only) was not complete. (E, F) After 8 weeks, new bone regeneration was observed with PM (E) and PM/dMSCs (F). Sections of representative implants are shown from each group. The sections were stained with hematoxylin & eosin. Original magnification ×100 (inner square, ×25) for all photographs. (G) 2 weeks control group, (H) 4 weeks control group, (I) 8 weeks control group, (J) 2 weeks PM group, (K) 4 weeks PM group, (L) 8 weeks PM group, (M) 2 weeks PM/dMSCs group, (N) 4 weeks PM/dMSCs group, and (O) 8 weeks PM/dMSCs group. See also Color Insert.

"injectable TEB,"[17–20] which had been established by the tissue engineering concept.[5]

PRP, one of scaffolds, which is a mixture of growth factors and an autologous modification of the fibrin glue, is believed to result in early consolidation and graft mineralization.[21] The use of PRP is based on the premise that the large numbers of platelets found in PRP release significant quantities of mitogenic polypeptides, such as platelet-derived growth factors (PDGF), transforming growth factor-β (TGF-β), and insulin-like growth factor-I (IGF-I). The potential effects of PDGF include the stimulation of mitogenesis of marrow stem cells and the stimulation of angiogenesis.[22] TGF-β has been shown to stimulate chemotaxis and mitogenesis of osteoblast precursors, stimulate the deposition of a collagen matrix for connective tissue healing and bone formation, and inhibit osteoclast formation and bone resorption.[23]

Before clinical application, we investigated the ability of bone regeneration by using the dog mandible model. The implanted regions were collected at two, four, and eight weeks and processed and decalcified for histology. The result showed that the cortical continuity was never restored and the cavities were invaded by a fibrous tissue and less new bone formation was seen in the PRP group (scaffold only). On the other hand, cavities filled with the TEB resulted in new bone formation even after two weeks, with a tubular pattern at eight weeks and abundant vascularization. This pattern reflected a normal bone macrostructure with a well-differentiated marrow cavity compared with cavities filled with autogenous particulate cancellous bone and marrow (PCBM), which showed space by PCBM resorption (Fig. 40.6).[18]

So the TEB-implanted group vascularized well. Ideally, the scaffold should be resorbed at a rate commensurate with new bone formation, within a few weeks. This makes it very different from most hydroxyapatite, β-TCP ceramics,[8] or coral scaffolds[24] that virtually do not degrade during the first few weeks of implantation. Presumably, the disappearance of the TEB left in place induced bone tissue formation, which then self-organized according to the surrounding environment. And the TEB performed better, suggesting a positive influence of PRP for the MSCs.

After the confirmation based on this series of experimental studies, it was performed on a human with the TEB for alveolar bone augmentation and simultaneous implant installation.[19,20,25]

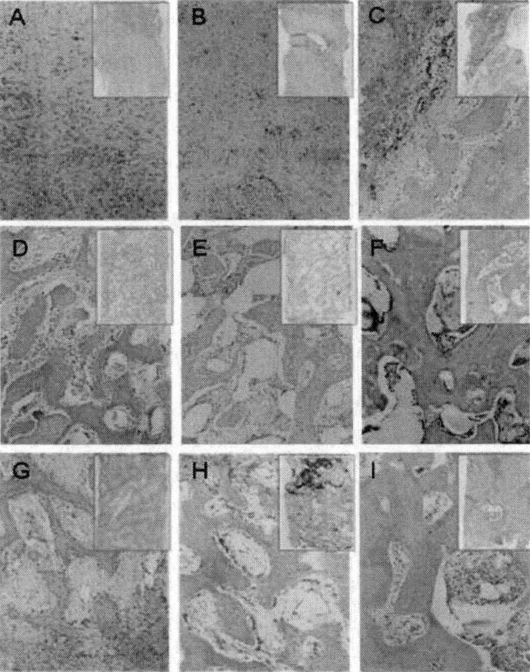

Figure 40.6. Histological evaluation of PRP, PCBM, and TEB (MSCs/PRP) implantations at each time point.[18] Bone defects were prepared with a trephine bar. Sections of representative implants are shown from each group. The sections were stained with hematoxylin & eosin. Original magnification ×200 (inner square, ×40) for all photographs. (A) 2 weeks PRP group, (B) 4 weeks PRP group, (C) 8 weeks PRP group, (D) 2 weeks PCBM group, (E) 4 weeks PCBM group, (F) 8 weeks PCBM group, (G) 2 weeks TEB group, (H) 4 weeks TEB group, and (I) 8 weeks TEB group. See also Color Insert.

40.3 Clinical Application

40.3.1 *Preparation and Clinical Application of MSCs, PRP, and Injectable TEB*

MSCs were isolated from the patients and incubated according to the reported method (Fig. 40.7A).[19,20,25] In cultures, MSCs were trypsinized and used for clinical treatment.

Figure 40.7. Injectable TEB preparation and application.[25] (A) Cultured MSCs, (B) prepared PRP, (C) the clinical application form of TEB, and (D) live cells (green color) and nonactive cells (red color) in TEB after implantation. See also Color Insert.

PRP, extracted one day before surgery, was isolated in a collection bag containing an anticoagulant, citrated by centrifugation, and stored at 22°C in a conventional shaker until used (Fig. 40.7B). Powdered human thrombin (5,000 units; Yoshitomi Co., Japan) was dissolved in 10% calcium chloride in a separate sterile cup. TEB was prepared as described previously.[19,20,25] Briefly, two syringes, one syringe containing MSCs, PRP, and air and the other containing a thrombin/calcium chloride mixture, were connected with a T connector, and the plungers of the syringes were pushed and pulled alternately, allowing air bubbles to go and return between the two syringes. Within 5 to 30 seconds, the contents achieved a gel-like consistency because thrombin affected the polymerization of fibrin to produce an insoluble gel (Fig. 40.7C). It is found that in the TEB gel, the grafted cells are live (Fig. 40.7D). Probably the PRP scaffold for MSCs would encourage MSCs adhesion, proliferation, and differentiation to elicit bone formation.

In the clinical application case of bone regeneration for dental implant, these results showed that injectable TEB would stably predict the success of bone formation and dental implants (Fig. 40.8). And it is that no adverse effects and remarkable bone absorption were seen in the follow-up time for four years.

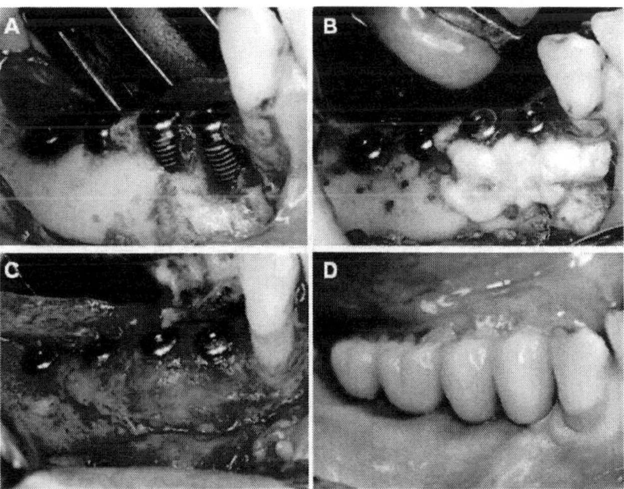

Figure 40.8. Clinical application case of bone regeneration with TEB for dental implant. (A) It is found that alveolar bone for dental implant insertion is insufficient. Therefore the implant thread is exposed. (B) TEB is applied for the thread exposure area. (C) It is found that the bone is regenerated after TEB implantation. (D) Final prosthesis view. See also Color Insert.

40.4 Conclusions and Outlook

This article has focused almost on the use of biodegradable scaffolds as the graft material for effective bone regeneration applied by tissue engineering and the regenerative medicine concept.[5] The result showed that they would be useful as a scaffold for bone regeneration and could be used in clinical application. The application is successful, and these cases would be observed and maintained in the future.

Moreover, the technology would reduce the patient burden and provide minimally invasive cell therapy for patients. It might be a clinical alternative to autogenous bone grafts, which are associated with attendant donor site morbidity.[17–20]

Acknowledgments

The authors wish to thank Drs. Minoru Ueda, Jae Seong Boo, Ryotaro Ozawa, Kenji Ito, Sayaka Nakamura, Ryoko Yoshimi, Kenji Hara, and Eri Umemura, members of the Department of Oral and Maxillofacial Surgery, Nagoya University Graduate School of Medicine, and Shuguang Zhang of the Center for Biomedical Engineering at Massachusetts Institute of Technology for their assistance and contributions to this study. We also thank ArBlast Co. Ltd., Kobe, Japan, and 3D Matrix Co. Ltd., Tokyo, Japan, for their help. This work was supported in part by Grants-in-Aid for Scientific Research (No. 20659297, 21390507) from the Japan Society for the Promotion of Science.

References

1. U. Bilkay, M. Alper, N. Celik, O. Erdem, H. Kerem, C. Ozek, O. Zekioglu, Y. Delen, E. Songur, and A. Cagdas. *J. Craniofac. Surg.*, **585** (2004).
2. H. Yoshimoto, Y. M. Shin, H. Terai, and J. P. Vacanti. *Biomaterials*, **2077** (2003).
3. W.-J. Li, C. T. Laurencin, E. J. Caterson, R. S. Tuan, and F. K. Ko. *J. Biomed. Mater. Res.*, **613** (2002).
4. R. H. Hardy, R. Kania, S. Verliac, A. Lortat-Jacob, and J. Benoit. *Eur. J. Orthop. Surg. Traumatol.*, **63** (1997).

5. R. Langer and J. P. Vacanti, *Science*, **920** (1993).
6. E. Lavik and R. Langer, *Appl. Microbiol. Biotechnol.*, **1** (2004).
7. S. P. Bruder, A. A. Kurth, M. Shea, W. C. Hayes, N. Jaiswal, and S. Kadiyala. *J. Orthop. Res.*, **155** (1998).
8. R. Cancedda, G. Bianchi, A. Derubeis, and R. Quarto. *Stem Cells*, **610** (2003).
9. T. Yoshikawa, H. Ohgushi, and S. Tamai *J. Biomed. Mater. Res.*, **481** (1996).
10. J. S. Boo, Y. Yamada, Y. Okazaki, Y. Hibino, K. Okada, K. Hata, T. Yoshikawa, Y. Sugiura, and M. Ueda, *J. Craniofac. Surg.*, **231** (2002).
11. N. Zagris. *Micron*, **427** (2001).
12. H. Yokoi, T. Kinoshita, and S. Zhang, *Proc. Natl. Acad. Sci. U S A*, **8414** (2005).
13. R. Yoshimi, Y. Yamada, K. Ito, S. Nakamura, A. Abe, T. Nagasaka, K. Okabe, T. Kohgo, S. Baba, and M. Ueda, *J. Craniofac. Surg.*, **1523** (2009).
14. T. C. Holmes, S. de Lacalle, X. Su, G. Liu, A. Rich, and S. Zhang. *Proc. Natl. Acad. Sci. U S A*, **6728** (2000).
15. J. Kisiday, M. Jin, B. Kurtz, H. Hung, C. Semino, S. Zhang, and A. J. Grodzinsky. *Proc. Natl. Acad. Sci. U S A*, **9996** (2002).
16. S. Zhang, T. C. Holmes, C. M. DiPersio, R. O. Hynes, X. Su, and A. Rich. *Biomaterals*, **1385** (1995).
17. Y. Yamada. J. S. Boo, R. Ozawa, T. Nagasaka, Y. Okazaki, K. Hata, and M. Ueda, *J. Cranio-Maxill. Surg.*, **27** (2003).
18. Y. Yamada, M. Ueda, T. Naiki, M. Takahashi, K. Hata, and T. Nagasaka, *Tissue Eng.*, **955** (2004).
19. Y. Yamada, M. Ueda, H. Hibi, and T. Nagasaka, *Cell Transplant*, **343** (2004).
20. Y. Yamada, M. Ueda, H. Hibi, and S. Baba, *Int. J. Periodont. Rest.*, **363** (2006).
21. R. E. Marx, *Clinics. Plast. Surg.*, **377** (1994).
22. R. E. Marx and A. K. Gard, In O.T. Jensen, Ed., *The Sinus Bone Graft* (Chicago: Quintessence, 1999), p. 183.
23. S. Mohan and D. Baylink, *Clin. Orthop. Rel. Res.*, **263** (1991).
24. H. Petite, V. Viateau, W. Bensaid, A. Meunier, C. de Pollak, and C. Bourguignon, *Nat. Biotech.*, **959** (2000).
25. Y. Yamada, S. Nakamura, K. Ito, T. Kohgo, H. Hibi, T. Nagasaka, and M. Ueda, *Tissue Eng.*, **1699** (2008).

Chapter 41

AN EFFICIENT *EX VIVO* EXPANSION OF ADULT MESENCHYMAL STEM CELLS IN SCAFFOLDS

Eui Kyun Park,[a*] Hong-In Shin,[a] and Shin-Yoon Kim[b]

[a]*Department of Oral Pathology and Regenerative Medicine, School of Dentistry,*
[b]*Department of Orthopaedic Surgery, School of Medicine, Kyungook National University 188-1 Samduk 2-ga, Jung-gu, Daegu, Korea*
*epark@knu.ac.kr

Tissue engineering utilizing mesenchymal stem cells (MSCs) in combination with scaffolds presents promising alternative therapies for bone defects. Because of their low numbers in adult tissues, most clinical applications of MSCs depend on isolation followed by *ex vivo* expansion. The development of methods to obtain sufficient quantities of MSCs, and to improve their safety and functionality, could dramatically facilitate their use in clinical applications in regenerative medicine. A number of papers have demonstrated stimulatory effects of growth factors on proliferation and differentiation of MSCs in culture. It is important to understand the biological effects of factors that regulate MSC activity. This review will discuss the effects of growth factors on MSCs and their mechanisms of action with respect to the efficient *ex vivo* expansion of MSCs for regeneration of hard tissue such as bone.

Handbook of Intelligent Scaffolds for Tissue Engineering and Regenerative Medicine
Edited by Gilson Khang

www.panstanford.com

41.1 Introduction

Currently, adult MSCs are used in a wide array of regenerative therapies. Some of the therapeutic uses for MSCs include tissue engineering using scaffolds, cell replacement for treatment of diseases related to functionally impaired cells, and delivery and targeting of MSCs to regions of interest by manipulating cell surfaces with specific antibodies or proteins. A growing number of clinical trials are currently being conducted to evaluate the therapeutic efficacy and safety of MSCs (www.clinicaltrials.gov).[1] In particular, there is a great deal of attention focused on the development of tissue engineering therapies for the treatment of bone and cartilage tissue defects that utilize MSCs, either alone or in combination with a scaffold.[1–5] In order to establish successful tissue engineering therapies, the growth characteristics of MSCs, as well as the ways in which they respond to various biomaterials and bioactive factors, need to be profoundly understood. Of prime importance is how to obtain sufficient quantities of MSCs and how to improve the safety and functionality of MSCs for clinical applications in regenerative medicine.

The characteristics of MSCs, such as their strong adherence to culture plates and fibroblast-like morphology, facilitate the isolation of cells from tissues. MSCs have been identified in and can be isolated from a variety of tissues in the human body, including bone marrow, blood, adipose tissue, trabecular bone, muscle, and dermis.[6,7] Although the initial isolation of MSCs can be achieved by plastic adhesion, the obtained population of cells is heterogeneous in both size and morphology,[8] and contains a mixture of cells at different commitment stages of the mesodermal lineage.[9] To isolate a homogenous MSC population, clonal expansion by limited dilution is often used. Another isolation method that utilizes antibodies against cell surface molecules has also been developed, but the lack of a specific marker in MSCs makes their isolation difficult. Currently, cell surface markers such as CD29, CD44, CD49a, CD71, CD73, CD90, CD105, CD146, CD166, CD271, and STRO-1 are used to identify MSCs.[9–11] However, it is still necessary to establish a specific cell surface marker for the selection of a homogenous population of MSCs and for use in tracking MSCs after transplantation.

In addition to their morphological characteristics, MSCs have the potential for extensive self-renewal, multilineage differentiation, and immunomodulation.[9,12,13] Under appropriate culture conditions, MSCs are capable of differentiating into cell from multiple lineages, such as osteoblasts, adipocytes, chondrocytes, and myoblasts.[7,13–15] A variety of growth factors and transcription factors are involved in lineage-specific differentiation and transdifferentiation of MSCs. The mechanisms that underlie lineage-specific differentiation are extensively reviewed elsewhere.[16]

The immunomodulatory effect of MSCs has also been recently demonstrated. MSCs were shown to modulate the functions of innate and adaptive immune cells and thus suppress inflammatory responses and modulate their survival.[9] MSCs can also suppress the proliferation, differentiation, and cytotoxicity of resting natural killer (NK) cells.[9,17,18] Similarly, proliferation, interferon-γ production, and cytotoxicity of preactivated NK cells is decreased in the presence of MSCs.[9,17–22] It is noteworthy that MSCs are only susceptible to cytokine-activated but not resting NK cells.[18,20,22] The inhibitory effects of MSCs on the differentiation of monocytes and CD34+ hematopoietic progenitors into immature myeloid dendritic cells (DCs) have been demonstrated. In addition, MSCs are able to suppress tumor necrosis factor production by DCs and increase interleukin-10 (IL-10) production by plasmacytoid DCs.[23–26] MSCs are also able to inhibit *in vitro* T cell proliferation and function in both naive and memory T cells.[27–30] Therefore, MSCs appear to have significant immunosuppressive potential.[30]

Ex vivo expansion is one of the most substantial challenges in the use of MSCs for clinical applications, particularly with regard to tissue engineering strategies for large lesions in bone. Therapeutic use of MSCs still depends on isolation and *ex vivo* expansion of MSCs because of their low prevalence in adult tissues.[31] Numerous factors can modulate the proliferative capacity of MSCs. They include donor age, plating density, serum type, and the presence of various cytokines and growth factors.[31] So far, several growth factors such as fibroblast growth factor (FGF), epidermal growth factor (EGF), platelet-derived growth factor (PDGF), transforming growth factor (TGF), insulin-like growth factor (IGF), hepatocyte growth factor (HGF), Wnt, and others have been shown to regulate the proliferative potential of MSCs.[7,31–42] MSC proliferation can be further potentiated by a combination of growth factors or of growth factors with glucocorticoids such as dexamethasone.[43–46] Although the mechanisms that govern enhanced proliferation of MSCs in response to growth factors and glucocorticoids are still largely unknown, a recent study has shown that Src kinase may be involved in the enhanced proliferation of MSCs.[43] Recently, alternative culture methods that exclude animal-derived culture constituents, such as fetal bovine serum (FBS), have also been proposed. In these systems, umbilical cord blood and platelet lysates are substituted for FBS for the expansion of MSCs.[1,47]

Collectively, plasticity, ease of isolation, immunomodulatory potential, and ready availability make MSCs attractive for tissue engineering therapy applications.[7] However, an effective technique for the *ex vivo* expansion of MSCs that maintains their multilineage differentiation potential is needed.

41.2 The Use of Growth Factors and Glucocorticoids for the Propagation of Adult MSCs

41.2.1 *Growth Factors*

The therapeutic application of MSCs is relatively restricted by a low yield of cells from adult tissues.[48,49] In order to obtain a sufficient number of MSCs for tissue engineering therapies, culture conditions need to be thoroughly considered. Serum in a growth medium contains a wide variety of nutrients, hormones, growth factors, vitamins, and microelements that affect the growth and stem cell potential of MSCs.[50–52] Of these, growth factors have the greatest potential to influence the cellular characteristics of MSCs. Therefore, an understanding of the effects exerted by various growth factors and the molecular mechanisms that underlie the proliferative potential of MSCs may promote the expansion of their therapeutic applications.

41.2.1.1 Fibroblast growth factors

The FGF family is comprised of 22 growth factors that regulate multiple biological processes, including bone and cartilage formation.[53] Individual FGFs bind to specific receptors or heparin-like proteoglycans[54] and activate signaling pathways in the cell, such as the Ras/extracellular-regulated kinase (ERK) and the phosphatidylinositol 3-kinase (PI3K)/protein kinase B/Akt (AKT) pathways.[55] Among the members of the FGF family, FGF-2 exerts the strongest influence on the self-renewal and growth capacity of MSCs. FGF-2 can accelerate the proliferation of bone marrow MSCs *in vitro*.[56–58] However, the effect of FGF-2 on MSC proliferation is variable, depending on culture condition and dosage.[59] A low dose of FGF-2 (3 ng/mL) induces proliferation of MSCs, but higher doses of FGF-2 (30 ng/mL) do not. In addition, FGF-2-stimulated MSC proliferation is more efficient at low cell density than at high cell density.[58] Therefore, stimulation with low doses of FGF-2 seems to be effective for the *in vitro* expansion of MSCs cultured at a low cell density. FGF also appears to induce commitment of MSCs toward an early stage of the osteoblast lineage. FGF-2-expanded MSCs can induce higher mineralization *in vitro* and increased bone formation *in vivo*.[56] In MSCs that have been cultured for multiple passages, the addition of FGF-2 can help to preserve a normal growth rate[60] and also induces the expression of FGF in MSCs, which allows them to maintain their self-renewal potential *in vitro*.[61] These results suggest that FGF-2 can induce the commitment of MSCs to an early or immature stage of the osteoblast lineage, but those MSCs still retain their potential for self-renewal.

In addition to FGF-2, other members of the FGF family, including FGF-1, FGF-4, and FGF-8, may also stimulate the growth of MSCs[40,62,42]

41.2.1.2 Epidermal growth factor

EGF has been implicated in the regulation of various cellular responses. The binding of EGF to its receptor (EGFR) activates numerous intracellular signaling molecules such as ERK, PI3K/AKT, and others,[63,64] leading to migration, adhesion, proliferation, and survival in various cell types[64–67] EGFRs are expressed on the surface of bone marrow MSCs.[68,69] Tatama *et al.* have shown that stimulation of immortalized human MSCs and primary porcine MSCs with soluble EGF can strongly induce phosphorylation of EGFR, ERK, AKT, and phospholipase C-γ.[41] EGF also partially promotes proliferation in human primary bone marrow MSCs.[70] Consequently, EGF is able to promote proliferation and migration of MSCs without affecting their multipotency.[41] In addition, a significant increase in clonogenic growth of MSCs, as assessed by colony forming unit-fibroblast (CFU-F) assay, occurs in response to EGF.[56] These results support that EGF can significantly promote proliferation of MSCs and could be used as a growth factor for *ex vivo* expansion.[41] On the other hand, Kratchmarova *et al.* have shown that EGF can promote osteoblastic differentiation of immortalized human MSCs.[71] The activation of intracellular signaling molecules was analyzed in response to EGF as an osteogenic factor and PDGF as a negative control. Tyrosine phosphorylation of PI3K exclusively occurred in response to PDGF, indicating the importance of PI3K in mediating the differential effects of these two growth factors on the osteogenic differentiation of MSCs. Furthermore, wortmannin, a PI3K inhibitor, can drive PDGF-mediated signals toward osteogenic differentiation.[71] Therefore, it will be interesting to see whether the differential effects exerted on MSCs by growth factors can be controlled by the use of chemical inhibitors of signaling molecules.

The effects of surface modifications elicited by use of a scaffold in combination with EGF have been demonstrated by Griffith *et al.* Tethering of EGF to the scaffold surface allows EGF to bind to the EGFR but inhibits internalization of the ligand[72] in a manner that resembles a matrix-embedded EGFR ligand.[41,73] In this study, surface-tethered EGF promoted both cell spreading and survival more strongly than soluble EGF. MSC death induced by Fas ligand, a proinflammatory cytokine, was also reduced by EGF tethering. These results demonstrate that tethered EGF offers not only a protective effect against inflammatory reactions to scaffolds but also a stimulatory effect on the proliferation of MSCs.[41]

41.2.1.3 Platelet-derived growth factor

PDGF exists as four subtypes (PDGF-A, -B, -C, and –D) and acts as homo- or heterodimers to stimulate their specific receptors, either PDGFRα or β.[74] The activated PDGFRs, which are tyrosine phosphorylated in their cytoplasmic domains, recruit adaptor proteins and activate several signaling pathways, such as Ras-ERK, PI3K/Akt, and phospholipase Cγ.[75] Consequently, PDGFs are able to potently modulate biological responses, such as proliferation of mesenchymal cells and the regulation of wound-healing processes.[76]

PDGFRs are found to be expressed in MSCs,[77] and binding of the PDGF-BB homodimer to its receptor strongly induces the proliferation and migration of MSCs.[41,71,78] Inhibition of PDGFRβ by imatinib mesylate suppresses proliferation of human MSCs *in vitro*,[41,78] suggesting that PDGFRβ signaling may mediate PDGF-induced proliferation of MSCs. Recently, a study conducted in PDGFRβ-impaired mice demonstrated that the mitogenic and migratory responses of MSCs to PDGF are dramatically reduced, while osteogenic differentiation is enhanced.[79] Therefore, PDGFRβ signaling appears to elicit opposing responses with regard to proliferation and osteoblast lineage differentiation in MSCs. In comparison to the effects mediated by PDGFRβ, the influence of PDGFRα on osteogenic differentiation is very subtle.[79] The stimulatory effect of PDGF on the proliferation of MSCs is demonstrated by the clonogenic growth of bone marrow MSCs. Both PDGF-BB and EGF show a strong ability to enhance clonogenic growth of CFU-F under serum-deprived culture conditions.[80] In addition, although a combination of growth factors, such as EGF, FGF-2, and PDGF, does not induce a significant increase in colony numbers, dose-dependent increases in mean colony diameter suggest a synergistic effect on cell proliferation.[80] Interestingly, the enhanced colony formation of MSCs by PDGF that occurs in a serum-free medium is not observed in a medium containing 10% serum.[56] Therefore, PDGF, PDGF-BB in particular, is able to induce proliferative and migratory responses but strongly inhibits osteogenic differentiation of MSCs.

41.2.1.4 Other growth factors

Other growth factors have also been implicated in regulation of the proliferative potential of MSCs. BMP-2, a member of the TGF-β superfamily, exerts a suppressive effect on MSC proliferation. BMP-2 antagonizes Wnt3a signaling and inhibits the proliferation of mouse bone marrow MSCs through the interaction of the BMP receptor, Smad, with Dishevelled-1, which is a component of the Wnt-signaling pathway.[81] Conversely, TGF-β1 induces the proliferation of human MSCs, through both Smad3-dependent nuclear accumulation of β-catenin and activation of the Smad-independent Ras-ERK signaling pathway in MSCs.[82,83] Therefore, TGF-β signaling may regulate MSC proliferation through crosstalk with the Wnt signaling pathway.[84]

IGF-1 is potentially capable of stimulating proliferation of MSCs,[85,86] but this effect is not reliably reproducible.[87] Proliferation of MSCs has also been shown to be influenced by HGF, TGF-α, tumor necrosis factor-α, IGF-2, stem cell factor, and a number of other growth factors.[7,86,88,89]

41.2.2 *Glucocorticoids*

Glucocorticoids play an important role in bone metabolism by regulating the activities of osteoblasts and osteoclasts. In particular, glucocorticoids regulate osteoblast and adipocyte lineage commitment.[90–92] Transgenic mice that overexpress 11β-hydroxysteroid-dehydrogenase type 2 (11βHSD2), an enzyme that catalyzes the conversion of active cortisol to inactive cortisone,[93–95] reveal phenotypic impairments in bone, such as vertebral osteopenia, reduced femoral cortical bone area and thickness, delayed cranial bone formation and suture closure, and impaired mineralized nodule formation of primary calvarial cells.[95,96] These phenotypic impairments appear to be due to a defect in Wnt signaling, which normally stimulates osteoblast lineage commitment of MSCs.[92,96] These observations underscore the important role that glucocorticoids play during normal bone development.

Glucocorticoids are commonly prescribed drugs used under a variety of circumstances, including organ transplantation, neoplastic diseases, and autoimmune disorders. Currently, various forms of glucocorticoids with varying half-lives, potentencies, and receptor affinities have been developed for clinical use. These drugs include dexamethasone, betamethasone, hydrocortisone, and prednisone.[97] Unfortunately, administration of excess amounts of glucocorticoids frequently results in detrimental effects on the skeleton, namely glucocorticoid-induced osteoporosis (GIO).[98,99] Glucocorticoids reduce bone formation by decreasing osteoblast proliferation and terminal differentiation, as well as by promoting apoptosis of osteocytes, resulting in the development of GIO.[100,101]

In an *in vitro* MSC culture system, glucocorticoids act as both osteogenic and adipogenic factors. The bifunctional effects of glucocorticoids are dependent on dosage: at high concentrations (about 10^{-6} M), MSCs respond to dexamethasone by differentiating into the adipocyte lineage, while at lower doses (about 10^{-8} M), MSCs differentiate into the osteoblast lineage. Under conditions favorable for differentiation into the osteoblast lineage, dexamethasone decreases MSC proliferation and triggers osteogenesis through the expression of multiple osteogenic markers, including alkaline phosphatase (ALP), osteocalcin, and many other osteoblast-related genes.[102–106] In the case of MSCs obtained from adipose tissue, treatment with glucocorticoids, such as dexamethasone, betamethasone, hydrocortisone, and prednisone, has a negligible effect on proliferation.[43] In addition to growth suppression, glucocorticoids also inhibit the terminal differentiation of MSCs and promote apoptosis of osteoblasts and mature osteocytes.[100,101] Concisely, glucocorticoids can inhibit proliferation of MSCs and influence the differentiation of MSCs into either the osteoblast or adipocyte lineage, depending on the dosage.

41.2.3 *Combination of Growth Factors and Steroids*

While synthetic glucocorticoids are suppressors of MSC growth and inducers of MSC differentiation, growth factors are essential for optimal growth of MSCs *in vitro*. However, when these two factors are combined, a significant enhancement of both MSC growth as well as MSC differentiation occurs.

Quarto *et al.* have demonstrated exciting results regarding the combination of growth factors and glucocorticoids in human bone marrow MSCs.[45] The osteogenic and proliferative potentials of human bone marrow MSCs are dramatically increased in response to dexamethasone combined with FGF-2. Both CFU-F size and proliferation of MSCs are significantly higher in the presence of combined FGF-2 and dexamethasone,[44–46] suggesting that a combination of cell-proliferative factors and osteogenic factors amplifies the proliferation potential in bone marrow MSCs. The combination of FGF-2 with other glucocorticoids, such as betamethasone, dexamethasone, hydrocortisone, and prednisolone, reveals that hydrocortisone and prednisolone do not exert strong cooperative effects on MSC proliferation but that betamethasone and dexamethasone are able to stimulate their proliferation.[43]

In addition, while dexamethasone alone increases levels of ALP, a key enzyme that mediates the mineralization process, a combination of FGF-2 and dexamethasone down-regulates ALP levels by 56%. This indicates that FGF-2 signaling can partially suppress dexamethasone-induced osteogenic differentiation. This indicates that in the presence of FGF-2, dexamethasone cannot further the differentiation of MSCs and thus maintains them at an early stage of commitment. Interestingly, after subcutaneous implantation into nude mice, MSCs that had been expanded in culture with a combination of FGF-2 and dexamethasone were able to form more bone than those expanded with either FGF-2 or dexamethasone alone. Therefore, osteogenic-specific priming of MSCs to the immature stage appears to induce enhanced bone formation. Enhanced proliferation and osteoblast differentiation in response to FGF-2 and dexamethasone are not restricted to human bone marrow MSCs; these effects have been observed in other tissue- and species-derived MSCs, such as rat bone marrow MSCs,[107] equine MSCs,[108] and human adipose tissue–derived MSCs.[43]

At the same time, a combination of FGF-2 and dexamethasone at the dosage used for induction of proliferation and osteoblast differentiation can also induce adipocyte lineage commitment.[43,44] FGF-2 and dexamethasone induce expression of ALP as well as peroxisome proliferator-activated receptor γ (PPARγ), an adipocyte marker in adipose-derived MSCs,[43] and enhance the formation of lipid droplets.[43,44] These results support that FGF-2 and dexamethasone can induce both osteoblast and adipocyte lineage commitment of MSCs. Under conditions in which PPARγ has been suppressed by an antagonist, the combination of FGF-2 and dexamethasone further stimulates proliferation and osteoblast lineage differentiation *in vitro*.[109] It will be interesting to determine whether the combination of FGF-2, dexamethasone, and PPARγ antagonist (or other adipogenesis inhibitors) can promote bone formation *in vivo*.

Although the mechanism underlying the enhanced proliferation and osteogenic differentiation induced by FGF-2 and dexamethasone is largely unknown, Src kinase is likely to be involved in this event. In response to FGF-2 and dexamethasone, phosphorylation of c-Jun N-terminal kinase, ERK, and p38 MAPK is decreased, while stimulatory tyrosine phosphorylation of Src kinase is increased, suggesting the involvement of a Src-dependent pathway in the enhanced proliferation of MSCs. Moreover, Src family kinase inhibitors substantially reduce FGF-2 and dexamethasone-induced proliferation and osteoblast lineage differentiation of MSCs.[43]

Gronthos and Simmons have demonstrated that a combination of growth factors can also stimulate the growth of MSCs. Simultaneous addition of PDGF and EGF resulted in dose-dependent increases in mean colony diameter, indicating enhanced growth of MSCs. Therefore, EGF/EGFR and PDGF/PDGFR signaling may cooperate to induce a synergistic induction of MSC proliferation.[80] In another study by Ng *et al.*, the combination of PDGF, TGF-β, and FGF was shown to be sufficient to induce MSC growth in a serum-free medium for up to five passages. Inhibiting any one of these pathways reduced MSC growth, demonstrating the importance of these three factors for the growth of MSCs.[110] These results are particularly significant, because elimination of animal-derived culture supplements is one of the challenges facing effective MSC expansion *in vitro*.

41.3 Growth of MSCs in Scaffolds

In order to regenerate damaged tissues, MSC-loaded scaffolds can be either directly transplanted or cultured for an appropriate length of time, both to ensure proper adhesion of MSCs to the scaffold and to grow MSCs to the optimal numbers within a scaffold. To ensure MSC adhesion and expansion, MSC-loaded scaffolds can be incubated for anywhere from an hour to several days under conditions favorable for MSC growth. In two-dimensional (2D) culture conditions, growth factors, alone or in combination with dexamethasone, dramatically enhance proliferation of MSCs. In a three-dimensional (3D) porous ceramic scaffold, the combination of FGF-2 and dexamethasone also significantly stimulates both MSC proliferation and osteoblast lineage commitment. Seeding of bone marrow MSCs in a porous biphasic calcium phosphate (BCP) scaffold, which is composed of hydroxyapatite and β-tricalcium phosphate (6:4), followed by incubation with FGF-2 and dexamethasone for 7 or 14 days, results in enhanced growth of MSCs compared to control (Fig. 41.1). Under

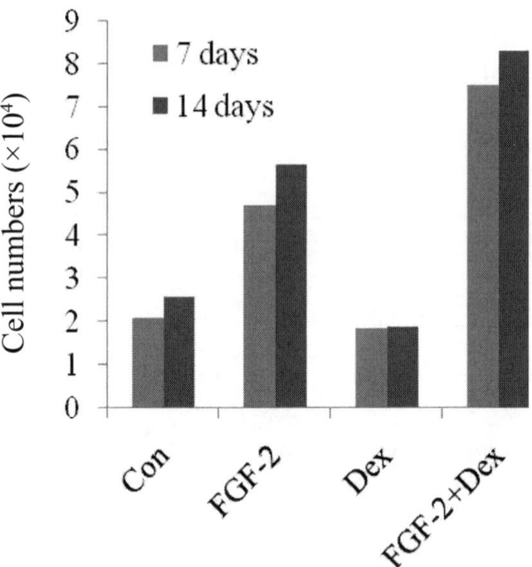

Figure 41.1. Growth of MSCs in a porous BCP scaffold. Bone marrow MSCs (2×10^5 cells in a 50 μL medium) were seeded in BCP ceramics ($5\times5\times2$ mm) by a static method and treated with or without FGF-2 (1 ng/mL), Dex (10^{-8} M), or a combination of FGF-2 and Dex for 7 and 14 days. Cell numbers were measured using a hematocytometer. *Abbreviation*: Dex, dexamethasone. See also Color Insert.

these conditions, FGF-2 alone is able to stimulate substantial growth of MSCs. However, the combination of FGF-2 and dexamethasone induces a much higher increase in MSC proliferation. As is also observed in 2D culture, dexamethasone alone arrests MSC growth in the 3D scaffold. These results suggest that, at least in the case of FGF-2 and dexamethasone, growth stimulation of MSCs can be reproduced in both 2D culture and 3D scaffold culture conditions.

Stimulation of MSC-loaded BCP scaffolds with FGF-2 and dexamethasone for 7 days, followed by stimulation with osteogenic medium (50 mg/mL ascorbic acid, 10^{-2} M β-glycerophosphate, and 10^{-8} M dexamethasone) for an additional 14 days, significantly induced initiation of differentiation toward the osteoblast lineage as assessed by ALP staining (Fig. 41.2). These results demonstrate that osteoblast commitment is also increased in MSCs grown in a scaffold with FGF-2 and dexamethasone. The enhancement of osteogenic differentiation by FGF-2 and dexamethasone may be in part due to the high proliferation rate of MSCs in 3D culture. Because inhibitors of Src kinase suppress both proliferation and osteogenic differentiation, and the degree of osteogenic differentiation in 2D culture is dependent on cell number,[43] the enhanced proliferation rate of MSCs in 3D may contribute to the osteogenic differentiation of MSCs expanded with FGF-2 and dexamethasone. It is also possible that osteoblast commitment may contribute to enhanced osteogenic

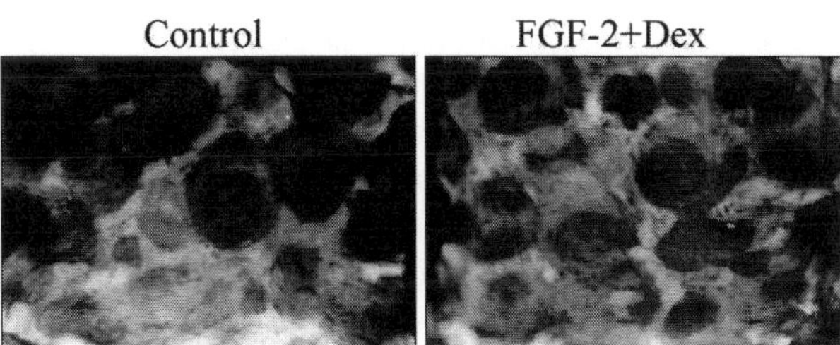

Figure 41.2. ALP staining of MSCs treated with FGF-2 and Dex in a BCP scaffold. Bone marrow MSCs (2×10^5 cells) were seeded onto a porous BCP ceramic scaffold, cultured with or without a combination of FGF-2 (1 ng/mL) and Dex (10^{-8} M) for 7 days, and then treated with an osteogenic cocktail for an additional 14 days. Cells committed to the osteoblast lineage were identified by ALP staining.

differentiation because MSCs expanded with FGF-2 and dexamethasone express ALP.

Because the effect of FGF-2 on MSC proliferation is more prominent at low cell density than at high density, enhanced bone formation can be achieved by seeding of low numbers of MSCs into a scaffold followed by culture for a given period in the presence of FGF-2 and dexamethasone prior to use.

41.4 Conclusions and Outlook

This article has focused on the *ex vivo* expansion of MSCs for their use in the therapeutic regeneration of bone defects. In addition, several factors that modulate the proliferative potential of MSCs, such as donor age, plating density, serum type, and the presence of growth factors,[31] were also discussed. Of these, growth factors have the strongest effect on MSC proliferation. In particular, FGF-2, EGF, and PDGF, either alone or in combination with glucocorticoids such as dexamethasone, strongly induce the proliferation of MSCs. Thus, a combination of growth factors and glucocorticoids may provide the optimal conditions for the expansion of MSCs *in vitro*.

However, other aspects of *ex vivo* expansion of MSCs with growth factors and glucocorticoids can induce both osteoblast and adipocyte lineage commitments, as revealed in articles by Lee and Locklin.[43,44] Because osteogenic differentiation occurs at the expense of adipocyte lineage commitment, any molecules that suppress adipogenic differentiation can potentiate the osteoblast lineage commitment of MSCs, while maintaining their proliferative potential. In fact, administration of a PPARγ antagonist during MSC expansion with FGF-2 and dexamethasone significantly stimulated both proliferation and osteoblast differentiation.[109] Examining the effects of other suppressors of adipogenic differentiation may provide relevant information about *ex vivo* expansion, osteoblast commitment of MSCs, and *in vivo* bone formation.

Because expansion of MSCs under culture conditions free of animal-derived serum is also critical for their therapeutic application,[111,112] it is of great importance to use a combination of growth factors or growth factors with glucocorticoids to ensure

MSC survival and proliferation. Successful culture of MSCs in an animal serum-free medium supplemented with growth factors has been reported by Ng and Gronthos.[80] These MSCs were able to propagate, at least to limited passages. These culture conditions may also be combined with other FBS substitutes, such as platelet lysates, as recently described by Schallmoser.[1]

The ultimate goal for the *ex vivo* expansion of MSCs is to produce safe and sufficient MSCs for therapeutic application. In order to achieve this, a combination of growth factors that exert a stimulatory effect on MSC proliferation when used in combination with other factors, such as autologous serum, needs to be evaluated in MSCs derived from large numbers of donors. On the basis of the advancements made thus far, it seems likely that the ability to expand MSCs under therapeutically safe and effective culture conditions will be in our hands in the near future.

Acknowledgments

Research described in this chapter was supported by a grant of the Korea Healthcare technology R&D Project, Ministry for Health, Welfare & Family Affair, Republic of Korea (A080840 and A091224). I would like to thank Dr. So Young Chun, Dr. Young Ae Choi, and Ms. Jiwon Lim for preparation of the early drafts and Ms. Hye-Jung Noh for technical assistance.

References

1. K. Schallmoser, E. Rohde, A. Reinisch, C. Bartmann, D. Thaler, C. Drexler, A. C. Obenauf, G. Lanzer, W. Linkesch, and D. Strunk, *Tissue Eng. Part C Methods*, **185** (2008).
2. A. Atala, *J. Tissue Eng. Regen. Med.*, **83** (2007).
3. A. I. Caplan, *Tissue Eng.*, **1198** (2005).
4. A. G. Mikos, S. W. Herring, P. Ochareon, J. Elisseeff, H. H. Lu, R. Kandel, F. J. Schoen, M. Toner, D. Mooney, A. Atala, M. E. Van Dyke, D. Kaplan, and G. Vunjak-Novakovic, *Tissue Eng.*, **3307** (2006).
5. J. T. Vilquin and P. Rosset, *Regen. Med.*, **589** (2006).

6. A. I. Caplan and S. P. Bruder, *Trends Mol. Med.*, **259** (2001).
7. L. Ling, V. Nurcombe, and S. M. Cool, *Gene*, **1** (2009).
8. R. A. Musina, E. S. Bekchanova, and G. T. Sukhikh, *Bull. Exp. Biol. Med.*, **504** (2005).
9. A. Uccelli, L. Moretta, and V. Pistoia, *Nat. Rev. Immunol.*, **726** (2008).
10. B. Sacchetti, A. Funari, S. Michienzi, S. Di Cesare, S. Piersanti, I. Saggio, E. Tagliafico, S. Ferrari, P. G. Robey, M. Riminucci, and P. Bianco, *Cell*, **324** (2007).
11. K. Stewart, P. Monk, S. Walsh, C. M. Jefferiss, J. Letchford, and J. N. Beresford, *Cell Tissue Res.*, **281** (2003).
12. J. N. Beresford, *Clin. Orthop. Relat. Res.*, **270** (1989).
13. A. I. Caplan, *J. Orthop. Res.*, **641** (1991).
14. M. F. Pittenger, A. M. Mackay, S. C. Beck, R. K. Jaiswal, R. Douglas, J. D. Mosca, M. A. Moorman, D. W. Simonetti, S. Craig, and D. R. Marshak, *Science*, **143** (1999).
15. D. J. Prockop, *Science*, **71** (1997).
16. S. Muruganandan, A. A. Roman, and C. J. Sinal, *Cell Mol. Life Sci.*, **236** (2009).
17. G. M. Spaggiari, A. Capobianco, H. Abdelrazik, F. Becchetti, M. C. Mingari, and L. Moretta, *Blood*, **1327** (2008).
18. G. M. Spaggiari, A. Capobianco, S. Becchetti, M. C. Mingari, and L. Moretta, *Blood*, **1484** (2006).
19. S. Aggarwal and M. F. Pittenger, *Blood*, **1815** (2005).
20. A. Poggi, C. Prevosto, A. M. Massaro, S. Negrini, S. Urbani, I. Pierri, R. Saccardi, M. Gobbi, and M. R. Zocchi, *J. Immunol.*, **6352** (2005).
21. Z. Selmani, A. Naji, I. Zidi, B. Favier, E. Gaiffe, L. Obert, C. Borg, P. Saas, P. Tiberghien, N. Rouas-Freiss, E. D. Carosella, and F. Deschaseaux, *Stem Cells*, **212** (2008).
22. P. A. Sotiropoulou, S. A. Perez, A. D. Gritzapis, C. N. Baxevanis, and M. Papamichail, *Stem Cells*, **74** (2006).
23. X. X. Jiang, Y. Zhang, B. Liu, S. X. Zhang, Y. Wu, X. D. Yu, and N. Mao, *Blood*, **4120** (2005).
24. Y. P. Li, S. Paczesny, E. Lauret, S. Poirault, P. Bordigoni, F. Mekhloufi, O. Hequet, Y. Bertrand, J. P. Ou-Yang, J. F. Stoltz, P. Miossec, and A. Eljaafari, *J. Immunol.*, **1598** (2008).
25. A. J. Nauta, A. B. Kruisselbrink, E. Lurvink, R. Willemze, and W. E. Fibbe, *J. Immunol.*, **2080** (2006).

26. R. Ramasamy, H. Fazekasova, E. W. Lam, I. Soeiro, G. Lombardi, and F. Dazzi, *Transplantation*, **71** (2007).
27. M. Di Nicola, C. Carlo-Stella, M. Magni, M. Milanesi, P. D. Longoni, P. Matteucci, S. Grisanti, and A. M. Gianni, *Blood*, **3838** (2002).
28. M. Krampera, S. Glennie, J. Dyson, D. Scott, R. Laylor, E. Simpson, and F. Dazzi, *Blood*, **3722** (2003).
29. R. Meisel, A. Zibert, M. Laryea, U. Gobel, W. Daubener, and D. Dilloo, *Blood*, **4619** (2004).
30. P. A. Sotiropoulou, S. A. Perez, M. Salagianni, C. N. Baxevanis, and M. Papamichail, *Stem Cells*, **462** (2006).
31. K. Ksiazek, *Rejuven. Res.*, **105** (2009).
32. G. Forte, M. Minieri, P. Cossa, D. Antenucci, M. Sala, V. Gnocchi, R. Fiaccavento, F. Carotenuto, P. De Vito, P. M. Baldini, M. Prat, and P. Di Nardo, *Stem Cells*, **23** (2006).
33. W. Huang, V. La Russa, A. Alzoubi, and P. Schwarzenberger, *Stem Cells*, **1512** (2006).
34. T. Ito, R. Sawada, Y. Fujiwara, Y. Seyama, and T. Tsuchiya, *Biochem. Biophys. Res. Commun.*, **108** (2007).
35. S. M. Kosmacheva, M. V. Volk, T. A. Yeustratenka, I. N. Severin, and M. P. Potapnev, *Bull. Exp. Biol. Med.*, **141** (2008).
36. M. Krampera, A. Pasini, A. Rigo, M. T. Scupoli, C. Tecchio, G. Malpeli, A. Scarpa, F. Dazzi, G. Pizzolo, and F. Vinante, *Blood*, **59** (2005).
37. D. Li, G. Y. Wang, B. H. Dong, Y. C. Zhang, Y. X. Wang, and B. C. Sun, *Cells Tissues Organs*, **169** (2007).
38. L. Longobardi, L. O'Rear, S. Aakula, B. Johnstone, K. Shimer, A. Chytil, W. A. Horton, H. L. Moses, and A. Spagnoli, *J. Bone Mineral Res.*, **626** (2006).
39. J. Pons, Y. Huang, J. Arakawa-Hoyt, D. Washko, J. Takagawa, J. Ye, W. Grossman, and H. Su, *Biochem. Biophys. Res. Commun.*, **419** (2008).
40. F. L. Quito, J. Beh, O. Bashayan, C. Basilico, and R. S. Basch, *Blood*, **1282** (1996).
41. K. Tamama, V. H. Fan, L. G. Griffith, H. C. Blair, and A. Wells, *Stem Cells*, **686** (2006).
42. M. P. Valta, T. Hentunen, Q. Qu, E. M. Valve, A. Harjula, J. A. Seppanen, H. K. Vaananen, and P. L. Harkonen, *Endocrinology*, **2171** (2006).
43. Lee SY, Lim J, Khang G, S. Y, C. PH, K. SS, C. S. S. HI, K. SY, and P. EK, *Tissue Eng. Part A*, (2009).

44. R. M. Locklin, R. O. Oreffo, and J. T. Triffitt, *Cell Biol. Int.*, **185** (1999).
45. A. Muraglia, I. Martin, R. Cancedda, and R. Quarto, *Bone*, **131S** (1998).
46. S. Pri-Chen, S. Pitaru, F. Lokiec, and N. Savion, *Bone*, **111** (1998).
47. J. Jung, N. Moon, J. Y. Ahn, E. J. Oh, M. Kim, C. S. Cho, J. C. Shin, and I. H. Oh, *Stem Cells Dev.*, **559** (2009).
48. H. Castro-Malaspina, R. E. Gay, G. Resnick, N. Kapoor, P. Meyers, D. Chiarieri, S. McKenzie, H. E. Broxmeyer, and M. A. Moore, *Blood*, **289** (1980).
49. P. J. Simmons and B. Torok-Storb, *Blood*, **55** (1991).
50. N. Dainiak, *Prog. Clin. Biol. Res.*, **59** (1985).
51. S. A. Kuznetsov, M. H. Mankani, and P. G. Robey, *Transplantation*, **1780** (2000).
52. A. Shahdadfar, K. Fronsdal, T. Haug, F. P. Reinholt, and J. E. Brinchmann, *Stem Cells*, **1357** (2005).
53. D. M. Ornitz and P. J. Marie, *Genes Dev.*, **1446** (2002).
54. C. J. Powers, S. W. McLeskey, and A. Wellstein, *Endocr. Relat. Cancer*, **165** (2000).
55. G. Xiao, D. Jiang, R. Gopalakrishnan, and R. T. Franceschi, *J. Biol. Chem.*, **36181** (2002).
56. I. Martin, A. Muraglia, G. Campanile, R. Cancedda, and R. Quarto, *Endocrinology*, **4456** (1997).
57. L. A. Solchaga, K. Penick, J. D. Porter, V. M. Goldberg, A. I. Caplan, and J. F. Welter, *J. Cell Physiol.*, **398** (2005).
58. S. Tsutsumi, A. Shimazu, K. Miyazaki, H. Pan, C. Koike, E. Yoshida, K. Takagishi, and Y. Kato, *Biochem. Biophys. Res. Commun.*, **413** (2001).
59. S. Hankemeier, M. Keus, J. Zeichen, M. Jagodzinski, T. Barkhausen, U. Bosch, C. Krettek, and M. Van Griensven, *Tissue Eng.*, **41** (2005).
60. L. J. Oliver, D. B. Rifkin, J. Gabrilove, M. J. Hannocks, and E. L. Wilson, *Growth Factors*, **231** (1990).
61. L. E. Zaragosi, G. Ailhaud, and C. Dani, *Stem Cells*, **2412** (2006).
62. C. van den Bos, J. D. Mosca, J. Winkles, L. Kerrigan, W. H. Burgess, and D. R. Marshak, *Hum. Cell*, **45** (1997).
63. S. G. Kennedy, A. J. Wagner, S. D. Conzen, J. Jordan, A. Bellacosa, P. N. Tsichlis, and N. Hay, *Genes Dev.*, **701** (1997).
64. P. P. Roux and J. Blenis, *Microbiol. Mol. Biol. Rev.*, **320** (2004).
65. V. H. Fan, K. Tamama, A. Au, R. Littrell, L. B. Richardson, J. W. Wright, A. Wells, and L. G. Griffith, *Stem Cells*, **1241** (2007).

66. K. J. Lembach, *Proc. Natl. Acad. Sci. U S A*, **183** (1976).
67. A. Wells, *Int. J. Biochem. Cell Biol.*, **637** (1999).
68. S. W. de Laat, J. Boonstra, L. H. Defize, W. Kruijer, P. T. van der Saag, L. G. Tertoolen, E. J. van Zoelen, and J. den Hertog, *Int. J. Dev. Biol.*, **681** (1999).
69. K. Satomura, A. R. Derubeis, N. S. Fedarko, K. Ibaraki-O'Connor, S. A. Kuznetsov, D. W. Rowe, M. F. Young, and P. Gehron Robey, *J. Cell Physiol.*, **426** (1998).
70. Shon MJ, Lee SY, Kim TH, Kim SY, Khang GS, Son YS, Kim SY, and P. EK, *Tissue Eng. Regener. Med.* **457** (2006).
71. I. Kratchmarova, B. Blagoev, M. Haack-Sorensen, M. Kassem, and M. Mann, *Science*, **1472** (2005).
72. P. R. Kuhl and L. G. Griffith-Cima, *Nat. Med.*, **1022** (1996).
73. C. S. Swindle, K. T. Tran, T. D. Johnson, P. Banerjee, A. M. Mayes, L. Griffith, and A. Wells, *J. Cell Biol.*, **459** (2001).
74. D. G. Gilbertson, M. E. Duff, J. W. West, J. D. Kelly, P. O. Sheppard, P. D. Hofstrand, Z. Gao, K. Shoemaker, T. R. Bukowski, M. Moore, A. L. Feldhaus, J. M. Humes, T. E. Palmer, and C. E. Hart, *J. Biol. Chem.*, **27406** (2001).
75. M. Tallquist and A. Kazlauskas, *Cytokine Growth Factor Rev.*, **205** (2004).
76. C. H. Heldin and B. Westermark, *Physiol. Rev.*, **1283** (1999).
77. J. J. Minguell, A. Erices, and P. Conget, *Exp. Biol. Med. (Maywood)*, **507** (2001).
78. F. Fierro, T. Illmer, D. Jing, E. Schleyer, G. Ehninger, S. Boxberger, and M. Bornhauser, *Cell Prolif.*, **355** (2007).
79. A. Tokunaga, T. Oya, Y. Ishii, H. Motomura, C. Nakamura, S. Ishizawa, T. Fujimori, Y. Nabeshima, A. Umezawa, M. Kanamori, T. Kimura, and M. Sasahara, *J. Bone Mineral Res.*, **1519** (2008).
80. S. Gronthos and P. J. Simmons, *Blood*, **929** (1995).
81. D. W. Burt and A. S. Law, *Prog. Growth Factor Res.*, **99** (1994).
82. H. Jian, X. Shen, I. Liu, M. Semenov, X. He, and X. F. Wang, *Genes Dev.*, **666** (2006).
83. K. Suzuki, M. C. Wilkes, N. Garamszegi, M. Edens, and E. B. Leof, *Cancer Res.*, **3673** (2007).
84. Z. Liu, Y. Tang, T. Qiu, X. Cao, and T. L. Clemens, *J. Biol. Chem.*, **17156** (2006).
85. Z. Q. Dai, R. Wang, S. K. Ling, Y. M. Wan, and Y. H. Li, *Cell Prolif.*, **671** (2007).

86. Z. Guo, J. Yang, X. Liu, X. Li, C. Hou, P. H. Tang, and N. Mao, *Chin. Med. J. (Engl.)*, **950** (2001).
87. S. Walsh, C. M. Jefferiss, K. Stewart, and J. N. Beresford, *Bone*, **80** (2003).
88. H. J. Liu, H. F. Duan, Z. Z. Lu, H. Wang, Q. W. Zhang, Z. Z. Wu, and L. S. Wang, *Zhongguo Shi Yan Xue Ye Xue Za Zhi*, **1044** (2005).
89. Q. R. Wang, Z. J. Yan, and N. S. Wolf, *Exp. Hematol.*, **341** (1990).
90. A. Herbertson and J. E. Aubin, *J. Bone Mineral Res.*, **285** (1995).
91. V. Shalhoub, D. Conlon, M. Tassinari, C. Quinn, N. Partridge, G. S. Stein, and J. B. Lian, *J. Cell Biochem*, **425** (1992).
92. H. Zhou, W. Mak, Y. Zheng, C. R. Dunstan, and M. J. Seibel, *J. Biol. Chem.*, **1936** (2008).
93. I. Kalajzic, Z. Kalajzic, M. Kaliterna, G. Gronowicz, S. H. Clark, A. C. Lichtler, and D. Rowe, *J. Bone Mineral Res.*, **15** (2002).
94. L. B. Sher, J. R. Harrison, D. J. Adams, and B. E. Kream, *Calcif. Tissue Int.*, **118** (2006).
95. L. B. Sher, H. W. Woitge, D. J. Adams, G. A. Gronowicz, Z. Krozowski, J. R. Harrison, and B. E. Kream, *Endocrinology*, **922** (2004).
96. H. Zhou, W. Mak, R. Kalak, J. Street, C. Fong-Yee, Y. Zheng, C. R. Dunstan, and M. J. Seibel, *Development*, **427** (2009).
97. B. J. Feldman, *Pediatr. Res.*, **249** (2009).
98. S. Khosla, E. G. Lufkin, S. F. Hodgson, L. A. Fitzpatrick, and L. J. Melton, 3rd, *Bone*, **551** (1994).
99. S. Pierotti, L. Gandini, A. Lenzi, and A. M. Isidori, *J. Steroid Biochem. Mol. Biol.*, **292** (2008).
100. C. A. O'Brien, D. Jia, L. I. Plotkin, T. Bellido, C. C. Powers, S. A. Stewart, S. C. Manolagas, and R. S. Weinstein, *Endocrinology*, **1835** (2004).
101. R. S. Weinstein, R. L. Jilka, A. M. Parfitt, and S. C. Manolagas, *J. Clin. Invest.*, **274** (1998).
102. U. D. Akavia, I. Shur, G. Rechavi, and D. Benayahu, *BMC Genomics*, **95** (2006).
103. P. L. Chang, H. C. Blair, X. Zhao, Y. W. Chien, D. Chen, A. B. Tilden, Z. Chang, X. Cao, O. M. Faye-Petersen, and P. Hicks, *Connect. Tissue Res.*, **67** (2006).
104. A. Fried, D. Benayahu, and S. Wientroub, *J. Cell Physiol.*, **472** (1993).

105. S. A. Kuznetsov, P. H. Krebsbach, K. Satomura, J. Kerr, M. Riminucci, D. Benayahu, and P. G. Robey, *J. Bone Mineral Res.*, **1335** (1997).
106. I. Shur, F. Lokiec, I. Bleiberg, and D. Benayahu, *J. Cell Biochem.*, **547** (2001).
107. Y. Hori, S. Inoue, Y. Hirano, and Y. Tabata, *Tissue Eng.*, **995** (2004).
108. A. A. Stewart, C. R. Byron, H. C. Pondenis, and M. C. Stewart, *Am. J. Vet. Res.*, **1013** (2008).
109. Kim S. Y., Shin H. I., and P. EK., *Kor J. Oral Maxillofac. Pathol.*, **19** (2007).
110. F. Ng, S. Boucher, S. Koh, K. S. Sastry, L. Chase, U. Lakshmipathy, C. Choong, Z. Yang, M. C. Vemuri, M. S. Rao, and V. Tanavde, *Blood*, **295** (2008).
111. M. Korhonen, *Eur. J. Haematol.*, **167**; author reply **168** (2007).
112. N. Meuleman, T. Tondreau, A. Delforge, M. Dejeneffe, M. Massy, M. Libertalis, D. Bron, and L. Lagneaux, *Eur. J. Haematol.*, **309** (2006).

Chapter 42

NANOPARTICLES FOR BIOIMAGING IN REGENERATIVE MEDICINE

Dongwon Lee,* **John M. Rhee, and Gilson Khang**
Department of Polymer-Nano Science and Technology, Choubuk National University, 664-14, Dukjin, Jeonju, 561-756, Korea
*dlee@chonbuk.ac.kr

Hydrogen peroxide is a major reactive oxygen species (ROS) and plays a fundamental role as a second messenger in normal cellular signal transduction. The overexpression of hydrogen peroxide has been implicated in inflammatory responses and numerous life-threatening diseases. Therefore, hydrogen peroxide has great potential as a diagnostic biomarker of inflammatory response. Inflammation is considered one of the most important host reactions to implanted biomaterials, and therefore how implanted biomaterials influence inflammation is of central important in determining the outcome of scaffold-based regenerative medicine. Imaging of hydrogen peroxide in cells and tissues has been performed by using optical probes that can react with hydrogen peroxide with high sensitivity and specificity and emit light. Optical imaging allows us to watch the behavior of target biological molecules in real time, providing temporal and spatial information on biological molecules in cells and tissues. The development of optical probes for molecular imaging of hydrogen peroxide is now a

Handbook of Intelligent Scaffolds for Tissue Engineering and Regenerative Medicine
Edited by Gilson Khang

www.panstanford.com

very active research area. This review has special emphasis on the design, detection mechanism, and performance of fluorescent and luminescent probes for hydrogen peroxide.

42.1 Introduction

ROS is a key physiological regulator and an important biological messenger in signal transduction cascades.[1,2] ROS includes hydrogen peroxide, superoxide, hydroxyl radicals, and nitric oxide.[3,4] In particular, hydrogen peroxide plays a fundamental role as a second messenger in normal cellular signal transduction. Recently, new data has suggested that a controlled burst of hydrogen peroxide can serve beneficial roles for cell survival, growth, differentiation, and maintenance. However, hydrogen peroxide is considered to be a major ROS in living organisms and a common oxidative stress.[5,6] A peroxide burst in response to cell receptor stimulation affects several classes of essential signaling proteins that control cell proliferation and/or cell death, including mitogen-activated protein (MAP) kinase and nuclear factor κB (NF-κB).[1,7–10] Moreover, excess generation of hydrogen peroxide induces oxidative stress, and cellular mismanagement of oxidation-reduction chemistry triggers subsequent oxidative damages to tissue and organs.[7] Hydrogen peroxide is thus a feature of human aging and numerous human diseases, particularly when inflammation is prominent.[8] These include acute and chronic inflammatory processes, atherosclerosis, neurodegenerative disorders, diabetics, and ischemic stroke. Mitochondria of inflammatory cells, including macrophages, neutrophiles, and eosinophiles, are believed to be the major source of ROS at the site of inflammation.[3] Therefore, hydrogen peroxide has great potential as a diagnostic biomarker of inflammatory response.[11–13]

In scaffold-based regenerative medicine, implanted biomaterials ultimately develop complications—adverse interactions of the patient with implants—which constitute implant failure and may cause harm to the patient. Inflammation is the most important host reaction to implanted biomaterials involving numerous cell types and mediator signals. Therefore, how implanted biomaterials influence inflammation is of central importance in determining the outcome of a regenerative medicine and tissue engineering strategy. Monitoring of inflammation is also important in regenerative medicine to evaluate the physiological effects of implants and the repair process.[14–16] Inflammation may be monitored by imaging hydrogen peroxide using optical probes that can react with ROS and emit light. Several systems have been developed for imaging of ROS in cells and tissues, including small molecules, proteins, and nanoparticles.[5]

The use of optical probes for tracking and reporting functional information of biological molecules is a new and rapidly growing technology, known as optical molecular imaging.[17] Optical molecular imaging is capable of providing temporal and spatial information on target biological molecules in cellular systems *in vitro* and *in vivo*; that is, it is possible to watch the behavior of a target biological molecules in living organisms in real time using optical probes.[18–20] Molecular imaging has a wide range of applications in pharmaceutical and toxicological research. This also has great potential for detecting diseases and monitoring the disease progress. In general, optical imaging probes have two classes, fluorescent and luminescent. Fluorescent probes require excitation light to excite the fluorophore to a higher energy state, followed by emission at a longer wavelength (less energy state). Genetically encoded proteins such as green fluorescent proteins (GFP) and fluorescent dyes are widely utilized as fluorescent probes. Luminescent probes do not require excitation light. Light emission from the luminescent probes is the result of a chemical reaction generating high energy. Peroxalate chemiluminescence and firefly luciferase proteins are good examples of luminescent probes.[17]

The development of optical probes capable of detecting ROS with high sensitivity and specificity is now a very active research area.[1–3] This chapter will focus on the fluorescent and luminescent probes for imaging of ROS. Special emphasis will be paid to the design and detection mechanism of fluorescent and luminescent probes that have high sensitivity and specificity to hydrogen peroxide. The probes are classified and subdivided based on the light emission mechanism.

42.2 Fluorescent Probes for Imaging of Hydrogen Peroxide

Fluorescent probes are promising tools for revealing the biological roles in living organisms and a number of fluorescent probes have been developed for a long time (Fig. 42.1).

A traditional fluorescent dye, 2',7'-dichlorodihydrofluorescein (DCFH), has been extensively used to detect ROS. DCFH is oxidized by ROS easily to form 2',7'dichlorofluorescein (DCF) that can emit strong fluorescence. Amplex Red, or 10-acetyl-3,7-dihydroxyphenoxazine, is also used to detect ROS.[5] However, these compounds have serious limitations such as background fluorescence. They tend to be autoxidized to produce large background fluorescence in the absence of ROS. Although they have been used to detect ROS extensively, it remains controversial that they react with ROS and detect individual ROS.[21–23]

The demands for new fluorescent probes that are capable of detecting hydrogen peroxide with high sensitivity and specificity have been increasing because the significance of hydrogen

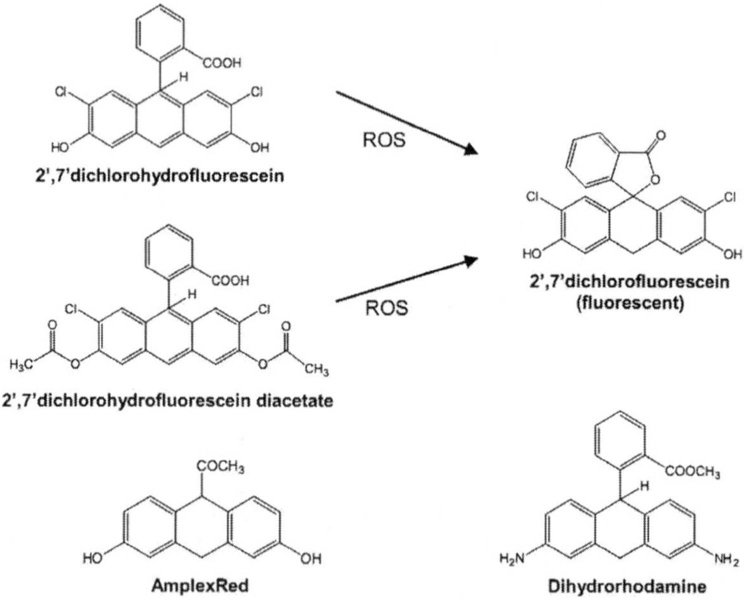

Figure 42.1. Representative and traditional fluorescent probes for ROS.

peroxide in physiology and numerous diseases is being increasingly recognized. The development of fluorescent probes for the selective and sensitive detection of hydrogen peroxide is now being energetically undertaken. Recently, several novel fluorescent probes for the selective detection of hydrogen peroxide have been developed, and some of them have been applied to the imaging of intracellular hydrogen peroxide. Most of these approaches for selective hydrogen peroxide imaging are based on the chemoselective deprotecting groups, which can be cleaved by hydrogen peroxide selectively.[1–3]

Maeda *et al.* synthesized pentafluorobenzenesulfonyl fluoresceins for the selective detection of hydrogen peroxide (Fig. 42.2). These probes are deprotected by hydrogen peroxide oxidation to release free fluorescent fluorescein. Unlike fluorescein with a dinitrobenzenesulfonate-leaving group, these probes selectively react with high selectivity to hydrogen peroxide, whose reaction rate constant is up to $\sim 250 M^{-1}s^{-1}$.[24] These probes were synthesized on the basis of the following rationale that the pentafluorobenzene ring enhances the reactivity of sulfonates toward hydrogen peroxide. They exhibited excellent selectivity to hydrogen peroxide over hydroxy radicals and superoxide. The detection limit of the probe (X=F) was reported to be 4.6 pmol.

Chang *et al.* designed and synthesized various fluorescent probes using chemoselective boronate deprotection. The major strategy of these optical probes involves exploiting the selective hydrogen peroxide–mediated transformation of arylboronates to phenols.[1–3] Upon the reaction with hydrogen peroxide, hydrolytic deprotection of the boronates produces the fluorescent fluorescien product. One of examples is shown in Fig. 42.3.

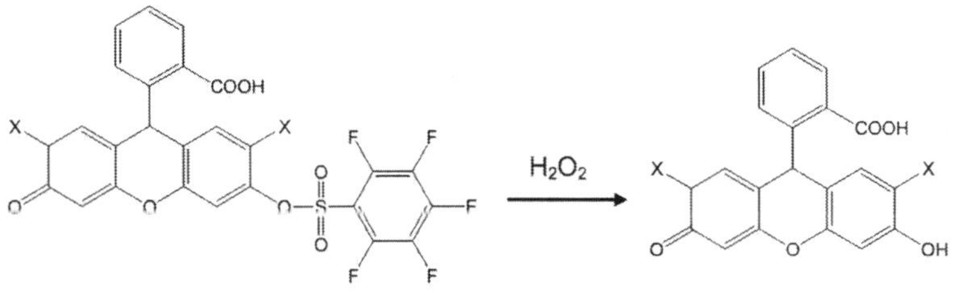

Figure 42.2. Hydrogen peroxide imaging based on benzenesulfonyl protection.

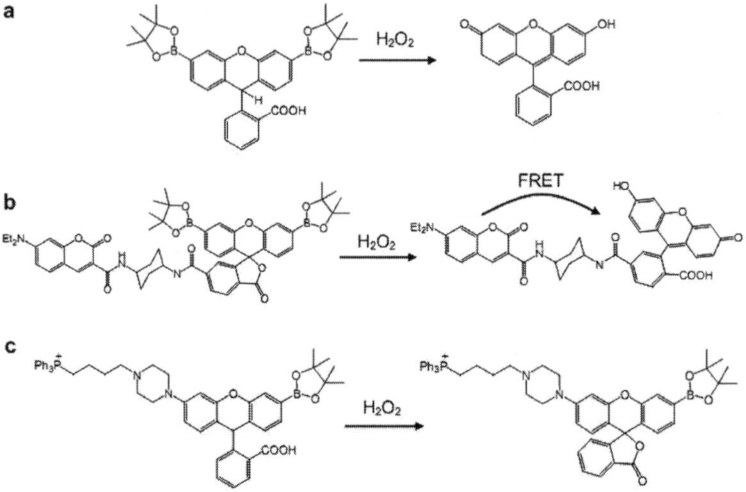

Figure 42.3. Boronate-protected fluorescent probes for imaging of hydrogen peroxide.

A rationally designed Peroxyfluor-1, in which installation of boronic ester groups renders the parent molecule nonfluorescent, reacts with hydrogen peroxide with excellent selectivity, leading to regeneration of strong fluorescence. A ratiometric fluorescent probe for hydrogen peroxide was also developed based on modulating fluorescence resonance energy transfer (FRET) in a two-fluorophore cassette comprised of coumarin- and boronate-protected fluorescein. Upon selective reaction with hydrogen peroxide to generate fluorescent fluorescein moiety, fluorescein shows a strong absorption in the coumarin emission region. The excitation of coumarin results in enhanced fluorescence from fluorescein by FRET. This ratiometric probe allows accurate and quantitative detection of hydrogen peroxide. The researchers also designed and developed a targetable fluorescent probe for imaging of mitochondrial hydrogen peroxide in living cells. The targetable fluorescent probe was designed to contain both mitochondrial-targeting moiety and a hydrogen peroxide–responsive element. The strategy of fluorescence imaging of mitochondrial hydrogen peroxide is based on the use of phosphonium head groups that have been used to deliver antioxidants to mitochondria. This probe was capable of imaging changes in the

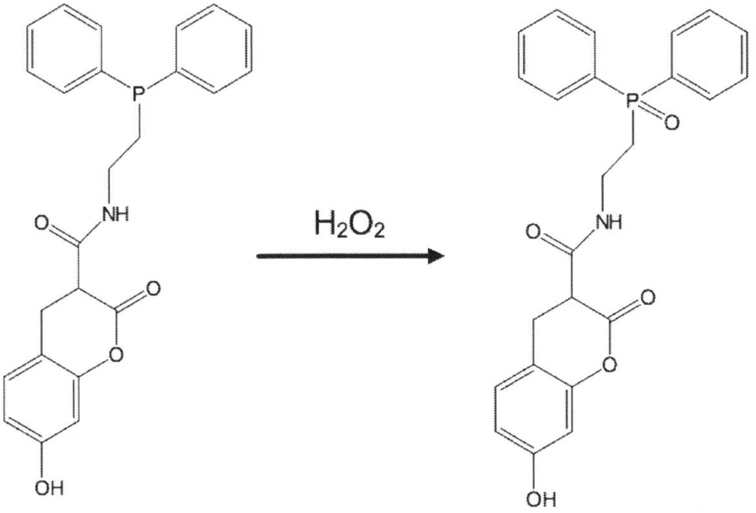

Figure 42.4. Fluorescent probes with diphenylphosphine moiety of hydrogen peroxide.

level of hydrogen peroxide within the mitochondria of various mammalian cells, as well as hydrogen peroxide elevation induced by an oxidant stress model of Parkinson's disease.

Another approach for the selective detection of hydrogen peroxide is based on the concept of photoinduced electron transfer (Fig. 42.4), in which 7-Hydroxy-2-oxy-*N*-(2-(diphenylphosphino) ethyl)-2H-choromene-3-carboxamide was rationally designed to consist of diphenylphosphine moiety and 7-hydroxycoumarin moiety.[25] With the diphenylphosphin moiety, 7-hydrocoumain had limited fluorescence. It produced increased fluorescence upon oxidation by hydrogen peroxide. However, it showed undesirably low reactivity toward hydrogen peroxide, fluorescence appearing over tens of minutes at 50 μM of hydrogen peroxide.

42.3 Luminescent Probes for Imaging of Hydrogen Peroxide

Peroxalate chemiluminescence is the most effective and sensitive chemiluminescence and is based on the reaction between an active oxyoxalate and an oxidant, typically hydrogen peroxide, in the presence of a fluorophore.[26] This reaction proceeds via a two-step process: first hydrogen peroxide reacts with the oxalate ester, generating the high-energy dioxetandione intermediate. This then excites the fluorescent dye, which upon relexation chemiluminesces and emits a photon.[27] Peroxalate chemiluminescence has been widely employed for the detection of hydrogen peroxide and fluorescent molecules due to its intrinsic nanomolar sensitivity and specificity to hydrogen peroxide.[28] Despite these appealing features, *in vivo* imaging of hydrogen peroxide based on peroxalate chemiluminescence was very challenging because of the instability of the peroxalate in water and its water insolubility. Moreover, the peroxalate chemiluminescence reaction requires biocompatible nanosized scaffolds that allow a three-component reaction in a biological environment.[11–13]

Recently, peroxalate chemiluminescence–based *in vivo* hydrogen peroxide imaging has been recently reported by Lee *et al.* (Fig. 42.5).

They hypothesized that solid nanoparticles composed of peroxalate polymers, which encapsulated fluorescent dyes, would chemiluminesce under aqueous conditions in the presence of hydrogen peroxide and also be suitable for *in vivo* imaging of hydrogen peroxide. Novel polymers were rationally designed, incorporating the peroxalate ester groups in the polymer backbone. They were synthesized from the reaction of oxalic chloride, hydroxybenzyl alcohol, and octanediol. The optimal composition of three monomers was chosen from the careful consideration of the reactivity to hydrogen peroxide and stability against water hydrolysis. Chemiluminescent nanoparticles were prepared by a single-emulsion method, and fluorescent dyes were encapsulated during the nanoparticles' formation. The peroxalate particles had a mean size of 550 nm.[11]

The peroxalate nanoparticles exhibited several attractive properties for *in vivo* imaging, such as tunable wavelength emission, excellent sensitivity, and high specificity for hydrogen peroxide over other ROS. As shown in Fig. 42.6, chemiluminescent intensity increased linearly with the concentration of hydrogen peroxide.

They could detect hydrogen peroxide as low as 250 nM. The chemiluminescent intensity of the peroxalate nanoparticles in the

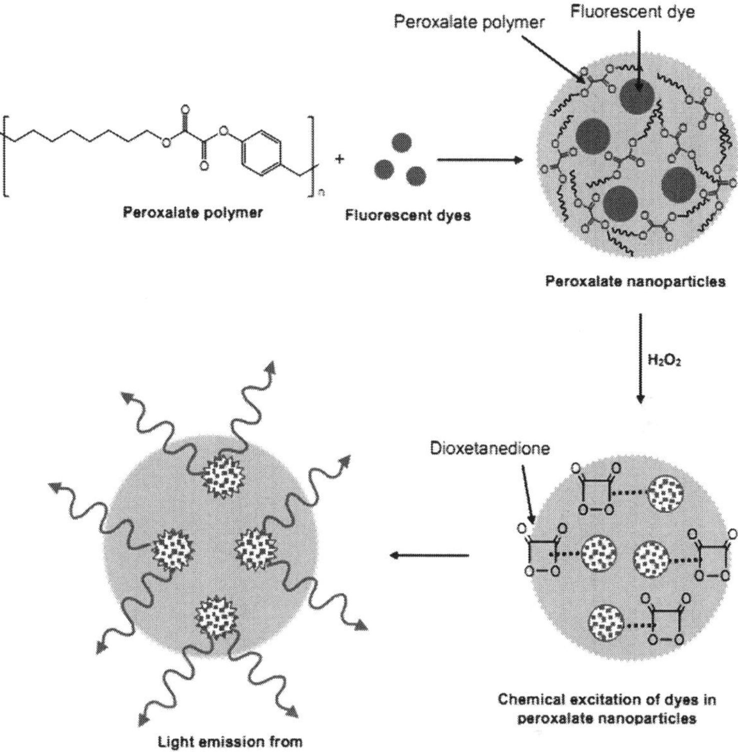

Figure 42.5. Chemiluminescent nanoparticles for the detection of hydrogen peroxide.

presence of hydrogen peroxide was almost 100 times higher that with either hydroxyl radicals or superoxide. The capability of imaging of *in vivo* hydrogen peroxide was performed by injecting peroxalate nanoparticles into the intraperitoneal cavity of mice after the induction of inflammation using lipopolysaccharide (LPS).

Figure 42.7 shows images chemiluminescence emitted from mice treated with peroxalate nanoparticles.

After the treatment of peroxalate nanoparticles, mice treated with LPS exhibited two times higher chemiluminescence intensity than mice treated with saline. The results clearly demonstrated that peroxalate nanoparticles composed of peroxalate polymers and

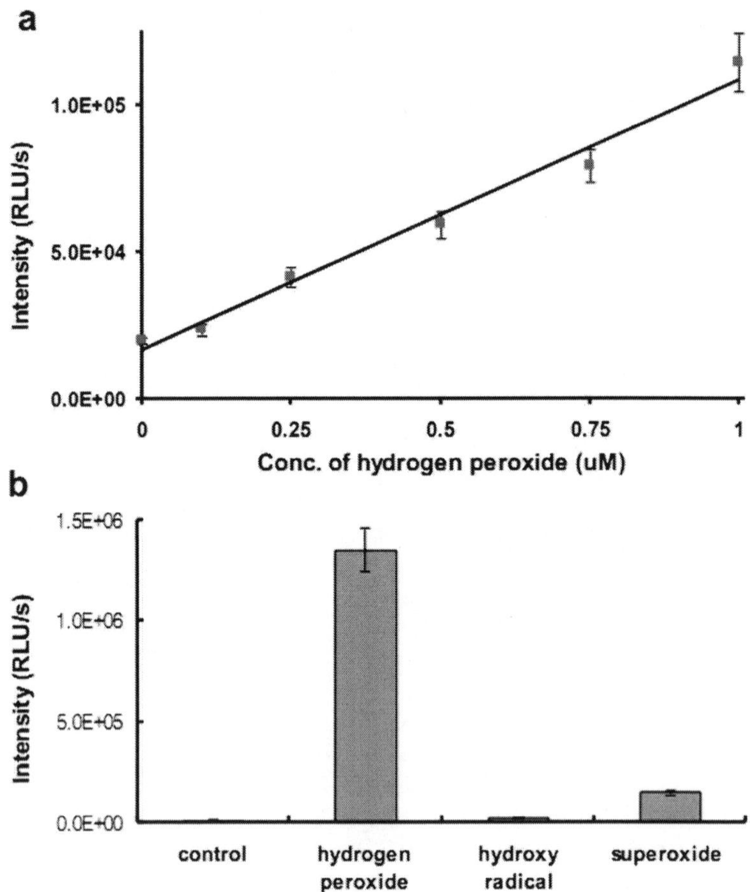

Figure 42.6. Detection of hydrogen peroxide with peroxalate nanoparticles. (a) Sensitivity and (b) specificity.

fluorescent dyes have great potential for imaging of hydrogen peroxide generated during inflammation-associated conditions.

Despite great promise for *in vivo* imaging of hydrogen peroxide using peroxalate polymer nanoparticles, their physicochemical properties such as large size and hydrophobic surface nature limit more biological applications. To overcome these limitations, peroxalate chemiluminescent micelles were developed. The micelles exploited the same concept of three-component peroxalate

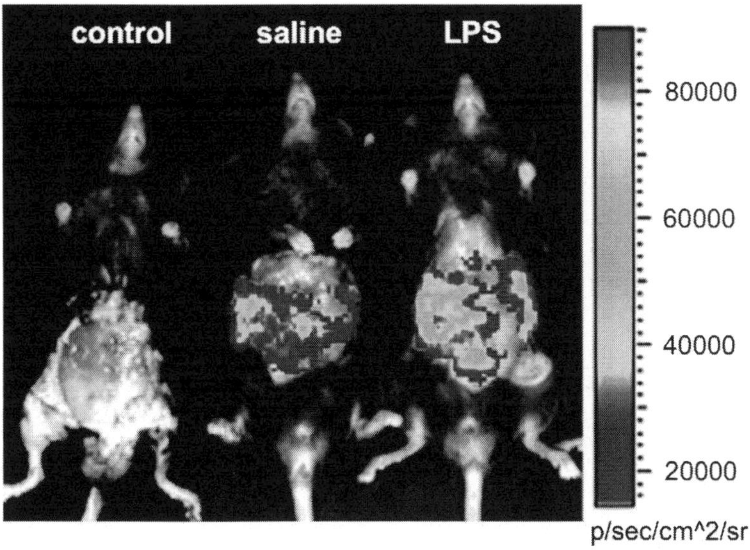

Figure 42.7. Detection of hydrogen peroxide using peroxalate chemiluminescent nanoparticles. After the induction of inflammation using LPS, 1 mg of peroxalate nanoparticles were injected into the inflammatory site and images were made for 1 min of acquisition time. See also Color Insert.

reactions employed in the peroxalate nanoparticles. Peroxalate compounds and fluorescent dyes were encapsulated in micelles composed of poly(ethylene glycol)-b-poly(caprolactone) (PEG PCL), as shown in Fig. 42.8.[13]

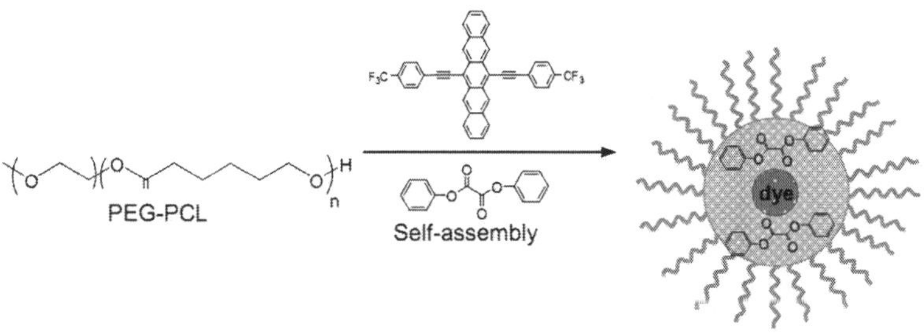

PEG-PCL

Self-assembly

dye

Figure 42.8. Chemiluminescent micelles composed of amphiphilic polymers, fluorescent dye, and peroxalate compounds.

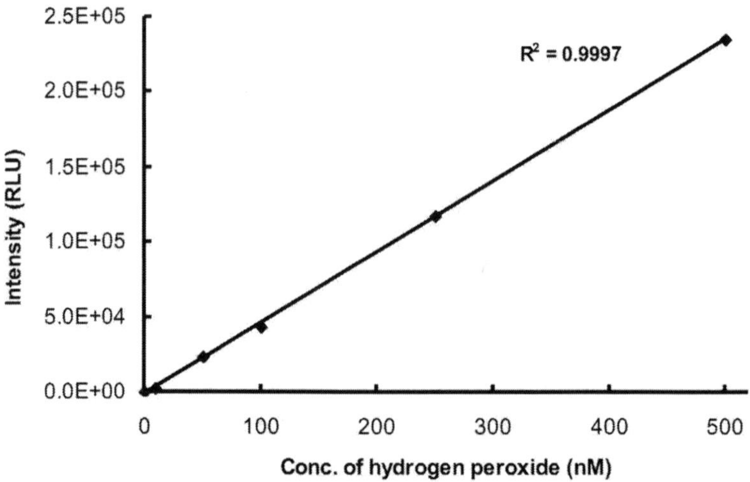

Figure 42.9. Detection of hydrogen peroxide using chemiluminescent micelles.

The chemiluminescent micelles have a mean size of 50 nm and possess a stealth PEG corona to enhance their circulation in the bloodstream. They were able to detect hydrogen peroxide at concentrations as low as 50 nM (Fig. 42.9). The significant enhancement of sensitivity is due mainly to the small size, which provides a larger surface area and enhances the diffusion of hydrogen peroxide into the micelles. The small size of chemiluminescent micelles should make them suitable for imaging extracellular hydrogen peroxide produced in inflammatory tissues as micelles with a mean size of 50 nm can extravasate into inflammatory tissues.[12]

42.4 Fluorescent Probes for Other ROS

Superoxide is known to be first generated in the respiratory chains and dismutes spontaneously to hydrogen peroxide.[8,29–30] The production and reaction of superoxide as a precursor to other ROS are of considerable importance to studies of cellular metabolism and pathogenesis of diverse cytotoxic phenomena. Therefore, there has been great interest to detect superoxide to elucidate the

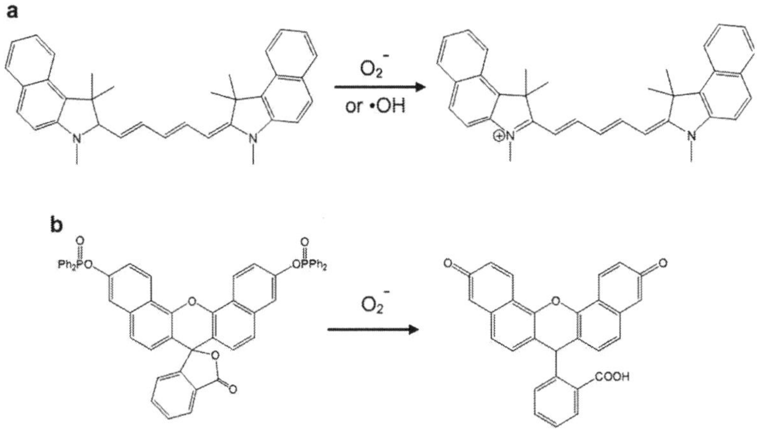

Figure 42.10. Fluorescent probes for detection of superoxide. (a) Cydrocyanine and (b) PNF-1.

mechanistic details and precise roles in pathology, aging, and diseases in living organisms. However, it is challenging to detect superoxide quantitatively because of its extremely short half-life.

Recently, a new class of fluorescent sensors has been developed that can image superoxide and hydroxyl radicals in cell cultures, tissue, and *in vivo* (Fig. 42.10a).[29] The fluorescent sensors were termed hydrocyanines, which were synthesized in one step from the commercially available cyanine dyes. Hydrocyanines are weakly fluorescent because of disrupted π-conjugations. However, after oxidation with either superoxide or hydroxyl radicals they exhibit dramatically increased fluorescence by regenerating their extended π-conjugations. Hydrocyanines have physicochemical properties needed for the detection of intracellular ROS. They are initially noncharged and membrane permeable. However, oxidation with either superoxide or hydroxyl radicals converts them into charged and membrane-impermeable molecules, which can accumulate in ROS-overexpressing cells. They have excellent stability, nanomolar sensitivity, and specificity to ROS. Furthermore, they were capable of imaging superoxide and hydroxy radical in cell cultures, tissue explants, and LPS-treated mice.

Xu *et al.* developed another fluorescent probe for imaging of superoxide based on a nucleophilic mechanism of superoxide to mediate deprotection of 4',9'-bis(diphenylphosphinyl) naphthofluorescein (PNF-1).[30] PNF-1 is a closed, nonfluorescent. After oxidation with superoxide, it rapidly regenerates into an open and fluorescent product. This probe exhibited a strong fluorescence response and high specificity to superoxide over other ROS and showed a detection limit of 0.1 nM (Fig. 42.10b).

42.5 Conclusions and Outlook

ROS, such as hydrogen peroxide and superoxide, plays a vital role in physiology and pathology. It is an essential oxygen metabolite for normal physiological functions and has been implicated in many inflammation-associated conditions. Therefore, there is great need for the development of methodology to image ROS in cells and tissues. Optical molecular imaging is a new and rapidly growing field, in which optical probes are used for tracking and reporting functional information of biological molecules. This chapter has focused almost exclusively on fluorescent and luminescent imaging for imaging of hydrogen peroxide and superoxide with high sensitivity and specificity. Special emphasis was made on the rational design and detection mechanism for hydrogen peroxide and superoxide. They are expected to have great potential for elucidating unknown physiological activities of ROS in inflammation-associated conditions.

Acknowledgments

This work was supported by a New Faculty Support Program of Chonbuk National University and a grant of the Korea Healthcare Technology R&D Project, Ministry of Health, Wealth & Family Affairs, Republic of Korea (A084304).

References

1. M. C. Y. Yang, *J. Am. Chem. Soc.*, **15392** (2004).
2. E. V. Miller, *J. Am. Chem. Soc.*, **16652** (2005).

3. B. C. Dickson, *J. Am. Chem. Soc.*, **9638** (2008).
4. C. W. Winterbourn, *Nat. Chem. Biol.*, **278** (2008).
5. N. Soh, *Anal. Bioanal. Chem.*, **532** (2006).
6. P. Wardman, *Free Radical Biol. Med.*, **995** (2007).
7. A. E. Albers, *J. Am. Chem. Soc.*, **9640** (2006).
8. M. Nagata, *Curr. Drug Targets-Inflamm. Aller.*, **503** (2005).
9. S. G. Rhee, *Nat. Chem. Biol.*, **244** (2007).
10. F. D. Virgilio, *Curr. Pharm. Des.*, **1647** (2004).
11. D. Lee, *Nat. Mater.*, **765** (2007).
12. D. Lee, *Int. J. Nanomed.*, **471** (2008).
13. M. Dasari, *J. Biomater. Mater. Res. Part A*, **561** (2009).
14. M. D. Kofron, *Adv. Funct. Mater.*, **1351** (2009).
15. N. P. Rhodes, *Biomaterials*, **51531** (2007).
16. E. E. Falco, *Pharm. Res.*, **2348** (2008).
17. W. W. Rice, *J. Biomed. Opt.*, **432** (2001).
18. J. V. Frangioni, *Curr. Opin. Chem. Biol.*, **626** (2003).
19. F. A. Jaffer, *J. Mole. Cellu. Cardio.*, **921** (2006).
20. V. Ntziachristos, *Eur. Radiol.*, **195** (2003).
21. E. Marchesi, *Free Radical Biol. Med.*, **148** (1999).
22. C. Rota, *Free Radical Biol. Med.*, **873** (1999).
23. M. Afzal, *Biochem. Biophys. Res. Comm.*, **619** (2003).
24. H. Maeda, *Angew. Chem. Int. Ed.*, **2389** (2004).
25. N. Soh, *Bioorg. Med. Chem.*, **1131** (2005).
26. J. Motoyoshiya, *J. Org. Chem.*, **7314** (2002).
27. C. Lee, *Synthe. Metals*, **97** (2002).
28. M. Tsunoda, *Anal. Chim. Acta*, **13** (2005).
29. K. Kundu, *Angew. Chem. Int. Ed.*, **299** (2009).
30. K. Xu, *Chem. Bio. Chem.*, **453** (2007).

Chapter 43

EFFECT OF SCAFFOLDS WITH BONE GROWTH FACTORS ON NEW BONE FORMATION

Hae-Ryong Song,[a*] Swee-Hin Teoh,[b] Jun-Ho Wang,[e] Hak-Jun Kim,[a] Ji-Hoon Bae,[a] Sung Eun Kim,[a] Jerry Chan,[c] Zhi-Yong Zhang,[b] and Chang-Wug Oh[d]

[a]*Rare Diseases Institute, Department of Orthopedic Surgery, College of Medicine, Korea University, Seoul, Korea*
[b]*Centre for Biomedical Materials and Applications (BIOMAT), Department of Mechanical Engineering, National University of Singapore*
[c]*Experimental Fetal Medicine Group, Department of Obstetrics and Gynaecology, KK Women's and Children's Hospital and National University of Singapore, Singapore*
[d]*Department of Orthopedic Surgery, Kyungpook National University Hospital, Daegu, Korea*
[e]*Department of Orthopedic Surgery, Samsung Medical Center, Seoul, Korea*
*songhae@korea.ac.kr

Bone tissue engineering (BTE) using scaffolds with growth factors/stem cells has received some success in clinical trails. Recently, osteogenic scaffolds using absorbable or nonabsorbable biomaterials have being investigated to repair bone defects. Scaffolds can provide the osteoconduction effect as well as a media to trap cells and allow them to proliferate and differentiate. Bioreactors provide a means to uniformly coat the scaffolds with cells

Handbook of Intelligent Scaffolds for Tissue Engineering and Regenerative Medicine
Edited by Gilson Khang

www.panstanford.com

and allow cells to proliferate better. Here we describe our work on preclinical animal trials applying the above concept to regenerate bone.

43.1 Introduction

New BTE methods are necessary to treat bone defects in orthopedic, plastic, and neurosurgical fields. Bone defect or bone shortening occurs after excision of bone tumor, open fracture, osteomyelitis, and pseudoarthrosis-associated congential diseases such as neurofibromatosis and hemimelia. There have been many methods to treat bone defects. These include autogenous bone grafting, allografting, vascularized fibular graft, and distraction osteogenesis using the Ilizarov method. There are limitations of these methods. Autogenous bone grafting has a limitation of donor site morbidity. Allografting has problems, including the follow-up resorption, infection transmission, or high incidence of refracture due to insufficient bone conduction and induction. Vascularizing fibular grafting has good vascularity, which can help control infection and promote new bone formation. However, it requires long surgical time and extensive experience to achieve good clinical results. Pioneered by Ilizarov, distraction osteogenesis is a remarkably effective technique for creating new bone to treat large bone defects resulting from trauma, tumor resection, or limb length discrepancies.[1–4] Distraction osteogenesis using the Ilizarov method can induce the periosteal new bone formation, and external fixator is utilized to apply 1 mm distraction per day. This technique is useful and can induce new bone formation without any bone grafting. However, one disadvantage of distraction osteogenesis is prolonged external fixation, such as large bone gaps or delay in bone formation in certain cases[5–7]—for instance, the formation of 1 cm new bone requires at least 1 month.

Recently, osteogenic scaffolds using absorbable or nonabsorbable biomaterials have being investigated to repair bone defects. Scaffolds can provide the osteoconduction effect as well as a media to trap cells and allow them to proliferate and differentiate. However, to achieve the osteoinduction effect, the incorporation of growth factors or stem cells is needed. Nonabsorbable materials such as hydroxyapatite or ceramic have been used with advantages of a rigid frame, which provides stability; however, these materials are radio-opaque. Therefore, it is difficult to differentiate the new bone formation from the ceramic scaffolds. In addition, the bioceramic is brittle and not suited for bone applications. A better approach is to use a composite of bioceramic and a resorbable polymer.

43.2 Bone Lengthening in Preclinical Animal Studies

Here, we report a few methods to accelerate the consolidation rate of distracted callus formation by the injection of calcium sulfate, cord blood (CB) stem cell, and human recombinant bone morphogenetic protein-2 (rhBMP-2).

43.2.1 *Calcium Sulfate in Tibial Lengthening*

Small monolateral external distraction devices (small bone distractor; U&I, Kyunggi- Do, Korea) were applied to the diaphysis of the tibia of New Zealand white rabbits (5–6 months old; 2.0–2.5 kg)

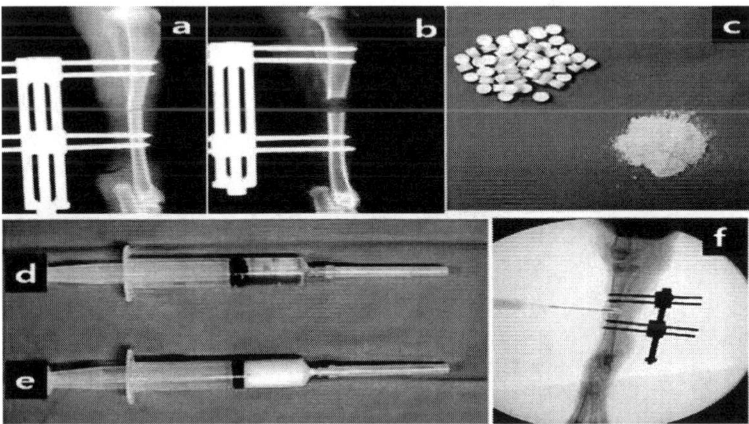

Figure 43.1. After lengthening of the rabbit tibia (a), lengthening was performed (b). The calcium sulfate pellet was made into the powder form (c). The mixed material of calcium sulfate and CMC was injected into group I (d), and only the CMC medium was injected into group II (e). Then, each material was injected using a fluoroscopic guide (f).

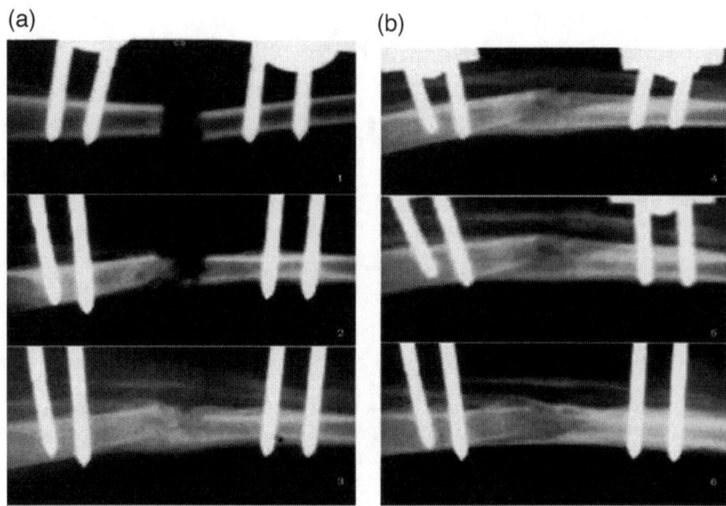

Figure 43.2. Radiographs of animals in experimental group I with calcium sulfate. There was a faint calcification shadow at the distraction gap at 2 weeks, marked calcification at 4 weeks, and bone bridging in the distracted area at 6 weeks.

(Fig. 43.1a). After a 3-day postsurgical waiting period, rabbits underwent distraction at a rate of 1 mm per 12 hours. Distraction lasted four days with the total lengthening of 8.0 mm. Rabbits were randomly divided into three groups. On day 7, group I was injected with calcium sulfate (Wright Medical Technology Inc, Arlington, Tennessee, USA) with a carboxy-methylcellulose (CMC) medium (200 mg calcium sulfate per 1 cm^3 of CMC) in the lengthening region under fluoroscopic guidance (Fig. 43.1b); group II was injected with a CMC medium alone as the control; and no injection was performed in group III. All the rabbits were euthanized 42 days postsurgery for further analysis.

There was early callus formation at week 4 in group I but not in groups II and III (Figs. 43.2–43.4). At week 6, all the rabbits in group I had a bony shadow with bridging of the proximal and distal ends (some rabbits showed mature bone union) in the distracted area. However, none of the rabbits in groups II and III showed bone union. Furthermore, the bone mineral density (BMD) analysis showed that at week 3, the percentage BMD of the experimental group I (16.1%)

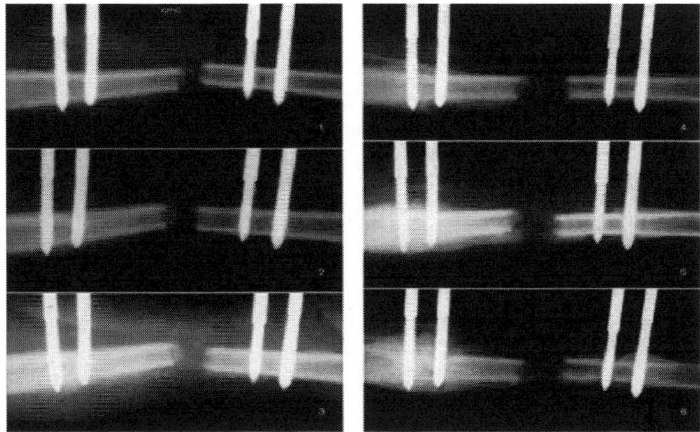

Figure 43.3. Radiographs of animals in control group II with the CMC medium. There was no callus formation at 2 weeks, minute calcification at 4 weeks, and immature callus formation at the end side of the cut part at 6 weeks.

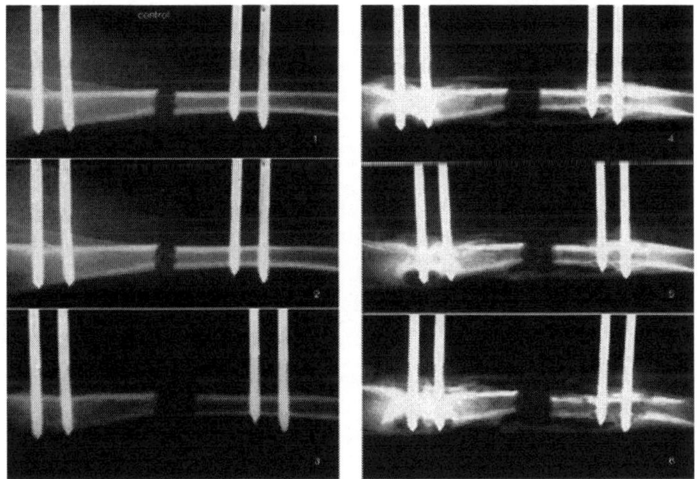

Figure 43.4. Radiographs of animals in control group III. There was no callus formation at 2 weeks, a little spotted calcification at 4 weeks, and irregular calcification in the distracted area at 6 weeks.

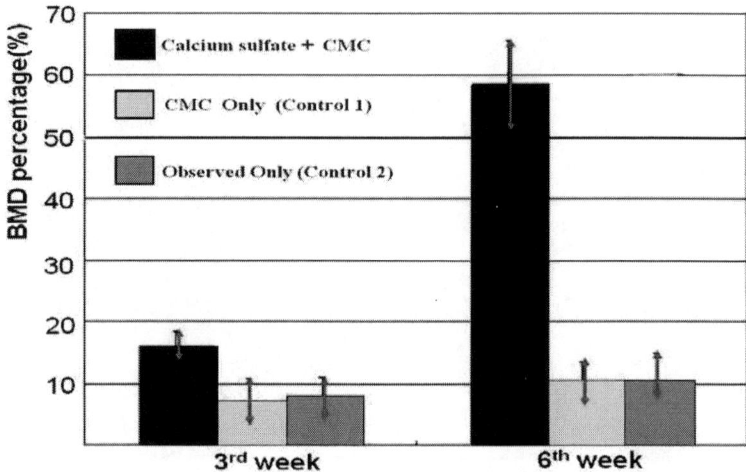

Figure 43.5. At 3 weeks and even 6 weeks, the percentage of BMD in group I increased significantly compared with control groups II and III. White box, group I; black box, group II; hatched box, group III. See also Color Insert.

was higher than the densities of the two control groups (group II, 6.9%; group III, 7.1%). At week 6, the percentage BMD of group I (58.4%) was significantly higher than that of group II (10.2%) and of group III (10.4%) ($P < 0.01$). However, there were no significant differences between the two control groups—groups II and III (Fig. 43.5).

From the hematoxylin and eosin (H&E) histological studies in the two control groups, the new bone formation around the cut end of the bone had abundant collagenous fibrous tissue with some portion of bony trabeculae in the distraction gap. However, without bridging between the two cut ends, the bone formation was incomplete in all control animals (Fig. 43.6). In all rabbits from group I with calcium sulfate, active new bone formation in the distracted area was noted. A similar trabecular pattern of bridging was also noted at the original cut bone ends, but this bridging was thin. The remaining calcium sulfate was nearly invisible. This study has shown that the implantation of calcium sulfate promotes bone maturation in the context of distraction osteogenesis.

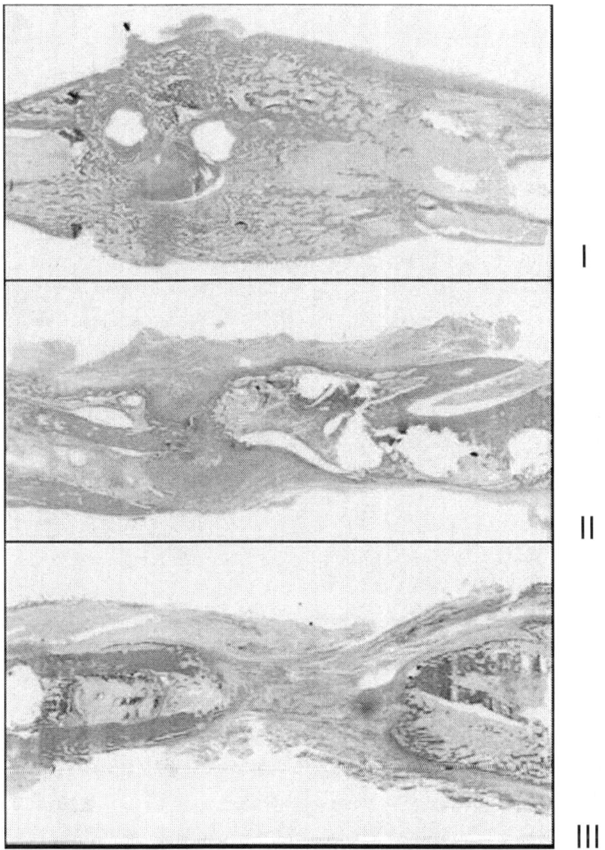

Figure 43.6. Histologic findings with H&E stain (x5) at 6 weeks. There was new trabecular bone with numerous osteoblasts found in the animals from group I with calcium sulfate, continuing the distraction gap. However, there was a small amount of new bone with abundant fibrous tissue found at the end side of the cut part in control group II with plain the CMC medium and in control group III with simple distraction.

43.2.2 *Cord Blood Stem Cells and rhBMP-2 in Tibial Lengthening*

Mesenchymal stem cells (MSCs) are multipotent cell types with a well-defined osteogenic differentiation pathway. Recently, it has been shown that the admixture of osteoblastic cells with collagen

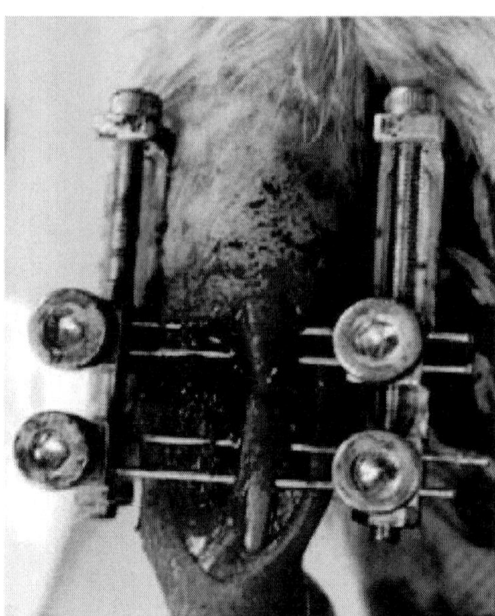

Figure 43.7. Photograph of osteotomy at a rat's tibia with an external fixator. See also Color Insert.

scaffolds enhanced consolidation of distraction osteogenesis experimentally. Also bone morphogenetic protein-2(BMP-2) has been revealed as mediator of tissue regeneration, especially regeneration of bone. However, few reports have been made about direct injection of CB stem cells and BMP-2 at the distraction osteogenesis site of bone. We designed a study to evaluate the effect of CB-derived MSCs (CB-MSCs) and BMP-2 on the healing of a rat model of distraction osteogenesis.

Nine-week-old male Sprague-Dawley rats weighing 250~300 g were used. A small bilateral external fixator was applied to the right tibia of the rats (Fig. 43.**7**). After 7 days, the lengthening of the tibia started at the rate of 0.25 mm per 12 hours for 10 days (total 5 mm lengthening). The rats were divided into control group (15 rats) and a CB-MSC-injected group (30 rats) at the end of the 5 mm distraction. The CB-MSC (Medipost, Seoul, Korea) group was also divided into an only CB-MSC group and an osteogenically primed CB-MSC and PRP-loaded group. Osteoinductive CB-MSCs were cultured in an osteogenic medium for three weeks, which was composed of minimum essential medium-α (MEM-α) supplemented with 10% fetal bovine serum (FBS), 0.1 μM dexamethasone (Sigma), 10 mm β-glycerol phosphate (Sigma), and 50 μM L-ascorbic acid 2-phosphate (Sigma). Platelet-rich plasma (PRP) was used as a carrier for both CB-MSC groups. The BMP-2 injected group(15 rats) was that E-coli derived human recombinant BMP-2s(rhBMP-2, Cowell Medi, Busan, Korea) were injected at distraction site. The control group was not injected with any material. The CB-MSC-injected group and the osteogenically primed CB-MSC group received an injection of 0.05 mL CB-MSC and osteogenically primed CB-MSC, respectively, at the distracted site. The BMP-2 injected group received 0.05 mL(0.1 mg/ml) at the distraction site.

In the osteogenically primed CB-MSC+PRP-injected group and the BMP-2 injected group, bone regeneration was evident in 100% of the animals, but in the control and the CB-MSC+PRP-injected group, bone regeneration was completed in 50% of the animals at four weeks and the osteogenically primed CB-MSC+PRP has positive effective bone regeneration at this early time point. However, more than 50% bone regeneration was also found in the CB-MSC+PRP-injected group from six weeks. Furthermore, similar amount of bone regeneration was estimated in the CB-MSC+PRP-injected group and the osteogenically primed CB-MSC+PRP-injected group and the BMP-2 injected group at eight weeks. The control group showed delayed bone regeneration in comparison to the CB-MSC+PRP-injected group and the osteogenically primed CB-MSC+PRP-injected group at eight weeks (Fig. 43.8). In the histological examination at eight weeks, new bone formation of the control group was visible at the distraction site, but new bone formation of the osteogenically primed CB-MSC-injected group and the BMP-2 injected group was more robust and had larger enchondral ossification layers than the control group and the CB-MSC-injected group (Fig. 43.9). The CB-MSC-injected group and the osteogenically primed CB-MSC+PRP-injected group also had more osteocytes than the control group; the size of the callus was compared through an image analyzer (Table 43.1).

From this study, osteogenically primed CB-MSCs and BMP-2 accelerated the consolidation of the distraction area in respect of

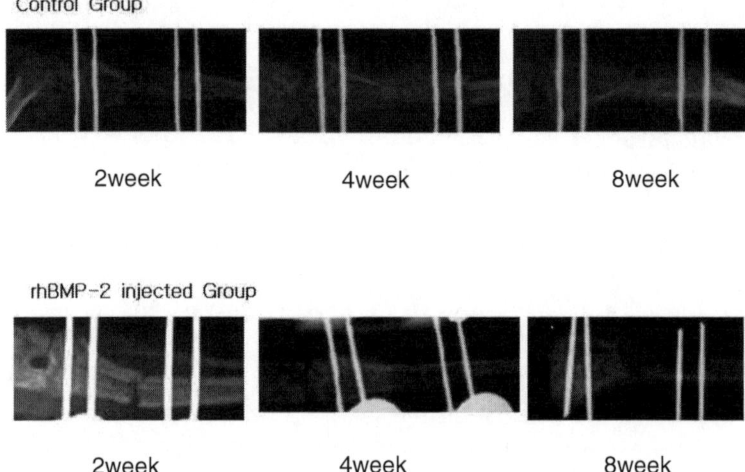

Figure 43.8. Serial radiographs of control group and rhBMP-2-injected group. Earlier consolidation was visible at the rhBMP-2-injected group than the control group.

radiology and histology. But osteogenically primed CB-MSCs and BMP-2s could not visible under radiography. A mixture of radio-opaque scaffolds with osteogenically primed CB-MSCs and MBP-2s could be helpful to identify the injected material and focus the injection area under the radiographic control.

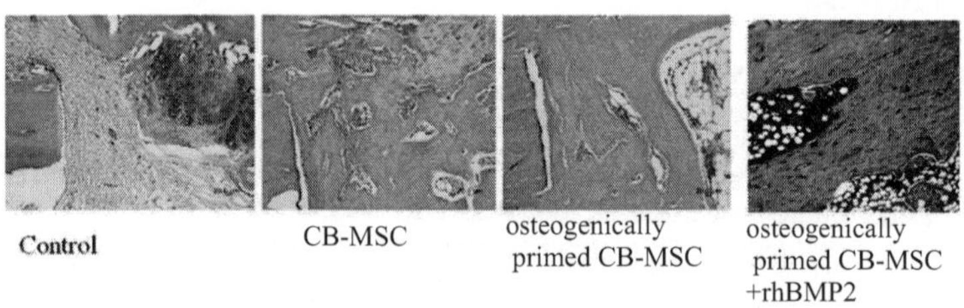

Figure 43.9. Histological photographs of control, CB-MSC, osteogenically primed CB-MSC and BMP-2 injected group at 8 weeks under x20 magnification. Bone formation and connection between the distraction callus and original bone was visible in osteogenically primed CB-MSC and BMP-2 injected group.

Table 43.1. Scores on the grading scale for histopathological evaluation.

	Histological Grade
Control	0.030±0.11
CB-MSC*	0.030±0.14
Osteoinductive CM-MSC**	0.048±0.33

*CB-MSC: cord blood mesenchymal stem cell
**osteoinductive CB-MSC: osteoinductive cord blood mesenchymal stem cell

43.3 Growth Factor–/Stem Cells–Mediated Scaffolds for Bone Tissue Engineering

Bone tissues have comparatively high self-regenerative capability. But this natural regenerative capacity is not sufficient to repair a large bone defect.[8] To treat large bone defects, allograft and autograft bone tissues have been used. However, the limited availability of and the risk of disease transmission have hampered the use of allografts. Tissue-engineered bone tissue has been studied as an alternative treatment for bone regeneration.[9] Tissue engineering (TE) approaches have the potential to overcome shortage of donor tissue.[10] TE strategies use specific combinations of cells, scaffolds, and bioactive factors for development of viable, three-dimensional (3D) tissue implants.[11] Here we described a few studies to use BTE scaffolds with or without growth factors and stem cells for bone defect healing.

43.3.1 *Use of Fibrin and Stem Cells for Bone Defect Healing in Rabbits*

MSCs, which can differentiate into osteoblasts, chondrocytes, and adipocytes,[12] can be easily isolated from bone marrow (BM) aspirates and expanded to large cell numbers *in vitro*. A challenge for bone regeneration in a scaffold system is how to provide the correct signal instructions to bring cells to the appropriate functional state *in vivo*.[11] Three-dimensional fibrin matrices used as cellular substrates *in vitro*. Fibrin has been utilized for application in the field of TE. BM stem cells can be embedded within fibrin gel scaffolds for transplanted cells.[13] The material acts instantaneously, thereby preventing cell loss. Fibrin glue is considered to be biocompatible and biodegradable because of a physiological component, commercially available.[14] Another advantage to use fibrin is that it can dissolve over time due to fibrinolysis.[15] Here we carried out a study to investigate the ability of bone formation with use of human BM stem cells, PRP, and a fibrin glue mixture. We then evaluated the potential of fibrin scaffolds *in vivo* for bone regeneration.

Eighteen New Zealand white rabbits (male, 2,500–3,000 g) were used in this study. A 6 mm diameter, 6 mm deep critical-sized defect was created on one tibia of each rabbit.[16] The rabbits were randomly divided into three groups. In the first group (defect), the defect in the tibia was left empty. In the second group (fibrin glue scaffold only), the contained defects were first made on the tibias and then manufactured scaffolds were implanted into the defects. In the third group (fibrin/human BM stem cells composite scaffold), the contained defects were created on the tibias, as described, and then prepared fibrin/human BM stem cells composite scaffolds were implanted into the defects.[17–19] The skin was sutured with 3/0 silk sutures. To evaluate the viability of human BM stem cells in a fibrin glue scaffold, live/dead staining was carried out at two, four, and six hours, which stained the live cells in green fluorescence and dead cells in red fluorescence (Fig. 43.10). There were no statistically significant differences in cell survival six hours after making a fibrin/marrow stem cells composite scaffold. The viability of BM

Figure 43.10. Cell viability was tested using live/dead kit at 2, 4 and 6 hours after a fibrin/human BM stem cells composite scaffold. The human BM stem cells were separated into two groups: live cells showed green fluorescence and dead cells showed red fluorescence. See also Color Insert.

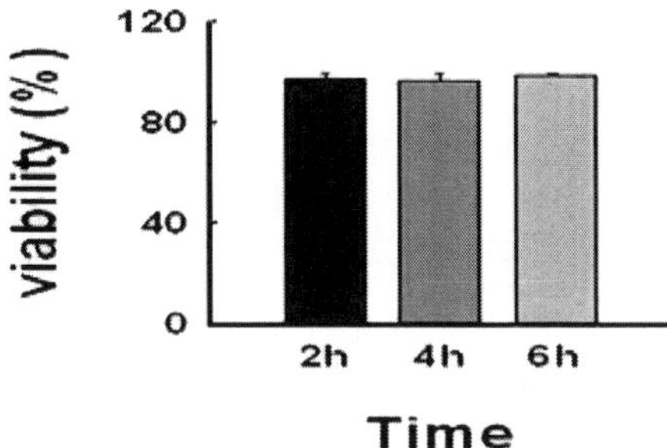

Figure 43.11. The viability of the human BM stem cells in a fibrin glue scaffold. The viability of human BM stem cells in a fibrin glue scaffold was measured 99.2 ± 0.5% at 6 hours (97.9 ± 1.7 at 2 hours; 97.3 ± 2.2 at 4 hours; 99.2 ± 0.5 at 6 hours). See also Color Insert.

stem cells in a fibrin glue scaffold was measured up to 90% average for six hours (Fig. 43.11).

To assay the osteogenic differentiation potential *in vitro*, a human BM stem cells/fibrin composite scaffold was cultured in an osteogenic medium for three weeks. Osteogenic differentiation in a section of the fibrin/BM stem cells composite scaffold was observed (Fig. 43.12a). Calcium deposition followed by osteogenic differentiation appeared (Fig. 43.12a). Also, osteogenic differentiation in the fibrin/human BM stem cells scaffold could be demonstrated by von Kossa staining (Fig. 43.12b). In this study, a fibrin glue scaffold provided a suitable environment for differentiation of osteoblasts *in vitro*.

Bone regeneration at the tibia defect site was assessed using X-rays 2, 4, 8, and 12 weeks after surgery (Fig. 43.13). X-ray evaluation showed there was no difference between the scaffold and the control group, comparing the degree of mineralization corresponding locations as well as the overall bone regeneration in the defects during 4, 8, and 12 weeks.[8,20] Judging from a macroscopic view, macroscopic examination of the tibia was performed at 4, 8, and 12 weeks (Fig. 43.14). All groups of bone defect were observed in the normal condition. New bone formation was not observed in the bottom of the defect.[21] And tissue adhesion occurred surrounding the tissue of the surgery regions. However the regions of adhesion were reduced gradually as time went by. H&E staining at 12

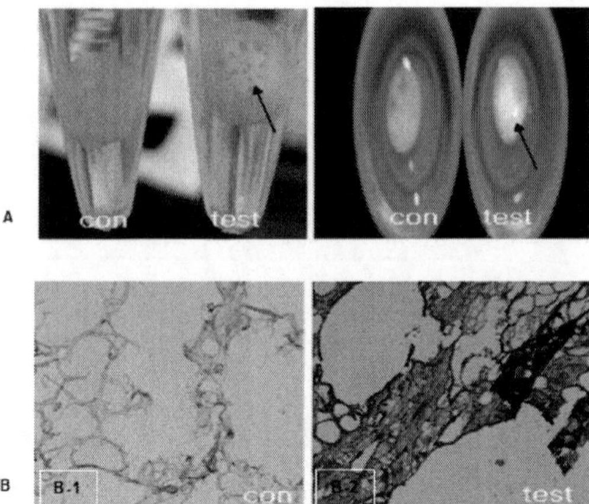

Figure 43.12. A The fibrin/human BM stem cells composite scaffold *in vitro* 12 days after induction of differentiation to judgment of differentiation ability to osteoblasts. Osteogenic differentiation was seen in test tube(A). Osteogenic differentiation in a section of fibrin/human BM stem cells composite scaffold was observed. (B-1) Fibrin/human BM stem cells composite scaffold in a normal growth medium. (B-2) Calcified matrix (black) is secreted by the human BM stem cells/fibrin composite. (con, control; test, experiment.) See also Color Insert.

weeks reduced the fibrin glue scaffold size, which was visible in all groups. The defect region was not filled with newly formed bone (Figs. 43.15a–c). That is, osteoid formation and woven bone structures were not seen. There were no significant differences among the groups. Also, a foreign-body reaction could be observed at the fibrin glue scaffold–host bone interface. But the foreign-body reaction was reduced gradually during 12 weeks.[22] The activity of BM stem cells seemed to be weak because of rejection.

To conclude, our study could not demonstrate the potential of bone regeneration in a fibrin glue scaffold *in vivo*. Toward a clinical application, further investigations have to be done using human BM stem cells, such as using immunosuppression agents. The performance of the composite constructs has to be assessed in a relevant implantation associated with a specific nutrient and mechanical situation.[23,24] The present approach will come to guide more mature research for TE using BM stem cells and scaffolds.

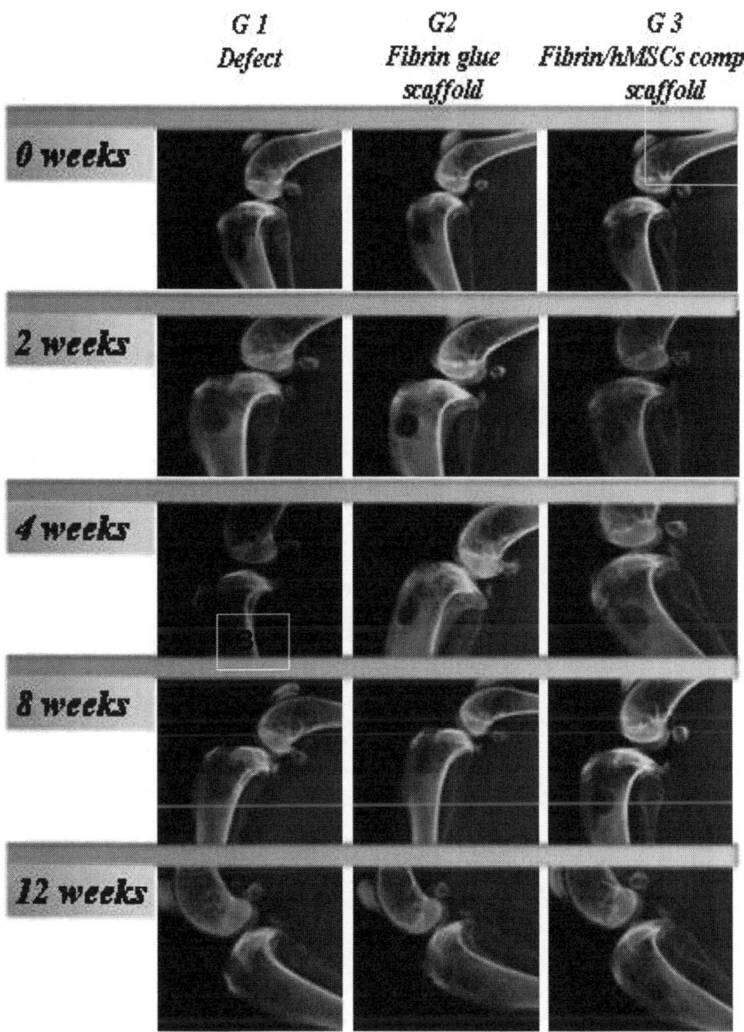

Figure 43.13. X-rays were obtained immediately after surgery and again after 2, 4, 8 and 12 weeks. G1 expresses group 1. Group 1 was not implanted with any kind of fibrin/human BM stem cells (sham operation). G2 expresses group 2. Group 2 was implanted with a fibrin glue scaffold in the defect region. G3 expresses group 3. Group 3 was implanted with fibrin/human BM stem cells composite scaffold in the defect region.

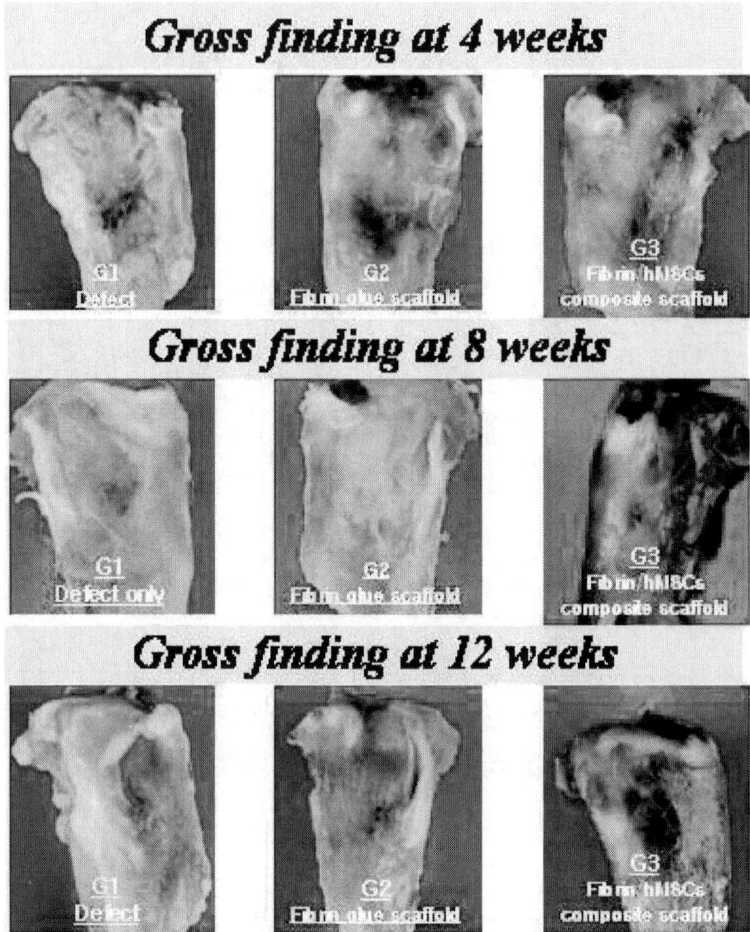

Figure 43.14. Gross finding examination of the defect group, the fibrin glue scaffold group, and the fibrin/human BM stem cells composite scaffold group at 4, 8 and 12 weeks after implantation into critical-sized contained bone defects of rabbits. Within the defects, mineralized tissues were not found. And tissue adhesion was observed at 4 weeks, so we could not confirm the defect region. But, as time went by, the adhesion region dwindled away. See also Color Insert.

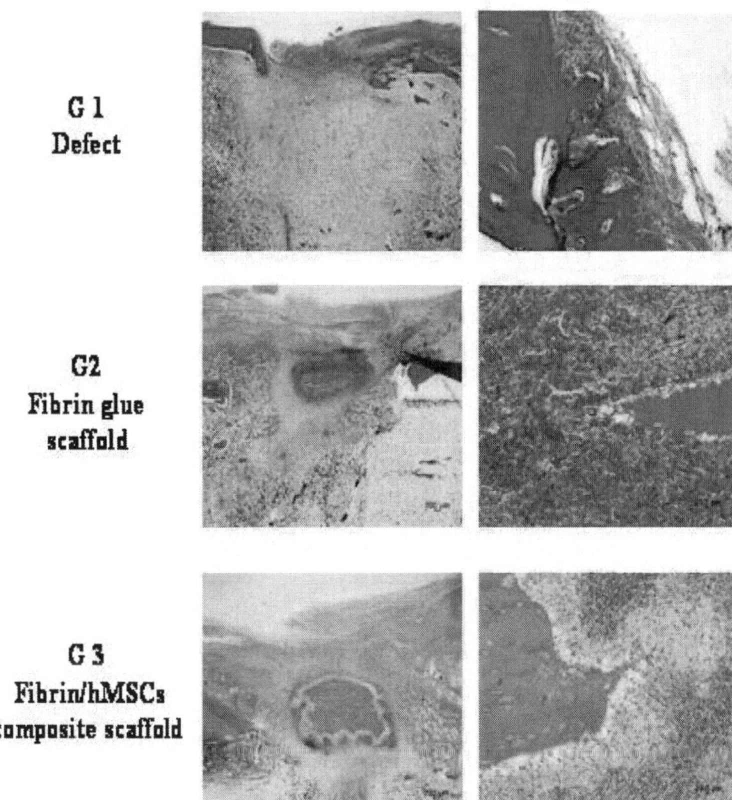

Figure 43.15a. H&E staining evaluation of specimen at 4 weeks after implantation. Group 1 was not implanted with any kind of fibrin/human BM stem cells (sham operation). Group 2 was with implanted a fibrin glue scaffold in the defect region. Group 3 was implanted with a fibrin/human BM stem cells composite scaffold in the defect region.

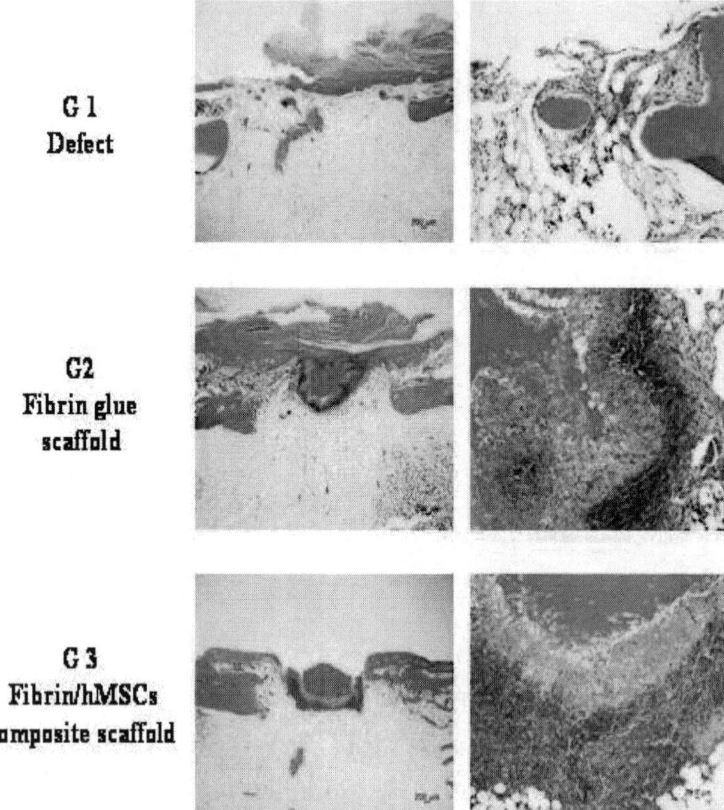

Figure 43.15b. H&E staining evaluation of specimen at 8 weeks after implantation. Group 1 was not implanted with any kind of fibrin/human BM stem cells (sham operation). Group 2 was implanted with a fibrin glue scaffold in the defect region. Group 3 was implanted with a fibrin/human BM stem cells composite scaffold in the defect region.

43.3.2 *Use of Bioreactors, Human Fetal Stem Cells, and 3D Scaffolds for Bone Tissue Engineering*

Cells are the central players for most biological process in the human body and play the essential role in the BTE. The selection of cell sources will eventually determine the clinical success of any BTE

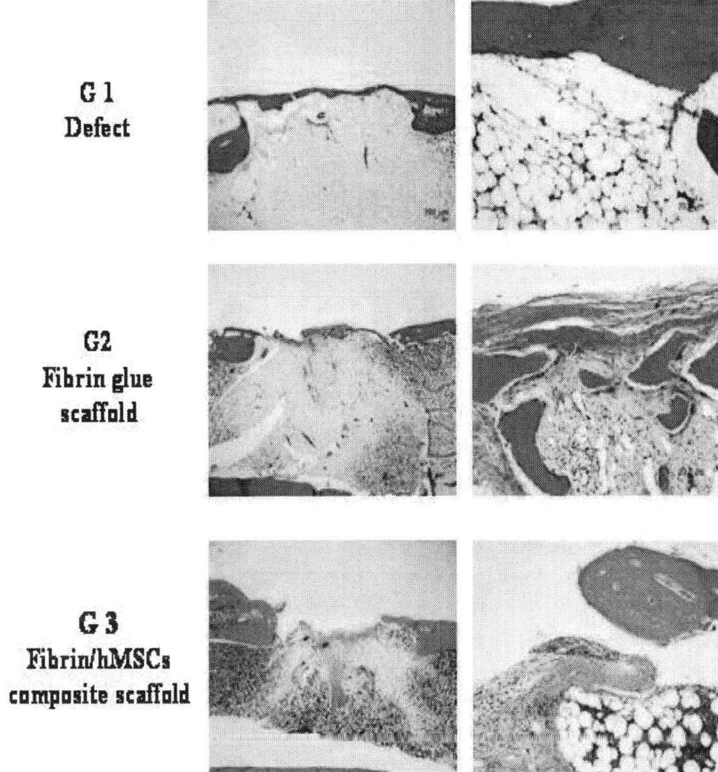

Figure 43.15c. H&E staining evaluation of specimen at 12 weeks after implantation. Group 1 was not implanted with any kind of fibrin/human BM stem cells (sham operation). Group 2 was implanted with a fibrin glue scaffold in the defect region. Group 3 was implanted with a fibrin/human BM stem cells composite scaffold in the defect region.

strategies. The ideal cellular source for a cell-based BTE approach should include the following characteristics: (1) no or low immunogenicity, (2) no tumorigenicity, (3) immediate availability, (4) availability in pertinent quantities, (5) a rapid cell proliferation rate, (6) a predictable and consistent osteogenic potential, and (7) controlled integration into the surrounding tissues.[25]

BM-MSCs, also known as marrow stromal cells or colony-forming units-fibroblast (CFU-F), were first identified from adult BM by Friedenstein *et al.* in the late 1960s.[26,27] Recently they have been extensively investigated and have demonstrated their great potential as an attractive and promising cell source for BTE applications because of their advantages: (1) MSCs can be easily isolated by the plastic adhesion method or other antibody selection techniques[28]; (2) the osteogenic differentiation pathway is quite well defined, and MSCs have been shown to generate greater bone formation than fresh BM in preclinical studies[29–31]; (3) MSCs and differentiated MSCs are reported to be nonimmunogenic and suitable for allogenic applications; and (4) cryostorage does not affect the osteogenic potential of MSCs, a condition that greatly facilitates their storage and clinical applications.[32] Besides these advantages, they are also found to play an essential role during bone remodeling and fracture healing *in vivo*. On the one hand, as the common progenitors for osteoblasts, osteocytes, and bone-lining cells, MSCs can affect bone remodeling via modulating their differentiation into these few types of osteogenic cells[33,34]; on the other hand, they are involved in bone remodeling through the regulation of the opposing action of osteoblasts and osteoclasts.[35] During bone fracture healing, MSCs can induce new bone formation by cellular differentiation into the osteoblasts and chondroblasts; moreover, MSCs were found to be able to create a regenerative microenvironment via their expression of a large spectrum of bioactive molecules.[35–37]

There are certain drawbacks that limit the clinical application of adult BM-derived MSCs, such as their low existing frequency in BM, high cellular senescence, limited proliferation capacity during the expansion,[38,39] and decreased osteogenic potential with ages.[40–42] As a result, efforts have been made to look for an alternative source of MSCs, and these efforts have led to the successful identification and isolation of MSCs with osteogenic potential from a diverse range of ontological and anatomical sources, including postnatal tissue such as adipose tissue,[43] periosteum,[44] trabecular bone,[45] synovial membrane,[46] and peripheral blood[7]; perinatal tissues such as the umbilical cord,[48] umbilical cord blood,[49,50] and amniotic fluid[51,52]; and prenatal tissue like fetal blood, BM, and liver.[53–56] While investigations into their basic biology, immunogenicity, and osteogenic potential have been reported, MSCs from different sources have not been systematically compared for BTE applications. Therefore, we performed a study to systematically compare human MSCs from different ontological and anatomical origins, including fetal BM, the umbilical cord, adult BM, and adult adipose tissue.[57] These four types of MSCs were compared head-to-head for their proliferation capacity and osteogenic potential mineralization *in vitro* under two different culture system settings, both a monolayer culture and a 3D polycaprolactone-tricalcium phosphate (PCL-TCP) scaffolds (manufactured by Osteopore Int Pte Ltd., Singapore) culture, and the *in vivo* ectopic bone formation capacity of different MSC-mediated scaffolds was investigated as well.

Our study unveiled the profound influence of the ontological and anatomical origin on cellular performance of MSCs for BTE applications. Human fetal MSCs (hfMSCs), derived from BM and an ontologically primitive origin, demonstrated their great potential as a superior cellular source for BTE applications. They demonstrated their primitiveness by expressing embryonic stem cell markers such as Oct-4 and Nanog.

In the monolayer culture, when compared with other MSCs, hfMSCs proliferated much faster (1.5–3.4x, $p<0.01$) with significantly higher self-renewal capacity (1.6–2.0x, $p<0.01$), as determined by the CFU assay (Fig. 43.16a). In the 3D PCL-TCP scaffold culture, Picogreen dsDNA quantification assay showed that hfMSCs proliferated much faster and saturated all the empty spaces in the scaffolds within 7 days, while it took 28 days to achieve cellular conference for other MSCs (Fig. 43.16b).

The hfMSCs can also proliferate for more than 100 population doublings before reaching cellular senescing, while human adult BM-derived MSCs (haMSC) can only sustain for 30 population doublings. Furthermore, hfMSCs have the highest osteogenic potential, both *in vitro* and *in vivo*, among the four types of MSCs, as assessed by von-Kossa staining, scanning electron microscopy (SEM) scanning ALP activity, and micro CT quantification of the ectopic bone formation *in vivo* (Fig. 43.17).[57] In addition, our study showed a lower HLA-1 expression in hfMSCs than other MSCs (55% vs. 95%–99%), indicating their lower immunogenicity. This is in consistence with a number of recent studies showing that hfMSCs have lower

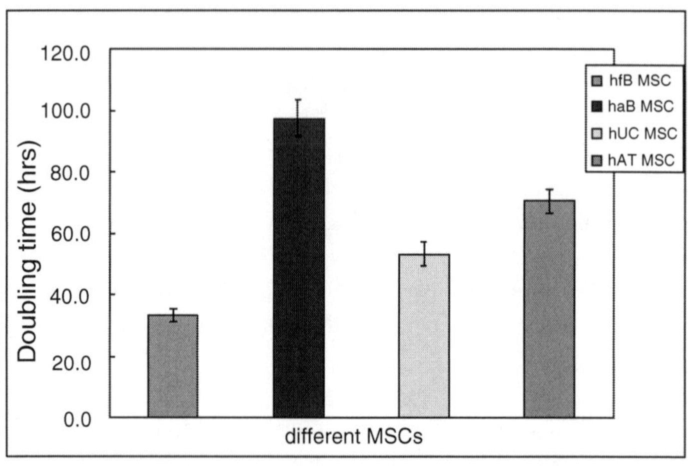

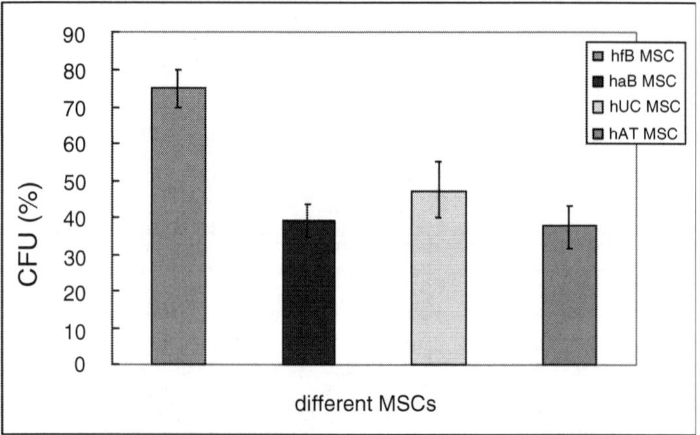

Figure 43.16a. Doubling time and CFUs of different MSCs. See also Color Insert.

immunogenicity than haMSCs.[58,59] In conclusion, because of their great proliferative capacity, high osteogenic potential both *in vitro* and *in vivo*, and low immunogenicity, hfMSCs can be a promising cellular source for BTE applications.

A challenging issue for the clinical translation of cell-based scaffold constructs is the maintenance of viability and proper

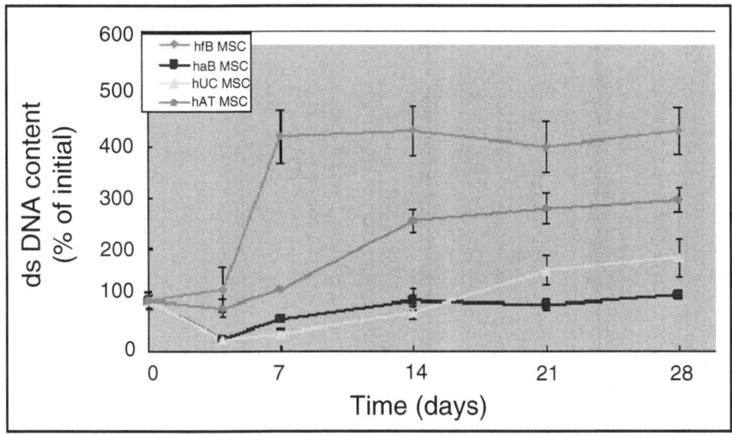

Figure 43.16b. dsDNA quantification by picogreen assay. See also Color Insert.

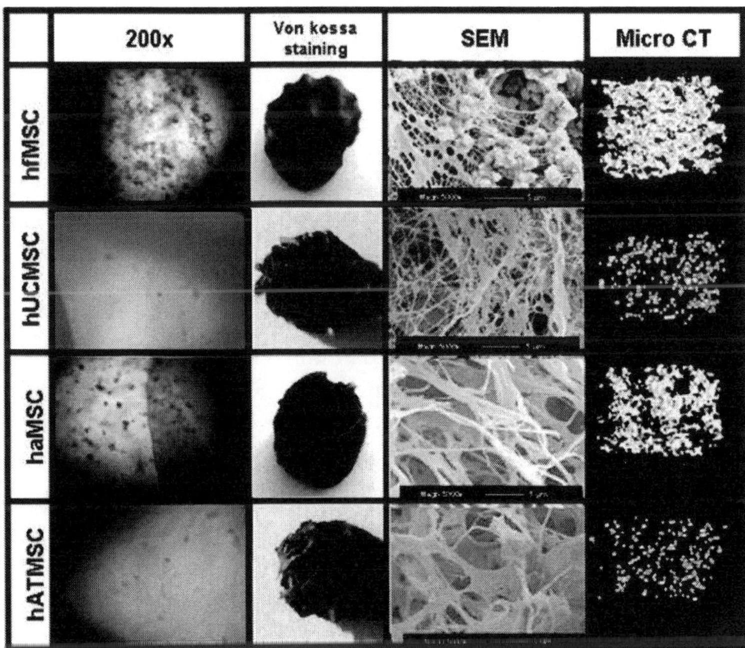

Figure 43.17. Mineralization of different MSC-mediated scaffolds by light microscopy, von Kossa staining, SEM, and micro CT. See also Color Insert.

predifferentiation prior to implantation, which has been shown to be very important to achieve the best therapeutic result. Under the *in vitro* culture condition, without the vasculature network, the traditional static culture cannot fulfill the special requirement of nutrient and oxygen supply for culturing large-dimension 3D scaffold constructs, because of the limited diffusion capacities and high demand in nutrient and gas exchange created by high cellular density. Many studies have demonstrated that the effective distance for nutrient and gas transport is only limited to a depth of a few hundred microns under the static culture.[60,61] Therefore, bioreactors have been widely used in the TE field because they can not only increase the mass transfer to maintain the viability of large 3D cellular scaffolds but also provide certain mechanical stimulation, which can promote the cell differentiation by triggering the mechanotransduction-signaling pathway.[62–64] The rotating wall vessel (RWV) bioreactor has been utilized for BTE and is known for its advanced characteristics of low shear, three dimensionality, and high mass transfer[65–69]; however, it has a problem that the single axial rotation may lead to the nonhomogenous distribution of the cell and extracellular matrix.[63,70,71] We developed a novel type of RWV bioreactor (biaxial rotating bioreactor), which can rotate simultaneously in two perpendicular axes and incorporate the perfusion system.[72,73] Computational simulation analysis has demonstrated that it can achieve manifold increases of fluid velocity with significant improvements of fluid transport through the scaffolds compared with the conventional uniaxial rotating bioreactor and static culture.[72] Our further laboratory investigation showed it maintained high cellular viability in the core of a thick scaffold, which was 2,000 mm from the surface, throughout a 28-day culture, while the traditional static culture resulted in the large necrotic core of the thick scaffold, as determined by floroscein diacetate/propidium iodine (FDA/PI) live/dead staining (Fig. 43.18). Compared to the static culture, the cellular scaffold under the biaxial rotating bioreactor culture proliferated faster and reached cellular confluence earlier (7 days vs. 28 days), with greater cellularity (2x, $p<0.01$). Furthermore, the biaxial bioreactor culture was associated with greater osteogenic induction, ALP expression (1.5x, $p<0.01$), calcium deposition (5.5x, $p<0.001$) and bony nodule formation on SEM, and *in vivo* ectopic bone formation in

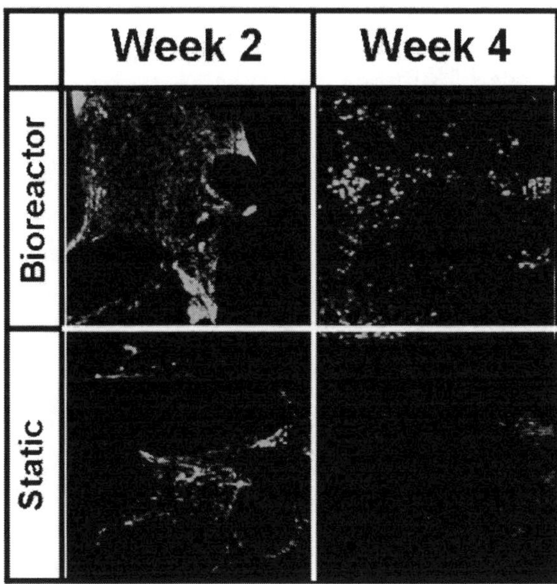

Figure 43.18. FDA/PI live/dead staining of the cellular scaffolds. The bioreactor culture led to complete cellular confluence in the scaffolds at week 2, while the static culture failed to do so. At week 4, the bioreactor culture maintained the high viability of cellular scaffolds, while the static culture resulted in large cellular necrosis (live cells were stained in green by FDA; dead cells were stained in red by PI). See also Color Insert.

immunodeficient mice (3.2x, $p<0.001$) compared with static-cultured scaffolds.[73] The use of the biaxial rotating bioreactor here allowed the maintenance of cellular viability beyond the limits of conventional diffusion, with increased proliferation and osteogenic differentiation both *in vitro* and *in vivo*, suggesting its utility for BTE applications.

On the basis of our expertise in scaffold fabrication, human fetal stem cell research, and bioreactor design, we conceptualized a BTE strategy by seeding hfMSCs onto a PCL-TCP scaffold, culturing and maturing cellular scaffolds into TE bone grafts. A critical-sized rat femoral defect model was selected to testify our concept and evaluate the effectiveness of our TE bone grafts for the defect healing. Results showed that the implantation of this TE bone graft can promote new bone formation and neovascularization and successfully bridge the defect within three months, while the treatment of empty scaffolds led to the nonunion of the critical-sized defect. Thus, this BTE strategy demonstrated a great potential for clinical applications.

43.3.3 *The Use of Scaffolds with or without Growth Factors and Cells for Clinical Trials*

There have been several clinical trials using stem cells with or without scaffolds for new bone formation. Kitoh *et al.* described that they could reduce the consolidation time of new bone after injection of culture-expanded BM stem cells with PRP as a scaffold into the distraction area in 17 patients. The healing index was 25 days per centimeter in the cell injection group, whereas the healing index was 36 days per centimeter in the control group. This method improved vascular invasion from surrounding muscle tissues at the site of distraction. The area without a sufficient vascular bed is not an ideal site for injection, such as the anteromedial aspect of the tibia. This method has the advantage of no limitation of the donor site compared with autografting because it is possible to keep a large number of viable cells during treatment. The disadvantage is that it takes four to six weeks for culturing stem cells before injection and secondary surgery is necessary.

Marcacci *et al.* tried to use stem cells with macroporous bioceramics for treatment of long bone defects in four patients. The age of patients ranged from 16 years to 41 years, and the sizes of bone defects ranged from 4 to 7 cm. External fixations were applied for stability. After six to seven years' follow-up, they found that there were no late fatigue fractures of the bioceramic implants, and there was bone union between the ceramics and the host bone. This method has the difficulty of evaluation of new bone formation inside the ceramics because the ceramics are visible as radio-opaque materials on radiographs and the callus could not be visible.

Jones *et al.* performed a randomized, controlled clinical trial using BMP-2 mixed with an allograft for reconstruction of a diaphyseal tibial fracture with cortical defects and compared the effect of this method with autogenous bone graft. The mean of bone defect length was 4 cm, ranging from 1 to 7 cm. They found that 87% of BMP-2 with the allograft group obtained fracture healing, whereas

67% of the autogenous graft group obtained fracture healing. This method suggested that addition of a growth factor to the allograft may replace the autogenous graft and overcome the disadvantages of the autogenous graft, such as the limitation of donor site, pain at the donor site, and scars. Even though many kinds of growth factors, stem cells, and scaffolds are being used to find the most efficient method for new bone formation in the clinical field, there is still no standard guideline regarding which stem cells or growth factors or scaffolds should be selected to reduce the healing period and which combination is the best among these biomaterials. The most important aspects are cost-effectiveness and easy accessibility clinically. Clinically, an absorbable scaffold that can be slightly visible during the initial stage of fracture healing on the radiograph is better than a nonabsorbable scaffold because the location of the scaffold should be identified on the radiograph to find out whether or not this scaffold is placed at the bone defect site. If the scaffold is used for the defect of long bones of the limbs, the scaffold can move due to gravity. During walking and exercising, incidence of displacement of the scaffold increases. Temporary visualization of the scaffold on conventional radiography requires the development of scaffolds with a different rate of combination of nonabsorable and absorbable scaffolds.

References

1. G. De Bastiani, R. Aldegheri, L. Renzi-Brivio, and G. Trivella, *J. Pediatr. Orthop.*, **7**, 129 (1987).
2. G. A. Ilizarov, *Clin. Orthop.*, **238**, 249 (1989).
3. G. A. Ilizarov, *Clin. Orthop.*, **239**, 263 (1989).
4. G. A. Ilizarov, *Clin. Orthop.*, **280**, 7 (1992).
5. J. Aronson and X. Shen, *Clin. Orthop.*, **301**, 25 (1994).
6. I. H. Choi, C. S. Sohn, C. Y. Chung, T. J. Cho, J. W. Lee, and D. Y. Lee, *Clin. Orthop.*, **368**, 240 (1999).
7. J. Fischgrund, D. Paley, and C. Suter, *Clin. Orthop.*, **301**, 31 (1994).
8. S. S. Kim, S. J. Gwak, and B. S. Kim, *J. Biomed. Mater. Res. A*, (2008).
9. S. Kim, S. S. Kim, S. H. Lee, Ahn S. Eun, S. J. Gwak, J. H. Song, B. S. Kim, and H. M. Chung, *Biomaterials* **29**, 1043 (2008).

10. R. M. Schek, J. M. Taboas, S. J. Segvich, S. J. Hollister, and P. H. Krebsach, *Tissue Eng.*, **10**, 1376 (2004).
11. T. C. Flanagan, C. Cornelissen, S. Koch, B. Tschoeke, J. S. Sachwen, T. Schmitz-Rode, and S. Jockenhoevel, *Biomaterials*, **28**, 3388 (2007).
12. J. W. Lee, Y. H. Kim, S. H. Kim, S. H. Han, and S. B. Hahn, *Yonsei Med J* **45**, **Suppl**, 41 (2004).
13. Y. E. Ju, P. A. Janmey, M. E. McCormick, E. S. Sawyer, and L. A. Flanagan, *Biomaterials* **28**, 2097 (2007).
14. D. Eyrich, H. Wiese, G. Maier, D. Skodacek, B. Appel, H. Sarhan, J. Tessmar, R. Staudenmaier, and M. Wenzel, *Tissue Eng*, **13**, 2207 (2007).
15. D. Eyrich, F. Brandl, B. Appel, H. Wiese, G. Maier, M. Wenzel, R. Staudenmaier, A. Goepferich, and T. Blunk, *Biomaterials*, **28(1)**, 55 (2007).
16. Y. Liu, X. Z. Shu, and G. D. Prestwich, *Tissue Eng.*, **12**, 3405 (2006).
17. B. Kreklau, M. Sittinger, M. B. Mensing, C. Voigt, G. Berger, G. R. Burmester, R. Rahmanzadeh, and U. Gross, *Biomaterials*, **20**, 1743 (1999).
18. S. D. Waldman, C. G. Spiteri, M. D. Grynpas, R. M. Pilliar, J. Hong, and R. A. Kandel,, *J. Bone Joint Surg. Am. 85-A*, **Suppl 2**, 101 (2003).
19. R. Katayama, S. Wakitani, N. Tsumaki, Y. Morita, I. Matsushita, R. Gejo, and T. Kimura, *Rheumatology. (Oxford)* **43**, 980 (2004).
20. M. R. Sarkar, P. Augat, S. J. Shefelbine, S. Schorlemmer, M. Huber-Lang, L. Claes, L. Kinzl, and A. Ignatius, *Biomaterials*, **27(9)**, 1817 (2006).
21. I. Nagura, H. Fujioka, T. Kokubu, T. Makino, Y. Sumi, and M. Kurosaka, *J. Bone Joint Surg. Br.*, **89**, 258 (2007).
22. U. Kneser, A. Voogd, J. Ohnolz, O. Buettner, L. Stangenberg, Y. H. Zhang, G. B. Stark, and D. J. Schaefer, *Cells Tissues Organs*, **179**, 158 (2005).
23. Linnes, P. Michael, Ratner, D. Buddy, Giachelli, and M. Cecilia, *Biomaterials*, **28**, 5298 (2007).
24. T. C. Flanagan, C. Cornelissen, S. Koch, B. Tschoeke, J. S. Sachweh, T. Schmitz-Rode, and S. Jockenhoevel, *Biomaterials*, **28**, 3388 (2007).
25. D. Logeart-Avramoglou, F. Anagnostou, R. Bizios, and H. Petite, *J. Cell Mol. Med.*, **9**, 72 (2005).
26. A. J. Friedenstein, S. I. Piatetzky, and K. V. Petrakova, *J. Embryol. Exp. Morphol.*, **16**, 381 (1966).
27. A. J. Friedenstein, K. V. Petrakova, A. I. Kurolesova, and G. P. Frolova, *Transplantation*, **6**, 230 (1968).

28. P. J. Simmons and B. Torok-Storb, *STRO-1. Blood*, **78**, 55 (1991).
29. K. Inoue, H. Ohgushi, T. Yoshikawa, M. Okumura, T. Sempuku, and S. Tamai, *J. Bone Miner. Res.*, **12**, 989 (1997).
30. A. Kahn, R. Gibbons, S. Perkins, and D. Gazit, *Clin. Orthop. Relat. Res.*, **69** (1995).
31. D. P. Pioletti, M. O. Montjovent, P. Y. Zambelli, and L. Applegate, *Swiss Med. Wkly.*, **136**, 557 (2006).
32. S. P. Bruder, N. Jaiswal, and S. E. Haynesworth, *J. Cell Biochem.*, **64**, 278 (1997).
33. J. A. Buckwalter, M. J. Glimcher, R. R. Cooper, and R. Recker, *Bone Biol.*, **45**, 387 (1996).
34. J. A. Buckwalter, M. J. Glimcher, R. R. Cooper, and R. Recker, *Bone Biol.*, **45** 371 (1996).
35. R. Bielby, E. Jones, and D. McGonagle, *Injury*, **38 Suppl 1**, 26 (2007).
36. A. I. Caplan, *J. Cell Physiol.*, **213**, 341 (2007).
37. A. I. Caplan, *Tissue Eng.*, **11**, 1198 (2005).
38. S. P. Bruder, N. Jaiswal, and S. E. Haynesworth, *J. Cell Biochem.*, **64**, 278 (1997).
39. D. G. Phinney, G. Kopen, W. Righter, S. Webster, N. Tremain, and D. J. Prockop, *J. Cell Biochem.*, **75**, 424 (1999).
40. S. C. Mendes, J. M. Tibbe, M. Veenhof, K. Bakker, S. Both, and P. P. Platenburg, *Tissue Eng.*, **8**, 911 (2002).
41. G. D'Ippolito G, P. C. Schiller, C. Ricordi, B. A. Roos, and G. A. Howard, *J. Bone Miner. Res.*, **14**, 1115 (1999).
42. S. M. Mueller and J. Glowacki, *J. Cell Biochem.*, **82**, 583–590 (2001).
43. P. A. Zuk, M. Zhu, H. Mizuno, J. Huang, J. W. Futrell, and A. J. Katz, *Tissue Eng.*, **7**, 211 (2001).
44. H. Nakahara, J. E. Dennis, S. P. Bruder, S. E. Haynesworth, D. P. Lennon, and A. I. Caplan, *Exp. Cell Res.*, **195**, 492 (1991).
45. Y. Sakaguchi, I. Sekiya, K. Yagishita, S. Ichinose, K. Shinomiya, and T. Muneta. *Blood*, **104**, 2728 (2004).
46. B. C. De and A. F. Dell, *Arthr. Rheum.*, **44**, 1928 (2001).
47. G. Z. Eghbali-Fatourechi, J. Lamsam, D. Fraser, D. Nagel, B. L. Riggs, and S. Khosla, *N. Engl. J. Med.*, **352**, 1959 (2005).
48. R. Sarugaser, D. Lickorish, D. Baksh, M. M. Hosseini, and J. E. Davies, *Stem Cells*, **23**, 220 (2005).
49. K. Bieback, S. Kern, H. Kluter, and H. Eichler, *Stem Cells*, **22**, 625 (2004).

50. O. K. Lee, T. K. Kuo, W. M. Chen, K. D. Lee, S. L. Hsieh, and T. H. Chen, *Blood*, **103**, 1669 (2004).
51. C. P. De, G. J. Bartsch, M. M. Siddiqui, T. Xu, C. C. Santos, and L. Perin, *Nat. Biotechnol.*, **25**, 100 (2007).
52. P. In't Anker, S. A. Scherjon, K. C. Kleijburg-van, W. A. Noort, F. H. Claas, R. Willemze, W. E. Fibbe, and H. H. Kanhai, *Blood*, **102**, 1548 (2003).
53. C. Campagnoli, I. A. Roberts, S. Kumar, P. R. Bennett, I. Bellantuono, and N. M. Fisk, *Blood*, **98**, 2396 (2001).
54. J. Chan, K. O' Donoghue, J. de Fuente, I. A. Roberts, S. Kumar, J. E. Morgan, and N. M. Fisk, *Stem Cells*, **23**, 93 (2005).
55. J. Chan, K. O' Donoghue, M. Gavina, Y. Torrente, N. Kennea, H. Mehmet, H. Stewart, D. J. Watt, J. E. Morgan, and N. M. Fisk, *Stem Cells*, **24**, 1879 (2006).
56. J. Chan, S. N. Waddington, D. K. O, H. Kurata, P. V. Guillot, and C. Gotherstrom, *Stem Cells*, **25**, 875 (2007).
57. Z. Y. Zhang, S. H. Teoh, M. S. Chong, J. T. Schantz, N. M. Fisk, and M. A. Choolani, *Stem Cells*, **27**, 126 (2009).
58. K. Le Blanc, *Cytotherapy*, **5**, 485 (2003).
59. C. Gotherstrom, O. Ringden, M. Westgren, C. Tammik, and B. K. Le, *Bone Marrow Transplant*, **32**, 265 (2003).
60. S. L. Ishaug, G. M. Crane, M. J. Miller, A. W. Yasko, M. J. Yaszemski, and A. G. Mikos, *J. Biomed. Mater. Res.*, **36**, 17 (1997).
61. I. Martin, B. Obradovic, L. E. Freed, and G. Vunjak-Novakovic, *Ann. Biomed. Eng.*, **27**, 656 (1999).
62. I. Martin, D. Wendt, and M. Heberer, *Trends Biotechnol.*, **22**, 80 (2004).
63. H. C. Chen and Y. C. Hu, *Biotechnol. Lett.*, **28**, 1415 (2006).
64. K. Bilodeau and D. Mantovani, *Tissue Eng.*, **12**, 2367 (2006).
65. X. Yu, E. A. Botchwey, E. M. Levine, S. R. Pollack, and C. T. Laurencin, *Proc. Natl. Acad. Sci. U S A*, **101**, 11203 (2004).
66. P. Eiselt, B. S. Kim, B. Chacko, B. Isenberg, M. C. Peters, and K. G. Greene, *Biotechnol. Prog.*, **14**, 134 (1998).
67. C. Granet, N. Laroche, L. Vico, C. Alexandre, and M. H. Lafage-Proust, *Med. Biol. Eng. Comput.*, **36**, 513 (1998).
68. B. J. Klement and B. S. Spooner, *J. Cell Biochem.*, **51**, 252 (1993).
69. G. Molnar, N. A. Schroedl, S. R. Gonda, and C. R. Hartzell, *In Vitro Cell Dev. Biol. Anim.*, **33**, 386 (1997).

70. A. S. Goldstein, T. M. Juarez, C. D. Helmke, M. C. Gustin, and A. G. Mikos, *Biomaterials*, **22**, 1279 (2001).
71. V. I. Sikavitsas, G. N. Bancroft, and A. G. Mikos, *J. Biomed. Mater. Res.*, **62**, 136 (2002).
72. H. Singh, S. H. Teoh, H. T. Low, and D. W. Hutmacher, *J. Biotechnol.*, **119**, 181 (2005).
73. Z. Y. Zhang, S. H. Teoh, W. S. Chong, T. T. Foo, Y. C Chng, and M. Choolani, *Biomaterials*, **30**, 2694 (2009).

Chapter 44

TEMPERATURE-RESPONSIVE CULTURE SURFACES FOR REGENERATIVE MEDICINE

Yoshikazu Kumashiro, Yoshikatsu Akiyama, Masayuki Yamato,* and Teruo Okano**

Institute of Advanced Biomedical Engineering and Science, Tokyo Women's Medical University, 8-1 Kawada-cho, Shinjuku-ku, Tokyo 162-8666, Japan
*myamato@abmes.twmu.ac.jp; **tokano@abmes.twmu.ac.jp

Cell-based sheet engineering using temperature-responsive cell culture surfaces was developed as a novel strategy for regenerative medicine to overcome the limits of conventional methods such as single-cell suspension injection and biodegradable polymer scaffolds. Here, we summarize the basic mechanisms of cell attachment and detachment and cell patterning on temperature-responsive cell culture surfaces to allow for rapid cell sheet fabrication and recovery. In addition, we show *in vivo* transplantations of cell sheets, with possible applications such as corneal reconstruction, periodontal ligament transplantation, esophageal reconstruction, and the sealing of lung air leaks.

44.1 Introduction

Tissue engineering utilizing biodegradable scaffolds has been pursued for its utility in the area of regenerative medicine.[1,2] Although early clinical treatment of cartilage has been reported,[3] clinical

Handbook of Intelligent Scaffolds for Tissue Engineering and Regenerative Medicine
Edited by Gilson Khang

www.panstanford.com

applications of biodegradable scaffold-based tissue engineering are extremely limited. In addition, it is possible that biodegradable scaffolds may not be ideal for the regeneration of cell-dense tissues, including myocardium and hepatic structures.[4–7] Although the degradation of biodegradable polymers, such as poly(lactic acid) and poly(glycolic acid), theoretically allows for proliferating cells to migrate into the spaces previously occupied by the degraded biomaterials, large amounts of extracellular matrix (ECM) components are often present within the scaffold structures. Additionally, significant inflammatory responses are commonly observed due to the degradation of the polymer materials. Although this inflammatory response can theoretically induce the recruitment of various leukocytes and possibly direct new blood vessel formation, host responses to the implanted scaffolds can also potentially damage the cells both within the[8] implanted constructs and in the surrounding host tissues. Based on these observations, tissue engineering techniques that can re-create adequate blood supplies are likely to be required to properly reconstruct tissues that possess the structure and function of the native organs.[9–11]

We have adopted an innovative approach to tissue engineering based on cell sheet technology.[12] Cell sheet technology consists mainly of a temperature-responsive culture dish, which allows for reversible cell adhesion to and detachment from the dish surface by controllable hydrophobicity of the surface.[13,14] This enables a noninvasive harvest of cultured cells as an intact monolayer cell sheet, including deposited ECM. The monolayer cell sheet can be collected simply by reducing the culture temperature to less than 32°C for less than one hour, without the need for any enzymes such as trypsin.[15] By using this technology, we can transplant cell sheets to host tissues without using biodegradable scaffolds (Fig. 44.1).

In this chapter, we focus on the following:

(1) The basic mechanism of cell attachment to and detachment from temperature-responsive cell culture surfaces
(2) Temperature-responsive cell culture surfaces that enable affinity control
(3) Applications for regenerative medicine
(4) Future aspects for regenerative medicine

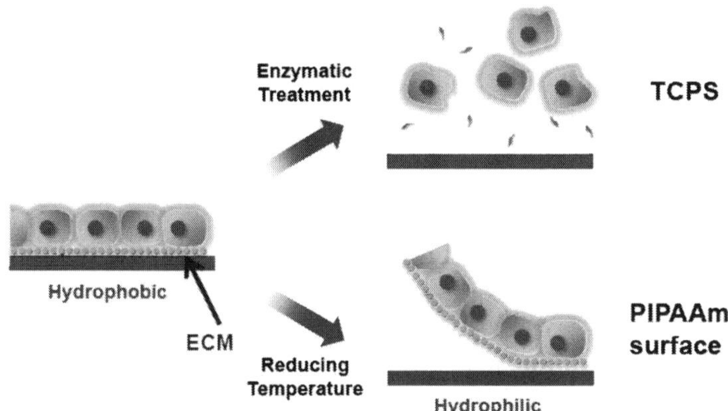

Figure 44.1. Cell sheet harvest from PIPAAm-grafted surfaces. Cells can adhere onto temperature-responsive surfaces at 37°C, whereas lowering temperature to 20°C facilitates the detachment of patterned cell sheets. In the case of tissue culture polystyrene (TCPS) dishes, enzymatic treatment, which is necessary for the harvesting of cells, seriously damages cell–cell connections and the deposited ECM. *Abbreviation*: PIPAAm poly(*N*-isopropylacrylamide). Reprinted with permission from Y. Kumashiro *et al.*, *Anal. Biomed. Eng.*, 38, 6, 1977. Copyright 2010.

44.2 The Basic Mechanism of Cell Attachment to and Detachment from Temperature-Responsive Surfaces

The temperature-responsive polymer poly(*N*-isopropylacrylamide) (PIPAAm) possesses a lower critical solution temperature (LCST) of 32°C in water. Below the LCST, PIPAAm is fully hydrated and soluble in aqueous solution, but it collapses extensively and becomes insoluble in the solution above the LCST. This unique feature of PIPAAm has been exploited for various biomedical applications, as have PIPAAm's hydrophobic interactions with amino acids and proteins.[16,17] By grafting polymer gels onto cell culture dishes using electron beam (EB) irradiation, temperature-dependent cell adhesion/detachment control can be achieved, with the development of temperature-responsive culture surfaces for tissue engineering and regenerative medicine applications. Confluently cultured cell monolayers on hydrophobized PIPAAm modified surfaces at 37°C can be detached as single-cell sheets by lowering the culture temperature to 20°C, at which point the modified surfaces become hydrophilic due to PIPAAm's hydration/dehydration transition at 32°C. In our preliminary studies, the graft amount of PIPAAm on the surface has a significant influence on cell adhesion behavior. Here, we focused on the correlation of the thickness of PIPAAm-grafted (covalently) layers on tissue culture polystyrene (TCPS) surfaces and cell adhesion/detachment behavior.[18] For this purpose, we utilized limited excimer laser ablation, time-of-flight secondary ion mass spectrometry (TOF-SIMS) images, and atomic force microscopy (AFM) methods to determine the thickness of PIPAAm-grafted layers.[19–21]

Two types of PIPAAm-grafted TCPS surfaces were evaluated with respect to the graft amount of PIPAAm on the surfaces and the surface thermo-responsive wettability behavior. The grafted polymer amounts on two PIPAAm-grafted surfaces were determined by attenuated total reflection Fourier transform infrared spectroscopy (ATR/FT-IR) measurements, which were 1.4 ± 0.1 μg/cm^2 ($n = 4$) and 2.9 ± 0.1 μg/cm^2 ($n = 4$), respectively. These two surface types are abbreviated as PIPAAm-1.4 and PIPAAm-2.9, respectively, hereafter. PIPAAm-grafted TCPS surfaces were ablated with a UV excimer laser until the TCPS region was exposed for a subsequent measurement of the graft layer thickness by means of AFM. The ablated domains were stained with a hydrophobic fluorescent dye, 1,1'-dioctadecyl-3,3,3',3'-tetramethylindocarbocyanine perchlorate (DiIC18), to elucidate whether the hydrophobic TCPS regions were exposed. With sharp contrast, the fluorescent dye selectively and clearly stained the ablated domains with an increasing number of laser shots. The fluorescent intensity for PIPAAm-1.4 is likely to be saturated with six laser shots. This result indicates that the TCPS region is successfully exposed by six times the laser ablation. Figure 44.2 shows AFM images and corresponding section profiles of the ablated domains for PIPAAm-1.4 (a) and PIPAAm-2.9 (b). The PIPAAm-1.4 surface was relatively smooth, and ablated areas were clearly observed. This is in sharp contrast with the relatively rough PIPAAm-2.9 surface. The root mean square (RMS) values of the PIPAAm-1.4 and PIPAAm-2.9 surfaces before laser ablation were approximately 7.7 and 16.5 nm, whereas those of the non-ablated areas shown in Fig. 44.2 were approximately 6.0 nm and 17.2 nm, respectively. The equivalent RMS values before and after laser ablation indicate that laser ablation with 10 and 20 mJ/cm^2

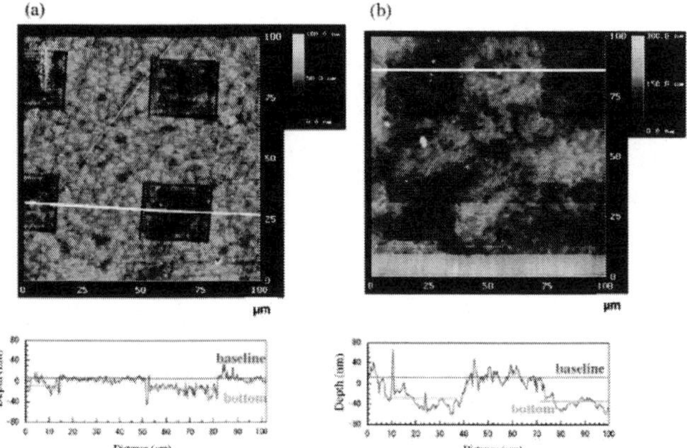

Figure 44.2. Cross-sectional profiles of the laser-ablated domains on PIPAAm-1.4 (a) and PIPAAm-2.9 surfaces (b). The laser fluence and the number of irradiations were 10 mJ/cm2 and 6 times for PIPAAm-1.4 (a) and 20 mJ/cm2 and 3 times for PIPAAm-2.9 (b), respectively. The baseline and the bottom were averages of the text data from the section profiles (see experimental section). The scan size was 100×100 mm^2. Reprinted with permission from Ref. 20. Copyright 2004 American Chemical Society.

did not contaminate the surfaces of PIPAAm-1.4 and PIPAAm-2.9 with ablated debris. Section profiles showed the depth to be 15.5 $\pm$ 7.2 nm, indicating the thickness of the grafted PIPAAm layer for PIPAAm-1.4. Likewise, the averaged thickness of PIPAAm-grafted layers for PIPAAm-2.9 was 29.3 $\pm$ 8.4 nm. Considering the amount and density of the grafted polymer on PIPAAm-1.4 and PIPAAm-2.9, the measured thicknesses of the grafted polymer layers are reasonable.

The cell attachment/detachment responses on the PIPAAm-grafted surfaces are the key factor for noninvasive recovery of cell sheets for further applications of tissue and organ reconstruction. Here, we examined cellular responses to these PIPAAm-grafted surfaces in terms of cell adhesion and detachment with temperature. Bovine endothelial cells (ECs) adhered and spread on the PIPAAm-1.4 dishes (Fig. 44.3), whereas they did not on PIPAAm-2.9 (Fig. 44.3). Adhered and proliferated cells on the PIPAAm-1.4 surfaces were detached from the surfaces by reducing the temperature below the PIPAAm's transition temperature for single-cell and/or monolayer-cell sheets, depending on the cell density of the surface.

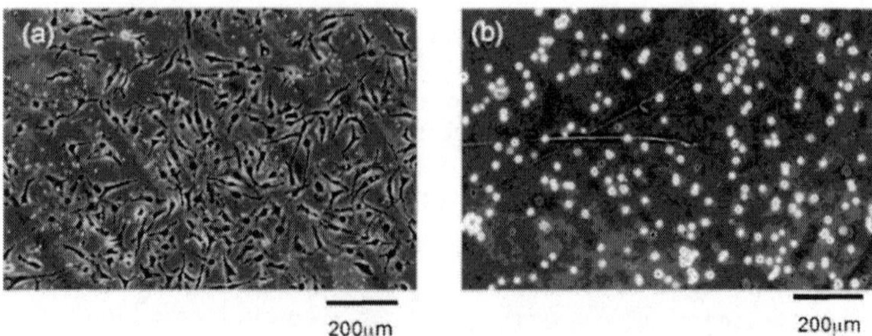

Figure 44.3. Phase-contrast microphotographs of bovine artery ECs cultured on PIPAAm-1.4 (a) and PIPAAm-2.9 (b) at 37°C for 1 day. Although cells can adhere and harvest onto PIPAAm-1.4 at 37°C, cells cannot attach onto PIPAAm-2.9 at 37°C. Reprinted with permission from Ref. 20. Copyright 2004 American Chemical Society.

In summary, the thickness and the amount of grafted PIPAAm layers have a significant influence on the surface properties in terms of temperature-responsive cell adhesion and detachment, and a thickness of approximately 20 nm is a key factor for the PIPAAm-grafted surface to achieve cell adhesion and detachment with temperature. The progressive dehydration of PIPAAm chains existing at the vicinity of hydrophobic TCPS may be important to the cell adhesion characteristics at 37°C. In contrast, the graft polymer at the outermost interfaces of PIPAAm-2.9 should be more polar at 37°C than PIPAAm-1.4, which inhibits cells from adhering.

44.3 Temperature-Responsive Cell Culture Surfaces that Enable Affinity Control

Specific interactions between biomolecules and their molecular recognition play important roles in biology and physiology. In particular, interactions between cell receptors and bioactive ECM proteins are necessary to maintain multiple cell functions. Appropriate chemical methodologies for biomaterial designs that regulate the bioactivities of proteins and cells are becoming increasingly popular.[22,23] Particularly, the control of specific interactions between bioactive molecules has been used to study, understand, and control molecular recognition–mediating biological processes.[24,25] The interactions of ligands and cell membrane receptors play multiple important roles in inducing cell spreading, proliferation, differentiation, and signal transduction.[26–31] The ECM elaborated by cells on surfaces constitutes a regulator of the cell adhesion process and cell differentiation and contributes to the mechanical properties of tissue. These regulatory effects of ECM are mediated through transmembrane proteins specialized in cell-substrate adhesion. The peptide sequence Arg-Gly-Asp (RGD) found in fibronectin, type I collagen, and other ECM proteins has been widely studied as an immobilized cell adhesion ligand specific to integrin-mediated cell adhesion.[32,33] The surface density and chemical presentation of these synthetic ligands and their effectiveness as surface modification agents in cell culture are correlated to their ability to enhance cell adhesion, which is particularly useful for applications in tissue engineering.[34–38] Temperature-responsive PIPAAm-grafted surfaces are modified with arginine–glycine–aspartic acid–serine (RGDS) peptides. Surface swelling, wettability, and elastic modulus of the grafted surface can be changed by temperature control. At temperatures greater than the LCST of PIPAAm, the grafted polymer chains collapse, and the immobilized peptide is therefore exposed to adherent cells. Below the LCST, hydrated polymer chains soften, expand, and swell, shielding immobilized RGDS from integrin access, limiting cell-surface attachment tension, and mechanically disrupting cell-surface contacts. Adherent cell populations are first rounded and then removed under this thermal treatment in the absence of serum proteins or trypsin treatment. Because PIPAAm does not have functional groups such as carboxyl, amino, or hydroxyl groups, these moieties have been introduced by co-polymerization in order to provide a means for the stable binding of peptides. Here, we have designed a functional co-monomer, 2-carboxyisopropylacrylamide (CIPAAm), because the use of conventional co-monomers such as acrylic acid (AAc) shifts the LCST to higher than 37°C.[8]

Poly(IPAAm-co-CIPAAm) was grafted onto a TCPS dish via electron-beam irradiation polymerization. Peptides were immobilized onto the grafted polymer chains via the carboxyl groups in CIPAAm using *N*-hydroxysuccinimide/1-ethyl-3-(3-dimethylaminopropyl) carbodiimide hydrochloride (NHS/EDC) chemistry.[39,40]

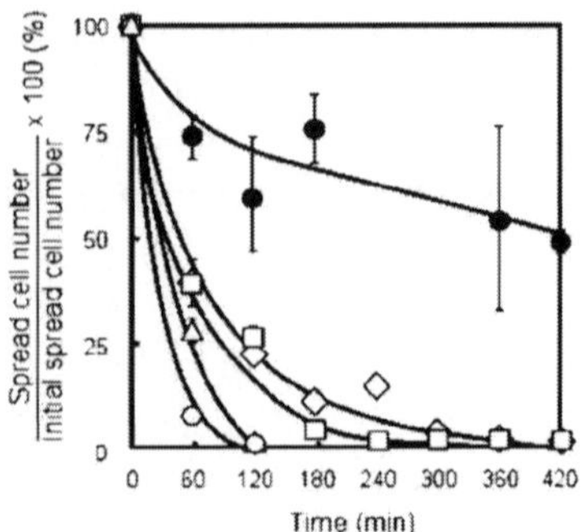

Figure 44.4. Competitive assay for HUVEC detachment from RGDS-immobilized temperature-responsive culture surfaces by soluble peptide addition. Soluble RGDS (100 mM) was added to cells on RGDS-IC(X) after a 24-h culture at 37°C without serum. CIPAAm compositions were 1 (open circle), 3 (triangle), 5 (square), and 10 mol% (diamond). As a comparison, cells cultured on TCPS for 24 h at 37°C with serum were detached at 20°C (closed circle). Reprinted with permission from Ref. 40. Copyright 2004 American Chemical Society.

Immobilized RGDS contributed to the enhancement of cell spreading, likely mediated by cell integrin. In contrast, the control arginine–glycine–glutamic acid–serine (RGES)-immobilized surfaces (RGES-IC(X)) did not demonstrate cell spreading enhancement against immobilized peptide density because they have no integrin biospecificity as a result of the substitution of glutamic acid (E) for aspartic acid (D). Figure 44.4 shows spread human vascular endothelial cell (HUVEC) numbers on various substrates after 6-hour and 24-hour cultures with or without serum at 37°C. Although the addition of dilute serum facilitates HUVEC spreading on TCPS, the RGDS modified poly(IPAAm-co-CIPAAm) surface dramatically promoted cell spreading under serum-free conditions. In contrast, little spreading improvement was observed on the RGDS-modified poly(IPAAm-co-AAc) surface on which peptides were immobilized via another co-monomer, AAc, in place of CIPAAm. As shown in Fig. 44.4, significant differences in HUVEC spreading on poly(IPAAm-*co*-CIPAAm)-grafted surfaces (CIPAAm; 1 mol % in feed) before and after RGDS immobilization after a 6-hour culture without serum at 37°C are readily evident.

To control the dissociation of specific interactions between cultured cell integrins and surface RGDS, spread HUVECs (1-day culture without serum at 37°C) were exposed to a lower temperature treatment at 20°C (Fig. 44.4). The detachment rates depended on the immobilized cell adhesive peptide densities represented by the monomer carboxylate group composition in each feed. Moreover, the detachment rates increased dramatically with the addition of higher concentrations of RGDS. Low temperature treatment at 20°C also promoted cell detachment from the RGDS-modified poly(IPAAm-co-CIPAAm) surface, for which the detachment rate also decreased with increasing cell adhesive peptide surface content.

Recently, more systematic chemical compositions achieved by using RGDS-modified poly(IPAAm-co-CIPAAm) systems have advanced cell attachment and detachment.[41]

By co-immobilizing a Pro–His–Ser–Arg–Asn (PHSRN) sequence on intelligent surfaces that enables the stable binding of RGDS to $\alpha 5\beta 1$ integrin, the synergistic roles of the PHSRN sequence have been investigated by studying time-dependent cell detachment from these surfaces. Notably, PHSRN has been found in FN and is thought to synergistically enhance the cell adhesive activity of the RGD sequence.[42–44] The synergistic site is located approximately 3.5 nm from the RGD loop in native FN, and its precise spatial positioning is thought to play a critical role in the observed synergistic activity, especially for downstream cell adhesion events involving focal adhesion kinase (FAK) phosphorylation.[11–16] In contrast, very little is known about the synergistic effects of PHSRN on binding strength. Therefore, here, we have developed a novel but simple protocol for the comparative study of cell adhesive strengths using RGDS- and PHSRN–immobilized intelligent surfaces. This approach provides a noninvasive method for examining time-dependent affinity changes between cells and peptides.

RGDS-immobilized, temperature-responsive polymer-grafted surfaces promote HUVEC adhesion and spreading by their biospecific activity in serum-free media. At temperatures below the grafted polymer's LCST, integrin-RGDS association decreases due to a loss of cell tension and surface anchoring, prompting cells to round and then detach (Fig. 44.5). Observed "on-off" control of specific integrin-RGDS binding achieved only by temperature regulation is

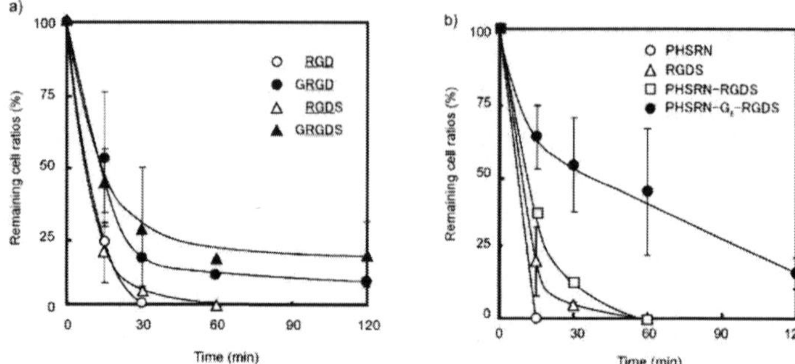

Figure 44.5. Detachment profiles for HUVECs seeded on the following peptide-immobilized poly(IPAAm-co-CIPAAm)-grafted surfaces as a function of time: (a) RGD, GRGD, RGDS, and GRGDS. (b) PHSRN, RGDS, PHSRN–RGDS, and PHSRN-G6-RGDS. The cells have been incubated at 37°C for 24 h in the absence of FBS and have been transferred to an incubator equipped with a cooling unit fixed at 20°C. The results are reported as the percentage of cells remaining on the surfaces. *Abbreviations*: FBS, fetal bovine serum; GRGD, glycine–arginine–glycin–aspartic aci; GRGDS, glycine–arginine–glycin–aspartic aci–serine. Reprinted with permission from Ref. 41. Copyright 2008 John Wiley & Sons, Inc.

very important for cell function regulation because cell surface integrins not only bind to cell adhesive ligands in ECM but also serve to produce subsequent cell signal transduction activity. Moreover, this approach facilitates a serum-free cell culture for the safety of clinical applications using cultured cells and noninvasive cell recovery (no enzymes) that help maintain original cell function, both of which are important to applications in tissue engineering.

44.4 Applications of Cell Sheet Technology

Cell sheet technology with a temperature-responsive culture dish allows for noninvasive harvesting of cultured cells as an intact monolayer cell sheet, including deposited ECM.[12] Using cell sheet engineering, we can completely avoid scaffolds, fixation, or sutures for conventional tissue engineering approaches using isolated cell injections and scaffold-based technologies, whose applicability is often limited. Direct transplantations have been applied to corneal epithelia, mucosal epithelia, periodontal ligament cells, bladder epithelia, and esophageal epithelia. In the following sections, specific applications of cell sheet technology are described.

44.5 Corneal Surface Reconstruction

The complete loss of limbal epithelial stem cells due to severe trauma (e.g., thermal and chemical burns) or eye disease (e.g., Stevens-Johnson syndrome, ocular pemphigoid) prompts adjacent conjunctival tissues to completely cover the cornea, followed by corneal vascularization and opacification with severe visual loss. Although corneal transplantation using donated eyes is the most common method of treatment, the high risk of graft rejection and a shortage of donor corneas remain significant inadequacies. From these aspects, cultured corneal epithelial cells treated with dispase or fibrin gel have been used for ocular surface reconstruction.[47–49] These methods require a carrier substrate to be surgically placed on host corneal stroma. In addition, infection, inflammation, and microtrauma due to incompletely biodegradable compounds remain controversial issues with regard to the use of the gel for tissue engineering.

In contrast, limbal epithelial stem cells can be isolated and cultured on temperature-responsive culture surfaces. To create functional bioengineered corneal epithelial sheets suitable for transplantation, we harvested human corneal limbal epithelial cells, containing corneal epithelial stem cells, from a US eye-bank cornea (Fig. 44.6).[50,51] These cells were seeded onto temperature-responsive culture surfaces on which mitomycin C (MMC)-treated 3T3 feeder cells had been plated one day previously. Corneal epithelial stem cells remained in the presence of feeder cell layers and other specialized conditions. Under our culture conditions, epithelial cell colonies grew to reach confluency within one to two weeks. After a few additional days in culture, a multilayered corneal epithelial sheet formed spontaneously, with cells on the sheet surface exhibiting a cobblestone morphology characteristic of epithelial cells. The corneal epithelial cell phenotype was confirmed by immunohistochemistry with an antikeratin 3 antibody specific for corneal epithelial cells. Bromodeoxyuridine (BrdU) uptake was

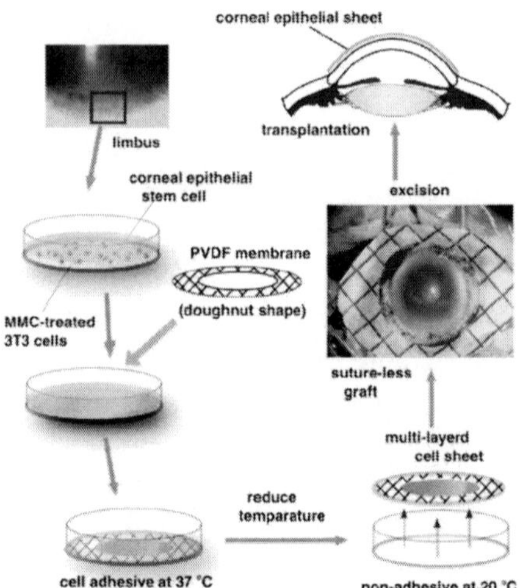

Figure 44.6. Corneal epithelial cell sheet transplantation. Limbal stem cells are collected and seeded on temperature-responsive culture surfaces. After 2 weeks in culture with an MMC-treated 3T3 feeder layer, all the multilayered cells are harvested as a cell sheet using a doughnutshaped PVDF membrane after release by temperature reduction. The cell sheet is grafted onto the corneal stroma without sutures. *Abbreviation*: PVDF, poly(vinylidene difluoride). Reprinted with permission from Ref. 50. Copyright 2004 Lippincott Williams & Wilkins, Inc.

observed in a small portion of cells. These multilayered corneal epithelial cells could be harvested as a transplantable cell sheet simply by reducing the culture temperature to 20°C, without the use of proteases or ethylenediaminetetraacetic acid (EDTA). Harvested sheets, flattened on their basal and apical surfaces, comprised three to five cell layers with small basal cells, flattened middle cells, and polygonal, flattened superficial cells. Such morphologic characteristics are similar to human corneal epithelium *in vivo* to some extent; however, they were less mature and differed in that the multilayer characteristics seemed to vary depending upon location and that the basal cells tended to be flattened, not cuboidal, especially in regions with fewer layers. Interestingly, only the flattened superficial cells expressed keratin 3, whereas BrdU uptake was detected in some basal cells. p63, which has recently been proposed as a corneal epithelial stem cell marker, was expressed continuously in the basal cells of the cell sheets. Immunostaining patterns of keratin 3 and p63 were highly similar to those in native limbal epithelium. These observations imply that epithelial stem cells were localized in the basal layer of the harvested cell sheets and that differentiated epithelial cells migrated to the surface layers as seen *in vivo*.

Conjunctival scar tissue on the cornea was surgically removed to expose the native, transparent corneal stroma. Then, cultured, autologous, multilayered corneal epithelial cell sheets were harvested by reducing the temperature and using a poly(vinylidene difluoride) (PVDF) membrane with an outer diameter of 30 mm and a 12-mm-diameter hole punched out of its center. The intact multilayered corneal sheet was placed directly over the bare corneal stroma. After five minutes of contact after placement, the sheet readily adhered to the underlying corneal stroma, self-stabilizing without any suturing. The PVDF membrane was then excised. Immediately after surgery, fluorescein staining showed that the corneal epithelial sheet covered the entire corneal surface homogeneously, reflecting a successful procedure. The next day, the graft remained firmly attached to the corneal surface, with some ocular surface inflammation seen. During postoperative healing, the ocular surface inflammation subsided, and corneal transparency was restored. During observation periods of up to 180 days, all bioengineered corneal sheet grafts remained stable at the initial placement site, covering the entire corneal surface. The corneal epithelium regenerated by corneal sheet graft healing exhibited a normal appearance, with slightly thin corneal epithelium, cuboidal basal cells, and flattened medial and superficial cells. All of the epithelial cell layers expressed keratin 3 in each of the two regenerated corneas examined by immunohistochemistry. To follow the transplanted corneal epithelial cells after grafting, transplanted cells were stained with a cell-tracer red dye one day before surgery. Later, the entire thickness of the regenerated corneal epithelium exhibited red fluorescence, indicating that the regeneration originated with the transplanted cells.

Clinical results have shown that the corneal surface remains clear with significantly improved visual acuity more than one year after the corneal epithelial cell sheet transplantation. Transplantation of corneal epithelial cell sheets also prevents the development of corneal haze after excimer laser keratectomy. We have also shown that autologous oral mucosal epithelial cell sheets can be used as an alternative to corneal epithelial cell sheets. Oral mucosal epithelium cell sheets can be harvested on temperature-responsive culture dishes and transplanted in the same manner as corneal epithelial sheets. The oral mucosal epithelial cell sheets more closely resemble the native corneal epithelium. Furthermore, transplanted oral mucosal epithelial cell sheets show modulation in their keratin expression profiles toward a corneal phenotype in a rabbit model. The direct interaction between the transplanted epithelial sheet and host stroma and the increased corneal transparency are the greatest benefits obtained by excluding the use of scaffolds or carrier substrates.

44.6 Periodontal Ligament Cell Sheets

Periodontitis is one of the most prevalent infectious diseases and is characterized by the destruction of tissues, such as alveolar bone, cementum, and the periodontal ligament, that surround and support the teeth. Recent studies have indicated that periodontitis is associated with not only oral diseases but also a number of systemic diseases such as adverse pregnancy outcomes, cardiovascular disease, stroke, pulmonary disease, and diabetes.[52] Although many surgical procedures have been applied to induce tissue regeneration, gingival epithelial tissues proliferate into the defect at a faster rate than the underlying mesenchymal tissues.[53] Thus, long junctional epithelium attachment to the dentin root surface is established, and the original form and functions of periodontal tissues cannot be restored. Therefore, the concept of guided tissue regeneration was established to promote the regeneration process.[54–58] We developed a periodontal ligament cell sheet that can regenerate cementum and periodontal ligaments in rats and canines.[59–63]

Human periodontal ligament cells cultured with ascorbic acid have been shown to enhance periodontal regeneration in an athymic rat model.[59] In addition, a subsequent study revealed that human periodontal ligament cells cultured in α-Minimum Essential Medium with osteoinductive reagents, which contained not only ascorbic acid but also β-glycerophosphate and dexamethasone, induced thicker cementum than that cultured in minimum essential medium-alpha (MEM-α) only, suggesting that osteoinductive reagents might be effective in creating thick cementum on denuded root surfaces.[63] Another study also demonstrated that three-layered human periodontal ligament cell sheets cultured in an osteoinductive medium regenerated thick cementum, including Sharpey's fibers, in a rat back model.[61]

In a canine model, we demonstrated periodontal regeneration using a monolayered canine periodontal ligament cell sheet reinforced with hyaluronan, although such periodontal tissue formation was not observed, except in one defect in the control group.[60] Histometric analysis revealed that the formation of new cementum in the experimental group was significantly higher than that in the control group. Its success rate was 60% (3 out of 5), and bone regeneration was limited. In order to improve the success rate, we transplanted three-layered PDL cell sheets supported with woven poly(glycolic acid) to dental root surfaces with three-wall periodontal defects in an autologous manner, and bone defects were filled with porous β-tricalcium phosphate.[62] The cell sheet transplantation regenerated both new bone and cementum connecting with well-oriented collagen fibers, whereas only limited bone regeneration was observed in the control group where cell sheet transplantation was eliminated. In detail, two-dimensional (2D) images of a micro–computer tomography (micro-CT) analysis demonstrated that almost 50% of the bone filling was observed after treatment with a poly(glycolic acid) sheet without periodontal ligament cells as a control group (Fig. 44.7a). In contrast, complete bone filling with an appropriate space of periodontal ligament was observed in the group treated with a poly(glycolic acid) sheet and a periodontal ligament cell sheet (Fig. 44.7b). Periodontal defects were still observed in a three-dimensional (3D) virtual model after a six-week healing period in the control group (Fig. 44.7c). In contrast, completely bone-filled alveolar ridges were seen in the experimental group (Fig. 44.7d). Based on the histometric analysis, there was no significant difference in the apical extension of the junctional epithelium between the experimental group and the control group. Ankylosis was not observed in any samples. Complete periodontal regeneration with both newly formed bone and cementum connecting with well-oriented collagen fibers was observed in all cases of the experimental group, whereas limited bone regeneration with poorly oriented periodontal fibers was seen in the control group.

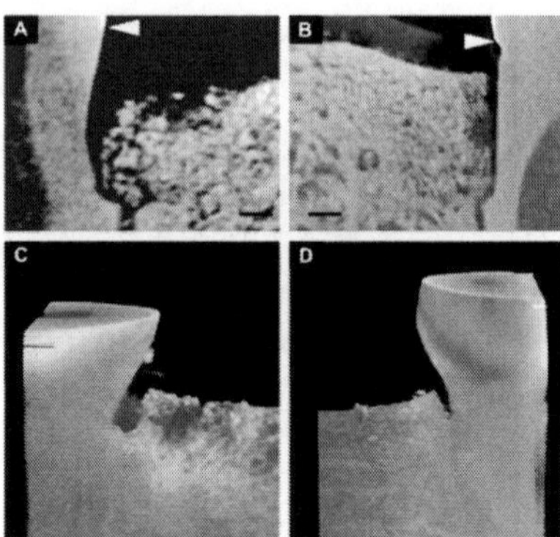

Figure 44.7. Micro-CT analysis of periodontal defects after 6-week period of healing. (a, b) 2D image of the 3-wall defect transplanted with bTCP (a) or three-layered PDL cell sheets/bTCP (b). Triangle indicates CEJ of bilateral mandibular first molars of dog. (bar, 1 mm) (c, d) 3D virtual model of the 3-wall defect transplanted with bTCP (c) or three-layered PDL cell sheets/bTCP (d). After the 6week healing period, micro-CT analysis was performed and images were reconstructed with VG max software. *Abbreviation*: CEJ, cement-enamel junction. Reprinted with permission from Ref. 62. Copyright 2009 Elsevier B.V.

In conclusion, transplantable multilayered periodontal ligament cell sheets were successfully fabricated with woven poly(glycolic acid), and its use combined with porous β-tricalcium phosphate induced true periodontal regeneration, including alveolar bone, cementum, and well-oriented fibers at the same time. Periodontal ligament cell sheet transplantation should prove useful for periodontal regeneration in clinical settings.

44.7 Endoscopic Esophageal Epithelial Transplantation

Endoscopic mucosal resection (EMR) has enabled the noninvasive removal of gastrointestinal tract cancers without the need for esophageal reconstruction.[64–67] Although EMR is a widely accepted technique, the size of treatment is limited, and the retrieval of multiple specimens, which can damage the tissues, is needed.[68] A novel method of endoscopic submucosal dissection (ESD) permits large *en bloc* resections without the size limitation, enabling the removal of large cancers using a single procedure.[69–74] However, severe inflammation causes esophageal scarring and stenosis, without any method to enhance wound healing.[71,74] We have developed a method combining ESD with the endoscopic transplantation of autologous oral mucosal epithelial cell sheets.[75] Detailed information is given as follows.

Cell sheets were noninvasively harvested with the use of a PVDF support membrane. Immediately after ESD, two autologous oral mucosal epithelial cell sheets were individually transplanted to the artificial esophageal ulceration using endoscopy ($n = 3$). An EMR tube was inserted in the esophagus of each animal, and the PVDF support membrane with an attached autologous oral mucosal epithelial cell sheet was grasped by endoscopic forceps and carefully maneuvered to the ulcer site through the EMR tube.[76] Cell sheets were then placed directly onto the ulcer sites, and gentle pressure was applied to the overlying PVDF support membranes using the EMR tube. After 10 minutes, the support membranes were simply removed by endoscopic forceps, with the cell sheets remaining on the ulcer wound beds. Animals receiving only ESD, without cell sheet transplantation ($n = 3$), were used as controls. The transplanted cell sheets were able to adhere and survive on the underlying muscle layers within the ulcer sites, providing an intact, stratified epithelium. Four weeks after surgery, complete wound healing, with no observable stenosis, was seen in the animals receiving autologous cell sheet transplantation (Fig. 44.8). In contrast, a noticeable

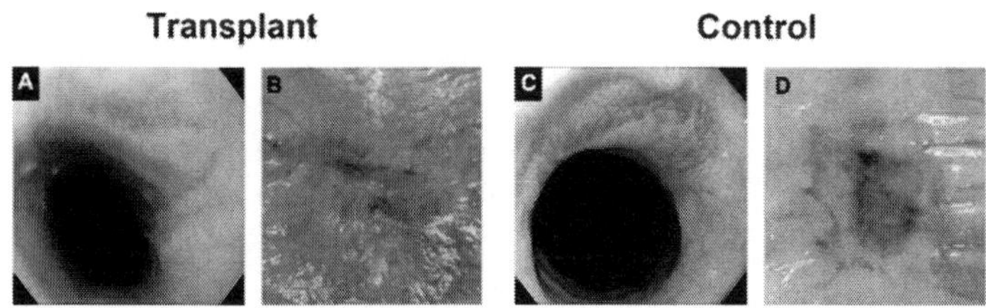

Figure 44.8. Transplantation of oral mucosal epithelial cell sheets promotes wound healing and reduces postoperative inflammation. Left and right panels represent transplant and control groups, respectively. (a, c) Endoscopic photographs taken 4 weeks postoperatively. (b, d) Macroscopic images of the esophageal sites receiving endoscopic submucosal dissection, after 4 weeks. Reprinted with permission from Ref. 75. Copyright 2006 BMJ Publishing Group.

fibrin mesh and host inflammation, consistent with the intermediate stages of wound healing, were observed in the control animals that received only ESD. After ESD, the cell sheets were attached directly to the esophageal ulcer beds using endoscopic forceps without suturing. The presence of an epithelial cell layer dramatically enhances post-ESD wound healing and reduces host inflammatory responses.

44.8 Sealing of Lung air Leaks

Air leaks caused by visceral pleural injury during lung resection are an unavoidable complication and can cause thoracic emphysema, leading to increased chest tube durations and prolonged hospital stays.[77–80] In these cases, the prevention of air leaks is a critical objective for patient management after pulmonary surgery. With improved surgical techniques such as manual suturing and electrocautery, complications due to intraoperative visceral pleural injuries have been slightly minimized.[81] Nevertheless, difficulty due to surgical constraints of the operations can still result in damage to the surrounding tissues, as well as failure to effectively close air leaks.

Current methods, including the use of various biological and synthetic sealants, are ineffective in the closure of intraoperative air leaks that often occur during cardiothoracic surgeries, resulting in a decreased quality of life for patients. We present the development of a novel lung air leak sealant using tissue-engineered cell sheets.[82,83] In contrast to previously used materials such as fibrin glue, these bioengineered cell sheets immediately and permanently seal air leaks in a dynamic fashion that allows for the extensive tissue contraction and expansion involved in respiration, without any postoperative recurrences. Additionally, we demonstrate that mesothelial cells migrate to cover the transplanted cell sheets, thereby confirming excellent biocompatibility and integration with the host tissues. Finally, we present the use of skin fibroblasts as an effective and readily available autologous cell source that can be easily applied.

References

1. R. Langer and J. Vacanti, *Science*, **260**, 920–926 (1993).
2. L. E. Freed *et al.*, *Nat. Biotech.*, **12**, 689–693 (1994).
3. S. Y. Chung, *Lancet*, **367**, 1215–1216 (2006).
4. T. Shimizu *et al.*, *Circul. Res.*, **90**, E40–E48 (2002).
5. T. Shimizu, M. Yamato, A. Kikuchi, and T. Okano, *Biomaterials*, **24**, 2309–2316 (2003).
6. S. Gupta *et al.*, *Hepatology*, **29**, 509–519 (1999).
7. K. Ohashi *et al.*, *Nat. Med.*, **13**, 880–885 (2007).
8. M. Ebara *et al.*, *Biomacromolecules*, **4**, 344–349 (2003).
9. R. E. Fillinger MF, R. A. Schwartz, D. E. Resetarits, A. M. Paskanik, D. Bruch, and B. CE, *J. Vasc. Surg.*, **11**, 556–566 (1990).
10. R. Langer and J. Vacanti, *Sci. Am.*, **280**, 86–89 (1999).
11. T. Shimizu *et al.*, *Faseb J.*, **20**, 708-+ (2006).
12. M. Yamato *et al.*, *Tissue Eng.*, **7**, 473–480 (2001).
13. N. Yamada *et al.*, *Rapid Commun.*, **11**, 571–576 (1990).
14. T. Okano, N. Yamada, M. Okuhara, H. Sakai, and Y. Sakurai, *Biomaterials*, **16**, 297–303 (1995).
15. M. Yamato, C. Konno, M. Utsumi, A. Kikuchi, and T. Okano, *Biomaterials*, **23**, 561–567 (2002).
16. K. Nagase *et al.*, *Langmuir*, **24**, 10981–10987 (2008).
17. K. Nagase *et al.*, *Langmuir*, **23**, 9409–9415 (2007).
18. M. Yamato *et al.*, *J. Biomed. Mater. Res.*, **67A**, 1065–1071 (2003).
19. Y. Akiyama, A. Kushida, M. Yamato, A. Kikuchi, and T. Okano, *Langmuir*, **23**, 796–802 (2007).
20. Y. Akiyama, A. Kikuchi, M. Yamato, and T. Okano, *Langmuir*, **20**, 5506–5511 (2004).
21. K. Fukumori *et al.*, *Acta Biomaterialia*, **5**, 470–476 (2009).
22. G. M. Whitesides, J. P. Mathias, and C. T. Seto, *Science*, **254**, 1312–1319 (1991).
23. D. J. Irvine, A. M. Mayes, and L. G. Griffith, *Biomacromolecules*, **2**, 85–94 (2000).
24. Z. Ding, R. B. Fong, C. J. Long, P. S. Stayton, and A. S. Hoffman, *Nature*, **411**, 59–62 (2001).
25. M. Strebelow, M. Schlosser, B. Ziegler, I. Rjasanowski, and M. Ziegler, *Diabetologia*, **42**, 661–670 (1999).

26. R. O. Hynes, *Cell*, **69**, 11–25 (1992).
27. S. Miyamoto, S. K. Akiyama, and K. M. Yamada, *Science*, **267**, 883–885 (1995).
28. M. E. Chicurel, C. S. Chen, and D. E. Ingber, *Curr. Opin. Cell Biol.*, **10**, 232–239 (1998).
29. D. E. Ingber, *J. Cell Sci.*, **116**, 1397–1408 (2003).
30. F. Grinnell, *Trends Cell Biol.*, **10**, 362–365 (2000).
31. F. Grinnell, M. Feld, and D. Minter, *Cell*, **19**, 517–525 (1980).
32. R. Pytela, M. D. Pierschbacher, M. H. Ginsberg, E. F. Plow, and E. Ruoslahti, *Science*, **231**, 1559–1562 (1986).
33. D. J. Leahy, W. A. Hendrickson, I. Aukhil, and H. P. Erickson, *Science*, **258**, 987–991 (1992).
34. S. P. Massia and J. A. Hubbell, *J. Cell Biol.*, **114**, 1089–1100 (1991).
35. B. T. Houseman, J. H. Huh, S. J. Kron, and M. Mrksich, *Nat. Biotechnol.*, **20**, 270–274 (2002).
36. L. Y. Koo, D. J. Irvine, A. M. Mayes, D. A. Lauffenburger, and L. G. Griffith, *J. Cell Sci.*, **115**, 1423–1433 (2002).
37. G. Maheshwari, G. Brown, D. A. Lauffenburger, A. Wells, and L. G. Griffith, *J. Cell Sci.*, **113**, 1677–1686 (2000).
38. L. G. Griffith and G. Naughton, *Science*, **295**, 1009–1014 (2002).
39. M. Ebara *et al.*, *Tissue Eng.*, **10**, 1125–1135 (2004).
40. M. Ebara *et al.*, *Biomacromolecules*, **5**, 505–510 (2004).
41. M. Ebara *et al.*, *Adv. Mater.*, **20**, 3034–3038 (2008).
42. D. J. Leahy, I. Aukhil, and H. P. Erickson, *Cell*, **84**, 155–164 (1996).
43. A. J. Garcia, J. E. Schwarzbauer, and D. Boettiger, *Biochemistry*, **41**, 9063–9069 (2002).
44. B. G. Keselowsky and D. M. Collard, *Proc. Natl. Acad. Sci. U S A*, **102**, 5953–5957 (2005).
45. S. M. Cutler and A. J. Garcia, *Biomaterials*, **24**, 1759–1770 (2003).
46. T. A. Petrie, J. R. Capadona, C. D. Reyes, and A. J. Garcia, *Biomaterials*, **27**, 5459–5470 (2006).
47. G. Pellegrini *et al.*, *Lancet*, **349**, 990–993 (1997).
48. P. Rama *et al.*, *Transplantation*, **72**, 1478–1485 (2001).
49. E. Di Iorio *et al.*, *Proc. Natl. Acad. Sci. U S A*, **102**, 9523–9528 (2005).
50. K. Nishida *et al.*, *Transplantation*, **77**, 379–385 (2004).
51. K. Nishida *et al.*, *New Engl. J. Med.*, **351**, 1187–1196 (2004).

52. B. L. Pihlstrom, B. S. Michalowicz, and N. W. Johnson, *Lancet*, **366**, 1809–1820 (2005).
53. M. Listgarten and M. Rosenberg, *J. Periodontol.*, **50**, 333–344 (1979).
54. B. M. Seo *et al.*, *Lancet*, **364**, 149–155 (2004).
55. K. Nagatomo *et al.*, *J. Periodont. Res.*, **41**, 303–310 (2006).
56. T. Nakahara *et al.*, *Tissue Eng.*, **10**, 537–544 (2004).
57. Y. Liu *et al.*, *Stem Cells*, **26**, 1065–1073 (2008).
58. W. Sonoyama *et al.*, *PLoS ONE*, **1**, e79 (2006).
59. M. Hasegawa, M. Yamato, A. Kikuchi, T. Okano, and I. Ishikawa, *Tissue Eng.*, **11**, 469–478 (2005).
60. T. Akizuki *et al.*, *J. Periodont. Res.*, **40**, 245–251 (2005).
61. M. G. Flores *et al.*, *J. Periodont. Res.*, **43**, 364–371 (2008).
62. T. Iwata *et al.*, *Biomaterials*, **30**, 2716–2723 (2009).
63. M. G. Flores *et al.*, *J. Clin. Periodontol.*, **35**, 1066–1072 (2008).
64. R. M. Soetikno, T. Gotoda, Y. Nakanishi, and N. Soehendra, *Gastrointest. Endosc.*, **57**, 567–579 (2003).
65. R. Soetikno, T. Kaltenbach, R. Yeh, and T. Gotoda, *J. Clin. Oncol.*, **23**, 4490–4498 (2005).
66. T. Gotoda, *Gastric Cancer*, **10**, 1–11 (2007).
67. T. Gotoda, H. Yamamoto, and R. M. Soetikno, *J. Gastroenterol.*, **41**, 929–942 (2006).
68. H. Inoue *et al.*, *Gastrointest. Endosc.*, **39**, 58–62 (1993).
69. T. Oyama *et al.*, *Clin. Gastroenterol. Hepatol.*, **3**, S67–S70 (2005).
70. H. Yamamoto, *Clin. Gastroenterol. Hepatol.*, **3**, S27–S29 (2005).
71. C. Katada *et al.*, *Gastrointest. Endosc.*, **57**, 165–169 (2003).
72. C. Katada, M. Muto, T. Manabe, A. Ohtsu, and S. Yoshida, *Gastrointest. Endosc.*, **61**, 219–225 (2005).
73. C. Lopes *et al.*, *Surg. Endosc.*, **21**, 820–824 (2007).
74. S. Seewald *et al.*, *Gastrointest. Endosc.*, **57**, 854–859 (2003).
75. T. Ohki *et al.*, *Gut*, **55**, 1704–1710 (2006).
76. H. Makuuchi, T. Yoshida, and C. Ell, *Endoscopy*, **36**, 1013–1018 (2004).
77. D. B. J. Loran, K. Woodside, R. J. Cerfolio, and J. B. Zwischenberger, *Chest Surg. Clin. N. Am.*, **12**, 477–488 (2002).
78. E. M. Toloza and D. H. Harpole, *Chest Surg. Clin. N. Am.*, **12**, 489–505 (2002).

79. I. Okereke, S. C. Murthy, J. M. Alster, E. H. Blackstone, and T. W. Rice, *Ann. Thorac. Surg.*, **79**, 1167–1173 (2005).
80. A. Brunelli *et al.*, *Ann Thorac Surg*, **77**, 1932-1937 (2004).
81. K. D. Murray, C. H. Ho, J. Y. J. Hsia, and A. G. Little, *Chest*, **122**, 2146–2149 (2002).
82. M. Kanzaki *et al.*, *Biomaterials*, **28**, 4294–4302 (2007).
83. M. Kanzaki *et al.*, *Eur. J. Cardio-Thorac. Surg.*, **34**, 864–869 (2008).

Chapter 45

CUSTOMIZED NANOCOMPOSITE SCAFFOLDS FABRICATED VIA SELECTIVE LASER SINTERING FOR BONE TISSUE ENGINEERING

Bin Duan and Min Wang*

Department of Mechanical Engineering, Faculty of Engineering, the University of Hong Kong, Pokfulam Road, Hong Kong

*memwang@hku.hk

In scaffold-based tissue engineering, controlling the macro- and micro-architecture of a scaffold and fulfilling a customized scaffold design with a complex anatomic shape are of great importance for the clinical success of a tissue engineering strategy. In order to achieve extensive and detailed control over scaffold architecture, solid free-form fabrication (SFF), also called "rapid prototyping (RP) technologies", is introduced into the tissue engineering field and have been developed for scaffold production. This chapter presents our investigations into the fabrication of three-dimensional (3D) calcium phosphate (Ca-P)/ poly(hydroxybutyrate-co-hydroxyvalerate) (PHBV) nanocomposite scaffolds through the use of selective laser sintering (SLS). A new strategy of using nanocomposite microspheres instead

Handbook of Intelligent Scaffolds for Tissue Engineering and Regenerative Medicine
Edited by Gilson Khang

www.panstanford.com

of dry-blended bioceramic-polymer powders as raw materials for SLS is adopted. A commercial SLS machine is modified to reduce the consumption of raw materials in scaffold fabrication. Based on designed scaffold models, nanocomposite scaffolds with well-defined architecture and pore structure can be fabricated via SLS. For using SLS-formed nanocomposite scaffolds as vehicles for the controlled release of biomolecules such as growth factors, both entrapment and attachment strategies are investigated. As a demonstration of the entrapment strategy, Ca-P/PHBV microspheres loaded with bovine serum albumin (BSA) are fabricated and sintered into 3D porous scaffolds. BSA-loaded nanocomposite scaffolds display controlled *in vitro* release behavior. However, there are problems in this biomolecule entrapment approach for SLS-formed scaffolds. Using the attachment strategy, recombinant human bone morphogenetic protein-2 (rhBMP-2) can be loaded onto surface-modified scaffolds through non-covalent binding. The scaffold surface modification by heparin not only provides a means to protect the loaded rhBMP-2 but also improves the sustained release behavior of rhBMP-2. The integration of advanced scaffold fabrication technology and nanocomposite and growth factor delivery to form multifunctional tissue engineering scaffolds holds promises for successful bone tissue regeneration.

45.1 Introduction

Orthopedic disorders and trauma are a serious and costly health issue and affect a large population of people around the world. In the United States alone, the cost for orthopedic health care exceeds US $28 billion per year, and this number is increasing due to an aging population.[1] Currently, various strategies are investigated by many researchers to maintain, repair, or improve bone tissue functions. Although bone can regenerate itself, when it comes to critical-size bone defects or bone tissue loss, bone grafts or bone graft substitutes are necessary for the replacement and regeneration of the fractured or missing bone. Current clinical treatments involve the use of autologous bone grafts, allogenous bone grafts, and bone graft substitutes such as biomedical polymers, bioceramics, and implantable metals. Although autografts remain as the golden standard in clinical practice and provide high bone formation rate, the availability of autografts is limited, especially for the repair of large bone defects, and the sacrifice of donor tissue may cause donor site morbidity with associated infection, pain, and hematoma.[2,3] Allogenous bone grafts (allografts), which are bone tissue that has been harvested from one individual and implanted into another individual of the same species, provide an alternative. However, for allografts, there are risks of immuno rejection and also disease transmission from the donor to the recipient.

With the increasing clinical demand for enhancing bone healing and avoiding donor site morbidity associated with harvesting bone grafts, bone tissue engineering has emerged as a promising new means for bone tissue repair. There are generally several strategies in tissue engineering, including cell-based tissue engineering, factor-based tissue engineering, and scaffold-based engineering, as illustrated in Fig. 45.1. For scaffold-based tissue engineering, cells and/or biomolecules are encapsulated within the matrices or seeded onto the surface of 3D scaffolds.[4,5] The biodegradable tissue engineering matrices or scaffolds, which maybe in the injectable, textile, or solid cellular form, serve as the extracellular matrix (ECM) to direct cell adhesion, proliferation, and

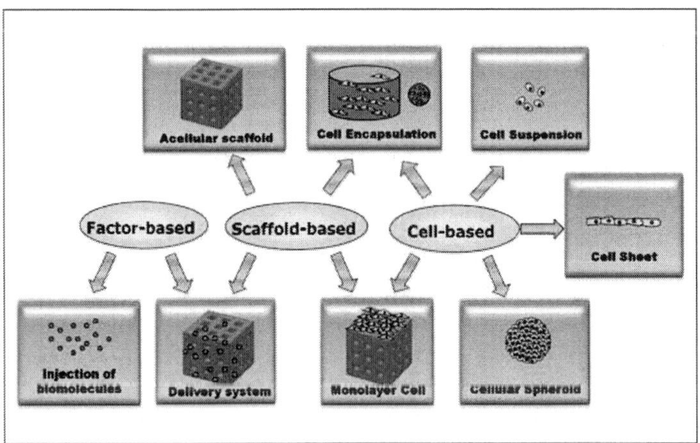

Figure 45.1. Schematic illustration of tissue engineering strategies.

differentiation and thus promote tissue regeneration. In scaffold-based tissue engineering, although ideal scaffolds have not yet been achieved (or actually, there are no such ideal scaffolds and each scaffold should be designed according to specific tissue regeneration requirements), certain minimum criteria, namely, biocompatibility, mechanical properties similar to those of the target tissue, biodegradability at a degradation rate commensurate with the tissue formation rate, etc., should be met for the scaffolds.

There are many methods for scaffold fabrication, including non-designed manufacturing techniques and designed manufacturing techniques. Using polymer solutions and sometimes with high pressure or high temperature, non-designed manufacturing techniques include solvent casting/particulate leaching, phase separation, gas foaming/high-pressure processing, melt molding, electrospinning, and combinations of these techniques (e.g., gas foaming plus particulate leaching, etc.). Designed manufacturing techniques, usually referred to as SFF or RP technologies in the general manufacturing sector of the industry, have attracted much attention recently and are investigated and developed for scaffold fabrication in the tissue engineering field.[6]

Solvent casting/particulate leaching is a simple and the most commonly used method for producing porous bone tissue engineering scaffolds.[7] This method involves dispersing granules of a water-soluble salt such as sodium chloride, sodium citrate, or sucrose in a polymer solution. The polymer solution with dispersed salt particles is then processed via either solvent casting or freeze drying in a mold of desired shape. After the solvent is removed, the salt particles are leached out, forming a highly porous structure.[8] The pore size of the scaffold can be controlled by selecting different sizes of the porogen and by changing the porogen/polymer ratio of the polymer solution. However, the porous scaffolds produced through solvent casting/particulate leaching are usually thin pieces and sometimes lack pore interconnectivity.

Phase separation occurs when, under certain conditions (usually induced by the change of solvent or temperature), a homogeneous multicomponent system becomes thermodynamically unstable and separates into more than one phase in order to lower the system free energy.[9] There are several methods to induce phase separation, forming polymer-rich and polymer-poor phases. Polymer solutions can be dispersed into a non-solvent, and this process will result in a phase inversion, which causes the polymer to precipitate.[10] Thermally induced phase separation processes are also utilized to fabricate biodegradable 3D polymer scaffolds. By carefully controlling the thermal gradient and supercooling temperature, different pore sizes and pore structures can be obtained.

Electrospinning is a highly versatile method for processing solutions or melts, mainly of polymers, into continuous fibers with diameters ranging from nanometers to submicrometers. In a typical electrospinning process, a polymer solution is contained in a syringe with a metal capillary connected to a high-voltage power supply as an electrode. The polymer solution is subjected to an electrical field, and the polymer jet travelling in the electrical field becomes thin fibers, which are deposited onto a conductive collector.[11] The electrospun mats consist of ultrafine fibers, which have high surface-to-volume ratios and porosity, mimicking the natural ECM of body tissues. Electrospun fibrous scaffolds have been employed for regenerating various tissues including skin, blood vessel, cartilage, bone, muscle, ligament, and nerve.[12–15]

Melt molding and particulate leaching techniques are often combined for scaffold fabrication. A polymer is mixed with a porogen prior to being loaded into a metal mold. The mold containing the mixture is then heated under pressure above the glass transition temperature (or melting temperature) of the polymer, forming a solid object. The molded product is subsequently immersed in a solvent for the selective dissolution of the porogen.[16] When the biodegradable polymers are in the form of microspheres, the term "microsphere sintering" is used.[17,18] Extrusion and injection molding are other melt-based processing techniques in this scaffold manufacturing route.[19]

Some of the non-designed manufacturing techniques can partially control the scaffold architecture by altering certain processing parameters, for example, porogen size, freezing temperature, organic solvent, and collection method for electrospun fibers.[20–22] Although the non-designed manufacturing techniques have attained much improvement and have been widely used in the tissue engineering field over the past two decades, their inherent limitations, such as manual intervention, inconsistent and inflexible processing procedures, use of cytotoxic solvent, and shape and dimension limitations for scaffolds produced, have restricted their scope of application.[23] Therefore, designed manufacturing techniques developed in recent years are promising alternative scaffold fabrication techniques for achieving extensive and detailed control over scaffold architecture and hence properties.

45.2 Application of Rapid Prototyping Technologies to Scaffold Fabrication

RP is a common name for a group of technologies that can generate a physical model directly from computer-aided design (CAD) data in a layer-by-layer manner, and each layer is the shape of the cross-section of the model at a specific level.[23] The data used to design and fabricate tissue engineering scaffolds is usually based on CAD systems, computer-based medical imaging technologies, or other computer-based technologies.[24,25] In a typical RP operation, computer-based data is first converted to an STL-type file format, "STL" being derived from "stereolithography," which is the oldest RP technology. Then the computer-generated 2D layers are created for building a solid model, starting from the bottom and proceeding upward. Each layer is glued or otherwise bonded to the previous layer, thus producing a solid model of the object according to the design. The advantages of RP techniques over conventional scaffold production methods include, but are not limited to, achieving well-defined external and internal architectures of the scaffolds, computer-controlled manufacturing processes, and high accuracy and reproducibility.

On the basis of their working principles, RP techniques can be divided into several categories: laser polymerization–based techniques, such as stereolithography apparatus (SLA); nozzle deposition–based techniques, such as fused deposition modeling (FDM); and powder-based techniques, such as 3D printing and SLS.[26,27] The techniques in each category are suitable for a specific form of raw materials and have their own advantages and disadvantages. More and newer RP techniques, such as low-temperature deposition manufacturing (LDM) and two-photon polymerization (TPP), have recently been investigated, developed, and commercialized for the fabrication of tissue engineering scaffolds utilizing different biomaterials, especially natural biodegradable polymers, as raw materials.

SLA is based on the use of an electromagnetic radiation source, typically ultraviolet (UV) radiation, to initiate photopolymerization of photopolymerizable resins. With the investigation and development of photopolymerizable biomaterials, SLA becomes increasingly popular for scaffold fabrication.[28,29] With the development of material science, some hydrophilic polymers can be modified by the addition of a photolabile group and thus become photolabile and cross-linkable. For example, some photoreactive and cross-linkable groups such as acrylates or methacrylates can be easily attached to poly(ethylene glycol) (PEG) and then cross-linked into a PEG hydrogel for use in tissue engineering and even for encapsulating cells in the presence of a cytocompatible photoinitiator.[30,31] Poly(propylene fumarate) (PPF), an unsaturated linear polyester that can be cross-linked by UV light and degraded by simple hydrolysis of the ester bonds into nontoxic products, is suitable for SLA and is a promising biodegradable material for bone tissue engineering.[32]

FDM developed by Stratasys Inc., USA, utilizes a temperature-controlled head to extrude thermoplastic materials to build solid objects layer by layer. FDM is constrained by the use of thermoplastic materials with good melt viscosity properties, and cells cannot be encapsulated into scaffolds during the FDM process due to the high temperatures applied. In the past, only a few non-resorbable polymeric materials, such as polyamide, acrylonitrile butadiene styrene (ABS), and other resins, could be used for FDM. Recently, poly(ε-caprolactone) (PCL) filaments were fabricated from PCL pellets, with a consistent filament diameter fitting for the FDM system for building fibrous PCL scaffolds.[33] Furthermore, biodegradable bioceramic-polymer composite scaffolds based on calcium phosphate/PCL were fabricated through FDM for bone tissue engineering.[34]

3D printing was first developed at the Massachusetts Institute of Technology (MIT) and has become perhaps one of the most widely investigated RP techniques for scaffold fabrication.[35] During

3D printing fabrication, a thin layer of powder is first spread over the building platform and then the inkjet print head prints or deposits a binder solution onto the powder bed. Many biodegradable polymers in the fine powder form can be processed into 3D scaffolds using the 3D printing technique. For synthetic poly(α-hydroxy esters) such as poly(L-lactic acid) (PLLA), poly(lactic-*co*-glycolic acid) (PLGA), and PCL, an organic solvent such as chloroform is used as the binder.[36,37] The particulate-leaching technique has also been combined with the 3D printing technique to create porous scaffolds using a PLGA mixture with NaCl particles.[38] Using an organic solvent as the binder could cause problems for cell incorporation in the 3D printing process. Therefore, a starch-based biomaterial composed of cornstarch, dextran, and gelatin was used for scaffold construction via 3D printing with water being the binder.[39]

LDM uses a computer-controlled nozzle to extrude a polymer solution in a layer-by-layer manner, and each layer of the deposited material is frozen on the platform. After the frozen objects are formed by the LDM system, they are freeze-dried in a freeze dryer to remove the solvent.[40] The scaffolds thus formed have both controlled micro- and macrostructures. Xu *et al.* used polyurethane (PU), an elastomer, to fabricate complex vascular systems via LDM.[41] An LDM system with two nozzles was later designed in order to extrude two types of biomaterials with different properties.[42] With this improved LDM system, a double-layer PU-collagen nerve conduit was produced for peripheral nerve regeneration. The hollow tubes achieved the optimal porosity of 75% and pore sizes of 15–25 μm for the external PU layer through the adjustment of polymer concentrations and orientation of filaments in the inner collagen layer.

TPP is based on the simultaneous absorption of two photons which induce chemical reactions between starter molecules and monomers within a transparent matrix.[26] An ultrashort pulse laser is needed to provide the high intensity, and the current capability of the TPP technique allows the generation of 3D porous structures at a resolution down to 100 nm.[43]

As an established member of the RP family, SLS employs a CO_2 laser to selectively sinter thin layers of powdered polymers or their composites, forming solid 3D objects. The schematic diagram for the operation of a commercial SLS machine is shown in Fig. 45.2.

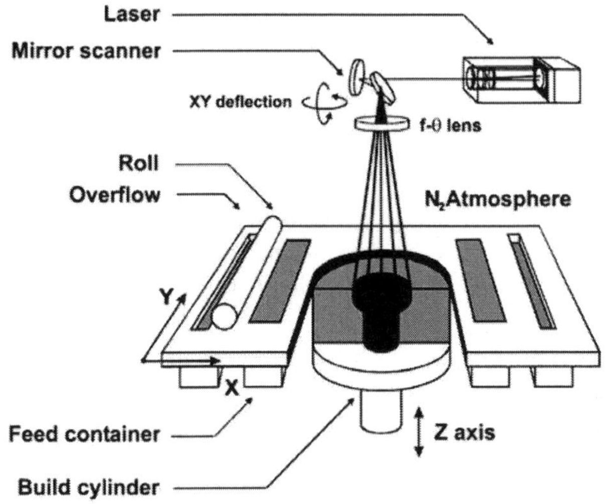

Figure 45.2. Schematic diagram of a commercial SLS machine.[44]

Theoretically, any material that can be processed into the powder form and sintered with high heat can be used in the SLS process. Therefore, investigations have been made by various groups to use biopolymers (both biodegradable and non-biodegradable polymers), ceramics, and composites in producing solid objects via SLS as bone graft substitutes and tissue engineering scaffolds. Rimell *et al.* reported the application of a simplified SLS apparatus to sinter ultrahigh molecular weight polyethylene.[45] Other powdered non-biodegradable polymers such as polyetheretherketone (PEEK) and high-density polyethylene (HDPE) have also been sintered via SLS.[46,47]

A scaffold made of a bioinert polymer (biodegradable or non-biodegradable) is not adequate for promoting bone tissue regeneration. Therefore, composite scaffolds based on non-degradable polymers and hydroxyapatite (HA), such as HA/PEEK and HA/HDPE, were constructed via SLS for potential bone tissue repair.[48–50] However, for bone tissue engineering, biodegradable materials are required in scaffold fabrication. Therefore, many researchers conducted research on composite scaffolds based on biodegradable polymers. In general, particulate HA is incorporated into composite scaffolds to make the scaffolds osteoconductive. A number of groups chose PCL as the matrix polymer due to its good biocompatibility.[51] Other biodegradable polymers, such as polyvinyl alcohol (PVA), PLLA, and PLGA (LA:GA=95:5), and their composites were also successfully processed into scaffolds by using the SLS technique.[47,52] Generally, dry-blending was used to mix polymer granules with bioceramic powders. The mixtures were then sintered to form scaffolds according to respective scaffold designs. Using dry-blending to make raw materials for RP techniques has advantages and disadvantages, and one obvious disadvantage is the inhomogeneous distribution of fine bioceramic particles in RP-produced scaffolds.

A new strategy for making good-quality composite scaffolds via SLS is to use composite microspheres instead of dry-blended bioceramic-polymer mixtures. Our previous investigations demonstrated the feasibility of this strategy by forming, via SLS, carbonated hydroxyapatite (CHAp)/PLLA composite scaffolds, which are osteoconductive and partially biodegradable, through the use of CHAp/PLLA nanocomposite microspheres, which were fabricated in a solid-in-oil-in-water (S/O/W) emulsion solvent evaporation process.[53] For obtaining osteoconductive and totally biodegradable composite scaffolds, microspheres composed of osteoconductive and totally resorbable Ca-P nanoparticles and a natural biodegradable polymer PHBV can be used.

45.3 Design of Scaffolds and the Nanocomposite Strategy

During the SLS process, the raw material in the particulate form is first loaded in the powder tanks (both left and right) of the SLS machine, and after preheating of the powder, the laser beam scans over the cross-section on the powder surface following the cross-sectional profiles of the design to heat up the powder to a preset temperature, causing the particles to fuse together to form a solid mass. When one layer of the cross-section is sintered, the piston in the left (or right) powder tank moves up and the piston in the build part moves down. Subsequently, the roller spreads another layer of powder over the previously sintered part. Relatively large quantities of powdered materials with appropriate particle sizes are

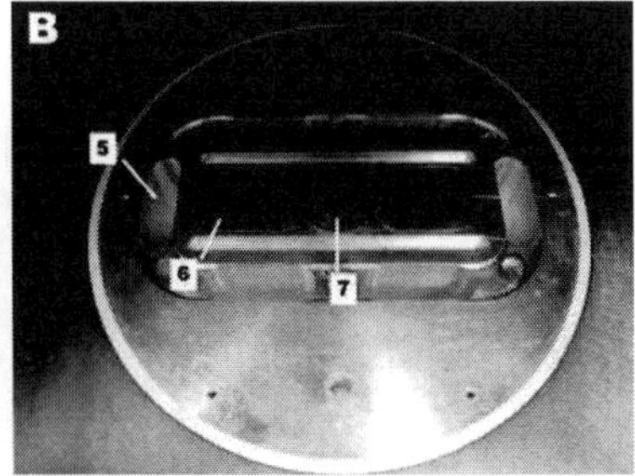

Figure 45.3. (a) Sinterstation® 2000 SLS machine; (b) miniature sintering platform for the modified Sinterstation® 2000 SLS machine. (1: left powder tank; 2: build part for original machine and miniature sintering platform for the modified machine; 3: roller; 4: left sensor; 5: recycle bin; 6: left powder tank for the modified machine; 7: build part for the modified machine.) See also Color Insert.

required for a commercial SLS machine in building (sintering) solid objects. A general view of the sintering section of a commercial Sinterstation® 2000 SLS machine (3D Systems, Valencia, CA, USA) is shown in Fig. 45.3a. A modification of this SLS machine was conducted in order to reduce the consumption of raw materials, that is, nanocomposite microspheres. As shown in Fig. 45.3b, a miniature sintering platform was designed and installed in the build part of the Sinterstation® 2000 SLS machine.[54] The miniature sintering platform consisted primarily of a miniature build part and two powder tanks similar to those of the commercial SLS machine but with much reduced sizes. The movement of the miniature build part was synchronized with the existing build part of Sinterstation® 2000 machine, and the two miniature powder tanks were driven by two additional stepping motors fixed within the miniature platform. In the sintering processes, small amounts of microspheres were fed into the miniature powder tanks with the original powder tanks being kept empty. Two sensors were installed to sense the roller positions, and the signals were fed back to a control panel, which could control the movement of miniature powder tanks. This modified SLS system only requires small amounts of particulate raw materials for scaffold fabrication and can sinter good-quality scaffolds in an effective and efficient manner. Similarly, for the same purpose of reducing raw material consumption, Wiria *et al.* modified their SLS system for scaffold manufacture.[55]

The SLS process requires raw materials to be in the powder form and of appropriate sizes. Therefore, for constructing PHBV-based tissue engineering scaffolds via SLS, PHBV microspheres and Ca-P/PHBV nanocomposite microspheres were fabricated prior to scaffold fabrication. Ca-P nanoparticles were firstly synthesized by rapid mixing of a $Ca(NO_3)_2 \cdot 4H_2O$ acetone solution with a $(NH_4)_2HPO_4$ aqueous solution.[56] With sizes in the range of 10–30 nm, the Ca-P nanoparticles obtained were amorphous, as shown in Fig. 45.4, and had a Ca:P molar ratio of about 1.5, which is similar to that of tricalcium phosphate, a commonly used bioactive and biodegradable bioceramic for bone tissue repair. PHBV microspheres were produced using the oil-in-water (O/W) emulsion solvent evaporation method. The polymer was dissolved in chloroform at 50°C and then added to an aqueous phase containing 1% PVA (w/v). The resulting emulsion was stirred at 600 rpm at room temperature until the solvent evaporated. The microspheres obtained were filtered, washed, and freeze-dried. Similarly, Ca-P/PHBV nanocomposite microspheres containing Ca-P nanoparticles were fabricated using the S/O/W emulsion solvent evaporation method.[56]

Both PHBV and Ca-P/PHBV microspheres were spherical in shape, and the surface of PHBV microspheres had a rough-wrinkled morphology (Fig. 45.4). The roughness of the PHBV microsphere surface was probably caused by the relatively high crystallinity of the PHBV polymer. The Ca-P/PHBV nanocomposite microspheres produced had an average diameter of 46.34 μm, as was measured by a particle sizer, which was smaller than that of PHBV microspheres (53.18 μm). The decrease in size and a narrower size distribution of Ca-P/PHBV nanocomposite microspheres were probably due to Ca-P nanoparticles, which may have acted as a pickering stabilizer[57] (co-emulsifier) during the fabrication of Ca-P/PHBV microspheres. Similarly, Fujii *et al.* fabricated HA nanoparticle–coated PLLA via a "pickering-type" emulsion route in the absence of any normally used surfactants.[58] Materials with particle sizes in the range of 10–150 μm are preferred for the SLS process. Therefore, both PHBV and Ca-P/PHBV microspheres were of suitable sizes for the SLS technology. The Ca-P content of Ca-P/PHBV nanocomposite microspheres was

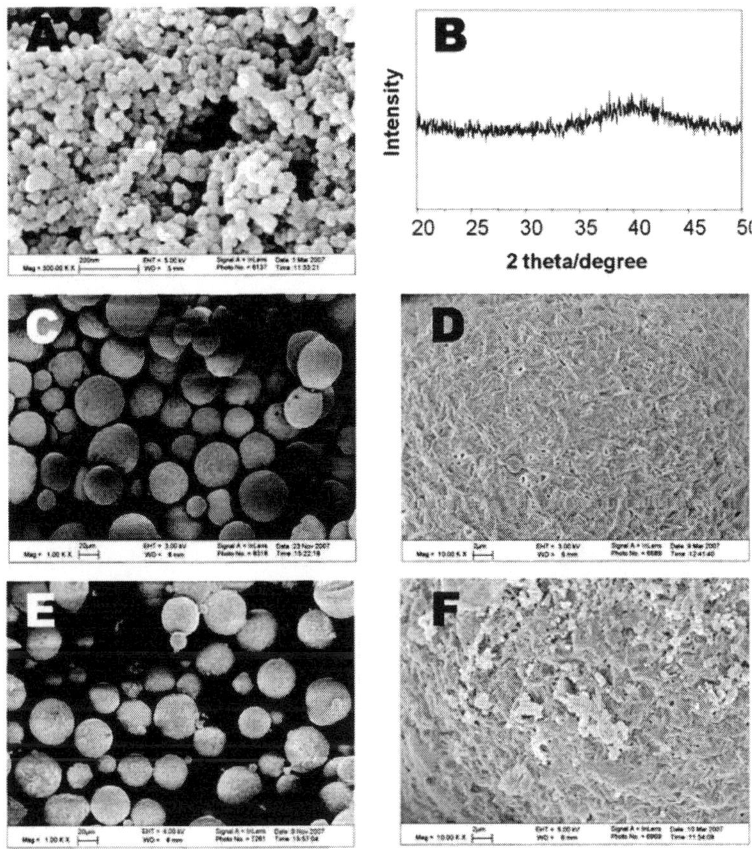

Figure 45.4. (a) SEM micrograph of Ca-P nanoparticles, (b) XRD pattern of Ca-P nanoparticles, (c, d) SEM micrographs of PHBV microspheres, (e, f) SEM micrographs of Ca-P/PHBV nanocomposite microspheres (c, e: general view; d, f close view). *Abbreviations*: SEM, scanning electron microscopy; XRD, X-ray diffraction.

12.9 wt% Ca-P (targeting at 15%), as was determined through thermalgravimetric analysis (TGA).

For fabricating scaffolds through SLS, scaffold models can be built using computer-based data that is obtained from medical imaging techniques such as computer tomography (CT) and magnetic resonance imaging (MRI) or designed using professional design software such as Solidworks®. Through careful design, accurate

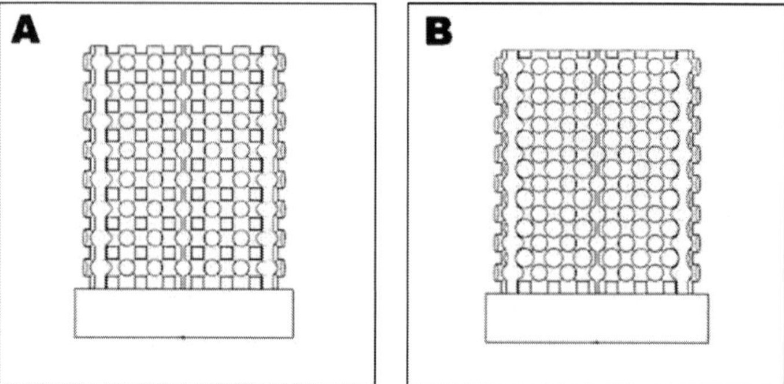

Figure 45.5. Front view of scaffold models: (a) scaffold with a square pore shape and (b) scaffold with a round pore shape.

scaffold models with a customized shape, complex architecture, and required pore size and porosity can be achieved. Figure 45.5 shows two scaffold models with different pore structures, which are designed by using Solidworks®. Different pore structures (with concave or convex pore surfaces) could have different effects on the cellular behavior.

45.4 Fabrication of Nanocomposite Scaffolds via SLS and Characteristics of the Scaffolds

Two porous scaffold models, namely, a bar-shaped scaffold model and a rod-shaped scaffold model, as shown in Fig. 45.6A,C, with 3D periodic architectures were designed using SolidWorks® in the current investigation.

To improve the quality of sintered scaffolds and facilitate scaffold handling, a solid base was incorporated in the scaffold design. The scaffold models consisted of a repeating array of struts. The bar-shaped model had a strut size of 0.5 mm and a pore size of 1.0 mm, and the rod-shaped model had a strut diameter of 1.0 mm and a pore size of 0.8 mm. Using the PHBV microspheres and Ca-P/PHBV nanocomposite microspheres, PHBV scaffolds and Ca-P/PHBV nanocomposite scaffolds, as shown in Fig. 45.6B,D, were successfully fabricated via SLS. The sintered scaffolds had good

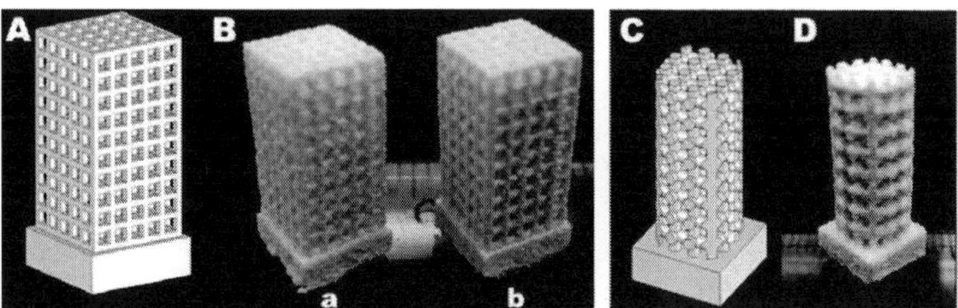

Figure 45.6. Bar-shaped and rod-shaped scaffold models and sintered scaffolds: (A) bar-shaped and (C) rod-shaped model in trimetric view; (B) sintered scaffolds, (a) PHBV scaffold and (b) Ca-P/PHBV nanocomposite scaffold, (D) sintered Ca-P/PHBV nanocomposite scaffold.

structure and handling stability, and hence loose microspheres entrapped in the sintered scaffolds could be easily removed. The macrostructure, including the height, width, and thickness, of sintered scaffolds could be well controlled by selecting carefully the values of various SLS parameters. However, it was not easy to achieve high accuracy for the pore size and strut size of scaffolds because of the growth effect in the SLS process as well as the limitation of resolution of the SLS machine used.[59]

The typical layer morphology of bar-shaped PHBV scaffolds and Ca-P/PHBV nanocomposite scaffolds is shown in Fig. 45.7. The SEM images indicated that the morphology and architecture of each layer of the scaffold were well preserved for both types of materials and the pores were clearly identified and comparable to the designed scaffold structure. The entrapped microspheres had been easily removed from the scaffolds by manual shaking. With the close examination of the strut surface, as shown in Fig. 45.7b,d, it could be seen that there was necking between adjacent microspheres and nearly intact microspheres without obvious fusion also existed. The presence of nearly intact microspheres could be explained as the sticking of microspheres onto the sintered strut due to very small melting of these microspheres caused by the heat generated during the SLS process. The porosity of the designed scaffold model was calculated to be 67.9% for the bar-shaped scaffold and 53.5% for the rod-shaped scaffold using the SolidWorks® software. For sintered scaffolds, the measured porosity values were $80.7 \pm 0.7\%$ for

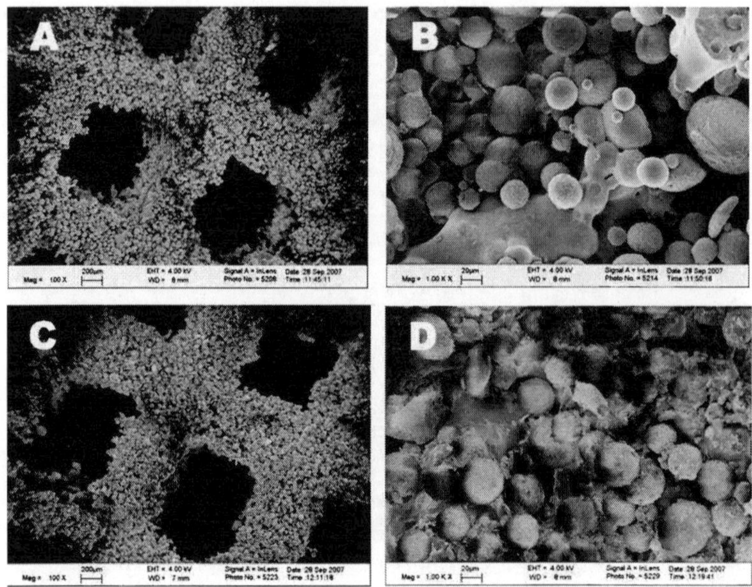

Figure 45.7. SEM micrographs of sintered scaffolds: (a) general view of PHBV scaffold, (b) close view of the strut surface of PHBV scaffold, (c) general view of Ca-P/PHBV scaffold, and (d) close view of the strut surface of Ca-P/PHBV scaffold.

bar-shaped nanocomposite scaffolds and 61.8±1.2% for rod-shaped nanocomposite scaffolds. It is obvious that the actual porosity values of all sintered scaffolds were higher than the theoretical porosity values, which is attributed to the presence of micropores on the strut surface of scaffolds.

The compressive strength and modulus of bar-shaped Ca-P/PHBV nanocomposite scaffolds were 0.24±0.02 MPa and 3.96±0.64 MPa, respectively, which were higher than those of PHBV scaffolds (0.19±0.02 MPa for compressive strength and 2.38±0.29 MPa for modulus). For rod-shaped Ca-P/PHBV nanocomposite scaffolds, the compressive strength and modulus increased to 2.20±0.18 MPa and 27.35±2.10 MPa, respectively. These results indicate that with increasing strut size and decreasing pore size, the mechanical properties of SLS-formed scaffolds could be improved. However, there is a limit for increasing the strut size and reducing the pore size for the simple reason of obtaining useful and usable tissue engineering scaffolds. The conventional norm of producing highly porous structures with interconnecting pores for tissue engineering has to be followed. And the pore size must be sufficiently large for cell migration.

45.5 Nanocomposite Scaffolds as Delivery Vehicles for Biomolecules

Despite the introduction of bioactive materials such as osteoconductive Ca-P nanoparticles into the polymer matrix, composite scaffolds alone may still not be sufficient to cause spontaneously bone healing and the regeneration of functional bone tissue, especially for large bone defects, due to the lack of osteoinductivity.[60] In order to further improve the bioactivity and functionality of scaffolds, biomolecules, especially growth factors, can be incorporated in bone tissue engineering scaffolds. Delivery vehicles for biomolecules can be made either by the attachment of biomolecules onto the surface of scaffolds, through non-covalent binding (surface adsorption, affinity binding, ionic complexation) or covalent binding (chemical conjugation), or by the physical entrapment of biomolecules inside scaffolds during the scaffold fabrication process.[61] The release rate and release kinetics of biomolecules from delivery vehicles are controlled by common release mechanisms, including diffusion, erosion, and swelling, subjected to the scaffold material, scaffold fabrication technique, scaffold architecture, and biomolecule incorporation method. In this chapter, our work on both entrapment and attachment of biomolecules within/onto sintered Ca-P/PHBV nanocomposite scaffolds is presented as examples of the two approaches.

For the encapsulation of biomolecules within sintered nanocomposite scaffolds, the biomolecules could be incorporated at the stage of making Ca-P/PHBV nanocomposite microspheres using a solid-in-water-in-oil-in-water (S/W/O/W) double-emulsion solvent evaporation method.[62] The double-emulsion method was chosen for the purpose of minimizing biomolecule denaturation, which could be caused by organic solvents that were used for making polymer solutions. Bovine serum albumin (BSA), instead of growth factors, was

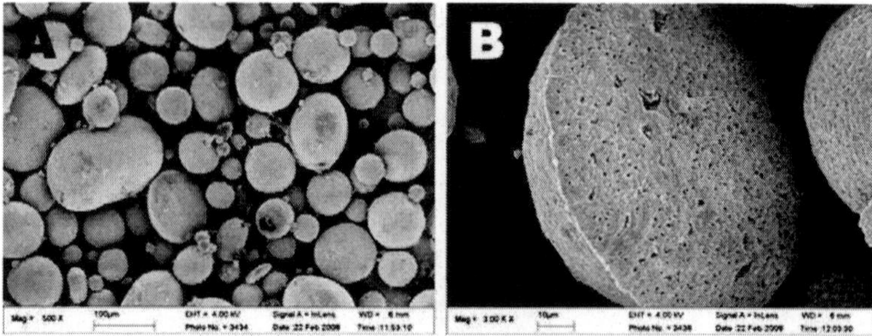

Figure 45.8. (a) Shape and morphology of BSA-loaded Ca-P/PHBV nanocomposite microspheres, and (b) cross-sectional view of BSA-loaded Ca-P/PHBV nanocomposite microsphere.

selected as a model biomolecule for the incorporation and release studies in this investigation because it is a well-characterized protein and is inexpensive for model studies. The surface morphology and internal structure of BSA-loaded Ca-P/PHBV nanocomposite microspheres were examined under SEM and are shown in Fig. 45.8. With the incorporation of BSA, small pores were observed on the cross sections of microspheres, as depicted in Fig. 45.8b, which were not observed in nanocomposite microspheres without BSA loading.

The actual BSA loading in Ca-P/PHBV nanocomposite microspheres, as was determined using a Micro BCA assay kit, was 6.13 ± 0.15 μg/mg, and the BSA encapsulation efficiency (EE) for Ca-P/PHBV nanocomposite microspheres was $24.51 \pm 0.60\%$. The low EE was mainly attributed to the inherent limited chemical and physical stability of BSA and relatively harsh microsphere fabrication process for BSA. The widely known reasons for low EE of biomolecules in polymer microspheres obtained via emulsion processes include (i) the use of organic solvents during the double-emulsion process, (ii) biomolecule elution from the inner water phase into the outer water phase during the solvent evaporation process, (iii) exposure of the biomolecules to high shear forces during emulsion preparation, and (iv) generation of organic-aqueous interfaces. These reasons could also be valid in the current investigation. In addition, the hydrophobility of PHBV due to its high crystallinity weakened the interaction between the PHBV polymer and BSA and thus further decreased the EE value.

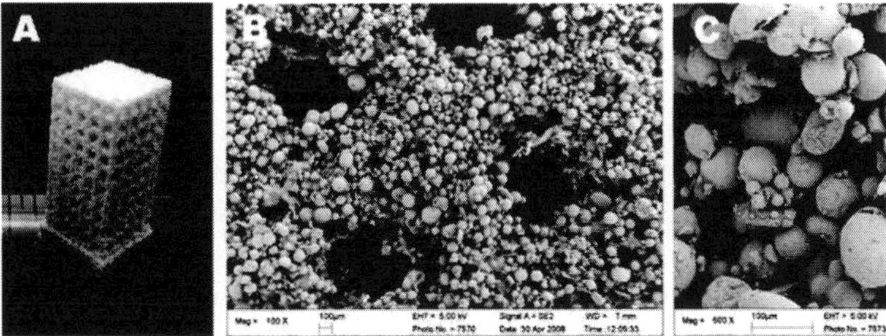

Figure 45.9. (a) BSA-loaded Ca-P/PHBV nanocomposite scaffold produced via SLS, (b) morphology of one layer of sintered scaffolds, and (c) close view of the strut surface.

The BSA-loaded Ca-P/PHBV microspheres were subsequently subjected to SLS for scaffold fabrication. The bar-shaped tetragonal porous scaffold model, as shown in Fig. 45.6a, was used in scaffold production. During the sintering process, the part bed temperature was fixed at 35°C, the scan speed at 1,257 mm/s, and the roller speed at 127 mm/s. The layer thickness was kept constant at 0.15 mm for all scaffolds according to the optimization results obtained in our previous SLS experiments. The laser power and scan spacing were set at 12.5 W and 0.10 mm/s, respectively. Fig. 45.9a shows BSA-loaded Ca-P/PHBV nanocomposite scaffolds produced via SLS. The typical layer morphology of BSA-loaded Ca-P/PHBV nanocomposite scaffold is presented as Fig. 45.9b. Once again, the SEM images indicated that the morphology of each layer of sintered scaffolds was well preserved and that the pores were clearly identified and comparable to the designed scaffold model. A close observation of the strut surface, as shown in Fig. 45.9c, revealed necking among adjacent microspheres. The loaded BSA in fused microspheres may lose some bioactivity because the microspheres were subjected to high thermal energy during sintering. The actual BSA loading level and EE for sintered Ca-P/PHBV scaffolds, as were calculated on the basis of the amount of initially applied BSA during microsphere fabrication, were 2.96 ± 0.15 μg/mg and 13.63 ± 0.71%, respectively. The low EE value was due not only to the denaturation of BSA during the laser sintering process but also to the low EE of Ca-P/PHBV nanocomposite microspheres fabricated. Experimental results showed that after

SLS, 55.63±2.89% of BSA encapsulated in the microspheres could maintain its bioactivity without denaturation.

For studying the *in vitro* release behavior of BSA, BSA-loaded microspheres and scaffolds were placed in centrifuge tubes, with each tube being filled with 5 mL of phosphate buffer saline (PBS, pH7.4) containing 0.05 wt% of sodium azide. The tubes were placed in a shaking water bath, which was maintained at 37°C and shaken horizontally at 30 rpm, for up to 28 days. At preset times, 2 mL of the supernatant was withdrawn from each tube after centrifugation to determine the amount of BSA released by using a Micro BCA assay kit and 2 mL fresh medium was replenished for each tube. Figure 45.10 shows *in vitro* release profiles of BSA from Ca-P/PHBV microspheres and scaffolds in relation to the total amount of encapsulated BSA and in relation to the carrier, respectively. As could be seen, an initial burst release occurred, followed by a low-rate release behavior for all test samples. After 28 days, approximately 85% cumulative release of BSA from Ca-P/PHBV nanocomposite scaffolds was achieved.

Differing from the traditional approach, which uses dry-blended polymeric particles with drug powders for SLS-formed scaffolds, in this investigation, the biomolecules were encapsulated inside PHBV-based microspheres and the microspheres were expected to protect biomolecules from high heat caused by SLS during scaffold fabrication and to control the release behavior of biomolecules *in vitro* and *in vivo*. However, although BSA could be successfully encapsulated within Ca-P/PHBV nanocomposite scaffolds and released *in vitro* in a relatively controlled manner, the EE in both BSA-loaded microspheres and scaffolds was relatively low. Therefore, other strategies need to be considered and investigated for achieving SLS-formed, multifunctional bone tissue engineering scaffolds.

Immobilizing biomolecules on the carrier surface and releasing them in a controlled manner has been appealing to many researchers and can be an effective way for the controlled delivery of growth factors from SLS-formed scaffolds. For Ca-P/PHBV nanocomposite scaffolds, this strategy requires surface modification of the scaffolds for introducing binding sites for growth factors.

The surface modification of sintered rod-shaped Ca-P/PHBV scaffolds was conducted in two steps: (1) physical entrapment of

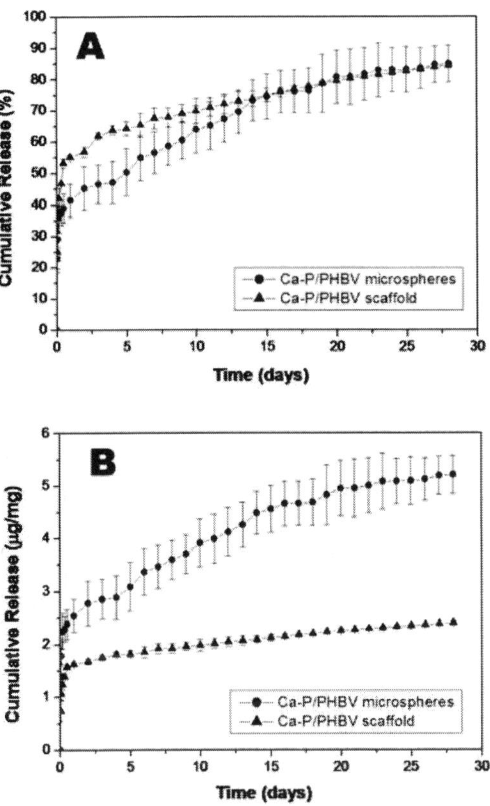

Figure 45.10. Cumulative release profiles of BSA from sintered Ca-P/PHBV scaffolds: (a) in relation to the total amount of encapsulated BSA and (b) in relation to the mass of the scaffold.

gelatin and (2) heparin immobilization,[59] as schematically illustrated in Fig. 45.11. For the physical entrapment of gelatin, gelatin was first dissolved in a miscible mixture of 2,2,2-trifluoroethanol (TFE) and distilled water (TFE:water = 30:70). TFE and water are solvent and non-solvent, respectively, for the PHBV matrix. Sintered Ca-P/PHBV scaffolds were immersed in the gelatin solution of 5 mg/mL concentration at room temperature for six hours. They were then removed for the gelatin solution and quickly soaked in cold distilled water. After this treatment, gelatin molecules were physically entrapped onto the surface of Ca-P/PHBV nanocomposite scaffolds.

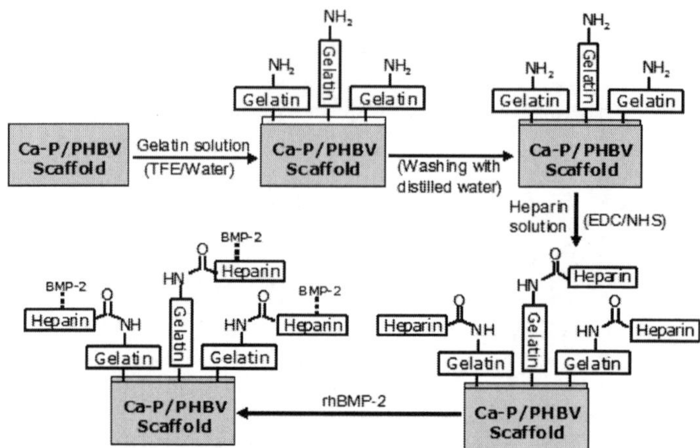

Figure 45.11. Illustration of physical entrapment of gelatin and immobilization of heparin on the surface of Ca-P/PHBV nanocomposite scaffolds and the binding strategy for rhBMP-2.

The amount of entrapped gelatin was measured by the BCA method, which is an effective way to determine the amount of the immobilized proteins on the surface of a biomaterial, and determined to be 1955.8±62.1 μg per scaffold.

For heparin immobilization, gelatin-entrapped Ca-P/PHBV nanocompoiste scaffolds were first prewetted in 2-(N-morpholino) ethanesulfonic acid (MES) buffer solution (0.1 M, pH5.6) for 30 minutes at room temperature. Two milligrams of heparin were dissolved in 1 mL MES buffer solution (0.1 M, pH5.6) containing 2 mg N-hydroxysuccinimide (NHS) and 1.2 mg 1-ethyl-3-(3-dimethylaminopropyl)-carbodiimide hydrochloride (EDC). After four-hour activation of the heparin solution at room temperature, the prewetted scaffolds were soaked in the activated heparin solution in MES buffer for another four hours and then extensively washed with PBS and dried overnight at room temperature. The amount of heparin conjugated to the surface of gelatin-entrapped nanocomposite scaffolds, as was determined using the toluidine blue method,[63] was 41.78±0.39 μg heparin per scaffold.

The gelatin entrapment improved the hydrophilicity of PHBV matrix and achieved a hydrophilicity-hydrophobicity balance without significantly affecting the surface morphology and mechanical properties of Ca-P/PHBV nanocomposite scaffolds. Moreover, it provided the scaffolds with plenty of amino and carboxyl groups (originated from the gelatin molecules), onto which other biomolecules or ligands can be covalently coupled. Based on carbodiimide chemistry, the entrapped gelatin provided amino groups for the conjugation of heparin. Heparin, a sulfated polysaccharide belonging to the glycosaminoglycans family, is known to have binding affinity with a number of growth factors and is capable of blocking the degradation of the growth factors and prolonging their release time.[64] Therefore, in many studies, heparin was immobilized onto the surface of various scaffolds for binding vascular endothelial growth factor (VEGF), bone morphogenetic protein (BMP), and basic fibroblast growth factor (bFGF).[65–67] Based on the specific affinity between heparin and growth factors, the bound growth factors would be released in a sustained manner and would improve angiogenesis or osteogenesis. The surface modification by entrapment of gelatin and immobilization of heparin significantly improved the wettability of Ca-P/PHBV nanocomposite scaffolds. However, the morphology and mechanical properties of Ca-P/PHBV nanocomposite scaffolds were not affected by this two-step surface modification process.[68]

For the incorporation of recombinant human bone morphogenetic protein (rhBMP)-2, 5 μg of rhBMP-2 was loaded onto each Ca-P/PHBV scaffold sample with or without surface modification by dripping 50 μL of rhBMP-2 solutions with 20 mM glacial acetic acid and 0.1% BSA. After incubation for one hour at room temperature, the rhBMP-2-loaded scaffolds were placed in centrifuge tubes and suspended in 2 mL of PBS (per tube) containing 0.1% BSA and 0.02% sodium azide. The amount of the released rhBMP-2 in the collected medium was determined at preset times by using a human BMP-2 enzyme-linked immunosorbent assay (ELISA) development kit. The *in vitro* release profiles of rhBMP-2 from Ca-P/PHBV nanocomposite scaffolds with and without surface modification are shown in Fig. 45.12. Both types of scaffolds showed an initial burst release with a subsequent sustained release over the 28-day test period. Approximately 2276.11$\pm$66.95 ng rhBMP-2 of cumulative release from surface-modified Ca-P/PHBV scaffolds was observed

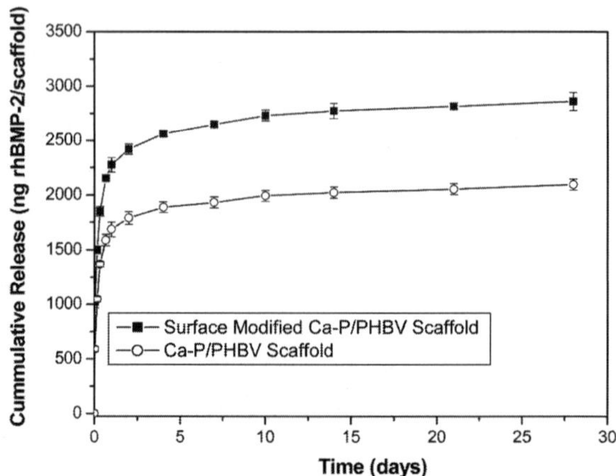

Figure 45.12. Cumulative release of rhBMP-2 from Ca-P/PHBV scaffolds with and without surface modification.

after the 1-day release time, while about 1686.71±66.07 ng rhBMP-2 of cumulative release was found for Ca-P/PHBV scaffolds without surface modification at the same release time.

The surface-modified Ca-P/PHBV nanocomposite scaffolds with heparin immobilization exhibited a higher initial burst release and delivered rhBMP-2 in a more sustained manner than scaffolds without surface modification. Similar sustained release behaviors for growth factors such as BMP-2, VEGF, bFGF, and platelet-derived growth factor (PDGF) were observed for heparin-conjugated microspheres or scaffold delivery systems.[65,69,70] After the 28-day test period, totally about 2860.79±82.54 ng rhBMP-2 was released from surface-modified Ca-P/PHBV scaffolds, which was significantly more than the amount released from Ca-P/PHBV scaffolds without surface modification (2095.90±51.48 ng).

Surface-modified Ca-P/PHBV nanocomposite scaffolds with immobilized heparin showed better sustained release behavior in the *in vitro* release process than scaffolds without surface modification, indicating good binding affinity between heparin and rhBMP-2 molecules. Furthermore, much more rhBMP-2 was finally released from surface modified scaffolds than from scaffolds without surface modification, with the protection of rhBMP-2 by the immobilized heparin being one of the contributing factors. Therefore, heparin immobilization on the surface of Ca-P/PHBV nanocomposite scaffolds not only provided a means to protect the loaded rhBMP-2 but also improved the sustained release behavior of rhBMP-2. The sustained release of growth factors in a control manner and at the desired concentration is very important for stimulating bone tissue regeneration.

45.6 Conclusions

This chapter focused on investigating the potentials of SLS in constructing multifunctional nanocomposite scaffolds for bone tissue engineering. It was demonstrated that 3D scaffolds with well-defined porous structures can be successfully produced through SLS by judiciously choosing the SLS parameters. Mimicking the hierarchical structure of bone, the composite approach can be adopted for making osteoconductive nanocomposite scaffolds for bone tissue regeneration. The new strategy of using nanocomposite microspheres instead of dry-blended bioceramic-polymer powders as raw materials for scaffold fabrication via SLS can ensure the quality of scaffolds and also a homogeneous distribution of osteoconductive bioceramic nanoparticles in the scaffolds. It may also provide an effective way for incorporating heat-insensitive drugs in scaffolds for their controlled *in vivo* release.

Multifunctionality is desired for tissue engineering scaffolds. The Ca-P/PHBV nanocomposite scaffolds formed through SLS in the current investigation can serve as delivery vehicles for biomolecules such as growth factors. Two strategies, namely, encapsulation of biomolecules within scaffolds and attachment of biomolecules onto the surface of scaffolds through non-covalent binding, can be employed. For the encapsulation of biomolecules within nanocomposite scaffolds, as a demonstration, BSA-loaded Ca-P/PHBV nanocomposite microspheres were first fabricated using the double-emulsion solvent evaporation method. The BSA-loaded nanocomposite microspheres were then processed via SLS into 3D porous scaffolds with good dimensional accuracy, while retaining the bioactivity of BSA. The BSA-loaded nanocomposite scaffolds displayed a controlled *in vitro* release of BSA, although the BSA loading level and EE were low for the scaffolds produced in the current investigation. For the attachment of biomolecules onto scaffolds through non-covalent binding, surface modification of Ca-P/PHBV nanocomposite scaffolds by physical entrapment of gelatin and heparin immobilization is needed in order to provide the binding site for the growth factor rhBMP-2. rhBMP-2 can be loaded onto surface-modified Ca-P/PHBV scaffolds, and the immobilized heparin not only provides a means to protect the loaded rhBMP-2 but also improves the sustained release behavior of rhBMP-2. The integration of advanced scaffold fabrication technology and nanocomposite and growth factor delivery can be a very effective strategy for bone tissue regeneration.

Acknowledgments

This work was supported by the University of Hong Kong (HKU) through a research grant in its Basic Research Programme and by the Hong Kong Research Grants Council (RGC) through GRF grants (HKU 7182/05E and 7181/09E). Assistance provided by staff and students in HKU's Department of Mechanical Engineering and Department of Orthopaedics and Traumatology is gratefully acknowledged. Prof. C.-S. Liu of East China University of Science and Technology, China, is thanked for providing rhBMP-2.

References

1. P. H. G. Chao, W. Grayson, and G. Vunjak-Novakovic, *J. Orthop. Sci.*, **12**, 398–404 (2007).
2. K. J. L. Burg, S. Porter, and J. F. Kellam, *Biomaterials*, **21**, 2347 (2000).
3. C. Laurencin, Y. Khan, and S. F. El-Amin, *Exp. Rev. Med. Dev.*, **3**, 49 (2006).
4. P. X. Ma, in Ed. J. I. Kroschwitz, *Encyclopedia of Polymer Science and Technol.* (John Wiley & Sons, Hoboken, NJ, 2004), p. 261.
5. L. G. Griffith and G. Naughton, *Science*, **295**, 1009 (2002).
6. S. J. Hollister, *Biofabrication*, **1**, 1 (2009).
7. L. Lu, S. J. Peter, M. D. Lyman, H. L. Lai, S. M. Leite, J. A. Tamada, S. Uyama, J. P. Vacanti, R. Langer, and A. G. Mikos, *Biomaterials*, **21**, 1837 (2000).
8. Q. Yang, L. Chen, X. Y. Shen, and Z. Q. Tan, *J. Macromol. Sci. Part B-Phys.*, **45**, 1171 (2006).

9. H. N. Wang, Y. B. Li, Y. Zuo, J. H. Li, S. S. Ma, and L. Cheng, *Biomaterials*, **28**, 3338 (2007).
10. A. J. Salgado, O. P. Coutinho, and R. L. Reis, *Macromol. Biosci.*, **4**, 743 (2004).
11. H. W. Tong and M. Wang, *J. Nanosci. Nanotechnol.*, **11**, 3834 (2006).
12. J. Lannutti, D. Reneker, T. Ma, D. Tomasko, and D. Farson, *Mater. Sci. Eng. C-Biomimetic Supramol. Sys.*, **27**, 504 (2007).
13. Y. Zuo, F. Yang, J. G. C. Wolke, Y. B. Li, and J. A. Jansen, *Acta Biomaterialia*, **6**, 1238 (2010).
14. Y. Yang, X. L. Zhu, W. G. Cui, X. H. Li, and Y. Jin, *Macromol. Mater. Eng.*, **294**, 611 (2009).
15. J. W. Xie, M. R. MacEwan, A. G. Schwartz, and Y. N. Xia, *Nanoscale*, **2**, 35 (2010).
16. S. H. Oh, S. G. Kang, and J. H. Lee, *J. Mater. Sci.-Mater. Med.*, **17**, 131 (2006).
17. T. Jiang, W. I. Abdel-Fattah, and C. T. Laurencin, *Biomaterials*, **27**, 4894 (2006).
18. J. L. Brown, L. S. Nair, and C. T. Laurencin, *J. Biomed. Mater. Res. Part B-Applied Biomaterials* **86B**, 396 (2008).
19. M. E. Gomes, A. S. Ribeiro, P. B. Malafaya, R. L. Reis, and A. M. Cunha, *Biomaterials*, **22**, 883 (2001).
20. X. H. Liu and P. X. Ma, *Biomaterials*, **30**, 4094 (2009).
21. Y. Cao, T. I. Croll, A. J. O'Connor, G. W. Stevens, and J. J. Cooper-White, *J. Biomater. Sci.-Polym. Ed.*, **17**, 369 (2006).
22. B. Sundaray, V. Subramanian, T. S. Natarajan, R. Z. Xiang, C. C. Chang, and W. S. Fann, *Appl. Phys. Lett.*, **84**, 1222 (2004).
23. K. F. Leong, C. M. Cheah, and C. K. Chua, *Biomaterials*, **24**, 2363 (2003).
24. S. F. Yang, K. F. Leong, Z. H. Du, and C. K. Chua, *Tissue Eng.*, **8**, 1 (2002).
25. B. Borah, G. J. Gross, T. E. Dufresne, T. S. Smith, M. D. Cockman, P. A. Chmielewski, M. W. Lundy, J. R. Hartke, and E. W. Sod, *Anat. Rec.*, **265**, 101 (2001).
26. S. M. Peltola, F. P. W. Melchels, D. W. Grijpma, and M. Kellomaki, *Ann. Med.*, **40**, 268 (2008).
27. S. J. Hollister, *Adv. Mater.*, **21**, 3330 (2009).
28. J. P. Fisher, D. Dean, P. S. Engel, and A. G. Mikos, *Ann. Rev. Mater. Res.*, **31**, 171 (2001).
29. F. P. W. Melchels, J. Feijen, and D. W. Grijpma, *Biomaterials*, **24**, 6121 (2010).

30. B. Dhariwala, E. Hunt, and T. Boland, *Tissue Eng.*, **10**, 1316 (2004).
31. K. Arcaute, B. K. Mann, and R. B. Wicker, *Ann. Biomed. Eng.*, **34**, 1429 (2006).
32. J. W. Lee, P. X. Lan, B. Kim, G. Lim, and D. W. Cho, *Microelectr. Eng.*, **84**, 1702 (2007).
33. I. Zein, D. W. Hutmacher, K. C. Tan, and S. H. Teoh, *Biomaterials*, **23**, 1169 (2002).
34. A. Yeo, B. Rai, E. Sju, J. J. Cheong, and S. H. Teoh, *J. Biomed. Mater. Res. - Part A*, **84**, 208 (2008).
35. E. M. Sachs, J. S. Haggerty, M. J. Cima, and P. A. Williams, US Patent No. 5204055 (1993).
36. W. S. Koegler and L. G. Griffith, *Biomaterials*, **25**, 2819 (2004).
37. B. Y. Tay, S. X. Zhang, M. H. Myint, F. L. Ng, M. Chandrasekaran, and L. K. A. Tan, *J. Mater. Proc. Technol.*, **182**, 117 (2007).
38. S. S. Kim, H. Utsunomiya, J. A. Koski, B. M. Wu, M. J.Cima, J. Sohn, K. Mukai, L. G. Griffith, and J. P. Vacanti, *Ann. Surg.*, **228**, 8 (1998).
39. C. X. F. Lam, X. M. Mo, S. H. Teoh, and D. W. Hutmacher, *Mater. Sci. & Eng. C-Biomimetic Supramol. Sys.*, **20**, 49 (2002).
40. X. H. Wang, Y. N. Yan, and R. J. Zhang, *Trends Biotechnol.*, **25**, 505 (2007).
41. W. Xu, X. H. Wang, Y. N. Yan, and R. J. Zhang, *J. Bioactive Compat. Polym.*, **23**, 103 (2008).
42. T. K. Cui, Y. N. Yan, R. J. Zhang, L. Liu, W. Xu, and X. H. Wang, *Tissue Eng.: Part C*, **15**, 1 (2009).
43. T. Weiss, G. Hildebrand, R. Schade, and K. Liefeith, *Eng. Life Sci.*, **9**, 384 (2009).
44. J. P. Kruth, P. Mercelis, J. Van Vaerenbergh, L. Froyen, and M. Rombouts, *Rapid Prototyp. J.*, **11**, 26 (2005).
45. J. T. Rimell and P. M. Marquis, *J. Biomed. Mater. Res.*, **53**, 414 (2000).
46. M. Schmidt, D. Pohle, and T. Rechtenwald, *Cirp Ann.-Manufac. Technol.*, **56**, 205 (2007).
47. K. H. Tan, C. K. Chua, K. F. Leong, C. M. Cheah, W. S. Gui, W. S. Tan, and F. E. Wiria, *Bio-Med. Mater. Eng.*, **5**, 113 (2005).
48. L. Hao, M. M. Savalani, Y. Zhang, K. E. Tanner, R. J. Heath, and R. A. Harris, *Proc. Royal Soc. A-Math. Phys. Eng. Sci.*, **463**, 1857 (2007).
49. K. H. Tan, C. K. Chua, K. F. Leong, C. M. Cheah, P. Cheang, M. S. Abu Bakar, and S. W. Cha, *Biomaterials*, **24**, 3115 (2003).

50. Y. Zhang, L. Hao, M. M. Savalani, R. A. Harris, L. Di Silvio, and K. E. Tanner, *J. Biomed. Mater. Res. Part A*, **91A**, 1018 (2009).
51. M. A. Woodruff and D. W. Hutmacher, *Progress in Polymer Science*, **35**, 1217 (2011).
52. R. L. Simpson, F. E. Wiria, A. A. Amis, C. K. Chua, K. F. Leong, U. N. Hansen, M. Chandraselkaran, and M. W. Lee, *J. Biomed. Mater. Res. Part B-Appl. Biomater.*, **84B**, 17 (2008).
53. W. Y. Zhou, M. Wang, W. L. Cheung, B. C. Gao, and D. M. Jia, *J. Mater. Sci.: Mater. Med.*, **19**, 103 (2008).
54. W. Y. Zhou, S. H. Lee, M. Wang, W. L. Cheung, and W. Y. Ip, *J. Mater. Sci.: Mater. Med.*, **19**, 2535 (2008).
55. F. E. Wiria, N. Sudarmadji, K. F. Leong, C. K. Chua, E. W. Chng, and C. C. Chan, *Rapid Prototyp. J.*, **16**, 90 (2010).
56. B. Duan, M. Wang, W. Y. Zhou, and W. L. Cheung, *Appl. Surf. Sci.*, **255**, 529 (2008).
57. G. Lagaly, M. Reese, and S. Abend, *Appl. Clay Sci.*, **14**, 83 (1999).
58. S. Fujii, M. Okada, H. Sawa, T. Furuzono, and Y. Nakamura, *Langmuir*, **25**, 9759 (2009).
59. B. Duan and M. Wang, *J. Royal Soc. Interface*, **7**, S615 (2010).
60. T. A. Holland and A. G. Mikos, *Adv. Biochem. Eng./Biotechnol.*, **102**, 161 (2006).
61. S. H. Lee and H. Shin, *Adv. Drug Deliv. Rev.*, **59**, 339 (2007).
62. B. Duan and M. Wang, *Polym. Degrad. Stab.*, **95**, 1655 (2010).
63. H. J. Chung, H. K. Kim, J. J. Yoon, and T. G. Park, *Pharm. Res.*, **23**, 1835 (2006).
64. I. Capila and R. J. Linhardt, *Angewandte Chmie - Int. Ed.*, **41**, 391 (2002).
65. Y. C. Ho, F. L. Mi, H. W. Sung, and P. L. Kuo, *Int. J. Pharm.*, **376**, 69 (2009).
66. D. Bezuidenhout, N. Davies, M. Black, C. Schmidt, A. Oosthuysen, and P. Zilla, *J. Biomater. Appl.*, **24**, 401 (2010).
67. L. Chen, Z. Q. He, B. Chen, M. J. Yang, Y. N. Zhao, W. J. Sun, Z. F. Xiao, J. Zhang, and J. W. Dai, *J. Mater. Sci.-Mater. Med.*, **21**, 309 (2010).
68. B. Duan, M. Wang, Z. Y. Li, W. C. Chan, and W. W. Lu, *Frontiers Mater. Sci.*, **5**, 57 (2011).
69. O. Jeon, S. J. Song, S. W. Kang, A. J. Putnam, and B. S. Kim, *Biomaterials*, **28**, 2763 (2007).
70. B. Sun, B. Chen, Y. N. Zhao, W. J. Sun, K. S. Chen, J. Zhang, Z. L. Wei, Z. F. Xiao, and J. W. Dai, *J. Biomed. Mater. Res. Part B-Appl. Biomater.*, **91B**, 366 (2009).

Index

Color Insert

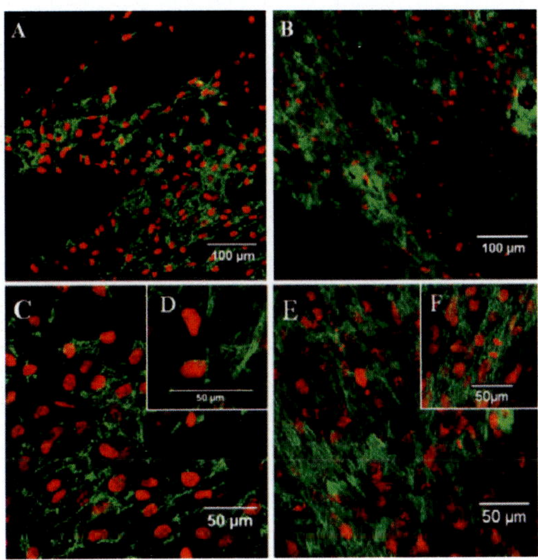

Figure 2.6

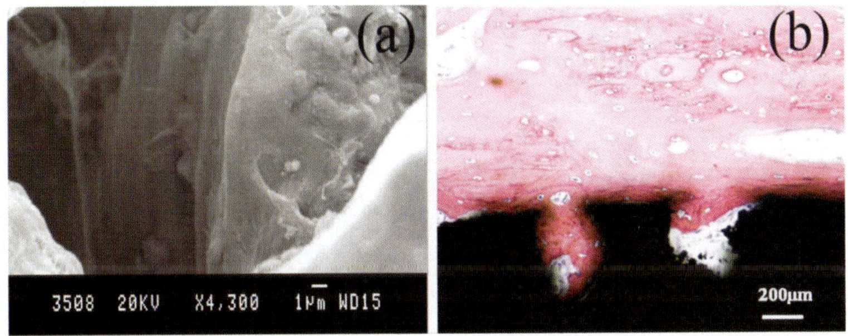

Figure 3.6

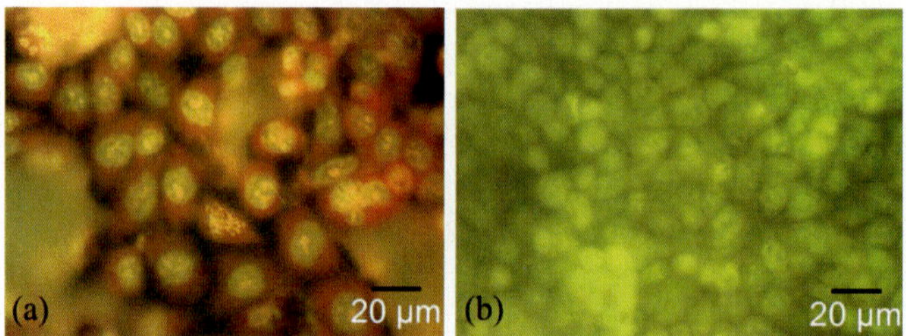

Figure 3.10

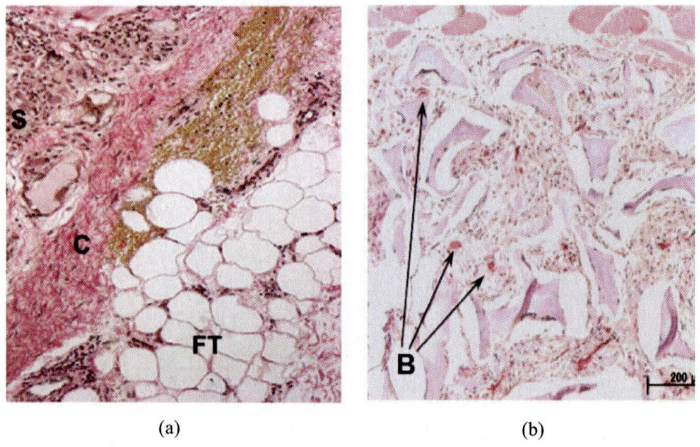

Figure 5.3

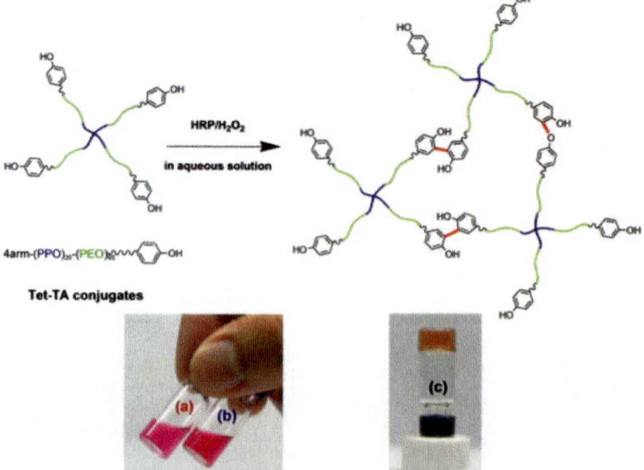

Figure 6.2

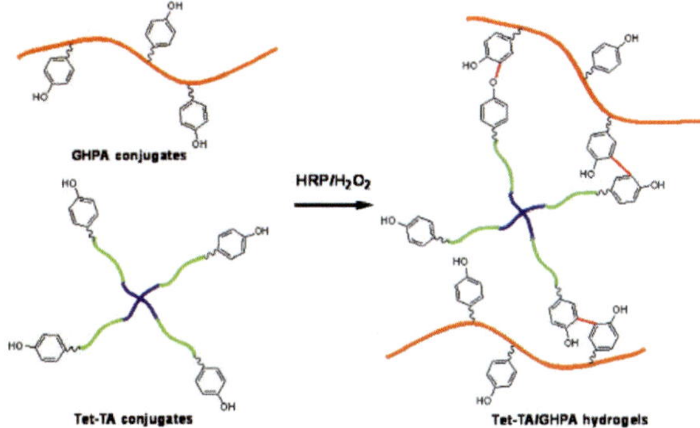

Figure 6.4

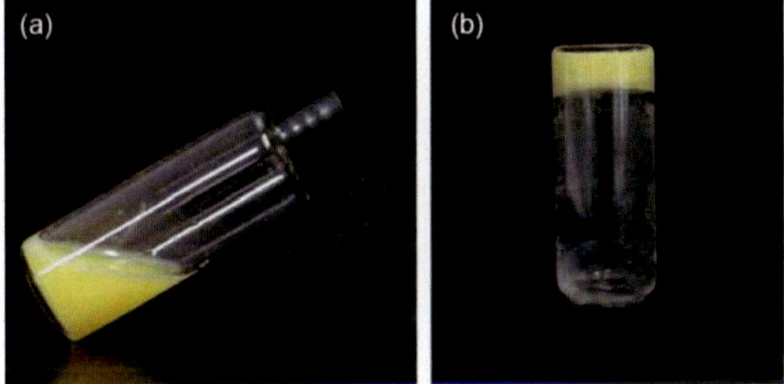

Figure 7.1

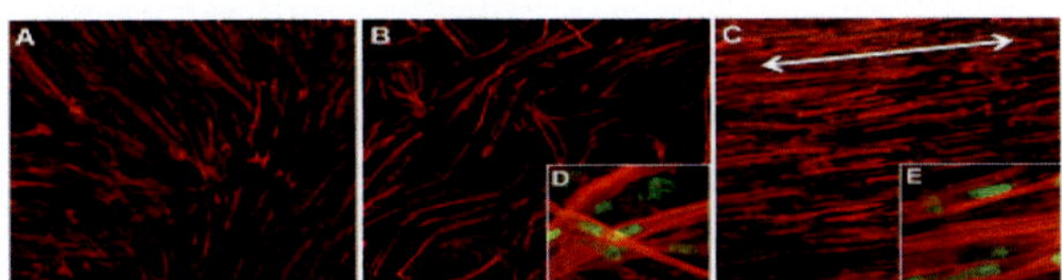

Figure 10.5

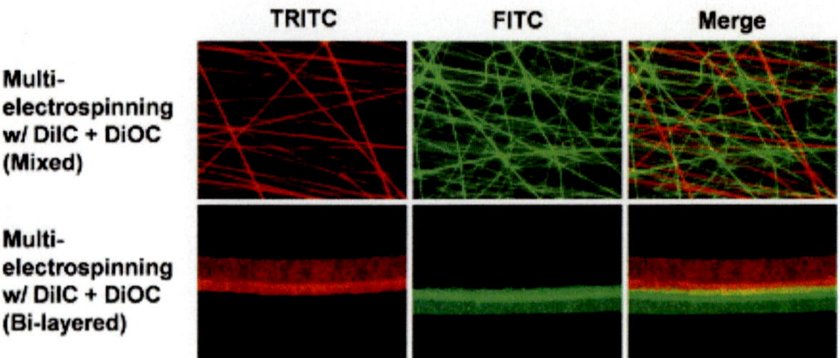

Figure 10.7

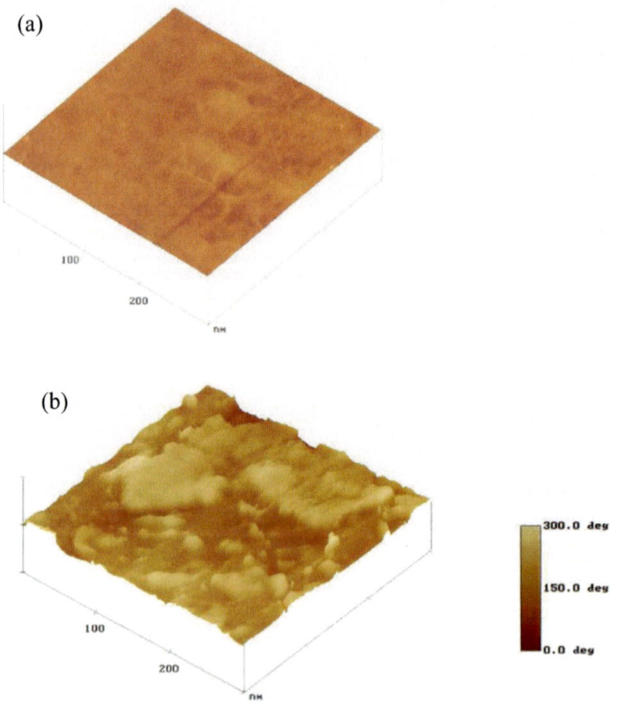

Figure 13.3

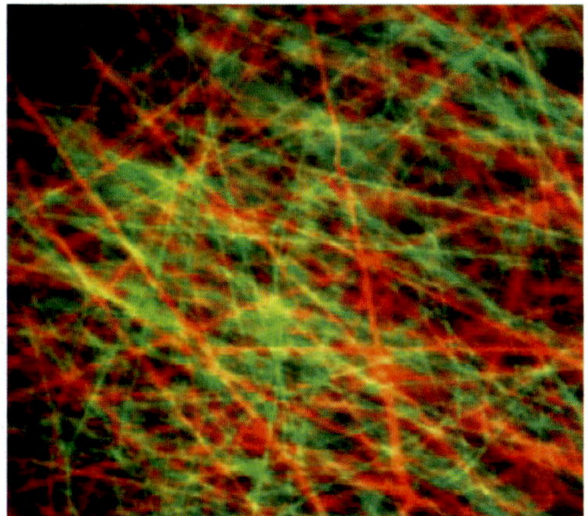

Figure 15.3

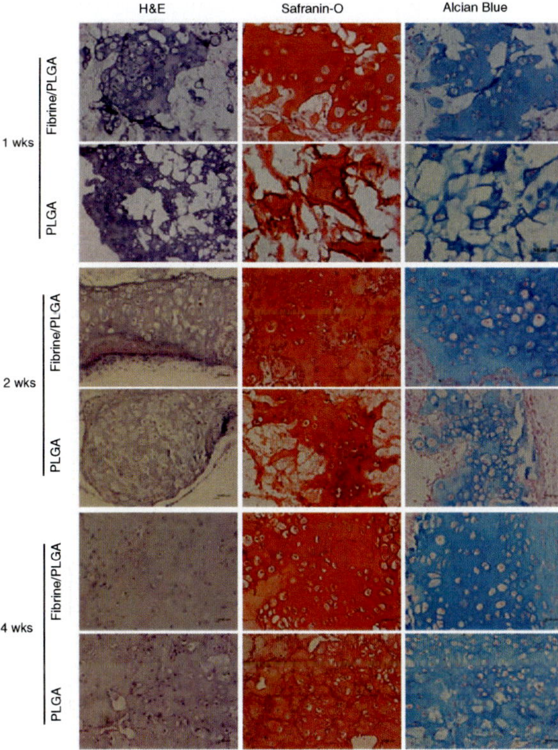

Figure 16.2

Collagen type II
Collagen type I
1 wks
Fibrine/PLGA
PLGA
2 wks
Fibrine/PLGA
PLGA
4 wks
Fibrine/PLGA
PLGA

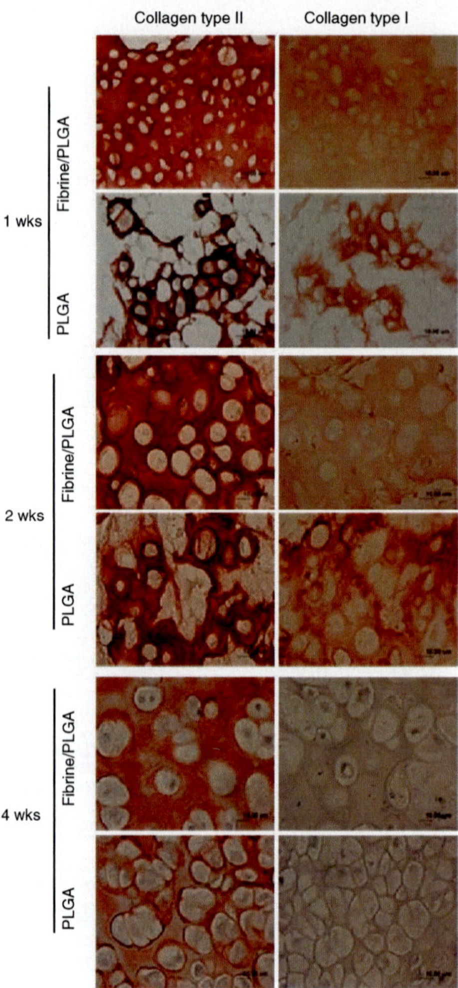

Figure 16.3

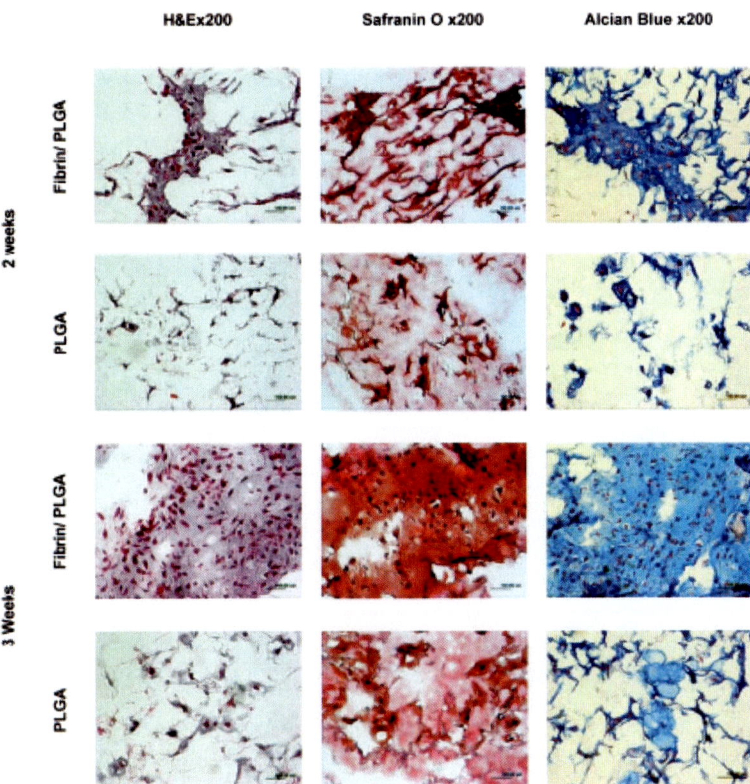

Figure 16.5

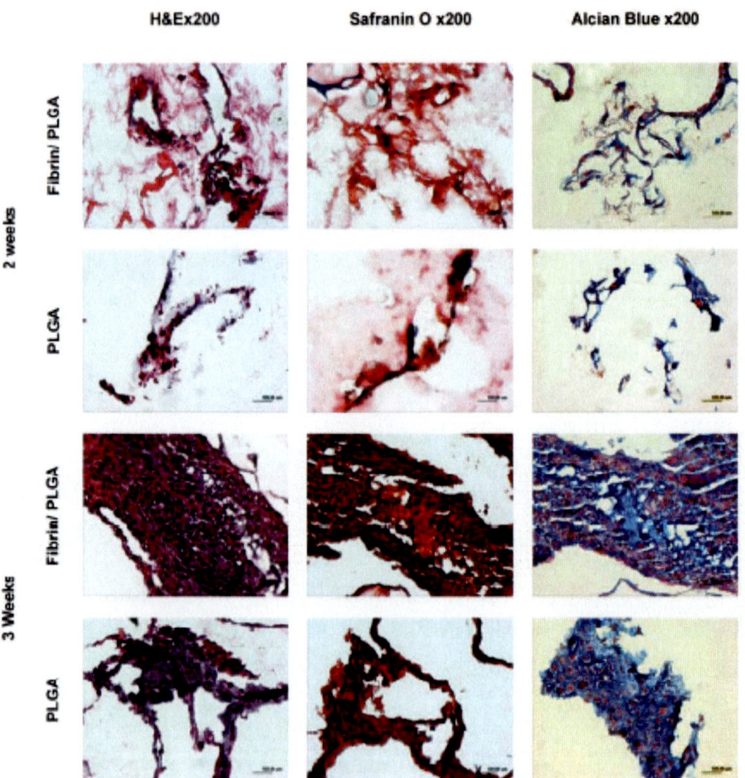

Figure 16.6

A
B
C
D
E
F
100 micrometer
50 micrometer
50 micrometer
50 micrometer
50 micrometer
50 micrometer

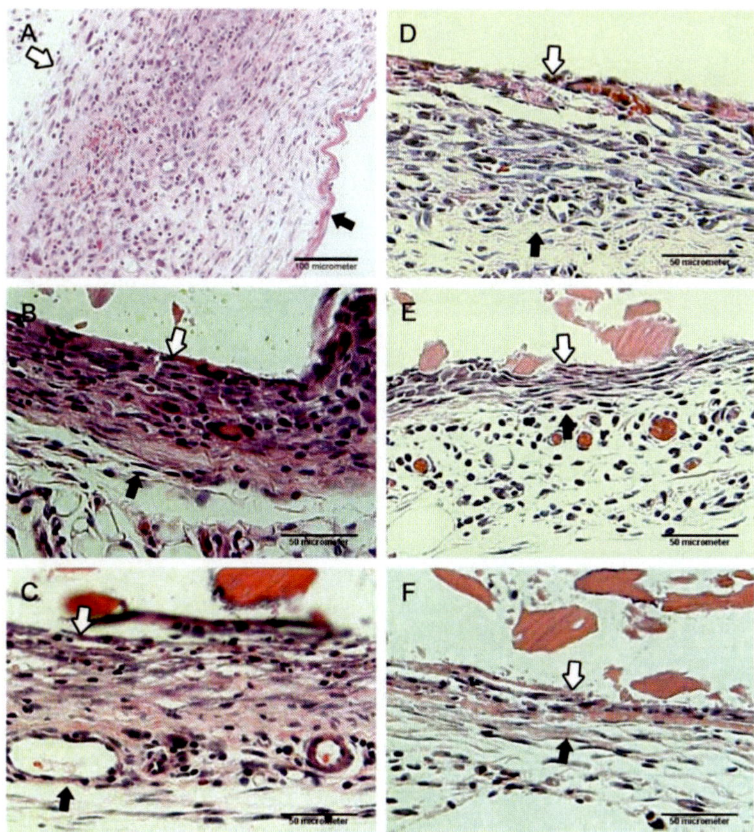

Figure 16.14

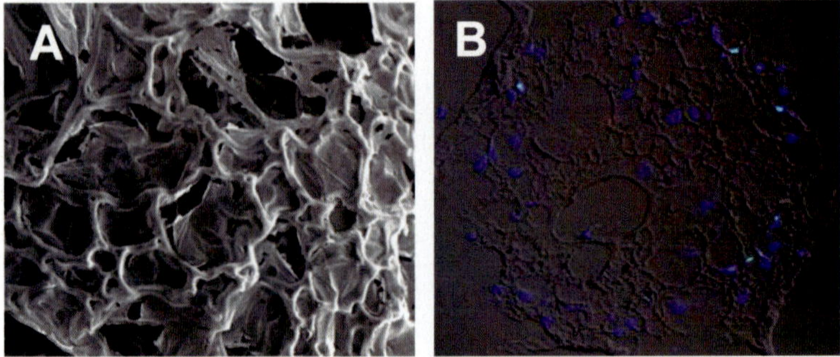

Figure 18.3

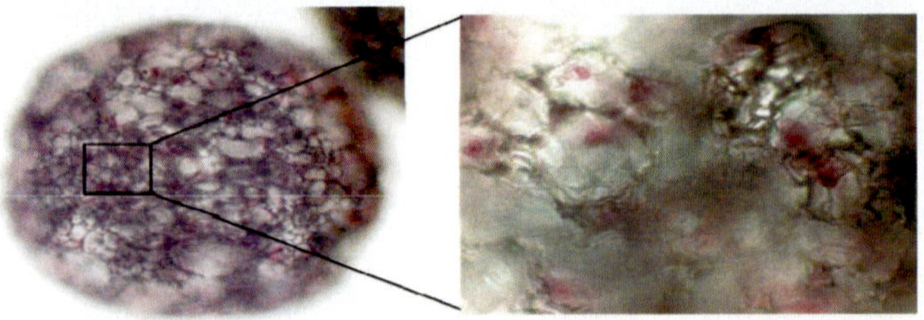

Figure 18.4

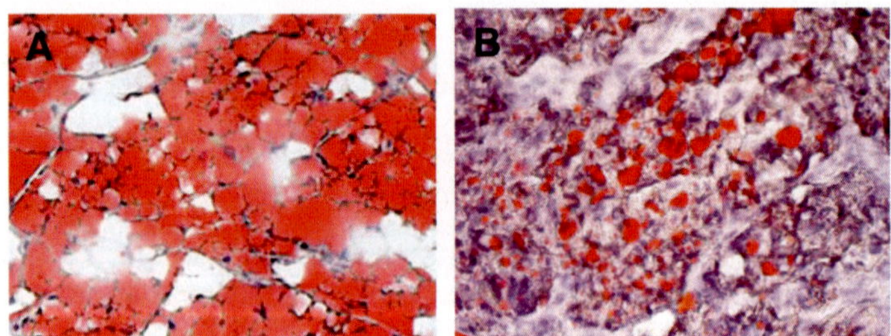

Figure 18.6

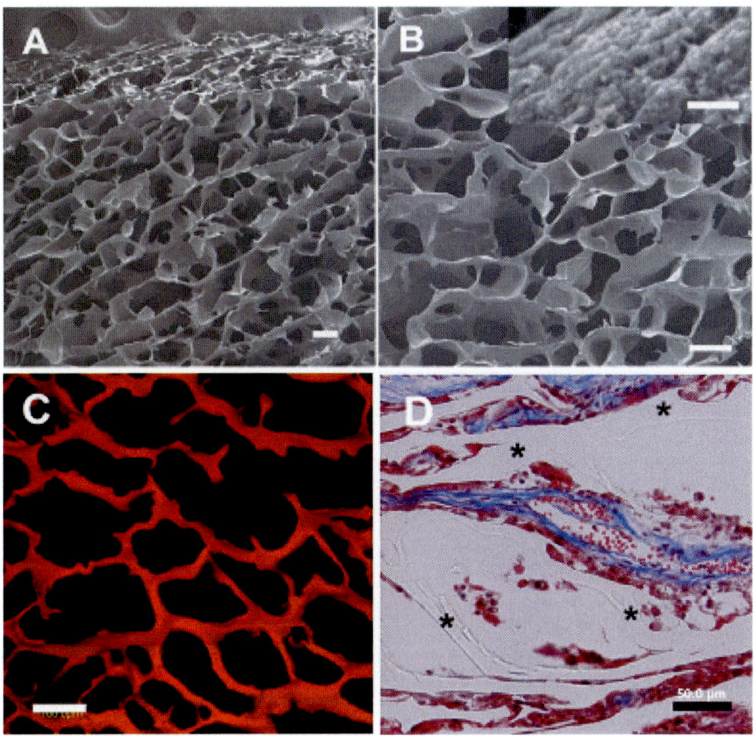

Figure 24.2

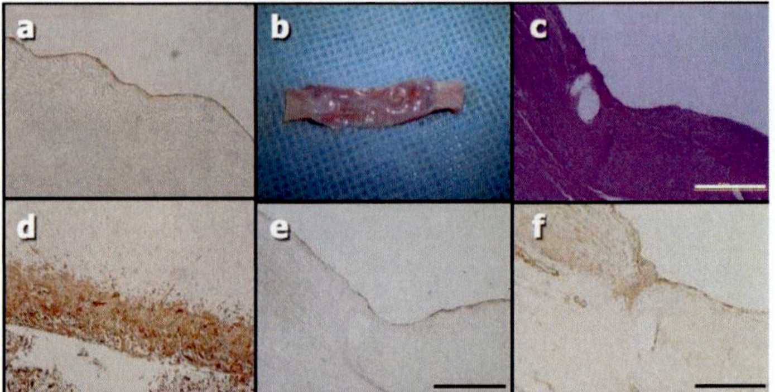

Figure 26.3

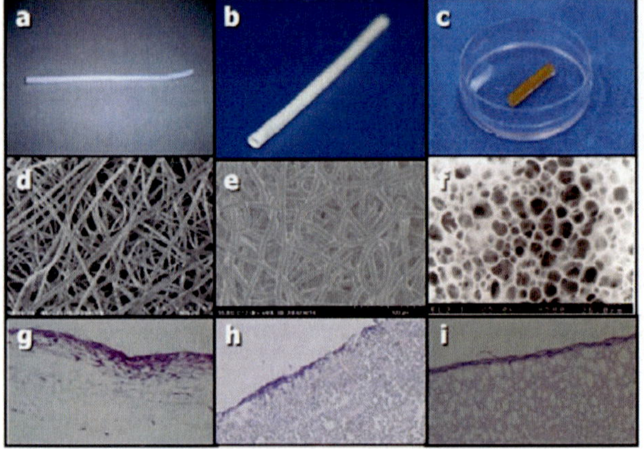

Figure 26.4

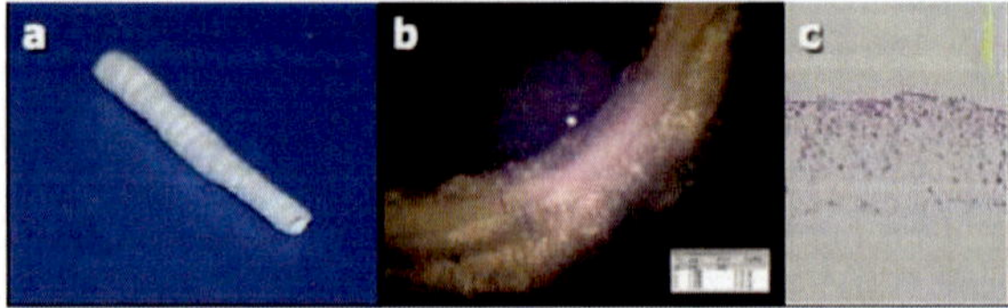

Figure 26.5

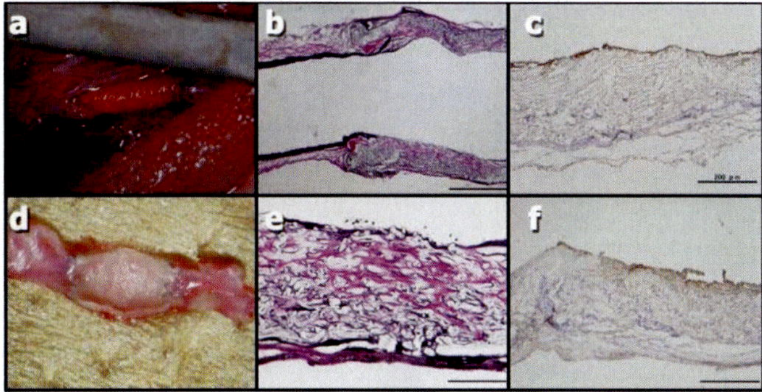

Figure 26.7

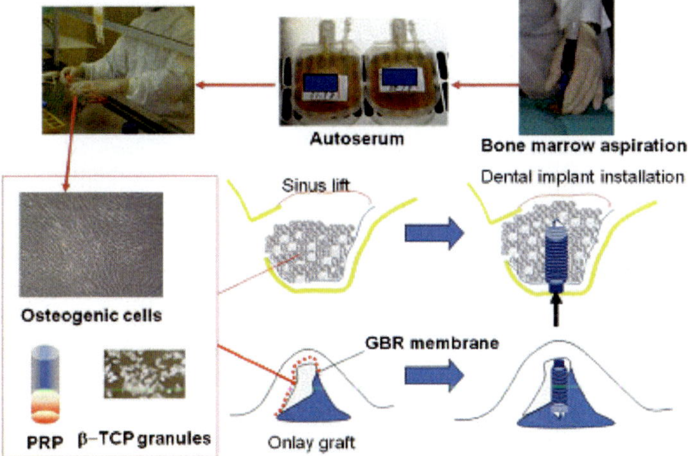

Figure 26.8

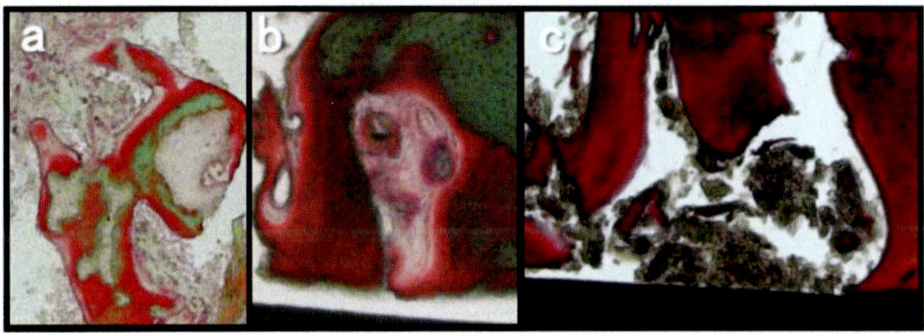

Figure 26.9

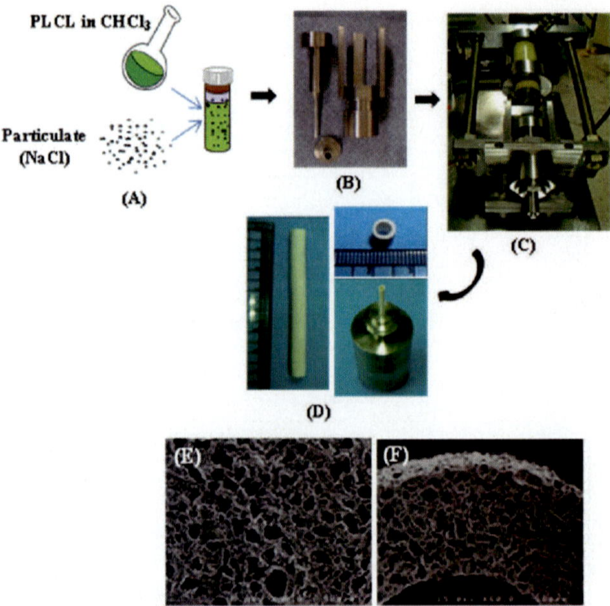

Figure 27.2

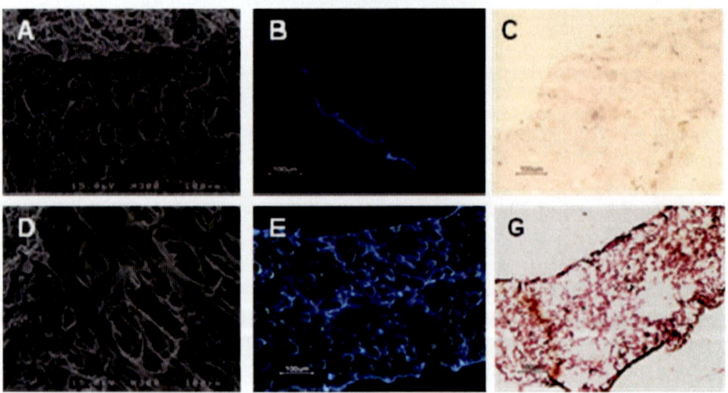

Figure 27.7

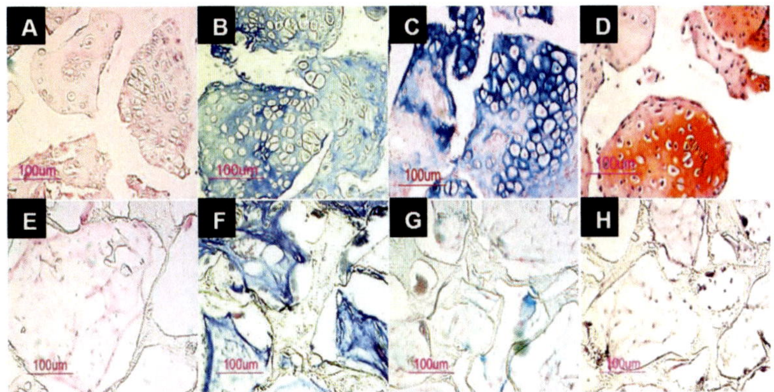

Figure 27.9

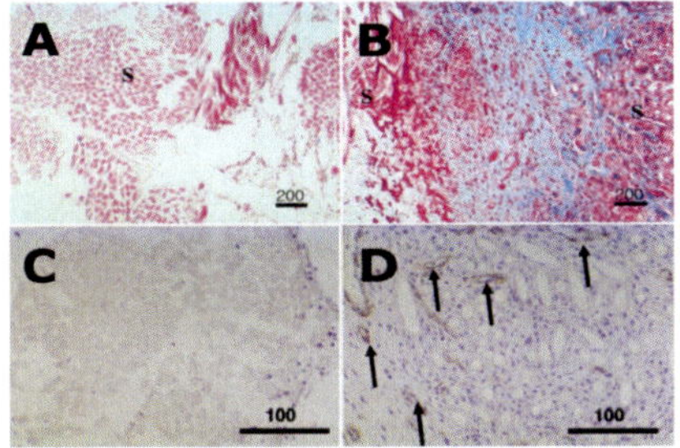

Figure 28.3

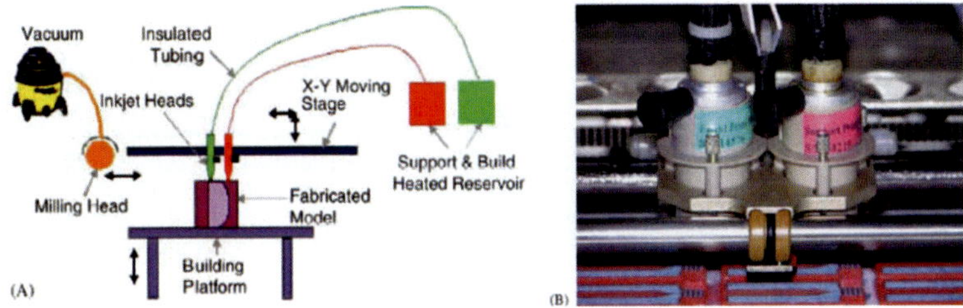

Figure 31.13

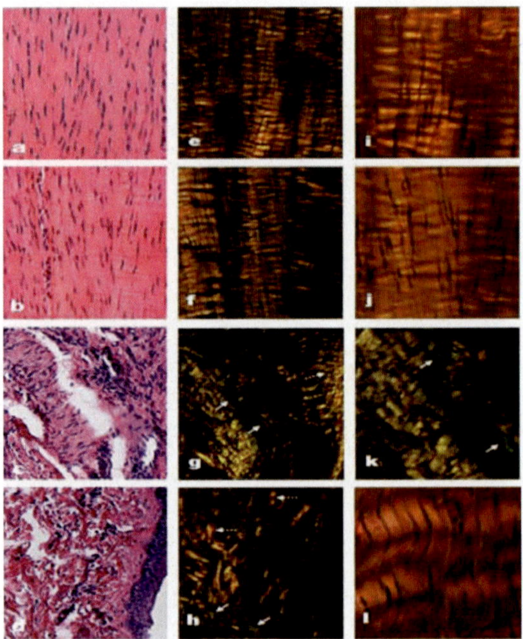

Figure 34.3

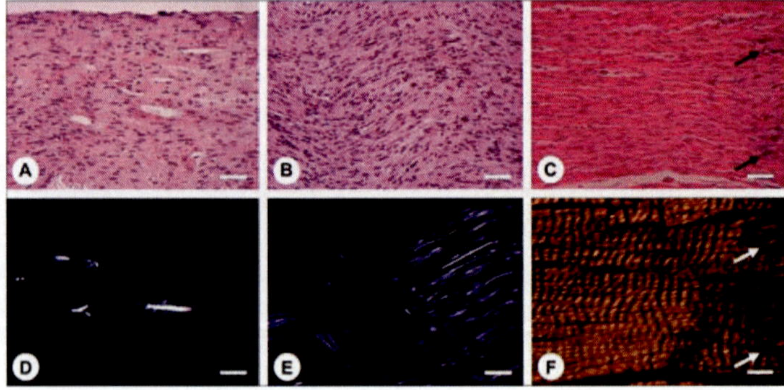

Figure 34.5

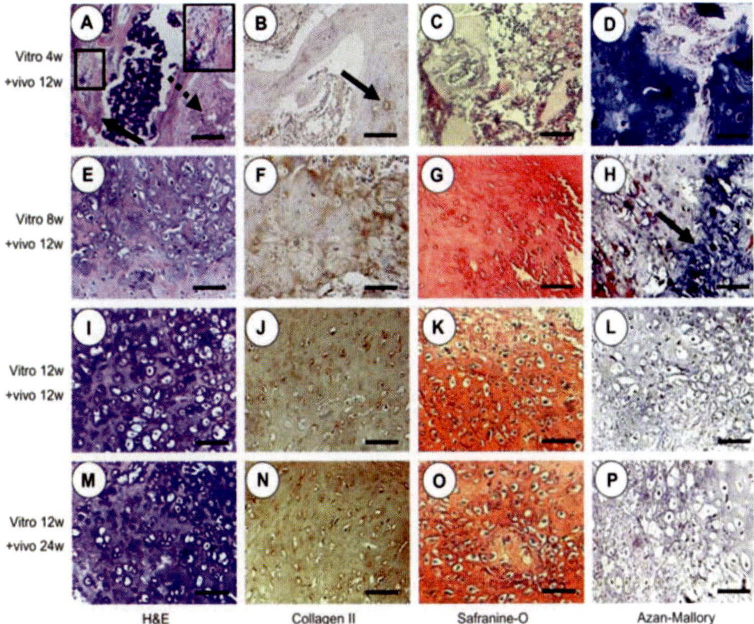

Figure 34.10

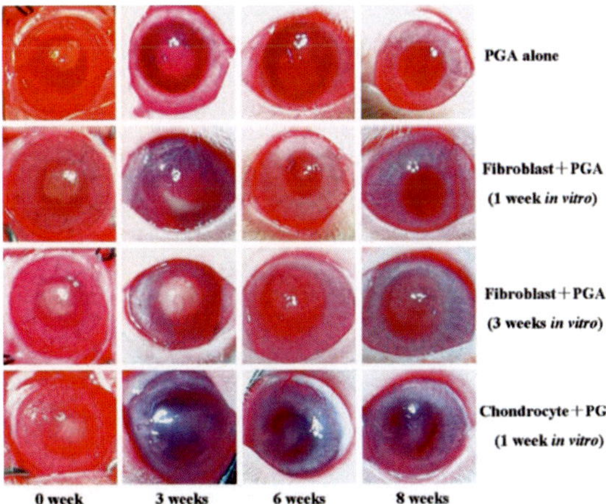

Figure 34.15

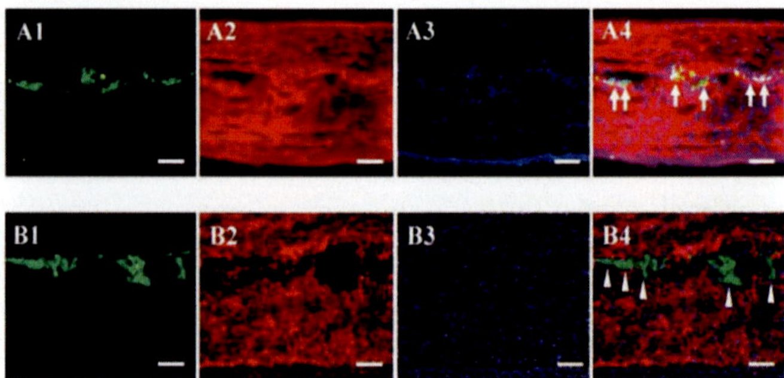

Figure 34.16

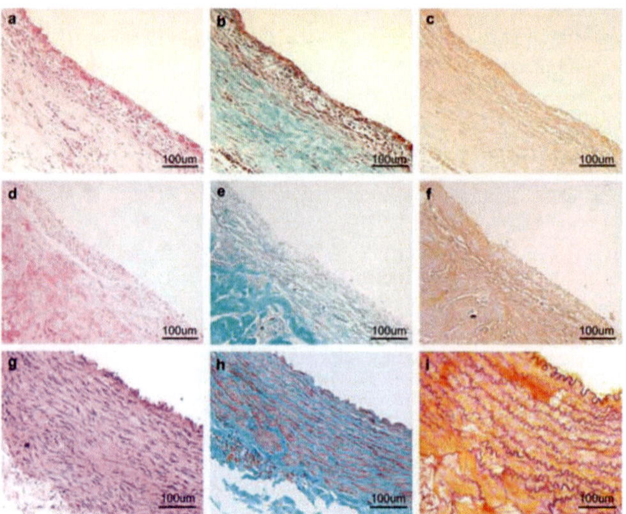

Figure 34.20

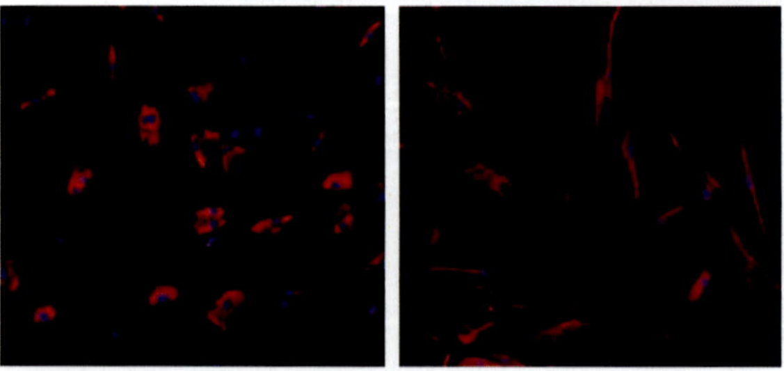

Figure 35.2

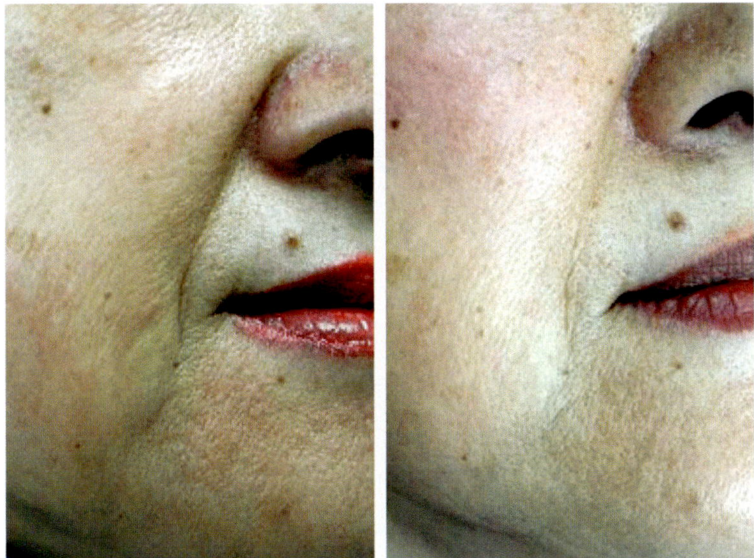

Figure 35.4

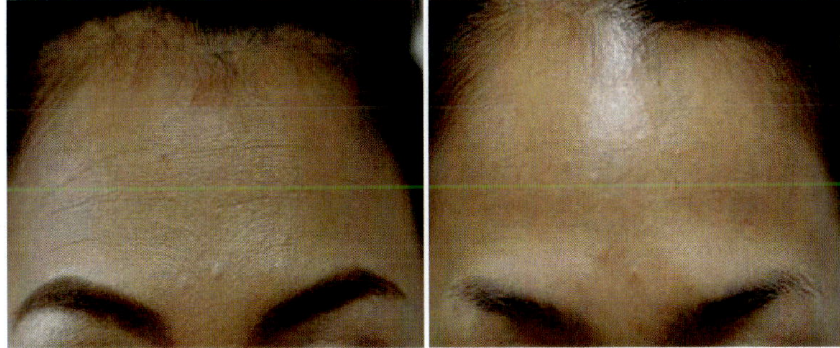

Figure 35.5

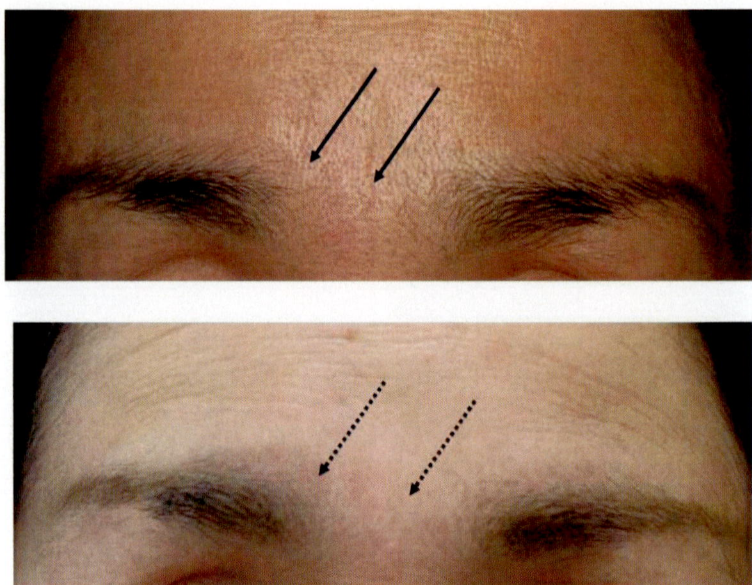

Figure 35.6

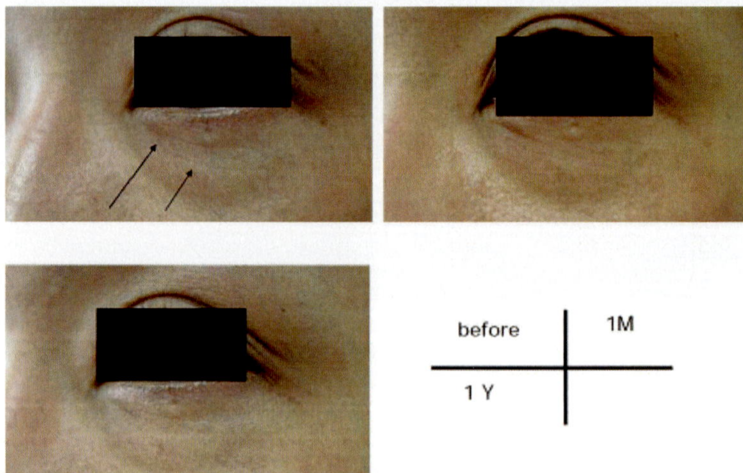

Figure 35.7

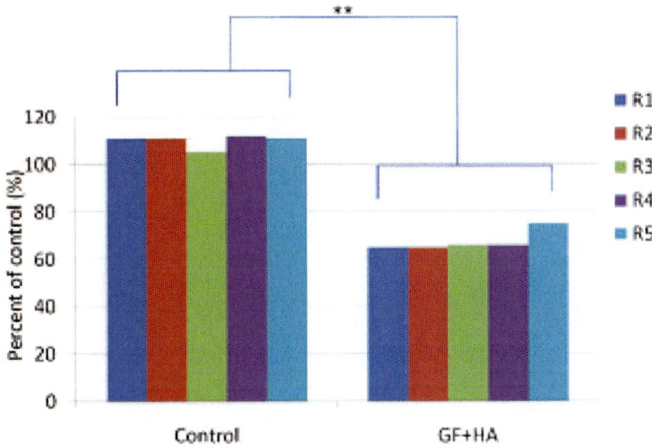

GF: Gingival Fibroblast
HA: Hyaluronic Acid

Figure 35.9

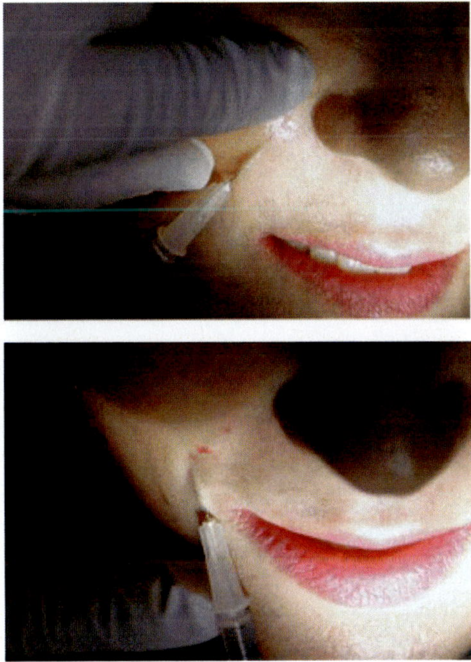

Figure 35.10

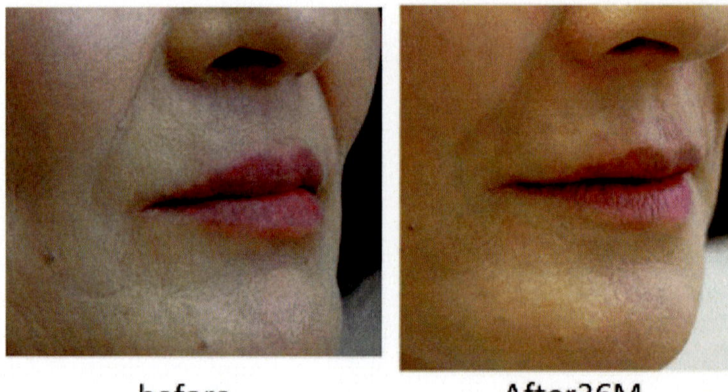

Figure 35.11

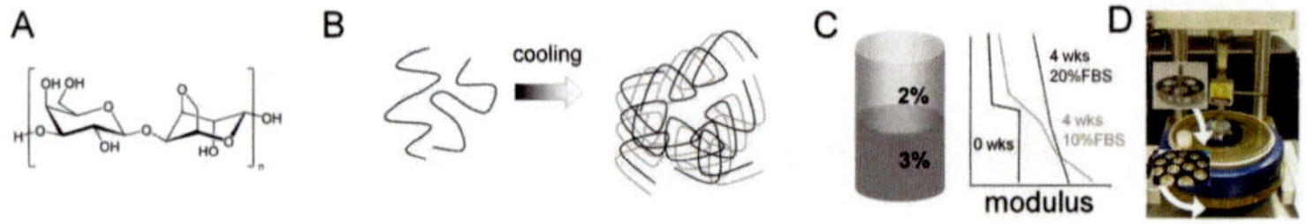

Figure 36.4

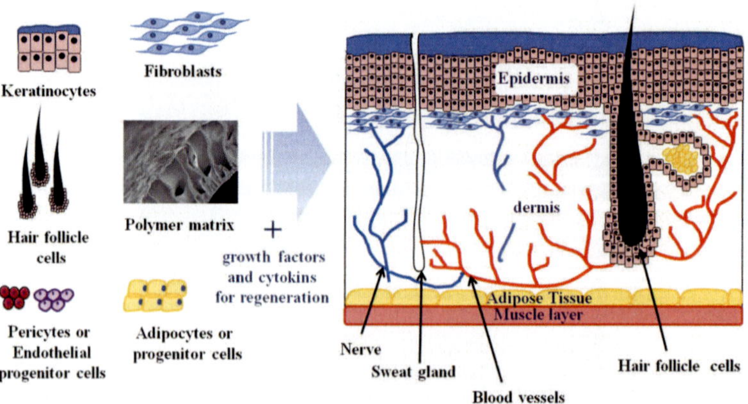

Figure 39.1

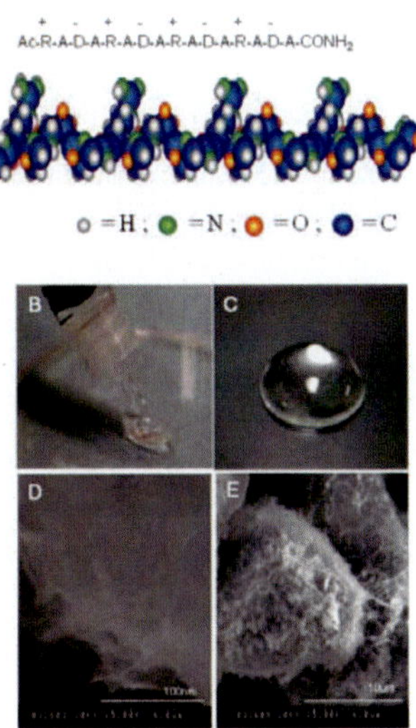

Figure 40.4

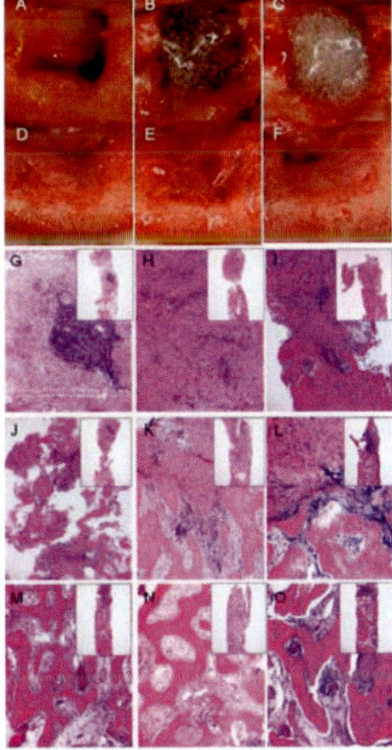

Figure 40.5

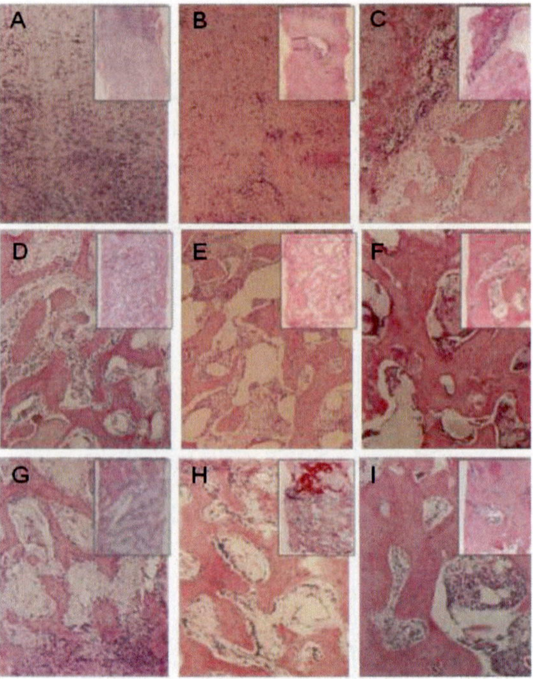

Figure 40.6

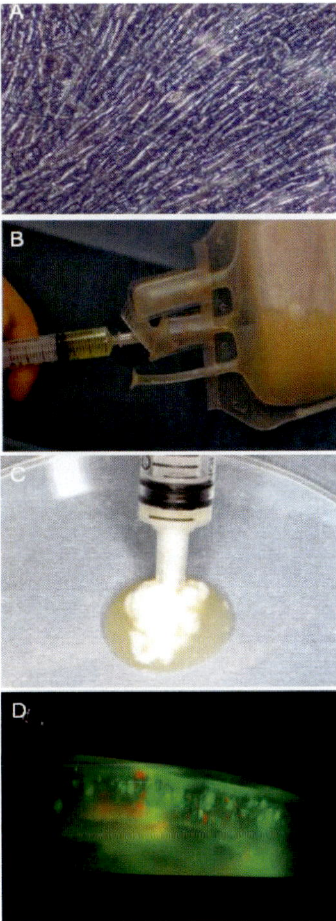

Figure 40.7

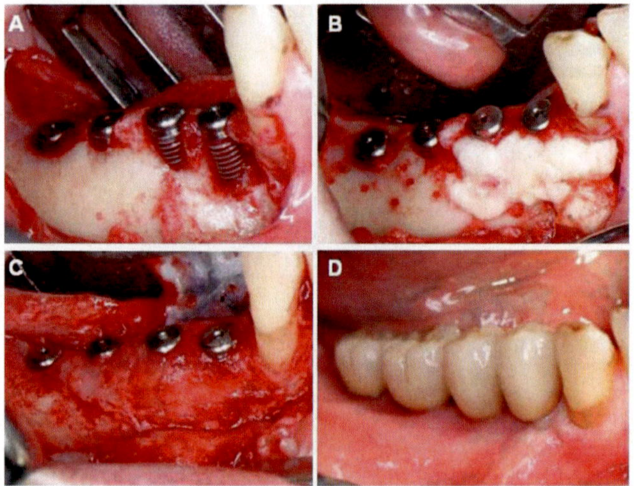

Figure 40.8

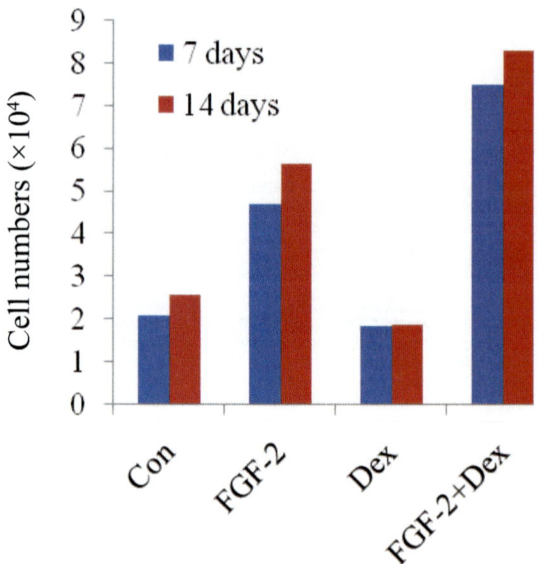

Figure 41.1

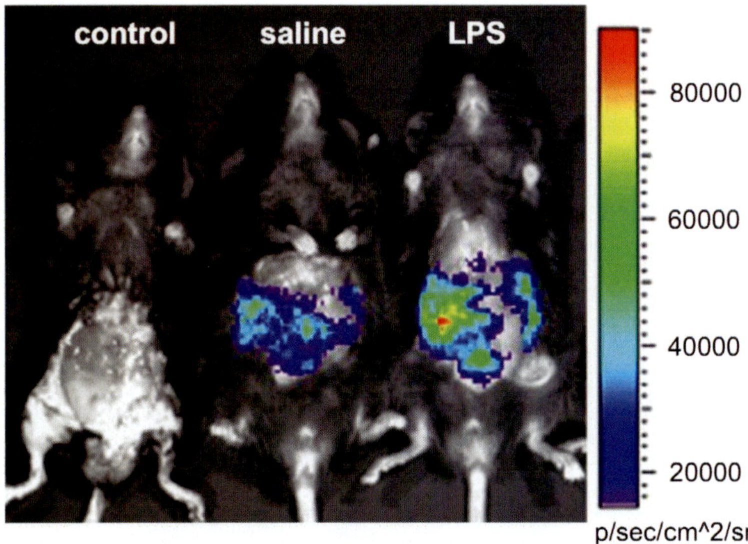

Figure 42.7

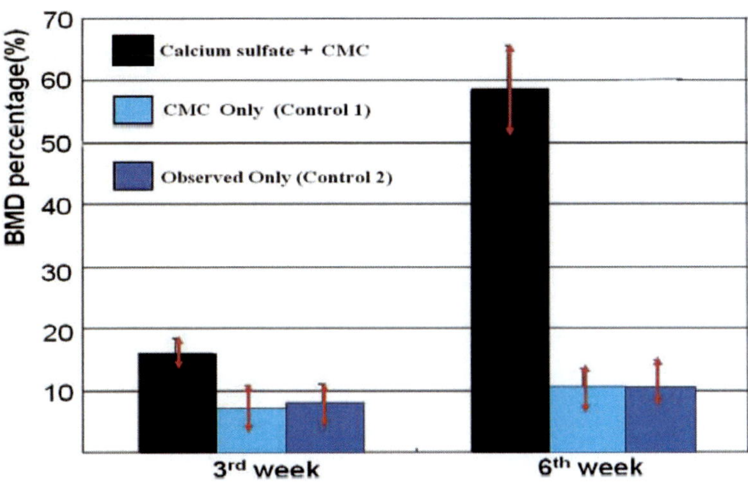

Figure 43.5

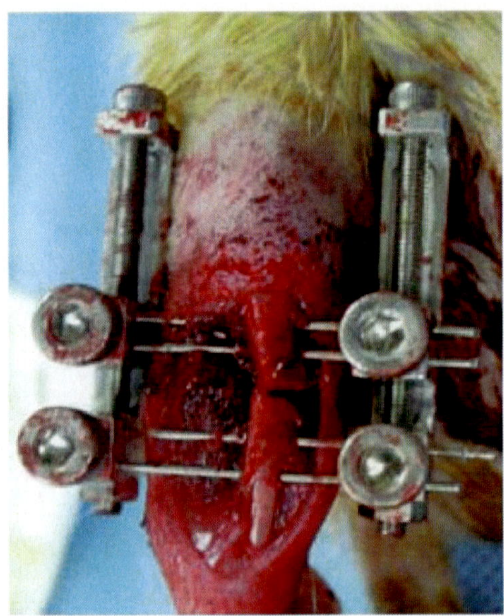

Figure 43.7

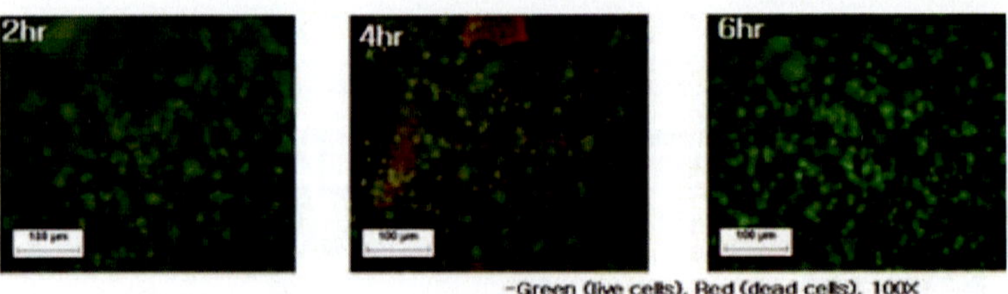

Figure 43.10

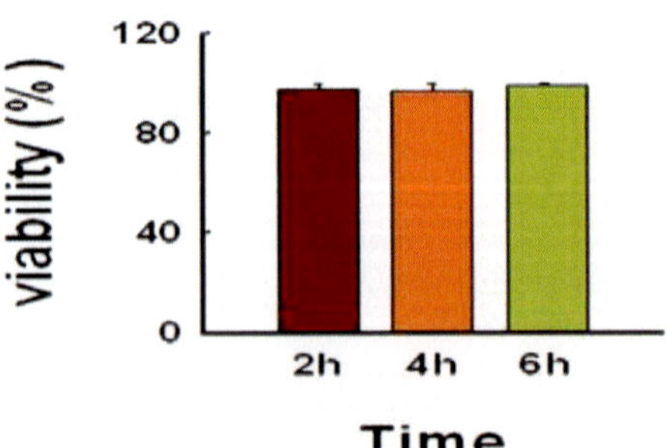

Figure 43.11

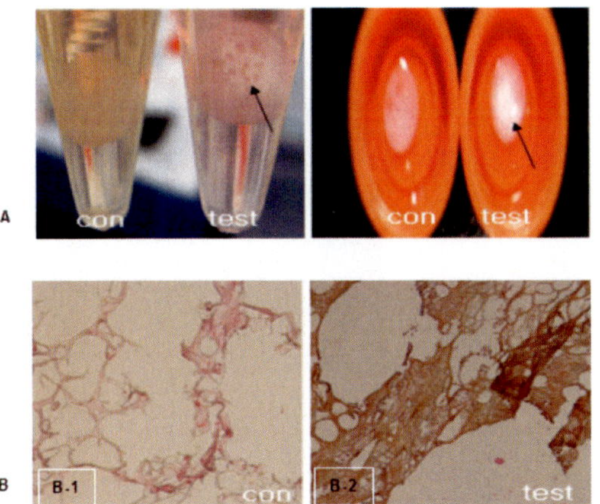

Figure 43.12

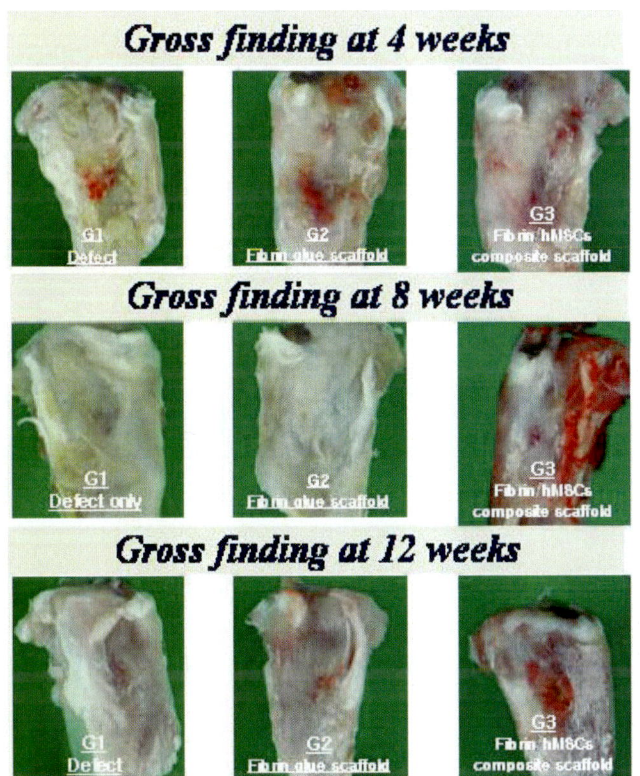

Figure 43.14

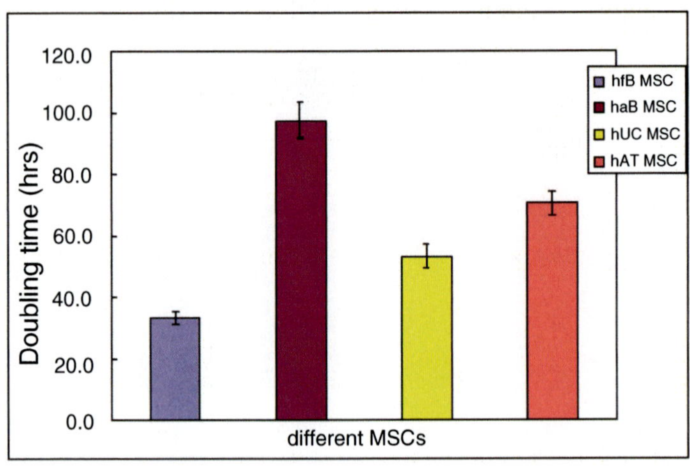

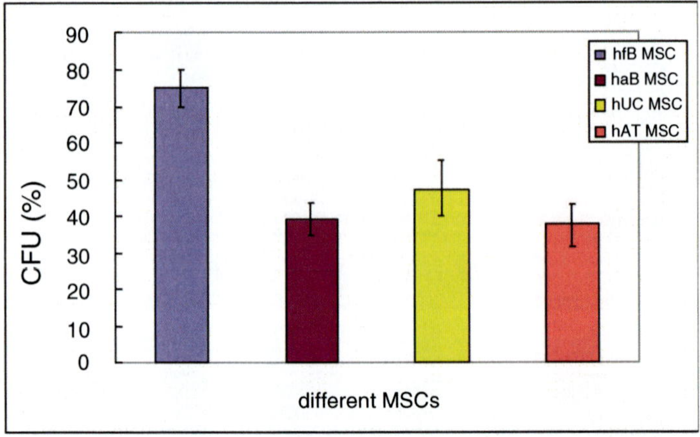

Figure 43.16a

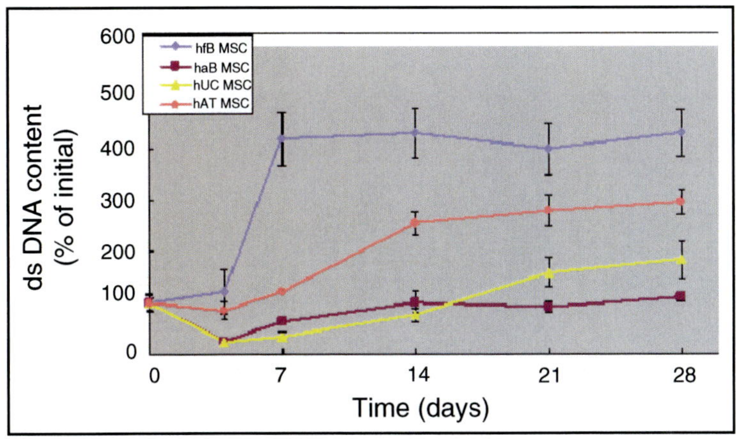

Figure 43.16b

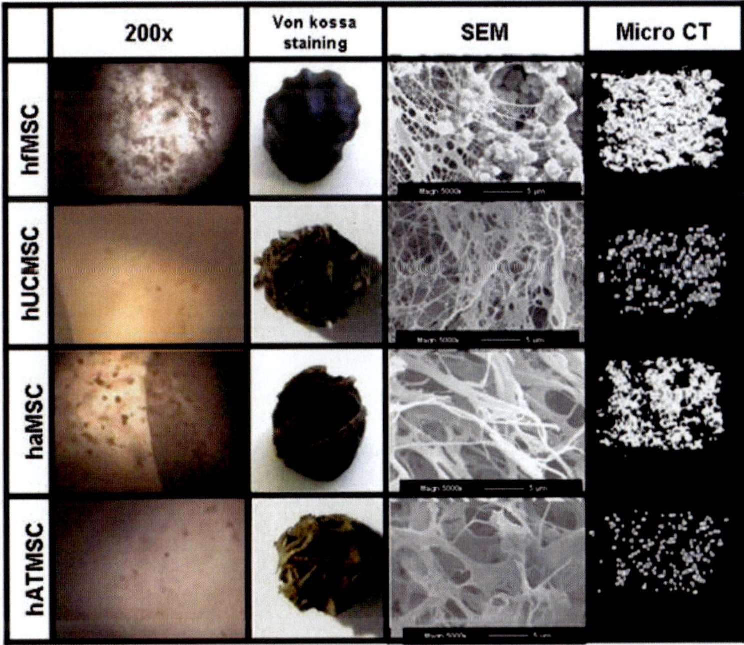

Figure 43.17

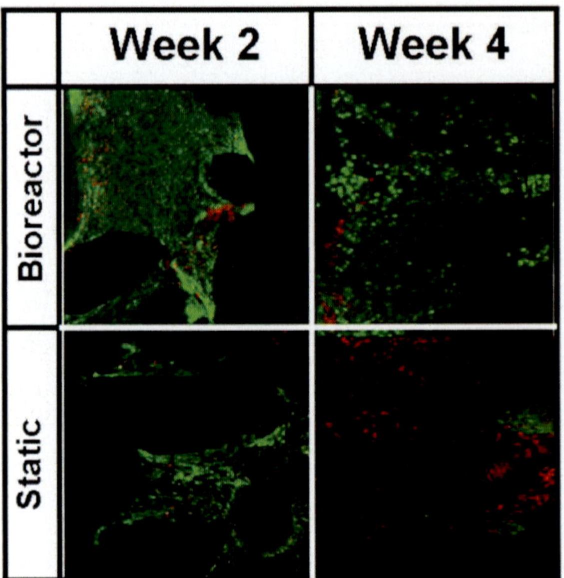

Figure 43.18

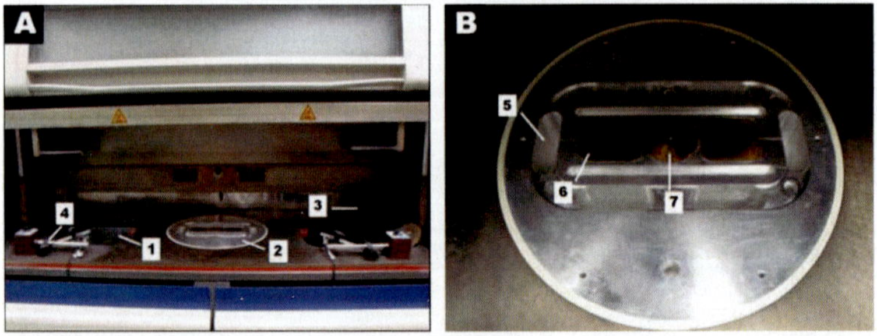

Figure 45.3